ISBN 978-0-656-33797-2
PIBN 10857710

FB &c Ltd, Dalton House, 60 Windsor Avenue, London, SW19 2RR.
Company number 08720141. Registered in England and Wales.

For support please visit www.forgottenbooks.com

VIEW OF THE TEMPLE OF SERAPIS AT PUZZUOLI IN 1836.

PRINCIPLES

OF

GEOLOGY;

OR,

THE MODERN CHANGES OF THE EARTH AND [illegible] INHABITANTS

CONSIDERED AS ILLUSTRATIVE OF GEOLOGY.

BY

SIR CHARLES LYELL, M.A. F.R.S.

[illegible]CE-PRESIDENT OF THE GEOLOGICAL [illegible]
ELEMENTARY GEOLOGY," "TRAVELS [illegible]
TO THE UNITED ST[illegible]," [illegible]

NEW AND ENTIRELY REVISED EDITION.

Illustrated with Maps, Plates, and [illegible]

PRINCIPLES

OF

GEOLOGY;

OR,

THE MODERN CHANGES OF THE EARTH AND ITS INHABITANTS

CONSIDERED AS ILLUSTRATIVE OF GEOLOGY.

BY

SIR CHARLES LYELL, M. A. F. R. S.

VICE-PRESIDENT OF THE GEOLOGICAL SOCIETY OF LONDON; AUTHOR OF "A MANUAL OF ELEMENTARY GEOLOGY," "TRAVELS IN NORTH AMERICA," "A SECOND VISIT TO THE UNITED STATES," ETC. ETC.

NEW AND ENTIRELY REVISED EDITION.

Illustrated with Maps, Plates, and Woodcuts.

"Verè scire est per causas scire."—BACON.

"The stony rocks are not primeval, but the daughters of Time."—LINNÆUS, *Syst. Nat.* ed. 5, *Stockholm*, 1748, p. 219.

"Amid all the revolutions of the globe, the economy of nature has been uniform, and her laws are the only things that have resisted the general movement. The rivers and the rocks, the seas and the continents have been changed in all their parts; but the laws which direct those changes, and the rules to which they are subject, have remained invariably the same."—PLAYFAIR, *Illustrations of the Huttonian Theory*, § 374.

"The inhabitants of the globe, like all the other parts of it, are subject to change. It is not only the individual that perishes, but whole species.

"A change in the animal kingdom seems to be a part of the order of Nature, and is visible in instances to which human power cannot have extended."—PLAYFAIR, *Illustrations of the Huttonian Theory*, § 413.

PREFACE TO THE NINTH EDITION.

The Principles of Geology in the first five editions embraced not only a view of the *modern changes* of the earth and its inhabitants, as set forth in the present work, but also some account of those monuments of analogous changes of *ancient* date, both in the organic and inorganic world, which it is the business of the geologist to interpret. The subject last mentioned, or "geology proper," constituted originally a fourth book, now omitted, the same having been enlarged into a separate treatise, first published in 1838, in one volume 12mo., and called "The Elements of Geology," afterwards recast in two volumes 12mo. in 1842, and again re-edited under the title of "Manual of Elementary Geology," in one volume 8vo. in 1851. The "Principles" and "Manual" thus divided, occupy, with one exception, to which I shall presently allude, very different ground. The "Principles" treat of such portions of the economy of existing nature, animate and inanimate, as are illustrative of Geology, so as to comprise an investigation of the permanent effects of causes now in action, which may serve as records to after ages of the present condition of the globe and its inhabitants. Such effects are the enduring monuments of the ever-varying state of the physical geography of the globe, the lasting signs of its destruction and renovation, and the memorials of the equally fluctuating condition of the organic world. They may be regarded, in short, as a symbolical language, in which the earth's autobiography is written.

In the "Manual of Elementary Geology," on the other hand, I have treated briefly of the component materials of the earth's crust, their arrangement and relative position, and their organic contents, which, when deciphered by aid of the key supplied

PREFACE TO THE NINTH EDITION.

The Principles of Geology in the first five editions embraced not only a view of the *modern changes* of the earth and its inhabitants, as set forth in the present work, but also some account of those monuments of analogous changes of *ancient* date, both in the organic and inorganic world, which it is the business of the geologist to interpret. The subject last mentioned, or "geology proper," constituted originally a fourth book, now omitted, the same having been enlarged into a separate treatise, first published in 1838, in one volume 12mo., and called "The Elements of Geology," afterwards recast in two volumes 12mo. in 1842, and again re-edited under the title of "Manual of Elementary Geology," in one volume 8vo. in 1851. The "Principles" and "Manual" thus divided, occupy, with one exception, to which I shall presently allude, very different ground. The "Principles" treat of such portions of the economy of existing nature, animate and inanimate, as are illustrative of Geology, so as to comprise an investigation of the permanent effects of causes now in action, which may serve as records to after ages of the present condition of the globe and its inhabitants. Such effects are the enduring monuments of the ever-varying state of the physical geography of the globe, the lasting signs of its destruction and renovation, and the memorials of the equally fluctuating condition of the organic world. They may be regarded, in short, as a symbolical language, in which the earth's autobiography is written.

In the "Manual of Elementary Geology," on the other hand, I have treated briefly of the component materials of the earth's crust, their arrangement and relative position, and their organic contents, which, when deciphered by aid of the key supplied by the study of the modern changes above alluded to, reveal to us the annals of a grand succession of past events—a series of revolutions which the solid exterior of the globe, and its living inhabitants, have experienced in times antecedent to the creation of man.

In thus separating the two works, however, I have retained in the "Principles" (book i.) the discussion of some matters which might fairly be regarded as common to both treatises; as for example, an historical sketch of the early progress of geology, followed by a series of preliminary essays to explain the facts and arguments which lead me to believe that the forces now operating upon and beneath the earth's surface may be the same, both in kind and degree, as those which at remote epochs have worked out geological changes. (See Analysis of Contents of this work, p. ix.)

If I am asked whether the "Principles" or the "Manual" should be studied first, I feel much the same difficulty in answering the question as if a student should inquire whether he ought to take up first a treatise on Chemistry, or one on Natural Philosophy, subjects sufficiently distinct, yet inseparably connected. On the whole, while I have endeavored to make each of the two treatises, in their present form, quite independent of the other, I would recommend the reader to study first the modern changes of the earth and its inhabitants as they are discussed in the present volume, proceeding afterwards to the classification and interpretation of the monuments of more remote ages.

CHARLES LYELL.

11 *Harley Street, London,* May 24, 1853.

Dates of the successive Editions of the "Principles" and "Elements" (or Manual) of Geology, by the Author.

Principles, 1st vol. in octavo, published in .... Jan. 1830.
————, 2d vol. do. do. .... Jan. 1832.
————, 1st vol. 2d edition in octavo........ 1832.
————, 2d vol. 2d edition do. .... Jan. 1833.
————, 3d vol. 1st edition do. .... May, 1833.
————, New edition (called the 3d) of the whole work in 4 vols. 12mo........ May, 1834.
————, 4th edition, 4 vols. 12mo........ June, 1835.
————, 5th do. do. do. .... Mar. 1837.
Elements, 1st edition in one vol. .... July, 1838.
Principles, 6th do. 3 vols. 12mo .... June, 1840.
Elements, 2d edition in 2 vols. 12mo........ July, 1841.
Principles, 7th edition in one vol. 8vo........ Feb. 1847.
————, 8th edition in one vol. 8vo........ May, 1850.
Manual of Elementary Geology (or "Elements," 3d edition) in one vol. 8vo........ Jan. 1851.
Manual, 4th edition, one vol. 8vo........ Jan. 1852.
Principles, 9th edition, now published in one vol. 8vo........ June, 1853

ANALYSIS OF THE CONTENTS

OF

THE PRINCIPLES OF GEOLOGY.

BOOK I. (CHAPTERS I. to XIII.)

HISTORICAL SKETCH OF THE PROGRESS OF GEOLOGY, WITH A SERIES OF ESSAYS TO SHOW THAT THE MONUMENTS OF THE ANCIENT STATE OF THE EARTH AND ITS INHABITANTS, WHICH THIS SCIENCE INTERPRETS, CAN ONLY BE UNDERSTOOD BY A PREVIOUS ACQUAINTANCE WITH TERRESTRIAL CHANGES NOW IN PROGRESS, BOTH IN THE ORGANIC AND INORGANIC WORLDS.

CHAPTER I.

CHAPTER II.

CHAPTER III.

CHAPTER IV.

CHAPTER V.

CHAPTER VI.

CHAPTERS VII. VIII.

CHAPTER IX.

CHAPTER X.

CHAPTER XI.

CHAPTER XII.

CHAPTER XIII.

BOOK II. (Chapters XIV. to XXXII.)

OBSERVED CHANGES IN THE INORGANIC WORLD NOW IN PROGRESS: FIRST, THE EFFECTS OF AQUEOUS CAUSES, SUCH AS RIVERS, SPRINGS, GLACIERS, WAVES, TIDES, AND CURRENTS; SECONDLY, OF IGNEOUS CAUSES, OR SUBTERRANEAN HEAT, AS EXHIBITED IN THE VOLCANO AND THE EARTHQUAKE.

CHAPTER XIV.

CHAPTER XV.

CHAPTER XVI.

CHAPTER XVII.

CHAPTER XVIII.

CHAPTERS XIX. XX. XXI.

CHAPTER XXII.

CHAPTERS XXIII. XXIV.

CHAPTER XXV.

CHAPTER XXVI.

CHAPTER XXVII.

CHAPTER XXVIII.

CHAPTER XXIX.

CHAPTER XXX.

CHAPTERS XXXI. XXXII.

BOOK III. (Chapters XXXIII to L.)

OBSERVED CHANGES OF THE ORGANIC WORLD NOW IN PROGRESS; FIRST, NATURE AND GEOGRAPHICAL DISTRIBUTION OF SPECIES, AND THEORIES RESPECTING THEIR CREATION AND EXTINCTION; SECONDLY, THE INFLUENCE OF ORGANIC BEINGS IN MODIFYING PHYSICAL GEOGRAPHY; THIRDLY, THE LAWS ACCORDING TO WHICH THEY ARE IMBEDDED IN VOLCANIC, FRESHWATER, AND MARINE DEPOSITS.

CHAPTERS XXXIII. XXXIV. XXXV. XXXVI.

CHAPTER XXXVII.

CHAPTER XXXVIII.

CHAPTER XXXIX.

CHAPTER XL.

LIST OF PLATES.

DIRECTIONS TO THE BINDER.

PRINCIPLES OF GEOLOGY.

BOOK I.

CHAPTER I.

Geology defined—Compared to History—Its relation to other Physical Sciences—Not to be confounded with Cosmogony.

Geology is the science which investigates the successive changes that have taken place in the organic and inorganic kingdoms of nature; it inquires into the causes of these changes, and the influence which they have exerted in modifying the surface and external structure of our planet.

By these researches into the state of the earth and its inhabitants at former periods, we acquire a more perfect knowledge of its present condition, and more comprehensive views concerning the laws now governing its animate and inanimate productions. When we study history, we obtain a more profound insight into human nature, by instituting a comparison between the present and former states of society. We trace the long series of events which have gradually led to the actual posture of affairs; and by connecting effects with their causes, we are enabled to classify and retain in the memory a multitude of complicated relations—the various peculiarities of national character—the different degrees of moral and intellectual refinement, and numerous other circumstances, which, without historical associations, would be uninteresting or imperfectly understood. As the present condition of nations is the result of many antecedent changes, some extremely remote, and others recent, some gradual, others sudden and violent; so the state of the natural world is the result of a long succession of events; and if we would enlarge our experience of the present economy of nature, we must investigate the effects of her operations in former epochs.

We often discover with surprise, on looking back into the chronicles of nations, how the fortune of some battle has influenced the fate of millions of our contemporaries, when it has long been forgotten by the mass of the population. With this remote event we may find inseparably connected the geographical boundaries of a great state, the language now spoken by the inhabitants, their peculiar manners, laws, and religious opinions. But far more astonishing and unexpected are the connections brought to light, when we carry back our researches into the history of nature. The form of a coast, the configuration of the interior of a country, the existence and extent of lakes, valleys, and mountains, can often be traced to the former prevalence of earthquakes and volcanoes in regions which have long been undisturbed. To these remote convulsions the present fertility of some districts, the sterile character of others, the elevation of land above the sea, the climate, and various peculiarities, may be distinctly referred. On the other hand, many distinguishing features of the surface may often be ascribed to the operation, at a remote era, of slow and tranquil causes—to the gradual deposition of sediment in a lake or in the ocean, or to the prolific increase of testacea and corals.

To select another example, we find in certain localities subterranean deposits of coal, consisting of vegetable matter, formerly drifted into seas and lakes. These seas and lakes have since been filled up, the lands whereon the forests grew have disappeared or changed their form, the rivers and currents which floated the vegetable masses can no longer be traced, and the plants belonged to species which for ages have passed away from the surface of our planet. Yet the commercial prosperity, and numerical strength of a nation, may now be mainly dependent on the local distribution of fuel determined by that ancient state of things.

Geology is intimately related to almost all the physical sciences, as history is to the moral. An historian should, if possible, be at once profoundly acquainted with ethics, politics, jurisprudence, the military art, theology; in a word, with all branches of knowledge by which any insight into human affairs, or into the moral and intellectual nature of man, can be obtained. It would be no less desirable that a geologist should be well versed in chemistry, natural philosophy, mineralogy, zoology, comparative anatomy, botany; in short, in every science relating to organic and inorganic nature. With these accomplishments, the historian and geologist would rarely fail to draw correct and philosophical conclusions from the various monuments transmitted to them of former occurrences. They would know to what combination of causes analogous effects were referable, and they would often be enabled to supply, by inference, information concerning many events unrecorded in the defective archives of former ages. But as such extensive acquisitions are scarcely within the reach of any individual, it is necessary that men who have devoted their lives to different departments should unite their efforts; and as the historian receives assistance from the antiquary, and from those who have cultivated different branches of moral and political science, so the geologist should avail himself of the aid of many naturalists, and particularly of those who have studied the fossil remains of lost species of animals and plants.

The analogy, however, of the monuments consulted in geology, and those available in history, extends no farther than to one class of historical monuments—those which may be said to be *undesignedly* commemorative of former events. The canoes, for example, and stone hatchets found in our peat bogs, afford an insight into the rude arts and manners of the earliest inhabitants of our island; the buried coin fixes the date of the reign of some Roman emperor; the ancient encampment indicates the districts once occupied by invading armies, and the former method of constructing military defences; the Egyptian mummies throw light on the art of embalming, the rites of sepulture, or the average stature of the human race in ancient Egypt. This class of memorials yields to no other in authenticity, but it constitutes a small part only of the resources on which the historian relies, whereas in geology it forms the only kind of evidence which is at our command. For this reason we must not expect to obtain a full and connected account of any series of events beyond the reach of history. But the testimony of geological monuments, if frequently imperfect, possesses at least the advantage of being free from all intentional misrepresentation. We may be deceived in the inferences which we draw, in the same manner as we often mistake the nature and import of phenomena observed in the daily course of nature; but our liability to err is confined to the interpretation, and, if this be correct, our information is certain.

It was long before the distinct nature and legitimate objects of geology were fully recognized, and it was at first confounded with many other branches of inquiry, just as the limits of history, poetry, and mythology were ill-defined in the infancy of civilization. Even in Werner's time, or at the close of the eighteenth century, geology appears to have been regarded as little other than a subordinate department of mineralogy; and Desmarest included it under the head of Physical Geography. But the most common and serious source of confusion arose from the notion, that it was the business of geology to discover the mode in which the earth originated, or, as some imagined, to study the effects of those cosmological causes which were employed by the Author of Nature to bring this planet out of a nascent and chaotic state into a more perfect and habitable condition. Hutton was the first who endeavored to draw a strong line of demarcation between his favorite science and cosmogony, for he declared that geology was in nowise concerned "with questions as to the origin of things."

An attempt will be made in the sequel of this work to demonstrate that geology differs as widely from cosmogony, as speculations concerning the mode of the first creation of man differ from history. But, before entering more at large on this controverted question, it will be desirable to trace the progress of opinion on this topic, from the earliest ages to the commencement of the present century.

CHAPTER II.

HISTORICAL SKETCH OF THE PROGRESS OF GEOLOGY.

Oriental Cosmogony—Hymns of the Vedas—Institutes of Menù—Doctrine of the successive destruction and renovation of the world—Origin of this doctrine—Common to the Egyptians—Adopted by the Greeks—System of Pythagoras—Of Aristotle—Dogmas concerning the extinction and reproduction of genera and species—Strabo's theory of elevation by earthquakes—Pliny—Concluding Remarks on the knowledge of the Ancients.

Oriental Cosmogony.—THE earliest doctrines of the Indian and Egyptian schools of philosophy agreed in ascribing the first creation of the world to an omnipotent and infinite Being. They concurred also in representing this Being, who had existed from all eternity, as having repeatedly destroyed and reproduced the world and all its inhabitants. In the sacred volume of the Hindoos, called the Ordinances of Menù, comprising the Indian system of duties religious and civil, we find a preliminary chapter treating of the Creation, in which the cosmogony is known to have been derived from earlier writings and traditions; and principally from certain hymns of high antiquity, called the Vedas. These hymns were first put together, according to Mr. Colebrooke,* in a connected series, about thirteen centuries before the Christian era, but they appear from internal evidence to have been written at various antecedent periods. In them, as we learn from the researches of Professor Wilson, the eminent Sanscrit scholar, two distinct philosophical systems are discoverable. According to one of them, all things were originally brought into existence by the sole will of a single First Cause, which existed from eternity; according to the other, there have always existed two principles, the one material, but without form, the other spiritual and capable of compelling "inert matter to develop its sensible properties." This development of matter into "individual and visible existences" is called creation, and is assigned to a subordinate agent, or the creative faculty of the Supreme Being embodied in the person of Brahma.

In the first chapter of the Ordinances of Menù above alluded to, we meet with the following passages relating to former destructions and renovations of the world:—

"The Being, whose powers are incomprehensible, having created me (Menù) and this universe, again became absorbed in the supreme spirit, changing the time of energy for the hour of repose.

"When that Power awakes, then has this world its full expansion; but when he slumbers with a tranquil spirit, then the whole system fades away. . . . . For while he reposes, as it were, embodied spirits

* Essays on the Philosophy of the Hindoos.

endowed with principles of action depart from their several acts, and the mind itself becomes inert."

The absorption of all beings into the Supreme essence is then described, and the Divine soul itself is said to slumber, and to remain for a time immersed in "the first idea, or in darkness." After which the text thus proceeds (verse fifty-seven), "Thus that immutable power by waking and reposing alternately, revivifies and destroys, in eternal succession, this whole assemblage of locomotive and immovable creatures."

It is then declared that there has been a long succession of *manwantaras*, or periods, each of the duration of many thousand ages, and—

"There are creations also, and destructions of worlds innumerable: the Being, supremely exalted, performs all this with as much ease as if in sport, again and again, for the sake of conferring happiness."*

No part of the Eastern cosmogony, from which these extracts are made, is more interesting to the geologist than the doctrine, so frequently alluded to, of the reiterated submersion of the land beneath the waters of a universal ocean. In the beginning of things, we are told, the First Sole Cause "with a thought created the waters," and then moved upon their surface in the form of Brahma the creator, by whose agency the emergence of the dry land was effected, and the peopling of the earth with plants, animals, celestial creatures, and man. Afterwards, as often as a general conflagration at the close of each manwantara had annihilated every visible and existing thing, Brahma, on awaking from his sleep, finds the whole world a shapeless ocean. Accordingly, in the legendary poems called the Puranas, composed at a later date than the Vedas, the three first Avatars or descents of the Deity upon earth have for their object to recover the land from the waters. For this purpose Vishnu is made successively to assume the form of a fish, a tortoise, and a boar.

Extravagant as may be some of the conceits and fictions which disfigure these pretended revelations, we can by no means look upon them as a pure effort of the unassisted imagination, or believe them to have been composed without regard to opinions and theories founded on the observation of Nature. In astronomy, for instance, it is declared that, at the North Pole, the year was divided into a long day and night, and that their long day was the northern, and their night the southern course of the sun; and to the inhabitants of the moon, it is said one day is equal in length to one month of mortals.† If such statements cannot be resolved into mere conjectures, we have no right to refer to mere chance the prevailing notion that the earth and its inhabitants had formerly undergone a succession of revolutions and aqueous catastrophes interrupted by long intervals of tranquillity.

Now there are two sources in which such a theory may have originated. The marks of former convulsions on every part of the surface of

* Institutes of Hindoo Law, or the Ordinances of Menû, from the Sanscrit, translated by Sir William Jones, 1796.

† Menû, Inst. c. i. 66, and 67.

our planet are obvious and striking. The remains of marine animals imbedded in the solid strata are so abundant, that they may be expected to force themselves on the attention of every people who have made some progress in refinement; and especially where one class of men are expressly set apart from the rest, like the ancient priesthoods of India and Egypt, for study and contemplation. If these appearances are once recognized, it seems natural that the mind should conclude in favor, not only of mighty changes in past ages, but of alternate periods of repose and disorder;—of repose, when the animals now fossil lived, grew, and multiplied—of disorder, when the strata in which they were buried became transferred from the sea to the interior of continents, and were uplifted so as to form part of high mountain-chains. Those modern writers, who are disposed to disparage the former intellectual advancement and civilization of Eastern nations, may concede some foundation of observed facts for the curious theories now under consideration, without indulging in exaggerated opinions of the progress of science; especially as universal catastrophes of the world, and exterminations of organic beings, in the sense in which they were understood by the Brahmins, are untenable doctrines.

We know that the Egyptian priests were aware, not only that the soil beneath the plains of the Nile, but that also the hills bounding the great valley, contained marine shells; and Herodotus inferred from these facts, that all lower Egypt, and even the high lands above Memphis, had once been covered by the sea.* As similar fossil remains occur in all parts of Asia hitherto explored, far in the interior of the continent as well as near the sea, they could hardly have escaped detection by some Eastern sages not less capable than the Greek historian of reasoning philosophically on natural phenomena.

We also know that the rulers of Asia were engaged in very remote eras in executing great national works, such as tanks and canals, requiring extensive excavations. In the fourteenth century of our era (in the year 1360), the removal of soil necessary for such undertakings brought to light geological facts, which attracted the attention of a people less civilized than were many of the older nations of the East. The historian Ferishta relates that fifty thousand laborers were employed in cutting through a mound, so as to form a junction between the rivers Selima and Sutlej; and in this mound were found the bones of elephants and men, some of them petrified, and some of them resembling bone. The gigantic dimensions attributed to the human bones show them to have belonged to some of the larger pachydermata.†

But, although the Brahmins, like the priests of Egypt, may have

* Herodot. Euterpe, 12.

† A Persian MS. copy of the historian Ferishta, in the library of the East India Company, relating to the rise and progress of the Mahomedan empire in India, was procured by Colonel Briggs from the library of Tippoo Sultan in 1799; which has been referred to at some length by Dr. Buckland. (Geol. Trans. 2d Series, vol. ii. part iii. p. 389.)

been acquainted with the existence of fossil remains in the strata, it is possible that the doctrine of successive destructions and renovations of the world, merely received corroboration from such proofs; and that it may have been originally handed down, like the religious traditions of most nations, from a ruder state of society. The system may have had its source, in part at least, in exaggerated accounts of those dreadful catastrophes which are occasioned by particular combinations of natural causes. Floods and volcanic eruptions, the agency of water and fire, are the chief instruments of devastation on our globe. We shall point out in the sequel the extent of many of these calamities, recurring at distant intervals of time, in the present course of nature; and shall only observe here, that they are so peculiarly calculated to inspire a lasting terror, and are so often fatal in their consequences to great multitudes of people, that it scarcely requires the passion for the marvellous, so characteristic of rude and half-civilized nations, still less the exuberant imagination of Eastern writers, to augment them into general cataclysms and conflagrations.

The great flood of the Chinese, which their traditions carry back to the period of Yaoù, something more than 2000 years before our era, has been identified by some persons with the universal deluge described in the Old Testament; but according to Mr. Davis, who accompanied two of our embassies to China, and who has carefully examined their written accounts, the Chinese cataclysm is therein described as interrupting the business of agriculture, rather than as involving a general destruction of the human race. The great Yu was celebrated for having "opened nine channels to draw off the waters," which "covered the low hills and bathed the foot of the highest mountains." Mr. Davis suggests that a great derangement of waters of the Yellow River, one of the largest in the world, might even now cause the flood of Yaou to be repeated, and lay the most fertile and populous plains of China under water. In modern times the bursting of the banks of an artificial canal, into which a portion of the Yellow River has been turned, has repeatedly given rise to the most dreadful accidents, and is a source of perpetual anxiety to the government. It is easy, therefore, to imagine how much greater may have been the inundation, if this valley was ever convulsed by a violent earthquake.*

Humboldt relates the interesting fact that, after the annihilation of a large part of the inhabitants of Cumana, by an earthquake in 1766, a season of extraordinary fertility ensued, in consequence of the great rains which accompanied the subterranean convulsions. "The Indians," he says, "celebrated, after the ideas of an antique superstition, by festivals and dancing, the destruction of the world and the approaching epoch of its regeneration."†

The existence of such rites among the rude nations of South Amer-

* See Davis on "The Chinese," published by the Soc. for the Diffus. of Use. Know. vol. i. pp. 137, 147.

† Humboldt et Bonpland, Voy. Relat. Hist. vol. i. p. 30.

ica is most important, as showing what effects may be produced by local catastrophes, recurring at distant intervals of time, on the minds of a barbarous and uncultivated race. I shall point out in the sequel how the tradition of a deluge among the Araucanian Indians may be explained, by reference to great earthquake-waves which have repeatedly rolled over part of Chili since the first recorded flood of 1590. (See chap. 29, Book II.) The legend also of the ancient Peruvians of an inundation many years before the reign of the Incas, in which only six persons were saved on a float, relates to a region which has more than once been overwhelmed by inroads of the ocean since the days of Pizarro. (Chap. 29, Book II.) I might refer the reader to my account of the submergence of a wide area in Cutch so lately as the year 1819, when a single tower only of the fort of Sindree appeared above the waste of waters (see Chap. 28, Book II.), if it were necessary, to prove how easily the catastrophes of modern times might give rise to traditionary narratives, among a rude people, of floods of boundless extent. Nations without written records, and who are indebted for all their knowledge of past events exclusively to oral tradition, are in the habit of confounding in one legend a series of incidents which have happened at various epochs; nor must we forget that the superstitions of a savage tribe are transmitted through all the progressive stages of society, till they exert a powerful influence on the mind of the philosopher. He may find, in the monuments of former changes on the earth's surface, an apparent confirmation of tenets handed down through successive generations, from the rude hunter, whose terrified imagination drew a false picture of those awful visitations of floods and earthquakes, whereby the whole earth as known to him was simultaneously devastated.

Egyptian Cosmogony.—Respecting the cosmogony of the Egyptian priests, we gather much information from writers of the Grecian sects, who borrowed almost all their tenets from Egypt, and amongst others that of the former successive destruction and renovation of the world.* We learn from Plutarch, that this was the theme of one of the hymns of Orpheus, so celebrated in the fabulous ages of Greece. It was brought by him from the banks of the Nile; and we even find in his verses, as in the Indian systems, a definite period assigned for the duration of each successive world.† The returns of great catastrophes were determined by the period of the Annus Magnus, or great year,—a cycle composed of the revolutions of the sun, moon, and planets, and terminating when these return together to the same sign whence they were supposed at some remote epoch to have set out. The duration of this great cycle was variously estimated. According to Orpheus, it was 120,000 years; according to others, 300,000; and by Cassander it was taken to be 360,000 years.‡

* Prichard's Egypt. Mythol. p. 177.

† Plut. de Defectu Oraculorum, cap. 12. Censorinus de Die Natali. See also Prichard's Egypt. Mythol. p. 182.

‡ Prichard's Egypt. Mythol. p. 182.

We learn particularly from the Timæus of Plato, that the Egyptians believed the world to be subject to occasional conflagrations and deluges, whereby the gods arrested the career of human wickedness, and purified the earth from guilt. After each regeneration, mankind were in a state of virtue and happiness, from which they gradually degenerated again into vice and immorality. From this Egyptian doctrine, the poets derived the fable of the decline from the golden to the iron age. The sect of Stoics adopted most fully the system of catastrophes destined at certain intervals to destroy the world. Those they taught were of two kinds;—the Cataclysm, or destruction by water, which sweeps away the whole human race, and annihilates all the animal and vegetable productions of nature; and the Ecpyrosis, or destruction by fire, which dissolves the globe itself. From the Egyptians also they derived the doctrine of the gradual debasement of man from a state of innocence. Towards the termination of each era, the gods could no longer bear with the wickedness of men, and a shock of the elements or a deluge overwhelmed them; after which calamity, Astrea again descended on the earth to renew the golden age.*

The connection between the doctrine of successive catastrophes and repeated deteriorations in the moral character of the human race is more intimate and natural than might at first be imagined. For, in a rude state of society, all great calamities are regarded by the people as judgments of God on the wickedness of man. Thus, in our own time, the priests persuaded a large part of the population of Chili, and perhaps believed themselves, that the fatal earthquake of 1822 was a sign of the wrath of Heaven for the great political revolution just then consummated in South America. In like manner, in the account given to Solon by the Egyptian priests, of the submersion of the island of Atlantis under the waters of the ocean, after repeated shocks of an earthquake, we find that the event happened when Jupiter had seen the moral depravity of the inhabitants.† Now, when the notion had once gained ground, whether from causes before suggested or not, that the earth had been destroyed by several general catastrophes, it would next be inferred that the human race had been as often destroyed and renovated. And since every extermination was assumed to be penal, it could only be reconciled with divine justice, by the supposition that man, at each successive creation, was regenerated in a state of purity and innocence.

A very large portion of Asia, inhabited by the earliest nations, whose traditions have come down to us, has been always subject to tremendous earthquakes. Of the geographical boundaries of these, and their effects, I shall speak in the proper place. Egypt has, for the most part, been exempt from this scourge, and the Egyptian doctrine of great catastrophes was probably derived in part, as before hinted, from early geological observations, and in part from Eastern nations.

Pythagorean Doctrines.—Pythagoras, who resided for more than

* Prichard's Egypt. Mythol. p. 193.

† Plato's Timæus.

twenty years in Egypt, and, according to Cicero, had visited the East, and conversed with the Persian philosophers, introduced into his own country, on his return, the doctrine of the gradual deterioration of the human race from an original state of virtue and happiness; but if we are to judge of his theory concerning the destruction and renovation of the earth from the sketch given by Ovid, we must concede it to have been far more philosophical than any known version of the cosmogonies of Oriental or Egyptian sects.

Although Pythagoras is introduced by the poet as delivering his doctrine in person, some of the illustrations are derived from natural events which happened after the death of the philosopher. But notwithstanding these anachronisms, we may regard the account as a true picture of the tenets of the Pythagorean school in the Augustan age; and although perhaps partially modified, it must have contained the substance of the original scheme. Thus considered, it is extremely curious and instructive; for we here find a comprehensive summary of almost all the great causes of change now in activity on the globe, and these adduced in confirmation of a principle of a perpetual and gradual revolution inherent in the nature of our terrestrial system. These doctrines, it is true, are not directly applied to the explanation of geological phenomena; or, in other words, no attempt is made to estimate what may have been in past ages, or what may hereafter be, the aggregate amount of change brought about by such never-ending fluctuations. Had this been the case, we might have been called upon to admire so extraordinary an anticipation with no less interest than astronomers, when they endeavor to define by what means the Samian philosopher came to the knowledge of the Copernican system.

Let us now examine the celebrated passages to which we have been adverting:*

"Nothing perishes in this world; but things merely vary and change their form. To be born, means simply that a thing begins to be something different from what it was before; and dying, is ceasing to be the same thing. Yet, although nothing retains long the same image, the sum of the whole remains constant." These general propositions are then confirmed by a series of examples, all derived from natural appearances, except the first, which refers to the golden age giving place to the age of iron. The illustrations are thus consecutively adduced.

1. Solid land has been converted into sea.

2. Sea has been changed into land. Marine shells lie far distant from the deep, and the anchor has been found on the summit of hills.

3. Valleys have been excavated by running water, and floods have washed down hills into the sea.†

* Ovid's Metamor. lib. 15.

† Eluvie mons est deductus in æquor, v. 267. The meaning of this last verse is somewhat obscure; but, taken with the context, may be supposed to allude to the abrading power of floods, torrents, and rivers.

4. Marshes have become dry ground.

5. Dry lands have been changed into stagnant pools.

6. During earthquakes some springs have been closed up, and new ones have broken out. Rivers have deserted their channels, and have been re-born elsewhere, as the Erasinus in Greece, and Mysus in Asia.

7. The waters of some rivers, formerly sweet, have become bitter; as those of the Anigris, in Greece, &c.*

8. Islands have become connected with the mainland by the growth of deltas and new deposits; as in the case of Antissa joined to Lesbos, Pharos to Egypt, &c.

9. Peninsulas have been divided from the main land, and have become islands, as Leucadia; and according to tradition, Sicily, the sea having carried away the isthmus.

10. Land has been submerged by earthquakes; the Grecian cities of Helice and Buris, for example, are to be seen under the sea, with their walls inclined.

11. Plains have been upheaved into hills by the confined air seeking vent; as at Trœzene in the Peloponnesus.

12. The temperature of some springs varies at different periods. The waters of others are inflammable.†

13. There are streams which have a petrifying power, and convert the substances which they touch into marble.

14. Extraordinary medicinal and deleterious effects are produced by the water of different lakes and springs.‡

15. Some rocks and islands, after floating and having been subject to violent movements, have at length become stationary and immovable; as Delos and the Cyanean Isles.§

16. Volcanic vents shift their position; there was a time when Etna was not a burning mountain, and the time will come when it will cease to burn. Whether it be that some caverns become closed up by the movements of the earth, and others opened, or whether the fuel is finally exhausted, &c., &c.

The various causes of change in the inanimate world having been thus enumerated, the doctrine of equivocal generation is next propounded, as illustrating a corresponding perpetual flux in the animate creation.‖

* The impregnation from new mineral springs, caused by earthquakes in volcanic countries, is perhaps here alluded to.

† That is probably an allusion to the escape of inflammable gas, like that in the district of Baku, west of the Caspian; at Pietramala, in the Tuscan Apennines; and several other places.

‡ Many of those described seem fanciful fictions, like the virtue still so commonly attributed to mineral waters.

§ Raspe, in a learned and judicious essay (De Novis Insulis, cap. 19), has made it appear extremely probable that all the traditions of certain islands in the Mediterranean having at some former time frequently shifted their positions, and at length become stationary, originated in the great change produced in their form by earthquakes and submarine eruptions, of which there have been modern examples in the new islands raised in the time of history. When the series of convulsions ended, the island was said to become fixed.

‖ It is not inconsistent with the Hindoo mythology to suppose that Pythagoras might have found in the East not only the system of universal and violent catas-

In the Egyptian and Eastern cosmogonies, and in the Greek version of them, no very definite meaning can, in general, be attached to the term "destruction of the world;" for sometimes it would seem almost to imply the annihilation of our planetary system, and at others a mere revolution of the surface of the earth.

Opinions of Aristotle.—From the works now extant of Aristotle, and from the system of Pythagoras, as above exposed, we might certainly infer that these philosophers considered the agents of change now operating in nature, as capable of bringing about in the lapse of ages a complete revolution; and the Stagyrite even considers occasional catastrophes, happening at distant intervals of time, as part of the regular and ordinary course of nature. The deluge of Deucalion, he says, affected Greece only, and principally the part called Hellas, and it arose from great inundations of rivers, during a rainy winter. But such extraordinary winters, he says, though after a certain period they return, do not always revisit the same places.*

Censorinus quotes it as Aristotle's opinion that there were general inundations of the globe, and that they alternated with conflagrations; and that the flood constituted the winter of the great year, or astronomical cycle, while the conflagration, or destruction by fire, is the summer, or period of greatest heat.† If this passage, as Lipsius supposes, be an amplification, by Censorinus, of what is written in "the Meteorics," it is a gross misrepresentation of the doctrine of the Stagyrite, for the general bearing of his reasoning in that treatise tends clearly in an opposite direction. He refers to many examples of changes now constantly going on, and insists emphatically on the great results which they must produce in the lapse of ages. He instances particular cases of lakes that had dried up, and deserts that had at length become watered by rivers and fertilized. He points to the growth of the Nilotic Delta since the time of Homer, to the shallowing of the Palus Mæotis within sixty years from his own time; and although, in the same chapter he says nothing of earthquakes, yet in others of the same treatise he shows himself not unacquainted with their effects.‡ He alludes, for example, to the upheaving of one of the Eolian islands previous to a volcanic eruption. "The changes of the earth," he says, "are so slow in comparison to the duration of our lives, that they are overlooked (λανθανει): and the migrations of people after great catastrophes, and their removal to other regions, cause the event to be forgotten."§

trophes and periods of repose in endless succession, but also that of periodical revolutions, effected by the continued agency of ordinary causes. For Brahma, Vishnu, and Siva, the first, second, and third persons of the Hindoo triad, severally represented the Creative, the Preserving, and the Destroying powers of the Deity. The coexistence of these three attributes, all in simultaneous operation, might well accord with the notion of perpetual but partial alterations finally bringing about a complete change. But the fiction expressed in the verses before quoted from Menû of eternal vicissitudes in the vigils and slumbers of Brahma seems accommodated to the system of great general catastrophes followed by new creations and periods of repose.

* Meteor. lib. i. cap. 12.

† De Die Nat.

‡ Lib. ii. cap. 14, 15, and 16.

§ Lib. ii. cap. 14, 15, and 16.

When we consider the acquaintance displayed by Aristotle, in his various works, with the destroying and renovating powers of Nature, the introductory and concluding passages of the twelfth chapter of his "Meteorics" are certainly very remarkable. In the first sentence he says, "The distribution of land and sea in particular regions does not endure throughout all time, but it becomes sea in those parts where it was land, and again it becomes land where it was sea: and there is reason for thinking that these changes take place according to a certain system, and within a certain period." The concluding observation is as follows:—"As time never fails, and the universe is eternal, neither the Tànais, nor the Nile, can have flowed forever. The places where they rise were once dry, and there is a limit to their operations; but there is none to time. So also of all other rivers; they spring up, and they perish; and the sea also continually deserts some lands and invades others. The same tracts, therefore, of the earth are not, some always sea, and others always continents, but every thing changes in the course of time."

It seems, then, that the Greeks had not only derived from preceding nations, but had also, in some slight degree, deduced from their own observations, the theory of periodical revolutions in the inorganic world: there is, however, no ground for imagining that they contemplated former changes in the races of animals and plants. Even the fact that marine remains were inclosed in solid rocks, although observed by some, and even made the groundwork of geological speculation, never stimulated the industry or guided the inquiries of naturalists. It is not impossible that the theory of equivocal generation might have engendered some indifference on this subject, and that a belief in the spontaneous production of living beings from the earth or corrupt matter, might have caused the organic world to appear so unstable and fluctuating, that phenomena indicative of former changes would not awaken intense curiosity. The Egyptians, it is true, had taught, and the Stoics had repeated, that the earth had once given birth to some monstrous animals, which existed no longer; but the prevailing opinion seems to have been, that after each great catastrophe the same species of animals were created over again. This tenet is implied in a passage of Seneca, where, speaking of a future deluge, he says, "Every animal shall be generated anew, and man free from guilt shall be given to the earth."*

An old Arabian version of the doctrine of the successive revolutions of the globe, translated by Abraham Ecchellensis,† seems to form a singular exception to the general rule, for here we find the idea of different genera and species having been created. The Gerbanites, a sect

* Omne ex integro animal generabitur, dabiturque terris homo inscius scelerum.—Quæst. Nat. iii. c. 29.

† This author was Regius Professor of Syriac and Arabic at Paris, where, in 1685, he published a Latin translation of many Arabian MSS. on different departments of philosophy. This work has always been considered of high authority.

of astronomers who flourished some centuries before the Christian era, taught as follows:—"That after every period of thirty-six thousand four hundred and twenty-five years, there were produced a pair of *every* species of animal, both male and female, from whom animals might be propagated and inhabit this lower world. But when a circulation of the heavenly orbs was completed, which is finished in that space of years, *other genera and species* of animals are propagated, as also of plants and other things, and the first order is destroyed, and so it goes on forever and ever."*

Theory of Strabo.—As we learn much of the tenets of the Egyptian and Oriental schools in the writings of the Greeks, so, many speculations of the early Greek authors are made known to us in the works of the Augustan and later ages. Strabo, in particular, enters largely, in the second book of his Geography, into the opinions of Eratosthenes and other Greeks on one of the most difficult problems in geology, viz., by what causes marine shells came to be plentifully buried in the earth at such great elevations and distances from the sea.

He notices, amongst others, the explanation of Xanthus the Lydian, who said that the seas had once been more extensive, and that they had afterwards been partially dried up, as in his own time many lakes, rivers, and wells in Asia had failed during a season of drought. Treating this conjecture with merited disregard, Strabo passes on to the hypothesis of Strato, the natural philosopher, who had observed that the quantity of mud brought down by rivers into the Euxine was so great, that its bed must be gradually raised, while the rivers still continue to pour in an undiminished quantity of water. He, therefore, conceived that, originally, when the Euxine was an inland sea, its level had by this means become so much elevated that it burst its barrier near Byzantium, and formed a communication with the Propontis; and this partial drainage, he supposed, had already converted the left side into marshy ground, and thus, at last, the whole would be choked up with soil. So, it was argued, the Mediterranean had once opened a passage for itself by the Columns of Hercules into the Atlantic; and perhaps the abundance of sea-shells in Africa, near the Temple of Jupiter Ammon, might also be the deposit of some former inland sea, which had at length forced a passage and escaped.

* Gerbanitæ docebant singulos triginta sex mille annos quadringentos, viginti quinque bina ex singulis animalium speciebus produci, marem scilicet ac feminam ex quibus animalia propagantur, huncque inferiorem incolunt orbem. Absolutâ autem cœlestium orbium circulatione, quæ illo annorum conficitur spatio, iterum alia producuntur animalium genera et species, quemadmodum et plantarum aliarumque rerum, et primus destruitur ordo, sicque in infinitum producitur.—Histor. Orient. Suppl. per Abrahamum Ecchellensem, Syrum Maronitam, cap. 7. et 8. ad calcem Chronici Orientali. Parisiis, e Typ. Regia. 1685, fol.

I have given the punctuation as in the Paris edition, there being no comma after quinque; but, at the suggestion of M. de Schlegel, I have referred the number twenty-five to the period of years, and not to the number of pairs of each species created at one time, as I had done in the two first editions. Fortis inferred that twenty-five new *species* only were created at a time; a construction which the passage will not admit. Mém. sur l'Hist. Nat. de l'Italie, vol. i. p. 202.

But Strabo rejects this theory, as insufficient to account for all the phenomena, and he proposes one of his own, the profoundness of which modern geologists are only beginning to appreciate. "It is not," he says, "because the lands covered by seas were originally at different altitudes, that the waters have risen, or subsided, or receded from some parts and inundated others. But the reason is, that the same land is sometimes raised up and sometimes depressed, and the sea also is simultaneously raised and depressed, so that it either overflows or returns into its own place again. We must, therefore, ascribe the cause to the ground, either to that ground which is under the sea, or to that which becomes flooded by it, but rather to that which lies beneath the sea, for this is more movable and, on account of its humidity, can be altered with greater celerity.* *It is proper,*" he observes in continuation, "*to derive our explanations from things which are obvious, and in some measure of daily occurrence, such as deluges, earthquakes, and volcanic eruptions,*† *and sudden swellings of the land beneath the sea*; for the last raise up the sea also; and when the same lands subside again, they occasion the sea to be let down. And it is not merely the small, but the large islands also, and not merely the islands, but the continents which can be lifted up together with the sea; and both large and small tracts may subside, for habitations and cities, like Bure, Bizoua, and many others, have been engulphed by earthquakes."

In another place, this learned geographer, in alluding to the tradition that Sicily had been separated by a convulsion from Italy, remarks, that at present the land near the sea in those parts was rarely shaken by earthquakes, since there were now open orifices whereby fire and ignited maters, and waters escape; but formerly, when the volcanoes of Etna, the Lipari Islands, Ischia, and others, were closed up, the imprisoned fire and wind might have produced far more vehement movements.‡ The doctrine, therefore, that volcanoes are safety-valves, and that the subterranean convulsions are probably most violent when first the volcanic energy shifts itself to a new quarter, is not modern.

We learn from a passage in Strabo,§ that it was a dogma of the Gaulish Druids that the universe was immortal, but destined to survive catastrophes both of fire and water. That this doctrine was communicated to them from the East, with much of their learning, cannot be doubted. Cæsar, it will be remembered, says that they made use of Greek letters in arithmetical computations.‖

* "Quod enim hoc attollitur aut subsidit, et vel inundat quædam loca, vel ab iis recedit, ejus rei causa non est, quod alia aliis sola humiliora sint aut altiora; sed quod idem solum modò attollitur modò deprimitur, simulque etiam modò attollitur modò deprimitur, mare: itaque vel exundat vel in suum redit locum."

Posteà, p. 88. "Restat, ut causam adscribamus solo, sive quod mari subest sive quod inundatur; potiùs tamen ei quod mari subest. Hoc enim multò est mobilius, et quod ob humiditatem celeriùs multari possit."—Strabo, Geog. Edit. Almelov. Amst. 1707, lib. 1.

† *Volcanic eruptions*, eruptiones flatuum, in the Latin translations, and in the original Greek, αναφυσηματα, gaseous eruptions? or *inflations* of land?—Ibid. p. 93.

‡ Strabo, lib. vi. p. 396.

§ Book iv.

‖ L. vi. ch. xiii.

Pliny.—This philosopher had no theoretical opinions of his own concerning changes of the earth's surface; and in this department, as in others, he restricted himself to the task of a compiler, without reasoning on the facts stated by him, or attempting to digest them into regular order. But his enumeration of the new islands which had been formed in the Mediterranean, and of other convulsions, shows that the ancients had not been inattentive observers of the changes which had taken place within the memory of man.

Such, then, appear to have been the opinions entertained before the Christian era, concerning the past revolutions of our globe. Although no particular investigations had been made for the express purpose of interpreting the monuments of ancient changes, they were too obvious to be entirely disregarded; and the observation of the present course of nature presented too many proofs of alterations continually in progress on the earth to allow philosophers to believe that nature was in a state of rest, or that the surface had remained, and would continue to remain unaltered. But they had never compared attentively the results of the destroying and reproductive operations of modern times with those of remote eras, nor had they ever entertained so much as a conjecture concerning the comparative antiquity of the human race, or of living species of animals and plants, with those belonging to former conditions of the organic world. They had studied the movements and positions of the heavenly bodies with laborious industry, and made some progress in investigating the animal, vegetable, and mineral kingdoms; but the ancient history of the globe was to them a sealed book, and, although written in characters of the most striking and imposing kind, they were unconscious even of its existence.

CHAPTER III.

HISTORY OF THE PROGRESS OF GEOLOGY—*continued.*

Arabian writers of the tenth century—Avicenna—Omar—Cosmogony of the Koran—Kazwini—Early Italian writers—Leonardo da Vinci—Fracastoro—Controversy as to the real nature of fossils—Attributed to the Mosaic deluge—Palissy—Steno—Scilla—Quirini—Boyle—Lister—Leibnitz—Hooke's Theory of Elevation by Earthquakes—Of lost species of animals—Ray—Physico-theological writers—Woodward's Diluvial Theory—Burnet—Whiston—Vallisneri—Lazzaro Moro—Generelli—Buffon—His theory condemned by the Sorbonne as unorthodox—His declaration—Targioni—Arduino—Michell—Catcott—Raspe Fuchsel—Fortis—Testa—Whitehurst—Pallas—Saussure.

Arabian writers.—AFTER the decline of the Roman empire, the cultivation of physical science was first revived with some success by the Saracens, about the middle of the eighth century of our era. The works of the most eminent classic writers were purchased at great expense from the Christians, and translated into Arabic; and Al Mamûn, son of the famous Harûn-al-Rashid, the contemporary of Charlemagne, received with marks of distinction, at his court at Bagdad, astronomers and men of learning from different countries. This caliph, and some of his successors, encountered much opposition and jealousy from the doctors of the Mahometan law, who wished the Moslems to confine their studies to the Koran, dreading the effects of the diffusion of a taste for the physical sciences.*

Avicenna.—Almost all the works of the early Arabian writers are lost. Amongst those of the tenth century, of which fragments are now extant, is a short treatise, "On the Formation and Classification of Minerals," by Avicenna, a physician, in whose arrangement there is considerable merit. The second chapter, "On the Cause of Mountains," is remarkable; for mountains, he says, are formed, some by essential, others by accidental causes. In illustration of the essential, he instances "a violent earthquake, by which land is elevated, and becomes a mountain;" of the accidental, the principal, he says, is excavation by water, whereby cavities are produced, and adjoining lands made to stand out and form eminences.†

Omar—Cosmogony of the Koran.—In the same century, also, Omar, surnamed "El Aalem," or "The Learned," wrote a work on "The Retreat of the Sea." It appears that on comparing the charts of his own time with those made by the Indian and Persian astronomers two thousand years before, he had satisfied himself that important changes had taken place since the times of history in the form of the coasts of Asia,

* Mod. Univ. Hist. vol. ii. chap. iv. section iii.

† Montes quandóque fiunt ex causa essentiali, quandóque ex causa accidentali. Ex essentiali causa, ut ex vehementi motu terræ elevatur terra, et fit mons. Accidentali, &c.—De Congelatione Lapidum, ed. Gedani, 1682.

and that the extension of the sea had been greater at some former periods. He was confirmed in this opinion by the numerous salt springs and marshes in the interior of Asia,—a phenomenon from which Pallas, in more recent times, has drawn the same inference.

Von Hoff has suggested, with great probability, that the changes in the level of the Caspian (some of which there is reason to believe have happened within the historical era), and the geological appearances in that district, indicating the desertion by that sea of its ancient bed, had probably led Omar to his theory of a general subsidence. But whatever may have been the proofs relied on, his system was declared contradictory to certain passages in the Koran, and he was called upon publicly to recant his errors; to avoid which persecution he went into voluntary banishment from Samarkand.*

The cosmological opinions expressed in the Koran are few, and merely introduced incidentally: so that it is not easy to understand how they could have interfered so seriously with free discussion on the former changes of the globe. The Prophet declares that the earth was created in two days, and the mountains were then placed on it; and during these, and two additional days, the inhabitants of the earth were formed; and in two more the seven heavens.† There is no more detail of circumstances; and the deluge, which is also mentioned, is discussed with equal brevity. The waters are represented to have poured out of an oven; a strange fable, said to be borrowed from the Persian Magi, who represented them as issuing from the oven of an old woman.‡ All men were drowned, save Noah and his family; and then God said, "O earth, swallow up thy waters; and thou, O heaven, withhold thy rain;" and immediately the waters abated.§

We may suppose Omar to have represented the desertion of the land by the sea to have been gradual, and that his hypothesis required a greater lapse of ages than was consistent with Moslem orthodoxy; for it is to be inferred from the Koran, that man and this planet were created at the same time; and although Mahomet did not limit expressly the antiquity of the human race, yet he gave an implied sanction to the Mosaic chronology, by the veneration expressed by him for the Hebrew Patriarchs.‖

* Von Hoff, Geschichte der Veränderungen der Erdoberfläche, vol. i. p. 406, who cites Delisle, bey Hismann Welt-und Völkergeschichte. Alte Geschichte 1ter theil, s. 234.—The Arabian persecutions for heretical dogmas in theology were often very sanguinary. In the same ages wherein learning was most in esteem, the Mahometans were divided into two sects, one of whom maintained that the Koran was increate, and had subsisted in the very essence of God from all eternity; and the other, the Motazalites, who, admitting that the Koran was instituted by God, conceived it to have been first made when revealed to the Prophet at Mecca, and accused their opponents of believing in two eternal beings. The opinions of each of these sects were taken up by diffent caliphs in succession, and the followers of each sometimes submitted to be beheaded, or flogged till at the point of death, rather than renounce their creed.—Mod. Univ. Hist. vol. ii. ch. iv.

† Koran, chap. xli.

‡ Sale's Koran, chap. xi. see note. § Ibid.

‖ Kossa, appointed master to the Caliph Al Mamûd, was author of a book en-

A manuscript work, entitled the "Wonders of Nature," is preserved in the Royal Library at Paris, by an Arabian writer, Mohammed Kazwini, who flourished in the seventh century of the Hegira, or at the close of the thirteenth century of our era.* Besides several curious remarks on aerolites, earthquakes, and the successive changes of position which the land and sea have undergone, we meet with the following beautiful passage which is given as the narrative of Kidhz, an allegorical personage:—"I passed one day by a very ancient and wonderfully populous city, and asked one of its inhabitants how long it had been founded. 'It is indeed a mighty city,' replied he; 'we know not how long it has existed, and our ancestors were on this subject as ignorant as ourselves.' Five centuries afterwards, as I passed by the same place, I could not perceive the slightest vestige of the city. I demanded of a peasant, who was gathering herbs upon its former site, how long it had been destroyed. 'In sooth a strange question!' replied he. 'The ground here has never been different from what you now behold it.'—'Was there not of old,' said I, 'a splendid city here?' —'Never,' answered he, 'so far as we have seen, and never did our fathers speak to us of any such.' On my return there 500 years afterwards, *I found the sea in the same place,* and on its shores were a party of fishermen, of whom I inquired how long the land had been covered by the waters? 'Is this a question,' said they, 'for a man like you? this spot has always been what it is now.' I again returned, 500 years afterwards, and the sea had disappeared; I inquired of a man who stood alone upon the spot, how long ago this change had taken place, and he gave me the same answer as I had received before. Lastly, on coming back again after an equal lapse of time, I found there a flourishing city, more populous and more rich in beautiful buildings, than the city I had seen the first time, and when I would fain have informed myself concerning its origin, the inhabitants answered me, 'Its rise is lost in remote antiquity: we are ignorant how long it has existed, and our fathers were on this subject as ignorant as ourselves.'"

Early Italian writers.—It was not till the earlier part of the sixteenth century that geological phenomena began to attract the attention of the Christian nations. At that period a very animated controversy sprang up in Italy, concerning the true nature and origin of marine shells, and other organized fossils, found abundantly in the strata of the peninsula. The celebrated painter Leonardo da Vinci, who in his youth had planned and executed some navigable canals in the north of Italy, was one of the first who applied sound reasoning to these subjects. The mud of rivers, he said, had covered and penetrated into the interior of fossil shells at a time when these were still at the bottom of the sea near the coast. "They tell us that these shells were formed in the

titled "The history of the Patriarchs and Prophets, *from the Creation of the World.*"—Mod. Univ. Hist. vol. ii. ch. iv.

* Translated by MM. Chezy and De Sacy, and cited by M. Elie de Beaumont, Ann. des Sci. Nat. 1832.

hills by the influence of the stars; but I ask where in the hills are the stars now forming shells of distinct ages and species? and how can the stars explain the origin of gravel, occurring at different heights and composed of pebbles rounded as if by the motion of running water; or in what manner can such a cause account for the petrifaction in the same places of various leaves, sea-weeds, and marine-crabs?"*

The excavations made in 1517, for repairing the city of Verona, brought to light a multitude of curious petrifactions, and furnished matter for speculation to different authors, and among the rest to Fracastoro,† who declared his opinion, that fossil shells had all belonged to living animals, which had formerly lived and multiplied where there exuviæ are now found. He exposed the absurdity of having recourse to a certain "plastic force," which it was said had power to fashion stones into organic forms; and with no less cogent arguments, demonstrated the futility of attributing the situation of the shells in question to the Mosaic deluge, a theory obstinately defended by some. That inundation, he observed, was too transient; it consisted principally of fluviatile waters; and if it had transported shells to great distances, must have strewed them over the surface, not buried them at vast depths in the interior of mountains. His clear exposition of the evidence would have terminated the discussion forever, if the passions of mankind had not been enlisted in the dispute; and even though doubts should for a time have remained in some minds, they would speedily have been removed by the fresh information obtained almost immediately afterwards, respecting the structure of fossil remains, and of their living analogues.

But the clear and philosophical views of Fracastoro were disregarded, and the talent and argumentative powers of the learned were doomed for three centuries to be wasted in the discussion of these two simple and preliminary questions: first, whether fossil remains had ever belonged to living creatures; and, secondly, whether, if this be admitted, all the phenomena could not be explained by the deluge of Noah. It had been the general belief of the Christian world down to the period now under consideration, that the origin of this planet was not more remote than a few thousand years; and that since the creation the deluge was the only great catastrophe by which considerable change had been wrought on the earth's surface. On the other hand, the opinion was scarcely less general, that the final dissolution of our system was an event to be looked for at no distant period. The era, it is true, of the expected millennium had passed away; and for five hundred years after the fatal hour when the annihilation of the planet

* See Venturi's extracts from Da Vinci's MMS. now in Library of Institute of France. They are not mentioned by Brocchi, and my attention was first called to them by Mr. Hallam. L. da Vinci died A. D. 1519.

† Museum Calceol.—See Brocchi's Discourse on the Progress of the Study of Fossil Conchology in Italy, where some of the following notices on Italian writers will be found more at large.

had been looked for, the monks remained in undisturbed enjoyment of rich grants of land bequeathed to them by pious donors, who, in the preamble of deeds beginning "appropinquante mundi termino"—— "appropinquante magno judicii die," left lasting monuments of the popular delusion.*

But although in the sixteenth century it had become necessary to interpret certain prophecies respecting the millennium more liberally, and to assign a more distant date to the future conflagration of the world, we find, in the speculations of the early geologists, perpetual allusion to such an approaching catastrophe; while in all that regarded the antiquity of the earth, no modification whatever of the opinions of the dark ages had been effected. Considerable alarm was at first excited when the attempt was made to invalidate, by physical proofs, an article of faith so generally received; but there was sufficient spirit of toleration and candor amongst the Italian ecclesiastics, to allow the subject to be canvassed with much freedom. They even entered warmly into the controversy themselves, often favoring different sides of the question; and however much we may deplore the loss of time and labor devoted to the defence of untenable positions, it must be conceded that they displayed far less polemic bitterness than certain writers who followed them "beyond the Alps," two centuries and a half later.

CONTROVERSY AS TO THE REAL NATURE OF FOSSIL ORGANIC REMAINS.

Mattioli—Falloppio.—The system of scholastic disputations, encouraged in the universities of the middle ages, had unfortunately trained men to habits of indefinite argumentation; and they often preferred absurd and extravagant propositions, because greater skill was required to maintain them; the end and object of these intellectual combats being victory, and not truth. No theory could be so far-fetched or fantastical as not to attract some followers, provided it fell in with popular notions; and as cosmogonists were not at all restricted, in building their systems, to the agency of known causes, the opponents of Fracastoro met his arguments by feigning imaginary causes, which differed from each other rather in name than in substance. Andrea Mattioli, for instance, an eminent botanist, the illustrator of Dioscorides, embraced the notion of Agricola, a skilful German miner, that a certain "materia pinguis," or "fatty matter," set into fermentation by heat, gave birth to fossil organic shapes. Yet Mattioli had come to the conclusion, from his own observations, that porous bodies, such as bones and shells, might be converted into stone, as being permeable to what he termed the "lapidifying juice." In like manner, Falloppio of Padua conceived that petrified shells were generated by fermentation in the spots where they are found, or that they had in some cases acquired

* In Sicily, in particular, the title-deeds of many valuable grants of land to the monasteries are headed by such preambles, composed by the testators about the period when the good King Roger was expelling the Saracens from that island.

their form from "the tumultuous movements of terrestrial exhalations." Although celebrated as a professor of anatomy, he taught that certain tusks of elephants, dug up in his time in Apulia, were mere earthy concretions; and, consistently with these principles, he even went so far as to consider it probable, that the vases of Monte Testaceo at Rome were natural impressions stamped in the soil.* In the same spirit, Mercati, who published, in 1574, faithful figures of the fossil shells preserved by Pope Sixtus V. in the Museum of the Vatican, expressed an opinion that they were mere stones, which had assumed their peculiar configuration from the influence of the heavenly bodies; and Olivi of Cremona, who described the fossil remains of a rich museum at Verona, was satisfied with considering them as mere "sports of nature."

Some of the fanciful notions of those times were deemed less unreasonable, as being somewhat in harmony with the Aristotelian theory of spontaneous generation, then taught in all the schools.† For men who had been taught in early youth, that a large proportion of living animals and plants was formed from the fortuitous concourse of atoms, or had sprung from the corruption of organic matter, might easily persuade themselves that organic shapes, often imperfectly preserved in the interior of solid rocks, owed their existence to causes equally obscure and mysterious.

Cardano, 1552.—But there were not wanting some who, during the progress of this century, expressed more sound and sober opinions. The title of a work of Cardano's, published in 1552, "De Subtilitate" (corresponding to what would now be called Transcendental Philosophy), would lead us to expect, in the chapter on minerals, many far-fetched theories characteristic of that age; but when treating of petrified shells, he decided that they clearly indicated the former sojourn of the sea upon the mountains.‡

Cesalpino—Majoli, 1597.—Cesalpino, a celebrated botanist, conceived that fossil shells had been left on the land by the retiring sea, and had concreted into stone during the consolidation of the soil;§ and in the following year (1597), Simeone Majoli‖ went still farther; and, coinciding for the most part with the views of Cesalpino, suggested that the shells and submarine matter of the Veronese, and other districts, might have been cast up upon the land by volcanic explosions, like those which gave rise, in 1538, to Monte Nuovo, near Puzzuoli. This hint seems to have been the first imperfect attempt to connect the position of fossil shells with the agency of volcanoes, a system afterwards more fully developed by Hooke, Lazzaro Moro, Hutton, and other writers.

Two years afterwards, Imperati advocated the animal origin of fossilized shells, yet admitted that stones could vegetate by force of "an internal principle;" and, as evidence of this, he referred to the teeth of fish and spines of echini found petrified.¶

* De Fossilib. pp. 109, 176. † Aristotle, On Animals, chaps. 1, 15.
‡ Brocchi, Con. Fos. Subap. Disc. sui Progressi. vol. i. p. 57.
§ De Metallicis. ‖ Dies Caniculares. ¶ Storia Naturale.

Palissy, 1580.—Palissy, a French writer on "The Origin of Springs from Rain-water,"and of other scientific works, undertook, in 1580, to combat the notions of many of his contemporaries in Italy, that petrified shells had all been deposited by the universal deluge. "He was the first," said Fontenelle, when, in the French Academy, he pronounced his eulogy, nearly a century and a half later, "who dared assert," in Paris, that fossil remains of testacea and fish had once belonged to marine animals.

Fabio Colonna.—To enumerate the multitude of Italian writers, who advanced various hypotheses, all equally fantastical, in the early part of the seventeenth century, would be unprofitably tedious; but Fabio Colonna deserves to be distinguished; for, although he gave way to the dogma, that all fossil remains were to be referred to the deluge of Noah, he resisted the absurd theory of Stelluti, who taught that fossil wood and ammonites were mere clay, altered into such forms by sulphureous waters and subterranean heat; and he pointed out the different states of shells buried in the strata, distinguishing between, first, the mere mould or impression; second, the cast or nucleus; and, thirdly, the remains of the shell itself. He had also the merit of being the first to point out that some of the fossils had belonged to marine and some to terrestrial testacea.*

Steno, 1669.—But the most remarkable work of that period was published by Steno, a Dane, once professor of anatomy at Padua, and who afterwards resided many years at the court of the Grand Duke of Tuscany. His treatise bears the quaint title of "De Solido intra Solidum naturaltier contento (1669)," by which the author intended to express, "On Gems, Crystals, and organic Petrifactions inclosed within solid Rocks." This work attests the priority of the Italian school in geological research; exemplifying at the same time the powerful obstacles opposed, in that age, to the general reception of enlarged views in the science. It was still a favorite dogma, that the fossil remains of shells and marine creatures were not of animal origin; an opinion adhered to by many from their extreme reluctance to believe, that the earth could have been inhabited by living beings before a great part of the existing mountains were formed. In reference to this controversy, Steno had dissected a shark recently taken from the Mediterranean, and had demonstrated that its teeth and bones were identical with many fossils found in Tuscany. He had also compared the shells discovered in the Italian strata with living species, pointed out their resemblance, and traced the various gradations from shells merely calcined, or which had only lost their animal gluten, to those petrifactions in which there was a perfect substitution of stony matter. In his division of mineral masses, he insisted on the secondary origin of those deposits in which the spoils of animals or fragments of older rocks were inclosed. He distinguished between marine formations and those of a fluviatile char-

* Osserv. sugli Animali aquat. e terrest. 1626.

acter, the last containing reeds, grasses, or the trunks and branches of trees. He argued in favor of the original horizontality of sedimentary deposits, attributing their present inclined and vertical position sometimes to the escape of subterranean vapors heaving the crust of the earth from below upwards, and sometimes to the falling in of masses overlying subterranean cavities.

He declared that he had obtained proof that Tuscany must successively have acquired six distinct configurations, having been twice covered by water, twice laid dry with a level, and twice with an irregular and uneven surface.* He displayed great anxiety to reconcile his new views with Scripture, for which purpose he pointed to certain rocks as having been formed before the existence of animals and plants: selecting unfortunately as examples certain formations of limestone and sandstone in his own country, now known to contain, though sparingly, the remains of animals and plants,—strata which do not even rank as the oldest part of our secondary series. Steno suggested that Moses, when speaking of the loftiest mountains as having been covered by the deluge, meant merely the loftiest of the hills then existing, which may not have been very high. The diluvian waters, he supposed, may have issued from the interior of the earth into which they had retired, when in the beginning the land was separated from the sea. These, and other hypotheses on the same subject, are not calculated to enhance the value of the treatise, and could scarcely fail to detract from the authority of those opinions which were sound and legitimate deductions from fact and observation. They have served, nevertheless, as the germs of many popular theories of later times, and in an expanded form have been put forth as original inventions by some of our contemporaries.

Scilla, 1670.—Scilla, a Sicilian painter, published, in 1670, a treatise, in Latin, on the fossils of Calabria, illustrated by good engravings. This work proves the continued ascendancy of dogmas often refuted; for we find the wit and eloquence of the author chiefly directed against the obstinate incredulity of naturalists as to the organic nature of fossil shells.† Like many eminent naturalists of his day, Scilla gave way to the popular persuasion, that all fossil shells were the effects and proofs of the Mosaic deluge. It may be doubted whether he was perfectly sincere, and some of his contemporaries who took the same course were certainly not so. But so eager were they to root out what they justly considered an absurd prejudice respecting the nature of organized fossils, that they seem to have been ready to make any concessions, in order to establish this preliminary point. Such a compromising policy was short-sighted, since it was to little purpose that the nature of the documents should at

* Sex itaque distinctas Etruriæ facies agnoscimus, dum bis fluida, bis plana, et sicca, bis aspera fuerit, &c.

† Scilla quotes the remark of Cicero on the story that a stone in Chios had been cleft open, and presented the head of Paniscus in relief:—"I believe," said the orator, "that the figure bore some resemblance to Paniscus, but not such that you would have deemed it sculptured by Scopas; for chance never perfectly imitates the truth."

length be correctly understood, if men were to be prevented from deducing fair conclusions from them.

Diluvial Theory.—The theologians who now entered the field in Italy, Germany, France, and England, were innumerable; and henceforward, they who refused to subscribe to the position, that all marine organic remains were proofs of the Mosaic deluge, were exposed to the imputation of disbelieving the whole of the sacred writings. Scarcely any step had been made in approximating to sound theories since the time of Fracastoro, more than a hundred years having been lost, in writing down the dogma that organized fossils were mere sports of nature. An additional period of a century and a half was now destined to be consumed in exploding the hypothesis, that organized fossils had all been buried in the solid strata by Noah's flood. Never did a theoretical fallacy, in any branch of science, interfere more seriously with accurate observation and the systematic classification of facts. In recent times, we may attribute our rapid progress chiefly to the careful determination of the order of succession in mineral masses, by means of their different organic contents, and their regular superposition. But the old diluvialists were induced by their system to confound all the groups of strata together instead of discriminating,—to refer all appearauces to one cause and to one brief period, not to a variety of causes acting throughout a long succession of epochs. They saw the phenomena only as they desired to see them, sometimes misrepresenting facts, and at other times deducing false conclusions from correct data. Under the influence of such prejudices, three centuries were of as little avail as a few years in our own times, when we are no longer required to propel the vessel against the force of an adverse current.

It may be well, therefore, to forewarn the reader, that in tracing the history of geology from the close of the seventeenth to the end of the eighteenth century, he must expect to be occupied with accounts of the retardation, as well as of the advance, of the science. It will be necessary to point out the frequent revival of exploded errors, and the relapse from sound to the most absurd opinions; and to dwell on futile reasoning and visionary hypothesis, because some of the most extravagant systems were invented or controverted by men of acknowledged talent. In short, a sketch of the progress of geology is the history of a constant and violent struggle of new opinions against doctrines sanctioned by the implicit faith of many generations, and supposed to rest on scriptural authority. The inquiry, therefore, although highly interesting to one who studies the philosophy of the human mind, is too often barren of instruction to him who searches for truths in physical science.

Quirini, 1676.—Quirini, in 1676,* contended, in opposition to Scilla, that the diluvian waters could not have conveyed heavy bodies to the summit of mountains, since the agitation of the sea never (as Boyle had

* De Testaceis fossilibus Mus. Septaliani.

demonstrated) extended to great depths;* and still less could the testacea, as some pretended, have lived in these diluvian waters; for "the duration of the flood was brief, and *the heavy rains must have destroyed the saltness of the sea!*" He was the first writer who ventured to maintain that the universality of the Mosaic cataclysm ought not to be insisted upon. As to the nature of petrified shells, he conceived that as earthy particles united in the sea to form the shells of mollusca, the same crystallizing process might be effected on the land; and that, in the latter case, the germs of the animals might have been disseminated through the substance of the rocks, and afterwards developed by virtue of humidity. Visionary as was this doctrine, it gained many proselytes even amongst the more sober reasoners of Italy and Germany; for it conceded that the position of fossil bodies could not be accounted for by the diluvial theory.

Plot—Lister, 1678.—In the mean time, the doctrine that fossil shells had never belonged to real animals maintained its ground in England, where the agitation of the question began at a much later period. Dr. Plot, in his "Natural History of Oxfordshire" (1677), attributed to a "plastic virtue latent in the earth" the origin of fossil shells and fishes; and Lister, to his accurate account of British shells, in 1678, added the fossil species, under the appellation of *turbinated and bivalve stones.* "Either," said he, "these were terriginous, or, if otherwise, the animals they so exactly represent *have become extinct.*" This writer appears to have been the first who was aware of the continuity over large districts of the principal groups of strata in the British series, and who proposed the construction of regular geological maps.†

Leibnitz, 1680.—The great mathematician Leibnitz published his "Protogœa" in 1680. He imagined this planet to have been originally a burning luminous mass, which ever since its creation has been undergoing refrigeration. When the outer crust had cooled down sufficiently to allow the vapors to be condensed, they fell, and formed a universal ocean, covering the loftiest mountains, and investing the whole globe. The crust, as it consolidated from a state of fusion, assumed a vesicular and cavernous structure; and being rent in some places, allowed the water to rush into the subterranean hollows, whereby the level of the primeval ocean was lowered. The breaking in of these vast caverns is supposed to have given rise to the dislocated and deranged position of the strata "which Steno had described," and the same disruptions com-

* The opinions of Boyle, alluded to by Quirini, were published a few years before, in a short article entitled "On the Bottom of the Sea." From observations collected from the divers of the pearl fishery, Boyle inferred that, when the waves were six or seven feet high above the surface of the water, there were no signs of agitation at the depth of fifteen fathoms; and that even during heavy gales of wind, the motion of the water was exceedingly diminished at the depth of twelve or fifteen feet. He had also learnt from some of his informants, that there were currents running in opposite directions at different depths.—Boyle's Works, vol. iii. p. 110. London, 1744.

† See Conybeare and Phillips, "Outlines of the Geology of England and Wales," p. 12.

municated violent movements to the incumbent waters, whence great inundations ensued. The waters, after they had been thus agitated, deposited their sedimentary matter during intervals of quiescence, and hence the various stony and earthy strata. "We may recognize, therefore," says Leibnitz, "a double origin of primitive masses, the one by refrigeration from igneous fusion, the other by concretion from aqueous solution."* By the repetition of similar causes (the disruption of the crust and consequent floods), alternations of new strata were produced, until at length these causes were reduced to a condition of quiescent equilibrium, and a more permanent state of things was established.†

Hooke, 1688.—The "Posthumous Works of Robert Hooke, M. D.," well known as a great mathematician and natural philosopher, appeared in 1705, containing "A Discourse of Earthquakes," which, we are informed by his editor, was written in 1668, but revised at subsequent periods.‡ Hooke frequently refers to the best Italian and English authors who wrote before his time on geological subjects; but there are no passages in his works implying that he participated in the enlarged views of Steno and Lister, or of his contemporary, Woodward, in regard to the geographical extent of certain groups of strata. His treatise, however, is the most philosophical production of that age, in regard to the causes of former changes in the organic and inorganic kingdoms of nature.

"However trivial a thing," he says, "a rotten shell may appear to some, yet these monuments of nature are more certain tokens of antiquity than coins or medals, since the best of those may be counterfeited or made by art and design, as may also books, manuscripts, and inscriptions, as all the learned are now sufficiently satisfied has often been actually practised," &c.; "and though it must be granted that it is very difficult to read them (the records of nature) and *to raise a chronology out of them*, and to state the intervals of the time wherein such or such catastrophes and mutations have happened, yet it is not impossible."§

Respecting the extinction of species, Hooke was aware that the fossil ammonites, nautili, and many other shells and fossil skeletons found in England, were of different species from any then known; but he doubted whether the species had become extinct, observing that the knowledge of naturalists of all the marine species, especially those inhabiting the deep sea, was very deficient. In some parts of his wri-

* Unde jam duplex origo intelligitur primorum corporum, una, cum ab ignis fusione refrigescerent, altera, cum reconcrescerent ex solutione aquarum.

† Redeunte mox simili causâ strata subinde alia aliis imponerentur, et facies teneri adhuc orbis sæpius novata est. Donec quiescentibus causis, atque æquilibratis, consistentior emergeret rerum status.—For an able analysis of the views of Leibnitz, in his Protogæa, see Mr. Conybeare's Report to the Brit. Assoc. on the Progress of Geological Science, 1832.

‡ Between the year 1688 and his death, in 1703, he read several memoirs to the Royal Society, and delivered lectures on various subjects, relating to fossil remains and the effects of earthquakes.

§ Posth. Works, Lecture, Feb. 29, 1688.

tings, however, he leans to the opinion that species had been lost; and in speculating on this subject, he even suggests that there might be some connection between the disappearance of certain kinds of animals and plants, and the changes wrought by earthquakes in former ages. Some species, he observes, with great sagacity, are "*peculiar to certain places*, and not to be found elsewhere. If, then, such a place had been swallowed up, it is not improbable but that those animate beings may have been destroyed with it; and this may be true both of aerial and aquatic animals; for those animated bodies, whether vegetables or animals, which were naturally nourished or refreshed by the air, would be destroyed by the water," &c.* Turtles, he adds, and such large ammonites as are found in Portland, seem to have been the productions of hotter countries; and it is necessary to suppose that England once lay under the sea within the torrid zone! To explain this and similar phenomena, he indulges in a variety of speculations concerning changes in the position of the axis of the earth's rotation, "a shifting of the earth's centre of gravity, analogous to the revolutions of the magnetic pole," &c. None of these conjectures, however, are proposed dogmatically, but rather in the hope of promoting fresh inquiries and experiments.

In opposition to the prejudices of his age, we find him arguing against the idea that nature had formed fossil bodies "for no other end than to play the mimic in the mineral kingdom;"—maintaining that figured stones were "really the several bodies they represent, or the mouldings of them petrified," and not, as some have imagined, 'a lusus naturæ,' sporting herself in the needless formation of useless beings."†

It was objected to Hooke, that his doctrine of the extinction of species derogated from the wisdom and power of the omnipotent Creator; but he answered, that, as individuals die, there may be some termination to the duration of a species; and his opinions, he declared, were not repugnant to Holy Writ: for the Scriptures taught that our system was degenerating, and tending to its final dissolution; "and as, when that shall happen, all the species will be lost, why not some at one time and some at another?"‡

But his principal object was to account for the manner in which shells

* Posth. Works, p. 327.

† Posth. Works, Lecture, Feb. 15, 1688. Hooke explained with considerable clearness the different modes wherein organic substances may become lapidified; and, among other illustrations, he mentions some silicified palm-wood brought from Africa, on which M. de la Hire had read a memoir to the Royal Academy of France (June, 1692), wherein he had pointed out, not only the tubes running the length of the trunk, but the roots at one extremity. De la Hire, says Hooke, also treated of certain trees found petrified in the "river that passes by Bakan, in the kingdom of *Ava*, and which has for the space of ten leagues the virtue of petrifying wood." It is an interesting fact that the silicified wood of the Irawadi should have attracted attention more than one hundred years ago. Remarkable discoveries have been made there in later times of fossil animals and vegetables, by Mr. Crawfurd and Dr. Wallich.—See Geol. Trans. vol. ii. part iii. p. 377, second series. De la Hire cites Father Duchatz, in the second volume of "Observations made in the Indies by the Jesuits."

‡ Posth. Works, Lecture, May 29, 1689.

had been conveyed into the higher parts of "the Alps, Apennines, and Pyrenean hills, and the interior of continents in general." These and other appearances, he said, might have been brought about by earthquakes, "which have turned plains into mountains, and mountains into plains, seas into land, and land into seas, made rivers where there were none before, and swallowed up others that formerly were, &c., &c.; and which, since the creation of the world, have wrought many great changes on the superficial parts of the earth, and have been the instruments of placing shells, bones, plants, fishes, and the like, in those places where, with much astonishment, we find them."* This doctrine, it is true, had been laid down in terms almost equally explicit by Strabo, to explain the occurrence of fossil shells in the interior of continents, and to that geographer, and other writers of antiquity, Hooke frequently refers; but the revival and development of the system was an important step in the progress of modern science.

Hooke enumerated all the examples known to him of subterranean disturbance, from "the sad catastrophe of Sodom and Gomorrah," down to the Chilian earthquake of 1646. The elevating of the bottom of the sea, the sinking and submersion of the land, and most of the inequalities of the earth's surface, might, he said, be accounted for by the agency of these subterranean causes. He mentions that the coast near Naples *was raised during the eruption of Monte Nuovo;* and that, in 1591, land rose in the island of St. Michael, during an eruption: and although it would be more difficult, he says, to prove, he does not doubt but that there had been as many earthquakes in the parts of the earth under the ocean, as in the parts of the dry land; in confirmation of which, he mentions the immeasurable depth of the sea near some volcanoes. To attest the extent of simultaneous subterranean movements, he refers to an earthquake in the West Indies, in the year 1690, where the space of earth raised, or "struck upwards," by the shock, exceeded, he affirms, the length of the Alps and Pyrenees.

Hooke's diluvial Theory.—As Hooke declared the favorite hypothesis of the day, "that marine fossil bodies were to be referred to Noah's flood," to be wholly untenable, he appears to have felt himself called upon to substitute a diluvial theory of his own, and thus he became involved in countless difficulties and contradictions. "During the great catastrophe," he said, "there might have been a changing of that part which was before dry land into sea by sinking, and of that which was sea into dry land by raising, and marine bodies might have been buried in sediment beneath the ocean, in the interval between the creation and the deluge."† Then follows a disquisition on the separation of the land from the waters, mentioned in Genesis; during which operation some places of the shell of the earth were forced outwards, and others pressed downwards or inwards, &c. His diluvial hypothesis very much resembled that of Steno, and was entirely opposed to the fundamental

* Posth. Works, p. 312.

† Posth. Works, p. 410.

principles professed by him, that he would explain the former changes of the earth *in a more natural manner* than others had done. When, in despite of this declaration, he required a former "crisis of nature," and taught that earthquakes had become debilitated, and that the Alps, Andes, and other chains, had been lifted up in a few months, he was compelled to assume so rapid a rate of change, that his machinery appeared scarcely less extravagant than that of his most fanciful predecessors. For this reason, perhaps, his whole theory of earthquakes met with undeserved neglect.

Ray, 1692.—One of his contemporaries, the celebrated naturalist, Ray, participated in the same desire to explain geological phenomena by reference to causes less hypothetical than those usually resorted to.* In his essay on "Chaos and Creation," he proposed a system, agreeing in its outline, and in many of its details, with that of Hooke; but his knowledge of natural history enabled him to elucidate the subject with various original observations. Earthquakes, he suggested, might have been the second causes employed at the creation, in separating the land from the waters, and in gathering the waters together into one place. He mentions, like Hooke, the earthquake of 1646, which had violently shaken the Andes for some hundreds of leagues, and made many alterations therein. In assigning a cause for the general deluge, he preferred a change in the earth's centre of gravity to the introduction of earthquakes. Some unknown cause, he said, might have forced the subterranean waters outwards, as was, perhaps, indicated by "the breaking up of the fountains of the great deep."

Ray was one of the first of our writers who enlarged upon the effects of running water upon the land, and of the encroachment of the sea upon the shores. So important did he consider the agency of these causes, that he saw in them an indication of the tendency of our system to its final dissolution; and he wondered why the earth did not proceed more rapidly towards a general submersion beneath the sea, when so much matter was carried down by rivers, or undermined in the sea-cliffs. We perceive clearly from his writings, that the gradual decline of our system, and its future consummation by fire, was held to be as necessary an article of faith by the orthodox, as was the recent origin of our planet. His discourses, like those of Hooke, are highly interesting, as attesting the familiar association in the minds of philosophers, in the age of Newton, of questions in physics and divinity. Ray gave an unequivocal proof of the sincerity of his mind, by sacrificing his preferment in the church, rather than take an oath against the Covenanters, which he could not reconcile with his conscience. His reputation, moreover, in the scientific world placed him high above the temptation of courting popularity, by pandering to the physico-theological taste of his age. It is, therefore,

* Ray's Physico-theological Discourses were of somewhat later date than Hooke's great work on earthquakes. He speaks of Hooke as one "whom for his learning and deep insight into the mysteries of nature he deservedly honored." —*On the Deluge*, chap. iv.

curious to meet with so many citations from the Christian fathers and prophets in his essays on physical science—to find him in one page proceeding, by the strict rules of induction, to explain the former changes of the globe, and in the next gravely entertaining the question, whether the sun and stars, and the whole heavens, shall be annihilated, together with the earth, at the era of the grand conflagration.

Woodward, 1695.—Among the contemporaries of Hooke and Ray, Woodward, a professor of medicine, had acquired the most extensive information respecting the geological structure of the crust of the earth. He had examined many parts of the British strata with minute attention; and his systematic collection of specimens, bequeathed to the University of Cambridge, and still preserved there as arranged by him, shows how far he had advanced in ascertaining the order of superposition. From the great number of facts collected by him, we might have expected his theoretical views to be more sound and enlarged than those of his contemporaries; but in his anxiety to accommodate all observed phenomena to the scriptural account of the Creation and Deluge, he arrived at most erroneous results. He conceived "the whole terrestrial globe to have been taken to pieces and dissolved at the flood, and the strata to have settled down from this promiscuous mass as any earthy sediment from a fluid."* In corroboration of these views he insisted upon the fact, that "marine bodies are lodged in the strata according to the order of their gravity, the heavier shells in stone, the lighter in chalk, and so of the rest."† Ray immediately exposed the unfounded nature of this assertion, remarking truly that fossil bodies "are often mingled, heavy with light, in the same stratum;" and he even went so far as to say, that Woodward "must have invented the phenomena for the sake of confirming his bold and strange hypothesis"‡—a strong expression from the pen of a contemporary.

Burnet, 1690.—At the same time Burnet published his "Theory of the Earth."§ The title is most characteristic of the age,—"The Sacred Theory of the Earth; containing an Account of the Original of the Earth, and of all the general Changes which it hath already undergone, or is to undergo, till the Consummation of all Things." Even Milton had scarcely ventured in his poem to indulge his imagination so freely in painting scenes of the Creation and Deluge, Paradise and Chaos. He explained why the primeval earth enjoyed a perpetual spring before the flood! showed how the crust of the globe was fissured by "the sun's rays," so that it burst, and thus the diluvial waters were let loose from a supposed central abyss. Not satisfied with these themes, he derived from the books of the inspired writers, and even from heathen authorities, prophetic views of the future revolutions of the globe, gave a most terrific description of the general conflagration, and proved that

* Essay towards a Natural History of the Earth, 1695. Preface. † Ibid.
‡ Consequences of the Deluge, p. 165.
§ First published in Latin between the years 1680 and 1690.

a new heaven and a new earth will rise out of a *second chaos*—after which will follow the blessed millennium.

The reader should be informed, that, according to the opinion of many respectable writers of that age, there was good scriptural ground for presuming that the garden bestowed upon our first parents was not on the earth itself, but above the clouds, in the middle region between our planet and the moon. Burnet approaches with becoming gravity the discussion of so important a topic. He was willing to concede that the geographical position of Paradise was not in Mesopotamia, yet he maintained that it was upon the earth, and in the southern hemisphere, near the equinoctial line. Butler selected this conceit as a fair mark for his satire, when, amongst the numerous accomplishments of Hudibras, he says,—

> "He knew the seat of Paradise,
> Could tell in what degree it lies;
> And, as he was disposed, could prove it
> Below the moon, or else above it."

Yet the same monarch, who is said never to have slept without Butler's poem under his pillow, was so great an admirer and patron of Burnet's book, that he ordered it to be translated from the Latin into English. The style of the "Sacred Theory" was eloquent, and the book displayed powers of invention of no ordinary stamp. It was, in fact, a fine historical romance, as Buffon afterwards declared; but it was treated as a work of profound science in the time of its author, and was panegyrized by Addison in a Latin ode, while Steele praised it in the "Spectator."

Whiston, 1696.—Another production of the same school, and equally characteristic of the time, was that of Whiston, entitled, "A New Theory of the Earth; wherein the Creation of the world in Six Days, the Universal Deluge, and the General Conflagration, as laid down in the Holy Scriptures, are shown to be perfectly agreeable to Reason and Philosophy." He was at first a follower of Burnet; but his faith in the infallibility of that writer was shaken by the declared opinion of Newton, that there was every presumption in astronomy against any former change in the inclination of the earth's axis. This was a leading dogma in Burnet's system, though not original, for it was borrowed from an Italian, Alessandro degli Alessandri, who had suggested it in the beginning of the fifteenth century, to account for the former occupatiou of the present continents by the sea. La Place has since strengthened the arguments of Newton, against the probability of any former revolution of this kind.

The remarkable comet of 1680 was fresh in the memory of every one when Whiston first began his cosmological studies; and the principal novelty of his speculations consisted in attributing the deluge to the near approach to the earth of one of these erratic bodies. Having ascribed an increase of the waters to this source, he adopted Woodward's theory, supposing all stratified deposits to have resulted from the "chaotic sediment of the flood." Whiston was one of the first who ventured to propose that the text of Genesis should be interpreted differently from its ordinary acceptation, so that the doctrine of the earth having existed long previous to the creation of man might no longer be regarded as unorthodox. He had the art to throw an air of plausibility over the most improbable parts of his theory, and seemed to be proceeding in the most sober manner, and, by the aid of mathematical demonstration, to the establishment of his various propositions. Locke pronounced a panegyric on his theory, commending him for having explained so many wonderful and before inexplicable things. His book, as well as Burnet's, was attacked and refuted by Keill. * Like all who introduced purely hypothetical causes to account for natural phenomena, Whiston retarded the progress of truth, diverting men from the investigation of the laws of sublunary nature, and inducing them to waste time in speculations on the power of comets to drag the waters of the ocean over the land—on the condensation of the vapors of their tails into water, and other matters equally edifying.

Hutchinson, 1724.—John Hutchinson, who had been employed by Woodward in making his collection of fossils, published afterwards, in 1724, the first part of his "Moses's Principia," wherein he ridiculed Woodward's hypothesis. He and his numerous followers were accustomed to declaim loudly against human learning; and they maintained that the Hebrew Scriptures, when rightly translated, comprised a perfect system of natural philosophy, for which reason they objected to the Newtonian theory of gravitation.

Celsius.—Andrea Celsius, the Swedish astronomer, published about this time his remarks on the gradual diminution and sinking of the waters in the Baltic, to which I shall have occasion to advert more particularly in the sequel (ch. 29).

Scheuchzer, 1708.—In Germany, in the mean time, Scheuchzer published his "Complaint and Vindication of the Fishes" (1708), "Piscium Querelæ et Vindiciæ," a work of zoological merit, in which he gave some good plates and descriptions of fossil fish. Among other conclusions he labored to prove that the earth had been remodelled at the deluge. Pluche, also, in 1732, wrote to the same effect; while Holbach, in 1753, after considering the various attempts to refer all the ancient formations to the flood of Noah, exposed the inadequacy of this cause.

Italian Geologists — Vallisneri.—I return with pleasure to the geologists of Italy, who preceded, as has been already shown, the naturalists of other countries in their investigations into the ancient history of the earth, and who still maintained a decided pre-eminence. They refuted and ridiculed the physico-theological systems of Burnet, Whiston,

* An Examination of Dr. Burnet's Theory, &c., 2d ed. 1734.

and Woodward;* while Vallisneri,† in his comments on the Woodwardian theory, remarked how much the interests of religion, as well as those of sound philosophy, had suffered by perpetually mixing up the sacred writings with questions in physical science. The works of this author were rich in original observations. He attempted the first general sketch of the marine deposits of Italy, their geographical extent, and most characteristic organic remains. In his treatise "On the Origin of Springs," he explained their dependence on the order, and often on the dislocations, of the strata, and reasoned philosophically against the opinions of those who regarded the disordered state of the earth's crust as exhibiting signs of the wrath of God for the sins of man. He found himself under the necessity of contending, in his preliminary chapter, against St. Jerome, and four other principal interpreters of Scripture, besides several professors of divinity, "that springs did not flow by subterranean siphons and cavities from the sea upwards, losing their saltness in the passage," for this theory had been made to rest on the infallible testimony of Holy Writ.

Although reluctant to generalize on the rich materials accumulated in his travels, Vallisneri had been so much struck with the remarkable continuity of the more recent marine strata, from one end of Italy to the other, that he came to the conclusion that the ocean formerly extended over the whole earth, and after abiding there for a long time, had gradually subsided. This opinion, however untenable, was a great step beyond Woodward's diluvian hypothesis, against which Vallisneri, and after him all the Tuscan geologists, uniformly contended, while it was warmly supported by the members of the Institute of Bologna.‡

Among others of that day, Spada, a priest of Grezzana, in 1737, wrote to prove that the petrified marine bodies near Verona were not diluvian.§ Mattani drew a similar inference from the shells of Volterra and other places; while Costantini, on the other hand, whose observations on the valley of the Brenta and other districts were not without value, undertook to vindicate the truth of the deluge, as also to prove that Italy had been peopled by the descendants of Japhet.‖

Moro, 1740.—Lazzaro Moro, in his work (published in 1740) "On the Marine Bodies which are found in the Mountains,"¶ attempted to apply the theory of earthquakes, as expounded by Strabo, Pliny, and other ancient authors, with whom he was familiar, to the geological phenomena described by Vallisneri.** His attention was awakened to

* Ramazzini even asserted, that the ideas of Burnet were mainly borrowed from a dialogue of one Patrizio; but Brocchi, after reading that dialogue, assures us that there was scarcely any other correspondence between these systems, except that both were equally whimsical.

† Dei Corpi Marini, Lettere critiche, &c. 1721.

‡ Brocchi, p. 28. § Ibid. p. 33. ‖ Ibid.

¶ Sui Crostacei ed altri Corpi Marini che si trovano sui Monti.

** Moro does not cite the works of Hooke and Ray; and although so many of his views were in accordance with theirs, he was probably ignorant of their writhe elevating power of subterranean forces by a remarkable phenomenon which happened in his own time, and which had also been noticed by Vallisneri in his letters. A new island rose in 1707 from deep water in the Gulf of Santorin, in the Mediterranean, during continued shocks of an earthquake, and, increasing rapidly in size, grew in less than a month to be half a mile in circumference, and about twenty-five feet above high-water mark. It was soon afterwards covered by volcanic ejections, but, when first examined. it was found to be a white rock, bearing on its surface living oysters and crustacea. In order to ridicule the various theories then in vogue, Moro ingeniously supposes the arrival on this new island of a party of naturalists ignorant of its recent origin. One immediately points to the marine shells, as proofs of the universal deluge; another argues that they demonstrate the former residence of the sea upon the mountains; a third dismisses them as mere *sports of nature;* while a fourth affirms that they were born and nourished within the rock in ancient caverns, into which salt water had been raised in the shape of vapor by the action of subterranean heat.

Moro pointed with great judgment to the *faults* and dislocations of the strata described by Vallisneri, in the Alps and other chains, in confirmation of his doctrine, that the continents had been heaved up by subterranean movements. He objected, on solid grounds, to the hypothesis of Burnet and of Woodward; yet he ventured so far to disregard the protest of Vallisneri, as to undertake the adaptation of every part of his own system to the Mosaic account of the creation. On the third day, he said, the globe was everywhere covered to the same depth by fresh water; and when it pleased the Supreme Being that the dry land should appear, volcanic explosions broke up the smooth and regular surface of the earth composed of primary rocks. These rose in mountain masses above the waves, and allowed melted metals and salts to ascend through fissures. The sea gradually acquired its saltness from volcanic exhalations, and, while it became more circumscribed in area, increased in depth. Sand and ashes ejected by volcanoes were regularly disposed along the bottom of the ocean, and formed the secondary strata, which in their turn were lifted up by earthquakes. We need not follow this author in tracing the progress of the creation of vegetables and animals on the other days of creation; but, upon the whole, it may be remarked, that few of the old cosmological theories had been conceived with so little violation of known analogies.

Generelli's illustrations of Moro, 1749.—The style of Moro was extremely prolix, and, like Hutton, who, at a later period, advanced many of the same views, he stood in need of an illustrator. The Scotch geologist was hardly more fortunate in the advocacy of Playfair, than was Moro in numbering amongst his admirers Cirillo Generelli, who, nine

tings, for they had not been translated. As he always refers to the Latin edition of Burnet, and a French translation of Woodward, we may presume that he did not read English.

years afterwards, delivered at a sitting of Academicians at Cremona a spirited exposition of his theory. This learned Carmelitan friar does not pretend to have been an original observer, but he had studied sufficiently to enable him to confirm the opinions of Moro by arguments from other writers; and his selection of the doctrines then best established is so judicious, that a brief abstract of them cannot fail to be acceptable, as illustrating the state of geology in Europe, and in Italy in particular, before the middle of the last century.

The bowels of the earth, says he, have carefully preserved the memorials of past events, and this truth the marine productions so frequent in the hills attest. From the reflections of Lazzaro Moro, we may assure ourselves that these are the effects of earthquakes in past times, which have changed vast spaces of sea into terra firma, and inhabited lands into seas. In this, more than in any other department of physics, are observations and experiments indispensable, and we must diligently consider facts. The land is known, wherever we make excavations, to be composed of different strata or soils placed one above the other, some of sand, some of rock, some of chalk, others of marl, coal, pummice, gypsum, lime, and the rest. These ingredients are sometimes pure, and sometimes confusedly intermixed. Within are often imprisoned different marine fishes, like dried mummies, and more frequently shells, crustacea, corals, plants, &c., not only in Italy, but in France, Germany, England, Africa, Asia, and America;—sometimes in the lowest, sometimes in the loftiest beds of the earth, some upon the mountains, some in deep mines, others near the sea, and others hundreds of miles distant from it. Woodward conjectured that these marine bodies might be found everywhere; but there are rocks in which none of them occur, as is sufficiently attested by Vallisneri and Marsilli. The remains of fossil animals consist chiefly of their more solid parts, and the most rocky strata must have been soft when such exuviæ were inclosed in them. Vegetable productions are found in different states of maturity, indicating that they were imbedded in different seasons. Elephants, elks, and other terrestrial quadrupeds, have been found in England and elsewhere, in superficial strata, never covered by the sea. Alternations are rare, yet not without example, of marine strata, with those which contain marshy and terrestrial productions. Marine animals are arranged in the subterraneous beds with admirable order, in distinct groups, oysters here, dentalia or corals there, &c., as now, according to Marsilli,* on the shores of the Adriatic. We must abandon the doctrine, once so popular, which denies that organized fossils were derived from living beings, and we cannot account for their present position by the ancient theory of Strabo, nor by that of Leibnitz, nor by the universal deluge, as explained by Woodward and others; "nor is it reasonable to call the Deity capriciously upon the stage, and to make him work miracles for the sake of confirming our preconceived hypothesis."

* Saggio fisico intorno alla Storia del Mare, part i. p. 24.

—"I hold in utter abomination, most learned Academicians! those systems which are built with their foundations in the air, and cannot be propped up without a miracle; and I undertake, with the assistance of Moro, to explain to you how these marine animals were transported into the mountains by natural causes."*

A brief abstract then follows of Moro's theory, by which, says Generelli, we may explain all the phenomena, as Vallisneri so ardently desired, "*without violence, without fictions, without hypothesis, without miracles.*"† The Carmelitan then proceeds to struggle against an obvious objection to Moro's system, considered as a method of explaining the revolutions of the earth, *naturally*. If earthquakes have been the agents of such mighty changes, how does it happen that their effects since the times of history have been so inconsiderable? This same difficulty had, as we have seen, presented itself to Hooke, half a century before, and forced him to resort to a former "crisis of nature:" but Generelli defended his position by showing how numerous were the accounts of eruptions and earthquakes, of new islands, and of elevations and subsidences of land, and yet how much greater a number of like events must have been unattested and unrecorded during the last six thousand years. He also appealed to Vallisneri as an authority to prove that the mineral masses containing shells, bore, upon the whole, but a small proportion to those rocks which were destitute of organic remains; and the latter, says the learned monk, might have been created as they now exist, *in the beginning*.

Generelli then describes the continual waste of mountains and continents, by the action of rivers and torrents, and concludes with these eloquent and original observations:—"Is it possible that this waste should have continued for six thousand, and *perhaps* a greater number of years, and that the mountains should remain so great, unless their ruins have been repaired? Is it credible that the Author of Nature should have founded the world upon such laws, as that the dry land should forever be growing smaller, and at last become wholly submerged beneath the waters? Is it credible that, amid so many created things, the mountains alone should daily diminish in number and bulk, without there being any repair of their losses? This would be contrary to that order of Providence which is seen to reign in all other things in the universe. Wherefore I deem it just to conclude, that the same cause which, in the beginning of time, raised mountains from the abyss, has down to the present day continued to produce others, in order to restore from time to time the losses of all such as sink down in different places, or are rent asunder, or in other way suffer disintegration. If this be admitted, we can easily understand why there should now be

* "Abbomino al sommo qualsivoglia sistema, che sia di pianta fabbricato in aria; massime quando è tale, che non possa sostenersi senza un miracolo," &c.—De' Crostacei e di altre Produz. del Mare, &c. 1749.

† "Senza violenze, senza finzioni, senza supposti, senza miracoli." De' Crostacei e di altre Produz. del Mare, &c. 1749.

found upon many mountains so great a number of crustacea and other marine animals."

In the above extract, I have not merely enumerated the opinions and facts which are confirmed by recent observation, suppressing all that has since proved to be erroneous, but have given a faithful abridgment of the entire treatise, with the omission only of Moro's hypothesis, which Generelli adopted, with all its faults and excellences. The reader will therefore remark, that although this admirable essay embraces so large a portion of the principal objects of geological research, it makes no allusion to the extinction of certain classes of animals; and it is evident that no opinions on this head had, at that time, gained a firm footing in Italy. That Lister and other English naturalists should long before have declared in favor of the loss of species, while Scilla and most of his countrymen hesitated, was perhaps natural, since the Italian museums were filled with fossil shells belonging to species of which a great portion did actually exist in the Mediterranean; whereas the English collectors could obtain no recent species from such of their own strata as were then explored.

The weakest point in Moro's system consisted in deriving *all* the stratified rocks from volcanic ejections; an absurdity which his opponents took care to expose, especially Vito Amici.* Moro seems to have been misled by his anxious desire to represent the formation of secondary rocks as having occupied an extremely short period, while at the same time he wished to employ known agents in nature. To imagine torrents, rivers, currents, partial floods, and all the operations of moving water, to have gone on exerting an energy many thousand times greater than at present, would have appeared preposterous and incredible, and would have required a hundred violent hypotheses; but we are so unacquainted with the true sources of subterranean disturbances, that their former violence may in theory be multiplied indefinitely, without its being possible to prove the same manifest contradiction or absurdity in the conjecture. For this reason, perhaps, Moro preferred to derive the materials of the strata from volcanic ejections, rather than from transportation by running water.

Marsilli.—Marsilli, whose work is alluded to by Generelli, had been prompted to institute inquiries into the bed of the Adriatic, by discovering, in the territory of Parma (what Spada had observed near Verona, and Schiavo in Sicily), that fossil shells were not scattered through the rocks at random, but disposed in regular order, according to certain genera and species.

Vitaliano Donati, 1750.—But with a view of throwing further light upon these questions, Donati, in 1750, undertook a more extensive investigation of the Adriatic, and discovered, by numerous soundings, that deposits of sand, marl, and tufaceous incrustations, most strictly analogous to those of the Subapennine hills, were in the act of accumulating there.

* Sui Testacei della Sicilia.

He ascertained that there were no shells in some of the submarine tracts, while in other places they lived together in families, particularly the genera Arca, Pecten, Venus, Murex, and some others. He also states that in divers localities he found a mass composed of corals, shells. and crustaceous bodies of different species, confusedly blended with earth, sand, and gravel. At the depth of a foot or more, the organic substances were entirely petrified and reduced to marble; at less than a foot from the surface, they approached nearer to their natural state; while at the surface they were alive, or, if dead, in a good state of preservation.

Baldassari.—A contemporary naturalist, Baldassari, had shown that the organic remains in the tertiary marls of the Siennese territory were grouped in families, in a manner precisely similar to that above alluded to by Donati.

Buffon, 1749.—Buffon first made known his theoretical views concerning the former changes of the earth, in his Natural History, published in 1749. He adopted the theory of an original volcanic nucleus, together with the universal ocean of Leibnitz. By this aqueous envelope the highest mountains were once covered. Marine currents then acted violently, and formed horizontal strata, by washing away solid matter in some parts, and depositing it in others; they also excavated deep submarine valleys. The level of the ocean was then depressed by the entrance of a part of its waters into subterranean caverns, and thus some land was left dry. Buffon seems not to have profited, like Leibnitz and Moro, by the observations of Steno, or he could not have imagined that the strata were generally horizontal, and that those which contain organic remains had never been disturbed since the era of their formation. He was conscious of the great power annually exerted by rivers and marine currents in transporting earthy materials to lower levels, and he even contemplated the period when they would destroy all the present continents. Although in geology he was not an original observer, his genius enabled him to render his hypothesis attractive; and by the eloquence of his style, and the boldness of his speculations, he awakened curiosity, and provoked a spirit of inquiry amongst his countrymen.

Soon after the publication of his "Natural History," in which was included his "Theory of the Earth," he received an official letter (dated January, 1751) from the Sorbonne, or Faculty of Theology in Paris, informing him that fourteen propositions in his works "were reprehensible, and contrary to the creed of the church." The first of these obnoxious passages, and the only one relating to geology, was as follows:—"The waters of the sea have produced the mountains and valleys of the land—the waters of the heavens, reducing all to a level, will at last deliver the whole land over to the sea, and the sea successively prevailing over the land, will leave dry new continents like those which we inhabit." Buffon was invited by the College, in very courteous terms, to send in an explanation, or rather a recantation of his unorthodox opinions. To this he submitted; and a general assembly of the Faculty having approved of his "Declaration," he was required to publish it in his next work. The document begins with these words:—"I declare that I had no intention to contradict the text of Scripture; that I believe most firmly all therein related about the creation, both as to order of time and matter of fact; and *I abandon every thing in my book respecting the foundation of the earth,* and, generally, all which may be contrary to the narration of Moses."*

The grand principle which Buffon was called upon to renounce was simply this,—that the present mountains and valleys of the earth are due to secondary causes, and that the same causes will in time destroy all the continents, hills, and valleys, and reproduce others like them." Now, whatever may be the defects of many of his views, it is no longer controverted that the present continents are of secondary origin. The doctrine is as firmly established as the earth's rotation on its axis; and that the land now elevated above the level of the sea will not endure forever, is an opinion which gains ground daily, in proportion as we enlarge our experience of the changes now in progress.

Targioni, 1751.—Targioni, in his voluminous "Travels in Tuscany, 1751 and 1754," labored to fill up the sketch of the geology of that region left by Steno sixty years before. Notwithstanding a want of arrangement and condensation in his memoirs, they contained a rich store of faithful observations. He has not indulged in many general views, but in regard to the origin of valleys, he was opposed to the theory of Buffon, who attributed them principally to submarine currents. The Tuscan naturalist labored to show that both the larger and smaller valleys of the Apennines were excavated by rivers and floods, caused by the bursting of the barriers of lakes, after the retreat of the ocean. He also maintained that the elephants and other quadrupeds, so frequent in the lacustrine and alluvial deposits of Italy, had inhabited that peninsula; and had not been transported thither, as some had conceived, by Hannibal or the Romans, nor by what they were pleased to term "a catastrophe of nature."

Lehman, 1756.—In the year 1756 the treatise of Lehman, a German mineralogist, and director of the Prussian mines, appeared, who also divided mountains into three classes: the first, those formed with the world, and prior to the creation of animals, and which contained no fragments of other rocks; the second class, those which resulted from the partial destruction of the primary rocks by a general revolution; and a third class, resulting from local revolutions, and in part from the deluge of Noah.

A French translation of this work appeared in 1759, in the preface of which, the translator displays very enlightened views respecting the operations of earthquakes, as well as of the aqueous causes.†

* Hist. Nat. tom. v. éd. de l'Imp. Royale, Paris, 1769.
† Essai d'une Hist. Nat. des Couches de la Terre, 1759.

Gesner, 1758.—In this year Gesner, the botanist, of Zurich, published an excellent treatise on petrifactions, and the changes of the earth which they testify.* After a detailed enumeration of the various classes of fossils of the animal and vegetable kingdoms, and remarks on the different states in which they are found petrified, he considers the geological phenomena connected with them; observing, that some, like those of Œningen, resembled the testacea, fish, and plants indigenous in the neighboring region;† while some, such as ammonites, gryphites, belemnites, and other shells, are either of unknown species, or found only in the Indian and other distant seas. In order to elucidate the structure of the earth, he gives sections, from Verenius, Buffon, and others, obtained in digging wells; distinguishes between horizontal and inclined strata; and, in speculating on the causes of these appearances, mentions Donati's examination of the bed of the Adriatic; the filling up of lakes and seas by sediment; the imbedding of shells now in progress; and many known effects of earthquakes, such as the sinking down of districts, or the heaving up of the bed of the sea, so as to form new islands, and lay dry strata containing petrifactions. The ocean, he says, deserts its shores in many countries, as on the borders of the Baltic; but the rate of recession has been so slow in the last 2000 years, that to allow the Apennines, whose summits are filled with marine shells, to emerge to their present height, would have required about 80,000 years,—a lapse of time ten times greater, or more, than the age of the universe. We must therefore refer the phenomenon to the command of the Deity, related by Moses, that "the waters should be gathered together in one place, and the dry land appear." Gesner adopted the views of Leibnitz, to account for the retreat of the primeval ocean: his essay displays much erudition; and the opinions of preceding writers of Italy, Germany, and England, are commented upon with fairness and discrimination.

Arduino, 1759.—In the year following, Arduino,‡ in his memoirs on the mountains of Padua, Vicenza, and Verona, deduced, from original observations, the distinction of rocks into primary, secondary, and tertiary, and showed that in those districts there had been a succession of submarine volcanic eruptions.

Michell, 1760.—In the following year (1760) the Rev. John Michell, Woodwardian Professor of Mineralogy at Cambridge, published in the Philosophical Transactions, an Essay on the Cause and Phenomena of Earthquakes.§ His attention had been drawn to this subject by the

* John Gesner published at Leyden, in Latin. † Part ii. chap. 9.

‡ Giornale del Criselini, 1759.

§ See a sketch of the History of English Geology, by Dr. Fitton, in Edinb. Rev. Feb. 1818; re-edited Lond. and Edinb. Phil. Mag. vols. i. and ii. 1832–3. Some of Michell's observations anticipate in so remarkable a manner the theories established forty years afterwards, that his writings would probably have formed an era in the science, if his researches had been uninterrupted. He held, however, his professorship only eight years, when his career was suddenly cut short by preferment to a benefice. From that time he appears to have been engaged in his clerical great earthquake of Lisbon in 1755. He advanced many original and philosophical views respecting the propagation of subterranean movements, and the caverns and fissures wherein steam might be generated. In order to point out the application of his theory to the structure of the globe, he was led to describe the arrangement and disturbance of the strata, their usual horizontality in low countries, and their contortions and fractured state in the neighborhood of mountain chains. He also explained, with surprising accuracy, the relations of the central ridges of older rocks to the "long narrow slips of similar earth, stones, and minerals," which are parallel to these ridges. In his generalizations, derived in great part from his own observations on the geological structure of Yorkshire, he anticipated many of the views more fully developed by later naturalists.

Catcott, 1761.—Michell's papers were entirely free from all physico-theological disquisitions, but some of his contemporaries were still earnestly engaged in defending or impugning the Woodwardian hypothesis. We find many of these writings referred to by Catcott, a Hutchinsonian, who published a "Treatise on the Deluge" in 1761. He labored particularly to refute an explanation offered by his contemporary, Bishop Clayton, of the Mosaic writings. That prelate had declared that the deluge "could not be literally true, save in respect to that part where Noah lived before the flood." Catcott insisted on the universality of the deluge, and referred to traditions of inundations mentioned by ancient writers, or by travellers, in the East Indies, China, South America, and other countries. This part of his book is valuable, although it is not easy to see what bearing the traditions have, if admitted to be authentic, on the Bishop's argument, since no evidence is adduced to prove that the catastrophes were contemporaneous events, while some of them are expressly represented by ancient authors to have occurred in succession.

Fortis—Odoardi, 1761.—The doctrines of Arduino, above adverted to, were afterwards confirmed by Fortis and Desmarest, in their travels in the same country; and they, as well as Baldassari, labored to complete the history of the Subapennine strata. In the work of Odoardi,* there was also a clear argument in favor of the distinct ages of the older Apennine strata, and the Subapennine formations of more recent origin. He pointed out that the strata of these two groups were *unconformable*, and must have been the deposits of different seas at distant periods of time.

Raspe, 1763.—A history of the new islands, by Raspe, a Hanove-

duties, and to have entirely discontinued his scientific pursuits, exemplifying the working of a system still in force at Oxford and Cambridge, where the chairs of mathematics, natural philosophy, chemistry, botany, astronomy, geology, mineralogy, and others, being frequently filled by clergymen, the reward of success disqualifies them, if they conscientiously discharge their new duties, from farther advancing the cause of science, and that, too, at the moment when their labors would naturally bear the richest fruits.

* Sui Corpi Marini del Feltrino, 1761.

rian, appeared in 1763, in Latin.* In this work, all the authentic accounts of earthquakes which had produced permanent changes on the solid parts of the earth were collected together and examined with judicious criticism. The best systems which had been proposed concerning the ancient history of the globe, both by ancient and modern writers, are reviewed; and the merits and defects of the doctrines of Hooke, Ray, Moro, Buffon, and others, fairly estimated. Great admiration is expressed for the hypothesis of Hooke, and his explanation of the origin of the strata is shown to have been more correct than Moro's, while their theory of the effects of earthquakes was the same. Raspe had not seen Michell's memoirs, and his views concerning the geological structure of the earth were perhaps less enlarged; yet he was able to add many additional arguments in favor of Hook's theory, and to render it, as he said, a nearer approach to what Hooke would have written had he lived in later times. As to the periods wherein all the earthquakes happened, to which we owe the elevation of various parts of our continents and islands, Raspe says he pretends not to assign their duration, still less to defend Hooke's suggestion, that the convulsions almost all took place during the deluge of Noah. He adverts to the apparent indications of the former tropical heat of the climate of Europe, and the changes in the species of animals and plants, as among the most obscure and difficult problems in geology. In regard to the islands raised from the sea, within the times of history or tradition, he declares that some of them were composed of strata containing organic remains, and that they were not, as Buffon had asserted, made of mere volcanic matter. His work concludes with an eloquent exhortation to naturalists to examine the isles which rose, in 1707, in the Grecian Archipelago, and, in 1720, in the Azores, and not to neglect such splendid opportunities of studying nature "in the act of parturition." That Hooke's writings should have been neglected for more than half a century, was matter of astonishment to Raspe; but it is still more wonderful that his own luminous exposition of that theory should, for more than another half century, have excited so little interest.

Fuchsel, 1762 and 1773.—Fuchsel, a German physician, published, in 1762, a geological description of the country between the Thuringerwald and the Hartz, and a memoir on the environs of Rudelstadt;† and afterwards, in 1773, a theoretical work on the ancient history of the earth and of man.‡ He had evidently advanced considerably beyond his predecessor Lehman, and was aware of the distinctness, both as to position and fossil contents, of several groups of strata of different ages, corresponding to the secondary formations now recognized by geologists

* De Novis e Mari Natis Insulis. Raspe was also the editor of the "Philosophical Works of Leibnitz. Amst. et Leipzig, 1765;" also author of "Tassie's Gems," and "Baron Munchausen's Travels."

† Acta Academiæ Electoralis Maguntinæ, vol. ii. Erfurt.

‡ This account of Fuchsel is derived from an excellent analysis of his memoirs by M. Keferstein. Journ. de Géologie, tom. ii. Oct. 1830.

in various parts of Germany. He supposed the European continents to have remained covered by the sea until the formation of the marine strata, called in Germany "muschelkalk," at the same time that the terrestrial plants of many European deposits, attested the existence of dry land which bordered the ancient sea; land which, therefore, must have occupied the place of the present ocean. The pre-existing continent had been *gradually* swallowed up by the sea, different parts having subsided in succession into subterranean caverns. All the sedimentary strata were originally horizontal, and their present state of derangement must be referred to subsequent oscillations of the ground.

As there were plants and animals in the ancient periods, so also there must have been men, but they did not all descend from one pair, but were created at various points on the earth's surface; and the number of these distinct birth-places was as great as are the original languages of nations.

In the writings of Fuchsel we see a strong desire manifested to explain geological phenomena as far as possible by reference to the agency of known causes; and although some of his speculations were fanciful, his views coincide much more nearly with those now generally adopted, than the theories afterwards promulgated by Werner and his followers.

Brander, 1766.—Gustavus Brander published, in 1766, his "Fossilia Hantoniensia," containing excellent figures of fossil shells from the more modern (or Eocene) marine strata of Hampshire. "Various opinions," he says in the preface, "had been entertained concerning the time when and how these bodies became deposited. Some there are who conceive that it might have been effected in a wonderful length of time by a gradual changing and shifting of the sea," &c. But the most common cause assigned is that of "the deluge." This conjecture, he says, even if the universality of the flood be not called in question, is purely hypothetical. In his opinion, fossil animals and testacea were, for the most part, of unknown species; and of such as were known, the living analogues now belonged to southern latitudes.

Soldani, 1780.—Soldani applied successfully his knowledge of zoology to illustrate the history of stratified masses. He explained that microscopic testacea and zoophytes inhabited the depths of the Mediterranean; and that the fossil species were, in like manner, found in those deposits wherein the fineness of their particles, and the absence of pebbles, implied that they were accumulated in a deep sea, or far from shore. This author first remarked the alternation of marine and freshwater strata in the Paris basin.*

Fortis—Testa, 1793.—A lively controversy arose between Fortis and another Italian naturalist, Testa, concerning the fish of Monte Bolca, in 1793. Their letters,† written with great spirit and elegance, show that they were aware that a large proportion of the Subapennine shells were

* Saggio orittografico, &c. 1780, and other Works.
† Lett. sui Pesci Fossili di Bolca. Milan, 1793.

identical with living species, and some of them with species now living in the torrid zone. Fortis proposed a somewhat fanciful conjecture, that when the volcanoes of the Vicentin were burning, the waters of the Adriatic had a higher temperature; and in this manner, he said, the shells of warmer regions may once have peopled their own seas. But Testa was disposed to think that these species of testacea were still common to their own and to equinoctial seas; for many, he said, once supposed to be confined to hotter regions, had been afterwards discovered in the Mediterranean.*

Cortesi—Spallanzani—Wallerius—Whitehurst.—While these Italian naturalists, together with Cortesi and Spallanzani, were busily engaged in pointing out the analogy between the deposits of modern and ancient seas, and the habits and arrangement of their organic inhabitants, and while some progress was making, in the same country, in investigating the ancient and modern volcanic rocks, some of the most original observers among the English and German writers, Whitehurst† and Wallerius, were wasting their strength in contending, according to the old Woodwardian hypothesis, that all the strata were formed by Noah's deluge. But Whitehurst's description of the rocks of Derbyshire was most faithful; and he atoned for false theoretical views, by providing data for their refutation.

Pallas—Saussure.—Towards the close of the eighteenth century, the idea of distinguishing the mineral masses on our globe into separate groups, and studying their relations, began to be generally diffused. Pallas and Saussure were among the most celebrated whose labors contributed to this end. After an attentive examination of the two great mountain chains of Siberia, Pallas announced the result, that the granitic rocks were in the middle, the schistose at their sides, and the limestones again on the outside of these; and this he conceived would prove a general law in the formation of all chains composed chiefly of primary rocks.‡

In his "Travels in Russia," in 1793 and 1794, he made many geological observations on the recent strata near the Wolga and the Caspian, and adduced proofs of the greater extent of the latter sea at no distant era in the earth's history. His memoir on the fossil bones of Siberia attracted attention to some of the most remarkable phenomena in geology. He stated that he had found a rhinoceros entire in the frozen soil, with its skin and flesh: an elephant, found afterwards in a mass

* This argument of Testa has been strengthened of late years by the discovery that dealers in shells had long been in the habit of selling Mediterranean species as shells of more southern and distant latitudes, for the sake of enhancing their price. It appears, moreover, from several hundred experiments made by that distinguished hydrographer, Capt. Smith, on the water within eight fathoms of the surface, that the temperature of the Mediterranean is on an average 3½° of Fahrenheit higher than the western part of the Atlantic ocean; an important fact, which in some degree may help to explain why many species are common to tropical latitudes and to the Mediterranean.

† Inquiry into the Original State and Formation of the Earth, 1778.

‡ Observ. on the Formation of Mountains. Act Petrop. ann. 1778, part i.

of ice on the shore of the North Sea, removed all doubt as to the accuracy of so wonderful a discovery.*

The subjects relating to natural history which engaged the attention of Pallas, were too multifarious to admit of his devoting a large share of his labors exclusively to geology. Saussure, on the other hand, employed the chief portion of his time in studying the structure of the Alps and Jura, and he provided valuable data for those who followed him. He did not pretend to deduce any general system from his numerous and interesting observations; and the few theoretical opinions which escaped from him, seem, like those of Pallas, to have been chiefly derived from the cosmological speculations of preceding writers.

CHAPTER IV.

HISTORY OF THE PROGRESS OF GEOLOGY—*continued.*

Werner's application of geology to the art of mining—Excursive character of his lectures—Enthusiasm of his pupils—His authority—His theoretical errors—Desmarest's Map and Description of Auvergne—Controversy between the Vulcanists and Neptunists—Intemperance of the rival sects—Hutton's Theory of the earth—His discovery of granite veins—Originality of his views—Why opposed—Playfair's illustrations—Influence of Voltaire's writings on geology—Imputations cast on the Huttonians by Williams, Kirwan, and De Luc—Smith's Map of England—Geological Society of London—Progress of the science in France—Growing importance of the study of organic remains.

Werner.—The art of mining has long been taught in France, Germany, and Hungary, in scientific institutions established for that purpose, where mineralogy has always been a principal branch of instruction.

Werner was named, in 1775, professor of that science in the "School of Mines," at Freyberg, in Saxony. He directed his attention not merely to the composition and external characters of minerals, but also to what he termed "geognosy," or the natural position of minerals in particular rocks, together with the grouping of those rocks, their geographical distribution, and various relations. The phenomena observed in the structure of the globe had hitherto served for little else than to furnish interesting topics for philosophical discussion; but when Werner pointed out their application to the practical purposes of mining, they were instantly regarded by a large class of men as an essential part of their professional education, and from that time the science was cultivated in Europe more ardently and systematically. Werner's mind was at once imaginative and richly stored with miscellaneous knowledge. He associated every thing with his favorite science, and in his excursive

* Nov. comm. Petr. XVII. Cuvier, Eloge de Pallas.

lectures, he pointed out all the economical uses of minerals, and their application to medicine; the influence of the mineral composition of rocks upon the soil, and of the soil upon the resources, wealth, and civilization of man. The vast sandy plains of Tartary and Africa, he would say, retained their inhabitants in the shape of wandering shepherds; the granitic mountains and the low calcareous and alluvial plains gave rise to different manners, degrees of wealth, and intelligence. The history even of languages, and the migration of tribes, had been determined by the direction of particular strata. The qualities of certain stones used in building would lead him to descant on the architecture of different ages and nations; and the physical geography of a country frequently invited him to treat of military tactics. The charm of his manners and his eloquence kindled enthusiasm in the minds of his pupils; and many, who had intended at first only to acquire a slight knowledge of mineralogy, when they had once heard him, devoted themselves to it as the business of their lives. In a few years, a small school of mines, before unheard of in Europe, was raised to the rank of a great university; and men already distinguished in science studied the German language, and came from the most distant countries to hear the great oracle of geology.*

Werner had a great antipathy to the mechanical labor of writing, and, with the exception of a valuable treatise on metalliferous veins, he could never be persuaded to pen more than a few brief memoirs, and those containing no development of his general views. Although the natural modesty of his disposition was excessive, approaching even to timidity, he indulged in the most bold and sweeping generalizations, and he inspired all his scholars with a most implicit faith in his doctrines. Their admiration of his genius, and the feelings of gratitude and friendship which they all felt for him, were not undeserved; but the supreme authority usurped by him over the opinions of his contemporaries, was eventually prejudicial to the progress of the science; so much so, as greatly to counterbalance the advantages which it derived from his exertions. If it be true that delivery be the first, second, and third requisite in a popular orator, it is no less certain, that to travel is of first, second, and third importance to those who desire to originate just and comprehensive views concerning the structure of our globe. Now Werner had not travelled to distant countries; he had merely explored a small portion of Germany, and conceived, and persuaded others to believe, that the whole surface of our planet, and all the mountain chains in the world, were made after the model of his own province. It became a ruling object of ambition in the minds of his pupils to confirm the generalizations of their great master, and to discover in the most distant parts of the globe his "universal formations," which he supposed had been each in succession simultaneously precipitated over the whole earth from a common menstruum, or "chaotic fluid." It now appears

* Cuvier, Eloge de Werner.

that the Saxon professor had misinterpreted many of the most important appearances even in the immediate neighborhood of Freyberg. Thus, for example, within a day's journey of his school, the porphyry, called by him primitive, has been found not only to send forth veins or dikes through strata of the coal formation, but to overlie them in mass. The granite of the Hartz mountains, on the other hand, which he supposed to be the nucleus of the chain, is now well known to traverse the other beds, as near Goslar; and still nearer Freyberg, in the Erzgebirge, the mica slate does not mantle round the granite as was supposed, but abuts abruptly against it. Fragments, also, of the greywacké slate, containing organic remains, have recently been found entangled in the granite of the Hartz, by M. de Seckendorf.*

The principal merit of Werner's system of instruction consisted in steadily directing the attention of his scholars to the constant relations of superposition of certain mineral groups; but he had been anticipated, as has been shown in the last chapter, in the discovery of this general law, by several geologists in Italy and elsewhere; and his leading divisions of the secondary strata were at the same time, and independently, made the basis of an arrangement of the British strata by our countryman, William Smith, to whose work I shall refer in the sequel.

Controversy between the Vulcanists and Neptunists.—In regard to basalt and other igneous rocks, Werner's theory was original, but it was also extremely erroneous. The basalts of Saxony and Hesse, to which his observations were chiefly confined, consisted of tabular masses capping the hills, and not connected with the levels of existing valleys, like many in Auvergne and the Vivarais. These basalts, and all other rocks of the same family in other countries, were, according to him, chemical precipitates from water. He denied that they were the products of submarine volcanoes; and even taught that, in the primeval ages of the world, there were no volcanoes. His theory was opposed, in a twofold sense, to the doctrine of the permanent agency of the same causes in nature; for not only did he introduce, without scruple, many imaginary causes supposed to have once effected great revolutions in the earth, and then to have become extinct, but new ones also were feigned to have come into play in modern times; and, above all, that most violent instrument of change, the agency of subterranean heat.

So early as 1768, before Werner had commenced his mineralogical studies, Raspe had truly characterized the basalts of Hesse as of igneous origin. Arduino, we have seen, had pointed out numerous varieties of trap-rock in the Vicentin as analogous to volcanic products, and as distinctly referable to ancient submarine eruptions. Desmarest, as before stated, had, in company with Fortis, examined the Vicentin in 1766, and confirmed Arduino's views. In 1772, Banks, Solander, and Troil compared the columnar basalt of Hecla with that of the Hebrides. Collini,

* I am indebted for this information partly to Messrs. Sedgwick and Murchison, who have investigated the country, and partly to Dr. Charles Hartmann, the translator of this work into German.

in 1774, recognized the true nature of the igneous rocks on the Rhine, between Andernach and Bonn. In 1775, Guettard visited the Vivarais, and established the relation of basaltic currents to lavas. Lastly, in 1779, Faujas published his description of the volcanoes of the Vivarais and Velay, and showed how the streams of basalt had poured out from craters which still remain in a perfect state.*

Desmarest.—When sound opinions had thus for twenty years prevailed in Europe concerning the true nature of the ancient trap-rocks, Werner by his simple dictum caused a retrograde movement, and not only overturned the true theory, but substituted for it one of the most unphilosophical that can well be imagined. The continued ascendancy of his dogmas on this subject was the more astonishing, because a variety of new and striking facts were daily accumulated in favor of the correct opinions previously entertained. Desmarest, after a careful examination of Auvergne, pointed out, first, the most recent volcanoes which had their craters still entire, and their streams of lava conforming to the level of the present river-courses. He then showed that there were others of an intermediate epoch, whose craters were nearly effaced, and whose lavas were less intimately connected with the present valleys; and, lastly, that there were volcanic rocks, still more ancient, without any discernible craters or scoriæ, and bearing the closest analogy to rocks in other parts of Europe, the igneous origin of which was denied by the school of Freyberg.†

Desmarest's map of Auvergne was a work of uncommon merit. He first made a trigonometrical survey of the district, and delineated its physical geography with minute accuracy and admirable graphic power. He contrived, at the same time, to express without the aid of colors, many geological details, including the different ages and sometimes even the structure, of the volcanic rocks, and distinguishing them from the fresh-water and the granitic. They alone who have carefully studied Auvergne, and traced the different lava streams from their craters to their termination,—the various isolated basaltic cappings,—the relation of some lavas to the present valleys,—the absence of such relations in others,—can appreciate the extraordinary fidelity of this elaborate work. No other district of equal dimensions in Europe exhibits, perhaps, so beautiful and varied a series of phenomena; and, fortunately, Desmarest possessed at once the mathematical knowledge required for the construction of a map, skill in mineralogy, and a power of original generalization.

Dolomieu—Montlosier.—Dolomieu, another of Werner's contemporaries, had found prismatic basalt among the ancient lavas of Etna; and, in 1784, had observed the alternations of submarine lavas and calcareous strata in the Val di Noto, in Sicily.‡ In 1790, also, he described similar phenomena in the Vicentin and in the Tyrol.§ Montlosier published,

* Cuvier, Eloge de Desmarest.

† Journ. de Phys. vol. xiii. p. 115; and Mém. de l'Inst., Sciences Mathémat. et Phys. vol. vi. p. 219.

‡ Journ. de Phys. tom. xxxv. p. 191. § Ib. tom. xxxvii. part ii. p. 200.

in 1788, an essay on the theories of volcanoes of Auvergne, combining accurate local observations with comprehensive views. Notwithstanding this mass of evidence the scholars of Werner were prepared to support his opinions to their utmost extent; maintaining, in the fulness of their faith, that even obsidian was an aqueous precipitate. As they were blinded by their veneration for the great teacher, they were impatient of opposition, and soon imbibed the spirit of a faction; and their opponents, the Vulcanists, were not long in becoming contaminated with the same intemperate zeal. Ridicule and irony were weapons more frequently employed than argument by the rival sects, till at last the controversy was carried on with a degree of bitterness almost unprecedented in questions of physical science. Desmarest alone, who had long before provided ample materials for refuting such a theory, kept aloof from the strife; and whenever a zealous Neptunist wished to draw the old man into an argument, he was satisfied with replying, "Go and see."*

Hutton, 1788.—It would be contrary to all analogy, in matters of graver import, that a war should rage with such fury on the Continent, and that the inhabitants of our island should not mingle in the affray. Although in England the personal influence of Werner was wanting to stimulate men to the defence of the weaker side of the question, they contrived to find good reason for espousing the Wernerian errors with great enthusiasm. In order to explain the peculiar motives which led many to enter, even with party feeling, into this contest, it will be necessary to present the reader with a sketch of the views unfolded by Hutton, a contemporary of the Saxon geologist. The former naturalist had been educated as a physician, but declining the practice of medicine, he resolved, when young, to remain content with the small indipendence inherited from his father, and thenceforth to give his undivided attention to scientific pursuits. He resided at Edinburgh, where he enjoyed the society of many men of high attainments, who loved him for the simplicity of his manners, and the sincerity of his character. His application was unwearied; and he made frequent tours through different parts of England and Scotland, acquiring considerable skill as a mineralogist, and consequently arriving at grand and comprehensive views in geology. He communicated the results of his observations unreservedly, and with the fearless spirit of one who was conscious that love of truth was the sole stimulus of his exertions. When at length he had matured his views, he published, in 1788, his "Theory of the Earth,"† and the same, afterwards more fully developed in a separate work, in 1795. This treatise was the first in which geology was declared to be in no way concerned about "questions as to the origin of things;" the first in which an attempt was made to dispense entirely with all hypothetical causes, and to explain the former changes of the earth's crust by reference exclusively to natural agents. Hutton

* Cuvier, Eloge de Desmarest. † Ed. Phil. Trans. 1788.

labored to give fixed principles to geology, as Newton had succeeded in doing to astronomy; but, in the former science, too little progress had been made towards furnishing the necessary data, to enable any philosopher, however great his genius, to realize so noble a project.

Huttonian theory.—"The ruins of an older world," said Hutton, "are visible in the present structure of our planet; and the strata which now compose our continents have been once beneath the sea, and were formed out of the waste of pre-existing continents. The same forces are still destroying, by chemical decomposition or mechanical violence, even the hardest rocks, and transporting the materials to the sea, where they are spread out, and form strata analogous to those of more ancient date. Although loosely deposited along the bottom of the ocean, they become afterwards altered and consolidated by volcanic heat, and then heaved up, fractured, and contorted."

Although Hutton had never explored any region of active volcanoes, he had convinced himself that basalt and many other trap-rocks were of igneous origin, and that many of them had been injected in a melted state through fissures in the older strata. The compactness of these rocks, and their different aspect from that of ordinary lava, he attributed to their having cooled down under the pressure of the sea; and in order to remove the objections started against this theory, his friend, Sir James Hall, instituted a most curious and instructive series of chemical experiments, illustrating the crystalline arrangement and texture assumed by melted matter cooled under high pressure.

The absence of stratification in granite, and its analogy, in mineral character, to rocks which he deemed of igneous origin, led Hutton to conclude that granite also must have been formed from matter in fusion; and this inference he felt could not be fully confirmed, unless he discovered at the contact of granite and other strata a repetition of the phenomena exhibited so constantly by the trap-rocks. Resolved to try his theory by this test, he went to the Grampians, and surveyed the line of junction of the granite and superincumbent stratified masses, until he found in Glen Tilt, in 1785, the most clear and unequivocal proofs in support of his views. Veins of red granite are there seen branching out from the principal mass, and traversing the black micaceous schist and primary limestone. The intersected stratified rocks are so distinct in color and appearance as to render the example in that locality most striking, and the alteration of the limestone in contact was very analogous to that produced by trap veins on calcareous strata. This verification of his system filled him with delight, and called forth such marks of joy and exultation, that the guides who accompanied him, says his biographer, were convinced that he must have discovered a vein of silver or gold.* He was aware that the same theory would not explain the origin of the primary schists, but these he called primary, rejecting the term primitive, and was disposed to consider them as sedimentary rocks al-

* Playfair's Works, vol. iv. p. 75.

tered by heat, and that they originated in some other form from the waste of previously existing rocks.

By this important discovery of granite veins, to which he had been led by fair induction from an independent class of facts, Hutton prepared the way for the greatest innovation of the systems of his predecessors. Vallisneri had pointed out the general fact that there were certain fundamental rocks which contained no organic remains, and which he supposed to have been formed before the creation of living beings. Moro, Generelli, and other Italian writers, embraced the same doctrine; and Lehman regarded the mountains called by him primitive, as parts of the original nucleus of the globe. The same tenet was an article of faith in the school of Freyberg; and if any one ventured to doubt the possibility of our being enabled to carry back our researches to the creation of the present order of things, the granitic rocks were triumphantly appealed to. On them seemed written, in legible characters, the memorable inscription—

> "Dinanzi a me non fur cose create
> Se non eterne;"*

and no small sensation was excited when Hutton seemed, with unhallowed hand, desirous to erase characters already regarded by many as sacred. "In the economy of the world," said the Scotch geologist, "I can find no traces of a beginning, no prospect of an end;" a declaration the more startling when coupled with the doctrine, that all past ages on the globe had been brought about by the slow agency of existing causes. The imagination was first fatigued and overpowered by endeavoring to conceive the immensity of time required for the annihilation of whole continents by so insensible a process; and when the thoughts had wandered through these interminable periods, no resting-place was assigned in the remotest distance. The oldest rocks were represented to be of a derivative nature, the last of an antecedent series, and that, perhaps, one of many pre-existing worlds. Such views of the immensity of past time, like those unfolded by the Newtonian philosophy in regard to space, were too vast to awaken ideas of sublimity unmixed with a painful sense of our incapacity to conceive a plan of such infinite extent. Worlds are seen beyond worlds immeasurably distant from each other, and, beyond them all, innumerable other systems are faintly traced on the confines of the visible universe.

The characteristic feature of the Huttonian theory was, as before hinted, the exclusion of all causes not supposed to belong to the present order of nature. But Hutton had made no step beyond Hooke, Moro, and Raspe, in pointing out in what manner the laws now governing subterranean movements might bring about geological changes, if sufficient time be allowed. On the contrary, he seems to have fallen far short of some of their views, especially when he refused to attribute any part

* "Before me things create were none, save things
Eternal."——Dante's *Inferno*, canto iii. Cary's Translation.

of the external configuration of the earth's crust to subsidence. He imagined that the continents were first gradually destroyed by aqueous degradation; and when their ruins had furnished materials for new continents, they were upheaved by violent convulsions. He therefore required alternate periods of general disturbance and repose; and such he believed had been, and would forever be, the course of nature.

Generelli, in his exposition of Moro's system, had made a far nearer approximation towards reconciling geological appearances with the state of nature as known to us; for while he agreed with Hutton, that the decay and reproduction of rocks were always in progress, proceeding with the utmost uniformity, the learned Carmelite represented the repairs of mountains by elevation from below to be effected by an equally constant and synchronous operation. Neither of these theories, considered singly, satisfies all the conditions of the great problem, which a geologist, who rejects cosmological causes, is called upon to solve; but they probably contain together the germs of a perfect system. There can be no doubt, that periods of disturbance and repose have followed each other in succession in every region of the globe; but it may be equally true, that the energy of the subterranean movements has been always uniform as regards the *whole earth.* The force of earthquakes may for a cycle of years have been invariably confined, as it is now, to large but determinate spaces, and may then have gradually shifted its position, so that another region, which had for ages been at rest, became in its turn the grand theatre of action.

Playfair's illustrations of Hutton.—The explanation proposed by Hutton, and by Playfair, the illustrator of his theory, respecting the origin of valleys and of alluvial accumulations, was also very imperfect. They ascribed none of the inequalities of the earth's surface to movements which accompanied the upheaving of the land, imagining that valleys in general were formed in the course of ages by the rivers now flowing in them; while they seem not to have reflected on the excavating and transporting power which the waves of the ocean might exert on land during its emergence.

Although Hutton's knowledge of mineralogy and chemistry was considerable, he possessed but little information concerning organic remains; they merely served him, as they did Werner, to characterize certain strata, and to prove their marine origin. The theory of former revolutions in organic life was not yet fully recognized; and without this class of proofs in support of the antiquity of the globe, the indefinite periods demanded by the Huttonian hypothesis appeared visionary to many; and some, who deemed the doctrine inconsistent with revealed truths, indulged very uncharitable suspicions of the motives of its author. They accused him of a deliberate design of reviving the heathen dogma of an "eternal succession," and of denying that this world ever had a beginning. Playfair, in the biography of his friend, has the following comment on this part of their theory:—"In the planetary motions, where geometry has carried the eye so far, both into the future and the past, we discover no mark either of the commencement or termination of the present order. It is unreasonable, indeed, to suppose that such marks should anywhere exist. The Author of Nature has not given laws to the universe, which, like the institutions of men, carry in themselves the elements of their own destruction. He has not permitted in His works any symptom of infancy or of old age, or any sign by which we may estimate either their future or their past duration. *He may put an end, as he no doubt gave a beginning,* to the present system, at some determinate period of time; but we may rest assured that this great catastrophe will not be brought about by the laws now existing, and that it is not indicated, by any thing which we perceive."*

The party feeling excited against the Huttonian doctrines, and the open disregard of candor and temper in the controversy, will hardly be credited by the reader, unless he recalls to his recollection that the mind of the English public was at that time in a state of feverish excitement. A class of writers in France had been laboring industriously for many years, to diminish the influence of the clergy, by sapping the foundations of the Christian faith; and their success, and the consequences of the Revolution, had alarmed the most resolute minds, while the imagination of the more timid was continually haunted by dread of innovation, as by the phantom of some fearful dream.

Voltaire.—Voltaire had used the modern discoveries in physics as one of the numerous weapons of attack and ridicule directed by him against the Scriptures. He found that the most popular systems of geology were accommodated to the sacred writings, and that much ingenuity had been employed to make every fact coincide exactly with the Mosaic account of the creation and deluge. It was, therefore, with no friendly feelings that he contemplated the cultivators of geology in general, regarding the science as one which had been successfully enlisted by theologians as an ally in their cause.† He knew that the majority of those who were aware of the abundance of fossil shells in the interior of continents, were still persuaded that they were proofs of the universal deluge; and as the readiest way of shaking this article of faith, he endeavored to inculcate skepticism as to the real nature of such shells, and to recall from contempt the exploded dogma of the sixteenth century, that they were sports of nature. He also pretended that vegetable impressions were not those of real plants.‡ Yet he was perfectly convinced that the shells had really belonged to living testacea, as may

* Playfair's Works, vol. iv. p. 55.

† In allusion to the theories of Burnet, Woodward, and other physico-theological writers, he declared that they were as fond of changes of scene on the face of the globe, as were the populace at a play. "Every one of them destroys and renovates the earth after his own fashion, as Descartes framed it: for philosophers put themselves without ceremony in the place of God, and think to create a universe with a word."—Dissertation envoyée a l'Academie de Boulogne, sur les Changemens arrivés dans notre Globe. Unfortunately, this and similar ridicule directed against the cosmogonists was too well deserved.

‡ See the chapter on "Des Pierres figurés."

be seen in his essay "On the formation of Mountains."* He would sometimes, in defiance of all consistency, shift his ground when addressing the vulgar; and, admitting the true nature of the shells collected in the Alps and other places, pretend that they were Eastern species, which had fallen from the hats of pilgrims coming from Syria. The numerous essays written by him on geological subjects were all calculated to strengthen prejudices, partly because he was ignorant of the real state of the science, and partly from his bad faith.† On the other hand, they who knew that his attacks were directed by a desire to invalidate Scripture, and who were unacquainted with the true merits of the question, might well deem the old diluvian hypothesis incontrovertible, if Voltaire could adduce no better argument against it than to deny the true nature of organic remains.

It is only by careful attention to impediments originating in extrinsic causes, that we can explain the slow and reluctant adoption of the simplest truths in geology. First, we find many able naturalists adducing the fossil remains of marine animals as proofs of an event related in Scripture. The evidence is deemed conclusive by the multitude for a century or more; for it favors opinions which they entertained before, and they are gratified by supposing them confirmed by fresh and unexpected proofs. Many who see through the fallacy have no wish to undeceive those who are influenced by it, approving the effect of the delusion, and conniving at it as a pious fraud; until, finally, an opposite party, who are hostile to the sacred writings, labor to explode the erroneous opinion, by substituting for it another dogma, which they know to be equally unsound.

The heretical Vulcanists were soon after openly assailed in England, by imputations of the most illiberal kind. We cannot estimate the malevolence of such a persecution, by the pain which similar insinuations might now inflict; for although charges of infidelity and atheism must always be odious, they were injurious in the extreme at that moment of political excitement; and it was better, perhaps, for a man's good reception in society, that his moral character should have been traduced, than that he should become a mark for these poisoned weapons.

I shall pass over the works of numerous divines, who may be excused for sensitiveness on points which then excited so much uneasiness in the public mind; and shall say nothing of the amiable poet Cowper,‡ who

* In that essay he lays it down, "that all naturalists are now agreed that deposits of shells in the midst of the continents are monuments of the continued occupation of these districts by the ocean." In another place also, when speaking of the fossil shells of Touraine, he admits their true origin.

† As an instance of his desire to throw doubt indiscriminately on all geological data, we may recall the passage where he says, that "the bones of a reindeer and hippopotamus discovered near Etempes did not prove, as some would have it, that Lapland and the Nile were once on a tour from Paris to Orleans, but merely that a lover of curiosities once preserved them in his cabinet."

‡ "Some drill and bore
The solid earth, and from the strata there
Extract a register, by which we learn

could hardly be expected to have inquired into the merit of doctrines in physics. But in the foremost ranks of the intolerant are found several laymen who had high claims to scientific reputation. Among these appears Williams, a mineral surveyor of Edinburgh, who published a "Natural History of the Mineral Kingdom," in 1789; a work of great merit, for that day, and of practical utility, as containing the best account of the coal strata. In his preface he misrepresents Hutton's theory altogether, and charges him with considering all rocks to be lavas of different colors and structure; and also with "warping every thing to support the eternity of the world."* He descants on the pernicious influence of such skeptical notions, as leading to downright infidelity and atheism, "and as being nothing less than to depose the Almighty Creator of the universe from his office."†

Kirwan—De Luc.—Kirwan, president of the Royal Academy of Dublin, a chemist and mineralogist of some merit, but who possessed much greater authority in the scientific world than he was entitled by his talents to enjoy, said, in the introduction to his "Geological Essays, 1799," "that *sound* geology *graduated* into religion, and was required to dispel certain systems of atheism or infidelity, of which they had had recent experience."‡ He was an uncompromising defender of the aqueous theory of all rocks, and was scarcely surpassed by Burnet and Whiston, in his desire to adduce the Mosaic writings in confirmation of his opinions.

De Luc, in the preliminary discourse to his Treatise on Geology,§ says, "The weapons have been changed by which revealed religion is attacked; it is now assailed by geology, and the knowledge of this science has become essential to theologians." He imputes the failure of former geological systems to their having been anti-Mosaical, and directed against a "sublime tradition." These and similar imputations, reiterated in the works of De Luc, seem to have been taken for granted by some modern writers: it is therefore necessary to state, in justice to the numerous geologists of different nations, whose works have been considered, that none of them were guilty of endeavoring, by arguments drawn from physics, to invalidate scriptural tenets. On the contrary, the majority of those who were fortunate enough "to discover the true causes of things," rarely deserved another part of the poet's panegyric, "*Atque metus omnes subjecit pedibus.*" The caution and even timid reserve, of many eminent Italian authors of the earlier period is very apparent; and there can hardly be a doubt, that they subscribed to certain dogmas, and particularly to the first diluvian theory, out of deference to popular prejudices, rather than from conviction. If they were guilty of dissimulation, we may feel regret, but must not blame their want of moral courage, reserving rather our condemnation for the intol-

That he who made it, and revealed its date
To Moses, was mistaken in its age."
The Task, book iii. "The Garden."

* P. 577. † P. 59. ‡ Introd. p. 2. § London, 1809.

erance of the times, and that inquisitorial power which forced Galileo to abjure, and the two Jesuits to disclaim the theory of Newton.*

Hutton answered Kirwan's attacks with great warmth, and with the indignation justly excited by unmerited reproach. "He had always displayed," says Playfair, "the utmost disposition to admire the beneficent design manifested in the structure of the world; and he contemplated with delight those parts of his theory which made the greatest additions to our knowledge of final causes." We may say with equal truth, that in no scientific works in our language can more eloquent passages be found, concerning the fitness, harmony, and grandeur of all parts of the creation, than in those of Playfair. They are evidently the unaffected expressions of a mind, which contemplated the study of nature, as best calculated to elevate our conceptions of the attributes of the First Cause. At any other time the force and elegance of Playfair's style must have insured popularity to the Huttonian doctrines; but by a singular coincidence, Neptunianism and orthodoxy were now associated in the same creed; and the tide of prejudice ran so strong, that the majority were carried far away into the chaotic fluid, and other cosmological inventions of Werner. These fictions the Saxon professor had borrowed with little modification, and without any improvement, from his predecessors. They had not the smallest foundation either in Scripture or in common sense, and were probably approved of by many as being so ideal and unsubstantial, that they could never come into violent collision with any preconceived opinions.

According to De Luc, the first essential distinction to be made between the various phenomena exhibited on the surface of the earth was, to determine which were the results of causes still in action, and which had been produced by causes that had ceased to act. The form and composition of the mass of our continents, he said, and their existence above the level of the sea, must be ascribed to causes no longer in action. These continents emerged, at no very remote period, on the sudden retreat of the ocean, the waters of which made their way into subterranean caverns. The formation of the rocks which enter into the crust of the earth began with the precipitation of granite from a primordial liquid, after which other strata containing the remains of or-

* In a most able article, by Mr. Drinkwater, on the "Life of Galileo," published in the "Library of Useful Knowledge," it is stated that both Galileo's work, and the book of Copernicus, "Nisi corrigatur" (for, with the omission of certain passages, it was sanctioned), were still to be seen on the forbidden list of the Index at Rome, in 1828. I was, however, assured in the same year, by Professor Scarpellini, at Rome, that Pius VII., a pontiff distinguished for his love of science, had procured a repeal of the edicts against Galileo and the Copernican system. He had assembled the Congregation; and the late Cardinal Toriozzi, assessor of the Sacred Office, proposed that they should wipe off this scandal from the church." The repeal was carried, with the dissentient voice of one Dominican only. Long before that time the Newtonian theory had been taught in the Sapienza, and all Catholic universities in Europe (with the exception, I am told, of Salamanca); but it was always required of professors, in deference to the decrees of the church, to use the term *hypothesis*, instead of theory. They now speak of the Copernican *theory*.

ganized bodies were deposited, till at last the present sea remained as the residuum of the primordial liquid, and no longer continued to produce mineral strata.*

William Smith, 1790.—While the tenets of the rival schools of Freyberg and Edinburgh were warmly espoused by devoted partisans, the labors of an individual, unassisted by the advantages of wealth or station in society, were almost unheeded. Mr. William Smith, an English surveyor, published his "Tabular View of the British Strata" in 1790, wherein he proposed a classification of the secondary formations in the West of England. Although he had not communicated with Werner, it appeared by this work that he had arrived at the same views respecting the laws of superposition of stratified rocks; that he was aware that the order of succession of different groups was never inverted; and that they might be identified at very distant points by their peculiar organized fossils.

From the time of the appearance of the "Tabular View," the author labored to construct a geological map of the whole of England; and with the greatest disinterestedness of mind, communicated the results of his investigations to all who desired information, giving such publicity to his original views, as to enable his contemporaries almost to compete with him in the race. The execution of his map was completed in 1815, and remains a lasting monument of original talent and extraordinary perseverance; for he had explored the whole country on foot, without the guidance of previous observers, or the aid of fellow-laborers, and had succeeded in throwing into natural divisions the whole complicated series of British rocks. D'Aubuisson, a distinguished pupil of Werner, paid a just tribute of praise to this remarkable performance, observing, that "what many celebrated mineralogists had only accomplished for a small part of Germany in the course of half a century, had been effected by a single individual for the whole of England."†

Werner invented a new language to express his divisions of rocks, and some of his technical terms, such as grauwacke, gneiss, and others, passed current in every country in Europe. Smith adopted for the most part English provincial terms, often of barbarous sound, such as gault, cornbrash, clunch clay; and affixed them to subdivisions of the British series. Many of these still retain their place in our scientific classifications, and attest his priority of arrangement.

MODERN PROGRESS OF GEOLOGY.

The contention of the rival factions of the Vulcanists and Neptunists had been carried to such a height, that these names had become terms of reproach; and the two parties had been less occupied in searching for truth, than for such arguments as might strengthen their own cause or serve to annoy their antagonists. A new school at last arose, whc

* Elementary Treatise on Geology. London, 1809. Translated by De la Fite
† See Dr. Fitton's Memoir, before cited, p. 57.

professed the strictest neutrality, and the utmost indifference to the systems of Werner and Hutton, and who resolved diligently to devote their labors to observation. The reaction, provoked by the intemperance of the conflicting parties, now produced a tendency to extreme caution. Speculative views were discountenanced, and, through fear of exposing themselves to the suspicion of a bias towards the dogmas of a party, some geologists became anxious to entertain no opinion whatever on the causes of phenomena, and were inclined to skepticism even where the conclusions deducible from observed facts scarcely admitted of reasonable doubt.

Geological Society of London.—But although the reluctance to theorize was carried somewhat to excess, no measure could be more salutary at such a moment than a suspension of all attempts to form what were termed "theories of the earth." A great body of new data were required; and the Geological Society of London, founded in 1807, conduced greatly to the attainment of this desirable end. To multiply and record observations, and patiently to await the result at some future period, was the object proposed by them; and it was their favorite maxim that the time was not yet come for a general system of geology, but that all must be content for many years to be exclusively engaged in furnishing materials for future generalizations. By acting up to these principles with consistency, they in a few years disarmed all prejudice, and rescued the science from the imputation of being a dangerous, or at best but a visionary pursuit.

A distinguished modern writer has with truth remarked, that the advancement of three of the main divisions of geological inquiry have during the last half century been promoted successively by three different nations of Europe,—the Germans, the English, and the French.* We have seen that the systematic study of what may be called mineralogical geology had its origin and chief point of activity in Germany, where Werner first described with precision the mineral characters of rocks. The classification of the secondary formations, each marked by their peculiar fossils, belongs, in a great measure, to England, where the labors before alluded to of Smith, and those of the most active members of the Geological Society of London, were steadily directed to these objects. The foundation of the third branch, that relating to the tertiary formations, was laid in France by the splendid work of Cuvier and Brongniart, published in 1808, "On the Mineral Geography and Organic Remains of the Neighborhood of Paris."

We may still trace, in the language of the science and our present methods of arrangement, the various countries where the growth of these several departments of geology was at different times promoted. Many names of simple minerals and rocks remain to this day German; while the European divisions of the secondary strata are in great part English, and are, indeed, often founded too exclusively on English types. Lastly, the subdivisions first established of the succession of strata in the Paris

* Whewell, British Critic, No. xvii. p. 187, 1831.

basin have served as normal groups to which other tertiary deposits throughout Europe have been compared, even in cases where this standard was wholly inapplicable.

No period could have been more fortunate for the discovery, in the immediate neighborhood of Paris, of a rich store of well-preserved fossils, than the commencement of the present century; for at no former era had Natural history been cultivated with such enthusiasm in the French metropolis. The labors of Cuvier in comparative osteology, and of Lamarck in recent and fossil shells, had raised these departments of study to a rank of which they had never previously been deemed susceptible. Their investigations had eventually a powerful effect in dispelling the illusion which had long prevailed concerning the absence of analogy between the ancient and modern state of our planet. A close comparison of the recent and fossil species and the inferences drawn in regard to their habits, accustomed the geologist to contemplate the earth as having been at successive periods the dwelling-place of animals and plants of different races, some terrestrial, and others aquatic—some fitted to live in seas, others in the waters of lakes and rivers. By the consideration of these topics, the mind was slowly and insensibly withdrawn from imaginary pictures of catastrophes and chaotic confusion, such as haunted the imagination of the early cosmogonists. Numerous proofs were discovered of the tranquil deposition of sedimentary matter, and the slow development of organic life. If many writers, and Cuvier himself in the number, still continued to maintain, that "the thread of induction was broken,"* yet, in reasoning by the strict rules of induction from recent to fossil species, they in a great measure disclaimed the dogma which in theory they professed. The adoption of the same generic, and, in some cases, even of the same specific, names for the exuviæ of fossil animals and their living analogues, was an important step towards familiarizing the mind with the idea of the identity and unity of the system in distant eras. It was an acknowledgment, as it were, that part at least of the ancient memorials of nature were written in a living language. The growing importance, then, of the natural history of organic remains may be pointed out as the characteristic feature of the progress of the science during the present century. This branch of knowledge has already become an instrument of great utility in geological classification, and is continuing daily to unfold new data for grand and enlarged views respecting the former changes of the earth.

When we compare the result of observations in the last fifty years with those of the three preceding centuries, we cannot but look forward with the most sanguine expectations to the degree of excellence to which geology may be carried, even by the labors of the present generation. Never, perhaps, did any science, with the exception of astronomy, unfold, in an equally brief period, so many novel and unexpected truths, and overturn so many preconceived opinions. The senses had for ages declared the earth to be at rest, until the astronomer taught that it was

* Discours sur les Révol. &c.

carried through space with inconceivable rapidity. In like manner was the surface of this planet regarded as having remained unaltered since its creation, until the geologist proved that it had been the theatre of reiterated change, and was still the subject of slow but never-ending fluctuations. The discovery of other systems in the boundless regions of space was the triumph of astronomy; to trace the same system through various transformations—to behold it at successive eras adorned with different hills and valleys, lakes and seas, and peopled with new inhabitants, was the delightful meed of geological research. By the geometer were measured the regions of space, and the relative distances of the heavenly bodies;—by the geologist myriads of ages were reckoned, not by arithmetical computation, but by a train of physical events—a succession of phenomena in the animate and inanimate worlds —signs which convey to our minds more definite ideas than figures can do of the immensity of time.

Whether our investigation of the earth's history and structure will eventually be productive of as great practical benefits to mankind as a knowledge of the distant heavens, must remain for the decision of posterity. It was not till astronomy had been enriched by the observations of many centuries, and had made its way against popular prejudices to the establishment of a sound theory, that its application to the useful arts was most conspicuous. The cultivation of geology began at a later period; and in every step which it has hitherto made towards sound theoretical principles, it had to contend against more violent prepossessions. The practical advantages already derived from it have not been inconsiderable; but our generalizations are yet imperfect, and they who come after us may be expected to reap the most valuable fruits of our labor. Meanwhile, the charm of first discovery is our own; and, as we explore this magnificent field of inquiry, the sentiment of a great historian of our times may continually be present to our minds, that "he who calls what has vanished back again into being, enjoys a bliss like that of creating."*

CHAPTER V.

PREJUDICES WHICH HAVE RETARDED THE PROGRESS OF GEOLOGY

Prepossessions in regard to the duration of past time—Prejudices arising from our peculiar position as inhabitants of the land—Of those occasioned by our not seeing subterranean changes now in progress—All these causes combine to make the former course of Nature appear different from the present—Objections to the doctrine, that causes similar in kind and energy to those now acting, have produced the former changes of the earth's surface, considered.

IF we reflect on the history of the progress of geology, as explained in the preceding chapters, we perceive that there have been great fluc-

* Niebuhr's Hist. of Rome, vol. i. p. 5. Hare and Thirlwall's translation.

tuations of opinion respecting the nature of the causes to which all former changes of the earth's surface are referable. The first observers conceived the monuments which the geologist endeavors to decipher to relate to an original state of the earth, or to a period when there were causes in activity, distinct, in kind and degree, from those now constituting the economy of nature. These views were gradually modified, and some of them entirely abandoned, in proportion as observations were multiplied, and the signs of former mutations more skilfully interpreted. Many appearances, which had for a long time been regarded as indicating mysterious and extraordinary agency, were finally recognized as the necessary result of the laws now governing the material world; and the discovery of this unlooked-for conformity has at length induced some philosophers to infer, that, during the ages contemplated in geology, there has never been any interruption to the agency of the same uniform laws of change. The same assemblage of general causes, they conceive, may have been sufficient to produce, by their various combinations, the endless diversity of effects, of which the shell of the earth has preserved the memorials; and, consistently with these principles, the recurrence of analogous changes is expected by them in time to come.

Whether we coincide or not in this doctrine, we must admit that the gradual progress of opinion concerning the succession of phenomena in very remote eras, resembles, in a singular manner, that which has accompanied the growing intelligence of every people, in regard to the economy of nature in their own times. In an early state of advancement, when a great number of natural appearances are unintelligible, an eclipse, an earthquake, a flood, or the approach of a comet, with many other occurrences afterwards found to belong to the regular course of events, are regarded as prodigies. The same delusion prevails as to moral phenomena, and many of these are ascribed to the intervention of demons, ghosts, witches, and other immaterial and supernatural agents. By degrees, many of the enigmas of the moral and physical world are explained, and, instead of being due to extrinsic and irregular causes, they are found to depend on fixed and invariable laws. The philosopher at last becomes convinced of the undeviating uniformity of secondary causes; and, guided by his faith in this principle, he determines the probability of accounts transmitted to him of former occurrences, and often rejects the fabulous tales of former times, on the ground of their being irreconcilable with the experience of more enlightened ages.

Prepossessions in regard to the duration of past time.—As a belief in the want of conformity in the causes by which the earth's crust has been modified in ancient and modern periods was, for a long time, universally prevalent, and that, too, amongst men who were convinced that the order of nature had been uniform for the last several thousand years, every circumstance which could have influenced their minds and given an undue bias to their opinions deserves particular attention.

Now the reader may easily satisfy himself, that, however undeviating the course of nature may have been from the earliest epochs, it was impossible for the first cultivators of geology to come to such a conclusion, so long as they were under a delusion as to the age of the world, and the date of the first creation of animate beings. However fantastical some theories of the sixteenth century may now appear to us,—however unworthy of men of great talent and sound judgment,—we may rest assured that, if the same misconception now prevailed in regard to the memorials of human transactions, it would give rise to a similar train of absurdities. Let us imagine, for example, that Champollion, and the French and Tuscan literati lately engaged in exploring the antiquities of Egypt, had visited that country with a firm belief that the banks of the Nile were never peopled by the human race before the beginning of the nineteenth century, and that their faith in this dogma was as difficult to shake as the opinion of our ancestors that the earth was never the abode of living beings until the creation of the present continents, and of the species now existing,—it is easy to perceive what extravagant systems they would frame, while under the influence of this delusion, to account for the monuments discovered in Egypt. The sight of the pyramids, obelisks, colossal statues, and ruined temples, would fill them with such astonishment, that for a time they would be as men spell-bound—wholly incapable of reasoning with sobriety. They might incline at first to refer the construction of such stupendous works to some superhuman powers of a primeval world. A system might be invented resembling that so gravely advanced by Manetho, who relates that a dynasty of gods originally ruled in Egypt, of whom Vulcan, the first monarch, reigned nine thousand years; after whom came Hercules and other demigods, who were at last succeeded by human kings.

When some fanciful speculations of this kind had amused their imaginations for a time, some vast repository of mummies would be discovered, and would immediately undeceive those antiquaries who enjoyed an opportunity of personally examining them; but the prejudices of others at a distance, who were not eye-witnesses of the whole phenomena, would not be so easily overcome. The concurrent report of many travellers would, indeed, render it necessary for them to accommodate ancient theories to some of the new facts, and much wit and ingenuity would be required to modify and defend their old positions. Each new invention would violate a greater number of known analogies; for if a theory be required to embrace some false principle, it becomes more visionary in proportion as facts are multiplied, as would be the case if geometers were now required to form an astronomical system on the assumption of the immobility of the earth.

Amongst other fanciful conjectures concerning the history of Egypt, we may suppose some of the following to be started. "As the banks of the Nile have been so recently colonized for the first time, the curious substances called mummies could never in reality have belonged to men. They may have been generated by some *plastic virtue* residing in the interior of the earth, or they may be abortions of Nature produced by her incipient efforts in the work of creation. For if deformed beings are sometimes born even now, when the scheme of the universe is fully developed, many more may have been 'sent before their time, scarce half made up,' when the planet itself was in the embryo state. But if these notions appear to derogate from the perfection of the Divine attributes, and if these mummies be in all their parts true representations of the human form, may we not refer them to the future rather than the past?—May we not be looking into the womb of Nature, and not her grave? May not these images be like the shades of the unborn in Virgil's Elysium—the archetypes of men not yet called into existence?"

These speculations, if advocated by eloquent writers, would not fail to attract many zealous votaries, for they would relieve men from the painful necessity of renouncing preconceived opinions. Incredible as such skepticism may appear, it has been rivalled by many systems of the sixteenth and seventeenth centuries, and among others by that of the learned Falloppio, who regarded the tusks of fossil elephants as earthy concretions, and the pottery or fragments of vases in the Monte Testaceo, near Rome, as works of nature, and not of art. But when one generation had passed away, and another, not compromised to the support of antiquated dogmas, had succeeded, they would review the evidence afforded by mummies more impartially, and would no longer controvert the preliminary question, that human beings had lived in Egypt before the nineteenth century: so that when a hundred years perhaps had been lost, the industry and talents of the philosopher would be at last directed to the elucidation of points of real historical importance.

But the above arguments are aimed against one only of many prejudices with which the earlier geologists had to contend. Even when they conceded that the earth had been peopled with animate beings at an earlier period than was at first supposed, they had no conception that the quantity of time bore so great a proportion to the historical era as is now generally conceded. How fatal every error as to the quantity of time must prove to the introduction of rational views concerning the state of things in former ages, may be conceived by supposing the annals of the civil and military transactions of a great nation to be perused under the impression that they occurred in a period of one hundred instead of two thousand years. Such a portion of history would immediately assume the air of a romance; the events would seem devoid of credibility, and inconsistent with the present course of human affairs. A crowd of incidents would follow each other in thick succession. Armies and fleets would appear to be assembled only to be destroyed, and cities built merely to fall in ruins. There would be the most violent transitions from foreign or intestine war to periods of profound peace, and the works effected during the years of disorder or tranquillity would appear alike superhuman in magnitude.

He who should study the monuments of the natural world under the influence of a similar infatuation, must draw a no less exaggerated picture of the energy and violence of causes, and must experience the same insurmountable difficulty in reconciling the former and present state of nature. If we could behold in one view all the volcanic cones thrown up in Iceland, Italy, Sicily, and other parts of Europe, during the last five thousand years, and could see the lavas which have flowed during the same period; the dislocations, subsidences, and elevations caused during earthquakes; the lands added to various deltas, or devoured by the sea, together with the effects of devastation by floods, and imagine that all these events had happened in one year, we must form most exalted ideas of the activity of the agents, and the suddenness of the revolutions. Were an equal amount of change to pass before our eyes in the next year, could we avoid the conclusion that some great crisis of nature was at hand? If geologists, therefore, have misinterpreted the signs of a succession of events, so as to conclude that centuries were implied where the characters imported thousands of years, and thousands of years where the language of Nature signified millions, they could not, if they reasoned logically from such false premises, come to any other conclusion than that the system of the natural world had undergone a complete revolution.

We should be warranted in ascribing the erection of the great pyramid to superhuman power, if we were convinced that it was raised in one day; and if we imagine, in the same manner, a continent or mountain-chain to have been elevated during an equally small fraction of the time which was really occupied in upheaving it, we might then be justified in inferring, that the subterranean movements were once far more energetic than in our own times. We know that during one earthquake the coast of Chili may be raised for a hundred miles to the average height of about three feet. A repetition of two thousand shocks, of equal violence, might produce a mountain-chain one hundred miles long, and six thousand feet high. Now, should one or two only of these convulsions happen in a century, it would be consistent with the order of events experienced by the Chilians from the earliest times; but if the whole of them were to occur in the next hundred years, the entire district must be depopulated, scarcely any animals or plants could survive, and the surface would be one confused heap of ruin and desolation.

One consequence of undervaluing greatly the quantity of past time, is the apparent coincidence which it occasions of events necessarily disconnected, or which are so unusual, that it would be inconsistent with all calculation of chances to suppose them to happen at one and the same time. When the unlooked-for association of such rare phenomena is witnessed in the present course of nature, it scarcely ever fails to excite a suspicion of the preternatural in those minds which are not firmly convinced of the uniform agency of secondary causes;—as if the death of some individual in whose fate they are interested happens to be accompanied by the appearance of a luminous meteor, or a comet, or the shock of an earthquake. It would be only necessary to multiply such coincidences indefinitely, and the mind of every philosopher would be disturbed. Now it would be difficult to exaggerate the number of physical events, many of them most rare and unconnected in their nature, which were imagined by the Woodwardian hypothesis to have happened in the course of a few months; and numerous other examples might be found of popular geological theories, which require us to imagine that a long succession of events happened in a brief and almost momentary period.

Another liability to error, very nearly allied to the former, arises from the frequent contact of geological monuments referring to very distant periods of time. We often behold, at one glance, the effects of causes which have acted at times incalculably remote, and yet there may be no striking circumstances to mark the occurrence of a great chasm in the chronological series of Nature's archives. In the vast interval of time which may really have elapsed between the results of operations thus compared, the physical condition of the earth may, by slow and insensible modifications, have become entirely altered; one or more races of organic beings may have passed away, and yet have left behind, in the particular region under contemplation, no trace of their existence.

To a mind unconscious of these intermediate events, the passage from one state of things to another must appear so violent, that the idea of revolutions in the system inevitably suggests itself. The imagination is as much perplexed by the deception, as it might be if two distant points in space were suddenly brought into immediate proximity. Let us suppose, for a moment, that a philosopher should lie down to sleep in some arctic wilderness, and then be transferred by a power, such as we read of in tales of enchantment, to a valley in a tropical country, where, on awaking, he might find himself surrounded by birds of brilliant plumage, and all the luxuriance of animal and vegetable forms of which Nature is so prodigal in those regions. The most reasonable supposition, perhaps, which he could make, if by the necromancer's art he were placed in such a situation, would be, that he was dreaming; and if a geologist form theories under a similar delusion, we cannot expect him to preserve more consistency in his speculations than in the train of ideas in an ordinary dream.

It may afford, perhaps, a lively illustration of the principle here insisted upon, if I recall to the reader's recollection the legend of the Seven Sleepers. The scene of that popular fable was placed in the two centuries which elapsed between the reign of the emperor Decius and the death of Theodosius the younger. In that interval of time (between the years 249 and 450 of our era) the union of the Roman Empire had been dissolved, and some of its fairest provinces overrun by the barbarians of the north. The seat of government had passed from Rome to Constantinople, and the throne from a pagan persecutor to a succession of Christian and orthodox princes. The genius of the empire had been humbled in the dust, and the altars of Diana and Hercules were on the point of being transferred to Catholic saints and martyrs. The legend relates, "that when Decius was still persecuting the Christians, seven noble youths of Ephesus concealed themselves in a spacious cavern in the side of an adjacent mountain, where they were doomed to perish by the tyrant, who gave orders that the entrance should be firmly secured with a pile of huge stones. They immediately fell into a deep slumber, which was miraculously prolonged, without injuring the powers of life, during a period of 187 years. At the end of that time the slaves of Adolius, to whom the inheritance of the mountain had descended, removed the stones to supply materials for some rustic edifice: the light of the sun darted into the cavern, and the Seven Sleepers were permitted to awake. After a slumber, as they thought, of a few hours, they were pressed by the calls of hunger, and resolved that Jamblichus, one of their number, should secretly return to the city to purchase bread for the use of his companions. The youth could no longer recognize the once familiar aspect of his native country, and his surprise was increased by the appearance of a large cross triumphantly erected over the principal gate of Ephesus. His singular dress and obsolete language confounded the baker, to whom he offered an ancient medal of Decius as the current coin of the empire; and Jamblichus, on the suspicion of a secret treasure, was dragged before the judge. Their mutual inquiries produced the amazing discovery, that two centuries were almost elapsed since Jamblichus and his friends had escaped from the rage of a pagan tyrant."*

This legend was received as authentic throughout the Christian world before the end of the sixth century, and was afterwards introduced by Mahomet as a divine revelation into the Koran, and from hence was adopted and adorned by all the nations from Bengal to Africa who professed the Mahometan faith. Some vestiges even of a similar tradition have been discovered in Scandinavia. "This easy and universal belief," observes the philosophical historian of the Decline and Fall, "so expressive of the sense of mankind, may be ascribed to the genuine merit of the fable itself. We imperceptibly advance from youth to age, without observing the gradual, but incessant, change of human affairs; and even, in our larger experience of history, the imagination is accustomed, by a perpetual series of causes and effects, to unite the most distant revolutions. But if the interval between two memorable eras could be instantly annihilated; if it were possible, after a momentary slumber of two hundred years, to display the new world to the eyes of a spectator who still retained a lively and recent impression of the old, his surprise and his reflections would furnish the pleasing subject of a philosophical romance."†

Prejudices arising from our peculiar position as inhabitants of the land.—The sources of prejudice hitherto considered may be deemed

* Gibbon, Decline and Fall, chap. xxxiii. † Id. Ibid.

peculiar for the most part to the infancy of the science, but others are common to the first cultivators of geology and to ourselves, and are all singularly calculated to produce the same deception, and to strengthen our belief that the course of nature in the earlier ages differed widely from that now established. Although these circumstances cannot be fully explained without assuming some things as proved, which it will be the object of another part of this work to demonstrate, it may be well to allude to them briefly in this place.

The first and greatest difficulty, then, consists in an habitual unconsciousness that our position as observers is essentially unfavorable, when we endeavor to estimate the nature and magnitude of the changes now in progress. In consequence of our inattention to this subject, we are liable to serious mistakes in contrasting the present with former states of the globe. As dwellers on the land, we inhabit about a fourth part of the surface; and that portion is almost exclusively a theatre of decay, and not of reproduction. We know, indeed, that new deposits are annually formed in seas and lakes, and that every year some new igneous rocks are produced in the bowels of the earth, but we cannot watch the progress of their formation; and as they are only present to our minds by the aid of reflection, it requires an effort both of the reason and the imagination to appreciate duly their importance. It is, therefore, not surprising that we estimate very imperfectly the result of operations thus invisible to us; and that, when analogous results of former epochs are presented to our inspection, we cannot immediately recognize the analogy. He who has observed the quarrying of stone from a rock, and has seen it shipped for some distant port, and then endeavors to conceive what kind of edifice will be raised by the materials, is in the same predicament as a geologist, who, while he is confined to the land, sees the decomposition of rocks, and the transportation of matter by rivers to the sea, and then endeavors to picture to himself the new strata which Nature is building beneath the waters.

Prejudices arising from our not seeing subterranean changes.—Nor is his position less unfavorable when, beholding a volcanic eruption, he tries to conceive what changes the column of lava has produced, in its passage upwards, on the intersected strata; or what form the melted matter may assume at great depths on cooling; or what may be the extent of the subterranean rivers and reservoirs of liquid matter far beneath the surface. It should, therefore, be remembered, that the task imposed on those who study the earth's history requires no ordinary share of discretion; for we are precluded from collating the corresponding parts of the system of things as it exists now, and as it existed at former periods. If we were inhabitants of another element—if the great ocean were our domain, instead of the narrow limits of the land, our difficulties would be considerably lessened; while, on the other hand, there can be little doubt, although the reader may, perhaps, smile at the bare suggestion of such an idea, that an amphibious being, who should possess our faculties, would still more easily arrive at sound theoretical opinions in geology, since he might behold, on the one hand, the decomposition of rocks in the atmosphere, or the transportation of matter by running water; and, on the other, examine the deposition of sediment in the sea, and the imbedding of animal and vegetable remains in new strata. He might ascertain, by direct observation, the action of a mountain torrent, as well as of a marine current; might compare the products of volcanoes poured out upon the land with those ejected beneath the waters; and might mark, on the one hand, the growth of the forest, and, on the other, that of the coral reef. Yet, even with these advantages, he would be liable to fall into the greatest errors, when endeavoring to reason on rocks of subterranean origin. He would seek in vain, within the sphere of his observation, for any direct analogy to the process of their formation, and would therefore be in danger of attributing them, wherever they are upraised to view, to some "primeval state of nature."

But if we may be allowed so far to indulge the imagination, as to suppose a being entirely confined to the nether world—some "dusky melancholy sprite," like Umbriel, who could "flit on sooty pinions to the central earth," but who was never permitted to "sully the fair face of light," and emerge into the regions of water and of air; and if this being should busy himself in investigating the structure of the globe, he might frame theories the exact converse of those usually adopted by human philosophers. He might infer that the stratified rocks, containing shells and other organic remains, were the oldest of created things, belonging to some original and nascent state of the planet. "Of these masses," he might say, "whether they consist of loose incoherent sand, soft clay, or solid stone, none have been formed in modern times. Every year some part of them are broken and shattered by earthquakes, or melted by volcanic fire; and when they cool down slowly from a state of fusion, they assume a new and more crystalline form, no longer exhibiting that stratified disposition and those curious impressions and fantastic markings, by which they were previously characterized. This process cannot have been carried on for an indefinite time, for in that case all the stratified rocks would long ere this have been fused and crystallized. It is therefore probable that the whole planet once consisted of these mysterious and curiously bedded formations at a time when the volcanic fire had not yet been brought into activity. Since that period there seems to have been a gradual development of heat; and this augmentation we may expect to continue till the whole globe shall be in a state of fluidity and incandescence."

Such might be the system of the Gnome at the very time that the followers of Leibnitz, reasoning on what they saw on the outer surface, might be teaching the opposite doctrine of gradual refrigeration, and averring that the earth had begun its career as a fiery comet, and might be destined hereafter to become a frozen mass. The tenets of the schools of the nether and of the upper world would be directly opposed to each other, for both would partake of the prejudices inevitably resulting from the continual contemplation of one class of phenomena to the exclusion of another. Man observes the annual decomposition of crystalline and igneous rocks, and may sometimes see their conversion into stratified deposits; but he cannot witness the reconversion of the sedimentary into the crystalline by subterranean fire. He is in the habit of regarding all the sedimentary rocks as more recent than the unstratified, for the same reason that we may suppose him to fall into the opposite error if he saw the origin of the igneous class only.

It was not an impossible contingency, that astronomers might have been placed at some period in a situation much resembling that in which the geologist seems to stand at present. If the Italians, for example, in the early part of the twelfth century, had discovered at Amalfi, instead of the pandects of Justinian, some ancient manuscripts filled with astronomical observations relating to a period of three thousand years, and made by some ancient geometers who possessed optical instruments as perfect as any in modern Europe, they would probably, on consulting these memorials, have come to a conclusion that there had been a great revolution in the solar and sidereal systems. "Many primary and secondary planets," they might say, "are enumerated in these tables, which exist no longer. Their positions are assigned with such precision that we may assure ourselves that there is nothing in their place at present but the blue ether. Where one star is visible to us, these documents represent several thousands. Some of those which are now single consisted then of two separate bodies, often distinguished by different colors, and revolving periodically round a common centre of gravity. There is nothing analogous to them in the universe at present; for they were neither fixed stars nor planets, but seem to have stood in the mutual relation of sun and planet to each other. We must conclude, therefore, that there has occurred, at no distant period, a tremendous catastrophe, whereby thousands of worlds have been annihilated at once, and some heavenly bodies absorbed into the substance of others."

When such doctrines had prevailed for ages, the discovery of some of the worlds, supposed to have been lost (the satellites of Jupiter, for example), by aid of the first rude telescope invented after the revival of science, would not dissipate the delusion, for the whole burden of proof would now be thrown on those who insisted on the stability of the system from a remote period, and these philosophers would be required to demonstrate the existence of *all* the worlds said to have been annihilated.

Such popular prejudices would be most unfavorable to the advancement of astronomy; for, instead of persevering in the attempt to improve their instruments, and laboriously to make and record observations, the greater number would despair of verifying the continued existence of the heavenly bodies not visible to the naked eye. Instead of confessing the extent of their ignorance, and striving to remove it by bringing to light new facts, they would indulge in the more easy and indolent employment of framing imaginary theories concerning catastrophes and mighty revolutions in the system of the universe.

For more than two centuries the shelly strata of the Subapennine hills afforded matter of speculation to the early geologists of Italy, and few of them had any suspicion that similar deposits were then forming in the neighboring sea. They were as unconscious of the continued action of causes still producing similar effects, as the astronomers, in the case above supposed, of the existence of certain heavenly bodies still giving and reflecting light, and performing their movements as of old. Some imagined that the strata, so rich in organic remains, instead of being due to secondary agents, had been so created in the beginning of things by the fiat of the Almighty. Others, as we have seen, ascribed the imbedded fossil bodies to some plastic power which resided in the earth in the early ages of the world. In what manner were these dogmas at length exploded? The fossil relics were carefully compared with their living analogues, and all doubts as to their organic origin were eventually dispelled. So, also, in regard to the nature of the containing beds of mud, sand, and limestone: those parts of the bottom of the sea were examined where shells are now becoming annually entombed in new deposits. Donati explored the bed of the Adriatic, and found the closest resemblance between the strata there forming, and those which constituted hills above a thousand feet high in various parts of the Italian peninsula. He ascertained by dredging that living testacea were there grouped together in precisely the same manner as were their fossil analogues in the inland strata; and while some of the recent shells of the Adriatic were becoming incrusted with calcareous rock, he observed that others had been newly buried in sand and clay, precisely as fossil shells occur in the Subapennine hills. This discovery of the identity of modern and ancient submarine operations was not made without the aid of artificial instruments, which, like the telescope, brought phenomena into view not otherwise within the sphere of human observation.

In like manner, the volcanic rocks of the Vicentin had been studied in the beginning of the last century; but no geologist suspected, before the time of Arduino, that these were composed of ancient submarine lavas. During many years of controversy, the popular opinion inclined to a belief that basalt and rocks of the same class had been precipitated from a chaotic fluid, or an ocean which rose at successive periods over the continents, charged with the component elements of the rocks in question. Few will now dispute that it would have been difficult to invent a theory more distant from the truth; yet we must cease to wonder that it gained so many proselytes, when we remember that its claims to probability arose partly from the very circumstance of its confirming the assumed want of analogy between geological causes and those now in action. By what train of investigations were geologists induced at length to reject these views, and to assent to the igneous origin of the trappean formations? By an examination of volcanoes now active, and by comparing their structure and the composition of their lavas with the ancient trap-rocks.

The establishment, from time to time, of numerous points of identification, drew at length from geologists a reluctant admission, that there was more correspondence between the condition of the globe at remote eras and now, and more uniformity in the laws which have regulated the changes of its surface, than they at first imagined. If, in this state of the science, they still despaired of reconciling every class of geological phenomena to the operations of ordinary causes, even by straining analogy to the utmost limits of credibility, we might have expected, at least, that the balance of probability would now have been presumed to incline towards the close analogy of the ancient and modern causes. But, after repeated experience of the failure of attempts to speculate on geological monuments, as belonging to a distinct order of things, new sects continued to persevere in the principles adopted by their predecessors. They still began, as each new problem presented itself, whether relating to the animate or inanimate world, to assume an original and dissimilar order of nature; and when at length they approximated, or entirely came round to an opposite opinion, it was always with the feeling, that they were conceding what they had been justified *à priori* in deeming improbable. In a word, the same men who, as natural philosophers, would have been most incredulous respecting any extraordinary deviations from the known course of nature, if reported to have happened *in their own time*, were equally disposed, as geologists, to expect the proofs of such deviations at every period of the past.

I shall proceed in the following chapters to enumerate some of the principal difficulties still opposed to the theory of the uniform nature and energy of the causes which have worked successive changes in the crust of the earth, and in the condition of its living inhabitants. The discussion of so important a question on the present occasion may appear premature, but it is one which naturally arises out of a review of the former history of the science. It is, of course, impossible to enter into such speculative topics, without occasionally carrying the novice beyond his depth, and appealing to facts and conclusions with which he will be unacquainted, until he has studied some elementary work on geology, but it may be useful to excite his curiosity, and lead him to study such works by calling his attention at once to some of the principal points of controversy.*

* In the earlier editions of this work, a fourth book was added on Geology Proper, or Systematic Geology, containing an account of the former changes of the animate and inanimate creation, brought to light by an examination of the crust of the earth. This I afterwards (in 1838) expanded into a separate publication called the Elements or Manual Geology, of which a fourth edition appeared December, 1851.

CHAPTER VI.

DOCTRINE OF THE DISCORDANCE OF THE ANCIENT AND MODERN CAUSES OF CHANGE CONTROVERTED.

Climate of the Northern Hemisphere formerly different—Direct proofs from the organic remains of the Italian strata—Proofs from analogy derived from extinct quadrupeds—Imbedding of animals in icebergs—Siberian mammoths—Evidence in regard to temperature, from the fossils of tertiary and secondary rocks—From the plants of the coal formation—Northern limit of these fossils—Whether such plants could endure the long continuance of an arctic night.

Climate of the Northern hemisphere formerly different.—PROOFS of former revolutions in climate, as deduced from fossil remains, have afforded one of the most popular objections to the theory which endeavors to explain all geological changes by reference to those now in progress on the earth. The probable causes, therefore, of fluctuations in climate, may first be treated of.

That the climate of the Northern hemisphere has undergone an important change, and that its mean annual temperature must once have more nearly resembled that now experienced within the tropics, was the opinion of some of the first naturalists who investigated the contents of the ancient strata. Their conjecture became more probable when the shells and corals of the older tertiary and many secondary rocks were carefully examined; for the organic remains of these formations were found to be intimately connected by generic affinity with species now living in warmer latitudes. At a later period, many reptiles, such as turtles, tortoises, and large saurian animals, were discovered in European formations in great abundance; and they supplied new and powerful arguments, from analogy, in support of the doctrine, that the heat of the climate had been great when our secondary strata were deposited. Lastly, when the botanist turned his attention to the specific determination of fossil plants, the evidence acquired still fuller confirmation; for the flora of a country is peculiarly influenced by temperature: and the ancient vegetation of the earth might have been expected more readily than the forms of animals, to have afforded conflicting proofs, had the popular theory been without foundation. When the examination of fossil remains was extended to rocks in the most northern parts of Europe and North America, and even to the Arctic regions, indications of the same revolution in climate were discovered.

It cannot be said, that in this, as in many other departments of geology, we have investigated the phenomena of former eras, and neglected those of the present state of things. On the contrary, since the first agitation of this interesting question, the accessions to our knowledge of living animals and plants have been immense, and have far surpassed all the data previously obtained for generalizing on the relation of certain types of organization to particular climates. The tropical and temperate zones of South America and of Australia have been explored; and, on close comparison, it has been found that scarcely any of the species of the animate creation in these extensive continents are identical with those inhabiting the old world. Yet the zoologist and botanist, well acquainted with the geographical distribution of organic beings in other parts of the globe, would have been able, if distinct groups of species had been presented to them from these regions, to recognize those which had been collected from latitudes within, and those which were brought from without the tropics.

Before I attempt to explain the probable causes of great vicissitudes of temperature on the earth's surface, I shall take a rapid view of some of the principal data which appear to support the popular opinions now entertained on the subject. To insist on the soundness of these inferences, is the more necessary, because some zoologists have undertaken to vindicate the uniformity of the laws of nature, not by accounting for former fluctuations in climate, but by denying the value of the evidence in their favor.*

Proofs from fossil shells in tertiary strata.—In Sicily, Calabria, and in the neighborhood of Naples, the fossil testacea of the most modern tertiary formations belong almost entirely to species now inhabiting the Mediterranean; but as we proceed northwards in the Italian peninsula we find in the strata called Subapennine an assemblage of fossil shells departing somewhat more widely from the type of the neighboring seas. The proportion of species identifiable with those now living in the Mediterranean is still considerable; but it no longer predominates, as in the South of Italy and part of Sicily, over the unknown species. Although occurring in localities which are removed several degrees farther from the equator (as at Sienna, Parma, Asti, &c.), the shells yield clear indications of a warmer climate. This evidence is of great weight, and is not neutralized by any facts of a conflicting character; such, for instance, as the association, in the same group, of individuals referable to species now confined to arctic regions. Whenever any of the fossil shells are identified with living species foreign to the Mediterranean, it is not in the Northern Ocean, but nearer the tropics, that they must be sought: on the other hand, the associated unknown species belong, for the most part, to *genera* which are now most largely developed in equinoctial regions, as, for example, the genera Cancellaria, Cassidaria, Pleurotoma, Conus, and Cypræa.

On comparing the fossils of the tertiary deposits of Paris and London with those of Bourdeaux, and these again with the more modern strata of Sicily, we should at first expect that they would each indicate a higher temperature in proportion as they are situated farther to the

* See two articles by the Rev. Dr. Fleming, in the Edinburgh New Phil. Journ. No. xii. p. 277, April, 1829; and No. xv. p. 65, Jan. 1830.

south. But the contrary is true; of the shells belonging to these several groups, whether freshwater or marine, some are of extinct, others of living species. Those found in the older, or Eocene, deposits of Paris and London, although six or seven degrees to the north of the Miocene strata at Bourdeaux, afford evidence of a warmer climate; while those of Bourdeaux imply that the sea in which they lived was of a higher temperature than that of Sicily, where the shelly strata were formed six or seven degrees nearer to the equator. In these cases the greater antiquity of the several formations (the Parisian being the oldest and the Sicilian the newest) has more than counterbalanced the influence which latitude would otherwise exert, and this phenomenon clearly points to a gradual and successive refrigeration of climate.

Siberian Mammoths.—It will naturally be asked, whether some recent geological discoveries bringing evidence to light of a colder, or as it has been termed "glacial epoch," towards the close of the tertiary periods throughout the northern hemisphere, does not conflict with the theory above alluded to, of a warmer temperature having prevailed in the eras of the Eocene, Miocene, and Pliocene formations. In answer to this inquiry, it may certainly be affirmed, that an oscillation of climate has occurred in times immediately antecedent to the peopling of the earth by man; but proof of the intercalation of a less genial climate at an era when nearly all the marine and terrestrial testacea had already become specifically the same as those now living, by no means rebuts the conclusion previously drawn, in favor of a warmer condition of the globe, during the ages which elapsed while the tertiary strata were deposited. In some of the most superficial patches of sand, gravel, and loam, scattered very generally over Europe, and containing recent shells, the remains of extinct species of land quadrupeds have been found, especially in places where the alluvial matter appears to have been washed into small lakes, or into depressions in the plains bordering ancient rivers. Similar deposits have also been lodged in rents and caverns of rocks, where they may have been swept in by land floods, or introduced by engulphed rivers during changes in the physical geography of these countries. The various circumstances under which the bones of animals have been thus preserved, will be more fully considered hereafter;* I shall only state here, that among the extinct mammalia thus entombed, we find species of the elephant, rhinoceros, hippopotamus, bear, hyæna, lion, tiger, monkey (macacus†), and many others; consisting partly of genera now confined to warmer regions.

It is certainly probable that when some of these quadrupeds abounded in Europe, the climate was milder than that now experienced. The hippopotamus, for example, is now only met with where the temperature of the water is warm and nearly uniform throughout the year, and

* Book iii. chaps. 46, 47, &c.

† Macacus pliocenus, Owen, Brit. Foss. Mam. Intr. p. 37, found with the extinct elephant, &c. in the modern freshwater beds at Grays Thurrock (Essex), in the valley of the Thames.

where the rivers are never frozen over. Yet when the great fossil species (*Hippopotamus major*, Cuv.) inhabited England, the testacea of our country were nearly the same as those now existing, and the climate cannot be supposed to have been very hot. The bones of this animal have lately been found by Mr. Strickland, together with those of a bear and other mammalia, at Cropthorn, near Evesham, in Worcestershire, in alluvial sand, together with twenty-three species of terrestrial and freshwater shells, all, with two exceptions, of British species. The bed of sand, containing the shells and bones, reposes on lias, and is covered with alternating strata of gravel, sand, and loam.*

The mammoth also appears to have existed in England when the temperature of our latitudes could not have been very different from that which now prevails; for remains of this animal have been found at North Cliff, in the county of York, in a lacustrine formation, in which all the land and freshwater shells, thirteen in number, can be identified with species and varieties now existing in that county. Bones of the bison, also, an animal now inhabiting a cold or temperate climate, have been found in the same place. That these quadrupeds, and the idigenous species of testacea associated with them, were all contemporary inhabitants of Yorkshire, has been established by unequivocal proof. The Rev. W. V. Vernon Harcourt caused a pit to be sunk to the depth of twenty-two feet through undisturbed strata, in which the remains of the mammoth were found imbedded, together with the shells, in a deposit which had evidently resulted from tranquil waters.†

In the valley of the Thames, as at Ilford and Grays, in Essex, bones of the elephant and rhinoceros occur in strata abounding in freshwater shells of the genera Unio, Cyclas, Paludina, Valvata, Ancylus, and others. These fossil shells belong for the most part to species now living in the same district, yet some few of them are extinct, as, for example, a species of Cyrena, a genus no longer inhabiting Europe, and now entirely restricted to warmer latitudes.

When reasoning on such phenomena, the reader must always bear in mind that the fossil individuals belonged to *species* of elephant, rhinoceros, hippopotamus, bear, tiger, and hyæna, distinct from those which now dwell within or near the tropics. Dr. Fleming, in a discussion on this subject, has well remarked that a near resemblance in form and osteological structure is not always followed, in the existing creation, by a similarity of geographical distribution; and we must therefore be on our guard against deciding too confidently, from mere analogy of anatomical structure, respecting the habits and physiological peculiarities of *species* now no more. "The zebra delights to roam over the tropical plains, while the horse can maintain its existence throughout an Iceland winter. The buffalo, like the zebra, prefers a high temperature, and cannot thrive even where the common ox prospers. The musk-ox, on the other hand, though nearly resembling the buffalo, prefers the stinted

* Geol. Proceedings, No. xxxvi. June, 1834.

† Phil. Mag., Sept. 1829, and Jan. 1830.

herbage of the arctic regions, and is able, by its periodical migrations, to outlive a northern winter. The jackal (*Canis aureus*) inhabits Africa, the warmer parts of Asia, and Greece; while the isatis (*Canis lagopus*) resides in the arctic regions. The African hare and the polar hare have their geographical distribution expressed in their trivial names;"* and different species of bears thrive in tropical, temperate, and arctic latitudes.

Recent investigations have placed beyond all doubt the important fact that a species of tiger, identical with that of Bengal, is common in the neighborhood of Lake Aral, near Sussac, in the forty-fifth degree of north latitude; and from time to time this animal is now seen in Siberia, in a latitude as far north as the parallel of Berlin and Hamburgh.† Humboldt remarks that the part of Southern Asia now inhabited by this Indian species of tiger is separated from the Himalaya by two great chains of mountains, each covered with perpetual snow,—the chain of Kuenlun, lat. 35° N., and that of Mouztagh, lat. 42°,—so that it is impossible that these animals should merely have made excursions from India, so as to have penetrated in summer to the forty-eighth and fifty-third degrees of north latitude. They must remain all the winter north of the Mouztagh, or Celestial mountains. The last tiger killed, in 1828, on the Lena, in lat. 52¼°, was in a climate colder than that of Petersburg and Stockholm.‡

We learn from Mr. Hodgson's account of the mammalia of Nepal, that the tiger is sometimes found at the very edge of perpetual snow in the Himalaya;§ and Pennant mentions that it is found among the snows of Mount Ararat in Armenia. The jaguar, also, has been seen in America, wandering from Mexico, as far north as Kentucky, lat. 37° N.,‖ and even as far as 42° S. in South America,—a latitude which corresponds to that of the Pyrenees in the northern hemisphere.¶ The range of the puma is still wider, for it roams from the equator to the Straits of Magellan, being often seen at Port Famine, in lat. 53° 38′ S.

A new species of panther (*Felis irbis*), covered with long hair, has been discovered in Siberia, evidently inhabiting, like the tiger, a region north of the Celestial Mountains, which are in lat. 42°.**

The two-horned African rhinoceros occurs without the tropics at the Cape of Good Hope, in lat. 34° 29′ S., where it is accompanied by the elephant, hippopotamus, and hyæna. Here the migration of all these species towards the south is arrested by the ocean; but if the continent had been prolonged still farther, and the land had been of moderate ele-

* Fleming, Ed. New Phil. Journ., No. xii. p. 282, 1829. The zebra, however, inhabits chiefly the extra-tropical parts of Africa.

† Humboldt, Fragmens de Géologie, &c., tome ii. p. 388. Ehrenberg, Ann. des Sci. Nat., tome xxi. p. 387.

‡ Ehrenberg, ibid. p. 390.

§ Journ. of Asiat. Soc., vol. i. p. 240.

‖ Rafinesque, Atlantic Journ., p. 18.

¶ Darwin's Journal of Travels in South America, &c., 1832 to 1836, in Voyage of H. M. S. Beagle, p. 159.

** Ehrenberg, ibid.

vation, it is very probable that they might have extended their range to a greater distance from the tropics.

Now, if the Indian tiger can range in our own times to the southern borders of Siberia, or skirt the snows of the Himalaya, and if the puma can reach the fifty-third degree of latitude in South America, we may easily understand how large species of the same *genera* may once have inhabited our temperate climates. The mammoth (*E. primigenius*), already alluded to, as occurring fossil in England, was decidedly different from the two existing species of elephants, one of which is limited to Asia, south of the 31° of N. lat., the other to Africa, where it extends, as before stated, as far south as the Cape of Good Hope. The bones of the great fossil species are very widely spread over Europe and North America; but are nowhere in such profusion as in Siberia, particularly near the shores of the Frozen Ocean. Are we, then, to conclude that this animal preferred a polar climate? If so, it may well be asked, by what food was it sustained, and why does it not still survive near the arctic circle?*

Pallas and other writers describe the bones of the mammoth as abounding throughout all the Lowlands of Siberia, stretching in a direction west and east, from the borders of Europe to the extreme point nearest America, and south and north, from the base of the mountains of Central Asia to the shores of the Arctic Sea. (See map, fig. 1.) Within this space, scarcely inferior in area to the whole of Europe, fossil ivory has been collected almost everywhere, on the banks of the Irtish, Obi, Yenesei, Lena, and other rivers. The elephantine remains do not occur in the marshes and low plains, but where the banks of the rivers present lofty precipices of sand and clay, from which circumstance Pallas very justly inferred that, if sections could be obtained, similar bones might be found in all the elevated lands intervening between the great rivers. Strahlenberg, indeed, had stated, before the time of Pallas, that wherever any of the great rivers overflowed and cut out fresh channels during floods, more fossil remains of the same kind were invariably disclosed.

As to the position of the bones, Pallas found them in some places imbedded together with marine remains; in others, simply with fossil wood, or lignite, such as, he says, might have been derived from carbonized peat. On the banks of the Yenesei, below the city of Krasnojarsk, in lat. 56°, he observed grinders, and bones of elephants, in strata of yel-

* The speculations which follow, on the ancient physical geography of Siberia, and its former fitness as a residence for the mammoth, were first given in their present form in my 4th edition, June, 1835. Recently Sir R. Murchison and his companions in their great work on the Geology of Russia, 1845 (vol. i. p. 497), have, in citing this chapter, declared that their investigations have led them to similar conclusions. Professor Owen, in his excellent History of British Fossil Mammalia, 1844, p. 261, *et seq.*, observes that the teeth of the mammoth differ from those of the living Asiatic or African elephant in having a larger proportion of dense enamel, which may have enabled it to subsist on the coarser ligneous tissues of trees and shrubs. In short, he is of opinion, that the structure of its teeth, as well as the nature of its epidermis and coverings, may have made it "a meet companion for the reindeer."

Fig. 1.

MAP OF SIBERIA.

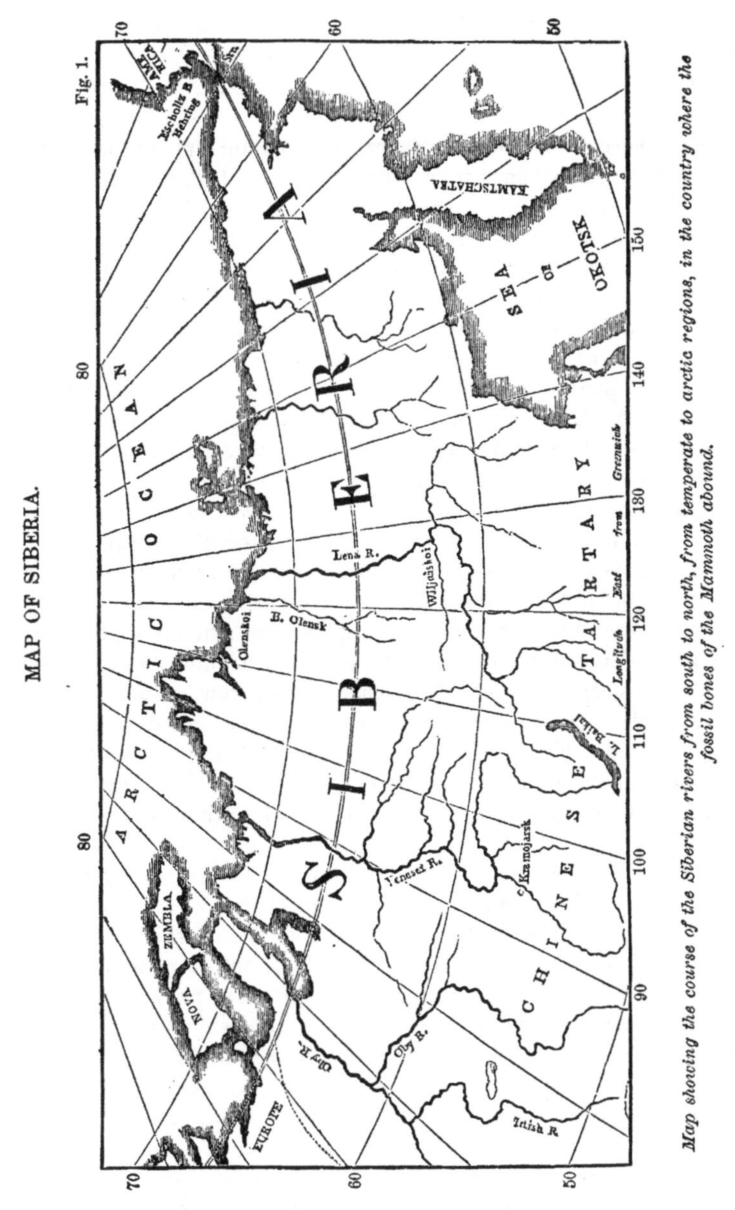

Map showing the course of the Siberian rivers from south to north, from temperate to arctic regions, in the country where the fossil bones of the Mammoth abound.

low and red loam, alternating with coarse sand and gravel, in which was also much petrified wood of the willow and other trees. Neither here nor in the neighboring country were there any marine shells, but merely layers of black coal.* But grinders of the mammoth were collected much farther down the same river, near the sea, in lat. 70°, mixed with *marine* petrifactions.† Many other places in Siberia are cited by Pallas,

* Pallas, Reise in Russ. Reiche, pp. 409, 410.
† Nov. Com. Petrop. vol. xvii. p. 584.

where sea shells and fishes' teeth accompany the bones of the mammoth, rhinoceros, and Siberian buffalo, or bison (*Bos priscus*). But it is not on the Obi nor the Yenesei, but on the Lena, farther to the east, where, in the same parallels of latitude, the cold is far more intense, that fossil remains have been found in the most wonderful state of preservation. In 1772, Pallas obtained from Wiljuiskoi, in lat. 64°, from the banks of the Wiljui, a tributary of the Lena, the carcass of a rhinoceros (*R. tichorhinus*), taken from the sand in which it must have remained congealed for ages, the soil of that region being always frozen to within a slight depth of the surface. This carcass was compared to a natural mummy, and emitted an odor like putrid flesh, part of the skin being still covered with black and gray hairs. So great, indeed, was the quantity of hair on the foot and head conveyed to St. Petersburg, that Pallas asked whether the rhinoceros of the Lena might not have been an inhabitant of the temperate regions of middle Asia, its clothing being so much warmer than that of the African rhinoceros.*

Professor Brandt, of St. Petersburg, in a letter to Baron Alex. Von Humboldt, dated 1846, adds the following particulars respecting this wonderful fossil relic:—"I have been so fortunate as to extract from cavities in the molar teeth of the Wiljui rhinoceros a small quantity of its half-chewed food, among which fragments of pine leaves, one-half of the seed of a polygonaceous plant, and very minute portions of wood with porous cells (or small fragments of coniferous wood), were still recognizable. It was also remarkable, on a close investigation of the head, that the blood-vessels discovered in the interior of the mass appeared filled, even to the capillary vessels, with a brown mass (coagulated blood), which in many places still showed the red color of blood."†

After more than thirty years, the entire carcass of a mammoth (or extinct species of elephant) was obtained in 1803, by Mr. Adams, much farther to the north. It fell from a mass of ice, in which it had been encased, on the banks of the Lena, in lat. 70°; and so perfectly had the soft parts of the carcass been preserved, that the flesh, as it lay, was devoured by wolves and bears. This skeleton is still in the museum of St. Petersburg, the head retaining its integument and many of the ligaments entire. The skin of the animal was covered, first, with black bristles, thicker than horse hair, from twelve to sixteen inches in length; secondly, with hair of a reddish brown color, about four inches long; and thirdly, with wool of the same color as the hair, about an inch in length. Of the fur, upwards of thirty pounds' weight were gathered from the wet sand-bank. The individual was nine feet high and sixteen feet long, without reckoning the large curved tusks: a size rarely surpassed by the largest living male elephants.‡

It is evident, then, that the mammoth, instead of being naked, like the

* Nov. Com. Petrop. vol. xvii. p. 591.
† Quart. Journ. Geol. Soc. Lond. vol. iv. p. 10, Memoirs.
‡ Journal du Nord, St. Petersburg, 1807.

living Indian and African elephants, was enveloped in a thick shaggy covering of fur, probably as impenetrable to rain and cold as that of the musk ox.* The species may have been fitted by nature to withstand the vicissitudes of a northern climate; and it is certain that, from the moment when the carcasses, both of the rhinoceros and elephant, above described, were buried in Siberia, in latitudes 64° and 70° N., the soil must have remained frozen, and the atmosphere nearly as cold as at this day.

The most recent discoveries made in 1843 by Mr. Middendorf, a distinguished Russian naturalist, and which he communicated to me in September, 1846, afford more precise information as to the climate of the Siberian lowlands, at the period when the extinct quadrupeds were entombed. One elephant was found on the Tas, between the Obi and Yenesei, near the arctic circle, about lat. 66° 30′ N., with some parts of the flesh in so perfect a state that the bulb of the eye is now preserved in the museum at Moscow. Another carcass, together with a young individual of the same species, was met with in the same year, 1843, in lat. 75° 15′ N., near the river Taimyr, with the flesh decayed. It was imbedded in strata of clay and sand, with erratic blocks, at about 15 feet above the level of the sea. In the same deposit Mr. Middendorf observed the trunk of a larch tree (*Pinus larix*), the same wood as that now carried down in abundance by the Taimyr to the Arctic Sea. There were also associated fossil shells of *living northern* species, and which are moreover characteristic of the drift or *glacial* deposits of Europe. Among these *Nucula pygmæa*, *Tellina calcarea*, *Mya truncata*, and *Saxicava rugosa* were conspicuous.

So fresh is the ivory throughout northern Russia, that, according to Tilesius, thousands of fossil tusks have been collected and used in turning; yet others are still procured and sold in great plenty. He declares his belief that the bones still left in northern Russia must greatly exceed in number all the elephants now living on the globe.

* Fleming, Ed. New Phil. Journ., No. xii. p. 285.

Bishop Heber informs us (Narr. of a Journey through the Upper Provinces of India, vol. ii. p. 166—219), that in the lower range of the Himalaya mountains, in the northeastern borders of the Delhi territory, between lat. 29° and 30°, he saw an Indian elephant of a small size, covered with shaggy hair. But this variety must be exceedingly rare; for Mr. Royle (late superintendent of the East India Company's Botanic Garden at Saharunpore) has assured me, that being in India when Heber's Journal appeared, and having never seen or heard of such elephants, he made the strictest inquiries respecting the fact, and was never able to obtain any evidence in corroboration. Mr. Royle resided at Saharunpore, lat. 30° N., upon *the extreme northern limits* of the range of the elephant. Mr. Everest also declares that he has been equally unsuccessful in finding any one aware of the existence of such a variety or breed of the animal, though one solitary individual was mentioned to him as having been seen at Delhi, with a good deal of long hair upon it. The greatest elevation, says Mr. E., at which the wild elephant is found in the mountains to the north of Bengal, is at a place called Nahun, about 4000 feet above the level of the sea, and in the 31st degree of N. lat., where the mean yearly temperature may be about 64° Fahrenheit, and the difference between winter and summer very great, equal to about 36° F., the month of January averaging 45°, and June, the hottest month, 81° F. (Everest on climate of Foss. Eleph., Journ. of Asiat. Soc., No. 25, p. 21.)

We are as yet ignorant of the entire geographical range of the mammoth; but its remains have been recently collected from the cliffs of frozen mud and ice on the east side of Behring's Straits, in Eschscholtz's Bay, in Russian America, lat. 66° N. As the cliffs waste away by the thawing of the ice, tusks and bones fall out, and a strong odor of animal matter is exhaled from the mud.*

On considering all the facts above enumerated, it seems reasonable to imagine that a large region in central Asia, including, perhaps, the southern half of Siberia, enjoyed, at no very remote period in the earth's history, a temperate climate, sufficiently mild to afford food for numerous herds of elephants and rhinoceroses, *of species distinct from those now living.* It has usually been taken for granted that herbivorous animals of large size require a very luxuriant vegetation for their support; but this opinion is, according to Mr. Darwin, completely erroneous:—"It has been derived," he says, "from our acquaintance with India and the Indian islands, where the mind has been accustomed to associate troops of elephants with noble forests and impenetrable jungles. But the southern parts of Africa, from the tropic of Capricorn to the Cape of Good Hope, although sterile and desert, are remarkable for the number and great bulk of the indigenous quadrupeds. We there meet with an elephant, five species of rhinoceros, a hippopotamus, a giraffe, the bos caffer, the elan, two zebras, the quagga, two gnus, and several antelopes. Nor must we suppose, that while the species are numerous, the individuals of each kind are few. Dr. Andrew Smith saw, in one day's march, in lat. 24° S., without wandering to any great distance on either side, about 150 rhinoceroses, with several herds of giraffes, and his party had killed, on the previous night, eight hippopotamuses. Yet the country which they inhabited was thinly covered with grass and bushes about four feet high, and still more thinly with mimosa-trees, so that the wagons of the travellers were not prevented from proceeding in a nearly direct line."†

In order to explain how so many animals can find support in this region, it is suggested that the underwood, of which their food chiefly consists, may contain much nutriment in a small bulk, and also that the vegetation has a rapid growth; for no sooner is a part consumed than its place, says Dr. Smith, is supplied by a fresh stock. Nevertheless, after making every allowance for this successive production and consumption, it is clear, from the facts above cited, that the quantity of food required by the larger herbivora is much less than we have usually imagined. Mr. Darwin conceives that the amount of vegetation supported at any one time by Great Britain may exceed, in a ten-fold ratio, the quantity existing on an equal area in the interior parts of Southern Africa.‡ It is remarked, moreover, in illustration of the small connec-

* See Dr. Buckland's description of these bones, Appen. to Beechy's Voy.

† Darwin, Journal of Travels in S. America, &c., 1832–36, in voyage of H. M. S. Beagle, p. 98. 2d Ed. London, 1845, p. 86.

‡ Darwin, Journal of Travels in S. America, &c., p. 99, 2d Ed. p. 85.

tion discoverable between abundance of food and the magnitude of indigenous mammalia, that while in the desert part of Southern Africa there are so many huge animals; in Brazil, where the splendor and exuberance of the vegetation are unrivalled, there is not a single wild quadruped of large size.*

It would doubtless be impossible for herds of mammoths and rhinoceroses to subsist, at present, throughout the year, even in the southern part of Siberia, covered as it is with snow during winter; but there is no difficulty in supposing a vegetation capable of nourishing these great quadrupeds to have once flourished between the latitudes 40° and 60° N.

Dr. Fleming has hinted, that "the kind of food which the existing species of elephant prefers, will not enable us to determine, or even to offer a probable conjecture, concerning that of the extinct species. No one acquainted with the gramineous character of the food of our fallow-deer, stag, or roe, would have assigned a lichen to the reindeer."

Travellers mention that, even now, when the climate of eastern Asia is so much colder than the same parallels of latitude farther west, there are woods not only of fir, but of birch, poplar, and alder, on the banks of the Lena, as far north as latitude 60°.

It has, moreover, been suggested, that as, in our own times, the northern animals migrate, so the Siberian elephant and rhinoceros may have wandered towards the north in summer. The musk oxen annually desert their winter quarters in the south, and cross the sea upon the ice, to graze for four months, from May to September, on the rich pasturage of Melville Island, in lat. 75°. The mammoths, without passing so far beyond the arctic circle, may nevertheless have made excursions, during the heat of a brief northern summer, from the central or temperate parts of Asia to the sixtieth parallel of latitude.

Now, in this case, the preservation of their bones, or even occasionally of their entire carcasses, in ice or frozen soil, may be accounted for, without resorting to speculations concerning sudden revolutions in the former state and climate of the earth's surface. We are entitled to assume, that, in the time of the extinct elephant and rhinoceros, the Lowland of Siberia was less extensive towards the north than now; for we have seen (p. 80) that the strata of this Lowland, in which the fossil bones lie buried, were originally deposited beneath the sea; and we know, from the facts brought to light in Wrangle's Voyage, in the years 1821, 1822, and 1823, that a slow upheaval of the land along the borders of the Icy Sea is now constantly taking place, similar to that experienced in part of Sweden. In the same manner, then, as the shores of the Gulf of Bothnia are extended, not only by the influx of sediment brought down by rivers, but also by the elevation and consequent drying up of the bed of the sea, so a like combination of causes may, in modern times, have been extending the low tract of land where marine

* Burchell, cited by Darwin, ibid. p. 101. 2d Ed. p. 87.

shells and fossil bones occur in Siberia.* Such a change in the physical geography of that region, implying a constant augmentation in the quantity of arctic land, would, according to principles to be explained in the next chapter, tend to increase the severity of the winters. We may conclude, therefore, that, before the land reached so far to the north, the temperature of the Siberian winter and summer was more nearly equalized; and a greater degree of winter's cold may, even more than a general diminution of the mean annual temperature, have finally contributed to the extermination of the mammoth and its contemporaries.

On referring to the map (p. 79), the reader will see how all the great rivers of Siberia flow at present from south to north, from temperate to arctic regions, and they are all liable, like the Mackenzie, in North America, to remarkable floods, in consequence of flowing in this direction. For they are filled with running water in their upper or southern course when completely frozen over for several hundred miles near their mouths, where they remain blocked up by ice for six months in every year. The descending waters, therefore, finding no open channel, rush over the ice, often changing their direction, and sweeping along forests and prodigious quantities of soil and gravel mixed with ice. Now the rivers of Siberia are among the largest in the world, the Yenesei having a course of 2500, the Lena of 2000 miles; so that we may easily conceive that the bodies of animals which fall into their waters may be transported to vast distances towards the Arctic Sea, and, before arriving there, may be stranded upon and often frozen into thick ice. Afterwards, when the ice breaks up, they may be floated still farther towards the ocean, until at length they become buried in fluviatile and submarine deposits near the mouths of rivers.

Humboldt remarks that near the mouths of the Lena a considerable thickness of frozen soil may be found at all seasons at the depth of a few feet; so that if a carcass be once imbedded in mud and ice in such a region and in such a climate, its putrefaction may be arrested for indefinite ages.† According to Prof. Von Baer of St. Petersburg, the ground is now frozen permanently to the depth of 400 feet, at the town of Yakutzt, on the western bank of the Lena, in lat. 62° N., 600 miles distant from the polar sea. Mr. Hedenstrom tells us that, throughout a wide area in Siberia, the boundary cliffs of the lakes and rivers consist of alternate layers of earthy materials and ice, in horizontal stratification;‡ and Mr. Middendorf informed us, in 1846, that, in his tour there three years before, he had bored in Siberia to the depth of seventy feet, and, after passing through much frozen soil mixed with ice, had come

* Since the above passage was first printed in a former edition, June, 1835, it has been shown by the observations of Sir R. Murchison, M. de Verneuil, and Count Keyserling, and more recently by M. Middendorf (see above, p. 81), that the Lowland of Siberia has actually been extended, since the *existing species* of shells inhabited the northern seas.

† Humboldt, Fragmens Asiatiques, tom. ii. p. 393.

‡ Reboul. Geol. de la Période Quaternaire, who cites Observ. sur la Sibérie, Bibl. Univ., Juillet, 1832.

down upon a solid mass of pure transparent ice, the thickness of which, after penetrating two or three yards, they did not ascertain. We may conceive, therefore, that even at the period of the mammoth, when the Lowland of Siberia was less extensive towards the north, and consequently the climate more temperate than now, the cold may still have been sufficiently intense to cause the rivers flowing in their present direction to sweep down from south to north the bodies of drowned animals, and there bury them in drift ice and frozen mud.

If it be true that the carcass of the mammoth was imbedded in pure ice, there are two ways in which it may have been frozen in. We may suppose the animal to have been overwhelmed by drift snow. I have been informed by Dr. Richardson, that, in the northern parts of America, comprising regions now inhabited by many herbivorous quadrupeds, the drift snow is often converted into permanent glaciers. It is commonly blown over the edges of steep cliffs, so as to form an inclined talus hundreds of feet high; and when a thaw commences, torrents rush from the land, and throw down from the top of the cliff alluvial soil and gravel. This new soil soon becomes covered with vegetation, and protects the foundation of snow from the rays of the sun. Water occasionally penetrates into the crevices and pores of the snow; but, as it soon freezes again, it serves the more rapidly to consolidate the mass into a compact iceberg. It may sometimes happen that cattle grazing in a valley at the base of such cliffs, on the borders of a sea or river, may be overwhelmed, and at length inclosed in solid ice, and then transported towards the polar regions. Or a herd of mammoths returning from their summer pastures in the north, may have been surprised, while crossing a stream, by the sudden congelation of the waters. The missionary Huc relates, in his travels in Thibet in 1846, that, after many of his party had been frozen to death, they pitched their tents on the banks of the Mouroui-Ousson (which lower down becomes the famous Blue River), and saw from their encampment "some black shapeless objects ranged in file across the stream. As they advanced nearer no change either in form or distinctness was apparent; nor was it till they were quite close, that they recognized in them a troop of the wild oxen, called Yak by the Thibetans.* There were more than fifty of them incrusted in the ice. No doubt they had tried to swim across at the moment of congelation, and had been unable to disengage themselves. Their beautiful heads, surmounted by huge horns, were still above the surface, but their bodies were held fast in the ice, which was so transparent that the position of the imprudent beasts was easily distinguishable; they looked as if still swimming, but the eagles and ravens had pecked out their eyes."†

The foregoing investigations, therefore, lead us to infer that the mammoth, and some other extinct quadrupeds fitted to live in high latitudes,

* Conjectured to be the wild stock of Bos grunniens.

† Recollections of a Journey through Tartary, Thibet, and China (ch. xv. p. 234), by M. Huc. Longman, 1852.

were inhabitants of Northern Asia at a time when the geographical conditions and climate of that continent were different from the present. But the age of this fauna was comparatively modern in the earth's history. It appears that when the oldest or eocene tertiary deposits were formed, a warm temperature pervaded the European seas and lands. Shells of the genus Nautilus and other forms characteristic of tropical latitudes; fossil reptiles, such as the crocodile, turtle, and tortoise; plants, such as palms, some of them allied to the cocoa-nut, the screw-pine, the custard-apple, and the acacia, all lead to this conclusion. This flora and fauna were followed by those of the miocene formation, in which indications of a southern, but less tropical climate are detected. Finally, the pliocene deposits, which come next in succession, exhibit in their organic remains a much nearer approach to the state of things now prevailing in corresponding latitudes. It was towards the close of this period that the seas of the northern hemisphere became more and more filled with floating icebergs often charged with erratic blocks, so that the waters and the atmosphere were chilled by the melting ice, and an arctic fauna enabled, for a time, to invade the temperate latitudes both of N. America and Europe. The extinction of a considerable number of land quadrupeds and aquatic mollusca was gradually brought about by the increasing severity of the cold; but many species survived this revolution in climate, either by their capacity of living under a variety of conditions, or by migrating for a time to more southern lands and seas. At length, by modifications in the physical geography of the northern regions, and the cessation of floating ice on the eastern side of the Atlantic, the cold was moderated, and a milder climate ensued, such as we now enjoy in Europe.*

Proofs from fossils in secondary and still older strata.—A great interval of time appears to have elapsed between the formation of the secondary strata, which constitute the principal portion of the elevated land in Europe, and the origin of the eocene deposits. If we examine the rocks from the chalk to the new red sandstone inclusive, we find many distinct assemblages of fossils entombed in them, all of unknown species, and many of them referable to genera and families now most abundant between the tropics. Among the most remarkable are reptiles of gigantic size; some of them herbivorous, others carnivorous, and far exceeding in size any now known even in the torrid zone. The genera are for the most part extinct, but some of them, as the crocodile and monitor, have still representatives in the warmer parts of the earth. Coral reefs also were evidently numerous in the seas of the same periods, composed of species often belonging to genera now characteristic of a tropical climate. The number of large chambered shells

* For an account of the more modern changes of the tertiary fauna and flora of the British Isles and adjoining countries, and particularly those facts which relate to the "glacial epoch," see an admirable essay by Prof. E. Forbes. Memoirs of Geol. Survey of Great Brit. vol. i. p. 336. London, 1846. To this important memoir I shall have frequent occasion to refer in the sequel.

also, including the nautilus, leads us to infer an elevated temperature; and the associated fossil plants, although imperfectly known, tend to the same conclusion, the Cycadeæ constituting the most numerous family.

But it is from the more ancient coal-deposits that the most extraordinary evidence has been supplied in proof of the former existence of a very different climate—a climate which seems to have been moist, warm, and extremely uniform, in those very latitudes which are now the colder, and in regard to temperature, the most variable regions of the globe. We learn from the researches of Adolphe Brongniart, Goeppert, and other botanists, that in the flora of the carboniferous era there was a great predominance of ferns, some of which were arborescent; as, for example, Caulopteris, Protopteris, and Psarronius; nor can this be accounted for, as some have supposed, by the greater power which ferns possess of resisting maceration in water.* This prevalence of ferns indicates a moist, equable, and temperate climate, and the absence of any severe cold; for such are the conditions which, at the present day, are found to be most favorable to that tribe of plants. It is only in the islands of the tropical oceans, and of the southern temperate zone, such as Norfolk Island, Otaheite, the Sandwich Islands, Tristan d'Acunha, and New Zealand, that we find any near approach to that remarkable preponderance of ferns which is characteristic of the Carboniferous flora. It has been observed that tree ferns and other forms of vegetation which flourished most luxuriantly within the tropics, extend to a much greater distance from the equator in the southern hemisphere than in the northern, being found even as far as 46° S. latitude in New Zealand. There is little doubt that this is owing to the more uniform and moist climate occasioned by the greater proportional area of sea. Next to ferns and pines, the most abundant vegetable forms in the coal formation are the Calamites, Lepidodendra, Sigillariæ, and Stigmariæ. These were formerly considered to be so closely allied to tropical genera, and to be so much greater in size than the corresponding tribes now inhabiting equatorial latitudes, that they were thought to imply an extremely hot, as well as humid and equable climate. But recent discoveries respecting the structure and relations of these fossil plants, have shown that they deviated so widely from all existing types in the vegetable world, that we have more reason to infer from this evidence a widely different climate in the Carboniferous era, as compared to that now prevailing, than a temperature extremely elevated.† Palms, if not entirely

* See a paper by Charles J. F. Bunbury, Esq., Journ. of Geol. Soc., London, No. 6, p. 88. 1846.

† The Calamites were formerly regarded by Adolphe Brongniart as belonging to the tribe of Equisetaceæ; but he is now inclined to refer them to the class of gymnogens, or gymnospermous exogens, which includes the Coniferæ and Cycadeæ. Lepidodendron appears to have been either a gigantic form of the lycopodium tribe, or, as Dr. Lindley thinks, intermediate between the lycopodia and the fir tribe. The Sigillariæ were formerly supposed by Ad. Brongniart, to be arborescent ferns; but the discovery of their internal structure, and of their leaves, has

wanting when the strata of the carboniferous group were deposited, appear to have been exceedingly rare.* The Coniferæ, on the other hand, so abundantly met with in the coal, resemble Araucariæ in structure, a family of the fir tribe, characteristic at present of the milder regions of the southern hemisphere, such as Chili, Brazil, New Holland, and Norfolk Island.

"In regard to the geographical extent of the ancient vegetation, it was not confined," says M. Brongniart, "to a small space, as to Europe, for example; for the same forms are met with again at great distances. Thus, the coal-plants of North America are, for the most part, identical with those of Europe, and all belong to the same genera. Some specimens, also, from Greenland, are referable to ferns, analogous to those of our European coal-mines."† The fossil plants brought from Melville Island, although in a very imperfect state, have been supposed to warrant similar conclusions;‡ and assuming that they agree with those of Baffin's Bay, mentioned by M. Brongniart, how shall we explain the manner in which such a vegetation lived through an arctic night of several months' duration?§

It may seem premature to discuss this question until the true nature of the fossil flora of the arctic regions has been more accurately determined; yet, as the question has attracted some attention, let us assume for a moment that the coal-plants of Melville Island are strictly analogous to those of the strata of Northumberland—would such a fact present an inexplicable enigma to the vegetable physiologist?

Plants, it is affirmed, cannot remain in darkness, even for a week,

since proved that they have no real affinity to ferns. According to the view now taken of their structure, their nearest allies in the recent world are the genera Cycas and Zamia; while Corda, on the other hand, maintains that they were closely related to the succulent euphorbias. Stigmaria is now generally admitted to have been merely the root of sigillaria. The scalariform vessels of these two genera are not conclusive in proving them to have a real affinity with ferns, as Mr. Brown has discovered the same structure of vessels in Myzodendron, a genus allied to the mistletoe; and Corda has lately shown that in two species of Stigmaria, hardly distinguishable by external characters, the vessels of the one are scalariform, and of the other dotted.

* Mr. Lindley endeavored formerly (1834) to show, in the "Fossil Flora," that Trigonocarpum Noeggerathii, a fruit found in the coal measures, has the true structure of a palm-fruit; but Ad. Brongniart has since inclined to regard it as cycadeous; nor is the French botanist satisfied that some specimens of supposed palm wood from the coal-mines of Radnitz in Bohemia, described by Corda, really belong to palms. On the other hand, Corda has proved Flabellaria borassifolia of Sternberg to be an exogenous plant, and Brongniart contends that it was allied to the Cycadeæ. See Tableau des Genres de Végétaux Fossiles. Paris, 1849.

† Prodrome d'une Hist. des Végét. Foss. p. 179. See also a late paper, Quart. Journ. of Geol. Soc. London, 1846, in which coal-plants of Alabama, lat. 33° N., collected by the author, are identified by Mr. Bunbury with British fossil species, showing the great southern extension of this flora.

‡ König, Journ. of Sci., vol. xv. p. 20. Mr. König informs me that he no longer believes any of these fossils to be tree ferns, as he at first stated, but that they agree generically with plants in our English coal-beds. The Melville Island specimens, now in the British Museum, are very obscure impressions.

§ Fossil Flora of Great Britain, by John Lindley and William Hutton, Esqrs., No. IV.

without serious injury, unless in a torpid state; and if exposed to heat and moisture they cannot remain torpid, but will grow, and must therefore perish. If, then, in the latitude of Melville Island, 75° N., a high temperature, and consequent humidity, prevailed at that period when we know the arctic seas were filled with corals and large multilocular shells, how could plants of tropical forms have flourished? Is not the bright light of equatorial regions as indispensable a condition of their well-being as the sultry heat of the same countries? and how could they annually endure a night prolonged for three months?*

Now, in reply to this objection, we must bear in mind, in the first place, that, so far as experiments have been made, there is every reason to conclude, that the range of intensity of light to which living plants can accommodate themselves is far wider than that of heat. No palms or tree ferns can live in our temperate latitudes without protection from the cold; but when placed in hot-houses they grow luxuriantly, even under a cloudy sky, and where much light is intercepted by the glass and frame-work. At St. Petersburg, in lat. 60° N., these plants have been successfully cultivated in hot-houses, although there they must exchange the perpetual equinox of their native regions, for days and nights which are alternately protracted to nineteen hours and shortened to five. How much farther towards the pole they might continue to live, provided a due quantity of heat and moisture were supplied, has not yet been determined; but St. Petersburg is probably not the utmost limit, and we should expect that in lat. 65° at least, where they would never remain twenty-four hours without enjoying the sun's light, they might still exist.

It should also be borne in mind, in regard to tree ferns, that they grow in the gloomiest and darkest parts of the forests of warm and temperate regions, even extending to nearly the 46th degree of south latitude in New Zealand. In equatorial countries, says Humboldt, they abound chiefly in the temperate, humid, and shady parts of *mountains*. As we know, therefore, that elevation often compensates for the effect of latitude in the geographical distribution of plants, we may easily understand that a class of vegetables, which grows at a certain height in the torrid zone, would flourish on the plains at greater distances from the equator, if the temperature, moisture, and other necessary conditions, were equally uniform throughout the year.

Nor must we forget that in all the examples above alluded to, we have been speaking of *living* species; but the coal-plants were of perfectly distinct species, nay, few of them except the ferns and pines can be referred to genera or even families of the existing vegetable kingdom. Having a structure, therefore, and often a form which appears to the botanist so anomalous, they may also have been endowed with a differ-

* Fossil Flora of Great Britain, by John Lindley and William Hutton, Esqrs. No. IV.

ent constitution, enabling them to bear a greater variation of circumstances in regard to light. We find that particular species of plants and tree ferns require at present different degrees of heat; and that some species can thrive only in the immediate neighborhood of the equator, others only a distance from it. In the same manner the *minimum* of *light*, sufficient for the now existing species, cannot be taken as the standard for all analogous tribes that may ever have flourished on the globe.

But granting that the extreme northern point to which a flora like that of the Carboniferous era could ever reach, may be somewhere between the latitudes of 65° and 70°, we should still have to inquire whether the vegetable remains might not have been drifted from thence, by rivers and currents, to the parallel of Melville Island, or still farther. In the northern hemisphere, at present, we see that the materials for future beds of lignite and coal are becoming amassed in high latitudes, far from the districts where the forests grew, and on shores where scarcely a stunted shrub can now exist. The Mackenzie, and other rivers of North America, carry pines with their roots attached for many hundred miles towards the north, into the Arctic Sea, where they are imbedded in deltas, and some of them drifted still farther by currents towards the pole.

Before we can decide on this question of transportation, we must know whether the fossil coal-plants occurring in high latitudes bear the marks of friction and of having decayed previously to fossilization. Many appearances in our English coal-fields certainly prove that the plants were not floated from great distances; for the outline of the stems of succulent species preserve their sharp angles, and others have their surfaces marked with the most delicate lines and streaks. Long leaves, also, are attached in many instances to the trunks or branches;* and leaves, we know, in general, are soon destroyed when steeped in water, although ferns will retain their forms after an immersion of many months.† It seems fair to presume, that most of the coal-plants grew upon the same land which supplied materials for the sandstones and conglomerates of the strata in which they are imbedded. The coarseness of the particles of many of these rocks attests that they were not borne from very remote localities, and that there was land therefore in the vicinity wasting away by the action of moving waters. The progress also of modern discovery has led to the very general admission of the doctrine that beds of coal have for the most part been formed of the remains of trees and plants that grew on the spot where the coal now exists; the land having been successively submerged, so that a covering of mud and sand was deposited upon accumulations of vegetable mater. That such has been the origin of some coal-seams is proved by the upright position of fossil trees, both in

* Fossil Flora, No. X.

† This has been proved by Mr. Lindley's experiments, ibid. No. XVII.

Europe and America, in which the roots terminate downwards in beds of coal.*

To return, therefore, from this digression,—the flora of the coal appears to indicate a uniform and mild temperature in the air, while the fossils of the contemporaneous mountain-limestone, comprising abundance of lamelliferous corals, large chambered cephalopods, and crinoidea, naturally lead us to infer a considerable warmth in the waters of the northern sea of the Carboniferous period. So also in regard to strata older than the coal, they contain in high northern latitudes mountain masses of corals which must have lived and grown on the spot, and large chambered univalves, such as Orthocerata and Nautilus, all seeming to indicate, even in regions bordering on the arctic circle, the former prevalence of a temperature more elevated than that now prevailing.

The warmth and humidity of the air, and the uniformity of climate, both in the different seasons of the year, and in different latitudes, appears to have been most remarkable when some of the oldest of the fossiliferous strata were formed. The approximation to a climate similar to that now enjoyed in these latitudes does not commence till the era of the formations termed tertiary; and while the different tertiary rocks were deposited in succession, from the eocene to the pliocene, the temperature seems to have been lowered, and to have continued to diminish even after the appearance upon the earth of a considerable number of the existing species, the cold reaching its maximum of intensity in European latitudes during the glacial epoch, or the epoch immediately antecedent to that in which all the species now contemporary with man were in being.

* I have treated of this subject in my Manual of Geology, and still more fully in my Travels in N. America, vol. ii. p. 178. For a full account of the facts at present known, and the theories entertained by the most eminent geologists and botanists on this subject, see Mr. Horner's Anniversary Address to the Geological Society of London, February, 1846. Consult also Sir H. de la Beche, on the formation of rocks in South Wales, Memoirs of Geol. Survey of Great Britain, 1846, p. 1 to 296.

CHAPTER VII.

FARTHER EXAMINATION OF THE QUESTION AS TO THE ASSUMED DISCORDANCE OF THE ANCIENT AND MODERN CAUSES OF CHANGE.

On the causes of vicissitudes in climate—Remarks on the present diffusion of heat over the globe—On the dependence of the mean temperature on the relative position of land and sea—Isothermal Lines—Currents from equatorial regions—Drifting of icebergs—Different temperature of Northern and Southern hemispheres—Combination of causes which might produce the extreme cold of which the earth's surface is susceptible—Conditions necessary for the production of the extreme of heat, and its probable effects on organic life.

*Causes of Vicissitudes in Climate.**—As the proofs enumerated in the last chapter indicate that the earth's surface has experienced great changes of climate since the deposition of the older sedimentary strata, we have next to inquire how such vicissitudes can be reconciled with the existing order of nature. The cosmogonist has availed himself of this, as of every obscure problem in geology, to confirm his views concerning a period when the planet was in a nascent or half-formed state, or when the laws of the animate and inanimate world differed essentially from those now established; and he has in this, as in many other cases, succeeded so far, as to divert attention from that class of facts which, if fully understood, might probably lead to an explanation of the phenomena. At first it was imagined that the earth's axis had been for ages perpendicular to the plane of the ecliptic, so that there was a perpetual equinox, and uniformity of seasons throughout the year;—that the planet enjoyed this "paradisiacal" state until the era of the great flood; but in that catastrophe, whether by the shock of a comet, or some other convulsion, it lost its equal poise, and hence the obliquity of its axis, and with that the varied seasons of the temperate zone, and the long nights and days of the polar circles.

When the progress of astronomical science had exploded this theory, it was assumed, that the earth at its creation was in a state of fluidity, and red-hot, and that ever since that era, it had been cooling down, contracting its dimensions, and acquiring a solid crust,—an hypothesis hardly less arbitrary, yet more calculated for lasting popularity; because, by referring the mind directly to the beginning of things, it requires no support from observation, nor from any ulterior hypothesis. But if, instead

* The theory proposed in this and the following chapters, to account for former fluctuations of climate at successive geological periods, agrees in every essential particular, and has indeed been reprinted almost verbatim from that published by me twenty years ago in the first edition of my Principles, 1830. It was referred to by Sir John F. W. Herschel in his Discourse on Natural Philosophy, published in 1830. In preceding works the gradual diminution of the earth's central heat was almost the only cause assigned for the acknowledged diminution of the superficial temperature of our planet.

of forming vague conjectures as to what might have been the state of the planet at the era of its creation, we fix our thoughts on the connection at present existing between climate and the distribution of land and sea; and then consider what influence former fluctuations in the physical geography of the earth must have had on superficial temperature, we may perhaps approximate to a true theory. If doubts and obscurities still remain, they should be ascribed to our limited acquaintance with the laws of Nature, not to revolutions in her economy;—they should stimulate us to farther research, not tempt us to indulge our fancies respecting the imaginary changes of internal temperature in an embryo world.

Diffusion of Heat over the Globe.—In considering the laws which regulate the diffusion of heat over the globe, we must be careful, as Humboldt well remarks, not to regard the climate of Europe as a type of the temperature which all countries placed under the same latitude enjoy. The physical sciences, observes this philosopher, always bear the impress of the places where they began to be cultivated; and as, in geology, an attempt was at first made to refer all the volcanic phenomena to those of the volcanoes in Italy, so in meteorology, a small part of the old world, the centre of the primitive civilization of Europe, was for a long time considered a type to which the climate of all corresponding latitudes might be referred. But this region, constituting only one-seventh of the whole globe, proved eventually to be the exception to the general rule. For the same reason, we may warn the geologist to be on his guard, and not hastily to assume that the temperature of the earth in the present era is a type of that which most usually obtains, since he contemplates far mightier alterations in the position of land and sea, at different epochs, than those which now cause the climate of Europe to differ from that of other countries in the same parallels.

It is now well ascertained that zones of equal warmth, both in the atmosphere and in the waters of the ocean, are neither parallel to the equator nor to each other.* It is also known that the *mean* annual temperature may be the same in two places which enjoy very different climates, for the seasons may be nearly uniform, or violently contrasted, so that the lines of equal winter temperature do not coincide with those of equal annual heat or isothermal lines. The deviations of all these lines from the same parallel of latitude are determined by a multitude of circumstances, among the principal of which are the position, direc-

* We are indebted to Baron Alex. von Humboldt for having first collected together the scattered data on which he founded an approximation to a true theory of the distribution of heat over the globe. Many of these data were derived from the author's own observations, and many from the works of M. Pierre Prevost, of Geneva, on the radiation of heat, and from other writers.—See Humboldt on Isothermal Lines, Mémoires d'Arcueil, tom. iii. translated in the Edin. Phil. Journ. vol. iii. July, 1820.

The map of Isothermal Lines, recently published by Humboldt and Dove (1848), supplies a large body of well-established data for such investigations, of which Mr. Hopkins has most ably availed himself in an essay "On the Causes which may have produced Changes in the earth's Superficial Temperature."—Q. Journ. Geol. Soc. 1852, p. 56.

tion, and elevation of the continents and islands, the position and depths of the sea, and the direction of currents and of winds.

On comparing the two continents of Europe and America, it is found that places in the same latitudes have sometimes a mean difference of temperature amounting to 11°, or even in a few cases to 17° Fahr.; and some places on the two continents, which have the same mean temperature, differ from 7° to 17° in latitude. Thus, Cumberland House, in North America, having the same latitude (54° N.) as the city of York in England, stands on the isothermal line of 32°, which in Europe rises to the North Cape, in lat. 71°, but its summer heat exceeds that of Brussels or Paris.* The principal cause of greater intensity of cold in corresponding latitudes of North America, as contrasted with Europe, is the connection of America with the polar circle, by a large tract of land, some of which is from three to five thousand feet in height; and, on the other hand, the separation of Europe from the arctic circle by an ocean. The ocean has a tendency to preserve everywhere a mean temperature, which it communicates to the contiguous land, so that it tempers the climate, moderating alike an excess of heat or cold. The elevated land, on the other hand, rising to the colder regions of the atmosphere, becomes a great reservoir of ice and snow, arrests, condenses, and congeals vapor, and communicates its cold to the adjoining country. For this reason, Greenland, forming part of a continent which stretches northward to the 82d degree of latitude, experiences under the 60th parallel a more rigorous climate than Lapland under the 72d parallel.

But if land be situated between the 40th parallel and the equator, it produces, unless it be of extreme height, exactly the opposite effect; for it then warms the tracts of land or sea that intervene between it and the polar circle. For the surface being in this case exposed to the vertical, or nearly vertical rays of the sun, absorbs a large quantity of heat, which it diffuses by radiation into the atmosphere. For this reason, the western parts of the old continent derive warmth from Africa, "which, like an immense furnace, distributes its heat to Arabia, to Turkey in Asia, and to Europe."† On the contrary, the northeastern extremity of Asia experiences in the same latitude extreme cold; for it has land on the north between the 60th and 70th parallel, while to the south it is separated from the equator by the Pacific Ocean.

In consequence of the more equal temperature of the waters of the ocean, the climate of islands and of coasts differs essentially from that of the interior of continents, the more maritime climate being characterized by mild winters, and more temperate summers; for the sea-breezes moderate the cold of winter, as well as the heat of summer. When, therefore, we trace round the globe those belts in which the mean annual temperature is the same, we often find great differences in climate; for there are *insular* climates in which the seasons are nearly equalized, and

* Sir J. Richardson's Appendix to Sir G. Bach's Journal, 1843—1845, p. 478.
† Malte-Brun, Phys. Geol. book xvii.

excessive climates, as they have been termed, where the temperature of winter and summer is strongly contrasted. The whole of Europe, compared with the eastern parts of America and Asia, has an insular climate. The northern part of China, and the Atlantic region of the United States, exhibit "excessive climates." We find at New York, says Humboldt, the summer of Rome and the winter of Copenhagen; at Quebec, the summer of Paris and the winter of Petersburg. At Pekin, in China, where the mean temperature of the year is that of the coasts of Brittany, the scorching heats of summer are greater than at Cairo, and the winters as rigorous as at Upsala.*

If lines be drawn round the globe through all those places which have the same winter temperature, they are found to deviate from the terrestrial parallels much farther than the lines of equal mean annual heat. The lines of equal winter in Europe, for example, are often curved so as to reach parallels of latitude 9° or 10° distant from each other, whereas the isothermal lines, or those passing through places having the same mean annual temperature, differ only from 4° to 5° in Europe.

Influence of currents and drift ice on temperature.—Among other influential causes, both of remarkable diversity in the mean annual heat, and of unequal division of heat in the different seasons, are the direction of currents and the accumulation and drifting of ice in high latitudes. The temperature of the Lagullas current is 10° or 12° Fahr. above that of the sea at the Cape of Good Hope; for it derives the greater part of its waters from the Mozambique channel, and south-east coast of Africa, and from regions in the Indian Ocean much nearer the line, and much hotter than the Cape.† An opposite effect is produced by the "equatorial" current, which crosses the Atlantic from Africa to Brazil, having a breadth varying from 160 to 450 nautical miles. Its waters are cooler by 3° or 4° Fahr. than those of the ocean under the line, so that it moderates the heat of the tropics.‡

But the effects of the Gulf stream on the climate of the North Atlantic Ocean are far more remarkable. This most powerful of known currents has its source in the Gulf or Sea of Mexico, which, like the Mediterranean and other close seas in temperate or low latitudes, is warmer than the open ocean in the same parallels. The temperature of the Mexican sea in summer is, according to Rennell, 86° Fahr., or at least 7° above that of the Atlantic in the same latitude.§ From this great reservoir or caldron of warm water, a constant current pours forth through the straits of Bahama at the rate of 3 or 4 miles an hour; it crosses the ocean in a northeasterly direction, skirting the great bank of Newfoundland, where it still retains a temperature of 8° above that of the surrounding sea. It reaches the Azores in about 78 days, after flowing nearly 3000 geographical miles, and from thence it some-

* On Isothermal Lines, &c.
† Rennell on Currents, p. 96. London, 1832.
‡ Ibid. p. 153.
§ Ibid. p. 25.

times extends its course a thousand miles farther, so as to reach the Bay of Biscay, still retaining an excess of 5° above the mean temperature of that sea. As it has been known to arrive there in the months of November and January, it may tend greatly to moderate the cold of winter in countries on the west of Europe.

There is a large tract in the centre of the North Atlantic, between the parallels of 33° and 45° N. lat., which Rennell calls the "recipient of the gulf water." A great part of it is covered by the weed called sargasso (*Sargassum bacciferum*), which the current floats in abundance from the Gulf of Mexico. This mass of water is nearly stagnant, is warmer by 7° or 10° than the waters of the Atlantic, and may be compared to the fresh water of a river overflowing the heavier salt water of the sea. Rennell estimates the area of the "recipient," together with that covered by the main current, as being 2000 miles in length from E. to W., and 350 in breadth from N. to S., which, he remarks, is a larger area than that of the Mediterranean. The heat of this great body of water is kept up by the incessant and quick arrivals of fresh supplies of warm water from the south; and there can be no doubt that the general climate of parts of Europe and America is materially affected by this cause.

It is considered probable by Scoresby that the influence of the Gulf stream extends even to the sea near Spitzbergen, where its waters may pass under those of melted ice; for it has been found that in the neighborhood of Spitzbergen, the water is warmer by 6° or 7° at the depth of one hundred and two hundred fathoms than at the surface. This might arise from the known law that fresh water passes the point of greatest density when cooled down below 40°, and between that and the freezing point expands again. The water of melted ice might be lighter, both as being fresh (having lost its salt in the decomposing process of freezing), and because its temperature is nearer the freezing point than the inferior water of the Gulf stream.

The great glaciers generated in the valleys of Spitzbergen, in the 79° of north latitude, are almost all cut off at the beach, being melted by the feeble remnant of heat still retained by the Gulf stream. In Baffin's Bay, on the contrary, on the west coast of Old Greenland, where the temperature of the sea is not mitigated by the same cause, and where there is no warmer under-current, the glaciers stretch out from the shore, and furnish repeated crops of mountainous masses of ice which float off into the ocean.* The number and dimensions of these bergs is prodigious. Captain Sir John Ross saw several of them together in Baffin's Bay aground in water fifteen hundred feet deep! Many of them are driven down into Hudson's Bay, and accumulating there, diffuse excessive cold over the neighboring continent; so that Captain Franklin reports, that at the mouth of Hayes' River, which

* Scoresby's Arctic Regions, vol. i. p. 208.—Dr. Latta's Observations on the Glaciers of Spitzbergen, &c. Edin. New Phil. Journ. vol. iii. p. 97.

lies in the same latitude as the north of Prussia or the south of Scotland, ice is found everywhere in digging wells, in summer, at the depth of four feet! Other bergs have been occasionally met with, at midsummer, in a state of rapid thaw, as far south as lat. 40° and longitude about 60° west, where they cool the water sensibly to the distance of forty or fifty miles around, the thermometer sinking sometimes 17°, or even 18°, Fahrenheit, in their neighborhood.* It is a well-known fact that every four or five years a large number of icebergs, floating from Greenland, double Cape Langaness, and are stranded on the west coast of Iceland. The inhabitants are then aware that their crops of hay will fail, in consequence of fogs which are generated almost incessantly; and the dearth of food is not confined to the land, for the temperature of the water is so changed that the fish entirely desert the coast.

Difference of climate of the Northern and Southern hemispheres.—When we compare the climate of the northern and southern hemispheres, we obtain still more instruction in regard to the influence of the distribution of land and sea on climate. The dry land in the southern hemisphere is to that of the northern in the ratio only of one to three, excluding from our consideration that part which lies between the pole and the 78° of south latitude, which has hitherto proved inaccessible. And whereas in the northern hemisphere, between the pole and the thirtieth parallel of north latitude, the land and sea occupy nearly equal areas, the ocean in the southern hemisphere covers no less than fifteen parts in sixteen of the entire space included between the antarctic circle and the thirtieth parallel of south latitude.

This great extent of sea gives a particular character to climates south of the equator, the winters being mild and the summers cool. Thus, in Van Dieman's Land, corresponding nearly in latitude to Rome, the winters are more mild than at Naples, and the summers not warmer than those at Paris, which is 7° farther from the equator.† The effects on animal and vegetable life are remarkable. Capt. King observed large shrubs of Fuchsia and Veronica, which in England are treated as tender plants, thriving and in full flower in Tierra del Fuego with the temperature at 36°. He states also that humming birds were seen sipping the sweets of the flowers "after two or three days of constant rain, snow, and sleet, during which time the thermometer had been at the freezing point." Mr. Darwin also saw parrots feeding on the seeds of a tree called the winter's bark, south of lat. 55°, near Cape Horn.‡

So the orchideous plants which are parasitical on trees, and are generally characteristic of the tropics, advance to the 38th and 42d degree of S. lat., and even beyond the 45th degree in New Zealand, where they were found by Forster. In South America also arborescent grasses abound in the dense forests of Chiloe, in lat. 42° S., where "they entwine the trees into one entangled mass to the height of thirty or forty

* Rennell on Currents, p. 95. † Humboldt on Isothermal Lines.
‡ Journ. of Travels in S. America, &c. p. 272.

feet above the ground. Palm-trees in the same quarter of the globe grow in lat. 37°, an arborescent grass very like a bamboo in 40°, and another closely allied kind, of great length, but not erect, even as far south as 45°."*

It has long been supposed that the general temperature of the southern hemisphere was considerably lower than that of the northern, and that the difference amounted to at least 10° Fahrenheit. Baron Humboldt, after collecting and comparing a great number of observations, came to the conclusion that even a much larger difference existed, but that none was to be observed within the tropics, and only a small difference as far as the thirty-fifth and fortieth parallel. Captain Cook was of opinion that the ice of the antarctic predominated greatly over that of the arctic region, that encircling the southern pole coming nearer to the equator by 10° than the ice around the north pole. All the recent voyages of discovery have tended to confirm this opinion, although Capt. Weddel penetrated, in 1823, three degrees farther south than Capt. Cook, reaching lat. 74° 15′ South, long. 34° 17′ West, and Sir James Ross, in 1842, arrived at lat. 78° 10′ S., as high a latitude, within three degrees, as the farthest point attained by Captain Parry in the arctic circle, or lat. 81° 12′ North.

The description given by ancient as well as modern navigators of the sea and land in high southern latitudes, clearly attests the greater severity of the climate as compared to arctic regions. In Sandwich Land, in lat. 59° S., or in nearly the same parallel as the north of Scotland, Capt. Cook found the whole country, from the summits of the mountains down to the very brink of the sea-cliffs, "covered many fathoms thick with everlasting snow," and this on the 1st of February, the hottest time of the year; and what is still more astonishing, in the island of S. Georgia, which is in the 54° south latitude, or the same parallel as Yorkshire, the line of perpetual snow descends to the level of the ocean.† When we consider this fact, and then recollect that the highest mountains in Scotland, which ascend to an elevation of nearly 5000 feet, and are four degrees farther to the north, do not attain the limit of perpetual snow on our side of the equator, we learn that latitude is one only of many powerful causes, which determine the climate of particular regions of the globe. Capt. Sir James Ross, in his exploring expedition in 1841–3, found that the temperature south of the 60th degree of latitude seldom rose above 32° Fahr. During the two summer months of the year 1841 (January and February) the range of the thermometer was between 11° and 32° Fahr.; and scarcely once rose above the freezing point. The permanence of snow in the southern hemisphere, is in this instance partly due to the floating ice, which chills the atmosphere and condenses the

* Darwin's travels in S. America, p. 271.

† Mr. Hopkins raises the question whether, in South Georgia, the descent of glaciers to the margin of the sea might not have been mistaken by Capt. Cook for the descent of the snow-line to the sea level. Quart. Journ. Geol. Soc. p. 85, 1852. The great navigator is generally very accurate, and there seem to be no observations of more recent date either to confirm or invalidate his statements.

vapor, so that in summer the sun cannot pierce through the foggy air. But besides the abundance of ice which covers the sea to the south of Georgia and Sandwich Land, we may also, as Humboldt suggests, ascribe the cold of those countries in part to the absence of land between them and the tropics.

If Africa and New Holland extended farther to the south, a diminution of ice would take place in consequence of the radiation of heat from these continents during summer, which would warm the contiguous sea and rarefy the air. The heated aerial currents would then ascend and flow more rapidly towards the south pole, and moderate the winter. In confirmation of these views, it is stated that the ice, which extends as far as the 68° and 71° of south latitude, advances more towards the equator whenever it meets an open sea; that is, where the extremities of the present continents are not opposite to it; and this circumstance seems explicable only on the principle above alluded to, of the radiation of heat from the lands so situated.

The cold of the antarctic regions was conjectured by Cook to be due to the existence of a large tract of land between the seventieth degree of south latitude and the pole. The justness of these and other speculations of that great navigator have since been singularly confirmed by the investigation made by Sir James Ross in 1841. He found Victoria Land, extending from 71° to 79° S. latitude, skirted by a great barrier of ice, the height of the land ranging from 4000 to 14,000 feet, the whole entirely covered with snow, except a narrow ring of black earth surrounding the huge crater of the active volcano of Mount Erebus, rising 12,400 feet above the level of the sea. The position of a mountainous territory of such altitude, so near the pole, and so obvious a source of intense cold, fully explains why Graham's and Enderby's Land, discovered by Captain Biscoe in 1831–2 (between lat. 64° and 68° S.), presented a most wintry aspect, covered even in summer with ice and snow, and nearly destitute of animal life. In corresponding latitudes of the northern hemisphere we not only meet with herds of wild herbivorous animals, but with land which man himself inhabits, and where he has even built ports and inland villages.*

The distance to which icebergs float from the polar regions on the opposite sides of the line is, as might have been anticipated, very different. Their extreme limit in the northern hemisphere is lat. 40°, as before mentioned, and they are occasionally seen in lat. 42° N., near the termination of the great bank of Newfoundland, and at the Azores, lat. 42° N., to which they are sometimes drifted from Baffin's Bay. But in the other hemisphere they have been seen, within the last few years, at different points off the Cape of Good Hope, between lat.

* After all these modern discoveries, the area still unexplored, within the antarctic circle, is more than double the area of Europe. The surface of the latter contains about 2,793,000 square geographical miles. The unexplored antarctic region, as calculated for me by Mr. Gardner, in 1840, equalled about 7,620,000 square miles.

Fig. 2.

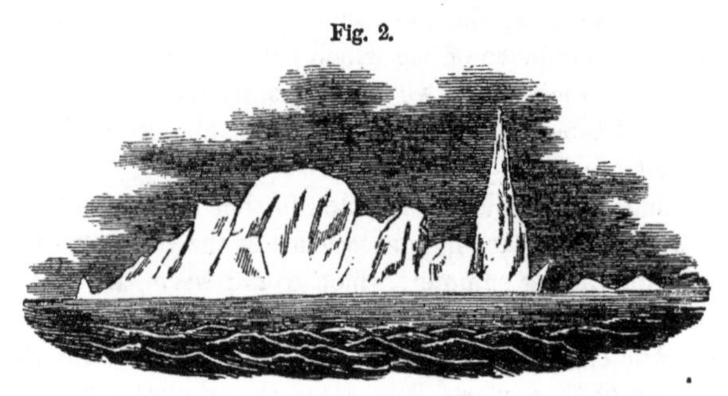

Iceberg seen off the Cape of Good Hope, April, 1829.
Lat. 39° 13′ S. Long. 48° 46′ E.

36° and 39°.* One of these (see fig. 2) was two miles in circumference, and 150 feet high, appearing like chalk when the sun was obscured, and having the lustre of refined sugar when the sun was shining on it. Others rose from 250 to 300 feet above the level of the sea, and were therefore of great volume below; since it is ascertained by experiments on the buoyancy of ice floating in sea-water, that for every cubic foot seen above, there must at least be eight cubic feet below water.† If ice islands from the north polar regions floated as far, they might reach Cape St. Vincent, and there, being drawn by the current that always sets in from the Atlantic through the Straits of Gibraltar, be drifted into the Mediterranean, so that the serene sky of that delightful region might soon be deformed by clouds and mists.

Before the amount of difference between the temperature of the two hemispheres was ascertained, it was referred by many astronomers to the precession of the equinoxes, or the acceleration of the earth's motion in its perihelium; in consequence of which the spring and summer of the southern hemisphere are now shorter, by nearly eight days, than those seasons north of the equator. But Sir J. Herschel reminds us that the excess of eight days in the duration of the sun's presence in the northern hemisphere is not productive of an excess of annual light and heat; since, according to the laws of elliptic motion, it is demonstrable that whatever be the ellipticity of the earth's orbit, the two hemispheres must receive *equal absolute quantities* of light and heat per annum, the proximity of the sun in perigee exactly compensating the effect of its swifter motion.‡ Humboldt, however, observes, that there must be a greater loss of heat by radiation in the southern hemi-

* On icebergs in low latitudes, by Capt. Horsburgh, by whom the sketch was made. Phil. Trans. 1830.

† Scoresby's Arctic Regions, vol. i. p. 234.

‡ This follows, observes Herschel, from a very simple theorem, which may be thus stated:—"The amount of heat received by the earth from the sun, while describing any part of its orbit, is proportional to the angle described round the sun's centre." So that if the orbit be divided into two portions by a line drawn *in any direction* through the sun's centre, the heat received in describing the two unequal segments of the eclipse so produced will be equal. Geol. Trans. vol. iii. part. ii. p. 298; second series.

sphere during a winter longer by eight days than that on the other side of the equator.*

Perhaps no very sensible effect may be produced by this source of disturbance; yet the geologist should bear in mind that to a certain extent it operates alternately on each of the two hemispheres for a period of upwards of 10,000 years, dividing unequally the times during which the annual supply of solar light and heat is received. This cause may sometimes tend to counterbalance inequalities of temperature resulting from other far more influential circumstances; but, on the other hand, it must sometimes tend to increase the extreme of deviation arising from particular combinations of causes.

But whatever may be at present the inferiority of heat in the temperate and frigid zones south of the line, it is quite evident that the cold would be far more intense if there happened, instead of open sea, to be tracts of elevated land between the 55th and 70th parallel; and, on the other hand, the cold would be moderated if there were more land between the line and the forty-fifth degree of south latitude.

Changes in the position of land and sea may give rise to vicissitudes in climate.—Having offered these brief remarks on the diffusion of heat over the globe in the present state of the surface, I shall now proceed to speculate on the vicissitudes of climate, which must attend those endless variations in the geographical features of our planet which are contemplated in geology. That our speculations may be confined within the strict limits of analogy, I shall assume, 1st, That the proportion of dry land to sea continues always the same. 2dly, That the volume of the land rising above the level of the sea is a constant quantity; and not only that its mean, but that its extreme height, is liable only to trifling variations. 3dly, That both the mean and extreme depth of the sea are invariable; and 4thly, It may be consistent with due caution to assume that the grouping together of the land in continents is a necessary part of the economy of nature; for it is possible that the laws which govern the subterranean forces, and which act simultaneously along certain lines, cannot but produce, at every epoch, continuous mountain-chains; so that the subdivision of the whole land into innumerable islands may be precluded.

If it be objected, that the maximum of elevation of land and depth of sea are probably not constant, nor the gathering together of all the land in certain parts, nor even perhaps the relative extent of land and water, I reply, that the arguments about to be adduced will be strengthened if, in these peculiarities of the surface, there be considerable deviations from the present type. If, for example, all other circumstances being the same, the land is at one time more divided into islands than at another, a greater uniformity of climate might be produced, the mean temperature remaining unaltered; or if, at another era, there were mountains higher than the Himalaya, these, when placed in

* On Isothermal Lines.

high latitudes, would cause a greater excess of cold. Or, if we suppose that at certain periods no chain of hills in the world rose beyond the height of 10,000 feet, a greater heat might then have prevailed than is compatible with the existence of mountains thrice that elevation.

However constant may be the relative proportion of sea and land, we know that there is annually some small variation in their respective geographical positions, and that in every century the land is in some parts raised, and in others depressed in level, and so likewise is the bed of the sea. By these and other ceaseless changes, the configuration of the earth's surface has been remodelled again and again, since it was the habitation of organic beings, and the bed of the ocean has been lifted up to the height of some of the loftiest mountains. The imagination is apt to take alarm when called upon to admit the formation of such irregularities in the crust of the earth, after it had once become the habitation of living creatures; but, if time be allowed, the operation need not subvert the ordinary repose of nature; and the result is in a general view insignificant, if we consider how slightly the highest mountain-chains cause our globe to differ from a perfect sphere. Chimborazo, though it rises to more than 21,000 feet above the sea, would be represented, on a globe of about six feet in diameter, by a grain of sand less than one-twentieth of an inch in thickness.

The superficial inequalities of the earth, then, may be deemed minute in quantity, and their distribution at any particular epoch must be regarded in geology as temporary peculiarities, like the height and outline of the cone of Vesuvius in the interval between two eruptions. But although, in reference to the magnitude of the globe, the unevenness of the surface is so unimportant, it is on the position and direction of these small inequalities that the state of the atmosphere, and both the local and general climate, are mainly dependent.

Before considering the effect which a material change in the distribution of land and sea must occasion, it may be well to remark, how greatly organic life may be affected by those minor variations, which need not in the least degree alter the general temperature. Thus, for example, if we suppose, by a series of convulsions, a certain part of Greenland to become sea, and, in compensation, a tract of land to rise and connect Spitzbergen with Lapland,—an accession not greater in amount than one which the geologist can prove to have occurred in certain districts bordering the Mediterranean, within a comparatively modern period,—this altered form of the land might cause an interchange between the climate of certain parts of North America and of Europe, which lie in corresponding latitudes. Many European species of plants and animals would probably perish in consequence, because the mean temperature would be greatly lowered; and others would fail in America, because it would there be raised. On the other hand, in places where the mean annual heat remained unaltered, some species which flourish in Europe, where the seasons are more uniform, would be unable to resist the greater heat of the North American summer, or the intenser cold of the winter; while others, now fitted by their habits for the great contrast of the American seasons, would not be fitted for the *insular* climate of Europe. The vine, for example, according to Humboldt, can be cultivated with advantage 10° farther north in Europe than in North America. Many plants endure severe frost, but cannot ripen their seeds without a certain intensity of summer heat and a certain quantity of light; others cannot endure a similar intensity either of heat or cold.

It is now established that many of the existing species of animals have survived great changes in the physical geography of the globe. If such species be termed modern, in comparison to races which preceded them, their remains, nevertheless, enter into submarine deposits many hundred miles in length, and which have since been raised from the deep to no inconsiderable altitude. When, therefore, it is shown that changes in the temperature of the atmosphere may be the consequence of such physical revolutions of the surface, we ought no longer to wonder that we find the distribution of existing species to be *local,* in regard to *longitude* as well as latitude. If all species were now, by an exertion of creative power, to be diffused uniformly throughout those zones where there is an equal degree of heat, and in all respects a similarity of climate, they would begin from this moment to depart more and more from their original distribution. Aquatic and terrestrial species would be displaced, as Hooke long ago observed, so often as land and water exchanged places; and there would also, by the formation of new mountains and other changes, be transpositions of climate, contributing, in the manner before alluded to, to the local extermination of species.*

If we now proceed to consider the circumstances required for a *general* change of temperature, it will appear, from the facts and principles already laid down, that whenever a greater extent of high land is collected in the polar regions, the cold will augment; and the same result will be produced when there is more sea between or near the tropics; while, on the contrary, so often as the above conditions are reversed, the heat will be greater. (See figs. 5 and 6, p. 111.) If this be admitted, it will follow, that unless the superficial inequalities of the earth be fixed and permanent, there must be never-ending fluctuations in the mean temperature of every zone; and that the climate of one era can no more be a type of every other, than is one of our four seasons of all the rest.

It has been well said, that the earth is covered by an ocean, in the midst of which are two great islands, and many smaller ones; for the whole of the continents and islands occupy an area scarcely exceeding one-fourth of the whole superficies of the spheroid. Now, according to this analogy, we may fairly speculate on the probability that there would not be usually, at any given epoch of the past, more than about one-fourth dry land in a particular region; as, for example, near the poles,

* A full consideration of the effect of changes in physical geography on the distribution and extinction of species is given in book iii.

or between them and the 75th parallels of N. and S. latitude. If, therefore, at present there should happen to be, in both these quarters of the globe, much *more* than this average proportion of land, some of it in the arctic region being above five thousand feet in height, and if in antarctic latitudes a mountainous country has been found varying from 4000 to 14,000 feet in height, this alone affords ground for concluding that, in the present state of things, the mean heat of the climate is below that which the earth's surface, in its more ordinary state, would enjoy. This presumption is heightened when we reflect on the results of the recent soundings made by Sir James Ross, in the Southern Ocean, and continued for four successive years, ending 1844, which seem to prove that the mean depth of the Atlantic and Pacific is as great as Laplace and other eminent astronomers had imagined;* for then we might look not only for more than two-thirds sea in the frigid zones, but for water of great depth, which could not readily be reduced to the freezing point. The same opinion is confirmed, when we compare the quantity of land lying between the poles and the 30th parallels of north and south latitude, with the quantity placed between those parallels and the equator; for, it is clear, that we have at present not only more than the usual degree of cold in the polar regions, but also less than the average quantity of heat within the tropics.

* For calculations founded on astronomical data, see Young's Nat. Phil., Lect. xlvii.; Mrs. Somerville's Connex. of Phys. Sci., sect. 14, p. 110. Laplace, endeavoring to estimate the probable depth of the sea from some of the phenomena of the tides, says of the ocean generally, "que sa profondeur moyenne est du même ordre que la hauteur moyenne des continens et des isles au-dessus de son niveau, hauteur qui ne surpasse pas mille mètres (3280 ft.)" Mec. Céleste, tom. xi. et Syst. du Monde. p. 254. The expression "du même ordre" admits in mathematical language of considerable latitude of signification, and does not mean that the depth of the water below the level of the sea corresponds exactly to the height of the land above it.

It appeared from the observations of Sir James Ross, communicated to me in 1849, by himself, and his fellow-voyager, Dr. Joseph Hooker, that in latitude 15° 3′ S., longitude 23° 14′ W. (the island of Trinidad, the nearest land, being 486 miles distant, and bearing S. 47 W.), they sounded with a weight of 76 lbs., and 4600 fathoms of line, which ran out to the very end, without finding bottom. Here therefore in mid-ocean the depth exceeded 27,600 feet. One of the shallowest soundings ever obtained in the open sea during the same survey, struck bottom with 2677 fathoms, or 16,062 feet, latitude 33° 21′ S., longitude 9° 4′ E. The surveyors arrived at the conclusion, that at a moderate distance from the shore, the depth of the great ocean always exceeds 4000 feet.

During the American survey in 1849, a much greater depth, or 5700 fathoms (34,200 feet), was sounded in the Atlantic by Lieut. Walsh, without reaching the bottom, in lat. 31° 59′ N., long. 58° 43′ W., or between the Bermudas and the Azores. But the deepest soundings yet published were taken Oct. 30th 1852, by Capt. Henry M. Denham, R. N., who reached bottom at 7706 fathoms (46,236 feet), lat. 36° 49′ S., long. 37° 6′ W., the nearest land being at the mouth of the River Plate. A weight of 9 lbs. was attached to the line, which was one-tenth of an inch in diameter; the day was calm, and the line took 9 hours 24 minutes to run out. When the bottom was struck the line was raised 50 fathoms, and then allowed to run out again. It struck at the same point as before, verifying the observations. Nevertheless some experienced surveyors have remarked that the experiment would have been more satisfactory had the weight been greater. The highest summits of the Himalaya are about 28,000 feet; the Pacific, according to this sounding, is probably at some points twice as deep as the Himalaya are high.

Position of land and sea which might produce the extreme of cold of which the earth's surface is susceptible.—To simplify our view of the various changes in climate, which different combinations of geographical circumstances may produce, we shall first consider the conditions neces-sary for bringing about the extreme of cold, or what would have been termed in the language of the old writers the winter of the "great year," or geological cycle, and afterwards, the conditions requisite to produce the maximum of heat, or the summer of the same year.

To begin with the northern hemisphere. Let us suppose those hills of the Italian peninsula and of Sicily, which are of comparatively mod-ern origin, and contain many fossil shells identical with living species, to subside again into the sea, from which they have been raised, and that an extent of land of equal area and height (varying from one to three thousand feet) should rise up in the Arctic Ocean between Siberia and the north pole. In speaking of such changes, I shall not allude to the manner in which I conceive it possible that they may be brought about, nor of the time required for their accomplishment—reserving for a future occasion, not only the proofs that revolutions of equal magnitude have taken place, but that analogous operations are still in gradual progress. The alteration now supposed in the physical geography of the northern regions, would cause additional snow and ice to accumulate where now there is usually an open sea; and the temperature of the greater part of Europe would be somewhat lowered, so as to resemble more nearly that of corresponding latitudes of North America: or, in other words, it might be necessary to travel about 10° farther south in order to meet with the same climate which we now enjoy. No compensation would be derived from the disappearance of land in the Mediterranean coun-tries; but the contrary, since the mean heat of the soil in those lati-tudes probably exceeds that which would belong to the sea, by which we imagine it to be replaced.

But let the configuration of the surface be still farther varied, and let some large district within or near the tropics, such as Brazil, with its plains and hills of moderate height, be converted into sea, while lands of equal elevation and extent rise up in the arctic circle. From this change there would, in the first place, result a sensible diminution of temperature near the tropic, for the Brazilian soil would no longer be heated by the sun; so that the atmosphere would be less warm, as also the neighboring Atlantic. On the other hand, the whole of Europe, Northern Asia, and North America, would be chilled by the enormous quantity of ice and snow, thus generated on the new arctic continent. If, as we have already seen, there are now some points in the southern hemisphere where snow is perpetual down to the level of the sea, in latitudes as low as central England, such might assuredly be the case throughout a great part of Europe, under the change of circumstances above supposed: and if at present the extreme range of drifted icebergs is the Azores, they might easily reach the equator after the assumed alteration. But to pursue the subject still farther, let the Himalaya mountains, with the whole of Hindostan, sink down, and their place be occupied by the Indian Ocean, while an equal extent of territory and mountains, of the same vast height, rise up between North Greenland and the Orkney Islands. It seems difficult to exaggerate the amount to which the climate of the northern hemisphere would then be cooled.*

But the refrigeration brought about at the same time in the southern hemisphere, would be nearly equal, and the difference of temperature between the arctic and equatorial latitudes would not be much greater than at present; for no important disturbance can occur in the climate of a particular region without its immediately affecting all other latitudes, however remote. The heat and cold which surround the globe are in a state of constant and universal flux and reflux. The heated and rarefied air is always rising and flowing from the equator towards the poles in the higher regions of the atmosphere; while in the lower, the colder air is flowing back to restore the equilibrium. That this circulation is constantly going on in the aerial currents is not disputed; it is often proved by the opposite course of the clouds at different heights, and the fact has been farther illustrated in a striking manner by two recent events. The trade wind continually blows with great force from the island of Barbadoes to that of St. Vincent; notwithstanding which, during the eruption of the volcano in the island of St. Vincent, in 1812, ashes fell in profusion from a great height in the atmosphere upon Barbadoes.† In like manner, during the great eruption of Sumbawa, in 1815, ashes were carried to the islands of Amboyna and Banda, which last is about 800 miles east from the site of the volcano. Yet the southeast monsoon was then at its height.‡ This apparent transposition of matter against the wind, confirmed the opinion of the existence of a counter-current in the higher regions, which had previously rested on theoretical conclusions only.

That a corresponding interchange takes place in the seas, is demonstrated, according to Humboldt, by the cold which is found to exist at great depths within the tropics; and, among other proofs, may be mentioned the mass of warmer water which the Gulf stream is constantly bearing northwards, while a cooler current flows *from* the north along the coast of Greenland and Labrador, and helps to restore the equilibrium.§

* Mr. Hopkins, reasoning on data furnished by Dove's Isothermal maps, has arrived at the very interesting conclusion, that both on Snowdon and the lower mountains of the West of Ireland the snow-line would descend to within 1000 feet of the sea level, and glaciers reach the sea, if we could simply assume the three following geographical changes:—

1st, The diversion of the Gulf stream from its present northerly course; 2dly, the depression of the existing land of Northern and Western Europe, to the amount of no more than 500 feet; and 3dly, a cold current from the North sweeping over the submerged area. Quart. Journ. Geol. Soc. 1852, p. 85.

† Daniell's Meteorological Essays, p. 103.

‡ Observed by J. Crawfurd, Esq.

§ In speaking of the circulation of air and water in this chapter, no allusion is made to the trade winds, or to irregularities in the direction of currents, caused

Currents of colder and therefore specifically heavier water pass from the poles towards the equator, which cool the inferior parts of the ocean; so that the heat of the torrid zone and the cold of the polar circle balance each other. The refrigeration, therefore, of the polar regions, resulting from the supposed alteration in the distribution of land and sea, would be immediately communicated to the tropics, and from them its influence would extend to the antarctic circle, where the atmosphere and the ocean would be cooled, so that ice and snow would augment. Although the mean temperature of higher latitudes in the southern hemisphere is, as before stated, for the most part, lower than that of the same parallels in the northern, yet, for a considerable space on each side of the line, the mean annual heat of the waters is found to be the same in corresponding parallels. If, therefore, by the new position of the land, the formation of icebergs had become of common occurrence in the northern temperate zone, and if these were frequently drifted as far as the equator, the same degree of cold which they generated would immediately be communicated as far as the tropic of Capricorn, and from thence to the lands or ocean to the south.

The freedom, then, of the circulation of heat and cold from pole to pole being duly considered, it will be evident that the mean temperature which may prevail at the same point at two distinct periods, may differ far more widely than that of any two points in the same parallels of latitude, at one and the same period. For the range of temperature, or in other words, the curvature of the isothermal lines in a given zone, and at a given period, must always be circumscribed within narrow limits, the climate of each place in that zone being controlled by the combined influence of the geographical peculiarities of all other parts of the earth. Whereas, if we compare the state of things at two distinct and somewhat distant epochs, a particular zone may at one time be under the influence of one class of disturbing causes, and at another time may be affected by an opposite combination. The lands, for example, to the north of Greenland cause the present climate of North America to be colder than that of Europe in the same latitudes; but the excess of cold is not so great as it would have been if the western hemisphere had been entirely isolated, or separated from the eastern like a distinct planet. For not only does the refrigeration produced by Greenland chill to a certain extent the atmosphere of northern and western Europe, but the mild climate of Europe reacts also upon North America, and moderates the chilling influence of the adjoining polar lands.

To return to the state of the earth after the changes above supposed, we must not omit to dwell on the important effects to which a wide expanse of perpetual snow would give rise. It is probable that nearly the whole sea, from the poles to the parallels of 45°, would be frozen over; for it is well known that the immediate proximity of land is not

by the rotary motion of the earth. These causes prevent the movements from being direct from north to south, or from south to north, but they do not affect the theory of a constant circulation.

essential to the formation and increase of field ice, provided there be in some part of the same zone a sufficient quantity of glaciers generated on or near the land, to cool down the sea. ' Captain Scoresby, in his account of the arctic regions, observes, that when the sun's rays "fall upon the snow-clad surface of the ice or land, they are in a great measure reflected, without producing any material elevation of temperature; but when they impinge on the black exterior of a ship, the pitch on one side occasionally becomes fluid while ice is rapidly generated at the other."*

Now field ice is almost always covered with snow;† and thus not only land as extensive as our existing continents, but immense tracts of sea in the frigid and temperate zones, might present a solid surface covered with snow, and reflecting the sun's rays for the greater part of the year. Within the tropics, moreover, where the ocean now predominates, the sky would no longer be serene and clear, as in the present era; but masses of floating ice would cause quick condensations of vapor, so that fogs and clouds would deprive the vertical rays of the sun of half their power. The whole planet, therefore, would receive annually a smaller portion of the solar influence, and the external crust would part, by radiation, with some of the heat which had been accumulated in it, during a different state of the surface. This heat would be dissipated in the spaces surrounding our atmosphere, which, according to the calculations of M. Fourier, have a temperature much inferior to that of freezing water.

After the geographical revolution above assumed, the climate of equinoctial lands might be brought at last to resemble that of the present temperate zone, or perhaps be far more wintry. They who should then inhabit such small isles and coral reefs as are now seen in the Indian Ocean and South Pacific, would wonder that zoophytes of large dimensions had once been so prolific in their seas; or if, perchance, they found the wood and fruit of the cocoa-nut tree or the palm silicified by the waters of some ancient mineral spring, or incrusted with calcareous matter, they would muse on the revolutions which had annihilated such genera, and replaced them by the oak, the chestnut, and the pine. With equal admiration would they compare the skeletons of their small lizards with the bones of fossil alligators and crocodiles more than twenty feet in length, which, at a former epoch, had multiplied between the tropics: and when they saw a pine included in an iceberg, drifted from latitudes which we now call temperate, they would be astonished at the proof thus afforded, that forests had once grown where nothing could be seen in their own times but a wilderness of snow.

If the reader hesitate to suppose so extensive an alteration of temperature as the probable consequence of geographical changes, confined to one hemisphere, he should remember how great are the local anomalies in climate now resulting from the peculiar distribution of land and sea

* See Scoreby's Arctic Regions, vol. i. p. 378. † Ibid. p. 320.

in certain regions. Thus, in the island of South Georgia, before mentioned (p. 98), Captain Cook found the everlasting snows descending to the level of the sea, between lat. 54° and 55° S.; no trees or shrubs were to be seen, and in summer a few rocks only, after a partial melting of the ice and snow, were scantily covered with moss and tufts of grass. If such a climate can now exist at the level of the sea in a latitude corresponding to that of Yorkshire in spite of all those equalizing causes before enumerated, by which the mixture of the temperatures of distant regions is facilitated throughout the globe, what rigors might we not anticipate in a winter generated by the transfer of the mountains of India to our arctic circle!

But we have still to contemplate the additional refrigeration which might be effected by changes in the relative position of land and sea in the southern hemisphere. If the remaining continents were transferred from the equatorial and contiguous latitudes to the south polar regions, the intensity of cold produced might, perhaps, render the globe uninhabitable. We are too ignorant of the laws governing the direction of subterranean forces, to determine whether such a crisis be within the limits of possibility. At the same time, it may be observed, that no distribution of land can well be imagined more irregular, or, as it were, capricious, than that which now prevails; for at present, the globe may be divided into two equal parts, in such a manner, that one hemisphere shall be almost entirely covered with water, while the other shall contain less water than land (see figs. 3 and 4);* and, what is still more extraordinary, on comparing the extratropical lands in the northern and southern hemispheres, the lands in the northern are found to be to those in the southern in the proportion of thirteen to one!† To imagine all the lands, therefore, in high, and all the sea in low latitudes, as delineated in fig. 6, p. 111, would scarcely be a more anomalous state of the surface.

Position of land and sea which might give rise to the extreme of heat.—Let us now turn from the contemplation of the winter of the "great year," and consider the opposite train of circumstances which would bring on the spring and summer. To imagine all the lands to be collected together in equatorial latitudes, and a few promontories only to project

* This is shown by projecting a map on the horizon of London, that is to say, by supposing the eye of the observer to be placed above that city, and to see from thence one half of the globe. For it so happens that from that point, and no other, we should behold the greatest possible quantity of land; and if we are then transferred to the opposite or antipodal point, we should see the greatest possible quantity of water. (See figs. 3 and 4.) A singular fact, first pointed out by Mr. James Gardner, namely, that only one twenty-seventh part of the dry land has any land opposite to it, is intimately connected with this excess of land in one of the two hemispheres above alluded to. Thus, in fig. 3, the land shaded black in part of China answers to that portion of the extremity of South America and Tierra del Fuego which is opposite or antipodal to it, whilst the dark spots in the northern and central parts of South America represent Borneo, Sumatra, and other antipodal islands in the Eastern Archipelago. See Gardner, Geol. Soc. Proceedings, 1833, vol. i. p. 488.

† Humboldt on Isothermal Lines

MAP showing the present unequal Distribution of LAND and WATER on the Surface of the GLOBE.

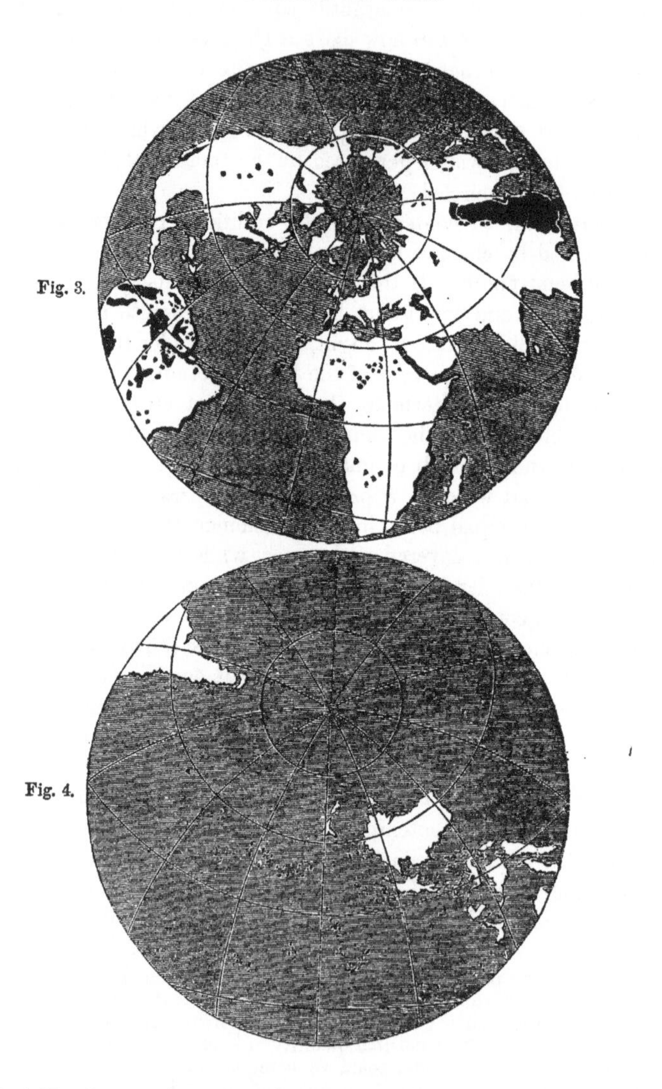

Fig. 3. Here London is taken as a centre, and we behold the greatest quantity of land existing in one hemisphere.

Fig. 4. Here the centre is the antipodal point to London, and we see the greatest quantity of water existing in one hemisphere.

The black shading expresses land having land opposite or antipodal to it.

MAPS showing the position of LAND and SEA which might produce the Extremes of HEAT and COLD in the Climates of the GLOBE.

Fig. 5.

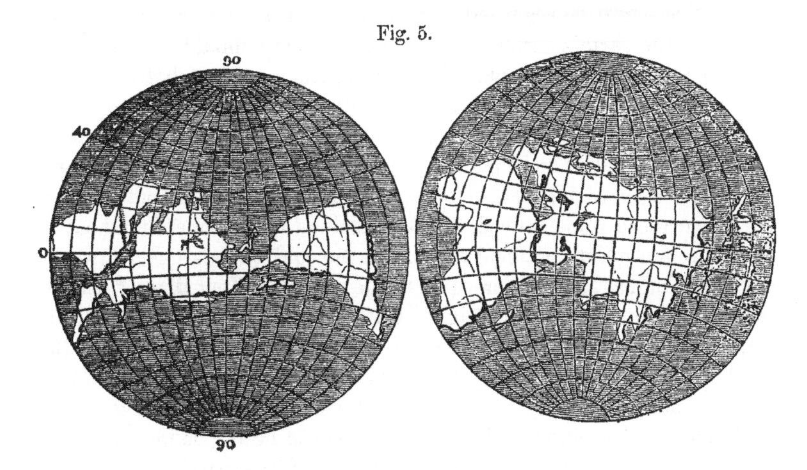

Extreme of Heat.

Fig. 6.

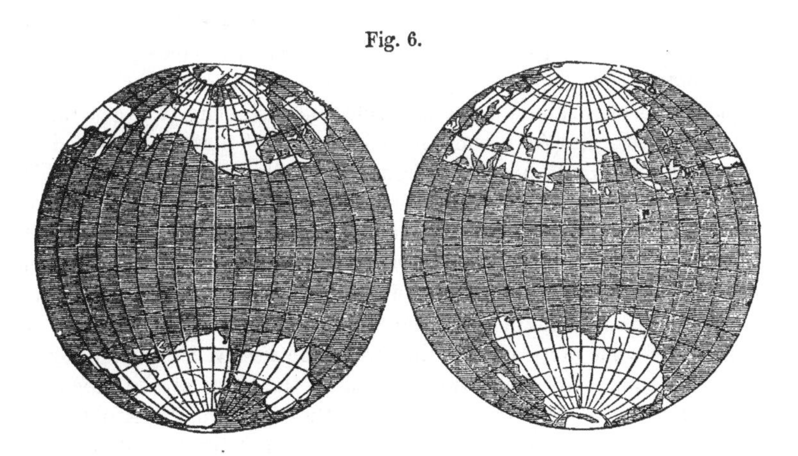

Extreme of Cold.

OBSERVATIONS.—These maps are intended to show that continents and islands having the same shape and relative dimensions as those now existing, might be placed so as to occupy either the equatorial or polar regions.

In fig. 5, scarcely any of the land extends from the equator towards the poles beyond the 30th parallel of latitude; and fig. 6, a very small proportion of it extends from the poles towards the Equator beyond the 40th parallel of latitude.

beyond the thirtieth parallel, as represented in the annexed maps (figs. 5 and 6), would be undoubtedly to suppose an extreme result of geological change. But if we consider a mere approximation to such a state of things, it would be sufficient to cause a general elevation of temperature. Nor can it be regarded as a visionary idea, that amidst the revolutions of the earth's surface, the quantity of land should, at certain periods, have been simultaneously lessened in the vicinity of both the poles, and increased within the tropics. We must recollect that even now it is necessary to ascend to the height of fifteen thousand feet in the Andes under the line, and in the Himalaya mountains, which are without the tropic, to seventeen thousand feet, before we reach the limit of perpetual snow. On the northern slope, indeed, of the Himalaya range, where the heat radiated from a great continent moderates the cold, there are meadows and cultivated land at an elevation equal to the height of Mont Blanc.* If then there were no arctic lands to chill the atmosphere, and freeze the sea, and if the loftiest chains were near the line, it seems reasonable to imagine that the highest mountains might be clothed with a rich vegetation to their summits, and that nearly all signs of frost would disappear from the earth.

When the absorption of the solar rays was in no region impeded, even in winter, by a coat of snow, the mean heat of the earth's crust would augment to considerable depths, and springs, which we know to be in general an index of the mean temperature of the climate, would be warmer in all latitudes. The waters of lakes, therefore, and rivers, would be much hotter in winter, and would be never chilled in summer by melted snow and ice. A remarkable uniformity of climate would prevail amid the archipelagoes of the temperate and polar oceans, where the tepid waters of equatorial currents would freely circulate. The general humidity of the atmosphere would far exceed that of the present period, for increased heat would promote evaporation in all parts of the globe. The winds would be first heated in their passage over the tropical plains, and would then gather moisture from the surface of the deep, till, charged with vapor, they arrived at extreme northern and southern regions, and there encountering a cooler atmosphere, discharged their burden in warm rain. If, during the long night of a polar winter, the snows should whiten the summits of some arctic islands, they would be dissolved as rapidly by the returning sun, as are the snows of Etna by the blasts of the sirocco.

We learn from those who have studied the geographical distribution of plants, that in very low latitudes, at present, the vegetation of small islands remote from continents has a peculiar character; the ferns and allied families, in particular, bearing a great proportion to the total number of other plants. Other circumstances being the same, the more remote the isles are from the continents, the greater does this proportion become. Thus, in the continent of India, and the tropical parts of New Holland, the proportion of ferns to the phænogamous plants is only as one to twenty-six; whereas, in the South-Sea Islands, it is as one to four, or even as one to three.†

* Humboldt, Tableaux de la Nature, tom. i. p. 112.

† Ad. Brongniart, Consid. Générales sur la Nat. de la Végét. &c. Ann. des Sciences Nat., Nov. 1828.

We might expect, therefore, in the summer of the "great year," or cycle of climate, that there would be a predominance of tree ferns and plants allied to genera now called tropical, in the islands of the wide ocean, while many forms now confined to arctic and temperate regions, or only found near the equator on the summit of the loftiest mountains, would almost disappear from the earth. Then might those genera of animals return, of which the memorials are preserved in the ancient rocks of our continents. The pterodactyle might flit again through the air, the huge iguanodon reappear in the woods, and the ichthyosaurs swarm once more in the sea. Coral reefs might be prolonged again beyond the arctic circle, where the whale and the narwal now abound; and droves of turtles might begin again to wander through regions now tenanted by the walrus and the seal.

But not to indulge too far in these speculations, I may observe, in conclusion, that however great, during the lapse of ages, may be the vicissitudes of temperature in every zone, it accords with this theory that the general climate should not experience any sensible change in the course of a few thousand years; because that period is insufficient to affect the leading features of the physical geography of the globe.

Notwithstanding the apparent uncertainty of the seasons, it is found that the mean temperature of particular localities is very constant, when observations made for a sufficient series of years are compared.

Yet there must be exceptions to this rule; and even the labors of man have, by the drainage of lakes and marshes, and the felling of extensive forests, caused such changes in the atmosphere as greatly to raise our conception of the more important influence of those forces to which, in certain latitudes, even the existence of land or water, hill or valley, lake or sea, must be ascribed. If we possessed accurate information of the amount of *local* fluctuation in climate in the course of twenty centuries, it would often, undoubtedly, be considerable. Certain tracts, for example, on the coast of Holland and of England consisted of cultivated land in the time of the Romans, which the sea, by gradual encroachments, has at length occupied. Here, at least, a slight alteration has been effected; for neither the distribution of heat in the different seasons, nor the mean annual temperature of the atmosphere investing the sea, is precisely the same as that which rests upon the land.

In those countries, also, where earthquakes and volcanoes are in full activity, a much shorter period may produce a sensible variation. The climate of the great table-land of Malpais in Mexico, must differ materially from that which prevailed before the middle of the last century; for, since that time, six mountains, the highest of them rising sixteen hundred feet above the plateau, have been thrown up by volcanic eruptions. It is by the repetition of an indefinite number of such local revolutions, and by slow movements extending simultaneously over wider areas, as will be afterwards shown, that a general change of climate may finally be brought about.

CHAPTER VIII.

ON FORMER CHANGES IN PHYSICAL GEOGRAPHY AND CLIMATE.

Geographical features of the northern hemisphere, at the period of the oldest fossiliferous strata—State of the surface when the mountain limestone and coal were deposited—Changes in physical geography, between the carboniferous period and the chalk—Abrupt transition from the secondary to the tertiary fossils—Accession of land, and elevation of mountain chains, after the consolidation of the secondary rocks—Explanation of Map, showing the area covered by sea, since the commencement of the tertiary period—Astronomical theories of the causes of variations in climate—Theory of the diminution of the supposed primitive heat of the globe.

In the sixth chapter, I stated the arguments derived from organic remains for concluding that in the period when the carboniferous strata were deposited, the temperature of the ocean and the air was more uniform in the different seasons of the year, and in different latitudes, than at present, and that there was a remarkable absence of cold as well as great moisture in the atmosphere. It was also shown that the climate had been modified more than once since that epoch, and that it had been reduced, by successive changes, more and more nearly to that now prevailing in the same latitudes. Farther, I endeavored, in the last chapter, to prove that vicissitudes in climate of no less importance may be expected to recur in future, if it be admitted that causes now active in nature have power, in the lapse of ages, to produce considerable variations in the relative position of land and sea. It remains to inquire whether the alterations, which the geologist can prove to have *actually taken place* at former periods, in the geographical features of the northern hemisphere, coincide in their nature, and in the time of their occurrence, with such revolutions in climate as might naturally have resulted, according to the meteorological principles already explained.

Period of the primary fossiliferous rocks.—The oldest system of strata which afford by their organic remains any evidence as to climate, or the former position of land and sea, are those formerly known as the *transition rocks*, or what have since been termed Lower Silurian or "primary fossiliferous" formations. These have been found in England, France, Germany, Sweden, Russia, and other parts of central and northern Europe, as also in the great Lake district of Canada and the United States. The multilocular or chambered univalves, including the Nautilus, and the corals, obtained from the limestones of these ancient groups, have been compared to forms now most largely developed in tropical seas. The corals, however, have been shown by M. Milne Edwards to differ generally from all living zoophytes; so that conclusions as to a warmer climate drawn from such remote analogies must be received with caution. Hitherto, few, if any, contemporaneous vegetable remains have been noticed; but such as are mentioned agree more nearly with the plants of the carboniferous era than any other, and would therefore imply a warm and humid atmosphere entirely free from intense cold throughout the year.

This absence or great scarcity of plants as well as of freshwater shells and other indications of neighboring land, coupled with the wide extent of marine strata of this age in Europe and North America, are facts which imply such a state of physical geography (so far at least as regards the northern hemisphere) as would, according to the principles before explained, give rise to such a moist and equable climate. (See p. 109, and fig. 5, p. 111.)

Carboniferous group.—This group comes next in the order of succession; and one of its principal members, the mountain limestone, was evidently a marine formation, as is shown by the shells and corals which it contains. That the ocean of that period was of considerable extent in our latitudes, we may infer from the continuity of these calcareous strata over large areas in Europe, Canada, and the United States. The same group has also been traced in North America, towards the borders of the arctic sea.*

There are also several regions in Scotland, and in the central and northern parts of England, as well as in the United States, where marine carboniferous limestones alternate with strata containing coal, in such a manner as to imply the drifting down of plants by rivers into the sea, and the alternate occupation of the same space by fresh and salt water.

Since the time of the earlier writers, no strata have been more extensively investigated, both in Europe and North America, than those of the ancient carboniferous group, and the progress of science has led to a general belief that a large portion of the purest coal has been formed, not, as was once imagined, by vegetable matter floated from a distance, but by plants which grew on the spot, and somewhat in the manner of peat on the spaces now covered by the beds of coal. The former existence of land in some of these spaces has been proved, as already stated, by the occurrence of numerous upright fossil trees, with their roots terminating downwards in seams of coal; and still more generally by the roots of trees (stigmariæ) remaining in their natural position in the clays which underlie almost every layer of coal.

As some nearly continuous beds of such coal have of late years been traced in North America, over areas 100 or 200 miles and upwards in diameter, it may be asked whether the large tracts of ancient land implied by this fact are not inconsistent with the hypothesis of the general prevalence of islands at the period under consideration? In reply, I may observe that the coal-fields must originally have been low alluvial grounds, resembling in situation the cypress-swamps of the Mississippi, or the sunderbunds of the Ganges, being liable like them to be inun-

* Sir J. Richardson, Proceedings of Geol. Soc. No. 7, p. 68, March, 1828.

dated at certain periods by a river or by the sea, if the land should be depressed a few feet. All the phenomena, organic and inorganic, imply conditions nowhere to be met with except in the deltas of large rivers. We have to account for an abundant supply of fluviatile sediment, carried for ages towards one and the same region, and capable of forming strata of mud and sand thousands of feet, or even fathoms, in thickness, many of them consisting of laminated shale, inclosing the leaves of ferns and other terrestrial plants. We have also to explain the frequent intercalations of root-beds, and the interposition here and there of brackish and marine deposits, demonstrating the occasional presence of the neighboring sea. But these forest-covered deltas could only have been formed at the termination of large hydrographical basins, each drained by a great river and its tributaries; and the accumulation of sediment bears testimony to contemporaneous denudation on a large scale, and, therefore, to a wide area of land, probably containing within it one or more mountain chains.

In the case of the great Ohio or Appalachian coal-field, the largest in the world, it seems clear that the uplands drained by one or more great rivers were chiefly to the eastward, or they occupied a space now filled by part of the Atlantic Ocean, for the mechanical deposits of mud and sand increase greatly in thickness and coarseness of material as we approach the eastern borders of the coal-field, or the southeast flanks of the Alleghany mountains, near Philadelphia. In that region numerous beds of pebbles, often of the size of a hen's egg, are seen to alternate with beds of pure coal.

But the American coal-fields are all comprised within the 30th and 50th degrees of north latitude; and there is no reason to presume that the lands at the borders of which they originated ever penetrated so far or in such masses into the colder and arctic regions, so as to generate a cold climate. In the southern hemisphere, where the predominance of sea over land is now the distinguishing geographical feature, we nevertheless find a large part of the continent of Australia, as well as New Zealand, placed between the 30th and 50th degrees of S. latitude. The two islands of New Zealand taken together, are between 800 and 900 miles in length, with a breadth in some parts of ninety miles, and they stretch as far south as the 46th degree of latitude. They afford, therefore, a wide area for the growth of a terrestrial vegetation, and the botany of this region is characterized by abundance of ferns, one hundred and forty species of which are already known, some of them attaining the size of trees. In this respect the southern shores of New Zealand in the 46th degree of latitude almost vie with tropical islands. Another point of resemblance between the Flora of New Zealand and that of the ancient carboniferous period is the prevalence of the fir tribe or of coniferous wood.

An argument of some weight in corroboration of the theory above explained respecting the geographical condition of the temperate and arctic latitudes of the northern hemisphere in the carboniferous period may also be derived from an examination of those groups of strata which immediately preceded the coal. The fossils of the Devonian and Silurian strata in Europe and North America have led to the conclusion, that they were formed for the most part in deep seas, far from land. In those older strata land plants are almost as rare as they are abundant or universal in the coal measures. Those ancient deposits, therefore, may be supposed to have belonged to an epoch when dry land had only just begun to be upraised from the deep; a theory which would imply the existence during the carboniferous epoch of islands, instead of an extensive continent, in the area where the coal was formed.

Such a state of things prevailing in the north, from the pole to the 30th parallel of latitude, if not neutralized by circumstances of a contrary tendency in corresponding regions south of the line, would give rise to a general warmth and uniformity of climate throughout the globe.

Changes in physical geography between the formation of the carboniferous strata and the chalk.—We have evidence in England that the strata of the ancient carboniferous group, already adverted to, were, in many instances, fractured and contorted, and often thrown into a vertical position, before the deposition of some even of the oldest known secondary rocks, such as the new red sandstone.

Fragments of the older formations are sometimes included in the conglomerates of the more modern; and some of these fragments still retain their fossil shells and corals, so as to enable us to determine the parent rocks from whence they were derived. There are other proofs of the disturbance at successive epochs of different secondary rocks before the deposition of others; and satisfactory evidence that, during these reiterated convulsions, the geographical features of the northern hemisphere were frequently modified, and that from time to time new lands emerged from the deep. The vegetation, during some parts of the period in question (from the lias to the chalk inclusive), when genera allied to Cycas and Zamia were abundant, appears to have approached to that of the larger islands of the equatorial zone; such, for example, as we now find in the West Indian archipelago.* These islands appear to have been drained by rivers of considerable size, which were inhabited by crocodiles and gigantic oviparous reptiles, both herbivorous and carnivorous, belonging for the most part to extinct genera. Of the contemporary inhabitants of the land we have as yet acquired but scanty information, but we know that there were flying reptiles, insects, and small mammifers, allied to the marsupial tribes.

A freshwater deposit, called the Wealden, occurs in the upper part of the secondary series of the south of England, which, by its extent and fossils, attests the existence in that region of a large river draining a continent or island of considerable dimensions. We know that this land was clothed with wood, and inhabited by huge terrestrial reptiles and birds. Its position so far to the north as the counties of Surrey and

* Ad. Brongniart, Consid. Générales sur la Nat. de la Végét. &c., Ann. des Sci. Nat., Nov. 1828.

Sussex, at a time when the mean temperature of the climate is supposed to have been much hotter than at present, may at first sight appear inconsistent with the theory before explained, that the heat was caused by the gathering together of all the great masses of land in low latitudes, while the northern regions were almost entirely sea. But it must not be taken for granted that the geographical conditions already described (p. 109, and fig. 5, p. 111) as capable of producing the extreme of heat were ever combined at any geological period of which we have yet obtained information. It is more probable, from what has been stated in the preceding chapters, that a slight approximation to such an extreme state of things would be sufficient; in other words, if most of the dry land were tropical, and scarcely any of it arctic or antarctic, a prodigious elevation of temperature must ensue, even though a part of some continents should penetrate far into the temperate zones.

Changes during the tertiary periods.—The secondary and tertiary formations of Europe, when considered separately, may be contrasted as having very different characters; the secondary appearing to have been deposited in open seas, the tertiary in regions where dry land, lakes, bays, and perhaps inland seas, abounded. The secondary series is almost exclusively marine; the tertiary, even the oldest part, contains lacustrine strata, and not unfrequently freshwater and marine beds alternating. In fact there is evidence of important geographical changes having occurred between the deposition of the cretaceous system, or uppermost of the secondary series, and that of the oldest tertiary group, and still more between the era of the latter and that of the newer tertiary formations. This change in the physical geography of Europe and North America was accompanied by an alteration no less remarkable in organic life, scarcely any *species* being common both to the secondary and tertiary rocks, and the fossils of the latter affording evidence of a different climate.

On the other hand, when we compare the tertiary formations of successive ages, we trace a gradual approximation in the imbedded fossils, from an assemblage in which extinct species predominate, to one where the species agree for the most part with those now existing. In other words, we find a gradual increase of animals and plants fitted for our present climates, in proportion as the strata which we examine are more modern. Now, during all these successive tertiary periods, there are signs of a great increase of land in European and North American latitudes. By reference to the map (Pl. 1), and its description, p. 121, the reader will see that about two-thirds of the present European lands have emerged since the earliest tertiary group originated. Nor is this the only revolution which the same region has undergone within the period alluded to, some tracts which were previously land having gained in altitude, others, on the contrary, having sunk below their former level.

That the existing lands were not all upheaved at once into their present position is proved by the most striking evidence. Several Italian geologists, even before the time of Brocchi, had justly inferred that the Apennines were elevated several thousand feet above the level of the Mediterranean before the deposition of the modern Subapennine beds which flank them on either side. What now constitutes the central calcareous chain of the Apennines must for a long time have been a narrow ridgy peninsula, branching off, at its northern extremity, from the Alps near Savona. This peninsula has since been raised from one to two thousand feet, by which movement the ancient shores, and, for a certain extent, the bed of the contiguous sea, have been laid dry, both on the side of the Mediterranean and the Adriatic.

The nature of these vicissitudes will be explained by the accompanying diagram, which represents a transverse section across the Italian peninsula. The inclined strata A are the disturbed formations of the

Fig. 7.

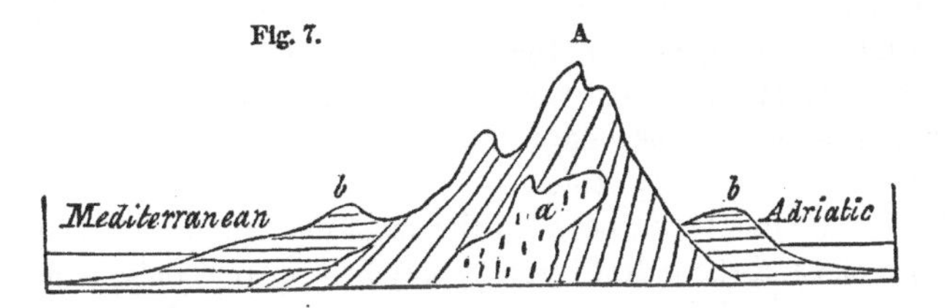

Apennines, into which the ancient igneous rocks *a* are supposed to have intruded themselves. At a lower level on each flank of the chain are the more recent shelly beds *b b*, which often contain rounded pebbles derived from the waste of contiguous parts of the older Apennine limestone. These, it will be seen, are horizontal, and lie in what is termed "unconformable stratification" on the more ancient series. They now constitute a line of hills of moderate elevation between the sea and the Apennines, but never penetrate to the higher and more ancient valleys of that chain.

The same phenomena are exhibited in the Alps on a much grander scale; those mountains being composed in some even of their higher regions of the newer secondary and oldest tertiary formations, while they are encircled by a great zone of more modern tertiary rocks both on their southern flank towards the plains of the Po, and on the side of Switzerland and Austria, and at their eastern termination towards Styria and Hungary.* This newer tertiary zone marks the position of former seas or gulfs, like the Adriatic, wherein masses of strata accumulated, some single groups of which are not inferior in thickness to the most voluminous of our secondary formations in England. Some even of these newer groups have been raised to the height of three or four thousand feet, and in proportion to their antiquity, they generally rise to greater heights, the older of them forming interior zones nearest to the central ridges of the Alps. We have already ascertained that the Alps gained accessions to their height and width at several successive peri-

* See a Memoir on the Alps, by Professor Sedgwick and Sir Rod. Murchison, Trans. of Geol. Soc. second ser. vol. iii. accompanied by a map.

ods, and that the last series of improvements occurred when the seas were inhabited by many existing species of animals.

We may imagine some future series of convulsions once more to heave up this stupendous chain, together with the adjoining bed of the sea, so that the mountains of Europe may rival the Andes in elevation; in which case the deltas of the Po, Adige, and Brenta, now encroaching upon the Adriatic, might be uplifted so as to form another exterior belt of considerable height around the southeastern flank of the Alps.

The Pyrenees, also, have acquired their present altitude, which in Mont Perdu exceeds eleven thousand feet, since the deposition of the nummulitic or Eocene division of the tertiary series. Some of the tertiary strata at the base of the chain are raised to the height of only a few hundred feet above the sea, and retain a horizontal position, without partaking in general in the disturbance to which the older series has been subjected; so that the great barrier between France and Spain was almost entirely upheaved in the interval between the deposition of certain groups of tertiary strata.

The remarkable break between the most modern of the known secondary rocks and the oldest tertiary, may be apparent only, and ascribable to the present deficiency of our information. Already the marles and green sand of Heers near Tongres, in Belgium, observed by M. Dumont, and the "pisolitic limestone" of the neighborhood of Paris, both intermediate in age between the Maestricht chalk and the lower Eocene strata, begin to afford us signs of a passage from one state of things to another. Nevertheless, it is far from impossible that the interval between the chalk and tertiary formations constituted an era in the earth's history, when the transition from one class of organic beings to another was, comparatively speaking, rapid. For if the doctrines above explained in regard to vicissitudes of temperature are sound, it will follow that changes of equal magnitude in the geographical features of the globe may at different periods produce very unequal effects on climate; and, so far as the existence of certain animals and plants depends on climate, the duration of species would be shortened or protracted, according to the rate at which the change of temperature proceeded.

For even if we assume that the intensity of the subterranean disturbing forces is uniform and capable of producing nearly equal amounts of alteration on the surface of the planet, during equal periods of time, still the rate of alteration in climate would be by no means uniform. Let us imagine the quantity of land between the equator and the tropic in one hemisphere to be to that in the other as thirteen to one, which, as before stated, represents the unequal proportion of the extra-tropical lands in the two hemispheres at present. (See figs. 3 and 4, p. 110.) Then let the first geographical change consist in the shifting of this preponderance of land from one side of the line to the other; from the southern hemisphere, for example, to the northern. Now this need not affect the *general* temperature of the earth. But if, at another epoch, we suppose a continuance of the same agency to transfer an equal volume of land from the torrid zone to the temperate and arctic regions of the northern and southern hemispheres, or into one of them, there might be so great a refrigeration of the mean temperature *in all latitudes*, that scarcely any of the pre-existing races of animals would survive; and, unless it pleased the Author of Nature that the planet should be uninhabited, new species, and probably of widely different forms, would then be substituted in the room of the extinct. We ought not, therefore, to infer that equal periods of time are always attended by an equal amount of change in organic life, since a great fluctuation in the mean temperature of the earth, the most influential cause which can be conceived in exterminating whole races of animals and plants, must, in different epochs, require unequal portions of time for its completion.

PLATE I. *Map showing the extent of surface in Europe which has at one period or another been covered by the sea since the commencement of the deposition of the older or Eocene Tertiary strata.*

THIS map will enable the reader to perceive at a glance the great extent of change in the physical geography of Europe, which can be proved to have taken place since some of the older tertiary strata began to be deposited. The proofs of submergence, during some part or other of this period, in all the districts distinguished by ruled lines, are of a most unequivocal character; for the area thus described is now covered by deposits containing the fossil remains of animals which could only have lived in salt water. The most ancient part of the period referred to cannot be deemed very remote, considered geologically; because the deposits of the Paris and London basins, and many other districts belonging to the older tertiary epoch, are newer than the greater part of the sedimentary rocks (those commonly called secondary and primary fossiliferous or paleozoic) of which the crust of the globe is composed. The species, moreover, of marine testacea, of which the remains are found in these older tertiary formations, are not entirely distinct from such as now live. Yet, notwithstanding the comparatively recent epoch to which this retrospect is carried, the variations in the distribution of land and sea depicted on the map form only a part of those which must have taken place during the period under consideration. Some approximation has merely been made to an estimate of the amount of *sea converted into land* in parts of Europe best known to geologists; but we cannot determine how much land has become sea during the same period; and there may have been repeated interchanges of land and water in the same places, changes of which no account is taken in the map, and respecting the amount of which little accurate information can ever be obtained.

I have extended the sea in some instances beyond the limits of the land now covered by tertiary formations, and marine drift, because other geological data have been obtained for inferring the submergence of these tracts after the deposition of the Eocene strata had begun. Thus, for example, there are good reasons for concluding that part of the chalk of England (the North and South Downs, for example, together with the intervening secondary tracts) continued beneath the sea until the oldest tertiary beds had begun to accumulate.

A strait of the sea separating England and Wales has also been introduced, on the evidence afforded by shells of existing species found in a deposit of gravel, sand, loam, and clay, called the northern drift, by Sir R. Murchison.* And Mr. Trimmer has discovered similar recent marine shells on the northern coast of North Wales, and on Moel Tryfane, near the Menai Straits, at the height of 1392 feet above the level of the sea!

Some raised sea-beaches, and drift containing marine shells, which I examined in 1843, between Limerick and Dublin, and which have been traced over other parts of Ireland by different geologists, have required an extension of the dark lines so as to divide that island into several. In improving this part of my map I have been especially indebted to the assistance of Mr. Oldham, who in 1843 announced to the British Association at Cork the fact that at the period when the drift or glacial beds were deposited, Ireland must have formed an archipelago such as is here depicted. A considerable part of Scotland might also have been represented in a similar manner as under water when the drift originated.

A portion of Brittany is divided into islands, because it is known to be covered with patches of marine tertiary strata chiefly miocene. When I examined these in 1830 and 1843, I convinced myself that the sea must have covered much larger areas than are now occupied by these small and detached deposits. The former connection of the White Sea and the Gulf of Finland is proved by the fact that a multitude of huge erratic blocks extend over the intervening space, and a large portion of Norway, Sweden, and Denmark, as well as Germany and Russia, are represented as sea, on the same evidence, strengthened by the actual occurrence of fossil sea-shells, of recent species, in the drift of various portions of those countries. The submergence of considerable areas under large bodies of fresh water, during the tertiary period, of which there are many striking geological proofs in Auvergne, and elsewhere, has not been expressed by ruled lines. They bear testimony to the former existence of neighboring lands, and a certain elevation of the areas where they occur above the level of the ocean; they are therefore left blank, together with all the space that cannot be demonstrated to have been part of the sea at some time or other, since the commencement of the Eocene epoch.

In compiling this map, which has been entirely recast since the first edition, I have availed myself of the latest geological maps of the British isles, and north of Europe; also of those published by the government surveyors of France, MM. de Beaumont and Dufresnoy; the map of

* See Proceedings of Geol. Soc. vol. ii. p. 334.

Germany and part of Europe, by Von Dechen, and that of Italy by M. Tchihatchoff (Berlin, 1842). Lastly, Sir R. Murchison's important map of Russia, and the adjoining countries, has enabled me to mark out not only a considerable area, previously little known, in which tertiary formations occur; but also a still wider expanse, over which the northern drift, and erratic blocks with occasional marine shells, are traceable. The southern limits of these glacial deposits in Russia and Germany indicate the boundary, so far as we can now determine it, of the northern ocean, at a period immediately antecedent to that of the human race.

I was anxious, even in the title of this map, to guard the reader against the supposition that it was intended to represent the state of the physical geography of part of Europe at any *one point of time*. The difficulty, or rather the impossibility, of restoring the geography of the globe as it may have existed at any former period, especially a remote one, consists in this, that we can only point out where part of the sea has been turned into land, and are almost always unable to determine what land may have become sea. All maps, therefore, pretending to represent the geography of remote geological epochs must be ideal. The map under consideration is not a restoration of a former state of things, at any particular moment of time, but a synoptical view of a certain amount of one kind of change (the conversion of sea into land) known to have been brought about within a given period.

It may be proper to remark that the vertical movements to which the land is subject in certain regions, occasion alternately the subsidence and the uprising of the surface; and that, by such oscillations at successive periods, a great area may have been entirely covered with marine deposits, although the whole may never have been beneath the waters at one time; nay, even though the relative proportion of land and sea may have continued unaltered throughout the whole period. I believe, however, that since the commencement of the tertiary period, the dry land in the northern hemisphere has been continually on the increase, both because it is now greatly in excess beyond the average proportion which land generally bears to water on the globe, and because a comparison of the secondary and tertiary strata affords indications, as I have already shown, of a passage from the condition of an ocean interspersed with islands to that of a large continent.

But supposing it were possible to represent all the vicissitudes in the distribution of land and sea that have occurred during the tertiary period, and to exhibit not only the actual existence of land where there was once sea, but also the extent of surface now submerged which may once have been land, the map would still fail to express all the important revolutions in physical geography which have taken place within the epoch under consideration. For the oscillations of level, as was before stated, have not merely been such as to lift up the land from below the water, but in some cases to occasion a rise of many thousand feet above the sea. Thus the Alps have acquired an additional altitude of 4000, and even in some places 10,000 feet; and the Apennines owe a considerable part of their present height to subterranean convulsions which have happened within the tertiary epoch.

On the other hand, some mountain chains may have been lowered during the same series of ages, in an equal degree, and shoals may have been converted into deep abysses.* Since this map was recast in 1847, geologists have very generally come to the conclusion that the nummulitic limestone, together with the overlying fucoidal grit and shale, called "Flysch," in the Alps, belongs to the older tertiary or Eocene group. As these nummulitic rocks enter into the structure of some of the most lofty and disturbed parts of the Alps, Apennines, Carpathians, Pyrenees, and other mountain chains, and form many of the elevated lands of Africa and Asia, their position almost implies the ubiquity of the post-Eocene ocean, not, indeed, by the simultaneous, but by the successive, occupancy of the whole ground by its waters.†

Concluding remarks on changes in physical geography.—The foregoing observations, it may be said, are confined chiefly to Europe, and therefore merely establish the increase of dry land in a space which constitutes but a small portion of the northern hemisphere; but it was stated in the preceding chapter, that the great Lowland of Siberia, lying chiefly between the latitudes 55° and 75° N. (an area nearly equal to all Europe), is covered for the most part by marine strata, which, from the account given by Pallas, and more recently by Sir R. Murchison, belongs to a period when all or nearly all the shells were of a species still living in the north. The emergence, therefore, of this area from the deep is, comparatively speaking, a very modern event, and must, as before remarked, have caused a great increase of cold throughout the globe.

Upon a review, then, of all the facts above enumerated, respecting the ancient geography of the globe as attested by geological monuments, there appear good grounds for inferring that changes of climate coincided with remarkable revolutions in the former position of sea and land. A wide expanse of ocean, interspersed with islands, seems to have pervaded the northern hemisphere at the periods when the Silurian and carboniferous rocks were formed, and a warm and very uniform temperature then prevailed. Subsequent modifications in climate accompanied the deposition of the secondary formations, when repeated changes were effected in the physical geography of our northern latitudes. Lastly, the refrigeration became most decided, and the climate most nearly assimilated to that now enjoyed, when the lands in Europe and northern Asia had attained their full extension, and the mountain chains their actual height.

Soon after the first publication of this theory of climate, an objection was made by an anonymous German critic in 1833 that there are no geological proofs of the prevalence at any former period of a temperature

* It may be observed, that the facts and inferences exhibited in this map bear not merely on the theory of climate above proposed, but serve also to illustrate the views explained in the third book respecting the migration of animals and plants and the gradual extinction of species.

† See Sir R. Murchison's Paper on the Alps, Quart. Journ. Geol. Soc. vol. v. and my Anniversary Address for 1850, ibid. vol. vi.

lower than that now enjoyed; whereas, if the causes above assigned were the true ones, it might reasonably have been expected that fossil remains would sometimes indicate colder as well as hotter climates than those now established.* In answer to this objection, I may suggest, that our present climates are probably far more distant from the extreme of possible heat than from its opposite extreme of cold. A glance at the map (fig. 6, p. 111) will show that all the existing lands might be placed between the 30th parallels of latitude on each side of the equator, and that even then they would by no means fill that space. In no other position would they give rise to so high a temperature. But the present geographical condition of the earth is so far removed from such a state of things, that the land lying between the poles and the parallels of 30, is in great excess; so much so that, instead of being to the sea in the proportion of 1 to 3, which is as near as possible the average general ratio throughout the globe, it is 9 to 23.† Hence it ought not to surprise us if, in our geological retrospect, embracing perhaps a small part only of a complete cycle of change in the terrestrial climates, we should happen to discover everywhere the signs of a higher temperature. The strata hitherto examined may have originated when the quantity of equatorial land was always decreasing and the land in regions nearer the poles augmenting in height and area, until at length it attained its present excess in high latitudes. There is nothing improbable in supposing that the geographical revolutions of which we have hitherto obtained proofs had this general tendency; and in that case the refrigeration must have been constant, although, for reasons before explained, the rate of cooling may not have been uniform.

It may, however, be as well to recall the reader's attention to what was before said of the indication brought to light of late years, of a considerable oscillation of temperature, in the period immediately preceding the human era. We have seen that on examining some of the most northern deposits, those commonly called the northern drift in Scotland, Ireland, and Canada, in which nearly all, in some cases, perhaps all, the fossil shells are of recent species, we discover the signs of a climate colder than that now prevailing in corresponding latitudes on both sides the Atlantic. It appears that an arctic fauna specifically resembling that of the present seas, extended farther to the south than now. This opinion is derived partly from the known habitations of the corresponding living species, and partly from the abundance of certain genera of shells

* Allgemeine Literatur Zeitung, No. cxxxix. July, 1833.

† In this estimate, the space within the antarctic circle is not taken into account: if included, it would probably add to the excess of dry land; for the late discoveries of Capt. Sir James Ross, who penetrated to lat. 78° 10′ S., confirm the conjecture of Captain Cook that the accumulation of antarctic ice implies the presence of a certain quantity of terra firma. The number of square miles on the surface of the globe are 148,522,000, the part occupied by the sea being 110,849,000, and that by land, 37,673,000; so that the land is very nearly to the sea as 1 part in 4. I am informed by Mr. Gardner that, according to a rough approximation, the land between the 30° N. lat. and the pole occupies a space *about equal to that of the sea*, and the land between the 30° S. lat. and the antarctic circle about one-sixteenth of that zone.

and the absence of others.* The date of the refrigeration thus inferred appears to coincide very nearly with the era of the dispersion of erratic blocks over Europe and North America, a phenomenon which will be ascribed in the sequel (ch. 16) to the cold then prevailing in the northern hemisphere. The force, moreover, of the German critic's objection has been since in a great measure destroyed, by the larger and more profound knowledge acquired in the last few years of the ancient carboniferous flora, which has led the ablest botanists to adopt the opinion, that the climate of the coal period was remarkable for its warmth, moisture, equability, and freedom from cold, rather than the intensity of its *tropical heat.* We are therefore no longer entitled to assume that there has been a constant and gradual decline in the absolute amount of heat formerly contained in the atmosphere and waters of the ocean, such as it was conjectured might have emanated from the incandescent central nucleus of a new and nearly fluid planet, before the interior had lost, by radiation into surrounding space, a great part of its original high temperature.

Astronomical causes of fluctuations in climate.—Sir John Herschel has lately inquired, whether there are any astronomical causes which may offer a possible explanation of the difference between the actual climate of the earth's surface, and those which formerly appear to have prevailed. He has entered upon this subject, he says, "impressed with the magnificence of that view of geological revolutions, which regards them rather as regular and necessary effects of great and general causes, than as resulting from a series of convulsions and catastrophes, regulated by no laws, and reducible to no fixed principles." Geometers, he adds, have demonstrated the absolute invariability of the mean distance of the earth from the sun; whence it would at first seem to follow, that the mean annual supply of light and heat derived from that luminary would be alike invariable: but a closer consideration of the subject will show, that this would not be a legitimate conclusion; but that on the contrary, the *mean* amount of solar radiation is dependent on the eccentricity of the earth's orbit, and therefore liable to variation.†

Now the eccentricity of the orbit, he continues, is actually diminishing, and has been so for ages beyond the records of history. In consequence, the ellipse is in a state of approach to a circle, and the annual average of solar heat radiated to the earth is actually on the *decrease.* So far this is in accordance with geological evidence, which indicates a general refrigeration of climate; but the question remains, whether the amount of diminution which the eccentricity may have ever undergone can be supposed sufficient to account for any sensible refrigeration. The calcula-

* See papers by Mr. Smith of Jordanhill, F. G. S., and the author, Proceedings Geol. Soc. No. 63, 1839, also that of Prof. E. Forbes, before cited, p. 86, note.

† The theorem is thus stated:—"The eccentricity of the orbit varying, the total quantity of heat received by the earth from the sun in one revolution is inversely proportional to the minor axis of the orbit. The major axis is invariable, and therefore, of course, the absolute length of the year: hence it follows that the mean annual average of heat will also be in the same inverse ratio of the minor axis."—Geol. Trans. second series, vol. iii. p. 295.

tions necessary to determine this point, though practicable, have never yet been made, and would be extremely laborious; for they must embrace all the perturbations which the most influential planets, Venus, Mars, Jupiter, and Saturn, would cause in the earth's orbit, and in each other's movements round the sun.

The problem is also very complicated, inasmuch as it depends not merely on the ellipticity of the earth's orbit, but on the assumed temperature of the celestial spaces beyond the earth's atmosphere; a matter still open to discussion, and on which M. Fourier and Sir J. Herschel have arrived at very different opinions. But if, says Herschel, we suppose an extreme case, as if the earth's orbit should ever become as eccentric as that of the planet Juno or Pallas, a great change of climate might be conceived to result, the winter and summer temperatures being sometimes mitigated, and at others exaggerated, in the same latitudes.

It is much to be desired that the calculations alluded to were executed, as even if they should demonstrate, as M. Arago thinks highly probable,* that the mean amount of solar radiation can never be materially affected by irregularities in the earth's motion, it would still be satisfactory to ascertain the point. Such inquiries, however, can never supersede the necessity of investigating the consequences of the varying position of continents, shifted as we know them to have been during successive epochs, from one part of the globe to the other.

Another astronomical hypothesis respecting the possible cause of secular variations in climate, has been proposed by a distinguished mathematician and philosopher, M. Poisson. He begins by assuming, 1st, that the sun and our planetary system are not stationary, but carried onward by a common movement through space; 2dly, that every point in space receives heat as well as light from innumerable stars surrounding it on all sides, so that if a right line of indefinite length be produced in any direction from such a point, it must encounter a star either visible or invisible to us. 3dly, He then goes on to assume, that the different regions of space, which in the course of millions of years are traversed by our system, must be of very unequal temperature, inasmuch as some of them must receive a greater, others a less, quantity of radiant heat from the great stellary inclosure. If the earth, he continues, or any other large body, pass from a hotter to a colder region, it would not readily lose in the second all the heat which it has imbibed in the first region, but retain a temperature increasing downwards from the surface, as in the actual condition of our planet.†

Now the opinion originally suggested by Sir W. Herschel, that our sun and its attendant planets were all moving onward through space, in the direction of the constellation Hercules, is very generally thought by eminent astronomers to be confirmed. But even if its reality be

* Ann. du Bur. des Long. 1834.

† Poisson, Théorie Mathémat. de la Chaleur, Comptes Rendus de l'Acad des Sci., Jan. 30, 1837.

no longer matter of doubt, conjectures as to its amount are still vague and uncertain; and great, indeed, must be the extent of the movement before this cause alone can work any material alteration in the terrestrial climates. Mr. Hopkins, when treating of this theory, remarked, that so far as we were acquainted with the position of the stars not very remote from the sun, they seem to be so distant from each other, that there are no points in space among them, where the intensity of radiating heat would be comparable to that which the earth derives from the sun, except at points very near to each star. Thus, in order that the earth should derive a degree of heat from stellar radiation comparable to that now derived from the sun, she must be in close proximity to some particular star, leaving the aggregate effect of radiation from the other stars nearly the same as at present. This approximation, however, to a single star could not take place consistently with the preservation of the motion of the earth about the sun, according to its present laws.

Suppose our sun should approach a star within the present distance of Neptune. That planet could no longer remain a member of the solar system, and the motions of the other planets would be disturbed in a degree which no one has ever contemplated as probable since the existence of the solar system. But such a star, supposing it to be no larger than the sun, and to emit the same quantity of heat, would not send to the earth much more than one-thousandth part of the heat which she derives from the sun, and would therefore produce only a very small change in terrestrial temperature.*

Variable splendor of stars.—There is still another astronomical suggestion respecting the possible causes of secular variations in the terrestrial climates which deserves notice. It has long been known that certain stars are liable to great and periodical fluctuations in splendor, and Sir J. Herschel has lately ascertained (Jan. 1840), that a large and brilliant star, called *alpha* Orionis, sustained, in the course of six weeks, a loss of nearly half its light. "This phenomenon," he remarks, "cannot fail to awaken attention, and revive those speculations which were first put forth by my father Sir W. Herschel, respecting the possibility of a change in the lustre of *our sun itself*. If there really be a community of nature between the sun and fixed stars, every proof that we obtain of the extensive prevalence of such periodical changes in those remote bodies, adds to the probability of finding something of the kind nearer home." Referring then to the possible bearing of such facts on ancient revolutions, in terrestrial climates, he says, that "it is a matter of observed fact, that many stars *have* undergone, in past ages, within the records of astronomical history, very extensive changes in apparent lustre, without a change of distance adequate to producing such an effect. If our sun were even *intrinsically* much brighter than at present, the mean temperature of the surface of our globe would, of course, be proportionally greater. I speak now not of periodical, but of secular

* Quart. Journ. Geol. Soc. 1852, p. 62.

changes. But the argument is complicated with the consideration of the possibly imperfect transparency of the celestial spaces, and with the cause of that imperfect transparency, which may be due to material non-luminous particles diffused irregularly in patches analogous to nebulæ, but of greater extent—to *cosmical clouds,* in short—of whose existence we have, I think, some indication in the singular and apparently capricious phenomena of temporary stars, and perhaps in the recent extraordinary sudden increase and hardly less sudden diminution of *η Argus*."*

More recently (1852) Schwabe has observed that the spots on the sun alternately increase and decrease in the course of every ten years, and Captain Sabine has pointed out that this variable obscuration coincides in time both as to its maximum and minimum with changes in all those terrestrial magnetic variations which are caused by the sun. Hence he infers that the period of alteration in the spots is a *solar magnetic period.* Assuming such to be the case, the variable light of some stars may indicate a similar phenomenon, or they may be stellar magnetic periods, differing only in the degree of obscuration and its duration. And as hitherto we have perceived no fluctuation in the heat received by the earth from the sun coincident with the solar magnetic period, so the fluctuations in the brilliancy of the stars may not perhaps be attended with any perceptible alteration in their power of radiating heat. But before we can speculate with advantage in this new and interesting field of inquiry, we require more facts and observations.

Supposed gradual diminution of the earth's primitive heat.—The gradual diminution of the supposed primitive heat of the globe has been resorted to by many geologists as the principal cause of alterations of climate. The matter of our planet is imagined, in accordance with the conjectures of Leibnitz, to have been originally in an intensely heated state, and to have been parting ever since with portions of its heat, and at the same time contracting its dimensions. There are, undoubtedly, good grounds for inferring from recent observation and experiment, that the temperature of the earth increases as we descend from the surface to that slight depth to which man can penetrate: but there are no positive proofs of a secular decrease of internal heat accompanied by contraction. On the contrary, La Place has shown, by reference to astronomical observations made in the time of Hipparchus, that in the last two thousand years at least there has been no sensible contraction of the globe by cooling; for had this been the case, even to an extremely small amount, the day would have been shortened, whereas its length has certainly not diminished during that period by $\frac{1}{300}$th of a second.

Baron Fourier, after making a curious series of experiments on the cooling of incandescent bodies, considers it to be proved mathematically, that the actual distribution of heat in the earth's envelope is precisely

* Proceedings Roy. Astronom. Soc. No. iii. Jan. 1840.

that which would have taken place if the globe had been formed in a medium of a very high temperature, and had afterwards been constantly cooled.* He contends, that although no contraction can be demonstrated to have taken place within the historical period (the operation being slow and the time of observation limited), yet it is no less certain that heat is annually passing out by radiation from the interior of the globe into the planetary spaces. He even undertook to demonstrate that the quantity of heat thus transmitted into space in the course of every century, through every square metre of the earth's surface, would suffice to melt a column of ice having a square metre for its base, and being three metres (or 9 feet 10 inches) high.

It is at the same time denied, that there is any assignable mode in which the heat thus lost by radiation can be again restored to the earth, and consequently the interior of our planet must, from the moment of its creation, have been subject to refrigeration, and is destined together with the sun and stars forever to grow colder. But I shall point out in the sequel (chapter 31) many objections to these views, and to the theory of the intense heat of the earth's central nucleus, and shall then inquire how far the observed augmentation of temperature, as we descend below the surface, may be referable to other causes unconnected with the supposed pristine fluidity of the entire globe.

CHAPTER IX.

THEORY OF THE PROGRESSIVE DEVELOPMENT OF ORGANIC LIFE AT SUCCESSIVE GEOLOGICAL PERIODS.

Theory of the progressive development of organic life—Evidence in its support inconclusive—Vertebrated animals, and plants of the most perfect organization, in strata of very high antiquity—Differences between the organic remains of successive formations—Comparative modern origin of the human race—The popular doctrine of successive development not established by the admission that man is of modern origin—Introduction of man, to what extent a change in the system.

Progressive development of organic life.—IN the preceding chapters I have considered whether revolutions in the general climate of the globe afford any just ground of opposition to the doctrine that the former changes of the earth which are treated of in geology belong to one uninterrupted series of physical events governed by ordinary causes. Against this doctrine some popular arguments have been derived from the great vicissitudes of the organic creation in times past; I shall

* See a Memoir on the Temperature of the Terrestrial Globe, and the Planetary Spaces, Ann. de Chimie et Phys. tom. xxvii. p. 136. Oct. 1824.

therefore proceed to the discussion of such objections, which have been thus formally advanced by the late Sir Humphrey Davy. "It is impossible," he affirms, "to defend the proposition, that the present order of things is the ancient and constant order of nature, only modified by existing laws: in those strata which are deepest, and which must, consequently, be supposed to be the earliest deposited, forms even of vegetable life are rare; shells and vegetable remains are found in the next order; the bones of fishes and oviparous reptiles exist in the following class; the remains of birds, with those of the same genera mentioned before, in the next order; those of quadrupeds of extinct species in a still more recent class; and it is only in the loose and slightly consolidated strata of gravel and sand, and which are usually called diluvian formations, that the remains of animals such as now people the globe are found, with others belonging to extinct species. But, in none of these formations, whether called secondary, tertiary, or diluvial, have the remains of man, or any of his works, been discovered; and whoever dwells upon this subject must be convinced, that the present order of things, and the comparatively recent existence of man as the master of the globe, is as certain as the destruction of a former and a different order, and the extinction of a number of living forms which have no types in being. In the oldest secondary strata there are no remains of such animals as now belong to the surface; and in the rocks, which may be regarded as more recently deposited, these remains occur but rarely, and with abundance of extinct species;—there seems, as it were, a gradual approach to the present system of things, and a succession of destructions and creations preparatory to the existence of man."*

In the above passages, the author deduces two important conclusions from geological data: first, that in the successive groups of strata, from the oldest to the most recent, there is a progressive development of organic life, from the simplest to the most complicated forms;—secondly, that man is of comparatively recent origin, and these conclusions he regards as inconsistent with the doctrine, "that the present order of things is the ancient and constant order of nature only modified by existing laws."

With respect, then, to the first of these propositions, we may ask whether the theory of the progressive development of animal and vegetable life, and their successive advancement from a simple to a more perfect state, has any secure foundation in fact? No geologists who are in possession of all the data now established respecting fossil remains, will for a moment contend for the doctrine in all its detail, as laid down by the distinguished philosopher to whose opinions we have referred: but naturalists, who are not unacquainted with recent discoveries, continue to defend it in a modified form. They say that in the first period of the world (by which they mean the earliest of which we have yet brought to light any memorials), the vegetation was characterized by a predominance of cryptogamic plants, while the animals

* Sir H. Davy, Consolations in Travel: Dialogue III. "The Unknown."

which coexisted were almost entirely confined to zoophytes, testacea, and a few fish. Plants of a less simple structure, coniferæ and cycadeæ, flourished largely in the next epoch, when oviparous reptiles began also to abound. Lastly, the terrestrial flora became most diversified and most perfect when the highest orders of animals, the mammalia and birds, were called into existence.

Now in the first place, it may be observed, that many naturalists are guilty of no small inconsistency in endeavoring to connect the phenomena of the earliest vegetation with a nascent condition of organic life, and at the same time to deduce from the numerical predominance of certain forms, the greater heat or uniformity of the ancient climate. The arguments in favor of the latter conclusion are without any force, unless we can assume that the rules followed by the Author of Nature in the creation and distribution of organic beings were the same formerly as now; and that, as certain families of animals and plants are now most abundant in, or exclusively confined to regions where there is a certain temperature, a certain degree of humidity, a certain intensity of light, and other conditions, so also analogous phenomena were exhibited at every former era.

If this postulate be denied, and the prevalence of particular families be declared to depend on a certain order of precedence in the introduction of different classes into the earth, and if it be maintained that the standard of organization was raised successively, we must then ascribe the numerical preponderance, in the earlier ages, of plants of simpler structure, *not to the heat*, or other climatal conditions, but to those different laws which regulate organic life in newly created worlds.

Before we can infer a warm and uniform temperature in high latitudes, from the presence of 250 species of ferns, some of them arborescent, accompanied by lycopadiacæ of large size, and araucariæ, we must be permitted to assume, that at all times, past, present, and future, a heated and moist atmosphere pervading the northern hemisphere has a tendency to produce in the vegetation a predominance of analogous forms.

It should moreover be borne in mind, when we are considering the question of development from a botanical point of view, that naturalists are by no means agreed as to the existence of an ascending scale of organization in the vegetable world corresponding to that which is very generally recognized in animals. "From the sponge to man," in the language of De Blainville, there may be a progressive chain of being, although often broken and imperfect; but if we seek to classify plants according to a linear arrangement, ascending gradually from the lichen to the lily or the rose, we encounter incomparably greater difficulties. Yet the doctrine of a more highly developed organization in the plants created at successive periods presupposes the admission of such a graduated scale.

We have as yet obtained but scanty information respecting the state of the terrestrial flora at periods antecedent to the coal. In the carboniferous epoch, about 500 species of fossil plants are enumerated by Adolphe Brongniart, which we may safely regard as a mere fragment of an ancient flora; since, in Europe alone, there are now no less than 11,000 living species. I have already hinted that the plants which produced coal were not drifted from a distance, but that nearly all of them grew on the spots where they became fossil. They appear to have belonged, as before explained (p. 115), to a peculiar class of *stations,*—to low level and swampy regions, in the deltas of large rivers, slightly elevated above the level of the sea. From the study, therefore, of such a vegetation, we can derive but little insight into the nature of the contemporaneous upland flora, still less of the plants of the mountainous or Alpine country; and if so, we are enabled to account for the apparent monotony of the vegetation, although its uniform character was doubtless in part owing to a greater uniformity of climate then prevailing throughout the globe. Some of the commonest trees of this period, such as the sigillariæ, which united the structure of ferns and of cycadeæ, departed very widely from all known living types. The coniferæ and ferns, on the contrary, were very closely allied to living genera. It is remarkable that none of the exogens of Lindley (dicotyledonous angiosperms of Brongniart), which comprise four-fifths of the living flora of the globe, and include all the forest trees of Europe except the fir-tribe, have yet been discovered in the coal measures, and a very small number—fifteen species only—of monocotyledons. If several of these last are true plants, an opinion to which Messrs. Lindley, Unger, Corda, and other botanists of note incline, the question whether any of the most highly organized plants are to be met with in ancient strata is at once answered in the affirmative. But the determination of these palms being doubtful, we have as yet in the coal no positive proofs either of the existence of the most perfect, or of the most simple forms of flowering or flowerless vegetation. We have no fungi, lichens, hepatici or mosses: yet this latter class may have been as fully represented then as now.

In the flora of the secondary eras, all botanists agree that palms existed, although in Europe plants of the family of zamia and cycas together with coniferæ predominated, and must have given a peculiar aspect to the flora. As only 200 or 300 species of plants are known in all the rocks ranging from the Trias to the Oolite inclusive, our data are too scanty as yet to affirm whether the vegetation of this second epoch was or was not on the whole of a simpler organization than that of our own times.

In the Lower Cretaceous formation, near Aix-la-Chapelle, the leaves of a great many dicotyledonous trees have lately been discovered by Dr. Debey, establishing the important fact of the coexistence of a large number of angiosperms with cycadeæ, and with that rich reptilian fauna comprising the ichthyosaur, plesiosaur, and pterodactyl, which some had supposed to indicate a state of the atmosphere unfavorable to a dicotyledonous vegetation.

The number of plants hitherto obtained from *tertiary* strata of different ages is very limited, but is rapidly increasing. They are referable to a much greater variety of families and classes than an equal number of fossil species taken from secondary or primary rocks, the angiosperms bearing the same proportion to the gymnosperms and acrogens as in the present flora of the globe. This greater variety may, doubtless, be partly ascribed to the greater diversity of stations in which the plants grew, as we have in this case an opportunity, rarely enjoyed in studying the secondary fossils, of investigating inland or lacustrine deposits accumulated at different heights above the sea, and containing the memorials of plants washed down from adjoining mountains.

In regard, then, to the strata from the cretaceous to the uppermost tertiary inclusive, we may affirm that we find in them all the principal classes of living plants, and during this vast lapse of time four or five complete changes in the vegetation occurred, yet no step whatever was made in advance at any of these periods by the addition of more highly organized species.

If we next turn to the fossils of the animal kingdom, we may inquire whether, when they are arranged by the geologists in a chronological series, they imply that beings of more highly developed structure and greater intelligence entered upon the earth at successive epochs, those of the simplest organization being the first created, and those more highly organized being the last.

Our knowledge of the Silurian fauna is at present derived entirely from rocks of marine origin, no fresh-water strata of such high antiquity having yet been met with. The fossils, however, of these ancient rocks at once reduce the theory of progressive development to within very narrow limits, for already they comprise a very full representation of the radiata, mollusca, and articulata proper to the sea. Thus, in the great division of radiata, we find asteriod and helianthoid zoophytes, besides crinoid and cystidean echinoderms. In the mollusca, between 200 and 300 species of cephalopoda are enumerated. In the articulata we have the crustaceans represented by more than 200 species of trilobites, besides other genera of the same class. The remains of fish are as yet confined to the upper part of the Silurian series; but some of these belong to placoid fish, which occupy a high grade in the scale of organization. Some naturalists have assumed that the earliest fauna was exclusively marine, because we have not yet found a single Silurian helix, insect, bird, terrestrial reptile or mammifer; but when we carry back our investigation to a period so remote from the present, we ought not to be surprised if the only accessible strata should be limited to deposits formed far from land, because the ocean probably occupied then, as now, the greater part of the earth's surface. After so many entire geographical revolutions, the chances are nearly three to one in favor of our finding that such small portions of the existing continents and islands as expose Silurian strata to view, should coincide in position with the ancient ocean rather than the land. We must not, therefore, too hastily infer, from the absence of fossil bones of mammalia in the older rocks, that the highest class of vertebrated animals did not exist in remoter ages. There are regions at present, in the Indian and Pacific Oceans, coextensive in area with the continents of Europe and North America, where we might dredge the bottom and draw up thousands of shells and corals, without obtaining one bone of a land quadruped. Suppose our mariners were to report, that, on sounding in the Indian Ocean near some coral reefs, and at some distance from the land, they drew up on hooks attached to their line portions of a leopard, elephant, or tapir, should we not be skeptical as to the accuracy of their statements? and if we had no doubt of their veracity, might we not suspect them to be unskilful naturalists? or, if the fact were unquestioned, should we not be disposed to believe that some vessel had been wrecked on the spot?

The casualties must always be rare by which land quadrupeds are swept by rivers far out into the open sea, and still rarer the contingency of such a floating body not being devoured by sharks or other predaceous fish, such as were those of which we find the teeth preserved in some of the carboniferous strata. But if the carcass should escape, and should happen to sink where sediment was in the act of accumulating, and if the numerous causes of subsequent disintegration should not efface all traces of the body, included for countless ages in solid rock, is it not contrary to all calculation of chances that we should hit upon the exact spot—that mere point in the bed of an ancient ocean, where the precious relic was entombed? Can we expect for a moment, when we have only succeeded, amidst several thousand fragments of corals and shells, in finding a few bones of *aquatic* or *amphibious* animals, that we should meet with a single skeleton of an inhabitant of the land?

Clarence, in his dream, saw, "in the slimy bottom of the deep,"

> —— a thousand fearful wrecks;
> A thousand men, that fishes gnaw'd upon:
> Wedges of gold, great anchors, heaps of pearl.

Had he also beheld, amid "the dead bones that lay scattered by," the carcasses of lions, deer, and the other wild tenants of the forest and the plain, the fiction would have been deemed unworthy of the genius of Shakspeare. So daring a disregard of probability and violation of analogy would have been condemned as unpardonable, even where the poet was painting those incongruous images which present themselves to a disturbed imagination during the visions of the night.

Until lately it was supposed that the old red sandstone, or Devonian rocks, contained no vertebrate remains except those of fish, but in 1850 the footprints of a chelonian, and in 1851 the skeleton of a reptile, allied both to the batrachians and lizards, were found in a sandstone of that age near Elgin in Scotland.* Up to the year 1844 it was laid

* Quart. Journ. Geol. Soc. 1852.

down as a received dogma in many works of high authority in geology, that reptiles were not created until after the close of the carboniferous epoch. In the course of that year, however, Hermann Von Meyer announced the discovery, in the coal measures of Rhenish Bavaria, of a reptile, called by him Apateon, related to the salamanders; and in 1847 three species of another genus, called archegosaurus by Goldfuss, were obtained from the coal of Saarbrück, between Treves and Strasburg. The footprints of a large quadruped, probably batrachian, had also been observed by Dr. King in the carboniferous rocks of Pennsylvania in 1844. The first example of the *bones* of a reptile in the Coal of North America was detected so lately as September, 1852, by Mr. G. W. Dawson and myself in Nova Scotia. These remains, referred by Messrs. Wyman and Owen to a perennibranchiate batrachian, were met with in the interior of an erect fossil tree, apparently a sigillaria. They seem clearly to have been introduced together with sediment into the tree, during its submergence and after it had decayed and was standing as a hollow cylinder of bark, this bark being now converted into coal.

When Agassiz, in his great work on fossil fish, described 152 species of ichthyolites from the Coal, he found them to consist of 94 placoids, belonging to the families of shark and ray, and 58 ganoids. One family of the latter he called "sauroid fish," including the megalicthys and holoptychius, often of great size, and all predaceous. Although true fish, and not intermediate between that class and reptiles, they seem to have been more highly organized than any living fish, reminding us of the skeletons of saurians by the close suture of their cranial bones, their large conical teeth, striated longitudinally, and the articulation of the spinous processes with the vertebræ. Among living species they are most nearly allied to the lepidosteus, or bony pike of the North American rivers. Before the recent progress of discovery above alluded to had shown the fallacy of such ideas, it was imagined by some geologists that this ichthyic type was the more highly developed, because it took the lead at the head of nature before the class of reptiles had been created. The confident assumption indulged in till the year 1844, that reptiles were first introduced into the earth in the Permian period, shows the danger of taking for granted that the date of the creation of any family of animals or plants in past time coincides with the age of the oldest stratified rock in which the geologist has detected its remains. Nevertheless, after repeated disappointments, we find some naturalists as much disposed as ever to rely on such negative evidence, and to feel now as sure that reptiles were not introduced into the earth till after the Silurian epoch, as they were in 1844, that they appeared for the first time at an era subsequent to the carboniferous.

Scanty as is the information hitherto obtained in regard to the articulata of the coal formation, we have at least ascertained that some insects winged their way through the ancient forests. In the ironstone of Coalbrook Dale, two species of coleoptera of the Linnæan genus curculio have been met with: and a neuropterous insect resembling a corydalis, together with another of the same order related to the phasmidæ. As an example of the insectivorous arachnidæ, I may mention the scorpion of the Bohemian coal, figured by Count Sternberg, in which even the eyes, skin, and minute hairs were preserved.* We need not despair, therefore, of obtaining eventually fossil representatives of all the principal orders of hexapods and arachnidæ in carboniferous strata.

Next in chronological order above the Coal comes the allied Magnesian Limestone, or Permian group, and the secondary formations from the Trias to the Chalk inclusive. These rocks comprise the monuments of a long series of ages in which reptiles of every variety of size, form, and structure peopled the earth; so that the whole period, and especially that of the Lias and Oolite, has been sometimes called "the age of reptiles." As there are now mammalia entirely confined to the land; others which, like the bat and vampire, fly in the air; others, again, of amphibious habits, frequenting rivers, like the hippopotamus, otter, and beaver; others exclusively aquatic and marine, like the seal, whale, and narwal; so in the early ages under consideration, there were terrestrial, winged, and aquatic reptiles. There were iguanodons walking on the land, pterodactyls winging their way through the air, monitors and crocodiles in the rivers, and ichthyosaurs and plesiosaurs in the ocean. It appears also that some of these ancient saurians approximated more nearly in their organization to the type of living mammalia than do any of the reptiles now existing.†

In the vast range of strata above alluded to, comprising the Permian, the Upper New Red Sandstone and Muschelkalk, the Lias, Oolite, Wealden, Green-sand, and Chalk, scarcely any well-authenticated instances of the occurrence of fossil birds in Europe are on record, and only two or three of fossil mammalia.

In regard to the absence of birds, they are usually wanting, for reasons afterwards to be explained (see chap. 47), in deposits of all ages, even in the tertiary periods, where we know that birds as well as land quadrupeds abounded. Some at least of the fossil remains formerly referred to this class in the Wealden (a great freshwater deposit below the chalk), have been recently shown by Mr. Owen to belong to pterodactyls.‡ But in North America still more ancient indications of the existence of the feathered tribe have been detected, the fossil foot-marks of a great variety of species, of various sizes, some larger than the ostrich, others smaller than the plover, having been observed. These bipeds have left marks of their footsteps on strata of an age decidedly intermediate between the Lias and the Coal.§

The examples of mammalia, above alluded to, are confined to the Trias and the Oolite. In the former, the evidence is as yet limited to two small molar teeth, described by Professor Plieninger in 1847, under the

* Buckland's Bridgewater Treatise, p. 409.
† Owen's Report on "British Fossil Reptiles, to Brit. Soc." 1841, p. 200.
‡ Quart. Journ. Geol. Soc. No. 6, p. 96.
§ See Hitchcock's Report on Geol. of Massachusetts, and Lyell's Travels in North America, chap. 12.

generic name of Microlestes. They were found near Stuttgart, and possess the double fangs so characteristic of mammalia.* The other fossil remains of the same class were derived from one of the inferior members of the oolitic series in Oxfordshire, and afford more full and satisfactory evidence, consisting of the lower jaws of three species of small quadrupeds about the size of a mole. Cuvier, when he saw one of them (during

Fig. 8. *Natural Size.*

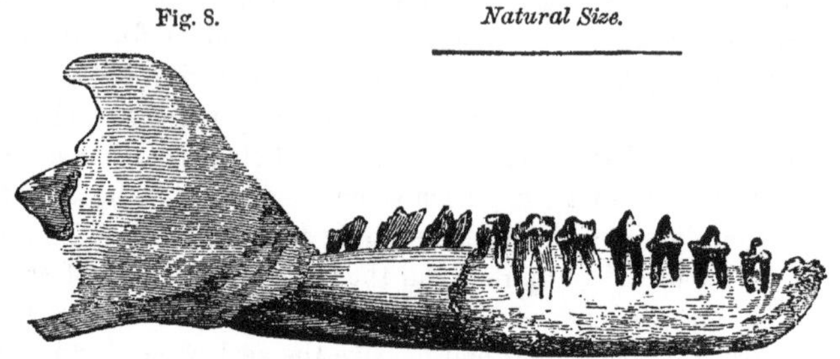

Thylacotherium Prevostii (*Valenciennes*). Amphitherium (*Owen*). Lower jaw, from the slate of Stonesfield, near Oxford.†

a visit to Oxford in 1818), referred it to the marsupial order, stating, however, that it differed from all known carnivora in having ten molar teeth in a row. Professor Owen afterwards pointed out that the jaw belonged to an extinct genus, having considerable affinity to a newly discovered Australian mammifer, the *Myrmecobius* of Waterhouse, which has nine molar teeth in the lower jaw. (Fig. 9.) A more perfect specimen enabled Mr. Owen in 1846 to prove that the inflection of the angular process of the lower jaw was not sufficiently marked to entitle the osteologist to infer that this quadruped was marsupial, as the process is not bent inwards in a greater degree than in the mole or hedgehog. Hence the genus amphitherium, of which there are two species from Stonesfield, must be referred to the ordinary or placental

Fig. 9.

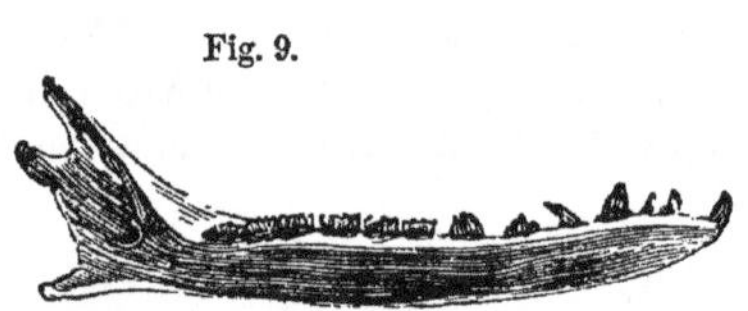

Myrmecobius fasciatus (*Waterhouse*). Recent from Swan River. Lower jaw of the natural size.‡

* See Manual of Geol. by the Author, index *Microlestes.*

† This figure (No. 8) is from a drawing by Professor C. Prevost, published Ann. des Sci. Nat. Avril, 1825. The fossil is a lower jaw, adhering by its inner side to the slab of oolite, in which it is sunk. The form of the condyle, or posterior process of the jaw, is convex, agreeing with the mammiferous type, and is distinctly seen, an impression of it being left on the stone, although in this specimen the bone is wanting. The anterior part of the jaw has been partially broken away, so that the double fangs of the molar teeth are seen fixed in their sockets, the form of the fangs being characteristic of the mammalia. Ten molars are preserved, and the place of an eleventh is believed to be apparent. The enamel of some of the teeth is well preserved.

‡ A colored figure of this small and elegant quadruped is given in the Trans. Zool. Soc. vol. ii. pl. 28. It is insectivorous, and was taken in a hollow tree, in a country abounding in ant-hills, ninety miles to the southeast of the mouth of Swan River in Australia.—It is the first living marsupial species known to have nine molar teeth in the lower jaw, and some of the teeth are widely separated from others, one of the peculiarities in the thylacotherium of Stonesfield, which at first induced M. Blainville to refer that creature to the class of reptiles.

type of insectivorous mammals, although it approximates in some points of structure to the myrmecobius and allied marsupials of Australia. The other contemporary genus, called phascolotherium, agrees much more nearly in osteological character and precisely in the number of the teeth with the opossums; and is believed to have been truly marsupial. (Fig. 10.)

Fig. 10.

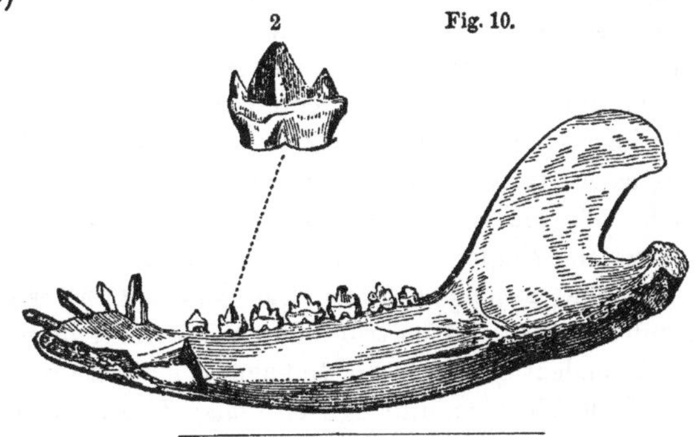

Natural size.

Phascolotherium Bucklandi, *Owen*. (*Syn.* Didelphis Bucklandi, *Brod.*)
Lower jaw, from Stonesfield.*

1. The jaw magnified twice in length. 2. The second molar tooth magnified six times.

The occurrence of these most ancient memorials of the mammiferous type, in so low a member of the oolitic series, while no other representatives of the same class (if we except the microlestes) have yet been found in any other of the inferior or superior secondary strata, is a striking fact, and should serve as a warning to us against hasty generalizations, founded solely on negative evidence. So important an exception to a general rule may be perfectly consistent with the conclusion, that a small number only of mammalia inhabited European latitudes when our secondary rocks were formed; but it seems fatal to the theory of progressive development, or to the notion that the order of precedence in the creation of animals, considered chronologically, has precisely coincided with the order in which they would be ranked according to perfection or complexity of structure.

It was for many years suggested that the marsupial order to which the fossil animals of Stonesfield were supposed exclusively to belong constitutes the lowest grade in the class Mammalia, and that this order, of which the brain is of more simple form, evinces an inferior degree of

* This figure (No. 10) was taken from the orignal, formerly in Mr. Broderip's collection, and now in the British Museum. It consists of the right half of a lower jaw, of which the inner side is seen. The jaw contains seven molar teeth, one canine, and three incisors; but the end of the jaw is fractured, and traces of the alveolus of a fourth incisor are seen. With this addition, the number of teeth would agree exactly with those of a lower jaw of a didelphis. The fossil is well preserved in a slab of oolitic structure containing shells of trigoniæ and other marine remains. Two or three other similar jaws, besides those above represented, have been procured from the quarries of Stonesfield.—See Broderip, Zool. Journ. vol. ii. p. 408. Owen, Proceedings Geol. Soc., November, 1838.

intelligence. If, therefore, in the oolitic period the marsupial tribes were the only warm-blooded quadrupeds which had as yet appeared upon our planet, the fact, it was said, confirmed the theory which teaches that the creation of the more simple forms in each division of the animal kingdom preceded that of the more complex. But on how slender a support, even if the facts had continued to hold true, did such important conclusions hang! The Australian continent, so far as it has been hitherto explored, contains no indigenous quadrupeds save those of the marsupial order, with the exception of a few small rodents, while some neighboring islands to the north, and even southern Africa, in the same latitude as Australia, abound in mammalia of every tribe except the marsupial. We are entirely unable to explain on what physiological or other laws this singular diversity in the habitations of living mammalia depends; but nothing is more clear than that the causes which stamp so peculiar a character on two different provinces of wide extent are wholly independent of time, or of the age or maturity of the planet.

The strata of the Wealden, although of a later date than the oolite of Stonesfield, and although filled with the remains of large reptiles, both terrestrial and aquatic, have not yielded as yet a single marsupial bone. Were we to assume on such scanty data that no warm-blooded quadrupeds were then to be found throughout the northern hemisphere, there would still remain a curious subject of speculation, whether the entire suppression of one important class of vertebrata, such as the mammiferous, and the great development of another, such as the reptilian, implies a departure from fixed and uniform rules governing the fluctuations of the animal world; such rules, for example, as appear from one century to another to determine the growth of certain tribes of plants and animals in arctic, and of other tribes in tropical regions.

In Australia, New Zealand, and many other parts of the southern hemisphere, where the indigenous land quadrupeds are comparatively few, and of small dimensions, the reptiles do not predominate in number or size. The deposits formed at the mouth of an Australian river, within the tropics, might contain the bones of only a few small marsupial animals, which, like those of Stonesfield, might hereafter be discovered with difficulty by geologists; but there would, at the same time, be no megalosauri and other fossil remains, showing that large saurians were plentiful on the land and in the waters at a time when mammalia were scarce. This example, therefore, would afford a very imperfect parallel to the state of the animal kingdom, supposed to have prevailed during the secondary periods, when a high temperature pervaded European latitudes.

It may nevertheless be advantageous to point to some existing anomalies in the geographical development of distinct classes of vertebrata which may be comparable to former conditions of the animal creation brought to light by geology. Thus in the arctic regions, at present, reptiles are small, and sometimes wholly wanting, where birds, large land quadrupeds, and cetacea abound. We meet with bears, wolves, foxes, musk oxen, and deer, walruses, seals, whales, and narwals, in regions of ice and snow, where the smallest snakes, efts, and frogs are rarely, if ever, seen.

A still more anomalous state of things presents itself in the southern hemisphere. Even in the temperate zone, between the latitudes 52° and 56° S., as, for example, in Tierra del Fuego, as well as in the woody region immediately north of the Straits of Magellan, and in the Falkland Islands, no reptiles of any kind are met with, not even a snake, lizard, or frog; but in these same countries we find the guanaco (a kind of llama), a deer, the puma, a large species of fox, many small rodentia, besides the seal and otter, together with the porpoise, whale, and other cetacea.

On what grand laws in the animal physiology these remarkable phenomena depend, cannot in the present state of science be conjectured; nor could we predict whether any opposite condition of the atmosphere, in respect to heat, moisture, and other circumstances, would bring about a state of animal life which might be called the converse of that above described, namely, a state in which reptiles of every size and order might abound, and mammalia disappear.

The nearest approximation to such a fauna is found in the Galapagos Archipelago. These islands, situated under the equator, and nearly 600 miles west of the coast of Peru, have been called "the land of reptiles," so great is the number of snakes, large tortoises, and lizards, which they support. Among the lizards, the first living species proper to the ocean has been discovered. Yet, although some of these islands are from 3000 to 4000 feet high, and one of them 75 miles long, they contain, with the exception of one small mouse, no indigenous mammifer. Even here, however, it is true that in the neighboring sea there are seals, and several kinds of cetacea.*

It may be unreasonable to look for a nearer analogy between the fauna now existing in any part of the globe, and that which we can show to have prevailed when our secondary strata were deposited, because we must always recollect that a climate like that now experienced at the equator, coexisting with the unequal days and nights of European latitudes, was a state of things to which there is now no counterpart on the globe. Consequently, the type of animal and vegetable existence required for such a climate might be expected to deviate almost as widely from that now established, as do the flora and fauna of our tropical differ from those of our arctic regions.

In the Tertiary strata.—The tertiary formations were deposited when the physical geography of the northern hemisphere had been entirely altered. Large inland lakes had become numerous, as in central France and other countries. There were gulfs of the sea, into which considerable rivers emptied themselves, and where strata like those of the Paris basin were accumulated. There were also formations in progress, in

* Darwin's Journal, chap. 19. Lyell's Manual of Geol. chap. 21, p. 279.

shallow seas not far from shore, such as are indicated by portions of the *Faluns* of the Loire, and the English *Crag*.

The proximity, therefore, of large tracts of dry land to the seas and lakes then existing, may, in a great measure, explain why the remains of land animals, so rare in the older strata, are not uncommon in these more modern deposits. Yet even these have sometimes proved entirely destitute of mammiferous relics for years after they had become celebrated for the abundance of their fossil testacea, fish, and reptiles. Thus the calcaire grossier, a marine limestone of the district round Paris, had afforded to collectors more than 1100 species of shells, besides many zoophytes, echinodermata, and the teeth of fish, before the bones of one or two land quadrupeds were met with in the same rock. The strata called London and Plastic clay in England have been studied for more than half a century, and about 400 species of shells, 50 or more of fish, besides several kinds of chelonian and saurian reptiles, were known before a single mammifer was detected. At length, in the year 1839, there were found in this formation the remains of a monkey, an opossum, a bat,* and a species of the extinct genus Hyracotherium, allied to the Peccary or hog tribe.

If we examine the strata above the London clay in England, we first meet with mammiferous remains in the Isle of Wight, in beds also belonging to the Eocene epoch, such as the remains of the Palæotherium, Anoplotherium, and other extinct quadrupeds, agreeing very closely with those first found by Cuvier, near Paris, in strata of the same age, and of similar freshwater origin.

In France we meet with another fauna, both conchological and mammalian in the Miocene "faluns" of the Loire; above which in the ascending series in Great Britain we arrive at the coralline crag of Suffolk, a marine formation which has yielded three or four hundred species of shells, very different from the Eocene testacea, and of which a large proportion, although a minority of the whole number, are recent, besides many corals, echini, foraminifera, and fish, but as yet no relic decidedly mammalian except the ear-bone of a whale.

In the shelly sand, provincially termed "Red Crag," in Suffolk, which immediately succeeds the coralline, constituting a newer member of the same tertiary group, about 250 species of shells have been recognized, of which a still larger proportion are recent. They are associated with numerous teeth of fish; but no signs of a warm-blooded quadruped had been detected until 1839, when the teeth of a leopard, a bear, a hog, and a species of ruminant, were found at Newbourn, in Suffolk, and since that time, several other genera of mammalia have been met with in the same formation, or in the Red Crag.†

Of a still newer date is the Norwich Crag, a fluvio marine deposit of the Pleiocene epoch, containing a mixture of marine, fluviatile, and land

* Taylor's Annals of Nat. Hist. Nov. 1839.

† See notice by the Author, and Professor Owen, Taylor's Annals of Nat. Hist. Nov. 1839.

shells, of which 90 per cent. or more are recent. These beds, since the time of their first investigation, have yielded a supply of mammalian bones of the genera mastodon, elephant, rhinoceros, pig, horse, deer, ox, and others, the bodies of which may have been washed down into the sea by rivers draining land, of which the contiguity is indicated by the occasional presence of terrestrial and freshwater shells.

Our acquaintance with the newer Pleiocene mammalia in Europe, South America, and Australia, is derived chiefly from cavern deposits, a fact which we ought never to forget if we desire to appreciate the superior facilities we enjoy for studying the more modern as compared to the more ancient terrestrial faunas. We know nothing of the fossil bones which must have been inclosed in the stalagmite of caverns in the older Pleiocene, or in the Miocene or Eocene epochs, much less can we derive any information respecting the inhabitants of the land from a similar source, when we carry back our inquiries to the Wealden or carboniferous epochs. We are as well assured that land and rivers then existed, as that they exist now; but it is evident that even a slight geographical revolution, accompanied by the submergence and denudation of land, would reduce to an extreme improbability the chance of our hitting on those minute points of space where caves may once have occurred in limestone rocks.

Fossil quadrumana.—Until within a few years (1836, 1837), not a single bone of any quadrumanous animal, such as the orang, ape, baboon, and monkey, had been discovered in a fossil state, although so much progress had been made in bringing to light the extinct mammalia of successive tertiary eras, both carnivorous and herbivorous. The total absence of these anthropomorphous tribes among the records of a former world, had led some to believe that the type of organization most nearly resembling the human, came so late in the order of creation, as to be scarcely, if at all, anterior to that of man. That such generalizations were premature, I endeavored to point out in the first edition of this work,* in which I stated that the bones of quadrupeds hitherto met with in tertiary deposits were chiefly those which frequent marshes, rivers, or the borders of lakes, as the elephant, rhinoceros, hippopotamus, tapir, hog, deer, and ox, while species which live in trees are extremely rare in a fossil state. I also hinted, that we had as yet no data for determining how great a number of the one kind we ought to find, before we have a right to expect a single individual of the other. Lastly, I observed that the climate of the more modern (or Post-Eocene) tertiary periods in England was not tropical, and that in regard to the London clay, of which the crocodiles, turtles, and fossil fruits implied a climate hot enough for the quadrumana, we had as yet made too little progress in ascertaining what were the Eocene pachydermata of England, to entitle us to expect to have discovered any quadrumana of the same date.

Since those remarks were first written, in 1829, a great number of extinct species have been added to our collections of tertiary mammalia

* See Principles of Geology, 1st ed. 1830, vol. i. p. 152.

from Great Britain and other parts of the world. At length, between the years 1836 and 1839, a few remains of quadrumana were found in France and England, India and Brazil. Those of India, belonging to more than one extinct species of monkey, were first discovered near the Sutlej, in lat. 30° N., in tertiary strata, of which the age is not yet determined; the Brazilian fossil, brought from the basin of the Rio das Velhas, about lat. 18° S., is referable to a form now peculiar in America, allied to the genus Callithrix, the species being extinct. The skull and other bones met with in the South of France belong to a gibbon, or one of the tailless apes, which stand next in the scale of organization to the orang. It occurred at Sansan, about forty miles west of Toulouse, in lat. 43° 40′ N., in freshwater strata, probably of the Miocene or middle tertiary period. Lastly, the English quadrumane first met with, occurred in a more ancient stratum than the rest, and at a point more remote from the equator. It belongs to the genus Macacus, is an extinct species, and was found in Suffolk, in lat. 52°,* in the London clay, the fossils of which, such as crocodiles, turtles, shells of the genus Nautilus, and many curious fruits, had already led geologists to the conclusion that the climate of that era (the Eocene) was warm and nearly tropical.

Some years later (in 1846) the jaw of another British species of fossil monkey, Macacus pliocenus, was announced by Mr. Owen as having been met with in the newer Pleiocene strata, on the banks of the Thames, at Grays, in Essex, accompanying the remains of hippopotamus, elephant, and other quadrupeds, and associated with freshwater and land shells, most of which are now inhabitants of the British Isles.†

When we consider the small area of the earth's surface hitherto explored geologically, and the new discoveries brought to light daily, even in the environs of great European capitals, we must feel that it would be rash to assume that the Lower Eocene deposits mark the era of the first creation of quadrumana. It would, however, be still

* The first quadrumanous fossils discovered in India were observed in 1836 in the Sewalik Hills, a lower range of the Himalayan Mountains, by Lieutenants Baker and Durond, by whom their osteological characters were determined (Journ. of Asiat. Soc. of Bengal, vol. v. p. 739), and in the year following, other fossils of the same class were brought to light and described by Capt. Cautley and Dr. Falconer. These were imbedded, like the former, in tertiary strata of conglomerate, sand, marl, and clay, in the Sub-Himalayan Mountains. (Ibid. vol. v. p. 379. Nov. 1836; and vol. vi. p. 354. May, 1837.)

The Brazilian quadrumane was found, with a great many other extinct species of animals, by a Danish naturalist, Dr. Lund, between the rivers Francisco and Velhas, in 1837.

The gibbon of the South of France was found by M. Lartet in the beginning of 1837, and determined by M. de Blainville. It occurred near Auch, in the department of Gers, about forty miles west of Toulouse, in freshwater marl, limestone, and sand. They were accompanied by the remains of the mastodon, dinotherium, palæotherium, rhinoceros, gigantic sloth, and other extinct quadrupeds. (Bulletin de la Soc. Geol. de France, tom. viii. p. 92.)

The British quadrumane was discovered in 1839, by Messrs. William Colchester and Searles Wood, at Kyson, near Woodbridge, in Suffolk, and was referred by Professor Owen to the genus Macacus. (Mag. of Nat. Hist. Sept. 1839. Taylor, Annals of Nat. Hist. No. xxiii. Nov. 1839.)

† Owen's Introduction to British Fossil Mammals, p. 46.

more unphilosophical to infer, as some writers have done, from a single extinct species of this family obtained in a latitude far from the tropics, that the Eocene quadrumana did not attain as high a grade of organization as they do in our own times. What would the naturalist know of the apes and orangs now contemporary with man, if our investigations were restricted to such northern latitudes as those where alone the geologist has hitherto found all the fossil quadrumana of Europe?

Cetacea.—The absence of Cetacea from rocks older than the Eocene has been frequently adduced as lending countenance to the theory of the very late appearance of the highest class of Vertebrata on the earth. Professor Sedgwick possesses in the Cambridge Museum a mass of anchylosed cervical vertebræ of a whale, which he found in drift clay near Ely, and which he has no doubt was washed out of the Kimmeridge clay, an upper member of the Oolite. According to Professor Owen, it exhibits well-marked specific characters, distinguishing it from all other known recent or fossil cetacea. Dr. Leidy, of Philadelphia, has lately described (1851) two species of cetacea of a new genus, which he has called Priscodelphinus from the green sand of New Jersey, which corresponds in age with the English Chalk or the cretaceous strata above the gault. The specimens consist of dorsal and cervical vertebræ.* Even in the Eocene strata of Europe, the discovery of cetaceans has never kept pace with that of land quadrupeds. The only instance cited in Great Britain is a species of Monodon, from the London clay, of doubtful authenticity as to its geological position. On the other hand, the gigantic Zeuglodon of North America occurs abundantly in the Middle Eocene strata of Georgia and Alabama, from which as yet no bones of land quadrupeds have been obtained.

In the present imperfect state then of our information, we can scarcely say more than that the cetacea seem to have been scarce in the secondary and primary periods. It is quite conceivable that when aquatic saurians, some of them carnivorous, like the Ichthyosaurus, were swarming in the sea, and when there were large herbivorous reptiles, like the Iguanodon, on the land, the class of reptiles may, to a certain extent, have superseded the cetacea, and discharged their functions in the animal economy.

That mammalia had been created long before the epoch of the Kimmeridge clay, is shown by the Microlestes of the Trias before alluded to, and by the Stonesfield quadrupeds from the Inferior Oolite. And we are bound to remember, whenever we infer the poverty of the flora or fauna of any given period of the past, from the small number of fossils occurring in ancient rocks, that it has been evidently no part of the plan of Nature to hand down to us a complete or systematic record of the former history of the animate world. We may have failed to discover a single shell, marine or freshwater, or a single coral or bone in certain sandstones, such as that of the valley of the Connecticut, where the footprints of bipeds and quadrupeds abound; but such failure may

* Proceedings of Acad. Nat. Sci. Philad. Dec. 9, 1851.

have arisen, not because the population of the land or sea was scanty at that era, but because in general the preservation of any relics of the animals or plants of former times is the exception to a general rule. Time so enormous as that contemplated by the geologist may multiply exceptional cases till they seem to constitute the rule, and so impose on the imagination as to lead us to infer the non-existence of creatures of which no monuments happen to remain. Professor Forbes has remarked, that few geologists are aware how large a proportion of all known species of fossils are founded on single specimens, while a still greater number are founded on a few individuals discovered in one spot. This holds true not only in regard to animals and plants inhabiting the land, the lake, and the river, but even to a surprising number of the marine mollusca, articulata, and radiata. Our knowledge, therefore, of the living creation of any given period of the past may be said to depend in a great degree on what we commonly call chance, and the casual discovery of some new localities rich in peculiar fossils may modify or entirely overthrow all our previous generalizations.

Upon the whole then we derive this result from a general review of the fossils of the successive tertiary strata, namely, that since the Eocene period, there have been several great changes in the land quadrupeds inhabiting Europe, probably not less than five complete revolutions, during which there has been no step whatever made in advance, no elevation in the scale of being; so that had man been created at the commencement of the Eocene era, he would not have constituted a greater innovation on the state of the animal creation previously established than now, when we believe him to have begun to exist at the close of the Pleiocene. The views, therefore, which I proposed in the first edition of this work, January, 1830, in opposition to the theory of progressive development, do not seem to me to require material modification, notwithstanding the large additions since made to our knowledge of fossil remains.

These views may be thus briefly stated. From the earliest period at which plants and animals can be proved to have existed, there has been a continual change going on in the position of land and sea, accompanied by great fluctuations of climate. To these ever-varying geographical and climatal conditions the state of the animate world has been unceasingly adapted. No satisfactory proof has yet been discovered of the gradual passage of the earth from a chaotic to a more habitable state, nor of any law of progressive development governing the extinction and renovation of species, and causing the fauna and flora to pass from an embryonic to a more perfect condition, from a simple to a more complex organization.

The principle of adaptation to which I have alluded, appears to have been analogous to that which now peoples the arctic, temperate, and tropical regions contemporaneously with distinct assemblages of species and genera, or which, independently of mere temperature, gives rise to a predominance of the marsupial or didelphous tribe of quadrupeds in

Australia, of the placental or monodelphous tribe in Asia and Europe, or which causes a profusion of reptiles without mammalia in the Galapagos Archipelago, and of mammalia without reptiles in Greenland.

Recent origin of man.—If, then, the popular theory of the successive development of the animal and vegetable world, from the simplest to the most perfect forms, rests on a very insecure foundation; it may be asked, whether the recent origin of man lends any support to the same doctrine, or how far the influence of man may be considered as such a deviation from the analogy of the order of things previously established, as to weaken our confidence in the uniformity of the course of nature.

Antecedently to investigation, we might reasonably have anticipated that the vestiges of man would have been traced back at least as far as those modern strata in which all the testacea and a certain number of the mammalia are of existing species, for of all the mammalia the human species is the most cosmopolite, and perhaps more capable than any other of surviving considerable vicissitudes in climate, and in the physical geography of the globe.

No inhabitant of the land exposes himself to so many dangers on the waters as man, whether in a savage or a civilized state;* and there is no animal, therefore, whose skeleton is so liable to become imbedded in lacustrine or submarine deposits; nor can it be said that his remains are more perishable than those of other animals; for in ancient fields of battle, as Cuvier has observed, the bones of men have suffered as little decomposition as those of horses which were buried in the same grave.† But even if the more solid parts of our species had disappeared, the impression of their form would have remained engraven on the rocks, as have the traces of the tenderest leaves of plants, and the soft integuments of many animals. Works of art, moreover, composed of the most indestructible materials, would have outlasted almost all the organic contents of sedimentary rocks. Edifices, and even entire cities, have, within the times of history, been buried under volcanic ejections, submerged beneath the sea, or engulfed by earthquakes; and had these catastrophes been repeated throughout an indefinite lapse of ages, the high antiquity of man would have been inscribed in far more legible characters on the framework of the globe than are the forms of the ancient vegetation which once covered the islands of the northern ocean, or of those gigantic reptiles which at still later periods peopled the seas and rivers of the northern hemisphere.‡

Dr. Prichard has argued that the human race have not always existed on the surface of the earth, because "the strata of which our continents are composed were once a part of the ocean's bed"—"mankind had a beginning, since we can look back to the period when the surface on which they lived began to exist."§ This proof, however, is insuffi-

* See ch. 48. † Ibid. ‡ Ibid.
§ Phys. Hist. of Mankind, vol. ii. p. 594.

cient, for many thousands of human beings now dwell in various quarters of the globe where marine species lived within the times of history, and, on the other hand, the sea now prevails permanently over large districts once inhabited by thousands of human beings. Nor can this interchange of sea and land ever cease while the present causes are in existence. Terrestrial species, therefore, might be older than the continents which they inhabit, and aquatic species of higher antiquity than the lakes and seas which they now people.

But so far as our interpretation of physical movements has yet gone, we have every reason to infer that the human race is extremely modern, even when compared to the larger number of species now our contemporaries on the earth, and we may, therefore, ask whether his creation can be considered as one step in a supposed progressive system, by which the organic world has advanced slowly from a more simple to a more complex and perfect state? If we concede, for a moment, the truth of the proposition, that the sponge, the cephalopod, the fish, the reptile, the bird, and the mammifer, have followed each other in regular chronological order, the creation of each class being separated from the other by vast intervals of time, should we be able to recognize, in man's entrance upon the earth, the last term of one and the same series of progressive developments?

In reply to this question it should first be observed, that the superiority of man depends not on those faculties and attributes which he shares in common with the inferior animals, but on his reason, by which he is distinguished from them. When it is said that the human race is of far higher dignity than were any pre-existing beings on the earth, it is the intellectual and moral attributes of our race, rather than the physical, which are considered; and it is by no means clear that the organization of man is such as would confer a decided pre-eminence upon him, if, in place of his reasoning powers, he was merely provided with such instincts as are possessed by the lower animals.

If this be admitted, it would not follow, even if there were sufficient geological evidence in favor of the theory of progressive development, that the creation of man was the last link in the same chain. For the sudden passage from an irrational to a rational animal, is a phenomenon of a distinct kind from the passage from the more simple to the more perfect forms of animal organization and instinct. To pretend that such a step, or rather leap, can be part of a regular series of changes in the animal world, is to strain analogy beyond all reasonable bounds.

Introduction of man, to what extent a change in the system.—But setting aside the question of progressive development, another and a far more difficult one may arise out of the admission that man is comparatively of modern origin. Is not the interference of the human species, it may be asked, such a deviation from the antecedent course of physical events, that the knowledge of such a fact tends to destroy all our confidence in the uniformity of the order of nature, both in regard to time past and future? If such an innovation could take place after the earth had been exclusively inhabited for thousands of ages by inferior animals, why should not other changes as extraordinary and unprecedented happen from time to time? If one new cause was permitted to supervene, differing in kind and energy from any before in operation, why may not others have come into action at different epochs? Or what security have we that they may not arise hereafter? And if such be the case, how can the experience of one period, even though we are acquainted with all the possible effects of the then existing causes, be a standard to which we can refer all natural phenomena of other periods?

Now these objections would be unanswerable, if adduced against one who was contending for the absolute uniformity throughout all time of the succession of sublunary events—if, for example, he was disposed to indulge in the philosophical reveries of some Egyptian and Greek sects, who represented all the changes both of the moral and material world as repeated at distant intervals, so as to follow each other in their former connection of place and time. For they compared the course of events on our globe to astronomical cycles; and not only did they consider all sublunary affairs to be under the influence of the celestial bodies, but they taught that on the earth, as well as in the heavens, the same identical phenomena recurred again and again in a perpetual vicissitude. The same individual men were doomed to be re-born, and to perform the same actions as before; the same arts were to be invented, and the same cities built and destroyed. The Argonautic expedition was destined to sail again with the same heroes, and Achilles with his Myrmidons to renew the combat before the walls of Troy.

> Alter erit tum Tiphys, et altera quæ vehat Argo
> Dilectos heroas; erunt etiam altera bella,
> Atque iterum ad Trojam magnus mittetur Achilles.*

The geologist, however, may condemn these tenets as absurd, without running into the opposite extreme, and denying that the order of nature has, from the earliest periods, been uniform in the same sense in which we believe it to be uniform at present, and expect it to remain so in future. We have no reason to suppose, that when man first became master of a small part of the globe, a greater change took place in its physical condition than is now experienced when districts, never before inhabited, become successively occupied by new settlers. When a powerful European colony lands on the shores of Australia, and introduces at once those arts which it has required many centuries to mature; when it imports a multitude of plants and large animals from the opposite extremity of the earth, and begins rapidly to extirpate many of the indigenous species, a mightier revolution is effected in a brief period than the first entrance of a savage horde, or their continued occupation

* Virgil, Eclog. iv. For an account of these doctrines, see Dugald Stewart's Elements of the Philosophy of the Human Mind, vol. ii. chap. ii. sect. 4, and Prichard's Egypt. Mythol. p. 177.

of the country for many centuries, can possibly be imagined to have produced. If there be no impropriety in assuming that the system is uniform when disturbances so unprecedented occur in certain localities, we can with much greater confidence apply the same language to those primeval ages when the aggregate number and power of the human race, or the rate of their advancement in civilization, must be supposed to have been far inferior. In reasoning on the state of the globe immediately before our species was called into existence, we must be guided by the same rules of induction as when we speculate on the state of America in the interval that elapsed between the introduction of man into Asia, the supposed cradle of our race, and the arrival of the first adventurers on the shores of the New World. In that interval, we imagine the state of things to have gone on according to the order now observed in regions unoccupied by man. Even now, the waters of lakes, seas, and the great ocean, which teem with life, may be said to have no immediate relation to the human race—to be portions of the terrestrial system of which man has never taken, nor ever can take possession; so that the greater part of the inhabited surface of the planet may still remain as insensible to our presence as before any isle or continent was appointed to be our residence.

If the barren soil around Sydney had at once become fertile upon the landing of our first settlers; if, like the happy isles whereof the poets have given such glowing descriptions, those sandy tracts had begun to yield spontaneously an annual supply of grain, we might then, indeed, have fancied alterations still more remarkable in the economy of nature to have attended the first coming of our species into the planet. Or if, when a volcanic island like Ischia was, for the first time, brought under cultivation by the enterprise and industry of a Greek colony, the internal fire had become dormant, and the earthquake had remitted its destructive violence, there would then have been some ground for speculating on the debilitation of the subterranean forces, when the earth was first placed under the dominion of man. But after a long interval of rest, the volcano bursts forth again with renewed energy, annihilates one half of the inhabitants, and compels the remainder to emigrate. The course of nature remains evidently unchanged; and, in like manner, we may suppose the general condition of the globe, immediately before and after the period when our species first began to exist, to have been the same, with the exception only of man's presence.

The modifications in the system of which man is the instrument do not, perhaps, constitute so great a deviation from previous analogy as we usually imagine; we often, for example, form an exaggerated estimate of the extent of our power in extirpating some of the inferior animals, and causing others to multiply; a power which is circumscribed within certain limits, and which, in all likelihood, is by no means exclusively exerted by our species.* The growth of human population can-

* See ch. 41.

not take place without diminishing the numbers, or causing the entire destruction, of many animals. The larger beasts of prey, in particular, give way before us; but other quadrupeds of smaller size, and innumerable birds, insects, and plants, which are inimical to our interests, increase in spite of us, some attacking our food, others our raiment and persons, and others interfering with our agricultural and horticultural labors. We behold the rich harvest which we have raised by the sweat of our brow, devoured by myriads of insects, and are often as incapable of arresting their depredations, as of staying the shock of an earthquake, or the course of a stream of lava.

A great philosopher has observed, that we can command nature only by obeying her laws; and this principle is true even in regard to the astonishing changes which are superinduced in the qualities of certain animals and plants by domestication and garden culture. I shall point out in the third book that we can only effect such surprising alterations by assisting the development of certain instincts, or by availing ourselves of that mysterious law of their organization, by which individual peculiarities are transmissible from one generation to another.*

It is probable from these and many other considerations, that as we enlarge our knowledge of the system, we shall become more and more convinced, that the alterations caused by the interference of man deviate far less from the analogy of those effected by other animals than is usually supposed.† We are often misled, when we institute such comparisons, by our knowledge of the wide distinction between the instincts of animals and the reasoning power of man; and we are apt hastily to infer, that the effects of a rational and irrational species, considered merely *as physical agents*, will differ almost as much as the faculties by which their actions are directed.

It is not, however, intended that a real departure from the antecedent course of physical events cannot be traced in the introduction of man. If that latitude of action which enables the brutes to accommodate themselves in some measure to accidental circumstances could be imagined to have been at any former period so great, that the operations of instinct were as much diversified as are those of human reason, it might, perhaps, be contended, that the agency of man did not constitute an anomalous deviation from the previously established order of things. It might then have been said, that the earth's becoming at a particular period the residence of human beings, was an era in the moral, not in the physical world—that our study and contemplation of the earth, and the laws which govern its animate productions, ought no more to be considered in the light of a disturbance or deviation from the system, than the discovery of the satellites of Jupiter should be regarded as a physical event affecting those heavenly bodies. Their influence in advancing the progress of science among men, and in aiding navigation and commerce, was accompanied by no reciprocal action

* See ch. 35. † See ch. 37, 38, 39, 41.

of the human mind upon the economy of nature in those distant planets; and so the earth might be conceived to have become, at a certain period, a place of moral discipline and intellectual improvement to man, without the slightest derangement of a previously existing order of change in its animate and inanimate productions.

The distinctness, however, of the human from all other species, considered merely as an efficient cause in the physical world, is real; for we stand in a relation to contemporary species of animals and plants widely different from that which other irrational animals can ever be supposed to have held to each other. We modify their instincts, relative numbers, and geographical distribution, in a manner superior in degree, and in some respects very different in kind from that in which any other species can affect the rest. Besides, the progressive movement of each successive generation of men causes the human species to differ more from itself in power at two distant periods, than any one species of the higher order of animals differs from another. The establishment, therefore, by geological evidence, of the first intervention of such a peculiar and unprecedented agency, long after other parts of the animate and inanimate world existed, affords ground for concluding that the experience during thousands of ages of all the events which may happen on this globe, would not enable a philosopher to speculate with confidence concerning future contingencies.

If, then, an intelligent being, after observing the order of events for an indefinite series of ages, had witnessed at last so wonderful an innovation as this, to what extent would his belief in the regularity of the system be weakened?—would he cease to assume that there was permanency in the laws of nature?—would he no longer be guided in his speculations by the strictest rules of induction? To these questions it may be answered, that, had he previously presumed to dogmatize respecting the absolute uniformity of the order of nature, he would undoubtedly be checked by witnessing this new and unexpected event, and would form a more just estimate of the limited range of his own knowledge, and the unbounded extent of the scheme of the universe. But he would soon perceive that no one of the fixed and constant laws of the animate or inanimate world was subverted by human agency, and that the modifications now introduced for the first time were the accompaniments of new and extraordinary circumstances, and those not of a *physical* but a *moral* nature. The deviation permitted would also appear to be as slight as was consistent with the accomplishment of the new *moral* ends proposed, and to be in a great degree temporary in its nature, so that, whenever the power of the new agent was withheld, even for a brief period, a relapse would take place to the ancient state of things; the domesticated animal, for example, recovering in a few generations its wild instinct, and the garden-flower and fruit-tree reverting to the likeness of the parent stock.

Now, if it would be reasonable to draw such inferences with respect to the future, we cannot but apply the same rules of induction to the past. We have no right to anticipate any modifications in the results of existing causes in time to come, which are not conformable to analogy, unless they be produced by the progressive development of human power, or perhaps by some other new relations which may hereafter spring up between the moral and material worlds. In the same manner, when we speculate on the vicissitudes of the animate and inanimate creation in former ages, we ought not to look for any anomalous results, unless where man has interfered, or unless clear indications appear of some other *moral* source of temporary derangement.

CHAPTER X.

SUPPOSED INTENSITY OF AQUEOUS FORCES AT REMOTE PERIODS.

Intensity of aqueous causes—Slow accumulation of strata proved by fossils—Rate of denudation can only keep pace with deposition—Erratics, and effects of ice—Deluges, and the causes to which they are referred—Supposed universality of ancient deposits.

Intensity of aqueous causes.—The great problem considered in the preceding chapters, namely, whether the former changes of the earth made known to us by geology, resemble in kind and degree those now in daily progress, may still be contemplated from several other points of view. We may inquire, for example, whether there are any grounds for the belief entertained by many, that the intensity both of aqueous and of igneous forces, in remote ages, far exceeded that which we witness in our own times.

First, then, as to aqueous causes: it has been shown, in our history of the science, that Woodward did not hesitate, in 1695, to teach that the entire mass of fossiliferous strata contained in the earth's crust had been deposited in a few months; and, consequently, as their mechanical and derivative origin was already admitted, the reduction of rocky masses into mud, sand, and pebbles, the transportation of the same to a distance, and their accumulation elsewhere in regular strata, were all assumed to have taken place with a rapidity unparalleled in modern times. This doctrine was modified by degrees, in proportion as different classes of organic remains, such as shells, corals, and fossil plants, had been studied with attention. Analogy led every naturalist to assume, that each full-grown individual of the animal or vegetable kingdom, had required a certain number of months or years for the attainment of maturity, and the perpetuation of its species by generation; and thus the first approach was made to the conception of a common standard of time, without which there are no means whatever of measuring the comparative rate at which any succession of events has taken place at two distinct periods. This standard consisted of the average duration of the lives of individuals of the same genera or families in the animal and vegetable kingdoms; and the multitude of fossils dispersed through successive strata implied the continuance of the same species for many generations. At length the idea that species themselves had had a limited duration, arose out of the observed fact that sets of strata of different ages contained fossils of distinct species. Finally, the opinion became general, that in the course of ages, one assemblage of animals and plants had disappeared after another again and again, and new tribes had started into life to replace them.

Denudation.—In addition to the proofs derived from organic remains, the forms of stratification led also, on a fuller investigation, to the belief that sedimentary rocks had been slowly deposited; but it was still supposed that *denudation,* or the power of running water, and the waves and currents of the ocean, to strip off superior strata, and lay bare the rocks below, had formerly operated with an energy wholly unequalled in our times. These opinions were both illogical and inconsistent, because deposition and denudation are parts of the same process, and what is true of the one must be true of the other. Their speed must be always limited by the same causes, and the conveyance of solid matter to a particular region can only keep pace with its removal from another, so that the aggregate of sedimentary strata in the earth's crust can never exceed in volume the amount of solid matter which has been ground down and washed away by running water. How vast, then, must be the spaces which this abstraction of matter has left vacant! how far exceeding in dimensions all the valleys, however numerous, and the hollows, however vast, which we can prove to have been cleared out by aqueous erosion! The evidences of the work of denudation are defective, because it is the nature of every destroying cause to obliterate the signs of its own agency; but the amount of reproduction in the form of sedimentary strata must always afford a true measure of the minimum of denudation which the earth's surface has undergone.

Erratics.—The next phenomenon to which the advocates of the excessive power of running water in times past have appealed, is the enormous size of the blocks called *erratic,* which lie scattered over the northern parts of Europe and North America. Unquestionably a large proportion of these blocks have been transported far from their original position, for between them and the parent rocks we now find, not unfrequently, deep seas and valleys intervening, or hills more than a thousand feet high. To explain the present situation of such travelled fragments, a deluge of mud has been imagined by some to have come from the north, bearing along with it sand, gravel, and stony fragments, some of them hundreds of tons in weight. This flood, in its transient passage over the continents, dispersed the boulders irregularly over hill, valley, and plain; or forced them along over a surface of hard rock, so as to polish it and leave it indented with parallel scratches and grooves—such markings as are still visible in the rocks of Scandinavia, Scotland, Canada, and many other countries.

There can be no doubt that the myriads of angular and rounded blocks above alluded to, cannot have been borne along by ordinary rivers or marine currents, so great is their volume and weight, and so clear are the signs, in many places, of time having been occupied in their successive deposition; for they are often distributed at various depths through heaps of regularly stratified sand and gravel. No waves of the sea raised by earthquakes, nor the bursting of lakes dammed up for a time by landslips or by avalanches of snow, can account for the observed facts; but I shall endeavor to show, in the next book, chap. 15,* that a combination of existing causes may have conveyed erratics into their present situations.

The causes which will be referred to are, first, the carrying power of ice, combined with that of running water; and second, the upward movement of the bed of the sea, converting it gradually into land. Without entering at present into any details respecting these causes, I may mention that the transportation of blocks by ice is now simultaneously in progress in the cold and temperate latitudes, both of the northern and southern hemisphere, as, for example, on the coasts of Canada and Gulf of St. Lawrence, and also in Chili, Patagonia, and the island of South Georgia. In those regions the uneven bed of the ocean is becoming strewed over with ice-drifted fragments, which have either stranded on shoals, or been dropped in deep water by melting bergs. The entanglement of boulders in drift-ice will also be shown to occur annually in North America, and these stones, when firmly frozen into ice, wander year after year from Labrador to the St. Lawrence, and reach points of the western hemisphere farther south than any part of Great Britain.

The general absence of erratics in the warmer parts of the equatorial regions of Asia, Africa, and America, confirms the same views. As to the polishing and grooving of hard rocks, it has lately been ascertained that glaciers give rise to these effects when pushing forward sand, pebbles, and rocky fragments, and causing them to grate along the bottom. Nor can there be any reasonable doubt that icebergs, when they run aground on the floor of the ocean, must imprint similar marks upon it.

It is unnecessary, therefore, to refer to deluges, or even to speculate on the former existence of a climate more severe than that now prevailing in the western hemisphere, to explain the geographical distribution of most of the European erratics.

Deluges.—As deluges have been often alluded to, I shall say something of the causes which may be supposed to give rise to these grand movements of water in addition to those already alluded to (p. 9). Geologists who believe that mountain-chains have been thrown up sud-

* See also Manual of Geology, ch. 11, 12.

denly at many successive epochs, imagine that the waters of the ocean may be raised by these convulsions, and then break in terrific waves upon the land, sweeping over whole continents, hollowing out valleys, and transporting sand, gravel, and erratics, to great distances. The sudden rise of the Alps or Andes, it is said, may have produced a flood even subsequently to the time when the earth became the residence of man. But it seems strange that none of the writers who have indulged their imaginations in conjectures of this kind, should have ascribed a deluge to the sudden conversion of part of the unfathomable ocean into a shoal rather than to the rise of mountain-chains. In the latter case, the mountains themselves could do no more than displace a certain quantity of atmospheric air, whereas, the instantaneous formation of the shoal would displace a vast body of water, which being heaved up to a great height might roll over and permanently submerge a large portion of a continent.

If we restrict ourselves to combinations of causes at present known, it would seem that the two principal sources of extraordinary inundations are, first, the escape of the waters of a large lake raised far above the sea; and, secondly, the pouring down of a marine current into lands depressed below the mean level of the ocean.

As an example of the first of these cases, we may take Lake Superior, which is more than 400 geographical miles in length and about 150 in breadth, having an average depth of from 500 to 900 feet. The surface of this vast body of fresh water is no less than 600 feet above the level of the ocean; the lowest part of the barrier which separates the lake on its southwest side from those streams which flow into the head waters of the Mississippi being about 600 feet high. If, therefore, a series of subsidences should lower any part of this barrier 600 feet, any subsequent rending or depression, even of a few yards at a time, would allow the sudden escape of vast floods of water into a hydrographical basin of enormous extent. If the event happened in the dry season, when the ordinary channels of the Mississippi and its tributaries are in a great degree empty, the inundation might not be considerable; but if in the flood-season, a region capable of supporting a population of many millions might be suddenly submerged. But even this event would be insufficient to cause a violent rush of water, and to produce those effects usually called diluvial; for the difference of level of 600 feet between Lake Superior and the Gulf of Mexico, when distributed over a distance of 1800 miles, would give an average fall of only four inches per mile.

The second case before adverted to is where there are large tracts of dry land beneath the mean level of the ocean. It seems, after much controversy, to be at length a settled point, that the Caspian is really 83 feet 6 inches lower than the Black Sea. As the Caspian covers an area about equal to that of Spain, and as its shores are in general low and flat, there must be many thousand square miles of country less than 83 feet above the level of that inland sea, and consequently depressed below the Black Sea and Mediterranean. This area includes the site of the populous city of Astrakhan and other towns. Into this region the ocean would pour its waters, if the land now intervening between the Sea of Azof and the Caspian should subside. Yet even if this event should occur, it is most probable that the submergence of the whole region would not be accomplished simultaneously, but by a series of minor floods, the sinking of the barrier being gradual.*

Supposed universality of ancient deposits.—The next fallacy which has helped to perpetuate the doctrine that the operations of water were on a different and grander scale in ancient times, is founded on the indefinite areas over which homogeneous deposits were supposed to extend. No modern sedimentary strata, it is said, equally identical in mineral character and fossil contents, can be traced continuously from one quarter of the globe to another. But the first propagators of these opinions were very slightly acquainted with the inconstancy in mineral composition of the ancient formations, and equally so of the wide spaces over which the same kind of sediment is now actually distributed by rivers and currents in the course of centuries. The persistency of character in the older series was exaggerated, its extreme variability in the newer was assumed without proof. In the chapter which treats of river-deltas and the dispersion of sediment by currents, and in the description of reefs of coral now growing over areas many hundred miles in length, I shall have opportunities of convincing the reader of the danger of hasty generalizations on this head.

In regard to the imagined universality of particular rocks of ancient date, it was almost unavoidable that this notion, when once embraced, should be perpetuated; for the same kinds of rock have occasionally been reproduced at successive epochs; and when once the agreement or disagreement in mineral character alone was relied on as the test of age, it followed that similar rocks, if found even at the antipodes, were referred to the same era, until the contrary could be shown.

* It has been suspected ever since the middle of the last century, that the Caspian was lower than the ocean, it being known that in Astrakhan the mercury in the barometer generally stands above thirty inches. In 1811, MM. Engelhardt and Parrot attempted to determine the exact amount of difference by a series of levellings and barometrical measurements across the isthmus at two different places near the foot of Mount Caucasus. The result of their operations led them to the opinion that the Caspian was more than 300 feet below the Black Sea. But the correctness of the observations having afterwards been called in question, M. Parrot revisited the ground in 1829 and 1830, and inferred from new levellings, that the mouth of the Don was between three and four feet lower than that of the Wolga; in other words, that the sea of Azof, which communicates with the Black Sea, was actually lower than the Caspian! Other statements, no less contradictory, having been made by other observers, the Russian government at length directed the Academy of St. Petersburg to send an expedition, in 1836, to decide the point by a trigonometrical survey, from which it appeared that the Caspian is 101 Russian, or 108 English, feet lower than the Black Sea. (For authorities, see Journ. Roy. Geograph. Soc. vol. viii. p. 135). Sir R. Murchison, however, concludes, in 1845, from the best Russian authorities, that the depression of the Caspian is only 83 feet 6 inches.

The measurements of Major Anthony Symonds, since confirmed by French authorities, make the Dead Sea to be 1200 feet below the Mediterranean.

Now it is usually impossible to combat such an assumption on geological grounds, so long as we are imperfectly acquainted with the order of superposition and the organic remains of these same formations. Thus, for example, a group of red marl and red sandstone, containing salt and gypsum, being interposed in England between the Lias and the Coal, all other red marls and sandstones, associated some of them with salt, and others with gypsum, and occurring not only in different parts of Europe, but in North America, Peru, India, the salt deserts of Asia, those of Africa—in a word, in every quarter of the globe, were referred to one and the same period. The burden of proof was not supposed to rest with those who insisted on the identity in age of all these groups—their identity in mineral composition was thought sufficient. It was in vain to urge as an objection the improbability of the hypothesis which implies that all the moving waters on the globe were once simultaneously charged with sediment of a red color.

But the rashness of pretending to identify, in age, all the red sandstones and marls in question, has at length been sufficiently exposed, by the discovery that, even in Europe, they belong decidedly to many different epochs. It is already ascertained, that the red sandstone and red marl containing the rock-salt of Cardona in Catalonia is newer than the Oolitic, if not more modern than the Cretaceous period. It is also known that certain red marls and variegated sandstones in Auvergne which are undistinguishable in mineral composition from the New Red Sandstone of English geologists, belong, nevertheless, to the Eocene period; and, lastly, the gypseous red marl of Aix, in Provence, formerly supposed to be a marine secondary group, is now acknowledged to be a tertiary freshwater formation. In Nova Scotia one great deposit of red marl, sandstone, and gypsum, precisely resembling in mineral character the "New Red" of England, occurs as a member of the Carboniferous group, and in the United States near the Falls of Niagara, a similar formation constitutes a subdivision of the Silurian series.*

Nor was the nomenclature commonly adopted in geology without its influence in perpetuating the erroneous doctrine of universal formations. Such names, for example, as Chalk, Green Sand, Oolite, Red Marl, Coal, and others, were given to some of the principal fossiliferous groups in consequence of mineral peculiarities which happened to characterize them in the countries where they were first studied. When geologists had at length shown, by means of fossils and the order of superposition, that other strata, entirely dissimilar in color, texture, and composition, were of contemporaneous date, it was thought convenient still to retain the old names. That these were often inappropriate was admitted; but the student was taught to understand them in no other than a chronological sense; so that the Chalk might not be a white cretaceous rock, but a hard dolomitic limestone, as in the Alps, or a brown sandstone or green marl, as in New Jersey, U. S. In like manner, the

* See Lyell's Travels in N. America, ch. 2 and 25.

Green Sand, it was said, might in some places be represented by red sandstone, red marl, salt, and gypsum, as in the north of Spain. . So the oolitic texture was declared to be rather an exception than otherwise to the general rule in rocks of the Oolitic period; and it often became necessary to affirm that no particle of carbonaceous matter could be detected in districts where the true Coal series abounded. In spite of every precaution the habitual use of this language could scarcely fail to instil into the mind of the pupil an idea that chalk, coal, salt, red marl, or the Oolitic structure were far more widely characteristic of the rocks of a given age than was really the case.

There is still another cause of deception, disposing us to ascribe a more limited range to the newer sedimentary formations as compared to the older, namely, the very general concealment of the newer strata beneath the waters of lakes and seas, and the wide exposure above waters of the more ancient. The Chalk, for example, now seen stretching for thousands of miles over different parts of Europe, has become visible to us by the effect, not of one, but of many distinct series of subterranean movements. Time has been required, and a succession of geological periods, to raise it above the waves in so many regions; and if calcareous rocks of the middle and upper tertiary periods have been formed, as homogeneous in mineral composition throughout equally extensive regions, it may require convulsions as numerous as all those which have occurred since the origin of the Chalk to bring them up within the sphere of human observation. Hence the rocks of more modern periods may appear partial, as compared to those of remoter eras, not because of any original inferiority in their extent, but because there has not been sufficient time since their origin for the development of a great series of elevatory movements.

In regard, however, to one of the most important characteristics of sedimentary rocks, their organic remains, many naturalists of high authority have maintained that the same species of fossils are more uniformly distributed through formations of high antiquity than in those of more modern date, and that distinct zoological and botanical provinces, as they are called, which form so striking a feature in the living creation, were not established at remote eras. Thus the plants of the Coal, the shells, the trilobites of the Silurian rocks, and the ammonites of the Oolite, have been supposed to have a wider geographical range than any living species of plants, crustaceans, or mollusks. This opinion seems in certain cases to be well founded, especially in relation to the plants of the Carboniferous epoch, owing probably to the more uniform temperature of the globe, at a time when the position of sea and land was less favorable to variations in climate, according to principles already explained in the seventh and eighth chapters. But a recent comparison of the fossils of North American rocks with those of corresponding ages in the European series, has proved that the terrestrial vegetation of the Carboniferous epoch is an exception to the general rule, and that the fauna and flora of the earth at successive periods, from the oldest Silurian to the newest Tertiary was as diversified as now. The shells, corals, and other classes of organic remains demonstrate the fact that the earth might then have been divided into separate zoological provinces, in a manner analogous to that observed in the geographical distribution of species now living.

CHAPTER XI.

ON THE SUPPOSED FORMER INTENSITY OF THE IGNEOUS FORCES.

Volcanic action at successive geological periods—Plutonic rocks of different ages—Gradual development of subterranean movements—Faults—Doctrine of the sudden upheaval of parallel mountain-chains—Objections to the proof of the suddenness of the upheaval, and the contemporaneousness of parallel chains—Trains of active volcanoes not parallel—As large tracts of land are rising or sinking slowly, so narrow zones of land may be pushed up gradually to great heights—Bending of strata by lateral pressure—Adequacy of the volcanic power to effect this without paroxysmal convulsions.

WHEN reasoning on the intensity of volcanic action at former periods, as well as on the power of moving water, already treated of, geologists have been ever prone to represent Nature as having been prodigal of violence and parsimonious of time. Now, although it is less easy to determine the relative ages of the volcanic than of the fossiliferous formations, it is undeniable that igneous rocks have been produced at all geological periods, or as often as we find distinct deposits marked by peculiar animal and vegetable remains. It can be shown that rocks commonly called trappean have been injected into fissures, and ejected at the surface, both before and during the deposition of the Carboniferous series, and at the time when the Magnesian Limestone, and when the Upper New Red Sandstone were formed, or when the Lias, Oolite, Green Sand, Chalk, and the several tertiary groups newer than the chalk, originated in succession. Nor is this all: distinct volcanic products may be referred to the subordinate divisions of each period, such as the Carboniferous, as in the county of Fife, in Scotland, where certain masses of contemporaneous trap are associated with the Lower, others with the Upper Coal measures. And if one of these masses is more minutely examined, we find it to consist of the products of a great many successive outbursts, by which scoriæ and lava were again and again emitted, and afterwards consolidated, then fissured, and finally traversed by melted matter, constituting what are called dikes.* As we enlarge, therefore, our knowledge of the ancient rocks formed by

* See Manual of Geology, chap. 29 to 33, inclusive.

subterranean heat, we find ourselves compelled to regard them as the aggregate effects of innumerable eruptions, each of which may have been comparable in violence to those now experienced in volcanic regions.

It may indeed be said that we have as yet no data for estimating the relative volume of matter simultaneously in a state of fusion at two given periods, as if we were to compare the columnar basalt of Staffa and its environs with the lava poured out in Iceland in 1783; but for this very reason it would be rash and unphilosophical to assume an excess of ancient as contrasted with modern outpourings of melted matter at particular periods of time.* It would be still more presumptuous to take for granted that the more deep-seated effects of subterranean heat surpassed at remote eras the corresponding effects of internal heat in our own times. Certain porphyries and granites, and all the rocks commonly called plutonic, are now generally supposed to have resulted from the slow cooling of materials fused and solidified under great pressure; and we cannot doubt that beneath existing volcanoes there are large spaces filled with melted stone, which must for centuries remain in an incandescent state, and then cool and become hard and crystalline when the subterranean heat shall be exhausted. That lakes of lava are continuous for hundreds of miles beneath the Chilian Andes, seems established by observations made in the year 1835.†

Now, wherever the fluid contents of such reservoirs are poured out successively from craters in the open air, or at the bottom of the sea, the matter so ejected may afford evidence by its arrangement of having originated at different periods; but if the subterranean residue after the withdrawal of the heat be converted into crystalline or plutonic rock, the entire mass may seem to have been formed at once, however countless the ages required for its fusion and subsequent refrigeration. As the idea that all the granite in the earth's crust was produced simultaneously, and in a primitive state of the planet, has now been universally abandoned; so the suggestion above adverted to, may put us on our guard against too readily adopting another opinion, namely, that each large mass of granite was generated in a brief period of time.

Modern writers indeed, of authority, seem more and more agreed that in the case of granitic rocks, the passage from a liquid or pasty to a solid and crystalline state must have been an extremely gradual process.

The doctrine so much insisted upon formerly, that crystalline rocks, such as granite, gneiss, mica-schist, quartzite, and others were produced in the greatest abundance in the earlier ages of the planet, and that their formation has ceased altogether in our own times, will be controverted in the next chapter.

Gradual development of subterranean movements.—The extreme violence of the subterranean forces in remote ages has been often inferred from the facts that the older rocks are more fractured and dislocated

* See ch. 26, *infrà*. † See ch. 27, *infrà*.

than the newer. But what other result could we have anticipated if the quantity of movement had been always equal in equal periods of time? Time must, in that case, multiply the derangement of strata in the ratio of their antiquity. Indeed the numerous exceptions to the above rule which we find in nature, present at first sight the only objection to the hypothesis of uniformity. For the more ancient formations remain in many places horizontal, while in others much newer strata are curved and vertical. This apparent anomaly, however, will be seen in the next chapter to depend on the irregular manner in which the volcanic and subterranean agency affect different parts of the earth in succession, being often renewed again and again in certain areas, while others remain during the whole time at rest.

That the more impressive effects of subterranean power, such as the upheaval of mountain-chains, may have been due to multiplied convulsions of moderate intensity rather than to a few paroxysmal explosions, will appear the less improbable when the gradual and intermittent development of volcanic eruptions in times past is once established. It is now very generally conceded that these eruptions have their source in the same causes as those which give rise to the permanent elevation and sinking of land; the admission, therefore, that one of the two volcanic or subterranean processes has gone on gradually, draws with it the conclusion that the effects of the other have been elaborated by successive and gradual efforts.

Faults.—The same reasoning is applicable to great *faults*, or those striking instances of the upthrow or downthrow of large masses of rock, which have been thought by some to imply tremendous catastrophes wholly foreign to the ordinary course of nature. Thus we have in England faults, in which the vertical displacement is between 600 and 3000 feet, and the horizontal extent thirty miles or more, the width of the fissures since filled up with rubbish varying from ten to fifty feet. But when we inquire into the proofs of the mass having risen or fallen suddenly on the one side of these great rents, several hundreds or thousands of feet above or below the rock with which it was once continuous on the other side, we find the evidence defective. There are grooves, it is said, and scratches on the rubbed and polished walls, which have often one common direction, favoring the theory that the movement was accomplished by a single stroke, and not by a series of interrupted movements. But, in fact, the striæ are not always parallel in such cases, but often irregular, and sometimes the stones and earth which are in the middle of the fault, or fissure, have been polished and striated by friction in different directions, showing that there have been slidings subsequent to the first introduction of the fragmentary matter. Nor should we forget that the last movement must always tend to obliterate the signs of previous trituration, so that neither its instantaneousness nor the uniformity of its direction can be inferred from the parallelism of the striæ that have been last produced.

When rocks have been once fractured, and freedom of motion communicated to detached portions of them, these will naturally continue to yield in the same direction, if the process of upheaval or of undermining be repeated again and again. The incumbent mass will always give way along the lines of least resistance, or where it was formerly rent asunder. Probably, the effects of reiterated movement, whether upward or downward, in a fault, may be undistinguishable from those of a single and instantaneous rise or subsidence; and the same may be said of the rising or falling of continental masses, such as Sweden or Greenland, which we know to take place slowly and insensibly.

Doctrine of the sudden upheaval of parallel mountain-chains.—The doctrine of the suddenness of many former revolutions in the physical geography of the globe has been thought by some to derive additional confirmation from a theory respecting the origin of mountain-chains, advanced in 1833 by a distinguished geologist, M. Elie de Beaumont. In several essays on this subject, the last published in 1852, he has attempted to establish two points; first, that a variety of independent chains of mountains have been thrown up suddenly at particular periods; and, secondly, that the contemporaneous chains thus thrown up, preserve a parallelism the one to the other.

These opinions, and others by which they are accompanied, are so adverse to the method of interpreting the history of geological changes which I have recommended in this work, that I am desirous of explaining the grounds of my dissent, a course which I feel myself the more called upon to adopt, as the generalizations alluded to are those of a skilful writer, and an original observer of great talent and experience. I shall begin, therefore, by giving a brief summary of the principal propositions laid down in the works above referred to.*

1st. M. de Beaumont supposes "that in the history of the earth there have been long periods of comparative repose, during which the deposition of sedimentary matter has gone on in regular continuity; and there have also been short periods of paroxysmal violence, during which that continuity was broken.

"2dly. At each of these periods of violence or 'revolution,' in the state of the earth's surface, a great number of mountain-chains have been formed suddenly.

"3dly. The chains thrown up by a particular revolution have one uniform direction, being parallel to each other within a few degrees of the compass, even when situated in remote regions; whilst the chains thrown up at different periods have, for the most part, different directions.

"4thly. Each 'revolution,' or 'great convulsion,' has fallen in with the date of another geological phenomenon; namely, 'the passage from

* Ann. des Sci. Nat., Septembre, Novembre, et Décembre, 1829. Revue Française, No. 15, May, 1830. Bulletin de la Société Géol. de France, p. 864, May, 1847. The latest edition of M. de Beaumont's theory will be found in the 12th vol. of the Dictionnaire Universel d'Hist. Nat. 1852, art. "Systèmes des Montagnes;" also the same printed separately.

one independent sedimentary formation to another,' characterized by a considerable difference in 'organic types.'

"5thly. There has been a recurrence of these paroxysmal movements from the remotest geological periods; and they may still be reproduced, and the repose in which we live may hereafter be broken by the sudden upthrow of another system of parallel chains of mountains.

"6thly. The origin of these chains depends not on partial volcanic action, or a reiteration of ordinary earthquakes, but on the secular refrigeration of the entire planet. For the whole globe, with the exception of a thin envelope, much thinner in proportion than the shell to an egg, is a fused mass, kept fluid by heat, but constantly cooling and contracting its dimensions. The external crust does not gradually collapse and accommodate itself century after century to the shrunken nucleus, subsiding as often as there is a slight failure of support, but it is sustained throughout whole geological periods, so as to become partially separated from the nucleus, until at last it gives way suddenly, cracking and falling in along determinate lines of fracture. During such a crisis the rocks are subjected to great lateral pressure, the unyielding ones are crushed, and the pliant strata bent, and are forced to pack themselves more closely into a smaller space, having no longer the same room to spread themselves out horizontally. At the same time, a large portion of the mass is squeezed upwards, because it is in the upward direction only that the excess in size of the envelope, as compared to the contracted nucleus, can find relief. This excess produces one or more of those folds or wrinkles in the earth's crust which we call mountain-chains.

"Lastly, some chains are comparatively modern; such as the Alps, which were partly upheaved after the middle tertiary period. The elevation of the Andes was much more recent, and was accompanied by the simultaneous outburst for the first time of 270 of the principal volcanoes now active.*

"The agitation of the waters of the ocean caused by this convulsion probably occasioned that transient and general deluge which is noticed in the traditions of so many nations."†

Several of the topics enumerated in the above summary, such as the cause of interruptions in the sedimentary series, will be discussed in the thirteenth chapter, and I shall now confine myself to what I conceive to be the insufficiency of the proofs adduced in favor of the suddenness of the upthrow, and the contemporaneousness of the origin of the parallel chains referred to. At the same time I may remark, that the great body of facts collected together by M. de Beaumont will always form a most valuable addition to our knowledge, tending as they do to confirm the doctrine that different mountain-chains have been formed in succession, and, as Werner first pointed out, that there are certain determinate lines of direction or strike in the strata of various countries.

* Système de Mont. p. 762.

† Ibid. pp. 761 and 773.

The following may serve as an analysis of the evidence on which the theory above stated depends. "We observe," says M. de Beaumont, "when we attentively examine nearly all mountain-chains, that the most recent rocks extend horizontally up to the foot of such chains, as we should expect would be the case if they were deposited in seas or lakes, of which these mountains have partly formed the shores; whilst the other sedimentary beds, tilted up, and more or less contorted, on the flanks of the mountains, rise in certain points even to their highest crests."* There are, therefore, in and adjacent to each chain, two classes of sedimentary rocks, the ancient and inclined beds, and the newer or horizontal. It is evident that the first appearance of the chain itself was an event "intermediate between the period when the beds now upraised were deposited, and the period when the strata were produced horizontally at its feet."

Fig. 11.

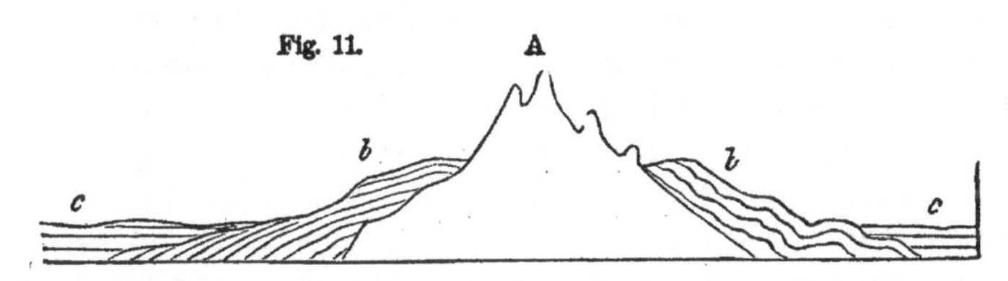

Thus the chain A assumed its present position after the deposition of the strata *b*, which have undergone great movements, and before the deposition of the group *c*, in which the strata have not suffered derangement.

If we then discover another chain B, in which we find not only the

Fig. 12.

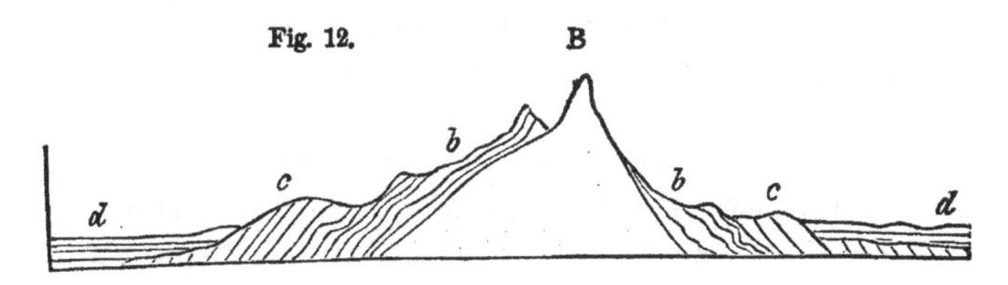

formation *b*, but the group *c* also, disturbed and thrown on its edges, we may infer that the latter chain is of subsequent date to A; for B must have been elevated *after* the deposition of *c*, and before that of the group *d*; whereas A had originated *before* the strata *c* were formed.

It is then argued, that in order to ascertain whether other mountain ranges are of contemporaneous date with A and B, or are referable to *distinct* periods, we have only to inquire whether the inclined and undisturbed sets of strata in each range correspond with or differ from those in the typical chain A and B.

Now all this reasoning is perfectly correct, so long as the period of time required for the deposition of the strata *b* and *c* is not made iden-

* Phil. Mag. and Annals, No. 58. New Series, p. 242.

tical in duration with the period of time during which the animals and plants found fossil in *b* and *c* may have flourished; for the latter, that is to say, the duration of certain groups of species, may have greatly exceeded, and probably did greatly exceed, the former, or the time required for the accumulation of certain local deposits, such as *b* and *c* (figs. 11 and 12). In order, moreover, to render the reasoning correct, due latitude must be given to the term contemporaneous; for this term must be understood to allude, not to a moment of time, but to the interval, whether brief or protracted, which elapsed between two events, namely, between the accumulation of the inclined and that of the horizontal strata.

But, unfortunately, no attempt has been made in the treatises under review to avoid this manifest source of confusion, and hence the very terms of each proposition are equivocal; and the possible length of some of the intervals is so vast, that to affirm that all the chains raised in such intervals were *contemporaneous* is an abuse of language.

In order to illustrate this argument, I shall select the Pyrenees as an example. Originally M. E. de Beaumont spoke of this range of mountains as having been uplifted suddenly (*à un seul jet*), but he has since conceded that in this chain, in spite of the general unity and simplicity of its structure, six, if not seven, systems of dislocation of different dates can be recognized.* In reference, however, to the latest, and by far the most important of these convulsions, the chain is said to have attained its present elevation at a certain epoch in the earth's history, namely, between the deposition of the chalk, or rocks of about that age, and that of certain tertiary formations "as old as the plastic clay;" for the chalk is seen in vertical, curved, and distorted beds on the flanks of the chain, as the beds *b*, fig. 11, while the tertiary formations rest upon them in horizontal strata at its base, as *c*, ibid.

The proof, then, of the extreme suddenness of the convulsion is supposed to be the shortness of the time which intervened between the formation of the chalk and the origin of certain tertiary strata.† Even if the interval were deducible within these limits, it might comprise an indefinite lapse of time. In strictness of reasoning, however, the author cannot exclude the Cretaceous or Tertiary periods from the possible duration of the interval during which the elevation may have taken place. For, in the first place, it cannot be assumed that the movement of upheaval took place after the close of the Cretaceous period; we can merely say, that it occurred after the deposition of certain strata of that period; secondly, although it were true that the event happened before the formation of all the tertiary strata now at the base of the Pyrenees, it would by no means follow that it preceded the whole Tertiary epoch.

The age of the strata, both of the inclined and horizontal series, may

* Système de Montagnes, 1852, p. 429.
† Phil. Mag. and Annals, No. 58. New series, p. 243.

have been accurately determined by M. De Beaumont, and still the upheaving of the Pyrenees may have been going on before the animals of the Chalk period, such as are found fossil in England, had ceased to exist, or when the Maestricht beds were in progress, or during the indefinite ages which may have elapsed between the extinction of the Maestricht animals and the introduction of the Eocene tribes, or during the Eocene epoch, or the rise may have been going on throughout one, or several, or all of these periods.

It would be a purely gratuitous assumption to say that the inclined cretaceous strata (*b*, fig. 11) on the flanks of the Pyrenees, were the very last which were deposited during the Cretaceous period, or that, as soon as they were upheaved, all or nearly all the species of animals and plants now found fossil in them were suddenly exterminated; yet, unless this can be affirmed, we cannot say that the Pyrenees were not upheaved during the Cretaceous period. Consequently, another range of mountains, at the base of which cretaceous rocks may lie in horizontal stratification, may have been elevated, like the chain A, fig. 12, during some part of the same great period.

There are mountains in Sicily two or three thousand feet high, the tops of which are composed of limestone, in which a large proportion of the fossil shells agree specifically with those now inhabiting the Mediterranean. Here, as in many other countries, the deposits now in progress in the sea must inclose shells and other fossils specifically identical with those of the rocks constituting the contiguous land. So there are islands in the Pacific where a mass of dead coral has emerged to a considerable altitude, while other portions of the mass remain beneath the sea, still increasing by the growth of living zoophytes and shells. The chalk of the Pyrenees, therefore, may at a remote period have been raised to an elevation of several thousand feet, while the species found fossil in the same chalk still continued to be represented in the fauna of the neighboring ocean. In a word, we cannot assume that the origin of a new range of mountains caused the Cretaceous period to cease, and served as the prelude to a new order of things in the animate creation.

To illustrate the grave objections above advanced, against the theory considered in the present chapter, let us suppose, that in some country three styles of architecture had prevailed in succession, each for a period of one thousand years; first the Greek, then the Roman, and then the Gothic; and that a tremendous earthquake was known to have occurred in the same district during one of the three periods—a convulsion of such violence as to have levelled to the ground all the buildings then standing. If an antiquary, desirous of discovering the date of the catastrophe, should first arrive at a city where several Greek temples were lying in ruins and half engulphed in the earth, while many Gothic edifices were standing uninjured, could he determine on these data the era of the shock? Could he even exclude any one of the three periods, and decide that it must have happened during one of the other two? Certainly not. He could merely affirm that it happened at some period after the introduction of the Greek style, and before the Gothic had fallen into disuse. Should he pretend to define the date of the convulsion with greater precision, and decide that the earthquake must have occurred after the Greek and before the Gothic period, that is to say, when the Roman style was in use, the fallacy in his reasoning would be too palpable to escape detection for a moment.

Yet such is the nature of the erroneous induction which I am now exposing. For as, in the example above proposed, the erection of a particular edifice is perfectly distinct from the period of architecture in which it may have been raised, so is the deposition of chalk, or any other set of strata, from the geological epochs characterized by certain fossils to which they may belong.

It is almost superfluous to enter into any farther analysis of the theory of parallelism, because the whole force of the argument depends on the accuracy of the data by which the contemporaneous or non-contemporaneous date of the elevation of two independent chains can be demonstrated. In every case, this evidence, as stated by M. de Beaumont, is equivocal, because he has not included in the possible interval of time between the depositions of the deranged and the horizontal formations, part of the periods to which each of those classes of formations are referable. Even if all the geological facts, therefore, adduced by the author were true and unquestionable, yet the conclusion that certain chains were or were not simultaneously upraised is by no means a legitimate consequence.

In the third volume of my first edition of the Principles, which appeared in April, 1833, I controverted the views of M. de Beaumont, then just published, in the same terms as I have now restated them. At that time I took for granted that the chronological date of the newest rocks entering into the disturbed series of the Pyrenees had been correctly ascertained. It now appears, however, that some of the most modern of those disturbed strata belong to the nummulitic formation, which are regarded by the majority of geologists as Eocene or older tertiary, an opinion not assented to by M. E. de Beaumont, and which I cannot discuss here without being led into too long a digression.*

Perhaps a more striking illustration of the difficulties we encounter, when we attempt to apply the theory under consideration even to the best known European countries, is afforded by what is called "The System of the Longmynds." This small chain, situated in Shropshire, is the third of the typical systems to which M. E. de Beaumont compares other mountain ranges corresponding in *strike* and structure. The date assigned to its upheaval is "after the unfossiliferous greywacke, or Cambrian strata, and before the Silurian." But Sir R. I. Murchison had shown in 1838, in his "Silurian System," and the British government surveyors, since that time, in their sections (about 1845), that the Longmynds and other chains of similar composition in North Wales are *post-*

* Système de Montagnes, 1852, p. 429.

Silurian. In all of them fossiliferous beds of the lower Silurian formation, or Llandeilo flags are highly inclined, and often vertical. In one limited region the Caradoc sandstone, a member of the lower Silurian, rests unconformably on the denuded edges of the inferior (or Llandeilo) member of the same group; whilst in some cases both of these sets of strata are upturned. When, therefore, so grave an error is detected in regard to the age of a typical chain, we are entitled to inquire with surprise, by what means nine other *parallel* chains in France, Germany, and Sweden, assumed to be "ante-Silurian," have been made to agree precisely in date with the Longmynds? If they are correctly represented as having been all deposited before the deposition of the Silurian strata, they cannot be contemporaneous with the Longmynds, and they only prove how little reliance can be placed on parallelism as a test of simultaneousness of upheaval. But in truth it is impossible, for reasons already given, to demonstrate that each of those nine chains coincide in date with one another, any more than with the Longmynds.

The reader will see in the sequel (chap. 31*) that Mr. Hopkins has inferred from astronomical calculations, that the solid crust of the earth cannot be less than 800 or 1000 miles thick, and may be more. Even if it be solid to the depth of 100 miles, such a thickness would be inconsistent with M. E. de Beaumont's hypothesis, which requires a shell not more than thirty miles thick, or even less. Mr. Hopkins admits that the exterior of the planet, though solid as a whole, may contain within it vast lakes or seas of lava. If so, the gradual fusion of rocks, and the expansive power of heat exerted for ages, as well as the subsequent contraction of the same during slow refrigeration, may perhaps account for the origin of mountain-chains, for these, as Dolomieu has remarked, are "far less important, proportionally speaking, than the inequalities on the surface of an egg-shell, which to the eye appears smooth." A "centripetal force" affecting the whole planet as it cools, seems a mightier cause than is required to produce wrinkles of such insignificant size.

In pursuing his investigations, M. E. de Beaumont has of late greatly multiplied the number of successive periods of instantaneous upheaval, admitting at the same time that occasionally new lines of upthrow have taken the direction of older ones.† These admissions render his views much more in harmony with the principles advocated in this work, but they impair the practical utility of parallelism considered as a chronological test; for no rule is laid down for limiting the interval, whether in time or space, which may separate two parallel lines of upheaval of different dates.‡

* For page, see Index, "Hopkins." † Art. Système de Montagnes, p. 775.

‡ M. E. de Beaumont in his later inquiries (Comptes rendus, Sept. 1850, and Systèmes des Montagnes) has come to the conclusion, that the principal mountain ranges, if prolonged, would intersect each other at certain angles, so as to produce a regular geometric arrangement, which he calls "a pentagonal network." This theory has been ably discussed and controverted by Mr. Hopkins, in his Anniversary Address as President of the Geol. Soc., Feb. 1853.

Among the various propositions above laid down (p. 164), it will be seen that the sudden rise of the Andes is spoken of as a modern event, but Mr. Darwin has brought together ample data in proof of the local persistency of volcanic action throughout a long succession of geological periods, beginning with times antecedent to the deposition of the oolitic and cretaceous formations of Chili, and continuing to the historical epoch. It appears that some of the parallel ridges which compose the Cordilleras, instead of being contemporaneous, were successively and slowly upheaved at widely different epochs. The whole range, after twice subsiding some thousands of feet, was brought up again by a slow movement in mass, during the era of the Eocene tertiary formations, after which the whole sank down once more several hundred feet, to be again uplifted to its present level by a slow and often interrupted movement.* In a portion of this latter period the "Pampean mud" was formed, in which the Megatherium mylodon and other extinct quadrupeds are buried. This mud contains in it recent species of shells, some of them proper to brackish water, and is believed by Mr. Darwin to be an estuary or delta deposit. M. A. d'Orbigny, however, has advanced an hypothesis referred to by M. E. de Beaumont, that the agitation and displacement of the waters of the ocean, caused by the elevation of the Andes, gave rise to a deluge, of which this Pampean mud, which rises sometimes to the height of 12,000 feet, is the result and monument.†

In studying many chains of mountains, we find that the strike or line of outcrop of continuous sets of strata, and the general direction of the chain, may be far from rectilinear. Curves forming angles of 20° or 30° may be found in the same range as in the Alleghanies; just as trains of active volcanoes and the zones throughout which modern earthquakes occur are often linear, without running in straight lines. Nor are all of these, though contemporaneous or belonging to our own epoch, by any means parallel, but some at right angles, the one to the other.

Slow upheaval and subsidence.—Recent observations have disclosed to us the wonderful fact, that not only the west coast of South America, but also other large areas, some of them several thousand miles in circumference, such as Scandinavia, and certain archipelagoes in the Pacific, are slowly and insensibly rising; while other regions, such as Greenland, and parts of the Pacific and Indian Oceans, in which atolls or circular coral islands abound, are as gradually sinking. That all the existing continents and submarine abysses may have originated in movements of this kind, continued throughout incalculable periods of time, is undeniable, and the denudation which the dry land appears everywhere to have suffered, favors the idea that it was raised from the deep by a succession of upward movements, prolonged throughout indefinite periods. For the action of waves and currents on land slowly emerging from the deep, affords the only power by which we can con-

* Darwin's Geology of South America, p. 248. London, 1846.
† Système de Montagnes, p. 748.

ceive so many deep valleys and wide spaces to have been denuded as those which are unquestionably the effects of running water.

But perhaps it may be said that there is no analogy between the slow upheaval of broad plains or table-lands, and the manner in which we must presume all mountain-chains, with their inclined strata, to have originated. It seems, however, that the Andes have been rising century after century, at the rate of several feet, while the Pampas on the east have been raised only a few inches in the same time. Crossing from the Atlantic to the Pacific, in a line passing through Mendoza, Mr. Darwin traversed a plain 800 miles broad, the eastern part of which has emerged from beneath the sea at a very modern period. The slope from the Atlantic is at first very gentle, then greater, until the traveller finds, on reaching Mendoza, that he has gained, almost insensibly, a height of 4000 feet. The mountainous district then begins suddenly, and its breadth from Mendoza to the shores of the Pacific is 120 miles, the average height of the principal chain being from 15,000 to 16,000 feet, without including some prominent peaks, which ascend much higher. Now all we require, to explain the origin of the principal inequalities of level here described, is to imagine, first, a zone of more violent movement to the west of Mendoza, and, secondly, to the east of that place, an upheaving force, which died away gradually as it approached the Atlantic. In short, we are only called upon to conceive, that the region of the Andes was pushed up four feet in the same period in which the Pampas near Mendoza rose one foot, and the plains near the shores of the Atlantic one inch. In Europe we have learnt that the land at the North Cape ascends about five feet in a century, while farther to the south the movements diminish in quantity first to a foot, and then, at Stockholm, to three inches in a century, while at certain points still farther south there is no movement.

But in what manner, it is asked, can we account for the great lateral pressure which has been exerted not only in the Andes, Alps, and other chains, but also on the strata of many low and nearly level countries? Do not the folding and fracture of the beds, the anticlinal and synclinal ridges and troughs, as they are called, and the vertical, and even sometimes the inverted position of the beds, imply an abruptness and intensity in the disturbing force wholly different in kind and energy to that which now rends the rocks during ordinary earthquakes? I shall treat more fully in the sequel (end of chap. 32) of the probable subterranean sources, whether of upward or downward movement, and of great lateral pressure; but it may be well briefly to state in this place that in our own times, as, for example, in Chili, in 1822, the volcanic force has overcome the resistance, and permanently uplifted a country of such vast extent that the weight and volume of the Andes must be insignificant in comparison, even if we indulge the most moderate conjectures as to the thickness of the earth's crust above the volcanic foci.

To assume that any set of strata with which we are acquainted are made up of such cohesive and unyielding materials, as to be able to resist a power of such stupendous energy, if its direction, instead of being vertical, happened to be oblique or horizontal, would be extremely rash. But if they could yield to a sideway thrust, even in a slight degree, they would become squeezed and folded to any amount if subjected for a sufficient number of times to the repeated action of the same force. We can scarcely doubt that a mass of rock several miles thick was uplifted in Chili in 1822 and 1835, and that a much greater volume of solid matter is upheaved wherever the rise of the land is very gradual, as in Scandinavia, the development of heat being probably, in that region, at a greater distance from the surface. If continents, rocked, shaken, and fissured, like the western region of South America, or very gently elevated, like Norway and Sweden, do not acquire in a few days or hours an additional height of several thousand feet, this can arise from no lack of mechanical force in the subterranean moving cause, but simply because the antagonist power, or the strength, toughness, and density of the earth's crust is insufficient to resist, so long, as to allow the volcanic energy an indefinite time to accumulate. Instead of the explosive charge augmenting in quantity for countless ages, it finds relief continuously, or by a succession of shocks of moderate violence, so as never to burst or blow up the covering of incumbent rock in one grand paroxysmal convulsion. Even in its most energetic efforts it displays an intermittent and mitigated intensity, being never permitted to lay a whole continent in ruins. Hence the numerous eruptions of lava from the same vent, or chain of vents, and the recurrence of similar earthquakes for thousands of years along certain areas or zones of country. Hence the numerous monuments of the successive ejection and injection of melted matter in ancient geological epochs, and the fissures formed in distinct ages, and often widened and filled at different eras.

Among the causes of lateral pressure, the expansion by heat of large masses of solid stone intervening between others which have a different degree of expansibility, or which happen not to have their temperature raised at the same time, may play an important part. But as we know that rocks have so often sunk down thousands of feet below their original level, we can hardly doubt that much of the bending of pliant strata, and the packing of the same into smaller spaces, has frequently been occasioned by subsidence. Whether the failure of support be produced by the melting of porous rocks, which, when fluid, and subjected to great pressure, may occupy less room than before, or which, by passing from a pasty to a crystalline condition, may, as in the case of granite, according to the experiments of Deville, suffer a contraction of 10 per cent., or whether the sinking be due to the subtraction of lava driven elsewhere to some volcanic orifice, and there forced outwards, or whether it be brought on by the shrinking of solid and stony masses during refrigeration, or by the condensation of gases, or any other imaginable cause, we have no reason to incline to the idea that the consequent geological changes are brought about so suddenly, as that large parts of continents are swallowed up at once in unfathomable subterranean abysses. If cavities be formed, they will be enlarged gradually, and as gradually filled. We read, indeed, accounts of engulphed cities and areas of limited extent which have sunk down many yards at once; but we have as yet no authentic records of the sudden disappearance of mountains, or the submergence or emergence of great islands. On the other hand, the creeps in coal mines* demonstrate that gravitation begins to act as soon as a moderate quantity of matter is removed even at a great depth. The roof sinks in, or the floor of the mine rises, and the bent strata often assume as regularly a curved and crumpled arrangement as that observed on a grander scale in mountain-chains. The absence, indeed, of chaotic disorder, and the regularity of the plications in geological formations of high antiquity, although not unfrequently adduced to prove the unity and instantaneousness of the disturbing force, might with far greater propriety be brought forward as an argument in favor of the successive application of some irresistible but moderated force, such as that which can elevate or depress a continent.

In conclusion, I may observe that one of the soundest objections to the theory of the sudden upthrow or downthrow of mountain-chains is this, that it provides us with too much force of one kind, namely, that of subterranean movement, while it deprives us of another kind of mechanical force, namely, that exerted by the waves and currents of the ocean, which the geologist requires for the denudation of land during its slow upheaval or depression. It may be safely affirmed that the quantity of igneous and aqueous action,—of volcanic eruption and denudation,—of subterranean movement and sedimentary deposition,—not only of past ages, but of one geological epoch, or even the fraction of an epoch, has exceeded immeasurably all the fluctuations of the inorganic world which have been witnessed by man. But we have still to inquire whether the time to which each chapter or page or paragraph of the earth's autobiography relates, was not equally immense when contrasted with a brief era of 3000 or 5000 years. The real point on which the whole controversy turns, is the relative amount of work done by mechanical force in given quantities of time, past and present. Before we can determine the relative intensity of the force employed, we must have some fixed standard by which to measure the time expended in its development at two distinct periods. It is not the magnitude of the effects, however gigantic their proportions, which can inform us in the slightest degree whether the operation was sudden or gradual, insensible or paroxysmal. It must be shown that a slow process could never in any series of ages give rise to the same results.

The advocate of paroxysmal energy might assume a uniform and fixed rate of variation in times past and present for the animate world, that is to say, for the dying-out and coming-in of species, and then en-

* See Lyell's Manual of Elementary Geology, ch. 5.

deavor to prove that the changes of the inanimate world have not gone on in a corresponding ratio. But the adoption of such a standard of comparison would lead, I suspect, to a theory by no means favorable to the pristine intensity of natural causes. That the present state of the organic world is not stationary, can be fairly inferred from the fact, that some species are known to have become extinct in the course even of the last three centuries, and that the exterminating causes always in activity, both on the land and in the waters, are very numerous; also, because man himself is an extremely modern creation; and we may therefore reasonably suppose that some of the mammalia now contemporary with man, as well as a variety of species of inferior classes, may have been recently introduced into the earth, to supply the places of plants and animals which have from time to time disappeared. But granting that some such secular variation in the zoological and botanical worlds is going on, and is by no means wholly inappreciable to the naturalist, still it is certainly far less manifest than the revolution always in progress in the inorganic world. Every year some volcanic eruptions take place, and a rude estimate might be made of the number of cubic feet of lava and scoriæ poured or cast out of various craters. The amount of mud and sand deposited in deltas, and the advance of new land upon the sea, or the annual retreat of wasting sea-cliffs, are changes the minimum amount of which might be roughly estimated. The quantity of land raised above or depressed below the level of the sea might also be computed, and the change arising from such movements in a century might be conjectured. Suppose the average rise of the land in some parts of Scandinavia to be as much as five feet in a hundred years, the present sea-coast might be uplifted 700 feet in fourteen thousand years; but we should have no reason to anticipate, from any zoological data hitherto acquired, that the molluscous fauna of the northern seas would in that lapse of years undergo any sensible amount of variation. We discover sea-beaches in Norway 700 feet high, in which the shells are identical with those now inhabiting the German Ocean; for the rise of land in Scandinavia, however insensible to the inhabitants, has evidently been rapid when compared to the rate of contemporaneous change in the testaceous fauna of the German Ocean. Were we to wait therefore until the mollusca shall have undergone as much fluctuation as they underwent between the period of the Lias and the Upper Oolite formations, or between the Oolite and Chalk, nay, even between any two of eight subdivisions of the Eocene series, what stupendous revolutions in physical geography ought we not to expect, and how many mountain-chains might not be produced by the repetition of shocks of moderate violence, or by movements not even perceptible by man!

Or, if we turn from the mollusca to the vegetable kingdom, and ask the botanist how many earthquakes and volcanic eruptions might be expected, and how much the relative level of land and sea might be altered, or how far the principal deltas will encroach upon the ocean, or the sea-cliffs recede from the present shores, before the species of

European forest-trees will die out, he would reply that such alterations in the inanimate world might be multiplied indefinitely before he should have reason to anticipate, by reference to any known data, that the existing species of trees in our forests would disappear and give place to others. In a word, the movement of the inorganic world is obvious and palpable, and might be likened to the minute-hand of a clock, the progress of which can be seen and heard, whereas the fluctuations of the living creation are nearly invisible, and resemble the motion of the hour-hand of a timepiece. It is only by watching it attentively for some time, and comparing its relative position after an interval, that we can prove the reality of its motion.*

CHAPTER XII.

DIFFERENCE IN TEXTURE OF THE OLDER AND NEWER ROCKS.

Consolidation of fossiliferous strata—Some deposits originally solid—Transition and slaty texture—Crystalline character of Plutonic and Metamorphic rocks—Theory of their origin—Essentially subterranean—No proofs that they were produced more abundantly at remote periods.

ANOTHER argument in favor of the dissimilarity of the causes operating at remote and recent eras has been derived by many geologists from the more compact, stony, and crystalline texture of the older as compared with the newer rocks.

Consolidation of strata.—This subject may be considered, first in reference to the fossiliferous strata; and, secondly, in reference to those crystalline and stratified rocks which contain no organic remains, such as gneiss and mica-schist. There can be no doubt that the former of these classes, or the fossiliferous, are generally more compact and stony in proportion as they are more ancient. It is also certain that a great part of them were originally in a soft and incoherent state, and that they have been since consolidated. Thus we find occasionally that shingle and sand have been agglutinated firmly together by a ferruginous or siliceous cement, or that lime in solution has been introduced, so as to bind together materials previously incoherent. Organic remains have sometimes suffered a singular transformation, as for ex-

* See the Author's Anniversary Address, Quart. Journ. Geol. Soc. 1850, vol. vi. p. 46, from which some of the above passages are extracted.

ple, where shells, corals, and wood are silicified, their calcareous or ligneous matter having been replaced by nearly pure silica. The constituents of some beds have probably set and become hard for the first time when they emerged from beneath the water.

But, on the other hand, we observe in certain formations now in progress, particularly in coral reefs, and in deposits from the waters of mineral springs, both calcareous and siliceous, that the texture of rocks may sometimes be stony from the first. This circumstance may account for exceptions to the general rule, not unfrequently met with, where solid strata are superimposed on others of a plastic and incoherent nature, as in the neighborhood of Paris, where the tertiary formations, consisting often of compact limestone and siliceous grit, are more stony than the subjacent chalk.

It will readily be understood, that the various solidifying causes, including those above enumerated, together with the pressure of incumbent rocks and the influence of subterranean heat, must all of them require time in order to exert their full power. If in the course of ages they modify the aspect and internal structure of stratified deposits, they will give rise to a general distinctness of character in the older as contrasted with the newer formations. But this distinctness will not be the consequence of any original diversity; they will be unlike, just as the wood in the older trees of a forest usually differs in texture and hardness from that of younger individuals of the same species.

Transition texture.—In the original classification of Werner, the highly crystalline rocks, such as granite and gneiss, which contain no organic remains, were called primary, and the fossiliferous strata secondary, while to another class of an age intermediate between the primary and secondary he gave the name of transition. They were termed transition because they partook in some degree in their mineral composition of the nature of the most crystalline rocks, such as gneiss and mica-schist, while they resembled the fossiliferous series in containing occasionally organic remains, and exhibiting evident signs of a mechanical origin. It was at first imagined, that the rocks having this intermediate texture had been all deposited subsequently to the series called primary, and before all the more earthy and fossiliferous formations. But when the relative position and organic remains of these transition rocks were better understood, it was perceived that they did not all belong to one period. On the contrary, the same mineral characters were found in strata of very different ages, and some formations occurring in the Alps, which several of the ablest scholars of Werner had determined to be transition, were ultimately ascertained, by means of their fossil contents and position, to be members of the Cretaceous, and even of the nummulitic or Eocene period. These strata had, in fact, acquired the *transition* texture from the influence of causes which, since their deposition, had modified their internal arrangement.

Texture and origin of Plutonic and metamorphic rocks.—Among the most singular of the changes superinduced on rocks, we have occasionally to include the slaty texture, the divisional planes of which sometimes intersect the true planes of stratification, and even pass directly through imbedded fossils. If, then, the crystalline, the slaty, and other modes of arrangement, once deemed characteristic of certain periods in the history of the earth, have in reality been assumed by fossiliferous rocks of different ages and at different times, we are prepared to inquire whether the same may not be true of the most highly crystalline state, such as that of gneiss, mica-schist, and statuary marble. That the peculiar characteristics of such rocks are really due to a variety of modifying causes has long been suspected by many geologists, and the doctrine has gained ground of late, although a considerable difference of opinion still prevails. According to the original Neptunian theory, all the crystalline formations were precipitated from a universal menstruum or chaotic fluid antecedently to the creation of animals and plants, the unstratified granite having been first thrown down so as to serve as a floor or foundation on which gneiss and other stratified rocks might repose. Afterwards, when the igneous origin of granite was no longer disputed, many conceived that a thermal ocean enveloped the globe, at a time when the first-formed crust of granite was cooling, but when it still retained much of its heat. The hot waters of this ocean held in solution the ingredients of gneiss, mica-schist, hornblende-schist, clay-slate, and marble, rocks which were precipitated, one after the other, in a crystalline form. No fossils could be inclosed in them, the high temperature of the fluid and the quantity of mineral matter which it held in solution, rendering it unfit for the support of organic beings.

It would be inconsistent with the plan of this work to enter here into a detailed account of what I have elsewhere termed the *metamorphic theory*;* but I may state that it is now demonstrable in some countries that fossiliferous formations, some of them of the age of the Silurian strata, as near Christiana in Norway, others belonging to the Oolitic period, as around Carrara in Italy, have been converted partially into gneiss, mica-schist, and statuary marble. The transmutation has been effected apparently by the influence of subterranean heat, acting under great pressure, or by chemical and electrical causes operating in a manner not yet understood, and which have been termed *Plutonic* action, as expressing, in one word, all the modifying causes which may be brought into play at great depths, and under conditions never exemplified at the surface. To this Plutonic action the fusion of granite itself in the bowels of the earth, as well as the superinducement of the metamorphic texture into sedimentary strata, must be attributed; and in accordance with these views the age of each metamorphic formation may be said to be twofold, for we have first to consider the period when it originated, as an aqueous deposit, in the form of mud, sand, marl, or limestone; secondly, the date at which it acquired a crystalline

* See Lyell's Manual of Elementary Geology.

texture. The same strata, therefore, may, according to this view, be very ancient in reference to the time of their deposition, and very modern in regard to the period of their assuming the metamorphic character.

No proofs that these crystalline rocks were produced more abundantly at remote periods.—Several modern writers, without denying the truth of the Plutonic or metamorphic theory, still contend that the crystalline and non-fossiliferous formations, whether stratified or unstratified, such as gneiss and granite, are essentially ancient as a class of rocks. They were generated, say they, most abundantly in the primeval state of the globe, since which time the quantity produced has been always on the decrease, until it became very inconsiderable in the Oolitic and Cretaceous periods, and quite evanescent before the commencement of the tertiary epoch.

Now the justness of these views depends almost entirely on the question whether granite, gneiss, and other rocks of the same order ever originated at the surface, or whether, according to the opinions above adopted, they are essentially subterranean in their origin, and therefore entitled to the appellation of *hypogene*. If they were formed superficially in their present state, and as copiously in the modern as in the more ancient periods, we ought to see a greater abundance of tertiary and secondary than of primary granite and gneiss; but if we adopt the hypogene theory before explained, their rapid diminution in volume among the visible rocks in the earth's crust in proportion as we investigate the formations of newer date, is quite intelligible. If a melted mass of matter be now cooling very slowly at the depth of several miles beneath the crater of an active volcano, it must remain invisible until great revolutions in the earth's crust have been brought about. So also if stratified rocks have been subjected to Plutonic action, and after having been baked or reduced to semi-fusion, are now cooling and crystallizing far under ground, it will probably require the lapse of many periods before they will be forced up to the surface and exposed to view, even at a single point. To effect this purpose there may be need of as great a development of subterranean movement as that which in the Alps, Andes, and Himalaya has raised marine strata containing ammonites to the height of 8000, 14,000, and 16,000 feet. By parity of reasoning we can hardly expect that any hypogene rocks of the tertiary periods will have been brought within the reach of human observation, seeing that the emergence of such rocks must always be so long posterior to the date of their origin, and still less can formations of this class become generally visible until so much time has elapsed as to confer on them a high relative antiquity. Extensive denudation must also combine with upheaval before they can be displayed at the surface throughout wide areas.

All geologists who reflect on subterranean movements now going on, and the eruptions of active volcanoes, are convinced that great changes are now continually in progress in the interior of the earth's crust far out of sight. They must be conscious, therefore, that the inaccessibility of the regions in which these alterations are taking place, compels them to remain in ignorance of a great part of the working of existing causes, so that they can only form vague conjectures in regard to the nature of the products which volcanic heat may elaborate under great pressure.

But when they find in mountain-chains of high antiquity, that what was once the interior of the earth's crust has since been forced outwards and exposed to view, they will naturally expect in the examination of those mountainous regions, to have an opportunity of gratifying their curiosity by obtaining a sight not only of the superficial strata of remote eras, but also of the contemporaneous nether-formed rocks. Having recognized, therefore, in such mountain-chains some ancient rocks of aqueous and volcanic origin, corresponding in character to superficial formations of modern date, they will regard any other class of ancient rocks, such as granite and gneiss, as the *residual phenomena* of which they are in search. These latter rocks will not answer the expectations previously formed of their probable nature and texture, unless they wear a foreign and mysterious aspect, and have in some places been fused or altered by subterranean heat; in a word, unless they differ wholly from the fossiliferous strata deposited at the surface, or from the lava and scoriæ thrown out by volcanoes in the open air. It is the total distinctness, therefore, of crystalline formations, such as granite, hornblende-schist, and the rest, from every substance of which the origin is familiar to us, that constitutes their claim to be regarded as the effects of causes now in action in the subterranean regions. They belong not to an order of things which has passed away; they are not the monuments of a primeval period, bearing inscribed upon them in obsolete characters the words and phrases of a dead language; but they teach us that part of the living language of nature, which we cannot learn by our daily intercourse with what passes on the habitable surface.

CHAPTER XIII.

UNIFORMITY IN THE SERIES OF PAST CHANGES IN THE ANIMATE AND INANIMATE WORLD.

Supposed alternate periods of repose and disorder—Observed facts in which this doctrine has originated—These may be explained by supposing a uniform and uninterrupted series of changes—Threefold consideration of this subject; first, in reference to the living creation, extinction of species, and origin of new animals and plants; secondly, in reference to the changes produced in the earth's crust by the continuance of subterranean movements in certain areas, and their transference after long periods to new areas; thirdly, in reference to the laws which govern the formation of fossiliferous strata, and the shifting of the areas of sedimentary deposition—On the combined influence of all these modes and causes of change in producing breaks and chasms in the chain of records—Concluding remarks on the identity of the ancient and present system of terrestrial changes.

Origin of the doctrine of alternate periods of repose and disorder.—It has been truly observed, that when we arrange the fossiliferous formations in chronological order, they constitute a broken and defective series of monuments: we pass without any intermediate gradations, from systems of strata which are horizontal to other systems which are highly inclined, from rocks of peculiar mineral composition to others which have a character wholly distinct,—from one assemblage of organic remains to another, in which frequently all the species, and most of the genera, are different. These violations of continuity are so common, as to constitute the rule rather than the exception, and they have been considered by many geologists as conclusive in favor of sudden revolutions in the inanimate and animate world. According to the speculations of some writers, there have been in the past history of the planet alternate periods of tranquillity and convulsion, the former enduring for ages, and resembling that state of things now experienced by man: the other brief, transient, and paroxysmal, giving rise to new mountains, seas, and valleys, annihilating one set of organic beings, and ushering in the creation of another.

It will be the object of the present chapter to demonstrate, that these theoretical views are not borne out by a fair interpretation of geological monuments. It is true that in the solid framework of the globe, we have a chronological chain of natural records, and that many links in this chain are wanting; but a careful consideration of all the phenomena will lead to the opinion that the series was originally defective,—that it has been rendered still more so by time—that a great part of what remains is inaccessible to man, and even of that fraction which is accessible, nine-tenths are to this day unexplored.

How the facts may be explained by assuming a uniform series of changes.—The readiest way, perhaps, of persuading the reader that we may dispense with great and sudden revolutions in the geological order of events, is by showing him how a regular and uninterrupted series of changes in the animate and inanimate world may give rise to such breaks in the sequence, and such unconformability of stratified rocks, as are usually thought to imply convulsions and catastrophes. It is scarcely necessary to state, that the order of events thus assumed to occur, for the sake of illustration, must be in harmony with all the conclusions legitimately drawn by geologists from the structure of the earth, and must be equally in accordance with the changes observed by man to be now going on in the living as well as in the inorganic creation. It may be necessary in the present state of science to supply some part of the assumed course of nature hypothetically; but if so, this must be done without any violation of probability, and always consistently with the analogy of what is known both of the past and present economy of our system. Although the discussion of so comprehensive a subject must carry the beginner far beyond his depth, it will also, it is hoped, stimulate his curiosity, and prepare him to read some elementary treatises on geology with advantage, and teach him the bearing on that science of the changes now in progress on the earth. At the same time it may enable him the better to understand the intimate connection between the second and third books of this work, the former of which is occupied with the changes in the inorganic, the latter with those of the organic creation.

In pursuance, then, of the plan above proposed, I shall consider in this chapter, first, what may be the course of fluctuation in the animate world; secondly, the mode in which contemporaneous subterranean movements affect the earth's crust; and, thirdly, the laws which regulate the deposition of sediment.

UNIFORMITY OF CHANGE CONSIDERED FIRST IN REFERENCE TO THE LIVING CREATION.

First, in regard to the vicissitudes of the living creation, all are agreed that the sedimentary strata found in the earth's crust are divisible into a variety of groups, more or less dissimilar in their organic remains and mineral composition. The conclusion universally drawn from the study and comparison of these fossiliferous groups is this, that at successive periods distinct tribes of animals and plants have inhabited the land and waters, and that the organic types of the newer formations are more analogous to species now existing, than those of more ancient rocks. If we then turn to the present state of the animate creation, and inquire whether it has now become fixed and stationary, we discover that, on the contrary, it is in a state of continual flux—that there are many causes in action which tend to the extinction of species, and which are conclusive against the doctrine of their unlimited durability. But natural history has been successfully cultivated for so short a period, that a few examples only of local, and perhaps but one or two of absolute, extirpation can as yet be proved, and these only where the interference of man has been conspicuous. It will nevertheless appear evident, from the facts and arguments detailed in the third book (from the thirty-seventh to the forty-second chapters, inclusive) that man is not the only exterminating agent; and that, independently of his intervention, the annihilation of species is promoted by the multiplication and gradual diffusion of every animal or plant. It will also appear, that every alteration in the physical geography and climate of the globe cannot fail to have the same tendency. If we proceed still farther, and inquire whether new species are substituted from time to time for those which die out, and whether there are certain laws appointed by the Author of Nature to regulate such new creations, we find that the period of human observation is as yet too short to afford data for determining so weighty a question. All that can be done is to show that the successive introduction of new species may be a constant part of the economy of the terrestrial system, without our having any right to expect that we should be in possession of direct proof of the fact. The appearance again and again of new species may easily have escaped detection, since the numbers of known animals and plants have augmented so rapidly within the memory of persons now living, as to have doubled in some classes, and quadrupled in others. It will also be remarked in the sequel (book iii. chap. 43), that it must always be more easy if species proceeded originally from single stocks, to prove that one which formerly abounded in a given district has ceased to be, than that another has been called into being for the first time. If, therefore, there be as yet only one or two unequivocal instances of extinction, namely, those of the dodo and solitaire (see ch. 41), it is scarcely reasonable as yet to hope that we should be cognizant of a single instance of the first appearance of a new species.

Recent origin of man, and gradual approach in the tertiary fossils of successive periods from an extinct to the recent fauna.—The geologist, however, if required to advance some fact which may lend countenance to the opinion that in the most modern times, that is to say, after the greater part of the existing fauna and flora were established on the earth, there has still been a new species superadded, may point to man himself as furnishing the required illustration—for man must be regarded by the geologist as a creature of yesterday, not merely in reference to the past history of the organic world, but also in relation to that particular state of the animate creation of which he forms a part. The comparatively modern introduction of the human race is proved by the absence of the remains of man and his works, not only from all strata containing a certain proportion of fossil shells of extinct species, but even from a large part of the newest strata, in which all the fossil individuals are referable to species still living.

To enable the reader to appreciate the full force of this evidence, I shall give a slight sketch of the information obtained from the newer strata, respecting fluctuations in the animate world, in times immediately antecedent to the appearance of man.

In tracing the series of fossiliferous formations from the more ancient to the more modern, the first deposits in which we meet with assemblages of organic remains, having a near analogy to the fauna of certain parts of the globe in our own time, are those commonly called tertiary. Even in the Eocene, or oldest subdivision of these tertiary formations, some few of the testacea belong to existing species, although almost all of them, and apparently all the associated vertebrata, are now extinct. These Eocene strata are succeeded by a great number of more modern deposits, which depart gradually in the character of their fossils from the Eocene type, and approach more and more to that of the living creation. In the present state of science, it is chiefly by the aid of shells that we are enabled to arrive at these results, for of all classes the testacea are the most generally diffused in a fossil state, and may be called the medals principally employed by nature, in recording the chronology of past events. In the Miocene deposits, which are next in succession to the Eocene, we begin to find a considerable number, although still a minority, of recent species, intermixed with some fossils common to the preceding epoch. We then arrive at the Pliocene strata, in which species now contemporary with man begin to preponderate, and in the newest of which nine-tenths of the fossils agree with species still inhabiting the neighboring sea.

In this passing from the older to the newer members of the tertiary system we meet with many chasms, but none which separate entirely, by a broad line of demarcation, one state of the organic world from another. There are no signs of an abrupt termination of one fauna and flora, and the starting into life of new and wholly distinct forms. Although we are far from being able to demonstrate geologically an insensible transition from the Eocene to the Miocene, or even from the latter to the recent fauna, yet the more we enlarge and perfect our general survey, the more nearly do we approximate to such a continuous series, and the more gradually are we conducted from times when many of the genera and nearly all the species were extinct, to those in which scarcely a single species flourished which we do not know to exist at present. Dr. A. Philippi, indeed, after an elaborate comparison of the fossil tertiary shells of Sicily with those now living in the Mediterranean, announces as the result of his examination that there are strata in that island, which attest a very gradual passage from a period, when only thirteen in a hundred of the shells were like the species now living in the sea, to an era when the recent species had attained a proportion of ninety-five in a hundred. There is therefore evidence, he says, in Sicily of this revolution in the animate world having been effected "without the intervention of any convulsion or abrupt changes, certain species having from time to time died out, and others having been introduced, until at length the existing fauna was elaborated."

It had often been objected that the evidence of fossil species occurring in two consecutive formations, was confined to the testacea or zoophytes, the characters of which are less marked and decisive than those afforded by the vertebrate animals. But Mr. Owen has lately insisted on the important fact, that not a few of the quadrupeds which now inhabit our island, and among others the horse, the ass, the hog, the smaller wild ox, the goat, the red deer, the roe, the beaver, and many of the diminutive rodents, are the same as those which once coexisted with the mammoth, the great northern hippopotamus, two kinds of rhinoceros, and other mammalia long since extinct. "A part," he observes, "and not the whole of the modern tertiary fauna has perished, and hence we may conclude that the cause of their destruction has not been a violent and universal catastrophe from which none could escape."*

Had we discovered evidence that man had come into the earth at a period as early as that when a large number of the fossil quadrupeds now living, and almost all the recent species of land, freshwater, and marine shells were in existence, we should have been compelled to ascribe a much higher antiquity to our species, than even the boldest speculations of the ethnologist require, for no small part of the great physical revolution depicted on the map of Europe (Pl. 3), before described, took place very gradually after the recent testacea abounded almost to the exclusion of the extinct. Thus, for example, in the deposits called the "northern drift," or the glacial formation of Europe and North America, the fossil marine shells can easily be identified with species either now inhabiting the neighboring sea, or living in the seas of higher latitudes. Yet they exhibit no memorials of the human race, or of articles fabricated by the hand of man. Some of the newest of these strata passing by the name of "raised beaches," occur at moderate elevations on the coast of England, Scotland, and Ireland. Other examples are met with on a more extended scale in Scandinavia, as at the height of 200 feet at Uddevalla in Sweden, and at twice that elevation, near Christiana, in Norway, also at an altitude of 600 or 700 feet in places farther north. They consist of beds of sand and clay, filling hollows in a district of granite and gneiss, and they must closely resemble the accumulations of shelly matter now in progress at the bottom of the Norwegian fiords. The rate at which the land is now rising in Scandinavia, is far too irregular in different places to afford a safe standard for estimating the minimum of time required for the upheaval

* Reports to Brit. Assoc. 1842, 1843, and Introd. to Brit. Foss. Mamm. p. 31. The conchological evidence respecting the British Miocene, Pliocene, and Pleistocene fossils, examined by Mr. Forbes, in the paper before cited, p. 88, note, bear out some of the most important conclusions of M. Deshayes, quoted by me in the first edition of the Principles, 1831, and the recent observations of Philippi in regard to the passage of species from one formation to another. I refer to these authorities more especially because this doctrine of a gradual transition has been opposed by some living naturalists of high distinction, among whom I may mention M. A. d'Orbigny and M. Agassiz. I have long been convinced that we must abandon many of the identifications formerly made of Eocene with recent shells; but some errors of this kind do not affect the general reasoning on the subject. See a discussion on this question, Quarterly Journ. of Geog. Soc., No. 5, p. 47 Feb. 1846.

of the fundamental granite, and its marine shelly covering, to the height of so many hundred feet; but according to the greatest average, of five or six feet in a century, the period required would be very considerable, and nearly the whole of it, as well as the antecedent epoch of submergence, seems to have preceded the introduction of man into these parts of the earth.

There are other post-tertiary formations of fluviatile origin, in the centre of Europe, in which the absence of human remains is perhaps still more striking, because, when formed, they must have been surrounded by dry land. I allude to the silt or *loess* of the basin of the Rhine, which must have gradually filled up the great valley of that river since the time when its waters, and the contiguous lands, were inhabited by the existing species of freshwater and terrestrial mollusks. Showers of ashes, thrown out by some of the last eruptions of the Eifel volcanoes, fell during the deposition of this fluviatile silt, and were interstratified with it. But these volcanoes became exhausted, the valley was re-excavated through the silt, and again reduced to its present form before the period of human history. The study, therefore, of this shelly silt reveals to us the history of a long series of events, which occurred after the testacea now living inhabited the land and rivers of Europe, and the whole terminated without any signs of the coming of man into that part of the globe.

To cite a still more remarkable example, we observe in Sicily a lofty table-land and hills, sometimes rising to the height of 3000 feet, capped with a limestone, in which from 70 to 85 per cent. of the fossil testacea are specifically identical with those now inhabiting the Mediterranean. These calcareous and other argillaceous strata of the same age are intersected by deep valleys which have been gradually formed by denudation, but have not varied materially in width or depth since Sicily was first colonized by the Greeks. The limestone, moreover, which is of so late a date in geological chronology, was quarried for building those ancient temples of Girgenti and Syracuse, of which the ruins carry us back to a remote era in human history. If we are lost in conjectures when speculating on the ages required to lift up these formations to the height of several thousand feet above the sea, how much more remote must be the era when the same rocks were gradually formed beneath the waters!

To conclude, it appears that, in going back from the recent to the Eocene period, we are carried by many successive steps from the fauna now contemporary with man to an assemblage of fossil species wholly different from those now living. In this retrospect we have not yet succeeded in tracing back a perfect transition from the recent to an extinct fauna; but there are usually so many species in common to the groups which stand next in succession as to show that there is no great chasm, no signs of a crisis when one class of organic beings was annihilated to give place suddenly to another. This analogy, therefore, derived from a period of the earth's history which can best be compared with the present state of things, and more thoroughly investigated than any other, leads to the conclusion that the extinction and creation of species, has been and is the result of a slow and gradual change in the organic world.

UNIFORMITY OF CHANGE CONSIDERED, SECONDLY, IN REFERENCE TO SUBTERRANEAN MOVEMENTS.

To pass on to another of the three topics before proposed for discussion, the reader will find, in the account given in the second book of the earthquakes recorded in history, that certain countries have, from time immemorial, been rudely shaken again and again, while others, comprising by far the largest part of the globe, have remained to all appearance motionless. In the regions of convulsion rocks have been rent asunder, the surface has been forced up into ridges, chasms have opened, or the ground throughout large spaces has been permanently lifted up above or let down below its former level. In the regions of tranquillity some areas have remained at rest, but others have been ascertained by a comparison of measurements, made at different periods, to have risen by an insensible motion, as in Sweden, or to have subsided very slowly, as in Greenland. That these same movements, whether ascending or descending, have continued for ages in the same direction has been established by geological evidence. Thus, we find both on the east and west coast of Sweden, that ground which formerly constituted the bottom of the Baltic and of the ocean has been lifted up to an elevation of several hundred feet above high-water mark. The rise within the historical period has not amounted to many yards, but the greater extent of antecedent upheaval is proved by the occurrence in inland spots, several hundred feet high, of deposits filled with fossil shells of species now living either in the ocean or the Baltic.

To detect proofs of slow and gradual subsidence must in general be more difficult; but the theory which accounts for the form of circular coral reefs and lagoon islands, and which will be explained in the last chapter of the third book, will satisfy the reader that there are spaces on the globe, several thousand miles in circumference, throughout which the downward movement has predominated for ages, and yet the land has never, in a single instance, gone down suddenly for several hundred feet at once. Yet geology demonstrates that the persistency of subterranean movements in one direction has not been perpetual throughout all past time. There have been great oscillations of level by which a surface of dry land has been submerged to a depth of several thousand feet, and then at a period long subsequent raised again and made to emerge. Nor have the regions now motionless been always at rest; and some of those which are at present the theatres of reiterated earthquakes have formerly enjoyed a long continuance of tranquillity. But although disturbances have ceased after having long prevailed, or have recommenced after a suspension for ages, there has been no universal disruption of the earth's crust or desolation of the surface since times the most remote. The non-occurrence of such a general convulsion is proved by the perfect horizontality now retained by some of the most ancient fossiliferous strata throughout wide areas.

Inferences derived from unconformable strata.—That the subterranean forces have visited different parts of the globe at successive periods, is inferred chiefly from the unconformability of strata belonging to groups of different ages. Thus, for example, on the borders of Wales and Shropshire we find the slaty beds of the ancient Silurian system curved and vertical, while the beds of the overlying carboniferous shale and sandstone are horizontal. All are agreed, that in such a case the older set of strata had suffered great dislocation before the deposition of the newer or carboniferous beds, and that these last have never since been convulsed by any movements of excessive violence. But the strata of the inferior group suffered only a local derangement, and rocks of the same age are by no means found everywhere in a curved or vertical position. In various parts of Europe, and particularly near Lake Wener in the south of Sweden, and in many parts of Russia, beds of the same Silurian system maintain the most perfect horizontality; and a similar observation may be made respecting limestones and shales of the like antiquity in the great lake district of Canada and the United States. They are still as flat and horizontal as when first formed; yet since their origin not only have most of the actual mountain-chains been uplifted, but the very rocks of which those mountains are composed have been formed.

It would be easy to multiply instances of similar unconformability in formations of other ages; but a few more will suffice. The coal measures before alluded to as horizontal on the borders of Wales are vertical in the Mendip Hills in Somersetshire, where the overlying beds of the New Red Sandstone are horizontal. Again, in the Wolds of Yorkshire the last mentioned sandstone supports on its curved and inclined beds the horizontal Chalk. The Chalk again is vertical on the flanks of the Pyrenees, and the tertiary strata repose unconformably upon it.

Consistency of local disturbances with general uniformity.—As almost every country supplies illustrations of the same phenomena, they who advocate the doctrine of alternate periods of disorder and repose may appeal to the facts above described, as proving that every district has been by turns convulsed by earthquakes and then respited for ages from convulsions. But so it might with equal truth be affirmed that every part of Europe has been visited alternately by winter and summer, although it has always been winter and always summer in some part of the planet, and neither of these seasons has ever reigned simultaneously over the entire globe. They have been always shifting about from place to place; but the vicissitudes which recur thus annually in a single spot are never allowed to interfere with the invariable uniformity of seasons throughout the whole planet.

So, in regard to subterranean movements, the theory of the perpetual uniformity of the force which they exert on the earth's crust is quite consistent with the admission of their alternate development and suspension for indefinite periods within limited geographical areas.

UNIFORMITY OF CHANGE CONSIDERED, THIRDLY, IN REFERENCE TO SEDIMENTARY DEPOSITION.

It now remains to speak of the laws governing the deposition of new strata. If we survey the surface of the globe we immediately perceive that it is divisible into areas of deposition and non-deposition, or, in other words, at any given time there are spaces which are the recipients, others which are not the recipients of sedimentary matter. No new strata, for example, are thrown down on dry land, which remains the same from year to year; whereas, in many parts of the bottom of seas and lakes, mud, sand, and pebbles are annually spread out by rivers and currents. There are also great masses of limestone growing in some seas, or in mid-ocean, chiefly composed of corals and shells.

No sediment deposited on dry land.—As to the dry land, so far from being the receptacle of fresh accessions of matter, it is exposed almost everywhere to waste away. Forests may be as dense and lofty as those of Brazil, and may swarm with quadrupeds, birds, and insects, yet at the end of ten thousand years one layer of black mould, a few inches thick, may be the sole representative of those myriads of trees, leaves, flowers, and fruits, those innumerable bones and skeletons of birds, quadrupeds, and reptiles, which tenanted the fertile region. Should this land be at length submerged, the waves of the sea may wash away in a few hours the scanty covering of mould, and it may merely impart a darker shade of color to the next stratum of marl, sand, or other matter newly thrown down. So also at the bottom of the ocean where no sediment is accumulating, sea-weed, zoophytes, fish, and even shells, may multiply for ages and decompose, leaving no vestige of their form or substance behind. Their decay, in water, although more slow, is as certain and eventually as complete as in the open air. Nor can they be perpetuated for indefinite periods in a fossil state, unless imbedded in some matrix which is impervious to water, or which at least does not allow a free percolation of that fluid, impregnated as it usually is, with a slight quantity of carbonic or other acid. Such a free percolation may be prevented either by the mineral nature of the matrix itself, or by the superposition of an impermeable stratum: but if unimpeded, the fossil shell or bone will be dissolved and removed, particle after particle, and thus entirely effaced, unless petrifaction or the substitution of mineral for organic matter happen to take place.

That there has been land as well as sea at all former geological periods, we know from the fact, that fossil trees and terrestrial plants are imbedded in rocks of every age. Occasionally lacustrine and fluviatile shells, insects, or the bones of amphibious or land reptiles, point to the same conclusion. The existence of dry land at all periods of the past implies, as before mentioned, the partial deposition of sediment, or its limitation to certain areas; and the next point to which I shall call the reader's attention, is the shifting of these areas from one region to another.

First, then, variations in the site of sedimentary deposition are brought about independently of subterranean movements. There is always a slight change from year to year, or from century to century. The sediment of the Rhone, for example, thrown into the Lake of Geneva, is now conveyed to a spot a mile and a half distant from that where it accumulated in the tenth century, and six miles from the point where the delta began originally to form. We may look forward to the period when this lake will be filled up, and then the distribution of the transported matter will be suddenly altered, for the mud and sand brought down from the Alps will thenceforth, instead of being deposited near Geneva, be carried nearly 200 miles southwards, where the Rhone enters the Mediterranean.

In the deltas of large rivers, such as those of the Ganges and Indus, the mud is first carried down for many centuries through one arm, and on this being stopped up it is discharged by another, and may then enter the sea at a point 50 or 100 miles distant from its first receptacle. The direction of marine currents is also liable to be changed by various accidents, as by the heaping up of new sand-banks, or the wearing away of cliffs and promontories.

But, secondly, all these causes of fluctuation in the sedimentary areas are entirely subordinate to those great upward or downward movements of land which have been already described as prevailing over large tracts of the globe. By such elevation or subsidence certain spaces are gradually submerged, or made gradually to emerge:—in the one case sedimentary deposition may be suddenly renewed after having been suspended for ages, in the other as suddenly made to cease after having continued for an indefinite period.

Causes of variation in mineral character of successive sedimentary groups.—If deposition be renewed after a long interval, the new strata will usually differ greatly from the sedimentary rocks previously formed in the same place, and especially if the older rocks have suffered derangement, which implies a change in the physical geography of the district since the previous conveyance of sediment to the same spot. It may happen, however, that, even when the inferior group is horizontal and conformable to the upper strata, these last may still differ entirely in mineral character, because since the origin of the older formation the geography of some distant country has been altered. In that country rocks before concealed may have become exposed by denudation; volcanoes may have burst out and covered the surface with scoriæ and lava, or new lakes may have been formed by subsidence; and other fluctuations may have occurred, by which the materials brought down from thence by rivers to the sea have acquired a distinct mineral character.

It is well known that the stream of the Mississippi is charged with sediment of a different color from that of the Arkansas and Red Rivers, which are tinged with red mud, derived from rocks of porphyry in "the far west." The waters of the Uruguay, says Darwin, draining a granitic country, are clear and black, those of the Parana, red.* The mud with which the Indus is loaded, says Burnes, is of a clayey hue, that of the Chenab, on the other hand, is reddish, that of the Sutlej is more pale.† The same causes which make these several rivers, sometimes situated at no great distance the one from the other, to differ greatly in the character of their sediment, will make the waters draining the same country at different epochs, especially before and after great revolutions in physical geography, to be entirely dissimilar. It is scarcely necessary to add, that marine currents will be affected in an analogous manner in consequence of the formation of new shoals, the emergence of new islands, the subsidence of others, the gradual waste of neighboring coasts, the growth of new deltas, the increase of coral reefs, and other changes.

Why successive sedimentary groups contain distinct fossils.—If, in the next place, we assume, for reasons before stated, a continual extinction of species and introduction of others into the globe, it will then follow that the fossils of strata formed at two distant periods on the same spot, will differ even more certainly than the mineral composition of the same. For rocks of the same kind have sometimes been reproduced in the same district after a long interval of time, whereas there are no facts leading to the opinion that species which have once died out have ever been reproduced. The submergence then of land must be often attended by the commencement of a new class of sedimentary deposits, characterized by a new set of fossil animals and plants, while the reconversion of the bed of the sea into land may arrest at once and for an indefinite time the formation of geological monuments. Should the land again sink, strata will again be formed; but one or many entire revolutions in animal or vegetable life may have been completed in the interval.

Conditions requisite for the original completeness of a fossiliferous series.—If we infer, for reasons before explained, that fluctuations in the animate world are brought about by the slow and successive removal and creation of species, we shall be convinced that a rare combination of circumstances alone can give rise to such a series of strata as will bear testimony to a gradual passage from one state of organic life to another. To produce such strata nothing less will be requisite than the fortunate coincidence of the following conditions: first, a never-failing supply of sediment in the same region throughout a period of vast duration; secondly, the fitness of the deposit in every part for the permanent preservation of imbedded fossils; and, thirdly, a gradual

* Darwin's Journal, p. 163. 2d. ed. p. 139.
† Journ. Roy. Geograph. Soc. vol. iii. p. 142.

subsidence to prevent the sea or lake from being filled up and converted into land.

It will appear in the chapter on coral reefs,* that, in certain parts of the Pacific and Indian Oceans, most of these conditions, if not all, are complied with, and the constant growth of coral, keeping pace with the sinking of the bottom of the sea, seems to have gone on so slowly, for such indefinite periods, that the signs of a gradual change in organic life might probably be detected in that quarter of the globe, if we could explore its submarine geology. Instead of the growth of coralline limestone, let us suppose, in some other place, the continuous deposition of fluviatile mud and sand, such as the Ganges and Brahmapootra have poured for thousands of years into the Bay of Bengal. Part of this bay, although of considerable depth, might at length be filled up before an appreciable amount of change was effected in the fish, mollusca, and other inhabitants of the sea and neighboring land. But, if the bottom be lowered by sinking at the same rate that it is raised by fluviatile mud, the bay can never be turned into dry land. In that case one new layer of matter may be superimposed upon another for a thickness of many thousand feet, and the fossils of the inferior beds may differ greatly from those entombed in the uppermost, yet every intermediate gradation may be indicated in the passage from an older to a newer assemblage of species. Granting, however, that such an unbroken sequence of monuments may thus be elaborated in certain parts of the sea, and that the strata happen to be all of them well adapted to preserve the included fossils from decomposition, how many accidents must still concur before these submarine formations will be laid open to our investigation! The whole deposit must first be raised several thousand feet, in order to bring into view the very foundation; and during the process of exposure the superior beds must not be entirely swept away by denudation.

In the first place, the chances are as three to one against the mere emergence of the mass above the waters, because three-fourths of the globe are covered by the ocean. But if it be upheaved and made to constitute part of the dry land, it must also, before it can be available for our instruction, become part of that area already surveyed by geologists; and this area comprehends perhaps less than a tenth of the whole earth. In this small fraction of land already explored, and still very imperfectly known, we are required to find a set of strata, originally of limited extent, and probably much lessened by subsequent denudation.

Yet it is precisely because we do not encounter at every step the evidence of such gradations from one state of the organic world to another, that so many geologists embrace the doctrine of great and sudden revolutions in the history of the animate world. Not content with simply availing themselves, for the convenience of classification, of

* Book iii. ch. 50.

those gaps and chasms which here and there interrupt the continuity of the chronological series, as at present known, they deduce, from the frequency of these breaks in the chain of records, an irregular mode of succession in the events themselves both in the organic and inorganic world. But, besides that some links of the chain which once existed are now clearly lost and others concealed from view, we have good reason to suspect that it was never complete originally. It may undoubtedly be said, that strata have been always forming somewhere, and therefore at every moment of past time nature has added a page to her archives; but, in reference to this subject, it should be remembered that we can never hope to compile a consecutive history by gathering together monuments which were originally detached and scattered over the globe. For as the species of organic beings contemporaneously inhabiting remote regions are distinct, the fossils of the first of several periods which may be preserved in any one country, as in America, for example, will have no connection with those of a second period found in India, and will therefore no more enable us to trace the signs of a gradual change in the living creation, than a fragment of Chinese history will fill up a blank in the political annals of Europe.

The absence of any deposits of importance containing recent shells in Chili, or anywhere on the western coast of South America, naturally led Mr. Darwin to the conclusion that "where the bed of the sea is either stationary or rising, circumstances are far less favorable than where the level is sinking to the accumulation of conchiferous strata of sufficient thickness and extension to resist the average vast amount of denudation."* An examination of the superficial clay, sand, and gravel of the most modern date in Norway and Sweden, where the land is also rising, would incline us to admit a similar proposition. Yet in these cases there has been a supply of sediment from the waste of the coast and the interior, especially in Patagonia and Chili. Nevertheless wherever the bottom of the sea has been continually elevated, the total thickness of sedimentary matter accumulating at depths suited to the habitation of most of the species of shells can never be great, nor can the deposits be thickly covered by superincumbent matter, so as to be consolidated by pressure. When they are upheaved, therefore, the waves on the beach will bear down and disperse the loose materials; whereas if the bed of the sea subsides slowly, a mass of strata containing abundance of such species as live at moderate depths may increase in thickness to any amount, and may extend over a broad area, as the water gradually encroaches on the land. If, then, at particular periods, as in the Miocene epoch, for example, both in Europe and North America, contemporaneous shelly deposits have originated, and have been preserved at very distant points, it may arise from the prevalence at that period of simultaneous subsidence throughout very wide areas. The absence in the same quarters of the globe of strata marking the

* Darwin's S. America, pp. 136, 139.

ages which immediately succeeded, may be accounted for by supposing that the level of the bed of the sea and the adjoining land was stationary or was undergoing slow upheaval.

How far some of the great violations of continuity which now exist in the chronological table of fossiliferous rocks, will hereafter be removed or lessened, must at present be mere matter of conjecture. The hiatus which exists in Great Britain between the fossils of the Lias and those of the Magnesian Limestone, is supplied in Germany by the rich fauna and flora of the Muschelkalk, Keuper, and Bunter Sandstein, which we know to be of a date precisely intermediate; those three formations being interposed in Germany between others which agree perfectly in their organic remains with our Lias and Magnesian Limestone. Until lately the fossils of the Coal-measures were separated from those of the antecedent Silurian group by a very abrupt and decided line of demarcation; but recent discoveries have brought to light in Devonshire, Belgium, the Eifel, and Westphalia, the remains of a fauna of an intervening period. This connecting link is furnished by the fossil shells, fish, and corals of the Devonian or Old Red Sandstone group, and some species of this newly intercalated fauna are found to be common to it and the subjacent Silurian rocks, while other species belong to it in common with the Coal-measures. We have also in like manner had some success of late years in diminishing the hiatus which still separates the Cretaceous and Eocene periods in Europe. Still we must expect, for reasons before stated, that some such chasms will forever continue to occur in some parts of our sedimentary series.

Consistency of the theory of gradual change with the existence of great breaks in the series.—To return to the general argument pursued in this chapter, it is assumed, for reasons above explained, that a slow change of species is in simultaneous operation everywhere throughout the habitable surface of sea and land; whereas the fossilization of plants and animals is confined to those areas where new strata are produced. These areas, as we have seen, are always shifting their position; so that the fossilizing process, by means of which the commemoration of the particular state of the organic world, at any given time, is affected, may be said to move about, visiting and revisiting different tracts in succession.

To make still more clear the supposed working of this machinery, I shall compare it to a somewhat analogous case that might be imagined to occur in the history of human affairs. Let the mortality of the population of a large country represent the successive extinction of species, and the births of new individuals the introduction of new species. While these fluctuations are gradually taking place everywhere, suppose commissioners to be appointed to visit each province of the country in succession, taking an exact account of the number, names, and individual peculiarities of all the inhabitants, and leaving in each district a register containing a record of this information. If, after the completion of one census, another is immediately made on the same plan, and then another, there will, at last, be a series of statistical documents in each province. When those belonging to any one province are arranged in chronological order, the contents of such as stand next to each other will differ according to the length of the intervals of time between the taking of each census. If, for example, there are sixty provinces, and all the registers are made in a single year, and renewed annually, the number of births and deaths will be so small, in proportion to the whole of the inhabitants, during the interval between the compiling of the two consecutive documents, that the individuals described in such documents will be nearly identical; whereas, if the survey of each of the sixty provinces occupies all the commissioners for a whole year, so that they are unable to revisit the same place until the expiration of sixty years, there will then be an almost entire discordance between the persons enumerated in two consecutive registers in the same province. There are, undoubtedly, other causes besides the mere quantity of time, which may augment or diminish the amount of discrepancy. Thus, at some periods a pestilential disease may have lessened the average duration of human life, or a variety of circumstances may have caused the births to be unusually numerous, and the population to multiply; or, a province may be suddenly colonized by persons migrating from surrounding districts.

These exceptions may be compared to the accelerated rate of fluctuation in the fauna and flora of a particular region, in which the climate and physical geography may be undergoing an extraordinary degree of alteration.

But I must remind the reader, that the case above proposed has no pretensions to be regarded as an exact parallel to the geological phenomena which I desire to illustrate; for the commissioners are supposed to visit the different provinces in rotation; whereas the commemorating processes by which organic remains become fossilized, although they are always shifting from one area to the other, are yet very irregular in their movements. They may abandon and revisit many spaces again and again before they once approach another district; and, besides this source of irregularity, it may often happen that, while the depositing process is suspended, denudation may take place, which may be compared to the occasional destruction by fire or other causes of some of the statistical documents before mentioned. It is evident that, where such accidents occur, the want of continuity in the series may become indefinitely great, and that the monuments which follow next in succession will by no means be equidistant from each other in point of time.

If this train of reasoning be admitted, the occasional distinctness of the fossil remains, in formations immediately in contact, would be a necessary consequence of the existing laws of sedimentary deposition and subterranean movement, accompanied by a constant mortality and renovation of species.

As all the conclusions above insisted on are directly opposed to opinions still popular, I shall add another comparison, in the hope of preventing any possible misapprehension of the argument. Suppose we had discovered two buried cities at the foot of Vesuvius, immediately superimposed upon each other, with a great mass of tuff and lava intervening, just as Portici and Resina, if now covered with ashes, would overlie Herculaneum. An antiquary might possibly be entitled to infer, from the inscriptions on public edifices, that the inhabitants of the inferior and older city were Greeks, and those of the modern towns Italians. But he would reason very hastily if he also concluded from these data that there had been a sudden change from the Greek to the Italian language in Campania. But if he afterwards found *three* buried cities, one above the other, the intermediate one being Roman, while, as in the former example, the lowest was Greek and the uppermost Italian, he would then perceive the fallacy of his former opinion, and would begin to suspect that the catastrophes by which the cities were inhumed might have no relation whatever to the fluctuations in the language of the inhabitants; and that, as the Roman tongue had evidently intervened between the Greek and Italian, so many other dialects may have been spoken in succession, and the passage from the Greek to the Italian may have been very gradual; some terms growing obsolete, while others were introduced from time to time.

If this antiquary could have shown that the volcanic paroxysms of Vesuvius were so governed as that cities should be buried one above the other, just as often as any variation occurred in the language of the inhabitants, then, indeed, the abrupt passage from a Greek to a Roman, and from a Roman to an Italian city, would afford proof of fluctuations no less sudden in the language of the people.

So, in Geology, if we could assume that it is part of the plan of Nature to preserve, in every region of the globe, an unbroken series of monuments to commemorate the vicissitudes of the organic creation, we might infer the sudden extirpation of species, and the simultaneous introduction of others, as often as two formations in contact are found to include dissimilar organic fossils. But we must shut our eyes to the whole economy of the existing causes, aqueous, igneous, and organic, if we fail to perceive *that such is not the plan of Nature.*

Concluding remarks on the identity of the ancient and present system of terrestrial changes.—I shall now conclude the discussion of a question with which we have been occupied since the beginning of the fifth chapter; namely, whether there has been any interruption, from the remotest periods, of one uniform system of change in the animate and inanimate world. We were induced to enter into that inquiry by reflecting how much the progress of opinion in Geology had been influenced by the assumption that the analogy was slight in kind, and still more slight in degree, between the causes which produced the former revolutions of the globe, and those now in every-day operation. It appeared clear that the earlier geologists had not only a scanty acquaintance with existing changes, but were singularly unconscious of the amount of their ignorance. With the presumption naturally inspired by this unconsciousness, they had no hesitation in deciding at once that time could never enable the existing powers of nature to work out changes of great magnitude, still less such important revolutions as those which are brought to light by Geology. They, therefore, felt themselves at liberty to indulge their imaginations in guessing at what *might be*, rather than inquiring *what is*; in other words, they employed themselves in conjecturing what might have been the course of nature at a remote period, rather than in the investigation of what was the course of nature in their own times.

It appeared to them more philosophical to speculate on the possibilities of the past, than patiently to explore the realities of the present; and having invented theories under the influence of such maxims, they were consistently unwilling to test their validity by the criterion of their accordance with the ordinary operations of nature. On the contrary, the claims of each new hypothesis to credibility appeared enhanced by the great contrast, in kind or intensity, of the causes referred to, and those now in operation.

Never was there a dogma more calculated to foster indolence, and to blunt the keen edge of curiosity, than this assumption of the discordance between the ancient and existing causes of change. It produced a state of mind unfavorable in the highest degree to the candid reception of the evidence of those minute but incessant alterations which every part of the earth's surface is undergoing, and by which the condition of its living inhabitants is continually made to vary. The student, instead of being encouraged with the hope of interpreting the enigmas presented to him in the earth's structure,—instead of being prompted to undertake laborious inquiries into the natural history of the organic world, and the complicated effects of the igneous and aqueous causes now in operation, was taught to despond from the first. Geology, it was affirmed, could never rise to the rank of an exact science,—the greater number of phenomena must forever remain inexplicable, or only be partially elucidated by ingenious conjectures. Even the mystery which invested the subject was said to constitute one of its principal charms, affording, as it did, full scope to the fancy to indulge in a boundless field of speculation.

The course directly opposed to this method of philosophizing consists in an earnest and patient inquiry, how far geological appearances are reconcilable with the effect of changes now in progress, or which may be in progress in regions inaccessible to us, and of which the reality is attested by volcanoes and subterranean movements. It also endeavors to estimate the aggregate result of ordinary operations multiplied by time, and cherishes a sanguine hope that the resources to be derived from observation and experiment, or from the study of nature such as she now is, are very far from being exhausted. For this reason all theories are rejected which involve the assumption of sudden and violent catastrophes and revolutions of the whole earth, and its inhabitants,—theories which are restrained by no reference to existing analogies, and in which a desire is manifested to cut, rather than patiently to untie, the Gordian knot.

We have now, at least, the advantage of knowing, from experience, that an opposite method has always put geologists on the road that leads to truth,—suggesting views which, although imperfect at first, have been found capable of improvement, until at last adopted by universal consent; while the method of speculating on a former distinct state of things and causes, has led invariably to a multitude of contradictory systems, which have been overthrown one after the other,—have been found incapable of modification,—and which have often required to be precisely reversed.

The remainder of this work will be devoted to an investigation of the changes now going on in the crust of the earth and its inhabitants. The importance which the student will attach to such researches will mainly depend in the degree of confidence which he feels in the principles above expounded. If he firmly believes in the resemblance or identity of the ancient and present system of terrestrial changes, he will regard every fact collected respecting the causes in diurnal action as affording him a key to the interpretation of some mystery in the past. Events which have occurred at the most distant periods in the animate and inanimate world, will be acknowledged to throw light on each other, and the deficiency of our information respecting some of the most obscure parts of the present creation will be removed. For as, by studying the external configuration of the existing land and its inhabitants, we may restore in imagination the appearance of the ancient continents which have passed away, so may we obtain from the deposits of ancient seas and lakes an insight into the nature of the subaqueous processes now in operation, and of many forms of organic life, which, though now existing, are veiled from sight. Rocks, also, produced by subterranean fire in former ages, at great depths in the bowels of the earth, present us, when upraised by gradual movements, and exposed to the light of heaven, with an image of those changes which the deep-seated volcano may now occasion in the nether regions. Thus, although we are mere sojourners on the surface of the planet, chained to a mere point in space, enduring but for a moment of time, the human mind is not only enabled to number worlds beyond the unassisted ken of mortal eye, but to trace the events of indefinite ages before the creation of our race, and is not even withheld from penetrating into the dark secrets of the ocean, or the interior of the solid globe; free, like the spirit which the poet described as animating the universe,

——— ire per omnes
Terrasque, tractusque maris, cœlumque profundum.

BOOK II.

CHANGES IN THE INORGANIC WORLD.

AQUEOUS CAUSES.

CHAPTER XIV.

Division of the subject into changes of the organic and inorganic world—Inorganic causes of change divided into aqueous and igneous—Aqueous causes first considered—Fall of rain—Recent rain-prints in mud—Destroying and transporting power of running water—Newly formed valleys in Georgia—Sinuosities of rivers—Two streams when united do not occupy a bed of double surface—Inundations in Scotland—Floods caused by landslips in the White Mountains—Bursting of a lake in Switzerland—Devastations caused by the Anio at Tivoli—Excavations in the lavas of Etna by Sicilian rivers—Gorge of the Simeto—Gradual recession of the cataract of Niagara.

Division of the subject.—GEOLOGY was defined to be the science which investigates the former changes that have taken place in the organic as well as in the inorganic kingdoms of nature. As vicissitudes in the inorganic world are most apparent, and as on them all fluctuations in the animate creation must in a great measure depend, they may claim our first consideration. The great agents of change in the inorganic world may be divided into two principal classes, the aqueous and the igneous. To the aqueous belong Rain, Rivers, Torrents, Springs, Currents, and Tides; to the igneous, Volcanoes, and Earthquakes. Both these classes are instruments of decay as well as of reproduction; but they may also be regarded as antagonist forces. For the aqueous agents are incessantly laboring to reduce the inequalities of the earth's surface to a level; while the igneous are equally active in restoring the unevenness of the external crust, partly by heaping up new matter in certain localities, and partly by depressing one portion, and forcing out another, of the earth's envelope.

It is difficult, in a scientific arrangement, to give an accurate view of the combined effects of so many forces in simultaneous operation; because, when we consider them separately, we cannot easily estimate either the extent of their efficacy, or the kind of results which they produce. We are in danger, therefore, when we attempt to examine the influence exerted singly by each, of overlooking the modifications which they produce on one another; and these are so complicated, that sometimes the igneous and aqueous forces co-operate to produce a joint effect, to which neither of them unaided by the other could give rise,—as when repeated earthquakes unite with running water to widen a valley; or when a thermal spring rises up from a great depth, and conveys the mineral ingredients with which it is impregnated from the interior of the earth to the surface. Sometimes the organic combine with the inorganic causes; as when a reef, composed of shells and corals, protects one line of coast from the destroying power of tides or currents, and turns them against some other point; or when drift timber, floated into a lake, fills a hollow to which the stream would not have had sufficient velocity to convey earthy sediment.

It is necessary, however, to divide our observations on these various causes, and to classify them systematically, endeavoring as much as possible to keep in view that the effects in nature are mixed and not simple, as they may appear in an artificial arrangement.

In treating, in the first place, of the aqueous causes, we may consider them under two divisions; first, those which are connected with the circulation of water from the land to the sea, under which are included all the phenomena of rain, rivers, glaciers, and springs; secondly, those which arise from the movements of water in lakes, seas, and the ocean, wherein are comprised the phenomena of waves, tides, and currents. In turning our attention to the former division, we find that the effects of rivers may be subdivided into, first, those of a destroying and transporting, and, secondly, those of a renovating nature; in the former are included the erosion of rocks and the transportation of matter to lower levels; in the renovating class, the formation of deltas by the influx of sediment, and the shallowing of seas; but these processes are so intimately related to each other, that it will not always be possible to consider them under their separate heads.

Fall of Rain.—It is well known that the capacity of the atmosphere to absorb aqueous vapor, and hold it in suspension, increases with every increment of temperature. This capacity is also found to augment in a higher ratio than the augmentation of the heat. Hence, as was first suggested by the geologist, Dr. Hutton, when two volumes of air, of different temperatures, both saturated with moisture, mingle together, clouds and rain are produced, for a mean degree of heat having resulted from the union of the two moist airs, the excess of vapor previously held in suspension by the warmer of the two is given out, and if it be in sufficient abundance is precipitated in the form of rain.

As the temperature of the atmosphere diminishes gradually from the equator towards the pole, the evaporation of water and the quantity of rain diminish also. According to Humboldt's computation, the average annual depth of rain at the equator is 96 inches, while in lat. 45° it is only 29 inches, and in lat. 60° not more than 17 inches. But there are so many disturbing causes, that the actual discharge, in any given locality, may deviate very widely from this rule. In England, for example, where the average fall at London is 24½ inches, as ascertained at the Greenwich Observatory, there is such irregularity in some districts, that while at Whitehaven, in Cumberland, there fell in 1849, 32 inches, the quantity of rain in Borrowdale, near Keswick (only 15 miles to the westward), was no less than 142 inches!* In like manner, in India, Colonel Sykes found by observations made in 1847 and 1848, that at places situated between 17° and 18° north lat., on a line drawn across the Western Ghauts in the Deccan, the fall of rain varied from 21 to 219 inches.† The annual average in Bengal is probably below 80 inches, yet Dr. G. Hooker witnessed at Churrapoonjee, in the year 1850, a fall of 30 inches in 24 hours, and in the same place during a residence of six months (from June to November) 530 inches! This occurred on the south face of the Khasia (or Garrow) mountains in Eastern Bengal (see map, Chap. XVIII.), where the depth during the whole of the same year probably exceeded 600 inches. So extraordinary a discharge of water, which, as we shall presently see, is very local, may be thus accounted for. Warm, southerly winds, blowing over the Bay of Bengal, and becoming laden with vapor during their passage, reach the low level delta of the Ganges and Brahmapootra, where the ordinary heat exceeds that of the sea, and where evaporation is constantly going on from countless marshes and the arms of the great rivers. A mingling of two masses of damp air of different temperatures probably causes the fall of 70 or 80 inches of rain, which takes place on the plains. The monsoon having crossed the delta, impinges on the Khasia mountains, which rise abruptly from the plain to a mean elevation of between 4000 and 5000 feet. Here the wind not only encounters the cold air of the mountains, but, what is far more effective as a refrigerating cause, the aerial current is made to flow upwards, and to ascend to a height of several thousand feet above the sea. Both the air and the vapor contained in it, being thus relieved of much atmospheric pressure, expand suddenly, and are cooled by rarefaction. The vapor is condensed, and about 500 inches of rain are thrown down annually, nearly twenty times as much as falls in Great Britain in a year, and almost all of it poured down in six months. The channel of every torrent and river is swollen at this season, and much sandstone horizontally stratified, and other rocks are reduced to sand and gravel by the flooded streams. So great is the superficial waste (or *denudation*), that what would otherwise be a rich and luxuriantly wooded region, is converted into a wild and barren moorland.

After the current of warm air has been thus drained of a large portion of its moisture, it still continues its northerly course to the opposite flank of the Khasia range, only 20 miles farther north, and here the fall of rain is reduced to 70 inches in the year. The same wind then blows northwards across the valley of the Brahmapootra, and at length arrives so dry and exhausted at the Bhootan Himalaya (lat. 28° N.), that those mountains, up to the height of 5000 feet, are naked and sterile, and all their outer valleys arid and dusty. The aerial current still con-

* Miller, Phil. Trans. 1851, p. 155.

† Phil. Trans. 1850, p. 354.

tinuing its northerly course and ascending to a higher region, becomes further cooled, condensation again ensues, and Bhootan, above 5000 feet, is densely clothed with vegetation.*

In another part of India, immediately to the westward, similar phenomena are repeated. The same warm and humid winds, copiously charged with aqueous vapor from the Bay of Bengal, hold their course due north for 300 miles across the flat and hot plains of the Ganges, till they encounter the lofty Sikkim mountains. (See map, Chap. XVIII.) On the southern flank of these they discharge such a deluge of rain that the rivers in the rainy season rise twelve feet in as many hours. Numerous landslips, some of them extending three or four thousand feet along the face of the mountains, composed of granite, gneiss, and slate, descend into the beds of streams, and dam them up for a time, causing temporary lakes, which soon burst their barriers. "Day and night," says Dr. Hooker, "we heard the crashing of falling trees, and the sound of boulders thrown violently against each other in the beds of torrents. By such wear and tear rocky fragments swept down from the hills are in part converted into sand and fine mud; and the turbid Ganges, during its annual inundation, derives more of its sediment from this source than from the waste of the fine clay of the alluvial plains below.†

On the verge of the tropics a greater quantity of rain falls annually than at the equator. Yet parts even of the tropical latitudes are entirely destitute of rain: Peru, for example, which owes its vegetation solely to rivers and nightly dews. In that country easterly winds prevail, blowing from the Pacific, and these being intercepted by the Andes, and cooled as they rise, are made to part with all their moisture before reaching the low region to the leeward. The desert zone of North Africa, between lat. 15° and 30° N., is another instance of a rainless region. Five or six consecutive years may pass in Upper Egypt, Nubia, and Dongola, or in the Desert of Sahara, without rain.

From the facts above mentioned, the reader will infer that in the course of successive geological periods there will be great variations in the quantity of rain falling in one and the same region. At one time there may be none whatever during the whole year; at another a fall of 100 or 500 inches; and these two last averages may occur on the two opposite flanks of a mountain-chain, not more than 20 miles wide. While, therefore, the valleys in one district are widened and deepened annually, they may remain stationary in another, the superficial soil being protected from waste by a dense covering of vegetation. This diversity depends on many geographical circumstances, but principally on the height of the land above the sea, the direction of the prevailing winds, and the relative position, at the time being, of the plains, hills, and the ocean, conditions all of which are liable in the course of ages to undergo a complete revolution.

* Hooker's Himalayan Journal, ined.

† Ibid.

Recent rain-prints.—When examining, in 1842, the extensive mud-flats of Nova Scotia, which are exposed at low tide on the borders of the Bay of Fundy, I observed not only the foot-prints of birds which had recently passed over the mud, but also very distinct impressions of rain-drops. A peculiar combination of circumstances renders these mud-flats admirably fitted to receive and retain any markings which may happen to be made on their surface. The sediment with which the waters are charged is extremely fine, being derived from the destruction of cliffs of red sandstone and shale, and as the tides rise fifty feet and upwards, large areas are laid dry for nearly a fortnight between the spring and neap tides. In this interval the mud is baked in summer by a hot sun, so that it solidifies and becomes traversed by cracks, caused by shrinkage. Portions of the hardened mud between these cracks may then be taken up and removed without injury. On examining the edges of each slab, we observe numerous layers, formed by successive tides, each layer being usually very thin, sometimes only one-tenth of an inch thick. When a shower of rain falls, the highest portion of the mud-covered flat is usually too hard to receive any impressions; while that recently uncovered by the tide near the water's edge is too soft. Between these areas a zone occurs, almost as smooth and even as a looking-glass, on which every drop forms a cavity of circular or oval form, and, if the shower be transient, these pits retain their shape permanently, being dried by the sun, and being then too firm to be effaced by the action of the succeeding tide, which deposits upon them a new layer of mud. Hence we often find, in splitting open a slab an inch or more thick, on the upper surface of which the marks of recent rain occur, that an inferior layer, deposited during some previous rise of the tide, exhibits on its under side perfect casts of rain-prints, which stand out in relief, the moulds of the same being seen on the layer below. But in some cases, especially in the more sandy layers, the markings have been somewhat blunted by the tide, and by several rain-prints having been joined into one by a repetition of drops falling on the same spot; in which case the casts present a very irregular and blistered appearance.

The finest examples which I have seen of these rain-prints were sent to me by Dr. Webster, from Kentville, on the borders of the Bay of Mines, in Nova Scotia. They were made by a heavy shower which fell on the 21st of July, 1849, when the rise and fall of the tides were at their maximum. The impressions (see fig. 13) consist of cup-shaped or hemispherical cavities, the average size of which is from one-eighth to one-tenth of an inch across, but the largest are fully half an inch in diameter, and one-tenth of an inch deep. The depth is chiefly below the general surface or plane of stratification, but the walls of the cavity consist partly of a prominent rim of sandy mud, formed of the matter which has been forcibly expelled from the pit. All the cavities having an oval form are deeper at one end, where they have also a higher rim, and all the deep ends have the same direction, showing towards which quarter the wind was blowing. Two or more drops are sometimes seen to have interfered with each other; in which case it is usually possible to determine which drop fell last, its rim being unbroken.

Fig. 13.

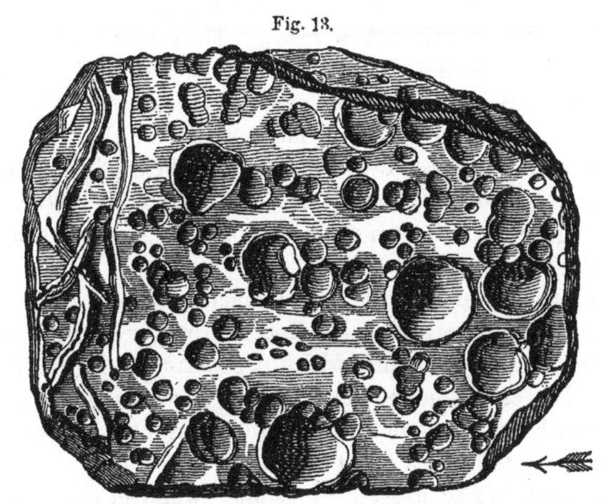

Recent rain-prints, formed July 21, 1849, at Kentville, Bay of Fundy, Nova Scotia. The arrow represents the direction of the shower.

On some of the specimens the winding tubular tracks of worms are seen, which have been bored just beneath the surface (see fig. 13, *left side*). They occasionally pass under the middle of a rain-mark, having been formed subsequently. Sometimes the worms have dived beneath the surface, and then reappeared. All these appearances, both of rain-prints and worm-tracks, are of great geological interest, as their exact counterparts are seen in rocks of various ages, even in formations of very high antiquity.* Small cavities, often corresponding in size to those produced by rain, are also caused by air-bubbles rising up through sand or mud; but these differ in character from rain-prints, being usually deeper than they are wide, and having their sides steeper. These, indeed, are occasionally vertical, or overarching, the opening at the top being narrower than the pit below. In their mode, also, of mutual interference they are unlike rain-prints.†

In consequence of the effects of mountains in cooling currents of moist air, and causing the condensation of aqueous vapor in the manner above described, it follows that in every country, as a general rule, the more elevated regions become perpetual reservoirs of water, which descends and irrigates the lower valleys and plains. The largest quantity of water is first carried to the highest region, and then made to descend by steep declivities towards the sea; so that it acquires superior velocity, and removes more soil, than it would do if the rain had been distributed over the plains and mountains equally in proportion to their relative

* See Manual of Geology, Index, *Rain-prints.*

† See Lyell on recent and fossil rains. Quart. Journ. Geol. Soc. 1851, vol. vii. p. 239.

areas. The water is also made by these means to pass over the greatest distances before it can regain the sea.

It has already been observed that in higher latitudes, where the atmosphere being colder is capable of holding less water in suspension, a diminished fall of rain takes place. Thus at St. Petersburg, the amount is only 16 inches, and at Uleaborg in the Gulf of Bothnia (N. lat. 65°), only 13½ inches, or less than half the average of England, and even this small quantity descends more slowly in the temperate zone, and is spread more equally over the year than in tropical climates. But in reference to geological changes, frost in the colder latitude acts as a compensating power in the disintegration of rocks, and the transportation of stones to lower levels.

Water when converted into ice augments in bulk more than one-twentieth of its volume, and owing to this property it widens the minute crevices (or *joints*) of rocks into which it penetrates. Ice also in various ways, as will be shown in the next chapter, gives buoyancy to mud and sand, even to huge blocks of stone, enabling rivers of moderate size and velocity to carry them to a great distance.

The mechanical force exerted by running water in undermining cliffs, and rounding off the angles of hard rock, is mainly due to the intermixture of foreign ingredients. Sand and pebbles, when hurried along by the violence of the stream, are thrown against every obstacle lying in their way, and thus a power of attrition is acquired, capable of wearing through the hardest siliceous stones, on which water alone could make no impression.

Newly formed valleys.—When travelling in Georgia and Alabama, in 1846, I saw in both those States the commencement of hundreds of valleys in places where the native forest had recently been removed. One of these newly formed gulleys or ravines is represented in the annexed woodcut (fig. 14), from a drawing which I made on the spot. It occurs three miles and a half due west of Milledgeville, the capital of Georgia, and is situated on the farm of Pomona, on the direct road to Macon.*

Twenty years ago, before the land was cleared, it had no existence; but when the trees of the forest were cut down, cracks three feet deep were caused by the sun's heat in the clay; and, during the rains, a sudden rush of water through the principal crack deepened it at its lower extremity, from whence the excavating power worked backwards, till, in the course of twenty years, a chasm, measuring no less than 55 feet in depth, 300 yards in length, and varying in width from 20 to 180 feet, was the result. The high road has been several times turned to avoid this cavity, the enlargement of which is still proceeding, and the old line of road may be seen to have held its course directly over what is now the wildest part of the ravine. In the perpendicular walls of this great chasm appear beds of clay and sand, red, white, yellow, and

* Lyell's Second Visit to the United States, 1846, vol. ii. p. 25.

Fig. 14.

Ravine on the farm of Pomona, near Milledgeville, Georgia, as it appeared January, 1846.
Excavated in twenty years, 55 feet deep, and 180 feet broad.

green, produced by the decomposition in situ of hornblendic gneiss, with layers and veins of quartz, which remain entire, to prove that the whole mass was once solid and crystalline.

I infer, from the rapidity of the denudation which only began here after the removal of the native wood, that this spot, elevated about 600 feet above the sea, has been always covered with a dense forest, from the remote time when it first emerged from the sea. The termination of the cavity on the right hand in the foreground is the head or upper end of the ravine, and in almost every case, such gulleys are lengthened by the streams cutting their way backwards. The depth at the upper end is often, as in this case, considerable, and there is usually at this point, during floods, a small cascade.

Sinuosities of rivers.—In proportion as such valleys are widened, sinuosities are caused by the deflection of the stream first to one side and then to the other. The unequal hardness of the materials through which the channel is eroded tends partly to give new directions to the lateral force of excavation. When by these, or by accidental shiftings of the alluvial matter in the channel, the current is made to cross its general line of descent, it eats out a curve in the opposite bank, or in the side of the hills bounding the valley, from which curve it is turned back again at an equal angle, so that it recrosses the line of descent, and gradually hollows out another curve lower down in the opposite bank, till the whole sides of the valley, or river bed, present a succession of salient and retiring angles. Among the causes of deviation from a straight course, by which torrents and rivers tend in mountainous regions to widen the valleys through which they flow, may be mentioned the confluence of lateral torrents, swollen irregularly at different seasons by partial storms, and discharging at different times unequal quantities of sand, mud, and pebbles, into the main channel.

When the tortuous flexures of a river are extremely great, as often happens in alluvial plains, the aberration from the direct line of descent may be restored by the river cutting through the isthmus which separates two neighboring curves. Thus in the annexed diagram, the extreme sinuosity of the river has caused it to return for a brief space

Fig. 15.

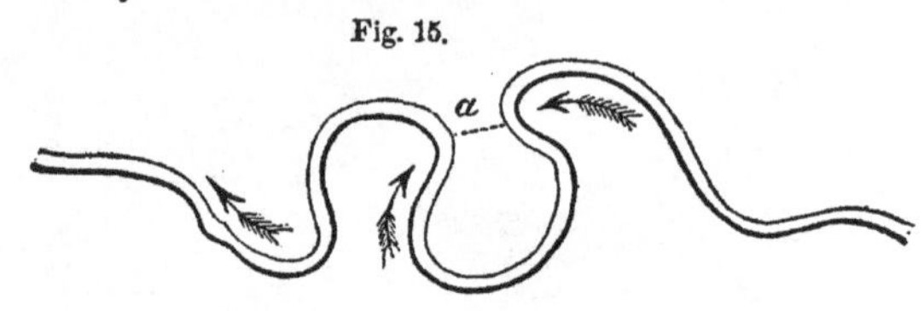

in a contrary direction to its main course, so that a peninsula is formed, and the isthmus (at *a*) is consumed on both sides by currents flowing in opposite directions. In this case an island is soon formed,—on either side of which a portion of the stream usually remains.

Transporting power of water.—In regard to the transporting power of water, we may often be surprised at the facility with which streams of a small size, and descending a slight declivity, bear along coarse sand and gravel; for we usually estimate the weight of rocks in air, and do not reflect on their comparative buoyancy when submerged in a denser fluid. The specific gravity of many rocks is not more than twice that of water, and very rarely more than thrice, so that almost all the fragments propelled by a stream have lost a third, and many of them a half, of what we usually term their weight.

It has been proved by experiment, in contradiction to the theories of the earlier writers on hydrostatics, to be a universal law, regulating the motion of running water, that the velocity at the bottom of the stream is everywhere less than in any part above it, and is greatest at the surface. Also that the superficial particles in the middle of the stream move swifter than those at the sides. This retardation of the lowest and lateral currents is produced by friction; and when the velocity is sufficiently great, the soil composing the sides and bottom gives way. A velocity of three inches per second at the bottom is ascertained to be sufficient to tear up fine clay,—six inches per second, fine sand,—twelve inches per second, fine gravel,—and three feet per second, stones of the size of an egg.*

When this mechanical power of running water is considered, we are prepared for the transportation before alluded to of large quantities of gravel, sand, and mud, by torrents which descend from mountainous regions. But a question naturally arises, How the more tranquil rivers of the valleys and plains, flowing on comparatively level ground, can remove the prodigious burden which is discharged into them by their numerous tributaries, and by what means they are enabled to convey the whole mass to the sea? If they had not this removing power, their channels would be annually choked up, and the valleys of the lower country, and plains at the base of mountain-chains, would be continually strewed over with fragments of rock and sterile sand. But this evil is prevented by a general law regulating the conduct of running water,—that two equal streams do not, when united, occupy a bed of double surface. Nay, the width of the principal river, after the junction of a tributary, sometimes remains the same as before, or is even lessened. The cause of this apparent paradox was long ago explained by the Italian writers, who had studied the confluence of the Po and its feeders in the plains of Lombardy.

The addition of a smaller river augments the velocity of the main stream, often in the same proportion as it does the quantity of water. Thus the Venetian branch of the Po swallowed up the Ferranese branch and that of Panaro without any enlargement of its own dimensions. The cause of the greater velocity is, first, that after the union of two rivers the water, in place of the friction of four shores, has only that of two to surmount; 2dly, because the main body of the stream being farther distant from the banks, flows on with less interruption; and lastly, because a greater quantity of water moving more swiftly, digs deeper into the river's bed. By this beautiful adjustment, the water which drains the interior country is made continually to occupy less room as it approaches the sea; and thus the most valuable part of our continents, the rich deltas and great alluvial plains, are prevented from being constantly under water.

River floods in Scotland, 1829.—Many remarkable illustrations of the power of running water in moving stones and heavy materials were afforded by the storm and floods which occurred on the 3d and 4th of August, 1829, in Aberdeenshire and other counties in Scotland. The elements during this storm assumed all the characters which mark the tropical hurricanes; the wind blowing in sudden gusts and whirlwinds, the lightning and thunder being such as is rarely witnessed in our climate, and heavy rain falling without intermission. The floods extended almost simultaneously, and with equal violence over that part

* Encyc. Brit. art. *Rivers*.

of the northeast of Scotland which would be cut off by two lines drawn from the head of Lochrannoch, one towards Inverness and the other to Stonehaven. The united line of the different rivers which were flooded, could not be less than from five to six hundred miles in length; and the whole of their courses were marked by the destruction of bridges, roads, crops, and buildings. Sir T. D. Lauder has recorded the destruction of thirty-eight bridges, and the entire obliteration of a great number of farms and hamlets. On the Nairn, a fragment of sandstone, fourteen feet long by three feet wide and one foot thick, was carried above 200 yards down the river. Some new ravines were formed on the sides of mountains where no streams had previously flowed, and ancient river-channels, which had never been filled from time immemorial, gave passage to a copious flood.*

The bridge over the Dee at Ballater consisted of five arches, having upon the whole a water-way of 260 feet. The bed of the river, on which the piers rested, was composed of rolled pieces of granite and gneiss. The bridge was built of granite, and had stood uninjured for twenty years; but the different parts were swept away in succession by the flood, and the whole mass of masonry disappeared in the bed of the river. "The river Don," observes Mr. Farquharson, in his account of the inundations, "has upon my own premises forced a mass of four or five hundred tons of stones, many of them two or three hundred pounds' weight, up an inclined plane, rising six feet in eight or ten yards, and left them in a rectangular heap, about three feet deep on a flat ground:—the heap ends abruptly at its lower extremity."†

The power even of a small rivulet, when swollen by rain, in removing heavy bodies, was exemplified in August, 1827, in the College, a small stream which flows at a slight declivity from the eastern watershed of the Cheviot Hills. Several thousand tons' weight of gravel and sand were transported to the plain of the Till, and a bridge, then in progress of building, was carried away, some of the arch-stones of which, weighing from half to three quarters of a ton each, were propelled two miles down the rivulet. On the same occasion, the current tore away from the abutment of a mill-dam a large block of greenstone-porphyry, weighing nearly two tons, and transported it to the distance of a quarter of a mile. Instances are related as occurring repeatedly, in which from one to three thousand tons of gravel are, in like manner, removed by this streamlet to still greater distances in one day.‡

Floods caused by landslips, 1826.—The power which running water may exert in the lapse of ages, in widening and deepening a valley, does not so much depend on the volume and velocity of the stream usually flowing in it, as on the number and magnitude of the obstructions which have, at different periods, opposed its free passage. If a torrent, however small, be effectually dammed up, the size of the valley

* Sir T. D. Lauder's Account of the Great Floods in Morayshire, August, 1829.
† Quarterly Jour. of Sci. &c. No. xii. New Series, p. 331.
‡ Culley, Proceed. Geol. Soc. 1829.

above the barrier, and its declivity below, and not the dimensions of the torrent, will determine the violence of the débâcle. The most universal source of local deluges, are landslips, slides, or avalanches, as they are sometimes called, when great masses of rock and soil, or sometimes ice and snow, are precipitated into the bed of a river, the boundary cliffs of which have been thrown down by the shock of an earthquake, or undermined by springs or other causes. Volumes might be filled with the enumeration of instances on record of these terrific catastrophes; I shall therefore select a few examples of recent occurrence, the facts of which are well authenticated.

Two dry seasons in the White Mountains, in New Hampshire (United States), were followed by heavy rains on the 28th August, 1826, when from the steep and lofty declivities which rise abruptly on both sides of the river Saco, innumerable rocks and stones, many of sufficient size to fill a common apartment, were detached, and in their descent swept down before them, in one promiscuous and frightful ruin, forests, shrubs, and the earth which sustained them. Although there are numerous indications on the steep sides of these hills of former slides of the same kind, yet no tradition had been handed down of any similar catastrophe within the memory of man, and the growth of the forest on the very spots now devastated, clearly showed that for a long interval nothing similar had occurred. One of these moving masses was afterwards found to have slid three miles, with an average breadth of a quarter of a mile. The natural excavations commenced generally in a trench a few yards in depth and a few rods in width, and descended the mountains, widening and deepening till they became vast chasms. At the base of these hollow ravines was seen a confused mass of ruins, consisting of transported earth, gravel, rocks, and trees. Forests of spruce-fir and hemlock, a kind of fir somewhat resembling our yew in foliage, were prostrated with as much ease as if they had been fields of grain; for, where they disputed the ground, the torrent of mud and rock accumulated behind, till it gathered sufficient force to burst the temporary barrier.

The valleys of the Amonoosuck and Saco presented, for many miles, an uninterrupted scene of desolation; all the bridges being carried away, as well as those over their tributary streams. In some places, the road was excavated to the depth of from fifteen to twenty feet; in others, it was covered with earth, rocks, and trees, to as great a height. The water flowed for many weeks after the flood, as densely charged with earth as it could be without being changed into mud, and marks were seen in various localities of its having risen on either side of the valley to more than twenty-five feet above its ordinary level. Many sheep and cattle were swept away, and the Willey family, nine in number, who in alarm had deserted their house, were destroyed on the banks of the Saco; seven of their mangled bodies were afterwards found near the river, buried beneath drift-wood and mountain ruins.* Eleven years

* Silliman's Journal, vol. xv. No. 2, p. 216. Jan. 1829.

after the event, the deep channels worn by the avalanches of mud and stone, and the immense heaps of boulders and blocks of granite in the river channel, still formed, says Professor Hubbard, a picturesque feature in the scenery.*

When I visited the country in 1845, eight years after Professor Hubbard, I found the signs of devastation still very striking; I also particularly remarked that although the surface of the bare granitic rocks had been smoothed by the passage over them of so much mud and stone, there were no continuous parallel and rectilinear furrows, nor any of the fine scratches or striæ which characterize *glacial* action. The absence of these is nowhere more clearly exemplified than in the bare rocks over which passed the great "Willey slide" of 1826.†

But the catastrophes in the White Mountains are insignificant, when compared to those which are occasioned by earthquakes, when the boundary hills, for miles in length, are thrown down into the hollow of a valley. I shall have opportunities of alluding to inundations of this kind, when treating expressly of earthquakes, and shall content myself at present with selecting an example of a flood due to a different cause.

Flood in the valley of Bagnes, 1818.—The valley of Bagnes is one of the largest of the lateral embranchments of the main valley of the Rhone, above the Lake of Geneva. Its upper portion was, in 1818, converted into a lake by the damming up of a narrow pass, by avalanches of snow and ice, precipitated from an elevated glacier into the bed of the river Dranse. In the winter season, during continued frost, scarcely any water flows in the bed of this river to preserve an open channel, so that the ice barrier remained entire until the melting of the snows in spring, when a lake was formed above, about half a league in length, which finally attained in some parts a depth of about two hundred feet, and a width of about seven hundred feet. To prevent or lessen the mischief apprehended from the sudden bursting of the barrier, an artificial gallery, seven hundred feet in length, was cut through the ice, before the waters had risen to a great height. When at length they accumulated and flowed through this tunnel, they dissolved the ice, and thus deepened their channel, until nearly half of the whole contents of the lake were slowly drained off. But at length, on the approach of the hot season, the central portion of the remaining mass of ice gave way with a tremendous crash, and the residue of the lake was emptied in half an hour. In the course of its descent, the waters encountered several narrow gorges, and at each of these they rose to a great height, and then burst with new violence into the next basin, sweeping along rocks, forests, houses, bridges, and cultivated land. For the greater part of its course the flood resembled a moving mass of rock and mud, rather than of water. Some fragments of granitic rocks, of enormous magnitude, and which from their dimensions, might be compared without exaggeration

* Silliman's Journal, vol. xxxiv. p. 115.

† See Lyell's Second Visit to the U. S. vol. i. p. 69.

to houses, were torn out of a more ancient alluvion, and borne down for a quarter of a mile. One of the fragments moved was sixty paces in circumference.* The velocity of the water, in the first part of its course, was thirty-three feet per second, which diminished to six feet before it reached the Lake of Geneva, where it arrived in six hours and a half, the distance being forty-five miles.†

This flood left behind it, on the plains of Martigny, thousands of trees torn up by the roots, together with the ruins of buildings. Some of the houses in that town were filled with mud up to the second story. After expanding in the plain of Martigny, it entered the Rhone, and did no farther damage; but some bodies of men, who had been drowned above Martigny, were afterwards found, at the distance of about thirty miles, floating on the farther side of the Lake of Geneva, near Vevay.

The waters, on escaping from the temporary lake, intermixed with mud and rock, swept along, for the first four miles, at the rate of above twenty miles an hour; and M. Escher, the engineer, calculated that the flood furnished 300,000 cubic feet of water every second—an efflux which is five times greater than that of the Rhine below Basle. Now, if part of the lake had not been gradually drained off, the flood would have been nearly double, approaching in volume to some of the largest rivers in Europe. It is evident, therefore, that when we are speculating on the excavating force which a river may have exerted in any particular valley, the most important question is, not the volume of the existing stream, nor the present levels of its channel, nor even the nature of the rocks, but the probability of a succession of floods at some period since the time when the valley may have been first elevated above the sea.

For several months after the débâcle of 1818, the Dranse, having no settled channel, shifted its position continually from one side to the other of the valley, carrying away newly-erected bridges, undermining houses, and continuing to be charged with as large a quantity of earthy matter as the fluid could hold in suspension. I visited this valley four months after the flood, and was witness to the sweeping away of a bridge, and the undermining of part of a house. The greater part of the ice-barrier was then standing, presenting vertical cliffs 150 feet high, like ravines in the lava-currents of Etna or Auvergne, where they are intersected by rivers.

Inundations, precisely similar, are recorded to have occurred at former periods in this district, and from the same cause. In 1595, for example, a lake burst, and the waters, descending with irresistible fury, destroyed the town of Martigny, where from sixty to eighty persons perished. In a similar flood, fifty years before, 140 persons were drowned.

Flood at Tivoli, 1826.—I shall conclude with one more example derived from a land of classic recollections, the ancient Tibur, and which,

* This block was measured by Capt. B. Hall, R. N.

† Inundation of the Val de Bagnes, in 1818, Ed. Phil. Journ., vol. i. p. 187, from memoir of M. Escher.

like all the other inundations above alluded to, occurred within the present century. The younger Pliny, it will be remembered, describes a flood on the Anio, which destroyed woods, rocks, and houses, with the most sumptuous villas and works of arts.* For four or five centuries consecutively, this "headlong stream," as Horace truly called it, has often remained within its bounds, and then, after so long an interval of rest, has at different periods inundated its banks again, and widened its channel. The last of these catastrophes happened 15th Nov. 1826, after heavy rains, such as produced the floods before alluded to in Scotland. The waters appear also to have been impeded by an artificial dike, by which they were separated into two parts, a short distance above Tivoli. They broke through this dike; and leaving the left trench dry, precipitated themselves, with their whole weight, on the right side. Here they undermined, in the course of a few hours, a high cliff, and widened the river's channel about fifteen paces. On this height stood the church of St. Lucia, and about thirty-six houses of the town of Tivoli, which were all carried away, presenting as they sank into the roaring flood, a terrific scene of destruction to the spectators on the opposite bank. As the foundations were gradually removed, each building, some of them edifices of considerable height, was first traversed with numerous rents, which soon widened into large fissures, until at length the roofs fell in with a crash, and then the walls sunk into the river, and were hurled down the cataract below.†

The destroying agency of the flood came within two hundred yards of the precipice on which the beautiful temple of Vesta stands; but fortunately this precious relic of antiquity was spared, while the wreck of modern structures was hurled down the abyss. Vesta, it will be remembered, in the heathen mythology, personified the stability of the earth; and when the Samian astronomer, Aristarchus, first taught that the earth revolved on its axis, and round the sun, he was publicly accused of impiety, "for removing the everlasting Vesta from her place." Playfair observed, that when Hutton ascribed instability to the earth's surface, and represented the continents which we inhabit as the theatre of incessant change and movement, his antagonists, who regarded them as unalterable, assailed him in a similar manner with accusations founded on religious prejudices.‡ We might appeal to the excavating power of the Anio as corroborative of one of the most controverted parts of the Huttonian theory; and if the days of omens had not gone by, the geologists who now worship Vesta might regard the late catastrophe as portentous. We may, at least, recommend the modern votaries of the goddess to lose no time in making a pilgrimage to her shrine, for the next flood may not respect the temple.

Excavation of rocks by running water.—The rapidity with which

* Lib. viii. Epist. 17.

† When at Tivoli, in 1829, I received this account from eye-witnesses of the event.

‡ Illustr. of Hutt. Theory, § 3, p. 147.

even the smallest streams hollow out deep channels in soft and destructible soils is remarkably exemplified in volcanic countries, where the sand and half-consolidated tuffs opposed but a slight resistance to the torrents which descend the mountain-side. After the heavy rains which followed the eruption of Vesuvius in 1824, the water flowing from the Atrio del Cavallo cut, in three days, a new chasm through strata of tuff and ejected volcanic matter, to the depth of twenty-five feet. I found the old mule-road, in 1828, intersected by this new ravine.

The gradual erosion of deep chasms through some of the hardest rocks, by the constant passage of running water, charged with foreign matter, is another phenomenon of which striking examples may be adduced. Illustrations of this excavating power are presented by many valleys in central France where the channels of rivers have been barred up by solid currents of lava, through which the streams have re-excavated a passage, to the depth of from twenty to seventy feet and upwards, and often of great width. In these cases there are decisive proofs that neither the sea, nor any denuding wave or extraordinary body of water, has passed over the spot since the melted lava was consolidated. Every hypothesis of the intervention of sudden and violent agency is entirely excluded, because the cones of *loose* scoriæ, out of which the lavas flowed, are oftentimes at no great elevation above the rivers, and have remained undisturbed during the whole period which has been sufficient for the hollowing out of such enormous ravines.

Recent excavation by the Simeto.—But I shall at present confine myself to examples derived from events which have happened since the time of history.

At the western base of Etna, a current of lava (A A, fig. 16), descending from near the summit of the great volcano, has flowed to the distance of five or six miles, and then reached the alluvial plain of the

Fig. 16.

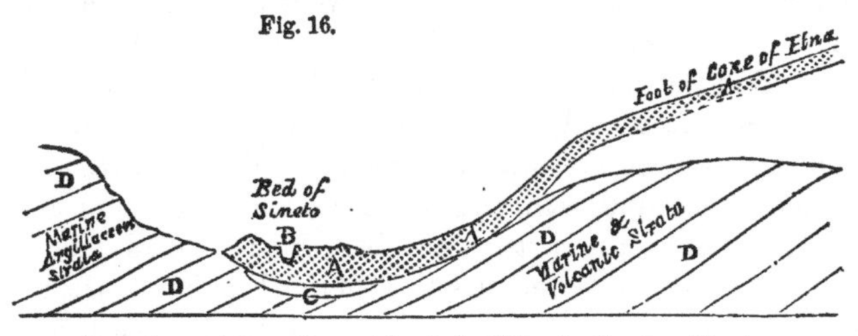

Recent excavation of lava at the foot of Etna by the river Simeto.

Simeto, the largest of the Sicilian rivers, which skirts the base of Etna, and falls into the sea a few miles south of Catania. The lava entered the river about three miles above the town of Aderno, and not only occupied its channel for some distance, but, crossing to the opposite side of the valley, accumulated there in a rocky mass. Gemmellaro gives the year 1603 as the date of the eruption.* The appearance of the

* Quadro Istorico dell' Etna, 1824.

current clearly proves, that it is one of the most modern of those of Etna; for it has not been covered or crossed by subsequent streams or ejections, and the olives which had been planted on its surface were all of small size, when I examined the spot in 1828, yet they were older than the natural wood on the same lava. In the course, therefore, of about two centuries, the Simeto has eroded a passage from fifty to several hundred feet wide, and in some parts from forty to fifty feet deep.

The portion of lava cut through is in no part porous or scoriaceous, but consists of a compact homogeneous mass of hard blue rock, somewhat inferior in weight to ordinary basalt, and containing crystals of olivine and glassy felspar. The general declivity of this part of the bed of the Simeto is not considerable; but, in consequence of the unequal waste of the lava, two water-falls occur at Passo Manzanelli, each about six feet in height. Here the chasm (B, fig. 16) is about forty feet deep, and only fifty broad.

The sand and pebbles in the river-bed consist chiefly of a brown quartzose sandstone, derived from the upper country; but the materials of the volcanic rock itself must have greatly assisted the attrition. This river, like the Caltabiano on the eastern side of Etna, has not yet cut down to the ancient bed of which it was dispossessed, and of which the probable position is indicated in the annexed diagram (C, fig. 16).

On entering the narrow ravine where the water foams down the two cataracts, we are entirely shut out from all view of the surrounding country; and a geologist who is accustomed to associate the characteristic features of the landscape with the relative age of certain rocks, can scarcely dissuade himself from the belief that he is contemplating a scene in some rocky gorge of a primary district. The external forms of the hard blue lava are as massive as any of the most ancient trap-rocks of Scotland. The solid surface is in some parts smoothed and almost polished by attrition, and covered in others with a white lichen, which imparts to it an air of extreme antiquity, so as greatly to heighten the delusion. But the moment we reascend the cliff the spell is broken; for we scarcely recede a few paces, before the ravine and river disappear, and we stand on the black and rugged surface of a vast current of lava, which seems unbroken, and which we can trace up nearly to the distant summit of that majestic cone which Pindar called "the pillar of heaven," and which still continues to send forth a fleecy wreath of vapor, reminding us that its fires are not extinct, and that it may again give out a rocky stream, wherein other scenes like that now described may present themselves to future observers.

Falls of Niagara.—The falls of Niagara afford a magnificent example of the progressive excavation of a deep valley in solid rock. That river flows over a flat table-land, in a depression of which Lake Erie is situated. Where it issues from the lake, it is nearly a mile in width, and 330 feet above Lake Ontario, which is about 30 miles distant. For the first fifteen miles below Lake Erie the surrounding country, comprising Upper Canada on the west, and the state of New York on the

Fig. 17.

Lake Erie.
The Falls.
Lime-
stone
Shale.
Lewiston.
Niagara River.
Queenstown.

east, is almost on a level with its banks, and nowhere more than thirty or forty feet above them.* (See fig. 17.). The river being occasionally interspersed with low wooded islands, and having sometimes a width of three miles, glides along at first with a clear, smooth, and tranquil current, falling only fifteen feet in as many miles, and in this part of its course resembling an arm of Lake Erie. But its character is afterwards entirely changed, on approaching the Rapids, where it begins to rush and foam over a rocky and uneven limestone bottom, for the space of nearly a mile, till at length it is thrown down perpendicularly 165 feet at the Falls. Here the river is divided into two sheets of water by an island, the largest cataract being more than a third of a mile broad, the smaller one having a breadth of six hundred feet. When the water has precipitated itself into an unfathomable pool, it rushes with great velocity down the sloping bottom of a narrow chasm, for a distance of seven miles. This ravine varies from 200 to 400 yards in width from cliff to cliff; contrasting, therefore, strongly in its breadth with that of the river above. Its depth is from 200 to 300 feet, and it intersects for about seven miles the table-land before described, which terminates suddenly at Queenstown in an escarpment or long line of inland cliff facing northwards, towards Lake Ontario. The Niagara, on reaching the escarpment and issuing from the gorge, enters the flat country, which is so nearly on a level with Lake Ontario, that there is only a fall of about four feet in the seven additional miles which intervene between Queenstown and the shores of that lake.

It has long been the popular belief that the Niagara once flowed in a shallow valley across the whole platform, from the present site of the Falls to the escarpment (called the Queenstown heights), where it is supposed that the cataract was first situated, and that the river has been slowly eating its way backwards through the rocks for the distance of seven miles. This hypothesis naturally suggests itself to every observer, who sees the narrowness of the gorge at its termination, and throughout its whole course, as far up as the Falls, above which point the river expands as before stated. The boundary cliffs of the ravine are usually perpendicular, and in many places undermined on one side by the impetuous stream. The uppermost rock of the table-land at the Falls consists of hard limestone (a member of the Silurian series), about ninety feet thick, beneath which lie soft shales of equal thickness, continually undermined by the action of the spray, which rises from the pool into which so large a body of water is projected, and is driven violently by gusts of wind against the base of the

* The reader will find in my Travels in North America, vol. i. ch. 2, a colored geological map and section of the Niagara district, also a bird's-eye view of the Falls and adjacent country, colored geologically, of which the first idea was suggested by the excellent original sketch given by Mr. Bakewell. I have referred more fully to these and to Mr. Hall's Report on the Geology of New York, as well as to the earlier writings of Hennepin and Kalm in the same work, and have speculated on the origin of the escarpment over which the Falls may have been originally precipitated. Vol. i. p. 32, and vol. ii. p. 93.

precipice. In consequence of this action, and that of frost, the shale disintegrates and crumbles away, and portions of the incumbent rock overhang 40 feet, and often when unsupported tumble down, so that the Falls do not remain absolutely stationary at the same spot, even for half a century. Accounts have come down to us, from the earliest period of observation, of the frequent destruction of these rocks, and the sudden descent of huge fragments in 1818 and 1828, are said to have shaken the adjacent country like an earthquake. The earliest travellers, Hennepin and Kalm, who in 1678 and 1751 visited the Falls, and published views of them, attest the fact, that the rocks have been suffering from dilapidation for more than a century and a half, and that some slight changes, even in the scenery of the cataract have been brought about within that time. The idea, therefore, of perpetual and progressive waste is constantly present to the mind of every beholder; and as that part of the chasm, which has been the work of the last hundred and fifty years resembles precisely, in depth, width, and character, the rest of the gorge which extends seven miles below, it is most natural to infer, that the entire ravine has been hollowed out in the same manner, by the recession of the cataract.

It must at least be conceded, that the river supplies an adequate cause for executing the whole task thus assigned to it, provided we grant sufficient time for its completion. As this part of the country was a wilderness till near the end of the last century, we can obtain no accurate data for estimating the exact rate at which the cataract has been receding. Mr. Bakewell, son of the eminent geologist of that name, who visited the Niagara in 1829, made the first attempt to calculate from the observations of one who had lived forty years at the Falls, and who had been the first settler there, that the cataract had during that period gone back about a yard annually. But after the most careful inquiries which I was able to make, during my visit to the spot in 1841–2, I came to the conclusion that the average of one foot a year would be a much more probable conjecture. In that case, it would have required thirty-five thousand years for the retreat of the Falls, from the escarpment of Queenstown to their present site. It seems by no means improbable that such a result would be no exaggeration of the truth, although we cannot assume that the retrograde movement has been uniform. An examination of the geological structure of the district, as laid open in the ravine, shows that at every step in the process of excavation, the height of the precipice, the hardness of the materials at its base, and the quantity of fallen matter to be removed, must have varied. At some points it may have receded much faster than at present, but in general its progress was probably slower, because the cataract, when it began to recede, must have had nearly twice its present height.

From observations made by me in 1841, when I had the advantage of being accompanied by Mr. Hall, state geologist of New York, and in 1842, when I re-examined the Niagara district, I obtained geological evidence of the former existence of an old river-bed, which, I have no doubt, indicates the original channel through which the waters once flowed from the Falls to Queenstown, at the height of nearly three hundred feet above the bottom of the present gorge. The geological monuments alluded to, consist of patches of sand and gravel, forty feet thick, containing fluviatile shells of the genera Unio, Cyclas, Melania, &c., such as now inhabit the waters of the Niagara above the Falls. The identity of the fossil species with the recent is unquestionable, and these freshwater deposits occur at the edge of the cliffs bounding the ravine, so that they prove the former extension of an elevated shallow valley, four miles below the falls, a distinct prolongation of that now occupied by the Niagara, in the elevated region intervening between Lake Erie and the Falls. Whatever theory be framed for the hollowing out of the ravine further down, or for the three miles which intervene between the whirlpool and Queenstown, it will always be necessary to suppose the former existence of a barrier of *rock*, not of loose and destructible materials, such as those composing the drift in this district, somewhere immediately below the whirlpool. By that barrier the waters were held back for ages, when the fluviatile deposit, 40 feet in thickness, and 250 feet above the present channel of the river, originated. If we are led by this evidence to admit that the cataract has cut back its way for four miles, we can have little hesitation in referring the excavation of the remaining three miles below to a like agency, the shape of the chasm being precisely similar.

There have been many speculations respecting the future recession of the Falls, and the deluge that might be occasioned by the sudden escape of the waters of Lake Erie, if the ravine should ever be prolonged 16 miles backwards. But a more accurate knowledge of the geological succession of the rocks, brought to light by the State Survey, has satisfied every geologist that the Falls would diminish gradually in height before they travelled back two miles, and in consequence of a gentle dip of the strata to the south, the massive limestone now at the top would then be at their base, and would retard, and perhaps put an effectual stop to, the excavating process.

CHAPTER XV.

TRANSPORTATION OF SOLID MATTER BY ICE.

Carrying power of river-ice—Rocks annually conveyed into the St. Lawrence by its tributaries—Ground-ice; its origin and transporting power—Glaciers—Theory of their downward movement—Smoothed and grooved rocks—The moraine unstratified—Icebergs covered with mud and stones—Limits of glaciers and icebergs—Their effects on the bottom when they run aground—Packing of coast-ice—Boulders drifted by ice on coast of Labrador—Blocks moved by ice in the Baltic.

THE power of running water to carry sand, gravel, and fragments of rock to considerable distances is greatly augmented in those regions where, during some part of the year, the frost is of sufficient intensity to convert the water, either at the surface or bottom of rivers, into ice.

This subject may be considered under three different heads:—first, the effect of surface-ice and ground-ice in enabling streams to remove gravel and stones to a distance; secondly, the action of glaciers in the transport of boulders, and in the polishing and scratching of rocks; thirdly, the floating off of glaciers charged with solid matter into the sea, and the drifting of icebergs and coast-ice.

River-ice.—Pebbles and small pieces of rock may be seen entangled in ice, and floating annually down the Tay in Scotland, as far as the mouth of that river. Similar observations might doubtless be made respecting almost all the larger rivers of England and Scotland; but there seems reason to suspect that the principal transfer from place to place of pebbles and stones adhering to ice goes on unseen by us under water. For although the specific gravity of the compound mass may cause it to sink, it may still be very buoyant, and easily borne along by a feeble current. The ice, moreover, melts very slowly at the bottom of running streams in winter, as the water there is often nearly at the freezing point, as will be seen from what will be said in the sequel of ground-ice.

As we traverse Europe in the latitudes of Great Britain, we find the winters more severe, and the rivers more regularly frozen over. M. Lariviere relates that, being at Memel on the Baltic in 1821, when the ice of the river Niemen broke up, he saw a mass of ice thirty feet long which had descended the stream, and had been thrown ashore. In the middle of it was a triangular piece of granite, about a yard in diameter, resembling in composition the red granite of Finland.*

* Consid. sur les Blocs Errat. 1829.

When rivers in the northern hemisphere flow from south to north, the ice first breaks up in the higher part of their course, and the flooded waters, bearing along large icy fragments, often arrive at parts of the stream which are still firmly frozen over. Great inundations are thus frequently occasioned by the obstructions thrown in the way of the descending waters, as in the case of the Mackenzie in North America, and the Irtish, Obi, Yenesei, Lena, and other rivers of Siberia. (See map, fig. 1, p. 79.) A partial stoppage of this kind lately occurred (Jan. 31, 1840) in the Vistula, about a mile and a half above the city of Dantzic, where the river, choked up by packed ice, was made to take a new course over its right bank, so that it hollowed out in a few days a deep and broad channel, many leagues in length, through a tract of sand-hills which were from 40 to 60 feet high.

In Canada, where the winter's cold is intense, in a latitude corresponding to that of central France, several tributaries of the St. Lawrence begin to thaw in their upper course, while they remain frozen over lower down, and thus large slabs of ice are set free and thrown upon the unbroken sheet of ice below. Then begins what is called the packing of the drifted fragments; that is to say, one slab is made to slide over another, until a vast pile is built up, and the whole being frozen together, is urged onwards by the force of the dammed up waters and drift-ice. Thus propelled, it not only forces along boulders, but breaks off from cliffs, which border the rivers, huge pieces of projecting rock. By this means several buttresses of solid masonry, which, up to the year 1836, supported a wooden bridge on the St. Maurice, which falls into the St. Lawrence, near the town of Trois Rivières, lat. 46° 20′, were thrown down, and conveyed by the ice into the main river; and instances have occurred at Montreal of wharfs and stone-buildings, from 30 to 50 feet square, having been removed in a similar manner. We learn from Captain Bayfield that anchors laid down within high-water mark, to secure vessels hauled on shore for the winter, must be cut out of the ice on the approach of spring, or they would be carried away. In 1834, the Gulnare's bower-anchor, weighing half a ton, was transported some yards by the ice, and so firmly was it fixed, that the force of the moving ice broke a chain-cable suited for a 10-gun brig, and which had rode the Gulnare during the heaviest gales in the gulf. Had not this anchor been cut out of the ice, it would have been carried into deep water and lost.*

The scene represented in the annexed plate (pl. 2), from a drawing by Lieutenant Bowen, R. N., will enable the reader to comprehend the incessant changes which the transport of boulders produces annually on the low islands, shores, and bed of the St. Lawrence above Quebec. The fundamental rocks at Richelieu Rapid, situated in lat. 46° N., are limestone and slate, which are seen at low-water to be covered with boulders of granite. These boulders owe their spheroidal form chiefly tc

* Capt. Bayfield, Geol. Soc. Proceedings, vol. ii. p. 223.

BOULDERS DRIFTED BY ICE ON SHORES OF THE ST. LAWRENCE

View taken by Lieut. Bowen, from the N. E., in the Spring of 1835, at Richelieu Rapid, lat. 46° N.

weathering, or action of frost, which causes the surface to exfoliate in concentric plates, so that all the more prominent angles are removed. At the point *a* is a cavity in the mud or sand of the beach, now filled with water, which was occupied during the preceding winter (1835) by the huge erratic *b*, a mass of granite, 70 tons' weight, found in the spring following (1836) at a distance of several feet from its former position. Many small islands are seen on the river, such as *c d*, which afford still more striking proofs of the carrying and propelling power of ice. These islets are never under water, yet every winter ice is thrown upon them in such abundance, that it *packs* to the height of 20, and even 30 feet, bringing with it a continual supply of large stones or boulders, and carrying away others; the greatest number being deposited, according to Lieutenant Bowen, on the edge of deep water. On the island *d*, on the left of the accompanying view, a lighthouse is represented, consisting of a square wooden building, which having no other foundation than the boulders, requires to be taken down every winter, and rebuilt on the reopening of the river.

These effects of frost, which are so striking on the St. Lawrence above Quebec, are by no means displayed on a smaller scale below that city, where the gulf rises and falls with the tide. On the contrary; it is in the estuary, between the latitudes 47° and 49°, that the greatest quantity of gravel and boulders of large dimensions are carried down annually towards the sea. Here the frost is so intense, that a dense sheet of ice is formed at low water, which, on the rise of the tide, is lifted up, broken, and thrown in heaps on the extensive shoals which border the estuary. When the tide recedes, this packed ice is exposed to a temperature sometimes 30° below zero, which freezes together all the loose pieces of ice, as well as the granitic and other boulders. The whole of these are often swept away by a high tide, or when the river is swollen by the melting of the snow in Spring. One huge block of granite, 15 feet long by 10 feet both in width and height, and estimated to contain 1500 cubic feet, was conveyed in this manner to some distance in the year 1837, its previous position being well known, as up to that time it had been used by Captain Bayfield as a mark for the surveying station.

Ground-ice.—When a current of cold air passes over the surface of a lake or stream it abstracts from it a quantity of heat, and the specific gravity of the water being thereby increased, the cooled portion sinks. This circulation may continue until the whole body of fluid has been cooled down to the temperature of 40° F., after which, if the cold increase, the vertical movement ceases, the water which is uppermost expands and floats over the heavier fluid below, and when it has attained a temperature of 32° Fahr. it sets into a sheet of ice. It should seem therefore impossible, according to this law of congelation, that ice should ever form at the bottom of a river; and yet such is the fact, and many speculations have been hazarded to account for so singular a phenomenon. M. Arago is of opinion that the mechanical action of a running stream produces a circulation by which the entire body of water is mixed up together, and cooled alike, and the whole being thus reduced to the freezing point, ice begins to form at the bottom for two reasons, first, because there is less motion there, and secondly, because the water is in contact with solid rock or pebbles which have a cold surface.* Whatever explanation we adopt, there is no doubt of the fact, that in countries where the intensity and duration of the cold is great, rivers and torrents acquire an increase of carrying power by the formation of what is called ground-ice. Even in the Thames we learn from Dr. Plott that pieces of this kind of ice, having gravel frozen on to their under side, rise up from the bottom in winter, and float on the surface. In the Siberian rivers, Weitz describes large stones as having been brought up from the river's bed in the same manner, and made to float.†

Glaciers.—In the temperate zone, the snow lies for months in winter on the summit of every high mountain, while in the arctic regions, a long summer's day of half a year's duration is insufficient to melt the snow, even on land just raised above the level of the sea. It is therefore not surprising, since the atmosphere becomes colder in proportion as we ascend in it, that there should be heights, even in tropical countries, where the snow never melts. The lowest limit to which the perpetual snow extends downwards, from the tops of mountains at the equator, is an elevation of not less than 16,000 feet above the sea; while in the Swiss Alps, in lat. 46° N. it reaches as low as 8,500 feet above the same level, the loftier peaks of the Alpine chain being from 12,000 to 15,000 feet high. The frozen mass augmenting from year to year would add indefinitely to the altitude of alpine summits, were it not relieved by its descent through the larger and deeper valleys to regions far below the general snow-line. To these it slowly finds its way in the form of rivers of ice, called glaciers, the consolidation of which is produced by pressure, and by the congelation of water infiltered into the porous mass, which is always undergoing partial liquefaction, and receiving in summer occasional showers of rain on its surface. In a day of hot sunshine, or mild rain, innumerable rills of pure and sparkling water run in icy channels along the surface of the glaciers, which in the night shrink, and come to nothing. They are often precipitated in bold cascades into deep fissures in the ice, and contribute together with springs to form torrents, which flow in tunnels at the bottom of the glaciers for many a league, and at length issue at their extremities, from beneath beautiful caverns or arches. The waters of these streams are always densely charged with the finest mud, produced by the grinding of rock and sand under the weight of the moving mass. (See fig. 18.)

* M. Arago, Annuaire, &c. 1833; and Rev. J. Farquharson, Phil. Trans. 1835, p. 329.

† Journ. of Roy. Geograph. Soc. vol. vi. p. 416.

Fig 18.

Glacier with medial and lateral moraines and with terminal cave.

The length of the Swiss glaciers is sometimes twenty miles, their width in the middle portion, where they are broadest, occasionally two or three miles; their depth or thickness sometimes more than 600 feet. When they descend steep slopes, and precipices, or are forced through narrow gorges, the ice is broken up, and assumes the most fantastic and picturesque forms, with lofty peaks and pinnacles, projecting above the general level. These snow-white masses are often relieved by a dark background of pines, as in the valley of Chamouni; and are not only surrounded with abundance of the wild rhododendron in full flower, but encroach still lower into the region of cultivation, and trespass on fields where the tobacco-plant is flourishing by the side of the peasant's hut.

The cause of glacier motion has of late been a subject of careful investigation and much keen controversy. Although a question of physics, rather than of geology, it is too interesting to allow me to pass it by without some brief mention. De Saussure, whose travels in the Alps are full of original observations, as well as sound and comprehensive general views, conceived that the weight of the ice might be sufficient to urge it down the slope of the valley, if the sliding motion were aided by the water flowing at the bottom. For this "gravitation theory" Charpentier, followed by Agassiz, substituted the hypothesis of dilatation. The most solid ice is always permeable to water, and penetrated by innumerable fissures and capillary tubes, often extremely minute. These tubes imbibe the aqueous fluid during the day, which freezes, it is said, in the cold of the night, and expands while in the act of congelation. The distension of the whole mass exerts an immense force, tending to propel the glacier in the direction of least resistance —"in other words, down the valley." This theory was opposed by Mr. Hopkins on mathematical and mechanical grounds, in several able papers. Among other objections, he pointed out that the friction of so enormous a body as a glacier on its bed is so great, that the vertical direction would always be that of least resistance, and if a considerable distension of the mass should take place, by the action of freezing, it would tend to increase its thickness, rather than accelerate its downward progress. He also contended (and his arguments were illustrated by many ingenious experiments), that a glacier can move along an extremely slight slope, solely by the influence of gravitation, owing to the constant dissolution of ice in contact with the rocky bottom, and the number of separate fragments into which the glacier is divided by fissures, so that freedom of motion is imparted to its several parts somewhat resembling that of an imperfect fluid. To this view Professor James Forbes objected, that gravitation would not supply an adequate cause for the sliding of solid ice down slopes having an inclination of no more than four or five degrees, still less would it explain how the glacier advances where the channel expands and contracts. The Mer de Glace in Chamouni, for example, after being 2000 yards wide, passes through a strait only 900 yards in width. Such a gorge, it is contended, would be choked up by the advance of any solid mass, even if it be broken up into numerous fragments. The same acute observer remarked, that water in the fissures and pores of glaciers cannot, and does not part with its latent heat, so as to freeze every night to a great depth, or far in the interior of the mass. Had the dilatation theory been true, the chief motion of the glacier would have occurred about sunset, when the freezing of the water must be greatest, and it had, in fact, been at first assumed by those who favored that hypothesis, that the mass moved faster at the sides, where the melting of ice was promoted by the sun's heat, reflected from boundary precipices.

Agassiz appears to have been the first to commence, in 1841, aided by a skilful engineer, M. Escher de la Linth, a series of exact measurements to ascertain the laws of glacier motion, and he soon discovered, contrary to his preconceived notions, that the stream of ice moved more slowly at the sides than at the centre, and faster in the middle region of the glacier than at its extremity.* Professor James Forbes, who had joined Mr. Agassiz during his earlier investigations in the Alps,

* See Système Glaciaire, by Agassiz, Guyot, and Desor, pp. 436, 437, 445. Mr. Agassiz, at p. 462, states that he published in the Deutsche Vierteljahrschrift for 1841, this result as to the central motion being greater than that of the sides, and was, therefore, the first to correct his own previous mistake.

undertook himself an independent series of experiments, which he followed up with great perseverance, to determine the laws of glacier motion. These he found to agree very closely with the laws governing the course of rivers, their progress being greater in the centre than at the sides, and more rapid at the surface than at the bottom. This fact was verified by carefully fixing a great number of marks in the ice, arranged in a straight line, which gradually assumed a beautiful curve, the middle part pointing down the glacier, and showing a velocity there, double or treble that of the lateral parts.* He ascertained that the rate of advance by night was nearly the same as by day, and that even the hourly march of the icy stream could be detected, although the progress might not amount to more than six or seven inches in twelve hours. By the incessant though invisible advance of the marks placed on the ice, "time," says Mr. Forbes, "was marked out as by a shadow on a dial, and the unequivocal evidence which I obtained, that even while walking on a glacier we are, day by day, and hour by hour, imperceptibly carried on by the resistless flow of the icy stream, filled me with admiration." (Travels in the Alps, p. 133.) In order to explain this remarkable regularity of motion, and its obedience to laws so strictly analogous to those of fluids, the same writer proposed the theory that the ice, instead of being solid and compact, is a viscous or plastic body, capable of yielding to great pressure, and the more so in proportion as its temperature is higher, and as it approaches more nearly to the melting point. He endeavors to show that this hypothesis will account for many complicated phenomena, especially for a ribboned or veined structure which is everywhere observable in the ice, and might be produced by lines of discontinuity, arising from the different rates at which the various portions of the semi-rigid glacier advance and pass each other. Many examples are adduced to prove that a glacier can model itself to the form of the ground over which it is forced, exactly as would happen if it possessed a certain ductility, and this power of yielding under intense pressure, is shown not to be irreconcilable with the idea of the ice being sufficiently compact to break into fragments, when the strain upon its parts is excessive; as where the glacier turns a sharp angle, or descends upon a rapid or convex slope. The increased velocity in summer is attributed partly to the greater plasticity of the ice, when not exposed to intense cold, and partly to the hydrostatic pressure of the water in the capillary tubes, which imbibe more of this liquid in the hot season.

On the assumption of the ice being a rigid mass, Mr. Hopkins attributed the more rapid motions in the centre to the unequal rate at which the broad stripes of ice, intervening between longitudinal fissures, advance; but besides that there are parts of the glacier where no such fissures exist, such a mode of progression, says Mr. Forbes, would cause the borders of large transverse rents or "crevasses," to be jagged like a

* J. Forbes. 8th Letter on Glaciers, Aug. 1844.

saw, instead of being perfectly even and straight-edged.* An experiment recently made by Mr. Christie, secretary to the Royal Society, appears to demonstrate that ice, under great pressure, possesses a sufficient degree of moulding and self-adapting power to allow it to be acted upon, as if it were a pasty substance. A hollow shell of iron an inch and a half thick, the interior being ten inches in diameter, was filled with water, in the course of a severe winter, and exposed to the frost, with the fuze-hole uppermost. A portion of the water expanded in freezing, so as to protrude a cylinder of ice from the fuze-hole; and this cylinder continued to grow inch by inch in proportion as the central nucleus of water froze. As we cannot doubt that an outer shell of ice is first formed, and then another within, the continued rise of the column through the fuze-hole must proceed from the squeezing of successive shells of ice concentrically formed, through the narrow orifice; and yet the protruded cylinder consisted of entire, and not fragmentary ice.†

The agency of glaciers in producing permanent geological changes consists partly in their power of transporting gravel, sand, and huge stones to great distances, and partly in the smoothing, polishing, and scoring of their rocky channels, and the boundary walls of the valleys through which they pass. At the foot of every steep cliff or precipice in high Alpine regions, a talus is seen of rocky fragments detached by the alternate action of frost and thaw. If these loose masses, instead of accumulating on a stationary base, happen to fall upon a glacier, they will move along with it, and, in place of a single heap, they will form in the course of years a long stream of blocks. If a glacier be 20 miles long, and its annual progression about 500 feet, it will require about two centuries for a block thus lodged upon its surface to travel down from the higher to the lower regions, or to the extremity of the icy mass. This terminal point remains usually unchanged from year to year, although every part of the ice is in motion, because the liquefaction by heat is just sufficient to balance the onward movement of the glacier, which may be compared to an endless file of soldiers, pouring into a breach, and shot down as fast as they advance.

The stones carried along on the ice are called in Switzerland the "moraines" of the glacier. There is always one line of blocks on each side or edge of the icy stream, and often several in the middle, where they are arranged in long ridges or mounds, often several yards high. (See fig. 18, p. 223.) The cause of these "medial moraines" was first explained by Agassiz, who referred them to the confluence of tributary glaciers.‡ Upon the union of two streams of ice, the right lateral moraine

* See Mr. Hopkins on Motion of Glaciers, Cambridge Phil. Trans. 1844, and Phil. Mag. 1845. Some of the late concessions of this author as to a certain plasticity in the mass, appear to me to make the difference between him and Professor Forbes little more than one of degree. (For the latest summary of Prof. Forbes' views, see Phil. Trans. 1846, pt. 2.)

† This experiment is cited by Mr. Forbes, Phil. Trans. 1846, p. 206; and I have conversed with Mr. Christie on the subject.

‡ Etudes sur les Glaciers, 1840.

of one of the streams comes in contact with the left lateral moraine of the other, and they afterwards move on together, in the centre, if the confluent glaciers are equal in size, or nearer to one side if unequal.

All sand and fragments of soft stone which fall through fissures and reach the bottom of the glaciers, or which are interposed between the glacier and the steep sides of the valley, are pushed along, and ground down into mud, while the larger and harder fragments have their angles worn off. At the same time the fundamental and boundary rocks are smoothed and polished, and often scored with parallel furrows, or with lines and scratches produced by hard minerals, such as crystals of quartz, which act like the diamond upon glass.* This effect is perfectly different from that caused by the action of water, or a muddy torrent forcing along heavy fragments; for when stones are fixed firmly in the ice, and pushed along by it under great pressure, in straight lines, they scoop out long rectilinear furrows or grooves parallel to each other.† The discovery of such markings at various heights far above the surface of the existing glaciers and for miles beyond their present terminations, affords geological evidence of the former extension of the ice beyond its present limits in Switzerland and other countries.

The moraine of the glacier, observes Charpentier, is entirely devoid of stratification, for there has been no sorting of the materials, as in the case of sand, mud, and pebbles, when deposited by running water. The ice transports indifferently, and to the same spots, the heaviest blocks and the finest particles, mingling all together, and leaving them in one confused and promiscuous heap wherever it melts.‡

Icebergs.—In countries situated in high northern latitudes, like Spitzbergen, between 70° and 80° N., glaciers, loaded with mud and rock, descend to the sea, and there huge fragments of them float off and become icebergs. Scoresby counted 500 of these bergs drifting along in latitudes 69° and 70° N., which rose above the surface from the height of 100 to 200 feet, and measured from a few yards to a mile in circumference.§ Many of them were loaded with beds of earth and rock of such thickness, that the weight was conjectured to be from 50,000 to 100,000 tons. Specimens of the rocks were obtained, and among them were granite, gneiss, mica-schist, clay-slate, granular felspar, and greenstone. Such bergs must be of great magnitude; because the mass of ice below the level of the water is about eight times greater than that above. Wherever they are dissolved, it is evident that the "moraine" will fall to the bottom of the sea. In this manner may submarine valleys, mountains, and platforms become strewed over with gravel, sand, mud, and scattered blocks of foreign rock, of a nature perfectly dissimilar from all in the vicinity, and which may have been transported across unfathomable abysses. If the bergs happen to melt in still water, so that the earthy and stony materials may fall tranquilly to the bottom,

* See Manual of Geol. ch. xi.

† Agassiz, Jam. Ed. New Phil. Journ. No. 54, p. 388.

‡ Charpentier, Ann. des Mines, tom. viii.; see also Papers by MM. Venetz and Agassiz.

§ Voyage in 1822, p. 233.

the deposit will probably be unstratified, like the terminal moraine of a glacier; but whenever the materials are under the influence of a current of water as they fall, they will be sorted and arranged according to their relative weight and size, and therefore more or less perfectly stratified.

In a former chapter it was stated that some ice islands have been known to drift from Baffin's Bay to the Azores, and from the South Pole to the immediate neighborhood of the Cape of Good Hope, so that the area over which the effects of moving ice may be experienced, comprehends a large portion of the globe.

We learn from Von Buch that the most southern point on the continent of Europe at which a glacier comes down to the sea is in Norway, in lat. 67° N.* But Mr. Darwin has shown, that they extend to the sea, in South America, in latitudes more than 20° nearer the equator than in Europe; as, for example, in Chili, where, in the Gulf of Penas, lat. 46° 40′ S., or the latitude of central France; and in Sir George Eyre's Sound, in the latitude of Paris, they give origin to icebergs, which were seen in 1834 carrying angular pieces of granite, and stranding them in fiords, where the shores were composed of clay-slate.† A large proportion, however, of the ice-islands seen floating both in the northern and southern hemispheres, are probably not generated by glaciers, but rather by the accumulation of coast ice. When the sea freezes at the base of a lofty precipice, the sheet of ice is prevented from adhering to the land by the rise and fall of the tide. Nevertheless, it often continues on the shore at the foot of the cliff, and receives accessions of drift snow blown from the land. Under the weight of this snow the ice sinks slowly if the water be deep, and the snow is gradually converted into ice by partial liquefaction and re-congelation. In this manner, islands of ice of great thickness and many leagues in length, originate, and are eventually blown out to sea by off-shore winds. In their interior are inclosed many fragments of stone which had fallen upon them from overhanging cliffs during their formation. Such floating icebergs are commouly flat-topped, but their lower portions are liable to melt in latitudes where the ocean at a moderate depth is usually warmer than the surface water and the air. Hence their centre of gravity changes continually, and they turn over and assume very irregular shapes.

In a voyage of discovery made in the antarctic regions in 1839, a dark-colored angular mass of rock was seen imbedded in an iceberg, drifting along in mid-ocean in lat. 61° S. That part of the rock which was visible was about 12 feet in height, and from 5 to 6 in width, but the dark color of the surrounding ice indicated that much more of the stone was concealed. A sketch made by Mr. Macnab, when the vessel was within a quarter of a mile of it, is now published.‡ This iceberg, one of many observed at sea on the same day, was between 250 and 300 feet high, and was no less than 1400 miles from any certainly known land. It is exceedingly improbable, says Mr. Darwin, in his notice of this phenom-

* Travels in Norway. † Darwin's Journal, p. 283.
‡ Journ. of Roy. Geograph. Soc. vol. ix. p. 526.

enon, that any land will hereafter be discovered within 100 miles of the spot, and it must be remembered that the erratic was still firmly fixed in the ice, and may have sailed for many a league farther before it dropped to the bottom.*

Captain Sir James Ross, in his antarctic voyage in 1841, 42, and 43, saw multitudes of icebergs transporting stones and rocks of various sizes, with frozen mud, in high southern latitudes. His companion, Dr. J. Hooker, informs me that he came to the conclusion that most of the southern icebergs have stones in them, although they are usually concealed from view by the quantity of snow which falls upon them.

In the account given by Messrs. Dease and Simpson, of their recent arctic discoveries, we learn that in lat. 71° N., long. 156° W., they found "a long low spit, named Point Barrow, composed of gravel and coarse sand, in some parts more than a quarter of a mile broad, which the pressure of the ice had forced up into numerous mounds, that, viewed from a distance, assumed the appearance of huge boulder rocks."†

This fact is important, as showing how masses of drift ice, when stranding on submarine banks, may exert a lateral pressure capable of bending and dislocating any yielding strata of gravel, sand, or mud. The banks on which icebergs occasionally run aground between Baffin's Bay and Newfoundland, are many hundred feet under water, and the force with which they are struck will depend not so much on the velocity as the momentum of the floating ice-islands. The same berg is often carried away by a change of wind, and then driven back again upon the same bank, or it is made to rise and fall by the waves of the ocean, so that it may alternately strike the bottom with its whole weight, and then be lifted up again until it has deranged the superficial beds over a wide area. In this manner the geologist may account, perhaps, for the circumstance that in Scandinavia, Scotland, and other countries where erratics are met with, the beds of sand, loam, and gravel are often vertical, bent, and contorted into the most complicated folds, while the underlying strata, although composed of equally pliant materials, are horizontal. But some of these curvatures of loose strata may also have been due to repeated alternations of layers of gravel and sand, ice and snow, the melting of the latter having caused the intercalated beds of indestructible matter to assume their present anomalous position.

There can be little doubt that icebergs must often break off the peaks and projecting points of submarine mountains, and must grate upon and polish their surface, furrowing or scratching them in precisely the same way as we have seen that glaciers act on the solid rocks over which they are propelled.‡

* Journ. of Roy. Geograph. Soc. vol. ix. p. 529.

† Ibid. vol. viii. p. 221.

‡ In my Travels in N. America, pp. 19, 23, &c., and Second Visit to the U. S., vol. i. ch. 2, also in my Manual of Geology, a more full account of the action of floating ice and coast-ice, and its bearing on geology, will be found.

To conclude: it appears that large stones, mud, and gravel are carried down by the ice of rivers, estuaries, and glaciers, into the sea, where the tides and currents of the ocean, aided by the wind, cause them to drift for hundreds of miles from the place of their origin. Although it will belong more properly to the seventh and eighth chapters to treat of the transportation of solid matter by the movements of the ocean, I shall add here what I have farther to say on this subject in connection with ice.

The saline matter which sea-water holds in solution, prevents its congelation, except where the most intense cold prevails. But the drifting of the snow from the land often renders the surface-water brackish near the coast, so that a sheet of ice is readily formed there, and by this means a large quantity of gravel is frequently conveyed from place to place, and heavy boulders also, when the coast-ice is packed into dense masses. Both the large and small stones thus conveyed usually travel in one direction like shingle-beaches, and this was observed to take place on the coast of Labrador and Gulf of St. Lawrence, between the latitudes 50° and 60° N., by Capt. Bayfield, during his late survey. The line of coast alluded to is strewed over for a distance of 700 miles with ice-borne boulders, often 6 feet in diameter, which are for the most part on their way from north to south, or in the direction of the prevailing current. Some points on this coast have been observed to be occasionally deserted, and then again at another season thickly bestrewed with erratics.

The accompanying drawing (fig. 19), for which I am indebted to

Fig. 19.

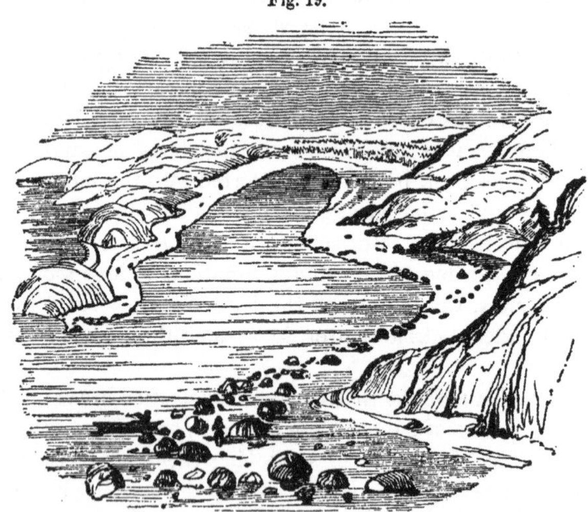

Boulders, chiefly of granite, stranded by ice on the coast of Labrador, between lat. 50° and 60° N. (Lieut. Bowen, R. N.)

Lieut. Bowen, R. N., represents the ordinary appearance of the Labrador coast, between the latitudes of 50° and 60° N. Countless blocks, chiefly granitic, and of various sizes, are seen lying between high and low-water mark. Capt. Bayfield saw similar masses carried by ice through the Straits of Belle Isle, between Newfoundland and the American continent, which he conceives may have travelled in the course of years from Baffin's Bay, a distance which may be compared in our hemisphere to the drifting of erratics from Lapland and Iceland as far south as Germany, France, and England.

It may be asked in what manner have these blocks been originally detached? We may answer that some have fallen from precipitous cliffs, others have been lifted up from the bottom of the sea, adhering by their tops to the ice, while others have been brought down by rivers and glaciers.

The erratics of North America are sometimes angular, but most of them have been rounded either by friction or decomposition. The granite of Canada, as before remarked (p. 221), has a tendency to concentric exfoliation, and scales off in spheroidal coats when exposed to the spray of the sea during severe frosts. The range of the thermometer in that country usually exceeds, in the course of the year, 100°, and sometimes 120° F.; and, to prevent the granite used in the buildings of Quebec from peeling off in winter, it is necessary to oil and paint the squared stones.

In parts of the Baltic, such as the Gulf of Bothnia, where the quantity of salt in the water amounts in general to one fourth only of that in the ocean, the entire surface freezes over in winter to the depth of 5 or 6 feet. Stones are thus frozen in, and afterwards lifted up about 3 feet perpendicularly on the melting of the snow in summer, and then carried by floating ice-islands to great distances. Professor Von Baer states, in a communication on this subject to the Academy of St. Petersburg, that a block of granite, weighing a million of pounds, was carried by ice during the winter of 1837–8 from Finland to the island of Hockland, and two other huge blocks were transported about the years 1806 and 1814 by packed ice on the south coast of Finland, according to the testimony of the pilots and inhabitants, one block having travelled about a quarter of a mile, and lying about 18 feet above the level of the sea.*

More recently Dr. Forchhammer has shown that in the Sound, the Great Belt, and other places near the entrance of the Baltic, ground-ice forms plentifully at the bottom and then rises to the surface, charged with sand and gravel, stones and sea-weed. Sheets of ice, also, with included boulders, are driven up on the coast during storms, and "packed" to a height of 50 feet. To the motion of such masses, but still more to that of the ground-ice, the Danish professor attributes the striation of rocky surfaces, forming the shores and bed of the sea, and he relates a striking fact to prove that large quantities of rocky fragments are annually carried by ice out of the Baltic. "In the year 1807," he says, "at the time of the bombardment of the Danish fleet, an English sloop-of-war, riding at anchor in the roads at Copenhagen, blew up. In

* Jam. Ed. New Phil. Journ. No. xlviii. p. 439.

1844, or thirty-seven years afterwards, one of our divers, known to be a trustworthy man, went down to save whatever might yet remain in the shipwrecked vessel. He found the space between decks entire, but covered with blocks from 6 to 8 cubic feet in size, and some of them heaped one upon the other. He also affirmed, that all the sunk ships which he had visited in the Sound, were in like manner strewed over with blocks."

Dr. Forchhammer also informs us, that during an intense frost in February, 1844, the Sound was suddenly frozen over, and sheets of ice, driven by a storm, were heaped up at the bottom of the Bay of Täarbeijk, threatening to destroy a fishing-village on the shore. The whole was soon frozen together into one mass, and forced up on the beach, forming a mound more than 16 feet high, which threw down the walls of several buildings. "When I visited the spot next day, I saw ridges of ice, sand, and pebbles, not only on the shore, but extending far out into the bottom of the sea, showing how greatly its bed had been changed; and how easily, where it is composed of rock, it may be furrowed and streaked by stones firmly fixed in the moving ice."*

CHAPTER XVI.

PHENOMENA OF SPRINGS.

Origin of Springs—Artesian wells—Borings at Paris—Distinct causes by which mineral and thermal waters may be raised to the surface—Their connection with volcanic agency—Calcareous springs—Travertin of the Elsa—Baths of San Vignone and of San Filippo, near Radicofani—Spheroidal structure in travertin—Lake of the Solfatara, near Rome—Travertin at Cascade of Tivoli—Gypseous, siliceous, and ferruginous springs—Brine springs—Carbonated springs—Disintegration of granite in Auvergne—Petroleum springs—Pitch lake of Trinidad.

Origin of springs.—The action of running water on the surface of the land having been considered, we may next turn our attention to what may be termed "the subterranean drainage," or the phenomena of springs. Every one is familiar with the fact, that certain porous soils, such as loose sand and gravel, absorb water with rapidity, and that the ground composed of them soon dries up after heavy showers. If a well be sunk in such soils, we often penetrate to considerable depths before we meet with water; but this is usually found on our approaching the lower parts of the formation, where it rests on some impervious bed; for here the water, unable to make its way downwards in a direct line, accumulates as in a reservoir, and is ready to ooze out into any opening which may be made, in the same manner as we see the salt water flow into, and fill, any hollow which we dig in the sands of the shore at low tide.

* Bulletin de la Soc. Géol. de France, 1847, tom. iv. pp. 1182, 1183.

The facility with which water can percolate loose and gravelly soils is clearly illustrated by the effect of the tides in the Thames between Richmond and London. The river, in this part of its course, flows through a bed of gravel overlying clay, and the porous superstratum is alternately saturated by the water of the Thames as the tide rises, and then drained again to the distance of several hundred feet from the banks when the tide falls, so that the wells in this tract regularly ebb and flow.

If the transmission of water through a porous medium be so rapid, we cannot be surprised that springs should be thrown out on the side of a hill, where the upper set of strata consist of chalk, sand, or other permeable substances, while the subjacent are composed of clay or other retentive soils. The only difficulty, indeed, is to explain why the water does not ooze out everywhere along the line of junction of the two formations, so as to form one continuous land-soak, instead of a few springs only, and these far distant from each other. The principal cause of this concentration of the waters at a few points is, first, the frequency of rents and fissures, which act as natural drains; secondly, the existence of inequalities in the upper surface of the impermeable stratum, which lead the water, as valleys do on the external surface of a country, into certain low levels and channels.

That the generality of springs owe their supply to the atmosphere is evident from this, that they become languid, or entirely cease to flow, after long droughts, and are again replenished after a continuance of rain. Many of them are probably indebted for the constancy and uniformity of their volume to the great extent of the subterranean reservoirs with which they communicate, and the time required for these to empty themselves by percolation. Such a gradual and regulated discharge is exhibited, though in a less perfect degree, in every great lake which is not sensibly affected in its level by sudden showers, but only slightly raised; so that its channel of efflux, instead of being swollen suddenly like the bed of a torrent, is enabled to carry off the surplus water gradually.

Much light has been thrown, of late years, on the theory of springs, by the boring of what are called by the French "Artesian wells," because the method has long been known and practised in Artois; and it is now demonstrated that there are sheets, and in some places currents of fresh water, at various depths in the earth. The instrument employed in excavating these wells is a large augur, and the cavity bored is usually from three to four inches in diameter. If a hard rock is met with, it is first triturated by an iron rod, and the materials being thus reduced to small fragments or powder, are readily extracted. To hinder the sides of the well from falling in, as also to prevent the spreading of the ascending water in the surrounding soil, a jointed pipe is introduced, formed of wood in Artois, but in other countries more commonly of metal. It frequently happens that, after passing through hundreds of feet of retentive soils, a water-bearing stratum is at length pierced, when the fluid immediately ascends to the surface, and flows over. The first rush of the water up the tube is often violent, so that for a time the water plays like a fountain, and then, sinking, continues to flow over tranquilly, or sometimes remains stationary at a certain depth below the orifice of the well. This spouting of the water in the first instance is probably owing to the disengagement of air and carbonic acid gas, for both of these have been seen to bubble up with the water.*

At Sheerness, at the mouth of the Thames, a well was bored on a low tongue of land near the sea, through 300 feet of the blue clay of London, below which a bed of sand and pebbles was entered, belonging, doubtless, to the plastic clay formation; when this stratum was pierced, the water burst up with impetuosity, and filled the well. By another perforation at the same place, the water was found at the depth of 328 feet below the surface clay; it first rose rapidly to the height of 189 feet, and then, in the course of a few hours, ascended to an elevation of eight feet above the level of the ground. In 1824 a well was dug at Fulham, near the Thames, at the Bishop of London's, to the depth of 317 feet, which, after traversing the tertiary strata, was continued through 67 feet of chalk. The water immediately rose to the surface, and the discharge was about 50 gallons per minute. In the garden of the Horticultural Society at Chiswick, the borings passed through 19 feet of gravel, 242½ feet of clay and loam, and 67½ feet of chalk, and the water then rose to the surface from a depth of 329 feet.† At the Duke of Northumberland's, above Chiswick, the borings were carried to the extraordinary depth of 620 feet, so as to enter the chalk, when a considerable volume of water was obtained, which rose four feet above the surface of the ground. In a well of Mr. Brooks, at Hammersmith, the rush of water from a depth of 360 feet was so great, as to inundate several buildings and do considerable damage; and at Tooting, a sufficient stream was obtained to turn a wheel, and raise the water to the upper stories of the houses.‡ In 1838, the total supply obtained from the chalk near London was estimated at six million gallons a day, and, in 1851, at nearly double that amount, the increase being accompanied by an average fall of no less than two feet a year in the level to which the water rose. The water stood commonly, in 1822, at high-water mark, and had sunk in 1851 to 45, and in some wells to 65 feet below high-water mark.§ This fact shows the limited capacity of the subterranean reservoir. In the last of three wells bored through the chalk at Tours, to the depth of several hundred feet, the water rose 32 feet above the level of the soil, and the discharge amounted to 300 cubic yards of water every twenty-four hours.‖

By way of experiment, the sinking of a well was commenced at Paris

* Consult J. Prestwich, Water-bearing Strata around London. 1851. (Van Voorst.)

† Sabine, Journ. of Sci. No. xxxiii. p. 72. 1824.

‡ Héricart de Thury, "Puits Forés," p. 49.

§ Prestwich, p. 69.

‖ Bull. de la Soc. Géol. de France, tom iii. p. 194.

in 1834, which had reached, in November, 1839, a depth of more than 1600 English feet, and yet no water ascended to the surface. The government were persuaded by M. Arago to persevere, if necessary, to the depth of more than 2000 feet; but when they had descended above 1800 English feet below the surface, the water rose through the tube (which was about ten inches in diameter), so as to discharge half a million of gallons of limpid water every twenty-four hours. The temperature of the water increased at the rate of 1° 8′ F. for every 101 English feet, as they went down, the result agreeing very closely with the anticipations of the scientific advisers of this most spirited undertaking.

Mr. Briggs, the British consul in Egypt, obtained water between Cairo and Suez, in a calcareous sand, at the depth of thirty feet; but it did not rise in the well.* But other borings in the same desert, of variable depth, between 50 and 300 feet, and which passed through alternations of sand, clay, and siliceous rock, yielded water at the surface.†

The rise and overflow of the water in Artesian wells is generally referred, and apparently with reason, to the same principle as the play of an artificial fountain. Let the porous stratum or set of strata, *a a*, rest on the impermeable rock *d*, and be covered by another mass of an impermeable nature. The whole mass *a a* may easily, in such a position, become saturated with water, which may descend from its higher and exposed parts—a hilly region to which clouds are attracted, and

Fig. 20.

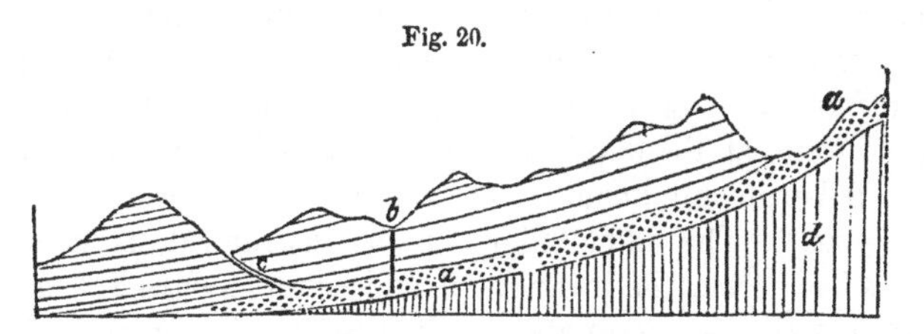

where rain falls in abundance. Suppose that at some point, as at *b*, an opening be made, which gives a free passage upwards to the waters confined in *a a*, at so low a level that they are subjected to the pressure of a considerable column of water collected in the more elevated portion of the same stratum. The water will then rush out, just as the liquid from a large barrel which is tapped, and it will rise to a height corresponding to the level of its point of departure, or, rather, to a height which balances the pressure previously exerted by the confined waters against the roof and sides of the stratum or reservoir *a a*. In like manner, if there happen to be a natural fissure *c*, a spring will be produced at the surface on precisely the same principle.

Among the causes of the failure of Artesian wells, we may mention those numerous rents and faults which abound in some rocks, and the

* Boué Résumé des Prog. de la Géol. en 1832, p. 184.
† Seventh Rep. Brit. Ass. 1837, p. 66.

deep ravines and valleys by which many countries are traversed; for, when these natural lines of drainage exist, there remains a small quantity only of water to escape by artificial issues. We are also liable to be baffled by the great thickness either of porous or impervious strata, or by the dip of the beds, which may carry off the waters from the adjoining high lands to some trough in an opposite direction, as when the borings are made at the foot of an escarpment where the strata incline inwards, or in a direction opposite to the face of the cliffs.

The mere distance of hills or mountains need not discourage us from making trials; for the waters which fall on these higher lands readily penetrate to great depths through highly inclined or vertical strata, or through the fissures of shattered rocks, and after flowing for a great distance, must often reascend and be brought up again by other fissures, so as to approach the surface in the lower country. Here they may be concealed beneath the covering of undisturbed horizontal beds, which it may be necessary to pierce in order to reach them. It should be remembered, that the course of waters flowing under ground bears but a remote resemblance to that of rivers on the surface, there being, in the one case, a constant descent from a higher to a lower level from the source of the stream to the sea; whereas, in the other, the water may at one time sink far below the level of the ocean, and afterwards rise again high above it.

Among other curious facts ascertained by aid of the borer, it is proved that in strata of different ages and compositions, there are often open passages by which the subterranean waters circulate. Thus, at St. Ouen, in France, five distinct sheets of water were intersected in a well, and from each of these a supply obtained. In the third water-bearing stratum, at the depth of 150 feet, a cavity was found in which the borer fell suddenly about a foot, and thence the water ascended in great volume.* The same falling of the instrument, as in a hollow space, has been remarked in England and other countries. At Tours, in 1830, a well was perforated quite through the chalk, when the water suddenly brought up, from a depth of 364 feet, a great quantity of fine sand, with much vegetable matter and shells. Branches of a thorn several inches long, much blackened by their stay in the water, were recognized, as also the stems of marsh plants, and some of their roots, which were still white, together with the seeds of the same in a state of preservation, which showed that they had not remained more than three or four months in the water. Among the seeds were those of the marsh plant *Galium uliginosum;* and among the shells, a freshwater species (*Planorbis marginatus*), and some land species, as *Helix rotundata* and *H. striata.* M. Dujardin, who, with others, observed this phenomenon, supposes that the waters had flowed from some valleys of Auvergne or the Vivarais since the preceding autumn.†

An analogous phenomenon is recorded at Reimke, near Bochum in

* H. de Thury, p. 295. † Bull. de la Soc. Géol de France, tom. i. p. 93.

Westphalia, where the water of an Artesian well brought up, from a depth of 156 feet, several small fish, three or four inches long, the nearest streams in the country being at a distance of some leagues.*

In both cases it is evident that water had penetrated to great depths, not simply by filtering through a porous mass, for then it would have left behind the shells, fish, and fragments of plants, but by flowing through some open channels in the earth. Such examples may suggest the idea that the leaky beds of rivers are often the feeders of springs.

MINERAL AND THERMAL SPRINGS.

Almost all springs, even those which we consider the purest, are impregnated with some foreign ingredients, which, being in a state of chemical solution, are so intimately blended with the water as not to affect its clearness, while they render it, in general, more agreeable to our taste, and more nutritious than simple rain-water. But the springs called mineral contain an unusual abundance of earthy matter in solution, and the substances with which they are impregnated correspond remarkably with those evolved in a gaseous form by volcanoes. Many of these springs are thermal, *i. e.*, their temperature is above the mean temperature of the place, and they rise up through all kinds of rock; as, for example, through granite, gneiss, limestone, or lava, but are most frequent in volcanic regions, or where violent earthquakes have occurred at eras comparatively modern.

The water given out by hot springs is generally more voluminous and less variable in quantity at different seasons than that proceeding from any others. In many volcanic regions, jets of steam, called by the Italians "stufas," issue from fissures, at a temperature high above the boiling point, as in the neighborhood of Naples, and in the Lipari Isles, and are disengaged unceasingly for ages. Now, if such columns of steam, which are often mixed with other gases, should be condensed before reaching the surface by coming in contact with strata filled with cold water, they may give rise to thermal and mineral springs of every degree of temperature. It is, indeed, by this means only, and not by hydrostatic pressure, that we can account for the rise of such bodies of water from great depths; nor can we hesitate to admit the adequacy of the cause, if we suppose the expansion of the same elastic fluids to be sufficient to raise columns of lava to the lofty summits of volcanic mountains. Several gases, the carbonic acid in particular, are disengaged in a free state from the soil in many districts, especially in the regions of active or extinct volcanoes; and the same are found more or less intimately combined with the waters of all mineral springs, both cold and thermal. Dr. Daubeny and other writers have remarked, not only that these springs are most abundant in volcanic regions, but that when remote from them, their site usually coincides with the position of some

* Bull. de la Soc. Géol. de France, tom. ii. p. 248.

great derangement in the strata; a fault, for example, or great fissure, indicating that a channel of communication has been opened with the interior of the earth at some former period of local convulsion. It is also ascertained that at great heights in the Pyrenees and Himalaya mountains hot springs burst out from granitic rocks, and they are abundant in the Alps also, these chains having all been disturbed and dislocated at times comparatively modern, as can be shown by independent geological evidence.

The small area of volcanic regions may appear, at first view, to present an objection to these views, but not so when we include earthquakes among the effects of igneous agency. A large proportion of the land hitherto explored by geologists can be shown to have been rent or shaken by subterranean movements since the oldest tertiary strata were formed. It will also be seen, in the sequel, that new springs have burst out, and others have had the volume of their waters augmented, and their temperature suddenly raised after earthquakes, so that the description of these springs might almost with equal propriety have been given under the head of "igneous causes," as they are agents of a mixed nature, being at once igneous and aqueous.

But how, it will be asked, can the regions of volcanic heat send forth such inexhaustible supplies of water? The difficulty of solving this problem would, in truth, be insurmountable, if we believed that all the atmospheric waters found their way into the basin of the ocean; but in boring near the shore we often meet with streams of fresh water at the depth of several hundred feet below the sea level; and these probably descend, in many cases, far beneath the bottom of the sea, when not artificially intercepted in their course. Yet, how much greater may be the quantity of salt water which sinks beneath the floor of the ocean, through the porous strata of which it is often composed, or through fissures rent in it by earthquakes. After penetrating to a considerable depth, this water may encounter a heat of sufficient intensity to convert it into vapor, even under the high pressure to which it would then be subjected. This heat would probably be nearest the surface in volcanic countries, and farthest from it in those districts which have been longest free from eruptions or earthquakes.

It would follow from the views above explained, that there must be a twofold circulation of terrestrial waters; one caused by solar heat, and the other by heat generated in the interior of our planet. We know that the land would be unfit for vegetation, if deprived of the waters raised into the atmosphere by the sun; but it is also true that mineral springs are powerful instruments in rendering the surface subservient to the support of animal and vegetable life. Their heat is said to promote the development of the aquatic tribes in many parts of the ocean, and the substances which they carry up from the bowels of the earth to the habitable surface, are of a nature and in a form which adapts them peculiarly for the nutrition of animals and plants.

As these springs derive their chief importance to the geologist from the quantity and quality of the earthy materials which, like volcanoes, they convey from below upwards, they may properly be considered in reference to the ingredients which they hold in solution. These consist of a great variety of substances; but chiefly salts with bases of lime, magnesia, alumine, and iron, combined with carbonic, sulphuric, and muriatic acids. Muriate of soda, silica, and free carbonic acid are frequently present; also springs of petroleum, or liquid bitumen, and of naphtha.

Calcareous springs.—Our first attention is naturally directed to springs which are highly charged with calcareous matter, for these produce a variety of phenomena of much interest in geology. It is known that rain-water collecting carbonic acid from the atmosphere has the property of dissolving the calcareous rocks over which it flows, and thus, in the smallest ponds and rivulets, matter is often supplied for the earthy secretions of testacea, and for the growth of certain plants on which they feed. But many springs hold so much carbonic acid in solution, that they are enabled to dissolve a much larger quantity of calcareous matter than rain-water; and when the acid is dissipated in the atmosphere, the mineral ingredients are thrown down, in the form of porous tufa or of more compact travertin.*

Auvergne.—Calcareous springs, although most abundant in limestone districts, are by no means confined to them, but flow out indiscriminately from all rock formations. In central France, a district where the primary rocks are unusually destitute of limestone, springs copiously charged with carbonate of lime rise up through the granite and gneiss. Some of these are thermal, and probably derive their origin from the deep source of volcanic heat, once so active in that region. One of these springs, at the northern base of the hill upon which Claremont is built, issues from volcanic peperino, which rests on granite. It has formed, by its incrustations, an elevated mound of travertin, or white concretionary limestone, 240 feet in length, and, at its termination, sixteen feet high and twelve wide. Another encrusting spring in the same department, situated at Chaluzet, near Pont Gibaud, rises in a gneiss country, at the foot of a regular volcanic cone, at least twenty miles from any calcareous rock. Some masses of tufaceous deposit, produced by this spring, have an oolitic texture.

Valley of the Elsa.—If we pass from the volcanic district of France to that which skirts the Apennines in the Italian peninsula, we meet with innumerable springs which have precipitated so much calcareous matter, that the whole ground in some parts of Tuscany is coated over with tufa and travertin, and sounds hollow beneath the foot.

In other places in the same country, compact rocks are seen descending the slanting sides of hills, very much in the manner of lava currents, except that they are of a white color and terminate abruptly when they reach the course of a river. These consist of a calcareous precipitate from springs, some of which are still flowing, while others have disap-

* See Glossary, "Tufa," "Travertin."

peared or changed their position. Such masses are frequent on the slope of the hills which bound the valley of the Elsa, one of the tributaries of the Arno, which flows near Colle, through a valley several hundred feet deep, shaped out of a lacustrine formation, containing fossil shells of existing species. I observed here that the travertin was unconformable to the lacustrine beds, its inclination according with the slope of the sides of the valley. One of the finest examples which I saw was at the Molino delle Caldane, near Colle. The Senà, and several other small rivulets which feed the Elsa, have the property of encrusting wood and herbs with calcareous stone. In the bed of the Elsa itself, aquatic plants, such as Charæ, which absorb large quantities of carbonate of lime, are very abundant.

Baths of San Vignone.—Those persons who have merely seen the action of petrifying waters in England, will not easily form an adequate conception of the scale on which the same process is exhibited in those regions which lie nearer to the active centres of volcanic disturbance. One of the most striking examples of the rapid precipitation of carbonate of lime from thermal waters, occurs in the hill of San Vignone in Tuscany, at a short distance from Radicofani, and only a few hundred yards from the high road between Sienna and Rome. The spring issues from near the summit of a rocky hill, about 100 feet in height. The top of

Fig. 21. *Baths of San Vignone.*

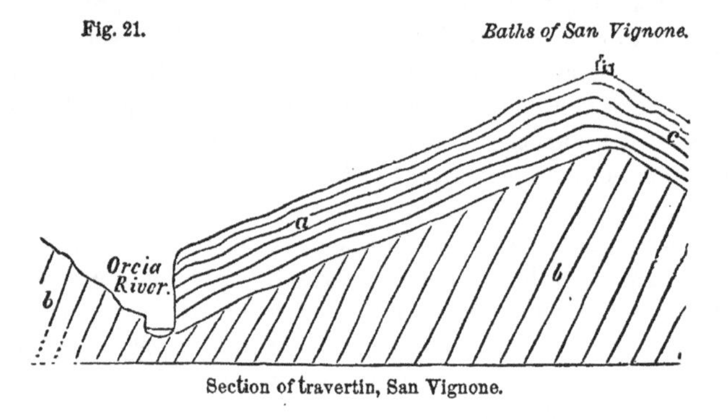

Section of travertin, San Vignone.

the hill stretches in a gently inclined platform to the foot of Mount Amiata, a lofty eminence, which consists in great part of volcanic products. The fundamental rock, from which the spring issues, is a black slate, with serpentine (*b b*, fig. 21), belonging to the older Apennine formation. The water is hot, has a strong taste, and, when not in very small quantity, is of a bright green color. So rapid is the deposition near the source, that in the bottom of a conduit-pipe for carrying off the water to the baths, and which is inclined at an angle of 30°, half a foot of solid travertin is formed every year. A more compact rock is produced where the water flows slowly; and the precipitation in winter, when there is least evaporation, is said to be more solid, but less in quantity by one-fourth, than in summer. The rock is generally white; some parts of it are compact, and ring to the hammer; others are cellular, and with such cavities as are seen in the carious part of bone or the siliceous millstone of the Paris basin. A portion of it also below the village of San Vignone consists of incrustations of long vegetable tubes, and may be called tufa. Sometimes the travertin assumes precisely the botryoidal and mammillary forms, common to similar deposits in Auvergne, of a much older date; and, like them, it often scales off in thin, slightly undulating layers.

A large mass of travertin (*c*, fig. 21) descends the hill from the point where the spring issues, and reaches to the distance of about half a mile east of San Vignone. The beds take the slope of the hill at about an angle of 6°, and the planes of stratification are perfectly parallel. One stratum, composed of many layers, is of a compact nature, and fifteen feet thick; it serves as an excellent building stone, and a mass of fifteen feet in length was, in 1828, cut out for the new bridge over the Orcia. Another branch of it (*a*, fig. 21) descends to the west, for 250 feet in length, of varying thickness, but sometimes 200 feet deep; it is then cut off by the small river Orcia, as some glaciers in Switzerland descend into a valley till their progress is suddenly arrested by a transverse stream of water.

The abrupt termination of the mass of rock at the river, where its thickness is undiminished, clearly shows that it would proceed much farther if not arrested by the stream, over which it impends slightly. But it cannot encroach upon the channel of the Orcia, being constantly undermined, so that its solid fragments are seen strewed amongst the alluvial gravel. However enormous, therefore, the mass of solid rock may appear which has been given out by this single spring, we may feel assured that it is insignificant in volume when compared to that which has been carried to the sea since the time when it began to flow. What may have been the length of that period of time we have no data for conjecturing. In quarrying the travertin, Roman tiles have been sometimes found at the depth of five or six feet.

Baths of San Filippo.—On another hill, not many miles from that last mentioned, and also connected with Mount Amiata, the summit of which is about three miles distant, are the celebrated baths of San Filippo. The subjacent rocks consist of alternations of black slate, limestone, and serpentine. There are three warm springs containing carbonate and sulphate of lime, and sulphate of magnesia. The water which supplies the baths falls into a pond, where it has been known to deposit a solid mass *thirty feet thick* in about *twenty years.** A manufactory of medallions in basso-relievo is carried on at these baths. The water is conducted by canals into several pits, in which it deposits travertin and crystals of sulphate of lime. After being thus freed from its grosser parts, it is conveyed by a tube to the summit of a small chamber, and made to fall through a space of ten or twelve feet. The current is broken in its descent by numerous crossed sticks, by which the

* Dr. Grosse on the Baths of San Filippo, Ed. Phil. Journ. vol. ii. p. 292.

spray is dispersed around upon certain moulds, which are rubbed lightly over with a solution of soap, and a deposition of solid matter like marble is the result, yielding a beautiful cast of the figures formed in the mould. The geologist may derive from these experiments considerable light, in regard to the high slope of the strata at which some semi-crystalline precipitations can be formed; for some of the moulds are disposed almost perpendicularly, yet the deposition is nearly equal in all parts.

A hard stratum of stone, about a foot in thickness, is obtained from the waters of San Filippo in four months; and, as the springs are powerful, and almost uniform in the quantity given out, we are at no loss to comprehend the magnitude of the mass which descends the hill, which is a mile and a quarter in length and the third of a mile in breadth, in some places attaining a thickness of 250 feet at least. To what length it might have reached it is impossible to conjecture, as it is cut off, like the travertin of San Vignone, by a small stream, where it terminates abruptly. The remainder of the matter held in solution is carried on probably to the sea.

Spheroidal structure in travertin.—But what renders this recent limestone of peculiar interest to the geologist, is the spheroidal form which it assumes, analogous to that of the cascade of Tivoli, afterwards to be described. (See fig. 22, p. 244.) The lamination of some of the concentric masses is so minute that sixty may be counted in the thickness of an inch, yet, notwithstanding these marks of gradual and successive deposition, sections are sometimes exhibited of what might seem to be perfect spheres. This tendency to a mammillary and globular structure arises from the facility with which the calcareous matter is precipitated in nearly equal quantities on all sides of any fragment of shell or wood or any inequality of the surface over which the mineral water flows, the form of the nucleus being readily transmitted through any number of successive envelopes. But these masses can never be perfect spheres, although they often appear such when a transverse section is made in any line not in the direction of the point of attachment. There are, indeed, occasionally seen small oolitic and pisolitic grains, of which the form is globular; for the nucleus, having been for a time in motion in the water, has received fresh accessions of matter on all sides.

In the same manner I have seen, on the vertical walls of large steam-boilers, the heads of nails or rivets covered by a series of enveloping crusts of calcareous matter, usually sulphate of lime; so that a concretionary nodule is formed, preserving a nearly globular shape, when increased to a mass several inches in diameter. In these, as in many travertins, there is often a combination of the concentric and radiated structure.

Campagna di Roma.—The country around Rome, like many parts of the Tuscan States already referred to, has been at some former period the site of numerous volcanic eruptions; and the springs are still copiously impregnated with lime, carbonic acid, and sulphuretted hydrogen. A hot spring was discovered about 1827, near Civita Vecchia, by Signor Riccioli, which deposits alternate beds of a yellowish travertin, and a white granular rock, not distinguishable, in hand specimens, either in grain, color, or composition, from statuary marble. There is a passage between this and ordinary travertin. The mass accumulated near the spring is in some places about six feet thick.

Lake of the Solfatara.—In the Campagna, between Rome and Tivoli, is the Lake of the Solfatara, called also Lago di Zolfo (lacus albula), into which flows continually a stream of tepid water from a smaller lake, situated a few yards above it. The water is a saturated solution of carbonic acid gas, which escapes from it in such quantities in some parts of its surface, that it has the appearance of being actually in ebullition. "I have found by experiment," says Sir Humphry Davy, "that the water taken from the most tranquil part of the lake, even after being agitated and exposed to the air, contained in solution more than its own volume of carbonic acid gas, with a very small quantity of sulphuretted hydrogen. Its high temperature, which is pretty constant at 80° of Fahr., and the quantity of carbonic acid that it contains, render it peculiarly fitted to afford nourishment to vegetable life. The banks of travertin are everywhere covered with reeds, lichen, confervæ, and various kinds of aquatic vegetables; and at the same time that the process of vegetable life is going on, the crystallizations of the calcareous matter, which is everywhere deposited, in consequence of the escape of carbonic acid, likewise proceed. There is, I believe, no place in the world where there is a more striking example of the opposition or contrast of the laws of animate and inanimate nature, of the forces of inorganic chemical affinity, and those of the powers of life." *

The same observer informs us that he fixed a stick in a mass of travertin covered by the water in the month of May, and in April following he had some difficulty in breaking, with a sharp-pointed hammer, the mass which adhered to the stick, and which was several inches in thickness. The upper part was a mixture of light tufa and the leaves of confervæ; below this was a darker and more solid travertin, containing black and decomposed masses of confervæ; in the inferior part the travertin was more solid, and of a gray color, but with cavities probably produced by the decomposition of vegetable matter.†

The stream which flows out of this lake fills a canal about nine feet broad and four deep, and is conspicuous in the landscape by a line of vapor which rises from it. It deposits calcareous tufa in this channel, and the Tiber probably receives from it, as well as from numerous other streams, much carbonate of lime in solution, which may contribute to the rapid growth of its delta. A large proportion of the most splendid edifices of ancient and modern Rome are built of travertin, derived from the quarries of Ponte Lucano, where there has evidently been a lake at a remote period, on the same plain as that already described.

* Consolations in Travel, pp. 123–125. † Ibid. p. 127.

Travertin of Tivoli.—In the same neighborhood the calcareous waters of the Anio incrust the reeds which grow on its banks, and the foam of the cataract of Tivoli forms beautiful pendant stalactites. On the sides of the deep chasm into which the cascade throws itself, there

Fig. 22.

Section of spheroidal concretionary Travertin under the Cascade of Tivoli.

is seen an extraordinary accumulation of horizontal beds of tufa and travertin, from four to five hundred feet in thickness. The section immediately under the temples of Vesta and the Sibyl, displays, in a precipice about four hundred feet high, some spheroids which are from *six to eight feet in diameter*, each concentric layer being about the eighth of an inch in thickness. The preceding diagram exhibits about fourteen feet of this immense mass, as seen in the path cut out of the rock in descending from the temple of Vesta to the Grotto di Nettuno. I have not attempted to express in this drawing the innumerable thin layers of which these magnificent spheroids are composed, but the lines given mark some of the natural divisions into which they are separated by minute variations in the size or color of the laminæ. The undulations also are much smaller in proportion to the whole circumference than in the drawing. The beds (*a a*) are of hard travertin and soft tufa; below them is a pisolite (*b*), the globules being of different sizes: underneath this appears a mass of concretionary travertin (*c c*), some of the spheroids being of the above-mentioned extraordinary size. In some places (as at *d*) there is a mass of amorphous limestone, or tufa, surrounded by concentric layers. At the bottom is another bed of pisolite (*b*), in which the small nodules are about the size and shape of beans, and some of them of filberts, intermixed with some smaller oolitic grains. In the tufaceous strata, wood is seen converted into a light tufa.

There can be little doubt that the whole of this deposit was formed in an extensive lake which existed when the external configuration of this country varied greatly from that now observed. The Anio throws itself into a ravine excavated in the ancient travertin, and its waters give rise to masses of calcareous stone, scarcely if at all distinguishable from the older rock. I was shown, in 1828, in the upper part of the travertin, the hollow left by a cart-wheel, in which the outer circle and the spokes had been decomposed, and the spaces which they filled left void. It seemed to me at the time impossible to explain the position of this mould without supposing that the wheel was imbedded before the lake was drained; but Sir R. Murchison suggests that it may have been washed down by a flood into the gorge in modern times, and then incrusted with calcareous tufa in the same manner as the wooden beam of the church of St. Lucia was swept down in 1826, and stuck fast in the Grotto of the Syren, where it still remains, and will eventually be quite imbedded in travertin.

I have already endeavored to explain (p. 241), when speaking of the travertin of San Filippo, how the spheroidal masses represented in figure 22 may have been formed.

Sulphureous and gypseus springs.—The quantity of other mineral ingredients wherewith springs in general are impregnated, is insignificant in comparison to lime, and this earth is most frequently combined with carbonic acid. But as sulphuric acid, and sulphuretted hydrogen are very frequently supplied by springs, gypsum may, perhaps, be deposited largely in certain seas and lakes. Among other gypseous precipitates at present known on the land, I may mention those of Baden, near Vienna, which feed the public bath. Some of these supply singly from 600 to 1000 cubic feet of water per hour, and deposit a fine powder, composed of a mixture of sulphate of lime with sulphur and muriate of lime.* The thermal waters of Aix, in Savoy, in passing through strata of Jurassic limestone, turn them into gypsum or sulphate of lime. In the Andes, at the Puenta del Inca, Lieutenant Brand found a thermal spring at the temperature of 91° Fahr., containing a large proportion of gypsum with carbonate of lime and other ingredi-

* C. Prevost, Essai sur la Constitution Physique du Bassin de Vienne, p. 10.

ents.* Many of the mineral springs of Iceland, says Mr. R. Bunsen, deposit gypsum,† and sulphureous acid gas escapes plentifully from them as from the volcanoes of the same island. It may, indeed, be laid down as a general rule, that the mineral substances dissolved in hot springs agree very closely with those which are disengaged in a gaseous form from the craters of active volcanoes.

Siliceous springs.—Azores.—In order that water should hold a very large quantity of silica in solution, it seems necessary that it should be raised to a high temperature.‡ The hot springs of the Valle das Fernas, in the island of St. Michael, rising through volcanic rocks, precipitate vast quantities of siliceous sinter. Around the circular basin of the largest spring, which is between twenty and thirty feet in diameter, alternate layers are seen of a coarser variety of sinter mixed with clay, including grass, ferns, and reeds, in different states of petrifaction. In some instances, alumina, which is likewise deposited from the hot waters, is the mineralizing material. Branches of the same ferns which now flourish in the island are found completely petrified, preserving the same appearance as when vegetating, except that they acquire an ash-gray color. Fragments of wood, and one entire bed from three to five feet in depth, composed of reeds now common in the island, have become completely mineralized.

The most abundant variety of siliceous sinter occurs in layers, from a quarter to half an inch in thickness, accumulated on each other often to the height of a foot and upwards, and constituting parallel, and for the most part horizontal, strata many yards in extent. This sinter has often a beautiful semi-opalescent lustre. A recent breccia is also in the act of forming, composed of obsidian, pumice, and scoriæ, cemented by siliceous sinter.§

Geysers of Iceland.—But the hot springs in various parts of Iceland, particularly the celebrated geysers, afford the most remarkable example of the deposition of silex.‖ The circular reservoirs into which the geysers fall, are lined in the interior with a variety of opal, and round the edges with sinter. The plants incrusted with the latter substance have much the same appearance as those incrusted with calcareous tufa in our own country. They consist of various grasses, the horse-tail (*Equisetum*), and leaves of the birch-tree, which are the most common of all, though no trees of this species now exist in the surrounding country. The petrified stems also of the birch occur in a state much resembling agatized wood.¶

By analysis of the water, Mr. Faraday has ascertained that the solution of the silex is promoted by the presence of the alkali, soda. He suggests that the deposition of silica in an insoluble state takes place partly because the water when cooled by exposure to the air is unable

* Travels across the Andes, p. 240. † Annalen der Chem. 1847.
‡ Daubeny on Volcanoes, p. 222.
§ Dr. Webster on the Hot Springs of Furnas, Ed. Phil. Journ. vol. vi. p. 306.
‖ See a cut of the Icelandic geyser, chap. 32.
¶ M. Robert, Bullétin de la Soc. Géol. de France, tom. vii. p. 11.

to retain as much silica as when it issues from the earth at a temperature of 180° or 190° Fahr.; and partly because the evaporation of the water decomposes the compound of silica and soda which previously existed. This last change is probably hastened by the carbonic acid of the atmosphere uniting with the soda. The alkali, when disunited from the silica, would readily be dissolved in and removed by running water.*

Mineral waters, even when charged with a small proportion of silica, as those of Ischia, may supply certain species of corals, sponges, and infusoria, with matter for their siliceous secretions; but there is little doubt that rivers obtain silex in solution from another and far more general source, namely, the decomposition of felspar. When this mineral, which is so abundant an ingredient in the hypogene and trappean rocks, has disintegrated, it is found that the residue, called porcelain clay, contains a small proportion only of the silica which existed in the original felspar, the other part having been dissolved and removed by water.†

Ferruginous springs.—The waters of almost all springs contain some iron in solution; and it is a fact familiar to all, that many of them are so copiously impregnated with this metal, as to stain the rocks or herbage through which they pass, and to bind together sand and gravel into solid masses. We may naturally, then, conclude that this iron, which is constantly conveyed from the interior of the earth into lakes and seas, and which does not escape again from them into the atmosphere by evaporation, must act as a coloring and cementing principle in the subaqueous deposits now in progress. Geologists are aware that many ancient sandstones and conglomerates are bound together or colored by iron.

Brine springs.—So great is the quantity of muriate of soda in some springs, that they yield one-fourth of their weight in salt. They are rarely, however, so saturated, and generally contain, intermixed with salt, carbonate and sulphate of lime, magnesia, and other mineral ingredients. The brine springs of Cheshire are the richest in our country; those of Northwich being almost saturated. Those of Barton also, in Lancashire, and Droitwich in Worcestershire, are extremely rich.‡ They are known to have flowed for more than 1000 years, and the quantity of salt which they have carried into the Severn and Mersey must be enormous. These brine springs rise up through strata of sandstone and red marl, which contain large beds of rock salt. The origin of the brine, therefore, may be derived in this and many other instances from beds of fossil salt; but as muriate of soda is one of the products of volcanic emanations and of springs in volcanic regions, the original source of salt may be as deep seated as that of lava.

Many springs in Sicily contain muriate of soda, and the "fiume salso,"

* Barrow's Iceland, p. 209.

† See Lyell's Manual of Elementary Geology; and Dr. Turner, Jam. Ed. New Phil. Journ. No. xxx. p. 246.

‡ L. Horner, Geol. Trans. vol. ii. p. 94.

in particular, is impregnated with so large a quantity, that cattle refuse to drink of it. A hot spring, rising through granite, at Saint Nectaire, in Auvergne, may be mentioned as one of many, containing a large proportion of muriate of soda, together with magnesia and other ingredients.*

Carbonated springs.—Auvergne.—Carbonic acid gas is very plentifully disengaged from springs in almost all countries, but particularly near active or extinct volcanoes. This elastic fluid has the property of decomposing many of the hardest rocks with which it comes in contact, particularly that numerous class in whose composition felspar is an ingredient. It renders the oxide of iron soluble in water, and contributes, as was before stated, to the solution of calcareous matter. In volcanic districts these gaseous emanations are not confined to springs, but rise up in the state of pure gas from the soil in various places. The Grotto del Cane, near Naples, affords an example, and prodigious quantities are now annually disengaged from every part of the Limagne d'Auvergne, where it appears to have been developed in equal quantity from time immemorial. As the acid is invisible, it is not observed, except an excavation be made, wherein it immediately accumulates, so that it will extinguish a candle. There are some springs in this district, where the water is seen bubbling and boiling up with much noise, in consequence of the abundant disengagement of this gas. In the environs of Pont-Gibaud, not far from Clermont, a rock belonging to the gneiss formation, in which lead-mines are worked, has been found to be quite saturated with carbonic acid gas, which is constantly disengaged. The carbonates of iron, lime, and manganese are so dissolved, that the rock is rendered soft, and the quartz alone remains unattacked.† Not far off is the small volcanic cone of Chaluzet, which once broke up through the gneiss, and sent forth a lava stream.

Supposed atmosphere of carbonic acid.—Prof. Bischoff in his history of volcanoes,‡ has shown what enormous quantities of carbonic acid gas are exhaled in the vicinity of the extinct craters of the Rhine (in the neighborhood of the Laacher-see, for example, and the Eifel), and also in the mineral springs of Nassau and other countries, where there are no immediate traces of volcanic action. It would be easy to calculate in how short a period the solid carbon, thus emitted from the interior of the earth in an invisible form, would amount to a quantity as great as could be obtained from the trees of a large forest, and how many thousand years would be required to supply the materials of a dense seam of pure coal from the same source. Geologists who favor the doctrine of the former existence of an atmosphere highly charged with carbonic acid, at the period of the ancient coal-plants, have not sufficiently reflected on the continual disengagement of carbon, which is taking place in a gaseous form from springs, as also in a free state from the ground and from

* Ann. de l'Auvergne, tome i. p. 234.
† Ann. Scient. de l'Auvergne, tome ii. June, 1829.
‡ Edinb. New Phil. Journ. Oct. 1839.

volcanic craters into the air. We know that all plants are now engaged in secreting carbon, and many thousands of large trees are annually floated down by great rivers, and buried in their alluvial deposits; but before we can assume that the quantity of carbon which becomes permanently locked up in the earth by such agency will bring about an essential change in the chemical composition of the atmosphere, we must be sure that the trees annually buried contain more carbon than is given out from the interior of the earth in the same lapse of time. Every large area covered by a dense mass of peat, bears ample testimony to the fact, that several million tons of carbon have been taken from the air, by the powers of vegetable life, and stored up in the earth's crust, a large quantity of oxygen having been at the same time set free; but we cannot infer from these circumstances, that the constitution of the atmosphere has been materially deranged, until we have data for estimating the rate at which dead animal and vegetable substances are daily putrefying,—organic remains and various calcareous rocks decomposing, and volcanic regions emitting fresh volumes of carbonic acid gas. That the ancient carboniferous period was one of vast duration all geologists are agreed; instead, therefore, of supposing an excess of carbonic acid in the air at that epoch, for the support of a peculiar flora, we may imagine Time to have multiplied the quantity of carbon given out annually by mineral springs, volcanic craters, and other sources, until the component elements of any given number of coal-seams had been evolved from below, without any variation taking place in the constitution of the atmosphere. It has been too common, in reasoning on this question, to compute the loss of carbon by the volume of coal stored up in the ancient strata, and to take no account of the annual gain, by the restoration of carbonic acid to the atmosphere, through the machinery above alluded to.*

Disintegrating effects of carbonic acid.—The disintegration of granite is a striking feature of large districts in Auvergne, especially in the neighborhood of Clermont. This decay was called by Dolomieu, "la maladie du granite;" and the rock may with propriety be said to have *the rot*, for it crumbles to pieces in the hand. The phenomenon may, without doubt, be ascribed to the continual disengagement of carbonic acid gas from numerous fissures.

In the plains of the Po, between Verona and Parma, especially at Villa Franca, south of Mantua, I observed great beds of alluvium, consisting chiefly of primary pebbles, percolated by spring-water, charged with carbonate of lime and carbonic acid in great abundance. They are for the most part incrusted with calc-sinter; and the rounded blocks of gneiss, which have all the outward appearance of solidity, have been so disintegrated by the carbonic acid as readily to fall to pieces.

The subtraction of many of the elements of rocks by the solvent power of carbonic acid, ascending both in a gaseous state and mixed with spring-water in the crevices of rocks, must be one of the most powerful

* See Lyell's Travels in N. America, vol. i. p. 150.

sources of those internal changes and rearrangements of particles so often observed in strata of every age. The calcareous matter, for example, of shells, is often entirely removed and replaced by carbonate of iron, pyrites, silex, or some other ingredient, such as mineral waters usually contain in solution. It rarely happens, except in limestone rocks, that the carbonic acid can dissolve all the constituent parts of the mass; and for this reason, probably, calcareous rocks are almost the only ones in which great caverns and long winding passages are found.

Petroleum springs.—Springs of which the waters contain a mixture of petroleum and the various minerals allied to it, as bitumen, naphtha, asphaltum, and pitch, are very numerous, and are, in many cases, undoubtedly connected with subterranean fires, which raise or sublime the more subtle parts of the bituminous matters contained in rocks. Many springs in the territory of Modena and Parma, in Italy, produce petroleum in abundance; but the most powerful, perhaps, yet known, are those on the Irawadi, in the Burman empire. In one locality there are said to be 520 wells, which yield annually 400,000 hogsheads of petroleum.*

Pitch lake of Trinidad.—Fluid bitumen is seen to ooze from the bottom of the sea, on both sides of the island of Trinidad, and to rise up to the surface of the water. Near Cape La Braye there is a vortex which, in stormy weather, according to Captain Mallet, gushes out, raising the water five or six feet, and covers the surface for a considerable space with petroleum, or tar; and the same author quotes Gumilla, as stating, in his "Description of the Orinoco," that about seventy years ago, a spot of land on the western coast of Trinidad, near half-way between the capital and an Indian village, sank suddenly, and was immediately replaced by a small lake of pitch, to the great terror of the inhabitants.†

It is probable that the great pitch lake of Trinidad owes its origin to a similar cause; and Dr. Nugent has justly remarked, that in that district all the circumstances are now combined from which deposits of pitch may have originated. The Orinoco has for ages been rolling down great quantities of woody and vegetable bodies into the surrounding sea, where, by the influence of currents and eddies, they may be arrested and accumulated in particular places. The frequent occurrence of earthquakes and other indications of volcanic action in those parts lend countenance to the opinion, that these vegetable substances may have undergone, by the agency of subterranean fire, those transformations and chemical changes which produce petroleum; and this may, by the same causes, be forced up to the surface, where, by exposure to the air, it becomes inspissated, and forms the different varieties of pure and earthy pitch, or asphaltum, so abundant in the island.‡

* Symes, Embassy to Ava, vol. ii. Geol. Trans. second series, vol. ii. part iii. p. 388.

† Dr. Nugent, Geol. Trans. vol. i. p. 69.

‡ Ibid. p. 67.

It may be stated generally, that a large portion of the finer particles and the more crystalline substances, found in sedimentary rocks of different ages, are composed of the same elements as are now held in solution by springs, while the coarser materials bear an equally strong resemblance to the pebbles and sedimentary matter carried down by torrents and rivers. It should also be remembered, that it is not only during inundations, when the muddy sediment is apparent, that rivers are busy in conveying solid matter to the sea, but that even when their waters are perfectly transparent, they are annually bearing along vast masses of carbon, lime, and silica to the ocean.

CHAPTER XVII.

REPRODUCTIVE EFFECTS OF RIVERS.

Lake deltas—Growth of the delta of the Upper Rhine in the Lake of Geneva—Computation of the age of deltas—Recent deposits in Lake Superior—Deltas of inland seas—Course of the Po—Artificial embankments of the Po and Adige—Delta of the Po, and other rivers entering the Adriatic—Rapid conversion of that gulf into land—Mineral characters of the new deposits—Marine delta of the Rhone—Various proofs of its increase—Stony nature of its deposits—Coast of Asia Minor—Delta of the Nile.

DELTAS IN LAKES.

I HAVE already spoken in the 14th chapter of the action of running water, and of the denuding power of rivers, but we can only form a just conception of the excavating and removing force exerted by such bodies of water, when we have the advantage of examining the reproductive effects of the same agents: in other words, of beholding in a palpable form the aggregate amount of matter, which they have thrown down at certain points in their alluvial plains, or in the basins of lakes and seas. Yet it will appear, when we consider the action of currents, that the growth of deltas affords a very inadequate standard by which to measure the entire carrying power of running water, since a considerable portion of fluviatile sediment is swept far out to sea.

Deltas may be divided into, first, those which are formed in lakes; secondly, those in island seas, where the tides are almost imperceptible; and, thirdly, those on the borders of the ocean. The most characteristic distinction between the lacustrine and marine deltas consists in the nature of the organic remains which become imbedded in their deposits; for, in the case of a lake, it is obvious that these must consist exclusively of such genera of animals as inhabit the land or the waters of a river or a lake; whereas, in the other case, there will be an admixture, and most frequently a predominance, of animals which inhabit salt water. In regard, however, to the distribution of inorganic matter, the deposits of lakes and seas are formed under very analogous circumstances.

Lake of Geneva.—Lakes exemplify the first reproductive operations in which rivers are engaged when they convey the detritus of rocks and the ingredients of mineral springs from mountainous regions. The accession of new land at the mouth of the Rhone, at the upper end of the Lake of Geneva, or the Leman Lake, presents us with an example of a considerable thickness of strata which have accumulated since the historical era. This sheet of water is about thirty-seven miles long, and its breadth is from two to eight miles. The shape of the bottom is very irregular, the depth having been found by late measurements to vary from 20 to 160 fathoms.* The Rhone, where it enters at the upper end, is turbid and discolored; but its waters, where it issues at the town of Geneva, are beautifully clear and transparent. An ancient town, called Port Vallais (Portus Valesiæ of the Romans), once situated at the water's edge, at the upper end, is now more than a mile and a half inland—this intervening alluvial tract having been acquired in about eight centuries. The remainder of the delta consists of a flat alluvial plain, about five or six miles in length, composed of sand and mud, a little raised above the level of the river, and full of marshes.

Sir Henry De la Beche found, after numerous soundings in all parts of the lake, that there was a pretty uniform depth of from 120 to 160 fathoms throughout the central region, and on approaching the delta, the shallowing of the bottom began to be very sensible at a distance of about a mile and three quarters from the mouth of the Rhone; for a line drawn from St. Gingoulph to Vevey gives a mean depth of somewhat less than 600 feet, and from that part of the Rhone, the fluviatile mud is always found along the bottom.† We may state, therefore, that the new strata annually produced are thrown down upon a slope about two miles in length; so that, notwithstanding the great depth of the lake, the new deposits are inclined at so slight an angle, that the dip of the beds would be termed, in ordinary geological language, horizontal.

The strata probably consist of alternations of finer and coarser particles; for, during the hotter months from April to August, when the snows melt, the volume and velocity of the river are greatest, and large quantities of sand, mud, vegetable matter, and drift-wood are introduced; but during the rest of the year, the influx is comparatively feeble, so much so, that the whole lake, according to Saussure, stands six feet lower. If, then, we could obtain a section of the accumulation formed in the last eight centuries, we should see a great series of strata, probably from 600 to 900 feet thick (the supposed original depth of the head of the lake), and nearly two miles in length, inclined at a very slight angle. In the mean time, a great number of

* De la Beche, Ed. Phil. Journ. vol. ii. p. 107. Jan. 1820.
† De la Beche, MS.

smaller deltas are growing around the borders of the lake, at the mouths of rapid torrents, which pour in large masses of sand and pebbles. The body of water in these torrents is too small to enable them to spread out the transported matter over so extensive an area as the Rhone does. Thus, for example, there is a depth of eighty fathoms within half a mile of the shore, immediately opposite the great torrent which enters east of Ripaille, so that the dip of the strata in that minor delta must be about four times as great as those deposited by the main river at the upper extremity of the lake.*

Chronological computations of the age of deltas.—The capacity of this basin being now ascertained, it would be an interesting subject of inquiry, to determine in what number of years the Leman Lake will be converted into dry land. It would not be very difficult to obtain the elements for such a calculation, so as to approximate at least to the quantity of time required for the accomplishment of the result. The number of cubic feet of water annually discharged by the river into the lake being estimated, experiments might be made in the winter and summer months, to determine the proportion of matter held in suspension or in chemical solution by the Rhone. It would be also necessary to allow for the heavier matter drifted along at the bottom, which might be estimated on hydrostatical principles, when the average size of the gravel and the volume and velocity of the stream at different seasons were known. Supposing all these observations to have been made, it would be more easy to calculate the future than the former progress of the delta, because it would be a laborious task to ascertain, with any degree of precision, the original depth and extent of that part of the lake which is already filled up. Even if this information were actually obtained by borings, it would only enable us to approximate within a certain number of centuries to the time when the Rhone began to form its present delta; but this would not give us the date of the origin of the Leman Lake in its present form, because the river may have flowed into it for thousands of years, without importing any sediment whatever. Such would have been the case, if the waters had first passed through a chain of upper lakes; and that this was actually the fact, seems indicated by the course of the Rhone between Martigny and the Lake of Geneva, and, still more decidedly, by the channels of many of its principal feeders.

If we ascend, for example, the valley through which the Dranse flows, we find that it consists of a succession of basins, one above the other, in each of which there is a wide expanse of flat alluvial lands, separated from the next basin by a rocky gorge, once perhaps the barrier of a lake. The river seems to have filled these lakes, one after the other, and to have partially cut through the barriers, some of which it is still gradually eroding to a greater depth. Before, therefore, we can pretend even to hazard a conjecture as to the era at which the principal delta of Lake Leman or any other delta commenced, we must be thoroughly acquaint-

* De la Beche, MS.

ed with the geographical features and geological history of the whole system of higher valleys which communicate with the main stream, and all the changes which they have undergone since the last series of convulsions which agitated and altered the face of the country.

Lake Superior.—Lake Superior is the largest body of freshwater in the world, being above 1700 geographical miles in circumference when we follow the sinuosities of its coasts, and its length, on a curved line drawn through its centre, being more than 400, and its extreme breadth above 150 geographical miles. Its surface is nearly as large as the whole of England. Its average depth varies from 80 to 150 fathoms; but, according to Captain Bayfield, there is reason to think that its greatest depth would not be overrated at 200 fathoms, so that its bottom is, in some parts, nearly 600 feet below the level of the Atlantic, its surface being about as much above it. There are appearances in different parts of this, as of the other Canadian lakes, leading us to infer that its waters formerly occupied a higher level than they reach at present; for at a considerable distance from the present shores, parallel lines of rolled stones and shells are seen rising one above the other, like the seats of an amphitheatre. These ancient lines of shingle are exactly similar to the present beaches in most bays, and they often attain an elevation of 40 or 50 feet above the present level. As the heaviest gales of wind do not raise the waters more than three or four feet, the elevated beaches have by some been referred to the subsidence of the lake at former periods, in consequence of the wearing down of its barrier; by others to the upraising of the shores by earthquakes, like those which have produced similar phenomena on the coast of Chili.

The streams which discharge their waters into Lake Superior are several hundred in number, without reckoning those of smaller size; and the quantity of water supplied by them is many times greater than that discharged at the Falls of St. Mary, the only outlet. The evaporation, therefore, is very great, and such as might be expected from so vast an extent of surface. On the northern side, which is encircled by primary mountains, the rivers sweep in many large boulders with smaller gravel and sand, chiefly composed of granitic and trap rocks. There are also currents in the lake in various directions, caused by the continued prevalence of strong winds, and to their influence we may attribute the diffusion of finer mud far and wide over great areas; for by numerous soundings made during Captain Bayfield's survey, it was ascertained that the bottom consists generally of a very adhesive clay, containing shells of the species at present existing in the lake. When exposed to the air, this clay immediately becomes indurated in so great a degree, as to require a smart blow to break it. It effervesces slightly with diluted nitric acid, and is of different colors in different parts of the lake; in one district blue, in another red, and in a third white, hardening into a substance resembling pipeclay.* From these statements, the geologist will

* Trans. of Lit. and Hist. Soc. of Quebec, vol. i. p. 5, 1829.

not fail to remark how closely these recent lacustrine formations in America resemble the tertiary argillaceous and calcareous marls of lacustrine origin in Central France. In both cases many of the genera of shells most abundant, as Limnea and Planorbis, are the same; and in regard to other classes of organic remains there must be the closest analogy, as I shall endeavor more fully to explain when speaking of the imbedding of plants and animals in recent deposits.

DELTAS OF INLAND SEAS.

Having thus briefly considered some of the lacustrine deltas now in progress, we may next turn our attention to those of inland seas.

Course of the Po.—The Po affords an instructive example of the manner in which a great river bears down to the sea the matter poured into it by a multitude of tributaries descending from lofty chains of mountains. The changes gradually effected in the great plain of Northern Italy, since the time of the Roman republic, are considerable. Extensive lakes and marshes have been gradually filled up, as those near Placentia, Parma, and Cremona, and many have been drained naturally by the deepening of the beds of rivers. Deserted river-courses are not unfrequent, as that of the Serio Morto, which formerly fell into the Adda, in Lombardy. The Po also itself has often deviated from its course, having after the year 1390 deserted part of the territory of Cremona, and invaded that of Parma; its old channel being still recognizable, and bearing the name of Po Morto. There is also an old channel of the Po in the territory of Parma, called Po Vecchio, which was abandoned in the twelfth century, when a great number of towns were destroyed.

Artificial embankments of Italian rivers.—To check these and similar aberrations, a general system of embankment has been adopted; and the Po, Adige, and almost all their tributaries, are now confined between high artificial banks. The increased velocity acquired by streams thus closed in, enables them to convey a much larger portion of foreign matter to the sea; and, consequently, the deltas of the Po and Adige have gained far more rapidly on the Adriatic since the practice of embankment became almost universal. But, although more sediment is borne to the sea, part of the sand and mud, which in the natural state of things would be spread out by annual inundations over the plain, now subsides in the bottom of the river-channels; and their capacity being thereby diminished, it is necessary, in order to prevent inundations in the following spring, to extract matter from the bed, and to add it to the banks of the river. Hence it happens that these streams now traverse the plain on the top of high mounds, like the waters of aqueducts, and at Ferrara the surface of the Po has become more elevated than the roofs of the houses.* The magnitude of these barriers is a subject

* Prony, see Cuvier, Disc. Prelim. p. 146.

of increasing expense and anxiety, it having been sometimes found necessary to give an additional height of nearly one foot to the banks of the Adige and Po in a single season.

The practice of embankment was adopted on some of the Italian rivers as early as the thirteenth century; and Dante, writing in the beginning of the fourteenth, describes, in the seventh circle of hell, a rivulet of tears separated from a burning sandy desert by embankments "like those which, between Ghent and Bruges, were raised against the ocean, or those which the Paduans had erected along the Brenta to defend their villas on the melting of the Alpine snows."

> Quale i Fiamminghi tra Guzzante e Bruggia,
> Temendo il fiotto che in ver lor s'avventa,
> Fanno lo schermo, perchè il mar si fuggia,
> E quale i Padovan lungo la Brenta,
> Per difender lor ville e lor castelli,
> Anzi che Chiarentana il caldo senta.—
>
> *Inferno*, Canto XV.

In the Adriatic, from the northern part of the Gulf of Trieste, where the Isonzo enters, down to the south of Ravenna, there is an uninterrupted series of recent accessions of land, more than 100 miles in length, which, within the last 2000 years, have increased from *two to twenty miles in breadth.* A line of sand-bars of great length has been formed nearly all along the western coast of this gulf, inside of which are lagunes, such as those of Venice, and the large lagune of Comacchio, 20 miles in diameter. Newly deposited mud brought down by the streams is continually lessening the depth of the lagunes, and converting part of them into meadows.* The Isonzo, Tagliamento, Piave, Brenta, Adige, and Po, besides many other inferior rivers, contribute to this advance of the coast-line and to the shallowing of the lagunes and the gulf.

Delta of the Po.—The Po and the Adige may now be considered as entering by one common delta, for two branches of the Adige are connected with arms of the Po, and thus the principal delta has been pushed out beyond those bars which separate the lagunes from the sea. The rate of the advance of this new land has been accelerated, as before stated, since the system of embanking the rivers became general, especially at that point where the Po and Adige enter. The waters are no longer permitted to spread themselves far and wide over the plains, and to leave behind them the larger portion of their sediment. Mountain torrents also have become more turbid since the clearing away of forests, which once clothed the southern flanks of the Alps. It is calculated that the mean rate of advance of the delta of the Po on the Adriatic between the years 1200 and 1600 was 25 yards or metres a year, whereas the mean annual gain from 1600 to 1804 was 70 metres.†

* See De Beaumont, Géologie Pratique, vol. i. p. 323, 1844.
† Prony, cited by Cuvier, Discours Prélimin.

Adria was a seaport in the time of Augustus, and had, in ancient times, given its name to the gulf; it is now about twenty Italian miles inland. Ravenna was also a seaport, and is now about four miles from the main sea. Yet even before the practice of embankment was introduced, the alluvium of the Po advanced with rapidity on the Adriatic; for Spina, a very ancient city, originally built in the district of Ravenna, at the mouth of a great arm of the Po, was, so early as the commencement of our era, eleven miles distant from the sea.*

But although so many rivers are rapidly converting the Adriatic into land, it appears, by the observations of M. Morlot, that since the time of the Romans, there has been a general subsidence of the coast and bed of this sea in the same region to the amount of five feet, so that the advance of the new-made land has not been so fast as it would have been had the level of the coast remained unaltered. The signs of a much greater depression anterior to the historical period have also been brought to light by an Artesian well, bored in 1847, to the depth of more than 400 feet, which still failed to penetrate through the modern fluviatile deposit. The auger passed chiefly through beds of sand and clay, but at four several depths, one of them very near the bottom of the excavation, it pierced beds of turf, or accumulations of vegetable matter, precisely similar to those now formed superficially on the extreme borders of the Adriatic. Hence we learn that a considerable area of what was once land has sunk down 400 feet in the course of ages.†

The greatest depth of the Adriatic, between Dalmatia and the mouths of the Po, is twenty-two fathoms; but a large part of the Gulf of Trieste and the Adriatic, opposite Venice, is less than twelve fathoms deep. Farther to the south, where it is less affected by the influx of great rivers, the gulf deepens considerably. Donati, after dredging the bottom, discovered the new deposits to consist partly of mud and partly of rock, the rock being formed of calcareous matter, incrusting shells. He also ascertained, that particular species of testacea were grouped together in certain places, and were becoming slowly incorporated with the mud or calcareous precipitates.‡ Olivi, also, found some deposits of sand, and others of mud, extending half way across the gulf; and he states that their distribution along the bottom was evidently determined by the prevailing current.§ It is probable, therefore, that the finer sediment of all the rivers at the head of the Adriatic may be intermingled by the influence of the current; and all the central parts of the gulf may be considered as slowly filling up with horizontal deposits, similar to those of the Subapennine hills, and containing many of the same species of shells. The Po merely introduces at present fine sand and mud, for it carries no pebbles farther than the spot where it joins the Trebia, west of Piacenza. Near the northern borders of the basin, the Isonzo,

* Brocchi, Conch. Foss. Subap. vol. i. p. 118.
† Archiac, Histoire des Progrés de la Géol. 1848, vol. ii. p. 232.
‡ Brocchi, Conch. Foss. Subap. vol. i. p. 39. § Ibid. vol. ii. p. 94.

Tagliamento, and many other streams, are forming immense beds of sand and some conglomerate; for here some high mountains of Alpine limestone approach within a few miles of the sea.

In the time of the Romans, the hot-baths of Monfalcone were on one of several islands of Alpine limestone, between which and the mainland, on the north, was a channel of the sea, about a mile broad. This channel is now converted into a grassy plain, which surrounds the islands on all sides. Among the numerous changes on this coast, we find that the present channel of the Isonzo is several miles to the west of its ancient bed, in part of which, at Ronchi, the old Roman bridge which crossed the Via Appia was lately found buried in fluviatile silt.

Marine delta of the Rhone.—The lacustrine delta of the Rhone in Switzerland has already been considered (p. 251), its contemporaneous marine delta may now be described. Scarcely has the river passed out of the Lake of Geneva before its pure waters are again filled with sand and sediment by the impetuous Arve, descending from the highest Alps, and bearing along in its current the granitic detritus annually brought down by the glaciers of Mont Blanc. The Rhone afterwards receives vast contributions of transported matter from the Alps of Dauphiny, and the primary and volcanic mountains of Central France; and when at length it enters the Mediterranean, it discolors the blue waters of that sea with a whitish sediment, for the distance of between six and seven miles, throughout which space the current of fresh water is perceptible.

Strabo's description of the delta is so inapplicable to its present configuration, as to attest a complete alteration in the physical features of the country since the Augustan age. It appears, however, that the head of the delta, or the point at which it begins to ramify, has remained unaltered since the time of Pliny, for he states that the Rhone divided itself at Arles into two arms. This is the case at present; one of the branches, the western, being now called Le Petit Rhône, which is again subdivided before entering the Mediterranean. The advance of the base of the delta, in the last eighteen centuries, is demonstrated by many curious antiquarian monuments. The most striking of these is the great and unnatural détour of the old Roman road from Ugernum to Beziers (*Bœterræ*) which went round by Nismes (*Nemausus*). It is clear that, when this was first constructed, it was impossible to pass in a direct line, as now, across the delta, and that either the sea or marshes intervened in a tract now consisting of terra firma.* Astruc also remarks, that all the places on low lands, lying to the north of the old Roman road between Nismes and Beziers, have names of Celtic origin, evidently given to them by the first inhabitants of the country; whereas, the places lying south of that road, towards the sea, have names of Latin derivation, and were clearly founded after the Roman language had been introduced.

* Mém. d'Astruc, cited by Von Hoff, vol. i. p. 288.

Another proof, also, of the great extent of land which has come into existence since the Romans conquered and colonized Gaul, is derived from the fact, that the Roman writers never mention the thermal waters of Balaruc in the delta, although they were well acquainted with those of Aix, and others still more distant, and attached great importance to them, as they invariably did to all hot springs. The waters of Balaruc, therefore, must have formerly issued under the sea—a common phenomenon on the borders of the Mediterranean; and on the advance of the delta they continued to flow out through the new deposits.

Among the more direct proofs of the increase of land, we find that Mese, described under the appellation of Mesua Collis by Pomponius Mela,* and stated by him to be nearly an island, is now far inland. Notre Dame des Ports, also, was a harbor in 898, but is now a league from the shore. Psalmodi was an island in 815, and is now two leagues from the sea. Several old lines of towers and sea-marks occur at different distances from the present coast, all indicating the successive retreat of the sea, for each line has in its turn become useless to mariners; which may well be conceived, when we state that the Tower of Tignaux, erected on the shore so late as the year 1737, is already a mile remote from it.†

By the confluence of the Rhone and the currents of the Mediterranean, driven by winds from the south, sand-bars are often formed across the mouths of the river; by these means considerable spaces become divided off from the sea, and subsequently from the river also, when it shifts its channels of efflux. As some of these lagoons are subject to the occasional ingress of the river when flooded, and of the sea during storms, they are alternately salt and fresh. Others, after being filled with salt water, are often lowered by evaporation till they become more salt than the sea; and it has happened, occasionally, that a considerable precipitate of muriate of soda has taken place in these natural salterns. During the latter part of Napoleon's career, when the excise laws were enforced with extreme rigor, the police was employed to prevent such salt from being used. The fluviatile and marine shells inclosed in these small lakes often live together in brackish water; but the uncongenial nature of the fluid usually produces a dwarfish size, and sometimes gives rise to strange varieties in form and color.

Captain Smyth in his survey of the coast of the Mediterranean, found the sea opposite the mouth of the Rhone, to deepen gradually from four to forty fathoms, within a distance of six or seven miles, over which the discolored fresh water extends; so that the inclination of the new deposits must be too slight to be appreciable in such an extent of section as a geologist usually obtains in examining ancient formations. When the wind blew from the southwest, the ships em-

* Lib. ii. c. v.

† Bouche, Chorographie et Hist. de Provence, vol. i. p. 23, cited by Von Hoff, vol. i. p. 290.

ployed in the survey were obliged to quit their moorings; and when they returned, the new sand-banks in the delta were found covered over with a great abundance of marine shells. By this means, we learn how occasional beds of drifted marine shells may become interstratified with freshwater strata at a river's mouth.

Stony nature of its deposits.—That a great proportion, at least, of the new deposit in the delta of the Rhone consists of *rock*, and not of loose incoherent matter, is perfectly ascertained. In the Museum at Montpelier is a cannon taken up from the sea near the mouth of the river, imbedded in a crystalline calcareous rock. Large masses, also, are continually taken up of an arenaceous rock, cemented by calcareous matter, including multitudes of broken shells of recent species. The observations lately made on this subject corroborate the former statement of Marsilli, that the earthy deposits of the coast of Languedoc form a stony substance, for which reason he ascribes a certain bituminous, saline, and glutinous nature to the substances brought down with sand by the Rhone.* If the number of mineral springs charged with carbonate of lime which fall into the Rhone and its feeders in different parts of France be considered, we shall feel no surprise at the lapidification of the newly deposited sediment in this delta. It should be remembered, that the fresh water introduced by rivers being lighter than the water of the sea, floats over the latter, and remains upon the surface for a considerable distance. Consequently it is exposed to as much evaporation as the waters of a lake; and the area over which the river-water is spread, at the junction of great rivers and the sea, may well be compared, in point of extent, to that of considerable lakes.

Now, it is well known, that so great is the quantity of water carried off by evaporation in some lakes, that it is nearly equal to the water flowing in; and in some inland seas, as the Caspian, it is quite equal. We may, therefore, well suppose that, in cases where a strong current does not interfere, the greater portion not only of the matter held mechanically in suspension, but of that also which is in chemical solution, may be precipitated at no great distance from the shore. When these finer ingredients are extremely small in quantity, they may only suffice to supply crustaceous animals, corals, and marine plants, with the earthy particles necessary for their secretions; but whenever it is in excess (as generally happens if the basin of a river lie partly in a district of active or extinct volcanoes), then will solid deposits be formed, and the shells will at once be included in a rocky mass.

Coast of Asia Minor.—Examples of the advance of the land upon the sea are afforded by the southern coast of Asia Minor. Admiral Sir F. Beaufort has pointed out in his survey the great alterations effected since the time of Strabo, where havens are filled up, islands joined to the mainland, and where the whole continent has increased many miles in extent. Strabo himself, on comparing the outline of the coast in his time with its ancient state, was convinced, like our country-

* Hist. Phys. de la Mer.

man; that it had gained very considerably upon the sea. The new-formed strata of Asia Minor consist *of stone,* not of loose incoherent materials. Almost all the streamlets and rivers, like many of those in Tuscany and the south of Italy, hold abundance of carbonate of lime in solution, and precipitate travertin, or sometimes bind together the sand and gravel into solid sandstones and conglomerates; every delta and sand-bar thus acquires solidity, which often prevents streams from forcing their way through them, so that their mouths are constantly changing their position.*

Delta of the Nile.—That Egypt was "the gift of the Nile," was the opinion of her priests before the time of Herodotus; and Rennell observes, that the "configuration and composition of the low lands leave no room for doubt that the sea once washed the base of the rocks on which the pyramids of Memphis stand, the *present* base of which is washed by the inundation of the Nile, at an elevation of 70 or 80 feet above the Mediterranean. But when we attempt to carry back our ideas to the remote period when the foundation of the delta was first laid, we are lost in the contemplation of so vast an interval of time."† Herodotus observes, "that the country round Memphis seemed formerly to have been an arm of the sea gradually filled by the Nile, in the same manner as the Meander, Achelous, and other streams, had formed deltas. Egypt, therefore, he says, like the Red Sea, was once a long narrow bay, and both gulfs were separated by a small neck of land. If the Nile, he adds, should by any means have an issue into the Arabian Gulf, it might choke it up with earth in 20,000 or even, perhaps, in 10,000 years; and why may not the Nile have filled a still greater gulf with mud in the space of time which has passed before our age?"‡

The distance between Memphis and the most prominent part of the delta in a straight line north and south, is about 100 geographical miles; the length of the base of the delta is more than 200 miles if we follow the coast between the ancient extreme eastern and western arms; but as these are now blocked up, that part only of Lower Egypt which intervenes between the Rosetta and Damietta branches, is usually called the delta, the coast line of which is about 90 miles in length. The bed of the river itself, says Sir J. G. Wilkinson, undergoes a gradual increase of elevation varying in different places, and always lessening in proportion as the river approaches the sea. "This increase of elevation in perpendicular height is much smaller in Lower than in Upper Egypt, and in the delta it diminishes still more; so that, according to an approximate calculation, the land about Elephantine, or the first cataract, lat. 24° 5′, has been raised nine feet in 1700 years; at Thebes, lat. 25° 43′, about seven feet; and at Heliopolis and Cairo, lat. 30°, about five feet ten inches. At Rosetta and the mouths of the Nile, lat. 31° 30′, the diminution in the perpendicular thickness of the deposit is lessened

* Karamania, or a brief Description of the Coast of Asia Minor, &c. London, 1817.

† Geog. Syst. of Herod. vol. ii. p. 107.

‡ Euterpe, XI.

in a much greater decreasing ratio than in the straitened valley of Central and Upper Egypt, owing to the great extent, east and west, over which the inundation spreads."*

For this reason the alluvial deposit does not cause the delta to protrude rapidly into the sea, although some ancient cities are now a mile or more inland, and the mouths of the Nile, mentioned by the earlier geographers, have been many of them silted up, and the outline of the coast entirely changed.

The bed of the Nile always keeps pace with the general elevation of the soil, and the banks of this river, like those of the Mississippi and its tributaries (see p. 265), are much higher than the flat land at a distance, so that they are seldom covered during the highest inundations. In consequence of the gradual rise of the river's bed, the annual flood is constantly spreading over a wider area, and the alluvial soil encroaches on the desert, covering, to the depth of six or seven feet, the base of statues and temples which the waters never reached 3000 years ago. Although the sands of the Libyan deserts have in some places been drifted into the valley of the Nile, yet these aggressions, says Wilkinson, are far more than counterbalanced by the fertilizing effect of the water which now reaches farther inland towards the desert, so that the number of square miles of arable soil is greater at present than at any previous period.

Mud of the Nile.—On comparing the different analyses which have been published of this mud, it will be found that it contains a large quantity of argillaceous matter, with much peroxide of iron, some carbonate of lime, and a small proportion of carbonate of magnesia. The latest and most careful analysis by M. Lassaigne shows a singularly close resemblance in the proportions of the ingredients of silica, alumina, iron, carbon, lime, and magnesia, and those observed in ordinary mica;† but a much larger quantity of calcareous matter is sometimes present.

In many places, as at Cairo, where artificial excavations have been made, or where the river has undermined its banks, the mud is seen to be thinly stratified, the upper part of each annual layer consisting of earth of a lighter color than the lower, and the whole separating easily from the deposit of the succeeding year. These annual layers are variable in thickness; but, according to the calculations of Girard and Wilkinson, the mean annual thickness of a layer at Cairo cannot exceed that of a sheet of thin pasteboard, and a stratum of two or three feet must represent the accumulation of a thousand years.

The depth of the Mediterranean is about twelve fathoms at a small distance from the shore of the delta; it afterwards increases gradually to 50, and then suddenly descends to 380 fathoms, which is, perhaps, the original depth of the sea where it has not been rendered shallower by fluviatile matter. We learn from Lieut. Newbold that nothing but the

* Journ. of Roy. Geograph. Soc. vol. ix. p. 432.

† Quart. Journ. Geol. Soc. vol. v.; Memoirs, p. 20; and Lassaigne, Journ. Pharm. t. v. p. 468.

finest and lightest ingredients reach the Mediterranean, where he has observed the sea discolored by them to the distance of 40 miles from the shore.* The small progress of the delta in the last 2000 years affords, perhaps, no measure for estimating its rate of growth when it was an inland bay, and had not yet protruded itself beyond the coast-line of the Mediterranean. A powerful current now sweeps along the shores of Africa, from the Straits of Gibraltar to the prominent convexity of Egypt, the western side of which is continually the prey of the waves; so that not only are fresh accessions of land checked, but ancient parts of the delta are carried away. By this cause, Canopus and some other towns have been overwhelmed; but to this subject I shall again refer when speaking of tides and currents.

CHAPTER XVIII.

REPRODUCTIVE EFFECTS OF RIVERS—*continued.*

Deltas formed under the influence of tides—Basin and delta of the Mississippi—Alluvial plain—River-banks and bluffs—Curves of the river—Natural rafts and snags—New lakes, and effects of earthquakes—Antiquity of the delta—Delta of the Ganges and Brahmapootra—Head of the delta and Sunderbunds—Islands formed and destroyed—Crocodiles—Amount of fluviatile sediment in the water—Artesian boring at Calcutta—Proofs of subsidence—Age of the delta—Convergence of deltas—Origin of existing deltas not contemporaneous—Grouping of strata and stratification in deltas—Conglomerates—Constant interchange of land and sea.

IN the last chapter several examples were given of the deltas of inland seas, where the influence of the tides is almost imperceptible. We may next consider those marine or oceanic deltas, where the tides play an important part in the dispersion of fluviatile sediment, as in the Gulf of Mexico, where they exert a moderate degree of force, and then in the Bay of Bengal, where they are extremely powerful. In regard to estuaries, which Rennel termed "negative deltas," they will be treated of more properly when our attention is specially turned to the operations of tides and currents (chapters 20, 21, and 22). In this case, instead of the land gaining on the sea at the river's mouth, the tides penetrate far inland beyond the general coast-line.

BASIN AND DELTA OF THE MISSISSIPPI.

Alluvial plain.—The hydrographical basin of the Mississippi displays, on the grandest scale, the action of running water on the surface of a vast continent. This magnificent river rises nearly in the forty-ninth

* Quart. Journ. Geol. Soc. 1848, vol. iv. p. 342.

parallel of north latitude, and flows to the Gulf of Mexico in the twenty-ninth—a course, including its meanders, of more than three thousand miles. It passes from a cold climate, where the hunter obtains his furs and peltries, traverses the temperate latitudes, and discharges its waters into the sea in the region of rice, the cotton plant, and the sugar-cane. From near its mouth at the Balize a steamboat may ascend for 2000 miles with scarcely any perceptible difference in the width of the river. Several of its tributaries, the Red River, the Arkansas, the Missouri, the Ohio, and others, would be regarded elsewhere as of the first importance, and, taken together, are navigable for a distance many times exceeding that of the main stream. No river affords a more striking illustration of the law before mentioned, that an augmentation of volume does not occasion a proportional increase of surface, nay, is even sometimes attended with a narrowing of the channel. The Mississippi is half a mile wide at its junction with the Missouri, the latter being also of equal width; yet the united waters have only, from their confluence to the mouth of the Ohio, a medial width of about half a mile. The junction of the Ohio seems also to produce no increase, but rather a decrease, of surface.* The St. Francis, White, Arkansas, and Red rivers are also absorbed by the main stream with scarcely any apparent increase of its width, although here and there it expands to a breadth of 1½, or even to 2 miles. On arriving at New Orleans, it is somewhat less than half a mile wide. Its depth there is very variable, the greatest at high water being 168 feet. The mean rate at which the whole body of water flows is variously estimated; according to Mr. Forshey the mean velocity of the current at the surface, somewhat exceeds 2¼ miles an hour when the water is at a mean height. For 300 miles above New Orleans the distance measured by the winding river is about twice as great as the distance in a right line. For the first 100 miles from the mouth the rate of fall is 1·80 inch per mile, for the second hundred 2 inches, for the third 2·30, for the fourth 2·57.

The alluvial plain of the Mississippi begins to be of great width below Cape Girardeau, 50 miles above the junction of the Ohio. At this junction it is about 50 miles broad, south of which it contracts to about 30 miles at Memphis, expands again to 80 miles at the mouth of the White River, and then, after various contractions and expansions, protrudes beyond the general coast-line, in a large delta, about 90 miles in width, from N. E. to S. W. Mr. Forshey estimates the area of the great plain as above defined at 31,200 square miles, with a circumference of about 3000 miles, exceeding the area of Ireland. If that part of this plain which lies below, or to the south of the branching off of the highest arm, called the Atchafalaya, be termed the delta, it constitutes less than half of the whole, being 14,000 square British miles in area. The delta may be said to be bounded on the east, west, and

* Flint's Geography, vol. i. p. 142. Lyell's Second Visit to the United States, vol. ii. chaps. 28 to 34.

south by the sea; on the north chiefly by the broad valley-plain which entirely resembles it in character as in origin. The east and west boundaries of the alluvial region above the head of the delta consists of cliffs or bluffs, which on the east side of the Mississippi are very abrupt, and are undermined by the river at many points. They consist, from Baton Rouge in Louisiana, where they commence, as far north as the borders of Kentucky, of geological formations newer than the cretaceous, the lowest being Eocene, and the uppermost consisting of loam, resembling the loess of the Rhine, and containing freshwater and land shells almost all of existing species. (See fig. 23.) These recent shells are associated with the bones of the mastodon, elephant, tapir, mylodon, horse, ox, and other quadrupeds, most of them of extinct species.

I have endeavored to show in my Second Visit to the United States, that this extensive formation of loam is either an ancient alluvial plain or a delta of the great river, formed originally at a lower level, and since upheaved, and partially denuded.

The Mississippi in that part of its course which is below the mouth of the Ohio, frequently washes the eastern bluffs, but never once comes in contact with the western. These are composed of similar formations; but I learn from Mr. Forshey that they rise up more gently from the alluvial plain (as at *a*, fig. 23). It is supposed that the

Fig. 23.
VALLEY OF THE MISSISSIPPI.

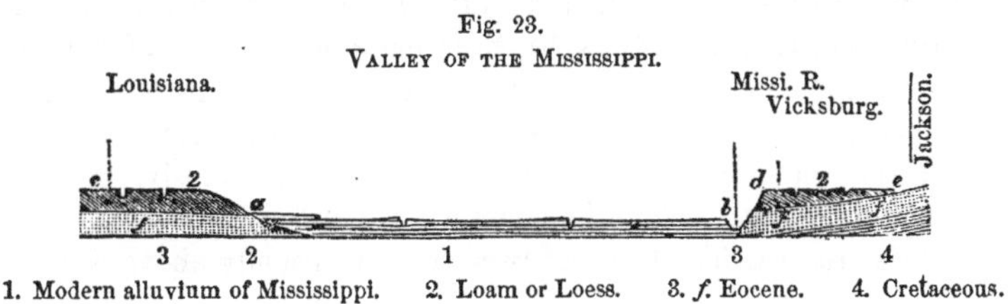

1. Modern alluvium of Mississippi. 2. Loam or Loess. 3. *f.* Eocene. 4. Cretaceous.

waters are thrown to the eastern side, because all the large tributary rivers entering from the west have filled that side of the great valley with their deltas, or with a sloping mass of clay and sand; so that the opposite bluffs are undermined, and the Mississippi is slowly but incessantly advancing eastward.*

Curves of the Mississippi.—The river traverses the plain in a meandering course, describing immense curves. After sweeping round the half of a circle, it is carried in a rapid current diagonally across the ordinary direction of its channel, to another curve of similar shape. Opposite to each of these, there is always a sand-bar, answering, in the convexity of its form, to the concavity of "the bend," as it is called.† The river, by continually wearing these curves deep, returns, like many other streams before described, on its own track, so that a vessel in some places, after sailing for twenty-five or thirty miles, is brought round again to within a mile of the place whence it started. When the waters approach so near to each other, it often happens at high floods

* Geograph. Descrip. of Louisiana, by W. Darby, Philadelphia, 1816, p. 102.
† Flint's Geography, vol. i. p. 152.

that they burst through the small tongue of land, and insulate a portion, rushing through what is called the "cut-off," so that vessels may pass from one point to another in half a mile to a distance which it previously required a voyage of twenty miles to reach. As soon as the river has excavated the new passage, bars of sand and mud are formed at the two points of junction with the old bend, which is soon entirely separated from the main river by a continuous mud-bank covered with wood. The old bend then becomes a semicircular lake of clear water, inhabited by large gar-fish, alligators, and wild fowl, which the steamboats have nearly driven away from the main river. A multitude of such crescent-shaped lakes, scattered far and wide over the alluvial plain, the greater number of them to the west, but some of them also eastward of the Mississippi, bear testimony of the extensive wanderings of the great stream in former ages. For the last two hundred miles above its mouth the course of the river is much less winding than above, there being only in the whole of that distance one great curve, that called the "English Turn." This great straightness of the stream is ascribed by Mr. Forshey to the superior tenacity of the banks, which are more clayey in this region.

The Mississippi has been incorrectly described by some of the earlier geographers, as a river running along the top of a long hill, or mound in a plain. In reality it runs in a valley, from 100 to 200 or more feet in depth, as *a, c, b*, fig. 24, its banks forming long strips of land parallel to the course of the main stream, and to the swamps *g, f*, and *d, e*, lying on each side. These extensive morasses, which are commonly well-wooded, though often submerged for months continuously, are rarely more than fifteen feet below the summit level of the banks. The banks themselves are occasionally overflowed, but are usually above water for

Fig. 24.

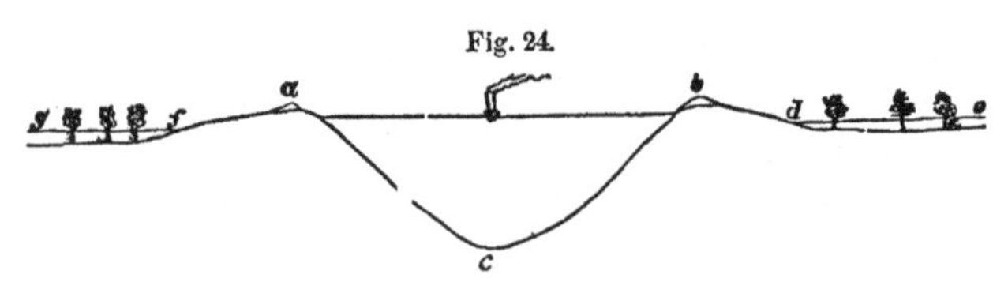

Section of channel, bank, levees (*a* and *b*), and swamps of Mississippi river.

a breadth of about two miles. They follow all the curves of the great river, and near New Orleans are raised artificially by embankments (or levees), *a b*, fig. 24, through which the river when swollen sometimes cuts a deep channel (or crevasse), inundating the adjoining low lands and swamps, and not sparing the lower streets of the great city.

The cause of the uniform upward slope of the river-bank above the adjoining alluvial plain is this: when the waters charged with sediment pass over the banks in the flood season, their velocity is checked among the herbage and reeds, and they throw down at once the coarser and more sandy matter with which they are charged. But the fine particles of mud are carried farther on, so that at the distance of about two miles, a thin film of fine clay only subsides, forming a stiff unctuous black soil, which gradually envelops the base of trees growing on the borders of the swamps.

Waste of the banks.—It has been said of a mountain torrent, that "it lays down what it will remove, and removes what it has laid down;" and in like manner the Mississippi, by the continual shifting of its course, sweeps away, during a great portion of the year, considerable tracts of alluvium, which were gradually accumulated by the overflow of former years, and the matter now left during the spring-floods will be at some future time removed. After the flood season, when the river subsides within its channel, it acts with destructive force upon the alluvial banks, softened and diluted by the recent overflow. Several acres at a time, thickly covered with wood, are precipitated into the stream; and large portions of the islands are frequently swept away.

"Some years ago," observes Captain Hall, "when the Mississippi was regularly surveyed, all its islands were numbered, from the confluence of the Missouri to the sea; but every season makes such revolutions, not only in the number, but in the magnitude and situation of these islands, that this enumeration is now almost obsolete. Sometimes large islands are entirely melted away; at other places they have attached themselves to the main shore, or, which is the more correct statement, the interval has been filled up by myriads of logs cemented together by mud and rubbish."*

Rafts.—One of the most interesting features in the great rivers of this part of America is the frequent accumulation of what are termed "rafts," or masses of floating trees, which have been arrested in their progress by snags, islands, shoals, or other obstructions, and made to accumulate, so as to form natural bridges, reaching entirely across the stream. One of the largest of these was called the raft of the Atchafalaya, an arm of the Mississippi, which was certainly at some former time the channel of the Red River, when the latter found its way to the Gulf of Mexico by a separate course. The Atchafalaya being in a direct line with the general direction of the Mississippi, catches a large portion of the timber annually brought down from the north; and the drift-trees collected in about thirty-eight years previous to 1816 formed a continuous raft, no less than ten miles in length, 220 yards wide, and eight feet deep. The whole rose and fell with the water, yet was covered with green bushes and trees, and its surface enlivened in the autumn by a variety of beautiful flowers. It went on increasing till about 1835, when some of the trees upon it had grown to the height of about sixty feet. Steps were then taken by the State of Louisiana to clear away the whole raft, and open the navigation, which was effected, not without great labor, in the space of four years.

The rafts on Red River are equally remarkable: in some parts of its course, cedar-trees are heaped up by themselves, and in other places,

* Travels in North America, vol. iii. p. 361.

pines. On the rise of the waters in summer hundreds of these are seen, some with their green leaves still upon them, just as they have fallen from a neighboring bank, others leafless, broken and worn in their passage from a far distant tributary: wherever they accumulate on the edge of a sand-bar they arrest the current, and soon become covered with sediment. On this mud the young willows and the poplars called cotton-wood spring up, their boughs still farther retarding the stream, and as the inundation rises, accelerating the deposition of new soil. The bank continuing to enlarge, the channel at length becomes so narrow that a single long tree may reach from side to side, and the remaining space is then soon choked up by a quantity of other timber.

"Unfortunately for the navigation of the Mississippi," observes Captain Hall, "some of the largest trunks, after being cast down from the position on which they grew, get their roots entangled with the bottom of the river, where they remain anchored, as it were, in the mud. The force of the current naturally gives their tops a tendency downwards, and, by its flowing past, soon strips them of their leaves and branches. These fixtures, called snags, or planters, are extremely dangerous to the steam-vessels proceeding up the stream, in which they lie like a lance in rest, concealed beneath the water, with their sharp ends pointed directly against the bows of the vessels coming up. For the most part these formidable snags remain so still that they can be detected only by a slight ripple above them, not perceptible to inexperienced eyes. Sometimes, however, they vibrate up and down, alternately showing their heads above the surface and bathing them beneath it."* So imminent, until lately, was the danger caused by these obstructions, that almost all the boats on the Mississippi were constructed on a particular plan, to guard against fatal accidents; but in the last ten years, by the aid of the power of steam and the machinery of a snag-boat, as it is called, the greater number of these trunks of trees have been drawn out of the mud.†

The prodigious quantity of wood annually drifted down by the Mississippi and its tributaries, is a subject of geological interest, not merely as illustrating the manner in which abundance of vegetable matter becomes, in the ordinary course of nature, imbedded in submarine and estuary deposits, but as attesting the constant destruction of soil and transportation of matter to lower levels by the tendency of rivers to shift their courses. Each of these trees must have required many years, some of them centuries, to attain their full size; the soil, therefore, whereon they grew, after remaining undisturbed for long periods, is ultimately torn up and swept away.

* Travels in North America, vol. iii. p. 362.

† "The boats are fitted," says Captain Hall, "with what is called a snag-chamber;—a partition formed of stout planks, which is calked, and made so effectually water-tight that the foremost end of the vessel is cut off as entirely from the rest of the hold as if it belonged to another boat. If the steam-vessel happen to run against a snag, and that a hole is made in her bow, under the surface, this chamber merely fills with water."—Travels in North America, vol. iii. p. 363.

It is also found in excavating at New Orleans, even at the depth of several yards below the level of the sea, that the soil of the delta contains innumerable trunks of trees, layer above layer, some prostrate, as if drifted, others broken off near the bottom, but remaining still erect, and with their roots spreading on all sides, as if in their natural position. In such situations they appeared to me to indicate a sinking of the ground, as the trees must formerly have grown in marshes above the sea-level. In the higher parts of the alluvial plain, for many hundred miles above the head of the delta, similar stools and roots of trees are also seen buried in stiff clay at different levels, one above the other, and exposed to view in the banks at low water. They point clearly to the successive growth of forests in the extensive swamps of the plain, where the ground was slowly raised, year after year, by the mud thrown down during inundations. These roots and stools belong chiefly to the deciduous cypress (*Taxodium distichum*), and other swamp-trees, and they bear testimony to the constant shifting of the course of the great river, which is always excavating land originally formed at some distance from its banks.

Formation of lakes in Louisiana.—Another striking feature in the basin of the Mississippi, illustrative of the changes now in progress, is the formation by natural causes of great lakes, and the drainage of others. These are especially frequent in the basin of the Red River in Louisiana, where the largest of them, called Bistineau, is more than *thirty miles* long, and has a medium depth of from *fifteen* to *twenty* feet. In the deepest parts are seen numerous cypress-trees, of all sizes, now dead, and most of them with their tops broken by the wind, yet standing erect under water. This tree resists the action of air and water longer than any other, and, if not submerged throughout the whole year, will retain life for an extraordinary period. Lake Bistineau, as well as Black Lake, Cado Lake, Spanish Lake, Natchitoches Lake, and many others, have been formed, according to Darby, by the gradual elevation of the bed of Red River, in which the alluvial accumulations have been so great as to raise its channel, and cause its waters, during the flood season, to flow up the mouths of many tributaries, and to convert parts of their courses into lakes. In the autumn, when the level of Red River is again depressed, the waters rush back, and some lakes become grassy meadows, with streams meandering through them.* Thus, there is a periodical flux and reflux between Red River and some of these basins, which are merely reservoirs, alternately emptied and filled, like our tide estuaries—with this difference, that in the one case the land is submerged for several months continuously, and in the other twice in every twenty-four hours. It has happened, in several cases, that a raft of timber or a bar has been thrown by Red River across some of the openings of these channels, and then the lakes become, like Bistineau, constant repositories of water. But, even in these cases, their level is liable to annual ele-

* Darby's Louisiana, p. 33.

vation and depression, because the flood of the main river, when at its height, passes over the bar; just as, where sand-hills close the entrance of an estuary on the Norfolk or Suffolk coast, the sea, during some high tide or storm, has often breached the barrier and inundated again the interior.

I am informed by Mr. Featherstonhaugh that the plains of the Red River and the Arkansas are so low and flat, that whenever the Mississippi rises thirty feet above its ordinary level, those great tributaries are made to flow back, and inundate a region of vast extent. Both the streams alluded to contain red sediment, derived from the decomposition of red porphyry; and since 1833, when there was a great inundation in the Arkansas, an immense swamp has been formed near the Mammelle mountain, comprising 30,000 acres, with here and there large lagoons, where the old bed of the river was situated; in which innumerable trees, for the most part dead, are seen standing, of cypress, cotton-wood, or poplar, the triple-thorned acacia, and others, which are of great size. Their trunks appear as if painted red for about fifteen feet from the ground; at which height a perfectly level line extends through the whole forest, marking the rise of the waters during the last flood.*

But most probably the causes above assigned for the recent origin of these lakes are not the only ones. Subterranean movements have altered, so lately as the years 1811–12, the relative levels of various parts of the basin of the Mississippi, situated 300 miles northeast of Lake Bistineau. In those years the great valley, from the mouth of the Ohio to that of the St. Francis, including a tract 300 miles in length, and exceeding in area the whole basin of the Thames, was convulsed to such a degree, as to create new islands in the river, and lakes in the alluvial plain. Some of these were on the left or east bank of the Mississippi, and were twenty miles in extent; as, for example, those named Reelfoot and Obion in Tennessee, formed in the channels or valleys of small streams bearing the same names.†

But the largest area affected by the great convulsion lies eight or ten miles to the westward of the Mississippi, and inland from the town of New Madrid, in Missouri. It is called "the sunk country," and is said

* Featherstonhaugh, Geol. Report, Washington, 1835, p. 84.

† Trees submerged in an upright position have been observed in other parts of N. America. Thus Captains Clark and Lewis found, about the year 1807, a forest of pines standing erect under water in the body of the Columbia river, which they supposed, from the appearance of the trees, to have been submerged only about twenty years. (Travels, &c. vol. ii. p. 241.) More lately (1835), the Rev. Mr. Parker observed on the same river (lat. 45° N., long. 121° W.) trees standing in their natural position in spots where the water was more than twenty feet deep. The tops of the trees had disappeared; but between high and low water-mark the trunks were only partially decayed; and the roots were seen through the clear water, spreading as they had grown in their native forest. (Tour beyond the Rocky Mountains, p. 132.) Some have inferred from these facts that a tract of land, more than twenty miles in length, must have subsided vertically; but Capt. Fremont, Dec. 1845 (Rep. of Explor. Exped. p. 195), satisfied himself that the submerged forests have been formed by immense land-slides from the mountains, which here closely shut in the river.

to extend along the course of the White Water and its tributaries, for a distance of between seventy and eighty miles north and south, and thirty miles or more east and west. Throughout this area, innumerable submerged trees, some standing leafless, others prostrate, are seen; and so great is the extent of lake and marsh, that an active trade in the skins of muskrats, mink, otters, and other wild animals, is now carried on there. In March, 1846, I skirted the borders of the "sunk country" nearest to New Madrid, passing along the Bayou St. John and Little Prairie, where dead trees of various kinds, some erect in the water, others fallen, and strewed in dense masses over the bottom, in the shallows, and near the shore, were conspicuous. I also beheld countless rents in the adjoining dry alluvial plains, caused by the movements of the soil in 1811–12, and still open, though the rains, frost, and river inundations, have greatly diminished their original depth. I observed, moreover, numerous circular cavities, called "sunk holes," from ten to thirty yards wide, and twenty feet or more in depth, which interrupt the general level of the plain. These were formed by the spouting out of large quantities of sand and mud during the earthquakes.*

That the prevailing changes of level in the delta and alluvial plain of the Mississippi have been caused by the subsidence, rather than the upheaval of land, appears to me established by the fact, that there are no protuberances of upraised alluvial soil, projecting above the level surface of the great plain. It is true that the gradual elevation of that plain, by new accessions of matter, would tend to efface every inequality derived from this source, but we might certainly have expected to find more broken ground between the opposite bluffs, had local upthrows of alluvial strata been of repeated occurrence.

Antiquity of the delta.—The vast size of the alluvial plain both above and below the head of the delta, or the branching off of the uppermost arm of the Atchafalaya, has been already alluded to. Its superficial dimensions, according to Mr. Forshey, exceed 30,000 square miles, nearly half of which belong to the true delta. The deposits consist partly of sand originally formed upon or near the banks of the river, and its tributaries, partly of gravel, swept down the main channel, of which the position has continually shifted, and partly of fine mud slowly accumulated in the swamps. The farther we descend the river towards its mouth, the finer becomes the texture of the sediment. The whole alluvial formation, from the base of the delta upwards, slopes with a very gentle inclination, rising about three inches in a mile from the level of the sea at the Balize, to the height of about 200 feet in a distance of about 800 miles.

That a large portion of this fluviatile deposit, together with the fluvio-marine strata now in progress near the Balize, consists of mud and sand with much vegetable matter intermixed, may be inferred from what has

* For an account of the "sunk country," shaken by the earthquake of 1811–12, see Lyell's Second Visit to the United States, ch. 33.

been said of the abundance of drift trees floated down every summer. These are seen matted together into a net-work around the extensive mud banks at the extreme mouths of the river. Every one acquainted with the geography of Louisiana is aware that the most southern part of the delta forms a long narrow tongue of land protruding for 50 miles into the Gulf of Mexico, at the end of which are numerous channels of discharge. This singular promontory consists simply of the river and its two low, flat banks, covered with reeds, young willows, and poplars. Its appearance answers precisely to that of the banks far in the interior, when nothing appears above water during inundations but the higher part of the sloping glacis or bank. In the one case we have the swamps or an expanse of freshwater with the tops of trees appearing above, in the other the bluish green surface of the Gulf of Mexico. An opinion has very commonly prevailed that this narrow promontory, the newest product of the river, has gained very rapidly upon the sea, since the foundation of New Orleans; but after visiting the Balize in 1846, in company with Dr. Carpenter, and making many inquiries of the pilots, and comparing the present outline of the coast with the excellent Spanish chart, published by Charlevoix 120 years before, we came to a different conclusion. The rate of permanent advance of the new land has been very slow, not exceeding perhaps one mile in a century. The gain may have been somewhat more rapid in former years, when the new strip of soil projected less far into the gulf, since it is now much more exposed to the action of a strong marine current. The tides also, when the waters of the river are low, enter into each opening, and scour them out, destroying the banks of mud and the sand-bars newly formed during the flood season.

An observation of Darby, in regard to the strata composing part of this delta, deserves attention. In the steep banks of the Atchafalaya, before alluded to, the following section, he says, is observable at low water:—first an upper stratum, consisting invariably of bluish clay, common to the banks of the Mississippi; below this a stratum of red ochreous earth, peculiar to Red River, under which the blue clay of the Mississippi again appears; and this arrangement is constant, proving, as that geographer remarks, that the waters of the Mississippi and the Red River occupied alternately, at some former periods, considerable tracts below their present point of union.* Such alternations are probably common in submarine spaces situated between two converging deltas; for, before the two rivers unite, there must almost always be a certain period when an intermediate tract will by turns be occupied and abandoned by the waters of each stream; since it can rarely happen that the season of highest flood will precisely correspond in each. In the case of the Red River and Mississippi, which carry off the waters from countries placed under widely distant latitudes, an exact coincidence in the time of greatest inundation is very improbable.

* Darby's Louisiana, p. 103.

The antiquity of the delta, or length of the period which has been occupied in the deposition of so vast a mass of alluvial matter, is a question which may well excite the curiosity of every geologist. Sufficient data have not yet been obtained to afford a full and satisfactory answer to the inquiry, but some approximation may already be made to the minimum of time required.

When I visited New Orleans, in February, 1846, I found that Dr. Riddell had made numerous experiments to ascertain the proportion of sediment contained in the waters of the Mississippi; and he concluded that the mean annual amount of solid matter was to the water as $\frac{1}{1245}$ in weight, or about $\frac{1}{3000}$ in volume.* From the observations of the same gentleman, and those of Dr. Carpenter and Mr. Forshey, an eminent engineer, to whom I have before alluded, the average width, depth, and velocity of the Mississippi, and thence the mean annual discharge of water were deduced. I assumed 528 feet, or the tenth of a mile, as the probable thickness of the deposit of mud and sand in the delta; founding my conjecture chiefly on the depth of the Gulf of Mexico, between the southern point of Florida and the Balize, which equals on an average 100 fathoms, and partly on some borings 600 feet deep in the delta, near Lake Pontchartrain, north of New Orleans, in which the bottom of the alluvial matter is said not to have been reached. The area of the delta being about 13,600 square statute miles, and the quantity of solid matter annually brought down by the river 3,702,758,400 cubic feet, it must have taken 67,000 years for the formation of the whole; and if the alluvial matter of the plain above be 264 feet deep, or half that of the delta,† it must have required 33,500 more years for its accumulation, even if its area be estimated as only equal to that of the delta, whereas it is in fact larger. If some deduction be made from the time here stated, in consequence of the effect of the drift-wood, which must have aided in filling up more rapidly the space above alluded to, a far more important allowance must be made on the other hand, for the loss of matter, owing to the finer particles of mud

* The calculations here given were communicated to the British Association, in a lecture which I delivered at Southampton in September, 1846. (See Athenæum Journal, Sept. 26, 1846, and Report of British Association, 1846, p. 117.) Dr. Riddell has since repeated his experiments on the quantity of sediment in the river at New Orleans without any material variation in the results.

Mr. Forshey, in a memoir on the Physics of the Mississippi, published in 1850, adopts Dr. Riddell's estimate for the quantity of mud, but takes 447,199 cubic feet per second as the average discharge of water for the year at Carrolton, nine miles above New Orleans, a result deduced from thirty years of observations. This being one-tenth more than I had assumed, would add a tenth to the sediment, and would diminish by one-eleventh the number of years required to accomplish the task above alluded to. "The cubic contents of sedimentary matter," says Forshey, "are equal to 4,083,333,333, and this sediment would annually cover twelve miles square one foot deep."

† The Mississippi is continually shifting its course in the great alluvial plain, cutting frequently to the depth of 100, and even sometimes to the depth of 250 feet. As the old channels become afterwards filled up, or in a great degree obliterated, this excavation alone must have given a considerable depth to the basin, which receives the alluvial deposit, and subsidences like those accompanying the earthquake of New Madrid in 1811–12 may have given still more depth.

not settling at the mouths of the river, but being swept out far to sea during the predominant action of the tides, and the waves in the winter months, when the current of fresh water is feeble. Yet however vast the time during which the Mississippi has been transporting its earthy burden to the ocean, the whole period, though far exceeding, perhaps, 100,000 years, must be insignificant in a geological point of view, since the bluffs or cliffs, bounding the great valley, and therefore older in date, and which are from 50 to 250 feet in perpendicular height, consist in great part of loam containing land, fluviatile, and lacustrine shells of species still inhabiting the same country. (See fig. 23, p. 265.)

Before we take leave of the great delta, we may derive an instructive lesson from the reflection that the new deposits already formed, or now accumulating, whether marine or freshwater, must greatly resemble in composition, and the general character of their organic remains, many ancient strata, which enter largely into the earth's structure. Yet there is no sudden revolution in progress, whether on the land or in the waters, whether in the animate or the inanimate world. Notwithstanding the excessive destruction of soil and uprooting of trees, the region which yields a never-failing supply of drift-wood is densely clothed with noble forests, and is almost unrivalled in its power of supporting animal and vegetable life. In spite of the undermining of many a lofty bluff, and the encroachments of the delta on the sea—in spite of the earthquake, which rends and fissures the soil, or causes areas more than sixty miles in length to sink down several yards in a few months, the general features of the district remain unaltered, or are merely undergoing a slow and insensible change. Herds of wild deer graze on the pastures, or browse upon the trees; and if they diminish in number, it is only where they give way to man and the domestic animals which follow in his train. The bear, the wolf, the fox, the panther, and the wild-cat, still maintain themselves in the fastnesses of the forests of cypress and gum-tree. The racoon and the opossum are everywhere abundant, while the musk-rat, otter, and mink still frequent the rivers and lakes, and a few beavers and buffaloes have not yet been driven from their ancient haunts. The waters teem with aligators, tortoises, and fish, and their surface is covered with millions of migratory waterfowl, which perform their annual voyage between the Canadian lakes and the shores of the Mexican Gulf. The power of man begins to be sensibly felt, and many parts of the wilderness to be replaced by towns, orchards, and gardens. The gilded steamboats, like moving palaces, stem the force of the current, or shoot rapidly down the descending stream, through the solitudes of the forests and prairies. Already does the flourishing population of the great valley far exceed that of the thirteen United States when first they declared their independence. Such is the state of a continent where trees and stones are hurried annually by a thousand torrents, from the mountains to the plains, and where sand and finer matter are swept down by a vast current to the sea, together with the wreck of countless forests and the bones of animals which perish in the inundations. When these materials reach the gulf, they do not render the waters unfit for aquatic animals; but on the contrary, the ocean here swarms with life, as it generally does where the influx of a great river furnishes a copious supply of organic and mineral matter. Yet many geologists, when they behold the spoils of the land heaped in successive strata, and blended confusedly with the remains of fishes, or interspersed with broken shells and corals; when they see portions of erect trunks of trees with their roots still retaining their natural position, and one tier of these preserved in a fossil state above another, imagine that they are viewing the signs of a turbulent instead of a tranquil and settled state of the planet. They read in such phenomena the proof of chaotic disorder and reiterated catastrophes, instead of indications of a surface as habitable as the most delicious and fertile districts now tenanted by man.

DELTA OF THE GANGES AND BRAHMAPOOTRA.

As an example of a still larger delta advancing upon the sea in opposition to more powerful tides, I shall next describe that of the Ganges and Brahmapootra (or Burrampooter). These, the two principal rivers of India, descend from the highest mountains in the world, and partially mingle their waters in the low plains of Hindostan, before reaching the head of the Bay of Bengal. The Brahmapootra, somewhat the larger of the two, formerly passed to the east of Dacca, even so lately as the beginning of the present century, pouring most of its waters into one of the numerous channels in the delta called "the Megna" By

Fig. 25.

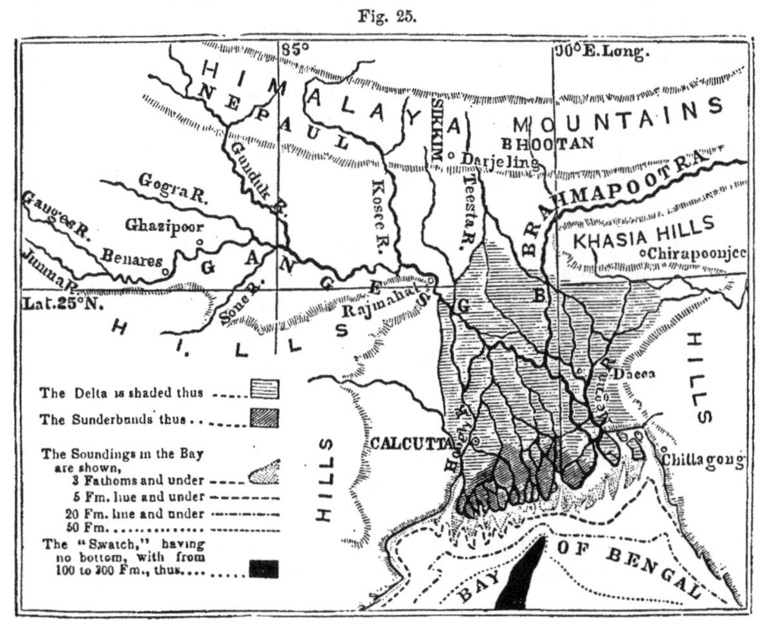

MAP of the DELTA of the GANGES and BRAHMAPOOTRA.

that name the main stream was always spoken of by Rennell and others in their memoirs on this region. But the main trunk now unites with an arm of the Ganges considerably higher up, at a point about 100 miles distant from the sea; and it is constantly, according to Dr. Hooker, working its way westward, having formerly, as may be seen by ancient maps, moved eastward for a long period.

The area of the delta of the combined rivers, for it is impossible now to distinguish what belongs to each, is considerably more than double that of the Nile, even if we exclude from the delta a large extent of low, flat, alluvial plain, doubtless of fluviatile origin, which stretches more than 100 miles to the hills west of Calcutta (see map, fig. 25), and much farther in a northerly direction beyond the head of the great delta. The head of a delta is that point where the first arm is given off. Above that point a river receives the waters of tributaries flowing from higher levels; below it, on the contrary, it gives out portions of its waters to lower levels, through channels which flow into adjoining swamps, or which run directly to the sea. The Mississippi, as before described, has a single head, which originated at an unknown period when the Red River joined it. In the great delta of Bengal there may be said to be two heads nearly equidistant from the sea, that of the Ganges (G, map, fig. 25), about 30 miles below Rajmahal, or 216 statute miles in a direct line from the sea, and that of the Brahmapootra (B), below Chirapoonjee, where the river issues from the Khasia mountains, a distance of 224 miles from the Bay of Bengal.

It will appear, by reference to the map, that the great body of fresh water derived from the two rivers enters the bay on its eastern side; and that a large part of the delta bordering on the sea is composed of a labyrinth of rivers and creeks, all filled with salt water, except those immediately communicating with the Hoogly, or principal arm of the Ganges. This tract alone, known by the name of the Woods, or Sunderbunds (more properly Soonderbuns), a wilderness infested by tigers and crocodiles, is, according to Rennell, equal in extent to the whole principality of Wales.*

On the sea-coast there are eight great openings, each of which has evidently, at some ancient period, served in its turn as the principal channel of discharge. Although the flux and reflux of the tide extend even to the heads of the delta when the rivers are low, yet, when they are periodically swollen by tropical rains, their volume and velocity counteract the tidal current, so that, except very near the sea, the ebb and flow become insensible. During the flood season, therefore, the Ganges and Brahmapootra almost assume in their delta, the character of rivers entering an inland sea; the movements of the ocean being then subordinate to the force of the rivers, and only slightly disturbing their operations. The great gain of the delta in height and area takes place during the inundations; and, during other seasons of the year, the

* Account of the Ganges and Burrampooter rivers, by Major Rennell, Phil. Trans. 1781.

ocean makes reprisals, scouring out the channels, and sometimes devouring rich alluvial plains.

Islands formed and destroyed.—Major R. H. Colebrooke, in his account of the course of the Ganges, relates examples of the rapid filling up of some of its branches, and the excavation of new channels, where the number of square miles of soil removed in a short time (the column of earth being 114 feet high) was truly astonishing. Forty square miles, or 25,600 acres, are mentioned as having been carried away, in one place, in the course of a few years.* The immense transportation of earthy matter by the Ganges and Brahmapootra is proved by the great magnitude of the islands formed in their channels during a period far short of that of a man's life. Some of these, many miles in extent, have originated in large sand-banks thrown up round the points at the angular turning of the rivers, and afterwards insulated by breaches of the streams. Others, formed in the main channel, are caused by some obstruction at the bottom. A large tree, or a sunken boat, is sometimes sufficient to check the current, and cause a deposit of sand, which accumulates till it usurps a considerable portion of the channel. The river then undermines its banks on each side, to supply the deficiency in its bed, and the island is afterwards raised by fresh deposits during every flood. In the great gulf below Luckipour, formed by the united waters of the Ganges and Megna, some of the islands, says Rennell, rival in size and fertility the Isle of Wight. While the river is forming new islands in one part, it is sweeping away old ones in others. Those newly formed are soon overrun with reeds, long grass, the Tamarix Indica, and other shrubs, forming impenetrable thickets, where the tiger, the rhinoceros, the buffalo, deer, and other wild animals, take shelter. It is easy, therefore, to perceive, that both animal and vegetable remains may occasionally be precipitated into the flood, and become imbedded in the sediment which subsides in the delta.

Three or four species of crocodile, of two distinct sub-genera, abound in the Ganges, and its tributary and contiguous waters; and Mr. H. T. Colebrooke informed me, that he had seen both forms in places far inland, many hundred miles from the sea. The Gangetic crocodile, or Gavial (in correct orthography, Garial), is confined to the fresh water, living exclusively on fish, but the commoner kinds, called Koomiah and Muggar, frequent both fresh and salt, being much larger and fiercer in salt and brackish water.† These animals swarm in the brackish water along the line of sand-banks, where the advance of the delta is most rapid. Hundreds of them are seen together in the creeks of the delta,

* Trans. of the Asiatic Society, vol. vii. p. 14.

† Cuvier referred the true crocodiles of the Ganges to a single species, *C. biporcatus*. But I learn from Dr. Falconer that there are three well-marked species, *C. biporcatus*, *C. palustris*, and *C. bombifrons*. *C. bombifrons* occurs in the northern branches of the Ganges, 1000 miles from Calcutta; *C. biporcatus* appears to be confined to the estuary; and *C. palustris*, to range from the estuary to the central parts of Bengal. The garial is found along with *C. bombifrons* in the north, and descends to the region of *C. biporcatus* in the estuary.

or basking in the sun on the shoals without. They will attack men and cattle, destroying the natives when bathing, and tame and wild animals which come to drink. "I have not unfrequently," says Mr. Colebrooke, "been witness to the horrid spectacle of a floating corpse seized by a crocodile with such avidity, that he half emerged above the water with his prey in his mouth." The geologist will not fail to observe how peculiarly the habits and distribution of these saurians expose them to become imbedded in the horizontal strata of fine mud, which are annually deposited over many hundred square miles in the Bay of Bengal. The inhabitants of the land, which happen to be drowned or thrown into the water, are usually devoured by these voracious reptiles; but we may suppose the remains of the saurians themselves to be continually entombed in the new formations. The number, also, of bodies of the poorer class of Hindoos thrown annually into the Ganges is so great, that some of their bones or skeletons can hardly fail to be occasionally enveloped in fluviatile mud.

It sometimes happens, at the season when the periodical flood is at its height, that a strong gale of wind, conspiring with a high spring-tide, checks the descending current of the river, and gives rise to most destructive inundations. From this cause, in 1763, the waters at Luckipour rose six feet above their ordinary level, and the inhabitants of a considerable district, with their houses and cattle, were totally swept away.

The population of all oceanic deltas are particularly exposed to suffer by such catastrophes, recurring at considerable intervals of time; and we may safely assume that such tragical events have happened again and again since the Gangetic delta was inhabited by man. If human experience and forethought cannot always guard against these calamities, still less cau the inferior animals avoid them; and the monuments of such disastrous inundations must be looked for in great abundance in strata of all ages, if the surface of our planet has always been governed by the same laws. When we reflect on the general order and tranquillity that reigns in the rich and populous delta of Bengal, notwithstanding the havoc occasionally committed by the depredations of the ocean, we perceive how unnecessary it is to attribute the imbedding of successive races of animals in older strata to extraordinary energy in the causes of decay and reproduction in the infancy of our planet, or to those general catastrophes and sudden revolutions so often resorted to.

Deposits in the delta.—The quantity of mud held in suspension by the waters of the Ganges and Brahmapootra is found, as might be expected, to exceed that of any of the rivers alluded to in this or the preceding chapters; for, in the first place, their feeders flow from mountains of unrivalled altitude, and do not clear themselves in any lakes, as does the Rhine in the Lake of Constance, or the Rhone in that of Geneva. And, secondly, their whole course is nearer the equator than that of the Mississippi, or any great river, respecting which careful experiments have been made, to determine the quantity of its water and earthy contents. The fall of rain, moreover, as we have before seen, is excessive on the southern flanks of the first range of mountains which rise from the plains of Hindostan, and still more remarkable is the quantity sometimes poured down in one day. (See above, p. 200.) The sea, where the Ganges and Brahmapootra discharge their main stream at the flood season, only recovers its transparency at the distance of from 60 to 100 miles from the delta; and we may take for granted that the current continues to transport the finer particles much farther south than where the surface water first becomes clear. The general slope, therefore, of the new strata must be extremely gentle. According to the best charts, there is a gradual deepening from four to about sixty fathoms, as we proceed from the base of the delta to the distance of about one hundred miles into the Bay of Bengal. At some few points seventy, or even one hundred, fathoms are obtained at that distance.

One remarkable exception, however, occurs to the regularity of the shape of the bottom. Opposite the middle of the delta, at the distance of thirty or forty miles from the coast, a deep submarine valley occurs, called the "swatch of no ground," about fifteen miles in diameter, where soundings of 180, and even 300, fathoms fail to reach the bottom. (See map, p. 275.) This phenomenon is the more extraordinary, since the depression runs north to within five miles of the line of shoals; and not only do the waters charged with sediment pass over it continually, but, during the monsoons, the sea, loaded with mud and sand, is beaten back in that direction towards the delta. As the mud is known to extend for eighty miles farther into the gulf, an enormous thickness of matter must have been deposited in "the swatch." We may conclude, therefore, either that the original depth of this part of the Bay of Bengal was excessive, or that subsidences have occurred in modern times. The latter conjecture is the less improbable, as the whole area of the delta has been convulsed in the historical era by earthquakes, and actual subsidences have taken place in the neighboring coast of Chittagong, while "the swatch" lies not far from the volcanic band which connects Sumatra, Barren Island, and Ramree.*

Opposite the mouth of the Hoogly river, and immediately south of Saugor Island, four miles from the nearest land of the delta, a new islet was formed about twenty years ago, called Edmonstone Island, on the centre of which a beacon was erected as a landmark in 1817. In 1818 the island had become two miles long and half a mile broad, and was covered with vegetation and shrubs. Some houses were then built upon it, and in 1820 it was used as a pilot station. The severe gale of 1823 divided it into two parts, and so reduced its size as to leave the beacon standing out in the sea, where, after remaining seven years, it was washed away. The islet in 1836 had been converted by successive storms into a sand-bank, half a mile long, on which a sea-mark was placed.

Although there is evidence of gain at some points, the general progress of the coast is very slow; for the tides, when the river water is low, are actively employed in removing alluvial matter. In the Sun-

* See below, ch. 22 and 29.

derbunds the usual rise and fall of the tides is no more than eight feet, but, on the east side of the delta, Dr. Hooker observed, in the winter of 1851, a rise of from sixty to eighty feet, producing among the islands at the mouths of the Megna and Fenny rivers, a lofty wave or "bore" as they ascend, and causing the river water to be ponded back, and then to sweep down with great violence when the tide ebbs. The bay for forty miles south of Chittagong is so fresh that neither algæ nor mangroves will grow in it. We may, therefore, conceive how effective may be the current formed by so great a volume of water in dispersing fine mud over a wide area. Its power is sometimes augmented by the agitation of the bay during hurricanes in the month of May. The new superficial strata consists entirely of fine sand and mud; such, at least, are the only materials which are exposed to view in regular beds on the banks of the numerous creeks. Neither here or higher up the Ganges, could Dr. Hooker discover any land or freshwater shells in sections of the banks, which in the plains higher up sometimes form cliffs eighty feet in height at low water. In like manner I have stated* that I was unable to find any buried shells in the delta or modern river cliffs of the Mississippi.

No substance so coarse as gravel occurs in any part of the delta of the Ganges and Brahmapootra, nor nearer the sea than 400 miles. Yet it is remarkable that the boring of an Artesian well at Fort William, near Calcutta, in the years 1835–1840, displayed, at the depth of 120 feet, clay and sand with pebbles. This boring was carried to a depth of 481 feet below the level of Calcutta, and the geological section obtained in the operation has been recorded with great care. Under the surface soil, at a depth of about ten feet, they came to a stiff blue clay about forty feet in thickness; below which was sandy clay, containing in its lower portion abundance of decayed vegetable matter, which at the bottom assumed the character of a stratum of black peat two feet thick. This peaty mass was considered as a clear indication (like the "dirt-bed" of Portland) of an ancient terrestrial surface, with a forest or Sunderbund vegetation. Logs and branches of a red-colored wood occur both above and immediately below the peat, so little altered that Dr. Wallich was able to identify them with the Soondri tree, *Heritiera littoralis*, one of the most prevalent forms at the base of the delta. Dr. Falconer tells me that similar peat has been met with at other points round Calcutta at the depth of nine feet and twenty-five feet. It appears, therefore, that there has been a sinking down of what was originally land in this region, to the amount of seventy feet or more perpendicular; for Calcutta is only a few feet above the level of the sea, and the successive peat-beds seem to imply that the subsidence of the ground was gradual or interrupted by several pauses. Below the vegetable mass they entered upon a stratum of yellowish clay about ten feet thick, containing horizontal layers of kunkar (or kankar), a nodular, concretionary, argillaceous limestone, met with abundantly at greater or

* Second Visit to the United States, vol. ii. p. 145.

less depths in all parts of the valley of the Ganges, over many thousand square miles, and always presenting the same characters, even at a distance of one thousand miles north of Calcutta. Some of this kunkar is said to be of very recent origin in deposits formed by river inundations near Saharanpoor. After penetrating 120 feet, they found loam containing water-worn fragments of mica-slate and other kinds of rock, which the current of the Ganges can no longer transport to this region. In the various beds pierced through below, consisting of clay, marl, and friable sandstone, with kunkar here and there intermixed, no organic remains of decidedly marine origin were met with. Too positive a conclusion ought not, it is true, to be drawn from such a fact, when we consider the narrow bore of the auger and its effect in crushing shells and bones. Nevertheless, it is worthy of remark, that the only fossils obtained in a recognizable state were of a fluviatile or terrestrial character. Thus, at the depth of 350 feet, the bony shell of a tortoise, or trionyx, a freshwater genus, was found in sand, resembling the living species of Bengal. From the same stratum, also, they drew up the lower half of the humerus of a ruminant, at first referred to a hyæna. It was the size and shape, says Dr. Falconer, of the shoulder-bone of the *Cervus porcinus*, or common hog-deer, of India. At the depth of 380 feet, clay with fragments of lacustrine shells was incumbent on what appears clearly to have been another "dirt-bed," or stratum of decayed wood, implying a period of repose of some duration, and a forest-covered land, which must have subsided 300 feet, to admit of the subsequent superposition of the overlying deposits. It has been conjectured that, at the time when this area supported trees, the land extended much farther out into the Bay of Bengal than now, and that in later times the Ganges, while enlarging its delta, has been only recovering lost ground from the sea.

At the depth of about 400 feet below the surface, an abrupt change was observed in the character of the strata, which were composed in great part of sand, shingle, and boulders, the only fossils observed being the vertebræ of a crocodile, shell of a trionyx, and fragments of wood very little altered, and similar to that buried in beds far above. These gravelly beds constituted the bottom of the section at the depth of 481 feet, when the operations were discontinued, in consequence of an accident which happened to the auger.

The occurrence of pebbles at the depths of 120 and 400 feet implies an important change in the geographical condition of the region around or near Calcutta. The fall of the river, or the general slope of the alluvial plain may have been formerly greater; or, before a general and perhaps unequal subsidence, hills once nearer the present base of the delta may have risen several hundred feet, forming islands in the bay, which may have sunk gradually, and become buried under fluviatile sediment.

Antiquity of the delta.—It would be a matter of no small scientific interest, if experiments were made to enable us to determine, with some degree of accuracy, the mean quantity of earthy matter discharged annually into the sea by the united waters of the Ganges and Brahmapootra. The Rev. Mr. Everest instituted, in 1831–2, a series of observations on the earthy matter brought down by the Ganges, at Ghazepoor, 500 miles from the sea. He found that, in 1831, the number of cubic feet of water discharged by the river per second at that place was, during the

Rains (4 months)	494,208
Winter (5 months)	71,200
Hot weather (3 months)	36,330

so that we may state in round numbers that 500,000 cubic feet per second flow down during the four months of the flood season, from June to September, and less than 60,000 per second during the remaining eight months.

The average quantity of solid matter suspended in the water during the rains was, by weight, $\frac{1}{428}$th part; but as the water is about one-half the specific gravity of the dried mud, the solid matter discharged is $\frac{1}{856}$th part in bulk, or 577 cubic feet per second. This gives a total of 6,082,041,600 cubic feet for the discharge in the 122 days of the rain. The proportion of sediment in the waters at other seasons was comparatively insignificant, the total amount during the five winter months being only 247,881,600 cubic feet, and during the three months of hot weather 38,154,240 cubic feet. The total annual discharge, then, would be 6,368,077,440 cubic feet.

This quantity of mud would in one year raise a surface of 228½ square miles, or a square space, each side of which should measure 15 miles, a height of one foot. To give some idea of the magnitude of this result, we will assume that the specific gravity of the dried mud is only one-half that of granite (it would, however, be more); in that case, the earthy matter discharged in a year would equal 3,184,038,720 cubic feet of granite. Now about 12½ cubic feet of granite weigh one ton; and it is computed that the great Pyramid of Egypt, if it were a solid mass of granite, would weigh about 600,000,000 tons. The mass of matter, therefore, carried down annually would, according to this estimate, more than equal in weight and bulk forty-two of the great pyramids of Egypt, and that borne down in the four months of the rains would equal forty pyramids. But if, without any conjecture as to what may have been the specific gravity of the mud, we attend merely to the weight of solid matter actually proved by Mr. Everest to have been contained in the water, we find that the number of tons weight which passed down in the 122 days of the rainy season was 339,413,760, which would give the weight of fifty-six pyramids and a half; and in the whole year 355,361,464 tons, or nearly the weight of sixty pyramids.

The base of the great Pyramid of Egypt covers eleven acres, and its perpendicular height is about five hundred feet. It is scarcely possible to present any picture to the mind which will convey an adequate conception of the mighty scale of this operation, so tranquilly and almost insensibly carried on by the Ganges, as it glides through its alluvial plain, even at a distance of 500 miles from the sea. It may, however, be stated, that if a fleet of more than eighty Indiamen, each freighted with about 1400 tons' weight of mud, were to sail down the river every hour of every day and night for four months continuously, they would only transport from the higher country to the sea a mass of solid matter equal to that borne down by the Ganges, even in this part of its course, in the four months of the flood season. Or the exertions of a fleet of about 2000 such ships going down daily with the same burden, and discharging it into the gulf, would be no more than equivalent to the operations of the great river.

The most voluminous current of lava which has flowed from Etna within historical times was that of 1669. Ferrara, after correcting Borelli's estimate, calculated the quantity of cubic yards of lava in this current at 140,000,000. Now, this would not equal in bulk one-fifth of the sedimentary matter which is carried down in a single year by the Ganges, past Ghazepoor, according to the estimate above explained; so that it would require five grand eruptions of Etna to transfer a mass of lava from the subterranean regions to the surface, equal in volume to the mud carried down in one year to that place.

Captain R. Strachey, of the Bengal Engineers, has remarked to me, not only that Ghazepoor, where Mr. Everest's observations were made, is 500 miles from the sea, but that the Ganges has not been joined there by its most important feeders. These drain upon the whole 750 miles of the Himalaya, and no more than 150 miles of that mountain-chain have sent their contributions to the main trunk at Ghazepoor. Below that place, the Ganges is joined by the Gogra, Gunduk, Khosee, and Teesta from the north, to say nothing of the Sone flowing from the south, one of the largest of the rivers which rise in the table-land of central India. (See map, fig. 25, p. 275.) Moreover the remaining 600 miles of the Himalaya comprise that eastern portion of the basin where the rains are heaviest. (See above, p. 200.) The quantity of water therefore carried down to the sea may probably be four or five times as much as that which passes Ghazepoor.

The Brahmapootra, according to Major Wilcox,* in the month of January, when it is near its minimum, discharges 150,000 cubic feet of water per second at Gwalpara, not many miles above the head of its delta. Taking the proportions observed at Ghazepoor at the different seasons as a guide, the probable average discharge of the Brahmapootra for the whole year may be estimated at about the same as that of the Ganges. Assuming this; and secondly, in order to avoid the risk of exaggeration, that the proportion of sediment in their waters is about a third less than Mr. Everest's estimate, the mud borne down to the Bay of Bengal in one year would equal 40,000 millions of cubic feet, or between six and seven times as much as that brought down to Ghazepoor, according to Mr. Everest's calculations in 1831, and ten times as much as that conveyed annually by the Mississippi to the Gulf of Mexico.

* Asiatic Researches, vol. xvii. p. 466.

Captain Strachey estimates the annually inundated portion of the delta at 250 miles in length by 80 in breadth, making an area of 20,000 square miles. The space south of this in the bay, where sediment is thrown down, may be 300 miles from E. to W. by 150 N. and S., or 45,000 square miles, which, added to the former, gives a surface of 65,000 square miles, over which the sediment is spread out by the two rivers. Suppose then the solid matter to amount to 40,000 millions of cubic feet per annum, the deposit, he observes, must be continued for forty-five years and three-tenths to raise the whole area a height of one foot, or 13,600 years to raise it 300 feet; and this, as we have seen, is much less than the thickness of the fluviatile strata actually penetrated, (and the bottom not reached) by the auger at Calcutta.

Nevertheless we can by no means deduce from these data alone, what will be the future rate of advance of the delta, nor even predict whether the land will gain on the sea, or remain stationary. At the end of 13,000 years the bay may be less shallow than now, provided a moderate depression, corresponding to that experienced in part of Greenland for many centuries shall take place (see chap. 30). A subsidence quite insensible to the inhabitants of Bengal, not exceeding two feet three inches in a century, would be more than sufficient to counterbalance all the efforts of the two mighty rivers to extend the limits of their delta. We have seen that the Artesian borings at Calcutta attest, what the vast depth of the "swatch" may also in all likelihood indicate, that the antagonist force of subsidence has predominated for ages over the influx of fluviatile mud, preventing it from raising the plains of Bengal, or from filling up a larger portion of the bay.

CONCLUDING REMARKS ON DELTAS.

Convergence of deltas.—If we possessed an accurate series of maps of the Adriatic for many thousand years, our retrospect would, without doubt, carry us gradually back to the time when the number of rivers descending from the mountains into that gulf by independent deltas was far greater in number. The deltas of the Po and the Adige, for instance, would separate themselves within the *recent* era, as, in all probability, would those of the Isonzo and the Torre. If, on the other hand, we speculate on future changes, we may anticipate the period when the number of deltas will greatly diminish; for the Po cannot continue to encroach at the rate of a mile in a hundred years, and other rivers to gain as much in six or seven centuries upon the shallow gulf, without new junctions occurring from time to time; so that Eridanus, "the king of rivers," will continually boast a greater number of tributaries. The Ganges and the Brahmapootra have perhaps become partially confluent in the same delta within the historical, or at least within the human era; and the date of the junction of the Red River and the Mississippi would, in all likelihood, have been known, if America had not been so recently discovered. The union of the Tigris and the Euphrates must undoubtedly have been one of the modern geographical changes of our Earth, for Col. Rawlinson informs me that the delta of those rivers has advanced two miles in the last sixty years, and is supposed to have encroached about forty miles upon the Gulf of Persia in the course of the last twenty-five centuries.

When the deltas of rivers, having many mouths, converge, a partial union at first takes place by the confluence of some one or more of their arms; but it is not until the main trunks are connected above the head of the common delta, that a complete intermixture of their joint waters and sediment takes place. The union, therefore, of the Po and Adige, and of the Ganges and Brahmapootra, is still incomplete. If we reflect on the geographical extent of surface drained by rivers such as now enter the Bay of Bengal, and then consider how complete the blending together of the greater part of their transported matter has already become, and throughout how vast a delta it is spread by numerous arms, we no longer feel so much surprise at the area occupied by some ancient formations of homogeneous mineral composition. But our surprise will be still farther lessened, when we afterwards inquire (ch. 21) into the action of tides and currents in disseminating sediment.

Age of existing deltas.—If we could take for granted, that the relative level of land and sea had remained stationary ever since all the existing deltas began to be formed—could we assume that their growth commenced at one and the same instant when the present continents acquired their actual shape—we might understand the language of geologists who speak of "the epoch of existing continents." They endeavor to calculate the age of deltas from this imaginary fixed period; and they calculate the gain of new land upon the sea, at the mouths of rivers, as having begun everywhere simultaneously. But the more we study the history of deltas, the more we become convinced that upward and downward movements of the land and contiguous bed of the sea have exerted, and continue to exert, an influence on the physical geography of many hydrographical basins, on a scale comparable in magnitude or importance to the amount of fluviatile deposition effected in an equal lapse of time. In the basin of the Mississippi, for example, proofs both of descending and ascending movements to a vertical amount of several hundred feet can be shown to have taken place since the existing species of land and freshwater shells lived in that region.*

The deltas also of the Po and Ganges have each, as we have seen (p. 257), when probed by the Artesian auger, borne testimony to a gradual subsidence of land to the extent of several hundred feet—old terrestrial surfaces, turf, peat, forest-land, and "dirt-beds," having been pierced at various depths. The changes of level at the mouth of the Indus in Cutch (see below, chap. 27), and those of New Madrid in the valley of the Mississippi (see p. 270, and chap. 27), are equally instructive, as demonstrating unceasing fluctuations in the levels of those areas into which running water is transporting sediment. If, therefore, the exact

* Lyell's Second Visit to the United States, vol. ii. chap. 34.

age of all modern deltas could be known, it is scarcely probable that we should find any two of them in the world to have coincided in date, or in the time when their earliest deposits originated.

Grouping of strata in deltas.—The changes which have taken place in deltas, even within the times of history, may suggest many important considerations in regard to the manner in which subaqueous sediment is distributed. With the exception of some cases hereafter to be noticed, there are some general laws of arrangement which must evidently hold good in almost all the lakes and seas now filling up. If a lake, for example, be encircled on two sides by lofty mountains, receiving from them many rivers and torrents of different sizes, and if it be bounded on the other sides, where the surplus waters issue, by a comparatively low country, it is not difficult to define some of the leading geological features which must characterize the lacustrine formation, when this basin shall have been gradually converted into dry land by the influx of sediment. The strata would be divisible into two principal groups: the *older* comprising those deposits which originated on the side adjoining the mountains, where numerous deltas first began to form; and the *newer* group consisting of beds deposited in the more central parts of the basin, and towards the side farthest from the mountains. The following characters would form the principal marks of distinction between the strata in each series:—The more ancient system would be composed, for the most part, of coarser materials, containing many beds of pebbles and sand, often of great thickness, and sometimes dipping at a considerable angle. These, with associated beds of finer ingredients, would, if traced round the borders of the basin, be seen to vary greatly in color and mineral composition, and would also be very irregular in thickness. The beds, on the contrary, in the newer group, would consist of finer particles, and would be horizontal, or very slightly inclined. Their color and mineral composition would be very homogeneous throughout large areas, and would differ from almost all the separate beds in the older series.

The following causes would produce the diversity here alluded to between the two great members of such lacustrine formations:—When the rivers and torrents first reach the edge of the lake, the detritus washed down by them from the adjoining heights sinks at once into deep water, all the heavier pebbles and sand subsiding near the shore. The finer mud is carried somewhat farther out, but not to the distance of many miles, for the greater part may be seen, as, for example, where the Rhone enters the Lake of Geneva, to fall down in clouds to the bottom, not far from the river's mouth. Thus alluvial tracts are soon formed at the mouths of every torrent and river, and many of these in the course of ages become of considerable extent. Pebbles and sand are then transported farther from the mountains; but in their passage they decrease in size by attrition, and are in part converted into mud and sand. At length some of the numerous deltas, which are all directed towards a common centre, approach near to each other; those of adjoining torrents become united, and each is merged, in its turn, in the delta of the largest river, which advances most rapidly into the lake, and renders all the minor streams, one after the other, its tributaries. The various mineral ingredients of all are thus blended together into one homogeneous mixture, and the sediment is poured out from a common channel into the lake.

As the average size of the transported particles decreases, while the force and volume of the main river augments, the newer deposits are diffused continually over a wider area, and are consequently more horizontal than the older. When at first there were many independent deltas near the borders of the basin, their separate deposits differed entirely from each other; one may have been charged, like the Arve where it joins the Rhone, with white sand and sediment derived from granite—another may have been black, like many streams in the Tyrol, flowing from the waste of decomposing rocks of dark slate—a third may have been colored by ochreous sediment, like the Red River in Louisiana—a fourth, like the Elsa in Tuscany, may have held much carbonate of lime in solution. At first they would each form distinct deposits of sand, gravel, limestone, marl, or other materials; but, after their junction, new chemical combinations and a distinct color would be the result, and the particles, having been conveyed ten, twenty, or a greater number of miles over alluvial plains, would become finer.

In those deltas where the tides and strong marine currents interfere, the above description would only be applicable, with certain modifications. If a series of earthquakes accompany the growth of a delta, and change the levels of the land from time to time, as in the region where the Indus now enters the sea, the phenomena will depart still more widely from the ordinary type. If, after a protracted period of rest, a delta sink down, pebbles may be borne along in shallow water near the foot of the boundary hills, so as to form conglomerates overlying the fine mud previously thrown into deeper water in the same area.

Causes of stratification in deltas.—The stratified arrangement, which is observed to prevail so generally in aqueous deposits, is most frequently due to variations in the velocity of running water, which cannot sweep along particles of more than a certain size and weight when moving at a given rate. Hence, as the force of the stream augments or decreases, the materials thrown down in successive layers at particular places are rudely sorted, according to their dimensions, form, and specific gravity. Where this cause has not operated, as where sand, mud, and fragments of rock are conveyed by a glacier, a confused heap of rubbish devoid of all stratification is produced.

Natural divisions are also occasioned in deltas, by the interval of time which separates annually the deposition of matter during the periodical rains, or melting of snow upon the mountains. The deposit of each year may acquire some degree of consistency before that of the succeeding year is superimposed. A variety of circumstances also give rise annually, or sometimes from day to day, to slight variations in color, fineness of the particles, and other characters, by which alternations of strata distinct in texture and mineral ingredients must be produced. Thus, for example, at one period of the year, drift-wood may be carried down, and, at another, mud, as was before stated to be the case in the delta of the Mississippi ; or at one time, when the volume and velocity of the stream are greatest, pebbles and sand may be spread over a certain area, over which, when the waters are low, fine matter or chemical precipitates are formed. During inundations, the turbid current of fresh water often repels the sea for many miles ; but when the river is low, salt water again occupies the same space. When two deltas are converging, the intermediate space is often, for reasons before explained, alternately the receptacle of different sediments derived from the converging streams (see p. 272). The one is, perhaps, charged with calcareous, the other with argillaceous matter; or one sweeps down sand and pebbles, the other impalpable mud. These differences may be repeated with considerable regularity, until a thickness of hundreds of feet of alternating beds is accumulated. The multiplication, also, of shells and corals in particular spots, and for limited periods, gives rise occasionally to lines of separation, and divides a mass which might otherwise be homogeneous into distinct strata.

An examination of the shell marl now forming in the Scotch lakes, or the sediment termed "warp," which subsides from the muddy water of the Humber and other rivers, shows that recent deposits are often composed of a great number of extremely thin layers, either even or slightly undulating, and preserving a general parallelism to the planes of stratification. Sometimes, however, the laminæ in modern strata are disposed diagonally at a considerable angle, which appears to take place where there are conflicting movements in the waters. In January, 1829, I visited, in company with Professor L. A. Necker, of Geneva, the confluence of the Rhone and Arve, when those rivers were very low, and were cutting channels through the vast heaps of débris thrown down from the waters of the Arve in the preceding spring. One of the sand-banks which had formed, in the spring of 1828, where the opposing currents of the two rivers neutralized each other, and caused a retardation in the motion, had been undermined; and the following is an exact representation of the arrangement of laminæ exposed in a vertical section. The length of the portion here seen is about twelve feet, and the height five. The strata A A consist of irregular alternations of pebbles and sand in undulating beds: below these are seams of very fine sand B B, some as thin as paper, others about a quarter of an inch thick. The strata C C are composed of layers of fine greenish-gray sand as thin as paper. Some of the inclined beds will be seen to be thicker at their upper, others at their lower extremity, the inclination of some being very considerable. These layers must have accumulated one on the other by lateral apposition, probably when one of the rivers was very gradually increasing or diminishing in velocity, so that the point of greatest retardation caused by their conflicting currents shifted slowly, allowing the sediment to be thrown down in successive layers on a sloping bank. The same phenomenon is exhibited in older strata of all ages.*

Fig. 26.

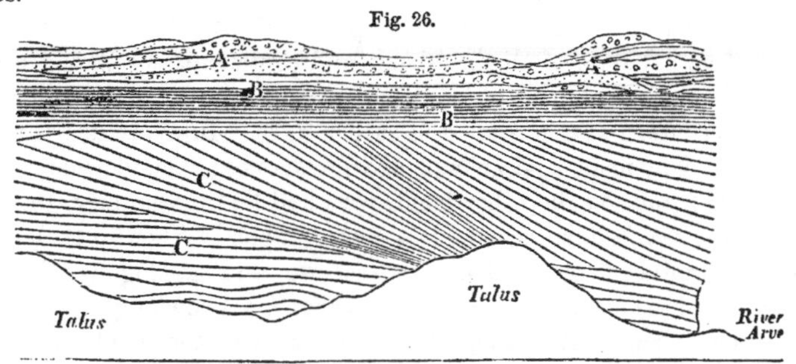

Section of a sand-bank in the bed of the Arve at its confluence with the Rhone, showing the stratification of deposits where currents meet.

If the bed of a lake or of the sea be sinking, whether at a uniform or an unequal rate, or oscillating in level during the deposition of sediment, these movements will give rise to a different class of phenomena, as, for example, to repeated alternations of shallow-water and deep-water deposits, each with peculiar organic remains, or to frequent repetitions of similar beds, formed at a uniform depth, and inclosing the same organic remains, and to other results too complicated and varied to admit of enumeration here.

Formation of conglomerates.—Along the base of the Maritime Alps, between Toulon and Genoa, the rivers, with few exceptions, are now forming strata of conglomerate and sand. Their channels are often several miles in breadth, some of them being dry, and the rest easily forded for nearly eight months in the year, whereas during the melting of the snow they are swollen, and a great transportation of mud and pebbles takes place. In order to keep open the main road from France to Italy, now carried along the sea-coast, it is necessary to remove annually great masses of shingle brought down during the flood season. A portion of the pebbles are seen in some localities, as near Nice, to form beds of shingle along the shore, but the greater part are swept into a deep sea. The small progress made by the deltas of minor rivers on this coast need not surprise us, when we recollect that there is sometimes a depth of two thousand feet at a few hundred yards from the beach, as near Nice. Similar observations might be made respecting a large proportion of the rivers in Sicily, and among others, respecting that which, immediately north of the port of Messina, hurries annually vast masses of granitic pebbles into the sea.

Constant interchange of land and sea.—I may here conclude my remarks on deltas, observing that, imperfect as is our information of the changes which they have undergone within the last three thousand years, they are sufficient to show how constant an interchange of sea

* See Manual of Geology by the Author.

and land is taking place on the face of our globe. In the Mediterranean alone, many flourishing inland towns, and a still greater number of ports, now stand where the sea rolled its waves since the era of the early civilization of Europe. If we could compare with equal accuracy the ancient and actual state of all the islands and continents, we should probably discover that millions of our race are now supported by lands situated where deep seas prevailed in earlier ages. In many districts not yet occupied by man, land animals and forests now abound where ships once sailed; and, on the other hand, we shall find, on inquiry, that inroads of the ocean have been no less considerable. When to these revolutions, produced by aqueous causes, we add analogous changes wrought by igneous agency, we shall, perhaps, acknowledge the justice of the conclusion of Aristotle, who declared that the whole land and sea on our globe periodically changed places.*

CHAPTER XIX.

DESTROYING AND TRANSPORTING EFFECTS OF TIDES AND CURRENTS.

Difference in the rise of tides—Lagullas and Gulf currents—Velocity of currents—Causes of currents—Action of the sea on the British coast—Shetland Islands—Large blocks removed—Isles reduced to clusters of rocks—Orkney isles—Waste of East coast of Scotland—and East coast of England—Waste of the cliffs of Holderness, Norfolk, and Suffolk—Sand-dunes, how far chronometers—Silting up of estuaries—Yarmouth estuary—Suffolk coast—Dunwich—Essex coast—Estuary of the Thames—Goodwin Sands—Coast of Kent—Formation of the Straits of Dover—South coast of England—Sussex—Hants—Dorset—Portland—Origin of the Chesil Bank—Cornwall—Coast of Brittany.

Although the movements of great bodies of water, termed tides and currents, are in general due to very distinct causes, their effects cannot be studied separately; for they produce, by their joint action, aided by that of the waves, those changes which are objects of geological interest. These forces may be viewed in the same manner as we before considered rivers, first, as employed in destroying portions of the solid crust of the earth and removing them to other places; secondly, as reproductive of new strata.

Tides.—It would be superfluous at the present day to offer any remarks on the cause of the tides. They are not perceptible in lakes or in most inland seas; in the Mediterranean even, deep and extensive as is that sea, they are scarcely sensible to ordinary observation, their effects being quite subordinate to those of the winds and currents. In some places, however, as in the Straits of Messina, there is an ebb and flow to the amount of two feet and upwards; at Naples and at the

* See p. 13.

Euripus, of twelve or thirteen inches; and at Venice, according to Rennell, of five feet.* In the Syrtes, also, of the ancients, two wide shallow gulfs, which penetrate very far within the northern coast of Africa, between Carthage and Cyrene, the rise is said to exceed five feet.†

In islands remote from any continent, the ebb and flow of the ocean is very slight, as at St. Helena, for example, where it is rarely above three feet.‡ In any given line of coast, the tides are greatest in narrow channels, bays, and estuaries, and least in the intervening tracts where the land is prominent. Thus, at the entrance of the estuary of the Thames and Medway, the rise of the spring tides is eighteen feet; but when we follow our eastern coast from thence northward, towards Lowestoff and Yarmouth, we find a gradual diminution, until at the places last mentioned, the highest rise is only seven or eight feet. From this point there begins again to be an increase, so that at Comer, where the coast again retires towards the west, the rise is sixteen feet; and towards the extremity of the gulf called "the Wash," as at Lynn and in Boston Deeps, it is from twenty-two to twenty-four feet, and in some extraordinary cases twenty-six feet. From thence again there is a decrease towards the north, the elevation at the Spurn Point being from nineteen to twenty feet, and at Flamborough Head and the Yorkshire coast from fourteen to sixteen feet.§

At Milford Haven in Pembrokeshire, at the mouth of the Bristol Channel, the tides rise thirty-six feet; and at King-Road near Bristol, forty-two feet. At Chepstow on the Wye, a small river which opens into the estuary of the Severn, they reach fifty feet, and sometimes sixty-nine, and even seventy-two feet. A current which sets in on the French coast, to the west of Cape La Hague, becomes pent up by Guernsey, Jersey, and other islands, till the rise of the tide is from twenty to forty-five feet, which last height it attains at Jersey, and at St. Malo, a seaport of Brittany. The tides in the Basin of Mines, at the head of the Bay of Fundy in Nova Scotia, rise to the height of seventy feet.

There are, however, some coasts where the tides seem to offer an exception to the rule above mentioned; for while there is scarcely any rise in the estuary of the Plata in S. America, there is an extremely high tide on the open coast of Patagonia, farther to the south. Yet even in this region the tides reach their greatest elevation (about fifty feet) in the Straits of Magellan, and so far at least they conform to the general rule.‖

Currents.—The most extensive and best determined system of currents, is that which has its source in the Indian Ocean under the influence of the trade winds; and which, after doubling the Cape of Good

* Geog. of Herod. vol. ii. p. 331.

† Ibid. p. 328.

‡ Romme, Vents et Courans, vol. ii. p. 2. Rev. F. Fallows, Quart. Journ. of Science, March, 1829.

§ The heights of these tides were given me by the late Captain Hewett, R. N.

‖ On the authority of Admiral Sir F. Beaufort, R. N.

Hope, inclines to the northward, along the western coast of Africa, then across the Atlantic, near the equator, where it is called the equatorial current, and is lost in the Caribbean Sea, yet seems to be again revived in the current which issues from the Gulf of Mexico. From thence it flows rapidly through the Straits of Bahama, taking the name of the Gulf Stream, and passing in a northeasterly direction, by the Banks of Newfoundland, towards the Azores.

We learn from the posthumous work of Rennell on this subject, that the Lagullas current, so called from the cape and bank of that name, is formed by the junction of two streams, flowing from the Indian Ocean; the one from the channel of Mozambique, down the southeast coast of Africa; the other from the ocean at large. The collective stream is from ninety to one hundred miles in breadth, and runs at the rate of from two and a half to more than four miles per hour. It is at length turned westward by the Lagullas bank, which rises from a sea of great depth to within one hundred fathoms of the surface. It must therefore be inferred, says Rennell, that the current here is more than one hundred fathoms deep, otherwise the main body of it would pass across the bank, instead of being deflected westward, so as to flow round the Cape of Good Hope. From this cape it flows northward, as before stated, along the western coast of Africa, taking the name of the South Atlantic current. It then enters the Bight, or Bay of Benin, and is turned westward, partly by the form of the coast there, and partly, perhaps, by the Guinea current, which runs from the north into the same great bay. From the centre of this bay proceeds the equatorial current already mentioned, holding a westerly direction across the Atlantic, which it traverses, from the coast of Guinea to that of Brazil, flowing afterwards by the shores of Guiana to the West Indies. The breadth of this current varies from 160 to 450 geographical miles, and its velocity is from twenty-five to seventy-nine miles per day, the mean rate being about thirty miles. The length of its whole course is about 4000 miles. As it skirts the coast of Guiana, it is increased by the influx of the waters of the Amazon and Orinoco, and by their junction acquires accelerated velocity. After passing the island of Trinidad it expands, and is almost lost in the Caribbean Sea; but there appears to be a general movement of that sea towards the Mexican Gulf, which discharges the most powerful of all currents through the Straits of Florida, where the waters run in the northern part with a velocity of four or five miles an hour, having a breadth of from thirty-five to fifty miles.*

The temperature of the Gulf of Mexico is 86° F. in summer, or 6° higher than that of the ocean, in the same parallel (25° N. lat.), and a large proportion of this warmth is retained, even where the stream reaches the 43° N. lat. After issuing from the Straits of Florida, the current runs in a northerly direction to Cape Hatteras, in North Carolina, about 35° N. lat., where it is more than seventy miles broad, and

* Consult the map of Currents by Capt. F. Beechy, R. N., Admiralty Manual, 1849, London.

still moves at the rate of seventy-five miles per day. In about the 40° N. lat., it is turned more towards the Atlantic by the extensive banks of Nantucket and St. George, which are from 200 to 300 feet beneath the surface of the sea; a clear proof that the current exceeds that depth. On arriving near the Azores, the stream widens, and overflows, as it were, forming a large expanse of warm water in the centre of the North Atlantic, over a space of 200 or 300 miles from north to south, and having a temperature of from 8° to 10° Fahr. above the surrounding ocean. The whole area, covered by the Gulf water, is estimated by Rennell at 2000 miles in length, and, at a mean, 350 miles in breadth; an area more extensive than that of the Mediterranean. The warm water has been sometimes known to reach the Bay of Biscay, still retaining five degrees of temperature above that of the adjoining ocean; and a branch of the Gulf current occasionally drifts fruits, plants, and wood, the produce of America and the West Indies, to the shores of Ireland and the Hebrides.

From the above statements we may understand why Rennell has characterized some of the principal currents as oceanic rivers, which he describes as being from 50 to 250 miles in breadth, and having a rapidity exceeding that of the largest navigable rivers of the continents, and so deep as to be sometimes obstructed, and occasionally turned aside, by banks, the tops of which do not rise within forty, fifty, or even one hundred fathoms of the surface of the sea.*

Greatest velocity of currents.—The ordinary velocity of the principal currents of the ocean is from one to three miles per hour; but when the boundary lands converge, large bodies of water are driven gradually into a narrow space, and then wanting lateral room, are compelled to raise their level. Whenever this occurs their velocity is much increased. The current which runs through the Race of Alderney, between the island of that name and the main land, has a velocity of about eight English miles an hour. Captain Hewett found that in the Pentland Firth, the stream, in ordinary spring tides, runs ten miles and a half an hour, and about thirteen miles during violent storms. The greatest velocity of the tidal current through the "Shoots" or New Passage, in the Bristol Channel, is fourteen English miles an hour; and Captain King observed, in his survey of the Straits of Magellan, that the tide ran at the same rate through the "First Narrows," and about eight geographical miles an hour, in other parts of those straits.

Causes of currents.—That movements of no inconsiderable magnitude should be impressed on an expansive ocean, by winds blowing for many months in one direction, may easily be conceived, when we observe the effects produced in our own seas by the temporary action of the same cause. It is well known that a strong southwest or northwest wind invariably raises the tides to an unusual height along the west coast of England and in the Channel; and that a northwest wind

* Rennell on Currents, p. 58.

of any continuance causes the Baltic to rise two feet and upwards above its ordinary level. Smeaton ascertained by experiment, that in a canal four miles in length, the water was kept up four inches higher at one end than at the other, merely by the action of the wind along the canal; and Rennell informs us that a large piece of water, ten miles broad, and generally only three feet deep, has, by a strong wind, had its waters driven to one side, and sustained so as to become six feet deep, while the windward side was laid dry.*

As water, therefore, he observes, when pent up so that it cannot escape, acquires a higher level, so, in a place *where it can escape,* the same operation produces a current; and this current will extend to a greater or less distance, according to the force by which it is produced. By the side of the principal oceanic currents, such as the Lagullas and the Gulf Stream, are parallel "counter-currents" running steadily in an opposite direction.

Currents flowing alternately in opposite directions are occasioned by the rise and fall of the tides. The effect of this cause is, as before observed, most striking in estuaries and channels between islands.

A third cause of oceanic currents is evaporation by solar heat, of which the great current setting through the Straits of Gibraltar into the Mediterranean is a remarkable example, and will be fully considered in the next chapter. A stream of colder water also flows from the Black Sea into the Mediterranean. It must happen in many other parts of the world that large quantities of water raised from one tract of the ocean by solar heat, are carried to some other where the vapor is condensed and falls in the shape of rain, and this, in flowing back again to restore equilibrium, will cause sensible currents.

These considerations naturally lead to the inquiry whether the level of those seas out of which currents flow, is higher than that of seas into which they flow. If not, the effect must be immediately equalized by under-currents or counter-currents. Arago is of opinion that, so far as observations have gone, there are no exact proofs of any such difference of level. It was inferred from the measurements of M. Lepére, that the level of the Mediterranean, near Alexandria, was lower by 26 feet 6 inches, than the Red Sea near Suez at low water, and about 30 feet lower than the Red Sea at the same place at high water,† but Mr. Robert Stevenson affirms, as the result of a more recent survey, that there is no difference of level between the two seas.‡

It was formerly imagined that there was an equal, if not greater, diversity in the relative levels of the Atlantic and Pacific, on the opposite sides of the Isthmus of Panama. But the levellings carried across that isthmus by Capt. Lloyd, in 1828, to ascertain the relative height of the Pacific Ocean at Panama, and of the Atlantic at the mouth of the river Chagres, have shown, that the difference of mean level be-

* Rennell on the Channel current.

† An. du Bureau des Long. 1836.

‡ Second Parliamentary Report on Steam Communication with India, July, 1851.

tween those oceans is not considerable, and, contrary to expectation, the difference which does exist is in favor of the greater height of the Pacific. According to this survey, the mean height of the Pacific is three feet and a half, or 3·52 above the Atlantic, if we assume the mean level of a sea to coincide with the mean between the extremes of the elevation and depression of the tides; for between the extreme levels of the greatest tides in the Pacific, at Panama, there is a difference of 27·44 feet; and at the usual spring tides 21·22 feet; whereas at Chagres this difference is only 1·16 feet, and is the same at all seasons of the year.

The tides, in short, in the Caribbean Sea are scarcely perceptible, not equalling those in some parts of the Mediterranean, whereas the rise is very high in the Bay of Panama; so that the Pacific is at high tide lifted up several feet above the surface of the Gulf of Mexico, and then at low water let down as far below it.* But astronomers are agreed that, on mathematical principles, the rise of the tidal wave above the mean level of a particular sea must be greater than the fall below it; and although the difference has been hitherto supposed insufficient to cause an appreciable error, it is, nevertheless, worthy of observation, that the error, such as it may be, would tend to reduce the small difference, now inferred, from the observations of Mr. Lloyd, to exist between the levels of the two oceans.

There is still another way in which heat and cold must occasion great movements in the ocean, a cause to which, perhaps, currents are principally due. Whenever the temperature of the surface of the sea is lowered, condensation takes place, and the superficial water, having its specific gravity increased, falls to the bottom, upon which lighter water rises immediately and occupies its place. When this circulation of ascending and descending currents has gone on for a certain time in high latitudes, the inferior parts of the sea are made to consist of colder or heavier fluid than the corresponding depths of the ocean between the tropics. If there be a free communication, if no chain of submarine mountains divide the polar from the equatorial basins, a horizontal movement will arise by the flowing of colder water from the poles to the equator, and there will then be a reflux of warmer superficial water from the equator to the poles. A well-known experiment has been adduced to elucidate this mode of action in explanation of the "trade winds."† If a long trough, divided in the middle by a sluice or partition, have one end filled with water and the other with quicksilver, both fluids will remain quiet so long as they are divided; but when the sluice is drawn up, the heavier fluid will rush along the bottom of the trough, while the lighter, being displaced, will rise, and, flowing in an opposite direction, spread itself at the top. In like manner the expansion and contraction of sea-water by heat and cold, have a tendency to set un-

* Phil. Trans. 1830, p. 59.

† See Capt. B. Hall, On Theory of Trade Winds, Fragments of Voy. second series, vol. i., and Appendix to Daniell's Meteorology.

der-currents in motion from the poles to the equator, and to cause counter-currents at the surface, which are impelled in a direction contrary to that of the prevailing trade winds. The geographical and other circumstances being very complicated, we cannot expect to trace separately the movements due to each cause, but must be prepared for many anomalies, especially as the configuration of the bed of the ocean must often modify and interfere with the course of the inferior currents, as much as the position and form of continents and islands alter the direction of those on the surface. Thus on sounding at great depths in the Mediterranean, Captains Berard and D'Urville have found that the cold does not increase in a high ratio as in the tropical regions of the ocean, the thermometer remaining fixed at about 55° F. between the depths of 1000 and 6000 feet. This might have been anticipated, as Captain Smyth in his survey had shown that the deepest part of the Straits of Gibraltar is only 1320 feet, so that a submarine barrier exists there which must prevent the influx of any under-current of the ocean cooled by polar ice.

Each of the four causes above mentioned, the wind, the tides, evaporation, and the expansion and contraction of water by heat and cold, may be conceived to operate independently of the others, and although the influence of all the rest were annihilated. But there is another cause, the rotation of the earth on its axis, which can only come into play when the waters have already been set in motion by some one or all of the forces above described, and when the direction of the current so raised happens to be from south to north, or from north to south.

The principle on which this cause operates is probably familiar to the reader, as it has long been recognized in the case of the trade winds. Without enlarging, therefore, on the theory, it will be sufficient to offer an example of the mode of action alluded to. When a current flows from the Cape of Good Hope towards the Gulf of Guinea, it consists of a mass of water, which, on doubling the Cape, in lat. 35°, has a rotatory velocity of about 800 miles an hour; but when it reaches the line, where it turns westward, it has arrived at a parallel where the surface of the earth is whirled round at the rate of 1000 miles an hour, or about 200 miles faster. If this great mass of water was transferred suddenly from the higher to the lower latitude, the deficiency of its rotatory motion, relatively to the land and water with which it would come into juxtaposition, would be such as to cause an apparent motion of the most rapid kind (of no less than 200 miles an hour) from east to west.

In the case of such a sudden transfer, the eastern coast of America, being carried round in an opposite direction, might strike against a large body of water with tremendous violence, and a considerable part of the continent might be submerged. This disturbance does not occur, because the water of the stream, as it advances gradually into new zones of the sea which are moving more rapidly, acquires by friction an accelerated velocity. Yet as this motion is not imparted instantaneously, thé fluid is unable to keep up with the full speed of the new surface over which it is successively brought. Hence, to borrow the language of Herschel, when he speaks of the trade winds, "it lags or hangs back, in a direction opposite to the earth's rotation, that is, from east to west,"* and thus a current, which would have run simply towards the north but for the rotation, may acquire a relative direction towards the west.

We may next consider a case where the circumstances are the converse of the above. The Gulf Stream flowing from about lat. 20° is at first impressed with a velocity of rotation of about 940 miles an hour, and runs to the lat. 40°, where the earth revolves only at the rate of 766 miles, or 174 miles slower. In this case a relative motion of an opposite kind may result; and the current may retain an excess of rotatory velocity, tending continually to deflect it eastward. Polar currents, therefore, or those flowing from high to low latitudes, are driven towards the eastern shores of continents, while tropical currents flowing towards the poles are directed against their western shores.

Thus it will be seen that currents depend, like the tides, on no temporary or accidental circumstances, but on the laws which preside over the motions of the heavenly bodies. But although the sum of their influence in altering the surface of the earth may be very constant throughout successive epochs, yet the points where these operations are displayed in fullest energy shift perpetually. The height to which the tides rise, and the violence and velocity of currents, depend in a great measure on the actual configuration of the land, the contour of a long line of continental or insular coast, the depth and breadth of channels, the peculiar form of the bottom of seas—in a word, on a combination of circumstances which are made to vary continually by many igneous and aqueous causes, and, amongst the rest, by the tides and currents themselves. Although these agents, therefore, of decay and reproduction are local in reference to periods of short duration, such as those which history embraces, they are nevertheless universal, if we extend our views to a sufficient lapse of ages.

Destroying and transporting power of currents.—After these preliminary remarks on the nature and causes of currents, their velocity and direction, we may next consider their action on the solid materials of the earth. We shall find that their efforts are, in many respects, strictly analogous to those of rivers. I have already treated in the third chapter, of the manner in which currents sometimes combine with ice, in carrying mud, pebbles, and large fragments of rock to great distances. Their operations are more concealed from our view than those of rivers, but extend over wider areas, and are therefore of more geological importance.

Waste of the British coasts.—Shetland Islands.—If we follow the

* Treatise on Astronomy, chap. 3.

eastern and southern shores of the British islands, from our Ultima Thule in Shetland to the Land's End in Cornwall, we shall find evidence of a series of changes since the historical era, very illustrative of the kind and degree of force exerted by tides and currents co-operating with the waves of the sea. In this survey we shall have an opportunity of tracing their joint power on islands, promontories, bays, and estuaries; on bold, lofty cliffs, as well as on low shores; and on every description of rock and soil, from granite to blown sand.

The northernmost group of the British islands, the Shetland, are composed of a great variety of rocks, including granite, gneiss, mica-slate, serpentine, greenstone, and many others, with some secondary rocks, chiefly sandstone and conglomerate. These islands are exposed continually to the uncontrolled violence of the Atlantic, for no land intervenes between their western shores and America. The prevalence, therefore, of strong westerly gales, causes the waves to be sometimes driven with irresistible force upon the coast, while there is also a current setting from the north. The spray of the sea aids the decomposition of the rocks, and prepares them to be breached by the mechanical force of the waves. Steep cliffs are hollowed out into deep caves and lofty arches; and almost every promontory ends in a cluster of rocks, imitating the forms of columns, pinnacles, and obelisks.

Drifting of large masses of rock.—Modern observations show that the reduction of continuous tracts to such insular masses is a process in which nature is still actively engaged. "The isle of Stenness," says Dr. Hibbert, "presents a scene of unequalled desolation. In stormy winters, huge blocks of stones are overturned, or are removed from their native beds, and hurried up a slight acclivity to a distance almost incredible. In the winter of 1802, a tabular-shaped mass, eight feet two inches by seven feet, and five feet one inch thick, was dislodged from its bed, and removed to a distance of from eighty to ninety feet. I measured the recent bed from which a block had been carried away the preceding winter (A. D. 1818), and found it to be seventeen feet and a half by seven feet, and the depth two feet eight inches. The removed mass had been borne to a distance of thirty feet, when it was shivered into thirteen or more lesser fragments, some of which were carried still farther, from 30 to 120 feet. A block, nine feet two inches by six feet and a half, and four feet thick, was hurried up the acclivity to a distance of 150 feet."*

At Northmavine, also, angular blocks of stone have been removed in a similar manner to considerable distances by the waves of the sea, some of which are represented in the annexed figure.

Effects of lightning.—In addition to numerous examples of masses detached and driven by the waves, tides, and currents from their place, some remarkable effects of lightning are recorded in these

* Descrip. of Shetland Islands, p. 527, Edin. 1822, to which work I am indebted for the following representations of rocks in the Shetland Isles.

Fig. 27.

Stony fragments drifted by the sea. Northmavine, Shetland.

isles. At Funzie, in Fetlar, about the middle of the last century, a rock of mica-schist, 105 feet long, ten feet broad, and in some places four feet thick, was in an instant torn by a flash of lightning from its bed, and broken into three large and several smaller fragments. One of these, twenty-six feet long, ten feet broad, and four feet thick, was simply turned over. The second, which was twenty-eight feet long, seventeen broad, and five feet in thickness, was hurled across a high point to the distance of fifty yards. Another broken mass, about forty feet long, was thrown still farther, but in the same direction, quite into the sea. There were also many smaller fragments scattered up and down.*

When we thus see electricity co-operating with the violent movements of the ocean in heaping up piles of shattered rocks on dry land and beneath the waters, we cannot but admit that a region which shall be the theatre, for myriads of ages, of the action of such disturbing causes, might present, at some future period, if upraised far above the bosom of the deep, a scene of havoc and ruin that may compare with any now found by the geologist on the surface of our continents.

In some of the Shetland Isles, as on the west of Meikle Roe, dikes, or veins of soft granite, have mouldered away; while the matrix in which they were inclosed, being of the same substance, but of a firmer texture, has remained unaltered. Thus, long narrow ravines, sometimes twenty feet wide, are laid open, and often give access to the waves. After describing some huge cavernous apertures into which the sea flows for 250 feet in Roeness, Dr. Hibbert, writing in 1822, enumerates other ravages of the ocean. "A mass of rock, the average dimensions of which may perhaps be rated at twelve or thirteen feet square, and four and a half or five in thickness, was first moved from its bed, about fifty years ago, to a distance of thirty feet, and has since been twice turned over."

Passage forced by the sea through porphyritic rocks.—"But the most sublime scene is where a mural pile of porphyry, escaping the process

* Dr. Hibbert, from MSS. of Rev. George Low, of Fetlar.

of disintegration that is devastating the coast, appears to have been left as a sort of rampart against the inroads of the ocean;—the Atlantic, when provoked by wintry gales, batters against it with all the force of real artillery—the waves having, in their repeated assaults, forced themselves an entrance. This breach, named the Grind of the Navir (fig. 28), is widened every winter by the overwhelming surge that, finding a

Fig. 28.

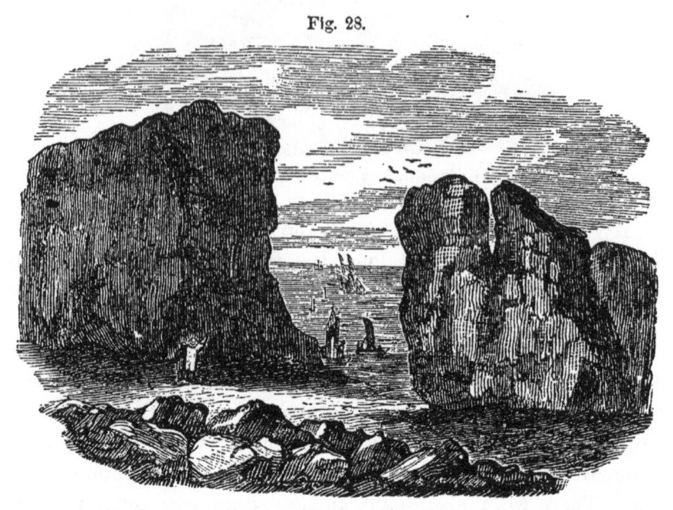

Grind of the Navir—passage forced by the sea through rocks of hard porphyry.

passage through it, separates large stones from its sides, and forces them to a distance of no less than 180 feet. In two or three spots, the fragments which have been detached are brought together in immense heaps, that appear as an accumulation of cubical masses, the product of some quarry."*

It is evident from this example, that although the greater indestructibility of some rocks may enable them to withstand, for a longer time, the action of the elements, yet they cannot permanently resist. There are localities in Shetland, in which rocks of almost every variety of mineral composition are suffering disintegration; thus the sea makes great inroads on the clay slate of Fitfel Head, on the serpentine of the Vord Hill in Fetlar, and on the mica-schist of the Bay of Triesta, on the east coast of the same island, which decomposes into angular blocks. The quartz rock on the east of Walls, and the gneiss and mica-schist of Garthness, suffer the same fate.

Destruction of islands.—Such devastation cannot be incessantly committed for thousands of years without dividing islands, until they become at last mere clusters of rocks, the last shreds of masses once continuous. To this state many appear to have been reduced, and innumerable fantastic forms are assumed by rocks adjoining these islands to which the name of Drongs is applied, as it is to those of similar shape in Feroe.

* Hibbert, p. 528.

The granite rocks (fig. 29), between Papa Stour and Hillswick Ness afford an example. A still more singular cluster of rocks is seen to

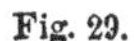

Fig. 29.

Granitic rocks named the Drongs, between Papa Stour and Hillswick Ness.

the south of Hillswick Ness (fig. 30), which presents a variety of forms as viewed from different points, and has often been likened to a small fleet of vessels with spread sails.* We may imagine that in the course

Fig. 30.

Granitic rocks to the south of Hillswick Ness, Shetland.

of time Hillswick Ness itself may present a similar wreck, from the unequal decomposition of the rocks whereof it is composed, consisting of gneiss and mica-schist traversed in all directions by veins of felspar-porphyry.

Midway between the groups of Shetland and Orkney is Fair Island, said to be composed of sandstone with high perpendicular cliffs. The current runs with such velocity, that during a calm, and when there is no swell, the rocks on its shores are white with the foam of the sea driven against them. The Orkneys, if carefully examined, would prob-

* Hibbert, p. 519.

ably illustrate our present topic as much as the Shetland group. The northeast promontory of Sanda, one of these islands, has been cut off in modern times by the sea, so that it became what is now called Start Island, where a lighthouse was erected in 1807, since which time the new strait has grown broader.

East coast of Scotland.—To pass over to the main land of Scotland, we find that in Inverness-shire there have been inroads of the sea at Fort George, and others in Morayshire, which have swept away the old town of Findhorn. On the coast of Kincardineshire, an illustration was afforded at the close of the last century, of the effect of promontories in protecting a line of low shore. The village of Mathers, two miles south of Johnshaven, was built on an ancient shingle beach, protected by a projecting ledge of limestone rock. This was quarried for lime to such an extent that the sea broke through, and in 1795 carried away the whole village in one night, and penetrated 150 yards inland, where it has maintained its ground ever since, the new village having been built farther inland on the new shore. In the bay of Montrose, we find the North Esk and the South Esk rivers pouring annually into the sea large quantities of sand and pebbles; yet they have formed no deltas, for the waves, aided by the current, setting across their mouths, sweep away all the materials. Considerable beds of shingle, brought down by the North Esk, are seen along the beach.

Proceeding southwards, we learn that at Arbroath, in Forfarshire, which stands on a rock of red sandstone, gardens and houses have been carried away since the commencement of the present century by encroachments of the sea. It had become necessary before 1828, to remove the lighthouses at the mouth of the estuary of the Tay, in the same county, at Button Ness, which were built on a tract of blown sand, the sea having encroached for three-quarters of a mile.

Force of waves and currents in estuaries.—The combined power which waves and currents can exert in *estuaries* (a term which I confine to bays entered both by rivers and the tides of the sea), was remarkably exhibited during the building of the Bell Rock Lighthouse, off the mouth of the Tay. The Bell Rock is a sunken reef, consisting of red sandstone, being from twelve to sixteen feet under the surface at high water, and about twelve miles from the mainland. At the distance of 100 yards, there is a depth, in all directions of two or three fathoms at low water. In 1807, during the erection of the lighthouse, six large blocks of granite, which had been landed on the reef, were removed by the force of the sea, and thrown over a rising ledge to the distance of twelve or fifteen paces; and an anchor, weighing about 22 cwt., was thrown up upon the rock.* Mr. Stevenson informs us moreover, that drift stones, measuring upwards of thirty cubic feet, or more than two tons' weight, have, during storms, been often thrown upon the rock from the deep water.†

* Account of Erection of Bell Rock Lighthouse, p. 163.
† Ed. Phil. Journ. vol. iii. p. 54, 1820.

Submarine forests.—Among the proofs that the sea has encroached on the land bordering the estuary of the Tay, Dr. Fleming has mentioned a submarine forest which has been traced for several miles along the northern shore of the county of Fife.* But subsequent surveys seem to have shown that the bed of peat containing tree-roots, leaves, and branches, now occurring at a lower level than the Tay, must have come into its present position by a general sinking of the ground on which the forest grew. The peat-bed alluded to is not confined, says Mr. Buist, to the present channel of the Tay, but extends far beyond it, and is covered by stratified clay from fifteen to twenty-five feet in thickness, in the midst of which, in some places, is a bed full of sea-shells.† Recent discoveries having established the fact that upward and downward movements have affected our island since the general coast-line had nearly acquired its present shape, we must hesitate before we attribute any given change to a single cause, such as the local encroachment of the sea upon low land.

On the coast of Fife, at St. Andrew's, a tract of land, said to have intervened between the castle of Cardinal Beaton and the sea, has been entirely swept away, as were the last remains of the Priory of Crail, in the same county, in 1803. On both sides of the Frith of Forth, land has been consumed; at North Berwick in particular, and at Newhaven, where an arsenal and dock, built in the reign of James IV., in the fifteenth century, has been overflowed.

East coast of England.—If we now proceed to the English coast, we find records of numerous lands having been destroyed in Northumberland, as those near Bamborough and Holy Island, and at Tynemouth Castle, which now overhangs the sea, although formerly separated from it by a strip of land. At Hartlepool, and several other parts of the coast of Durham composed of magnesian limestone, the sea has made considerable inroads.

Coast of Yorkshire.—Almost the whole coast of Yorkshire, from the mouth of the Tees to that of the Humber, is in a state of gradual dilapidation. That part of the cliffs which consist of lias, the oolite series, and chalk, decays slowly. They present abrupt and naked precipices, often 300 feet in height; and it is only at a few points that the grassy covering of the sloping talus marks a temporary relaxation of the erosive action of the sea. The chalk cliffs are worn into caves and needles in the projecting headland of Flamborough, where they are decomposed by the salt spray, and slowly crumble away. But the waste is most rapid between that promontory and Spurn Point, or the coast of Holderness, as it is called, a tract consisting of beds of clay, gravel, sand, and chalk rubble. The irregular intermixture of the argillaceous beds causes many springs to be thrown out, and this facilitates the undermining process, the waves beating against them, and a strong current

* Quart. Journ. of Sci. &c., No. xiii. *N. S.* March, 1830.
† Buist, Quart. Journ. of Agricult. No. xlv. p. 34, June, 1839.

setting chiefly from the north. The wasteful action is very conspicuous at Dimlington Height, the loftiest point in Holderness, where the beacon stands on a cliff 146 feet above high water, the whole being composed of clay, with pebbles scattered through it.* "For many years," says Professor Phillips, "the rate at which the cliffs recede from Bridlington to Spurn, a distance of thirty-six miles, has been found by measurement to equal on an average two and a quarter yards annually, which, upon thirty-six miles of coast, would amount to about thirty acres a year. At this rate, the coast, the mean height of which above the sea is about forty feet, has lost one mile in breadth since the Norman Conquest, and more than two miles since the occupation of York (Eboracum) by the Romans."† The extent of this denudation, as estimated by the number of cubic feet of matter removed annually, will be again spoken of in chapter 22.

In the old maps of Yorkshire, we find spots, now sand-banks in the sea, marked as the ancient sites of the towns and villages of Auburn, Hartburn, and Hyde. "Of Hyde," says Pennant, "only the tradition is left; and near the village of Hornsea, a street called Hornsea Beck has long since been swallowed."‡ Owthorne and its church have also been in great part destroyed, and the village of Kilnsea; but these places are now removed farther inland. The annual rate of encroachment at Owthorne for several years preceding 1830, is stated to have averaged about four yards. Not unreasonable fears are entertained that at some future time the Spurn Point will become an island, and that the ocean, entering into the estuary of the Humber, will cause great devastation.§ Pennant, after speaking of the silting up of some ancient ports in that estuary, observes, "But, in return, the sea has made most ample reprisals; the site, and even the very names of several places, once towns of note upon the Humber, are now only recorded in history; and Ravensper was at one time a rival to Hull (Madox, Ant. Exch. i. 422), and a port so very considerable in 1332, that Edward Baliol and the confederated English barons sailed from hence to invade Scotland; and Henry IV., in 1399, made choice of this port to land at, to effect the deposal of Richard II.; yet the whole of this has long since been devoured by the merciless ocean; extensive sands, dry at low water, are to be seen in their stead."‖

Pennant describes Spurn Head as a promontory in the form of a sickle, and says the land, for some miles to the north, was "perpetually preyed on by the fury of the German Sea, which devours whole acres at a time, and exposes on the shores considerable quantities of beautiful amber."

Lincolnshire.—The maritime district of Lincolnshire consists chiefly of lands that lie below the level of the sea, being protected by embank-

* Phillips's Geology of Yorkshire, p. 61.
† Rivers, Mountains, and Sea-coast of Yorkshire p. 122, 1853, London.
‡ Arctic Zoology, vol. i. p. 10, Introduction.
§ Phillips's Geol. of York. p. 60.
‖ Arct. Zool. vol. i. p. 13, Introd.

ments. Some of the fens were embanked and drained by the Romans; but after their departure the sea returned, and large tracts were covered with beds of silt, containing marine shells, now again converted into productive lands. Many dreadful catastrophes are recorded by incursions of the sea, whereby several parishes have been at different times overwhelmed.

Norfolk.—The decay of the cliffs of Norfolk and Suffolk is incessant. At Hunstanton, on the north, the undermining of the lower arenaceous beds at the foot of the cliff, causes masses of red and white chalk to be precipitated from above. Between Hunstanton and Weybourne, low hills, or dunes, of blown sand, are formed along the shore, from fifty to sixty feet high. They are composed of dry sand, bound in a compact mass by the long creeping roots of the plant called Marram (*Arundo arenaria*). Such is the present set of the tides, that the harbors of Clay, Wells, and other places are securely defended by these barriers; affording a clear proof that it is not the strength of the material at particular points that determines whether the sea shall be progressive or stationary, but the general contour of the coast.

The waves constantly undermine the low chalk cliffs, covered with sand and clay, between Weybourne and Sherringham, a certain portion of them being annually removed. At the latter town I ascertained, in 1829, some facts which throw light on the rate at which the sea gains upon the land. It was computed, when the present inn was built, in 1805, that it would require seventy years for the sea to reach the spot: the mean loss of land being calculated, from previous observations, to be somewhat less than one yard, annually. The distance between the house and the sea was fifty yards; but uo allowance was made for the slope of the ground being *from* the sea, in consequence of which the waste was naturally accelerated every year, as the cliff grew lower, there being at each succeeding period less matter to remove when portions of equal area fell down. Between the years 1824 and 1829, no less than seventeen yards were swept away, and only a small garden was then left between the building and the sea. There was, in 1829, a depth of twenty feet (sufficient to float a frigate) at one point in the harbor of that port, where, only forty-eight years before, there stood a cliff fifty feet high, with houses upon it! If once in half a century an equal amount of change were produced suddenly by the momentary shock of an earthquake, history would be filled with records of such wonderful revolutions of the earth's surface; but, if the conversion of high land into deep sea be gradual, it excites only local attention. The flagstaff of the Preventive Service station, on the south side of this harbor, was thrice removed inland between the years 1814 and 1829, in cousequence of the advance of the sea.

Farther to the south we find cliffs, composed, like those of Holderness before mentioned, of alternating strata of blue clay, gravel, loam, and fine sand. Although they sometimes exceed 300 feet in height, the havoc made on the coast is most formidable. The whole site of ancient

Cromer now forms part of the German Ocean, the inhabitants having gradually retreated inland to their present situation, from whence the sea still threatens to dislodge them. In the winter of 1825, a fallen mass was precipitated from near the lighthouse, which covered twelve acres, extending far into the sea, the cliffs being 250 feet in height.* The undermining by springs has sometimes caused large portions of the upper part of the cliffs, with houses still standing upon them, to give way, so that it is impossible, by erecting breakwaters at the base of the cliffs, permanently to ward off the danger.

On the same coast, says Mr. R. C. Taylor, the ancient villages of Shipden, Wimpwell, and Eccles have disappeared; several manors and large portions of neighboring parishes having, piece after piece, been swallowed up; nor has there been any intermission, from time immemorial, in the ravages of the sea along a line of coast twenty miles in length, in which these places stood.† Of Eccles, however, a monument still remains in the ruined tower of the old church, which is half buried in the dunes of sand within a few paces (60 ?) of the sea-beach (fig. 31). So early as 1605 the inhabitants petitioned James I. for a

Fig. 31.

Tower of the buried Church of Eccles, Norfolk, A. D. 1839.
The inland slope of the hills of blown sand is shown in this view, with the lighthouse of Hasborough in the distance.

reduction of taxes, as 300 acres of land, and all their houses, save fourteen, had then been destroyed by the sea. Not one half that number of acres now remains in the parish, and hills of blown sand now occupy the site of the houses which were still extant in 1605. When I visited the spot in 1839, the sea was fast encroaching on the sand-hills, and had laid open on the beach the foundations of a house fourteen yards square, the upper part of which had evidently been pulled down before it had been buried under sand. The body of the church has also been long buried, but the tower still remains visible.

* Taylor's Geology of East Norfolk, p. 32. † Ibid.

M. E. de Beaumont has suggested that sand-dunes in Holland and other countries may serve as natural chronometers, by which the date of the existing continents may be ascertained. The sands, he says, are continually blown inland by the force of the winds, and by observing the rate of their march we may calculate the period when the movement commenced.* But the example just given will satisfy every geologist that we cannot ascertain the starting-point of dunes, all coasts being liable to waste, and the shores of the Low Countries in particular, being not only exposed to inroads of the sea, but, as M. de Beaumont himself has well shown, having even in historical times undergone a change of level. The dunes may indeed, in some cases, be made use of as chronometers, to enable us to assign a minimum of antiquity to existing coast-lines; but this test must be applied with great caution, so variable is the rate at which the sands may advance into the interior.

Hills of blown sand, between Eccles and Winterton, have barred up and excluded the tide for many hundred years from the mouths of several small estuaries; but there are records of nine breaches, from 20 to 120 yards wide, having been made through these, by which immense damage was done to the low grounds in the interior. A few miles south of Happisburgh, also, are hills of blown sand, which extend to Yarmouth. These *dunes* afford a temporary protection to the coast, and an inland cliff about a mile long, at Winterton, shows clearly that at that point the sea must have penetrated formerly farther than at present.

Silting up of estuaries.—At Yarmouth, the sea has not advanced upon the sands in the slightest degree since the reign of Elizabeth. In the time of the Saxons, a great estuary extended as far as Norwich, which city is represented, even in the thirteenth and fourteenth centuries, as "situated on the banks of an arm of the sea." The sands whereon Yarmouth is built, first became firm and habitable ground about the year 1008, from which time a line of dunes has gradually increased in height and breadth, stretching across the whole entrance of the ancient estuary, and obstructing the ingress of the tides so completely, that they are only admitted by the narrow passage which the river keeps open, and which has gradually shifted several miles to the south. The ordinary tides at the river's mouth rise, at present, only to the height of three or four feet, the spring tides to about eight or nine.

By the exclusion of the sea, thousands of acres in the interior have become cultivated lands; and, exclusive of smaller pools, upwards of sixty freshwater lakes have been formed, varying in depth from fifteen to thirty feet, and in extent from one acre to twelve hundred.† The Yare, and other rivers, frequently communicate with these sheets of water; and thus they are liable to be filled up gradually with lacustrine and fluviatile deposits, and to be converted into land covered with

* De Beaumont, Géologie Pratique, p. 218.
† Taylor's Geology of East Norfolk, p. 10.

forests. Yet it must not be imagined, that the acquisition of new land fit for cultivation in Norfolk and Suffolk indicates any permanent growth of the eastern limits of our island to compensate its reiterated losses. No *delta* can form on such a shore.

Immediately off Yarmouth, and parallel to the shore, is a great range of sand-banks, the shape of which varies slowly from year to year, and often suddenly after great storms. Captain Hewitt, R. N., found in these banks, in 1836, a broad channel sixty-five feet deep, where there was only a depth of four feet during a prior survey in 1822. The sea had excavated to the depth of sixty feet in the course of fourteen years, or perhaps a shorter period. The new channel thus formed serves at present (1838), for the entrance of ships into Yarmouth Roads; and the magnitude of this change shows how easily a new set of the waves and currents might endanger the submergence of the land gained within the ancient estuary of the Yare.

That great banks should be thrown across the mouths of estuaries on our eastern coast, where there is not a large body of river-water to maintain an open channel, is perfectly intelligible, when we bear in mind that the marine current, sweeping along the coast, is charged with the materials of wasting cliffs, and ready to form a bar anywhere the instant its course is interrupted or checked by any opposing stream. The mouth of the Yare has been, within the last five centuries, diverted about four miles to the south. In like manner it is evident that, at some remote period, the river Alde entered the sea at Aldborough, until its ancient outlet was barred up and at length transferred to a point no less than ten miles distant to the southwest. In this case, ridges of sand and shingle, like those of Lowestoff Ness, which will be described by and by, have been thrown up between the river and the sea; and an ancient sea-cliff is to be seen now inland.

It may be asked why the rivers on our east coast are always deflected southwards, although the tidal current flows alternately from the south and north? The cause is to be found in the superior force of what is commonly called "the flood tide from the north," a tidal wave derived from the Atlantic, a small part of which passes eastward up the English Channel, and through the Straits of Dover and then northwards, while the principal body of water, moving much more rapidly in a more open sea, on the western side of Britain, first passes the Orkneys, and then turning, flows down between Norway and Scotland, and sweeps with great velocity along our eastern coast. It is well known that the highest tides on this coast are occasioned by a powerful northwest wind, which raises the eastern part of the Atlantic, and causes it to pour a greater volume of water into the German Ocean. This circumstance of a violent *off-shore* wind being attended with a rise of the waters, instead of a general retreat of the sea, naturally excites the wonder of the inhabitants of our coast. In many districts they look with confidence for a rich harvest of that valuable manure, the sea-weed, when the northwesterly gales prevail, and are rarely disappointed.

Coast of Suffolk.—The cliffs of Suffolk, to which we next proceed, are somewhat less elevated than those of Norfolk, but composed of similar alternations of clay, sand, and gravel. From Gorleston in Suffolk, to within a few miles north of Lowestoff, the cliffs are slowly undermined. Near the last-mentioned town, there is an inland cliff about sixty feet high, the sloping talus of which is covered with turf and heath. Between the cliff and the sea is a low flat tract of sand called the Ness, nearly three miles long, and for the most part out of reach of the highest tides. The point of the Ness projects from the

Fig. 32.

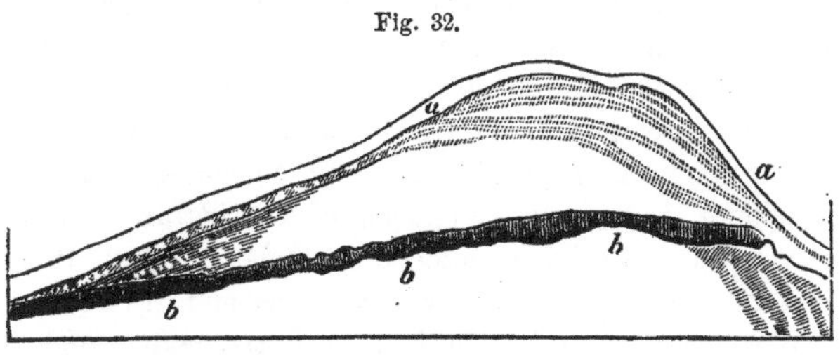

Map of Lowestoff Ness, Suffolk.*

a, a. The dotted lines express a series of sand and shingle, forming the extremity of the triangular space called the Ness.
b, b, b. The dark line represents the inland cliff on which the town of Lowestoff stands, between which and the sea is the Ness.

base of the original cliff to the distance of 660 yards. This accession of land, says Mr. Taylor, has been effected at distinct and distant intervals, by the influence of currents running between the land and a shoal about a mile off Lowestoff, called the Holm Sand. The lines of growth in the Ness are indicated by a series of concentric ridges or embankments inclosing limited areas, and several of these ridges have been formed within the observation of persons now living. A rampart of heavy materials is first thrown up to an unusual altitude by some extraordinary tide, attended with a violent gale. Subsequent tides extend the base of this high bank of shingle, and the interstices are then filled with sand blown from the beach. The Arundo and other marine plants by degrees obtain a footing; and creeping along the ridge, give solidity to the mass, and form in some cases a matted covering of turf. Meanwhile another mound is forming externally, which by the like process rises and gives protection to the first. If the sea forces its way through one of the external and incomplete mounds, the breach is soon repaired. After a while the marine plants within the areas inclosed by these embankments are succeeded by a better species of herbage affording good pasturage, and the sands become sufficiently firm to support buildings.

Destruction of Dunwich by the sea.—Of the gradual destruction of Dunwich, once the most considerable seaport on this coast, we have many authentic records. Gardner, in his history of that borough, pub-

* From Mr. R. C. Taylor's Mem., see Phil. Mag., Oct. 1827, p. 297.

lished in 1754, shows, by reference to documents, beginning with Doomsday Book, that the cliffs at Dunwich, Southwold, Eastern, and Pakefield, have been always subject to wear away. At Dunwich, in particular, two tracts of land which had been taxed in the eleventh century, in the time of King Edward the Confessor, are mentioned in the Conqueror's survey, made but a few years afterwards, as having been devoured by the sea. The losses, at a subsequent period, of a monastery,—at another of several churches,—afterwards of the old port,—then of four hundred houses at once,—of the church of St. Leonard, the high-road, town-hall, jail, and many other buildings, are mentioned, with the dates when they perished. It is stated that, in the sixteenth century, not one-quarter of the town was left standing; yet the inhabitants retreating inland, the name was preserved, as has been the case with many other ports when their ancient site has been blotted out. There is, however, a church of considerable antiquity still standing, the last of twelve mentioned in some records. In 1740, the laying open of the churchyard of St. Nicholas and St. Francis, in the sea-cliffs, is well described by Gardner, with the coffins and skeletons exposed to view—some lying on the beach, and rocked

"In cradle of the rude imperious surge."

Of these cemeteries no remains can now be seen. Ray also says, "that ancient writings make mention of a wood a mile and a half to the east of Dunwich, the site of which must at present be so far within the sea." * This city, once so flourishing and populous, is now a small village, with about twenty houses, and one hundred inhabitants.

There is an old tradition, "that the tailors sat in their shops at Dunwich, and saw the ships in Yarmouth Bay;" but when we consider how far the coast at Lowestoff Ness projects between these places, we cannot give credit to the tale, which, nevertheless, proves how much the inroads of the sea in times of old had prompted men of lively imagination to indulge their taste for the marvellous.

Gardner's description of the cemeteries laid open by the waves reminds us of the scene which has been so well depicted by Bewick,† and of which numerous points on the same coast might have suggested the idea. On the verge of a cliff, which the sea has undermined, are represented the unshaken tower and western end of an abbey. The eastern aisle is gone, and the pillars of the cloister are soon to follow. The waves have almost isolated the promontory, and invaded the cemetery, where they have made sport with the mortal relics, and thrown up a skull upon the beach. In the foreground is seen a broken tombstone, erected, as its legend tells, "to *perpetuate* the memory"—of one whose name is obliterated, as is that of the county for which he was "Custos Rotulorum." A cormorant is perched on the monument,

* Consequences of the Deluge, Phys. Theol. Discourses.
† History of British Birds, vol. ii. p. 220 ed. 1821.

defiling it, as if to remind some moralizer like Hamlet, of "the base uses" to which things sacred may be turned. Had this excellent artist desired to satirize certain popular theories of geology, he might have inscribed the stone to the memory of some philosopher who taught "the permanency of existing continents"—"the era of repose"—"the impotence of modern causes."

The incursions of the sea at Aldborough, were formerly very destructive, and this borough is known to have been once situated a quarter of a mile east of the present shore. The inhabitants continued to build farther inland, till they arrived at the extremity of their property, and then the town decayed greatly; but two sand-banks, thrown up at a short distance, now afford a temporary safeguard to the coast. Between these banks and the present shore, where the current now flows, the sea is twenty-four feet deep on the spot where the town formerly stood.

Essex.—Harwich is said to have owed its rise to the destruction of Orwell, a town which stood on the spot now called "the west rocks," and was overwhelmed by an inroad of the sea since the Conquest. Apprehensions have been entertained that the isthmus on which Harwich stands may at no remote period become an island, for the sea may be expected to make a breach near Lower Dover Court, where Beacon Cliff is composed of horizontal beds of London clay containing septaria. It had wasted away considerably between the years 1829 and 1838, at both which periods I examined this coast. In that short interval several gardens and many houses had been swept into the sea, and in April, 1838, a whole street was threatened with destruction. The advance of the sea is much accelerated by the traffic carried on in septaria, which are shipped off for cement as fast as they fall down upon the beach. These stones, if allowed to remain in heaps on the shore, would break the force of the waves and retard the conversion of the peninsula into an island, an event which might be followed by the destruction of the town of Harwich. Captain Washington, R. N., ascertained in 1847, that Beacon Cliff, above mentioned, which is about fifty feet high, had given way at the rate of forty feet in forty-seven years, between 1709 and 1756; eighty feet between 1756 and 1804; and three hundred and fifty feet between the latter period and 1841; showing a rapidly accelerated rate of destruction.*

Among other losses it is recorded that, since the year 1807, a field called the Vicar's Field, which belonged to the living of Harwich, has been overwhelmed;† and in the year 1820 there was a considerable space between the battery at Harwich, built in the beginning of the present century, and the sea; part of the fortification had been swept away in 1829, and the rest then overhung the water.

At Walton Naze, in the same county, the cliffs, composed of London clay, capped by the shelly sands of the crag, reach the height of about

* Tidal Harbor Commissioners' First Report, 1845, p. 176.
† On authority of Dr. Mitchell, F. G. S.

100 feet, and are annually undermined by the waves. The old churchyard of Walton has been washed away, and the cliffs to the south are constantly disappearing.

Kent.—Isle of Sheppey.—On the coast bounding the estuary of the Thames, there are numerous examples both of the gain and loss of land. The Isle of Sheppey, which is now about six miles long by four in breadth, is composed of London clay. The cliffs on the north, which are from sixty to eighty feet high, decay rapidly, fifty acres having been lost in twenty years, between 1810 and 1830. The church at Minster, now near the coast, is said to have been in the middle of the island in 1780; and if the present rate of destruction should continue, we might calculate the period, and that not a very remote one, when the whole island will be annihilated. On the coast of the mainland, to the east of Sheppey, is Herne Bay: a place still retaining the name of a bay, although it is no longer appropriate, as the waves and currents have swept away the ancient headlands. There was formerly a small promontory in the line of the shoals where the present pier is built, by which the larger bay was divided into two, called the Upper and Lower.*

Fig. 33.

View of Reculver Church, taken in the year 1781.
1. Isle of Sheppey. 2. Ancient chapel now destroyed. The cottage between this chapel and the cliff was demolished by the sea, in 1782.

Still farther east stands the church of Reculver, upon a cliff composed of clay and sand, about twenty-five feet high. Reculver (Regulvium) was an important military station in the time of the Romans, and appears, from Leland's account, to have been, so late as Henry VIII.'s reign, nearly one mile distant from the sea. In the "Gentleman's Magazine," there is a view of it, taken in 1781, which still represents a considerable space as intervening between the north wall of the churchyard and the cliff.† Sometime before the year 1780, the waves had reached the site of the ancient Roman camp or fortification, the walls of which

* On the authority of W. Gunnell, Esq., and W. Richardson, Esq., F. G. S.
† Vol. ii. New Ser. 1809, p. 801.

had continued for several years after they were undermined to overhang the sea, being firmly cemented into one mass. They were eighty yards nearer the sea than the church, and they are spoken of in the "Topographica Britannica," in the year 1780, as having recently fallen down. In 1804, part of the churchyard with some adjoining houses was washed away, and the ancient church, with its two spires, was dismantled and abandoned as a place of worship, but kept in repair as a landmark well known to mariners. I visited the spot in June, 1851, and saw human bones and part of a wooden coffin projecting from the cliff, near the top. The whole building would probably have been swept away long

Fig. 34.

Reculver Church, in 1834.

ere this, had not the force of the waves been checked by an artificial causeway of stones and large wooden piles driven into the sands on the beach to break the force of the waves.

Isle of Thanet.—The isle of Thanet was, in the time of the Romans, separated from the rest of Kent by a navigable channel, through which the Roman fleets sailed on their way to and from London. Bede describes this small estuary as being, in the beginning of the eighth century, three furlongs in breadth; and it is supposed that it began to grow shallow about the period of the Norman conquest. It was so far silted up in the year 1485, that an act was then obtained to build a bridge across it; and it has since become marsh land with small streams running through it. On the coast, Bedlam Farm, belonging to the hospital of that name, lost eight acres in the twenty years preceding

1830, the land being composed of chalk from forty to fifty feet above the level of the sea. It has been computed that the average waste of the cliff between the North Foreland and the Reculvers, a distance of about eleven miles, is not less than two feet per annum. The chalk cliffs on the south of Thanet, between Ramsgate and Pegwell Bay, have on an average lost three feet per annum for the last ten years (preceding 1830).

Goodwin Sands.—The Goodwin Sands lie opposite this part of the Kentish coast. They are about ten miles in length, and are in some parts three, and in others seven, miles distant from the shore; and, for a certain space, are laid bare at low water. That they are a remnant of land, and not "a mere accumulation of sea sand," as Rennell imagined,* may be presumed from the fact that, when the erection of a lighthouse on this shoal was in contemplation by the Trinity Board in the year 1817, it was found, by borings, that the bank consisted of fifteen feet of sand, resting on blue clay; and, by subsequent borings, the subjacent chalk has been reached. An obscure tradition has come down to us, that the estates of Earl Goodwin, the father of Harold, who died in the year 1053, were situated here, and some have conjectured that they were overwhelmed by the flood mentioned in the Saxon chronicle, *sub anno* 1099. The last remains of an island, consisting, like Sheppey, of clay, may perhaps have been carried away about that time.

Fig. 35.

Shakspeare's Cliff in 1836, seen from the northeast.

There are other records of waste in the county of Kent, as at Deal; and at Dover, where Shakspeare's Cliff, composed entirely of chalk, has suffered greatly, and continually diminishes in height, the slope of the hill being towards the land. (See fig. 35.) There was an immense landslip from this cliff in 1810, by which Dover was shaken as if by an

* Geog. of Herod. vol. ii. p. 326.

earthquake, and a still greater one in 1772.* We may suppose, therefore, that the view from the top of the precipice in the year 1600, when the tragedy of King Lear was written, was more "fearful and dizzy" than it is now. The best antiquarian authorities are agreed, that Dover Harbor was formerly an estuary, the sea flowing up a valley between the chalk hills. The remains found in different excavations confirm the description of the spot given by Cæsar and Antoninus, and there is clear historical evidence to prove that at an early period there was no shingle at all at Dover.†

Straits of Dover.—In proceeding from the northern parts of the German Ocean towards the Straits of Dover, the water becomes gradually more shallow, so that, in the distance of about two hundred leagues, we pass from a depth of 120 to that of 58, 38, 18, and even less than 2 fathoms. The shallowest part follows a line drawn between Romney Marsh and Boulogne. From this point the English Channel again deepens progressively as we proceed westward, so that the Straits of Dover may be said to part two seas.‡

Whether England was formerly united with France has often been a favorite subject of speculation. So early as 1605 our countryman Verstegan, in his "Antiquities of the English Nation," observed that many preceding writers had maintained this opinion, but without supporting it by any weighty reasons. He accordingly endeavors himself to confirm it by various arguments, the principal of which are, first, the proximity and identity of the composition of the opposite cliffs and shores of Albion and Gallia, which, whether flat and sandy, or steep and chalky, correspond exactly with each other; secondly the occurrence of a submarine ridge, called "our Lady's Sand," extending from shore to shore at no great depth, and which, from its composition, appears to be the original basis of the isthmus; thirdly, the identity of the noxious animals in France and England, which could neither have swum across, nor have been introduced by man. Thus no one, he says, would have imported wolves, therefore "these wicked beasts did of themselves pass over." He supposes the ancient isthmus to have been about six English miles in breadth, composed entirely of chalk and flint, and in some places of no great height above the sea-level. The operation of the waves and tides, he says, would have been more powerful when the straits were narrower, and even now they are destroying cliffs composed of similar materials. He suggests the possible co-operation of earthquakes; and when we consider how many submarine forests skirt the southern and eastern shores of England, and that there are raised beaches at many points above the sea-level, containing fossil shells of recent species, it seems reasonable to suppose that such up-

* Dodsley's Ann. Regist. 1772.

† See J. B. Redman on Changes of S. E. Coast of England, Proceed. Instit. Civil Engin. vol. ii. 1851, 1852.

‡ Stevenson, Ed. Phil. Journ. No. v. p. 45, and Dr. Fitton, Geol. Trans. 2d series, vol. iv. plate 9.

ward and downward movements, taking place perhaps as slowly as those now in progress in Sweden and Greenland, may have greatly assisted the denuding force of "the ocean stream," Ποταμοιο μεγα σθενος Ωκεανοιο.

Folkstone.—At Folkstone, the sea undermines the chalk and subjacent strata. About the year 1716 there was a remarkable sinking of a tract of land near the sea, so that houses became visible from certain points at sea, and from particular spots on the sea cliffs, from whence they could not be seen previously. In the description of this subsidence in the Phil. Trans. 1716, it is said, "that the land consisted of a solid stony mass (chalk), resting on wet clay (gault), so that it slid forwards towards the sea, just as a ship is launched on tallowed planks." It is also stated that, within the memory of persons then living, the cliff there had been washed away to the extent of ten rods.

Encroachments of the sea at Hythe are also on record; but between this point and Rye there has been a gain of land within the times of history; the rich level tract called Romney Marsh, or Dungeness, about ten miles in width and five in breadth, and formed of silt, having received great accession. It has been necessary, however, to protect it from the sea, from the earliest periods, by embankments, the towns of Lydd and Romney being the only parts of the marsh above the level of the highest tides.* Mr. Redman has cited numerous old charts and trustworthy authorities to prove that the average annual increase of the promontory of shingle called Dungeness amounted for two centuries, previous to 1844, to nearly six yards. Its progress, however, has fluctuated during that period; for between 1689 and 1794, a term of 105 years, the rate was as much as 8¼ yards per annum.† It is ascertained that the shingle is derived from the westward. Whether the pebbles are stopped by the meeting of the tide from the north flowing through the Straits of Dover, with that which comes up the Channel from the west, as was formerly held, or by the check given to the tidal current by the waters of the Rother, as some maintain, is still a disputed question.

Rye, situated to the south of Romney Marsh, was once destroyed by the sea, but it is now two miles distant from it. The neighboring town of Winchelsea was destroyed in the reign of Edward I., the mouth of the Rother stopped up, and the river diverted into another channel. In its old bed, an ancient vessel, apparently a Dutch merchantman, was found about the year 1824. It was built entirely of oak, and much blackened.‡ Large quantities of hazel-nuts, peat, and wood are found in digging in Romney Marsh.

South coast of England.—Westward of Hastings, or of St. Leonard's, the shore line has been giving way as far as Pevensey Bay, where formerly there existed a haven now entirely blocked up by shingle. The

* On the authority of Mr. J. Meryon, of Rye.
† Redman, ibid. see p. 315.
‡ Edin. Journ. of Sci. No. xix. p. 56.

degradation has equalled for a series of years seven feet per annum in some places, and several martello towers had in consequence, before 1851, been removed by the Ordnance.* At the promontory of Beachy Head a mass of chalk, three hundred feet in length, and from seventy to eighty in breadth, fell in the year 1813 with a tremendous crash; and similar slips have since been frequent.†

About a mile to the west of the town of Newhaven, the remains of an ancient intrenchment are seen on the brow of Castle Hill. This earth-work, supposed to be Roman, was evidently once of considerable extent and of an oval form, but the greater part has been cut away by the sea. The cliffs, which are undermined here, are high; more than one hundred feet of chalk being covered by tertiary clay and sand, from sixty to seventy feet in thickness. In a few centuries the last vestiges of the plastic clay formation on the southern borders of the chalk of the South Downs on this coast will probably be annihilated, and future geologists will learn, from historical documents, the ancient geographical boundaries of this group of strata in that direction. On the opposite side of the estuary of the Ouse, on the east of Newhaven harbor, a bed of shingle, composed of chalk flints derived from the waste of the adjoining cliffs, had accumulated at Seaford for several centuries. In the great storm of November, 1824, this bank was entirely swept away, and the town of Seaford inundated. Another great beach of shingle is now forming from fresh materials.

The whole coast of Sussex has been incessantly encroached upon by the sea from time immemorial; and, although sudden inundations only, which overwhelmed fertile or inhabited tracts, are noticed in history, the records attest an extraordinary amount of loss. During a period of no more than eighty years, there are notices of about *twenty* inroads, in which tracts of land of from twenty to *four hundred acres* in extent were overwhelmed at once, the value of the tithes being mentioned in the Taxatio Ecclesiastica.‡ In the reign of Elizabeth, the town of Brighton was situated on that tract where the chain pier now extends into the sea. In the year 1665, twenty-two tenements had been destroyed under the cliff. At that period there still remained under the cliff 113 tenements, the whole of which were overwhelmed in 1703 and 1705. No traces of the ancient town are now perceptible, yet there is evidence that the sea has merely resumed its ancient position at the base of the cliffs, the site of the whole town having been merely a beach abandoned by the ocean for ages.

Hampshire.—Isle of Wight.—It would be endless to allude to all the localities on the Sussex and Hampshire coasts where the land has given way; but I may point out the relation which the geological structure of the Isle of Wight bears to its present shape, as attesting that the

* Redman as cited, p. 315.
† Webster, Geol. Trans. vol. ii. p. 192, 1st series.
‡ Mantell, Geology of Sussex, p. 293.

coast owes its outline to the continued action of the sea. Through the middle of the island runs a high ridge of chalk strata, in a vertical position, and in a direction east and west. This chalk forms the projecting promontory of Culver Cliff on the east, and of the Needles on the west; while Sandown Bay on the one side, and Compton Bay on the other, have been hollowed out of the softer sands and argillaceous strata, which are inferior, in geological position, to the chalk.

The same phenomena are repeated in the Isle of Purbeck, where the line of vertical chalk forms the projecting promontory of Handfast Point; and Swanage Bay marks the deep excavation made by the waves in the softer strata, corresponding to those of Sandown Bay.

Hurst Castle bank—progressive motion of sea beaches.—Although the loose pebbles and grains of sand composing any given line of sea-beach are carried sometimes one way, sometimes another, they have, nevertheless, an ultimate motion in one particular direction.* Their progress, for example, on the south coast of England, is from west to east, which is owing partly to the action of the waves driven eastwards by the prevailing wind, and partly to the current, or the motion of the general body of water caused by the tides and winds. The force of the waves gives motion to pebbles which the velocity of the currents alone would be unable to carry forwards; but as the pebbles are finally reduced to sand or mud, by continual attrition, they are brought within the influence of a current; and this cause must determine the course which the main body of matter derived from wasting cliffs will eventually take.

It appears, from the observations of Mr. Palmer and others, that if a pier or groin be erected anywhere on our southern or southeastern coast to stop the progress of the beach, a heap of shingle soon collects on the western side of such artificial barriers. The pebbles continue to accumulate till they rise as high as the pier or groin, after which they pour over in great numbers during heavy gales.†

The western entrance of the Channel, called the Solent, is crossed for more than two-thirds of its width by the shingle-bank of Hurst Castle, which is about two miles long, seventy yards broad, and twelve feet high, presenting an inclined plane to the west. This singular bar consists of a bed of rounded chalk flints, resting on a submarine argillaceous base. The flints and a few other pebbles, intermixed, are derived from the waste of Hordwell, and other cliffs to the westward, where tertiary strata, capped with a covering of broken chalk flints, from five to fifty feet thick, are rapidly undermined. In the great storm of November, 1824, this bank of shingle was moved bodily forwards for forty yards towards the northeast; and certain piles, which served to mark the boundaries of two manors, were found after the storm on the opposite side of the bar. At the same time many acres of pasture land were

* See Palmer on Shingle Beaches, Phil. Trans. 1834, p. 568.

† Groins are formed of piles and wooden planks, or of fagots staked down and are used either to break the force of the waves, or to retain the beach.

covered by shingle, on the farm of Westover, near Lymington. But the bar was soon restored in its old position by pebbles drifted from the west; and it appears from ancient maps that it has preserved the same general outline and position for centuries.*

Mr. Austen remarks that, as a general rule, it is only when high tides concur with a gale of wind, that the sea reaches the base of cliffs so as to undermine them and throw down earth and stone. But the waves are perpetually employed in abrading and fashioning the materials already strewed over the beach. Much of the gravel and shingle is always travelling up and down, between high-water mark and a slight depth below the level of the lowest tides, and occasionally the materials are swept away and carried into deeper water. Owing to these movements every portion of our southern coast may be seen at one time or other in the condition of bare rock. Yet other beds of sand and shingle soon collect, and, although composed of new materials, invariably exbibit on the same spots precisely similar characters.†

The cliffs between Hurst Shingle Bar and Christchurch are undermined continually, the sea having often encroached for a series of years at the rate of a yard annually. Within the memory of persons now living, it has been necessary thrice to remove the coast-road farther inland. The tradition, therefore, is probably true, that the church of Hordwell was once in the middle of that parish, although now (1830) very near the sea. The promontory of Christchurch Head gives way slowly. It is the only point between Lymington and Poole Harbor, in Dorsetshire, where any hard stony masses occur in the cliffs. Five layers of large ferruginous concretions, somewhat like the septaria of the London clay, have occasioned a resistance at this point, to which we may ascribe this headland. In the mean time, the waves have cut deeply into the soft sands and loam of Poole Bay; and, after severe frosts, great landslips take place, which by degrees become enlarged into narrow ravines, or chines, as they are called, with vertical sides. One of these chines, near Boscomb, has been deepened twenty feet within a few years. At the head of each there is a spring, the waters of which have been chiefly instrumental in producing these narrow excavations, which are sometimes from 100 to 150 feet deep.

Isle of Portland.—The peninsulas of Purbeck and Portland are continually wasting away. In the latter, the soft argillaceous substratum (Kimmeridge clay) hastens the dilapidation of the superincumbent mass of limestone.

In 1655 the cliffs adjoining the principal quarries in Portland gave way to the extent of one hundred yards, and fell into the sea; and in December, 1734, a slide to the extent of 150 yards occurred on the east side of the isle, by which several skeletons buried between slabs of stone, were discovered. But a much more memorable occurrence of

* Redman as cited, p. 315.

† Rob. A. C. Austen on the Valley of the English Channel, Quart. Journ. G. S. vol. vi. p. 72.

this nature, in 1792, occasioned probably by the undermining of the cliffs, is thus described in Hutchin's History of Dorsetshire:—"Early in the morning the road was observed to crack: this continued increasing, and before two o'clock the ground had sunk several feet, and was in one continued motion, but attended with no other noise than what was occasioned by the separation of the roots and brambles, and now and then a falling rock. At night it seemed to stop a little, but soon moved again; and, before morning, the ground from the top of the cliff to the water-side had sunk in some places fifty feet perpendicular. The extent of ground that moved was about *a mile and a quarter* from north to south, and 600 yards from east to west."

Formation of the Chesil Bank.—Portland is connected with the mainland by the Chesil Bank, a ridge of shingle about seventeen miles in length, and, in most places, nearly a quarter of a mile in breadth. The pebbles forming this immense barrier are chiefly siliceous, all loosely thrown together, and rising to the height of from twenty to thirty feet above the ordinary high-water mark; and at the southeastern end, which is nearest the Isle of Portland, where the pebbles are largest, forty feet. The fundamental rocks whereon the shingle rests are found at the depth of a few yards only below the level of the sea. The formation of that part of the bar which attaches Portland to the mainland may have been due to an original shoal or reef, or to the set of the tides in the narrow channel, by which the course of the pebbles, which are always coming from the west, has been arrested. It is a singular fact that, throughout the Chesil Bank, the pebbles increase gradually in size as we proceed southeastward, or as we go farther from the quarter which supplied them. Had the case been reversed, we should naturally have attributed the circumstance to the constant wearing down of the pebbles by friction, as they are rolled along a beach seventeen miles in length. But the true explanation of the phenomenon is doubtless this: the tidal current runs strongest from west to east, and its power is greater in the more open channel or farther from the land. In other words its force increases southwards, and as the direction of the bank is from northwest to southeast, the size of the masses coming from the westward and thrown ashore must always be largest where the motion of the water is most violent. Colonel Reid states that all calcareous stones rolled along from the west are soon ground into sand, and in this form they pass round Portland Island.*

The storm of 1824 burst over the Chesil Bank with great fury, and the village of Chesilton, built upon its southern extremity, was overwhelmed, with many of the inhabitants. The same storm carried away part of the Breakwater at Plymouth, and huge masses of rock, from two to five tons in weight, were lifted from the bottom of the weather side, and rolled fairly to the top of the pile. One block of limestone,

* See Palmer on Motion of Shingle Beaches, Phil. Trans. 1834, p. 568; and Col. Sir W. Reid, Papers of Royal Engineers, 1838, vol. ii. p. 128.

weighing seven tons, was washed round the western extremity of the Breakwater, and carried 150 feet.* The propelling power is derived in these cases from the breaking of the waves, which run fastest in shallow water, and for a short space far exceed the most rapid currents in swiftness. It was in the same month, and also during a spring-tide, that a great flood is mentioned on the coasts of England, in the year 1099. Florence of Worcester says, "On the third day of the nones of Nov. 1099, the sea came out upon the shore and buried towns and men very many, and oxen and sheep innumerable." We also read in the Saxon Chronicle, for the year 1099, "This year eke on St. Martin's mass day, the 11th of Novembre, sprung up so much of the sea flood, and so myckle harm did, as no man minded that it ever afore did, and there was the ylk day a new moon."

South of the Bill, or southern point of Portland, is a remarkable shoal in the channel at the depth of seven fathoms, called "the Shambles," consisting entirely of rolled and broken shells of Purpura lapillus, Mytilus edulis, and other species now living. This mass of light materials is always in motion, varying in height from day to day, and yet the shoal remains constant.

Dorsetshire.—Devonshire.—At Lyme Regis, in Dorsetshire, the "Church Cliffs," as they are called, consisting of lias about one hundred feet in height, gradually fell away at the rate of one yard a year, from 1800 to 1829.†

An extraordinary landslip occurred on the 24th of December, 1839, on the coast between Lyme Regis and Axmouth, which has been described by the Rev. W. D. Conybeare, to whose kindness I am indebted for the accompanying section, fig. 36. The tract of downs ranging

Fig. 36.

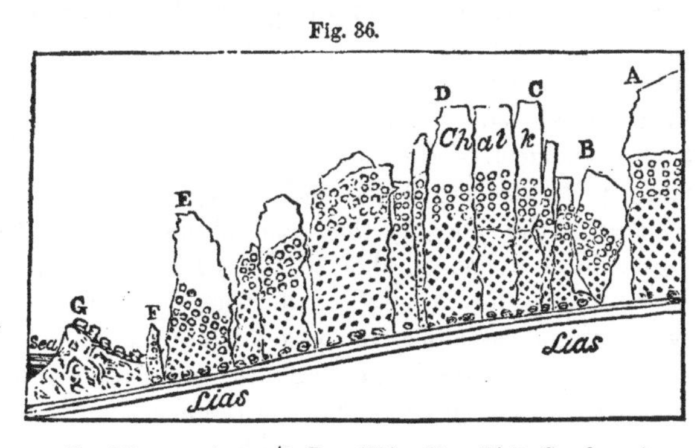

Landslip, near Axmouth, Dec. 1839. (Rev. W. D. Conybeare.)

A. Tract of Downs still remaining at their original level.
B. New ravine.
C, D. Sunk and fractured strip united to A, before the convulsion.
D, E. Bendon undercliff as before, but more fissured, and thrust forward about fifty feet, towards the sea.
F. Pyramidal crag, sunk from seventy to twenty feet in height.
G. New reef upheaved from the sea.

* De la Beche, Geolog. Manual, p. 82.

† According to the measurement of Carpenter of Lyme.

there along the coast is capped by chalk (*h*), which rests on sandstone, alternating with chert (*i*), beneath which is more than 100 feet of loose sand (*k*), with concretions at the bottom, and belonging like *i* to the green-sand formation; the whole of the above masses, *h, i, k*, reposing on retentive beds of clay (*l*), belonging to the lias, which shelves towards the sea. Numerous springs issuing from the loose sand (*k*), have gradually removed portions of it, and thus undermined the superstratum, so as to have caused subsidences at former times, and to have produced a line of undercliff between D and E. In 1839 an excessively wet season had saturated all the rocks with moisture, so as to increase the weight of the incumbent mass, from which the support had already been withdrawn by the action of springs. Thus the superstrata were precipitated into hollows prepared for them, and the adjacent masses of partially undermined rock, to which the movement was communicated, were made to slide down on a slippery basis of watery sand towards the sea. These causes gave rise to a convulsion, which began on the morning of the 24th of December, with a crashing noise; and, on the evening of the same day, fissures were seen opening in the ground, and the walls of tenements rending and sinking, until a deep chasm or ravine, B, was formed, extending nearly three-quarters of a mile in length, with a depth of from 100 to 150 feet, and a breadth exceeding 240 feet. At the bottom of this deep gulf lie fragments of the original surface thrown together in the wildest confusion. In consequence of lateral movements, the tract intervening between the new fissure and the sea, including the ancient undercliff, was fractured, and the whole line of sea-cliff carried bodily forwards for many yards. "A remarkable pyramidal crag, F, off Culverhole Point, which lately formed a distinguishing landmark, has sunk from a height of about seventy to twenty feet, and the main cliff, E, before more than fifty feet distant from this insulated crag, is now brought almost close to it. This motion of the sea-cliff has produced a farther effect, which may rank among the most striking phenomena of this catastrophe. The lateral pressure of the descending rocks has urged the neighboring strata, extending beneath the shingle of the shore, by their state of unnatural condensation, to burst upwards in a line parallel to the coast—thus an elevated ridge, G, more than a mile in length, and rising more than forty feet, covered by a confused assemblage of broken strata, and immense blocks of rock, invested with sea-weed and corallines, and scattered over with shells and star-fish, and other productions of the deep, forms an extended reef in front of the present range of cliffs."*

A full account of this remarkable landslip, with a plan, sections, and many fine illustrative drawings, was published by Messrs. Conybeare and Buckland,† from one of which the annexed cut has been reduced, fig. 37.

* Rev. W. D. Conybeare, letter dated Axminster, Dec. 31, 1839.
† London, J. Murray, 1840.

Fig. 87.

View of the Axmouth landslip from Great Bindon, looking westward to the Sidmouth hills, and estuary of the Exe. From an original drawing by Mrs. Buckland.

Cornwall.—Near Penzance, in Cornwall, there is a projecting tongue of land, called the "Green," formed of granitic sand, from which more than thirty acres of pasture land have been gradually swept away, in the course of the last two or three centuries.* It is also said that St. Michael's Mount, now an insular rock, was formerly situated in a wood, several miles from the sea; and its old Cornish name (Caraclowse in Cowse) signifies, according to Carew, the Hoar Rock in the wood.† Between the Mount and Newlyn there is seen under the sand, black vegetable mould, full of hazel-nuts, and the branches, leaves, roots, and trunks of forest-trees, all of indigenous species. This stratum has been traced seaward as far as the ebb permits, and many proofs of a submerged vegetable accumulation, with stumps of trees in the position in which they grew, have been traced, says Sir Henry De la Beche, round the shores of Devon, Cornwall, and Western Somerset. The facts not only indicate a change in the relative level of the sea and land, since the species of animals and plants were the same as those now living in this district; but, what is very remarkable, there seems evidence of the submergence having been effected, in part at least, since the country was inhabited by man.‡

A submarine forest occurring at the mouth of the Parret in Somersetshire, on the south side of the Bristol Channel, was described by Mr. L. Horner, in 1815, and its position attributed to subsidence. A bed of peat is there seen below the level of the sea, and the trunks of

* Boase, Trans. Royal Geol. Soc. of Cornwall, vol. ii. p. 129.
† Boase, ibid. vol. ii. p. 135.
‡ De la Beche's Report on the Geology of Devon, &c. chap. xiii.

large trees, such as the oak and yew, having their roots still diverging as they grew, and fixed in blue clay.*

Tradition of loss of land in Cornwall.—The oldest historians mention a tradition in Cornwall, of the submersion of the Lionnesse, a country said to have stretched from the Land's End to the Scilly Islands. The tract, if it existed, must have been thirty miles in length, and perhaps ten in breadth. The land now remaining on either side is from two hundred to three hundred feet high; the intervening sea about three hundred feet deep. Although there is no authentic evidence for this romantic tale, it probably originated in some former inroads of the Atlantic, accompanying, perhaps, a subsidence of land on this coast.†

West coast of England.—Having now brought together an ample body of proofs of the destructive operations of the waves, tides, and currents, on our eastern and southern shores, it will be unnecessary to enter into details of changes on the western coast, for they present merely a repetition of the same phenomena, and in general on an inferior scale. On the borders of the estuary of the Severn the flats of Somersetshire and Gloucestershire have received enormous accessions, while, on the other hand, the coast of Cheshire, between the rivers Mersey and Dee, has lost, since the year 1764, many hundred yards, and some affirm more than half a mile, by the advance of the sea upon the abrupt cliffs of red clay and marls. Within the period above mentioned several lighthouses have been successively abandoned.‡ There are traditions in Pembrokeshire§ and Cardiganshire‖ of far greater losses of territory than that which the Lionnesse tale of Cornwall pretends to commemorate. They are all important, as demonstrating that the earliest inhabitants were familiar with the phenomenon of incursions of the sea.

Loss of land on the coast of France.—The French coast, particularly that part of Brittany, where the tides rise to an extraordinary height, is the constant prey of the waves. In the ninth century many villages and woods are reported to have been carried away, the coast undergoing great change, whereby the hill of St. Michael was detached from the mainland. The parish of Bourgneuf, and several others in that neighborhood, were overflowed in the year 1500. In 1735, during a great storm, the ruins of Palnel were seen uncovered in the sea.¶

* Geol. Trans. 1st series, vol. iii. p. 383.
† Boase, vol. ii. p. 130.
‡ Stevenson, Jameson's Ed. New Phil. Journ. No. 8, p. 386.
§ Camden, who cites Gyraldus; also Ray, "On the Deluge," Phys. Theol. p. 228.
‖ Meyrick's Cardigan.
¶ Von Hoff, Geschichte, &c. vol. i. p. 49.

CHAPTER XX.

ACTION OF TIDES AND CURRENTS—*continued.*

Inroads of the sea at the mouths of the Rhine in Holland—Changes in the arms of the Rhine—Proofs of subsidence of land—Estuary of the Bies Bosch, formed in 1421—Zuyder Zee, in the 13th century—Islands destroyed—Delta of the Ems converted into a bay—Estuary of the Dollart formed—Encroachment of the sea on the coast of Sleswick—On shores of North America—Tidal wave, called the Bore—Influence of tides and currents on the mean level of seas—Action of currents in inland lakes and seas—Baltic—Cimbrian deluge—Straits of Gibraltar—No under-current there—Whether salt is precipitated in the Mediterranean—Waste of shores of Mediterranean.

Inroads of the sea at the mouths of the Rhine.—The line of British coast considered in the preceding chapter offered no example of the conflict of two great antagonist forces; the influx, on the one hand, of a river draining a large continent, and, on the other, the action of the waves, tides, and currents of the ocean. But when we pass over by the Straits of Dover to the Continent, and proceed northeastwards, we find an admirable illustration of such a contest, where the ocean and the Rhine are opposed to each other, each disputing the ground now occupied by Holland; the one striving to shape out an estuary, the other to form a delta. There was evidently a period when the river obtained the ascendancy, when the shape and perhaps the relative level of the coast and set of the tides were very different; but for the last two thousand years, during which man has witnessed and actively participated in the struggle, the result has been in favor of the ocean; the area of the whole territory having become more and more circumscribed; natural and artificial barriers having given away, one after another; and many hundred thousand human beings having perished in the waves.

Changes in the arms of the Rhine.—The Rhine, after flowing from the Grison Alps, copiously charged with sediment, first purifies itself in the Lake of Constance, where a large delta is formed; then swelled by the Aar and numerous other tributaries, it flows for more than six hundred miles towards the north; when, entering a low tract, it divides into two arms, about ten miles northeast of Cleves,—a point which must therefore be considered the head of its delta. (See *, map, fig. 8.) In speaking of the delta, I do not mean to assume that all that part of Holland which is comprised within the several arms of the Rhine can be called a delta in the strictest sense of the term; because some portion of the country thus circumscribed, as, for example, a part of Gelderland and Utrecht, consists of strata which may have been deposited in the sea before the Rhine existed. These older tracts may either have been raised like the Ullah Bund in Cutch, during the period when the

Fig. 38.

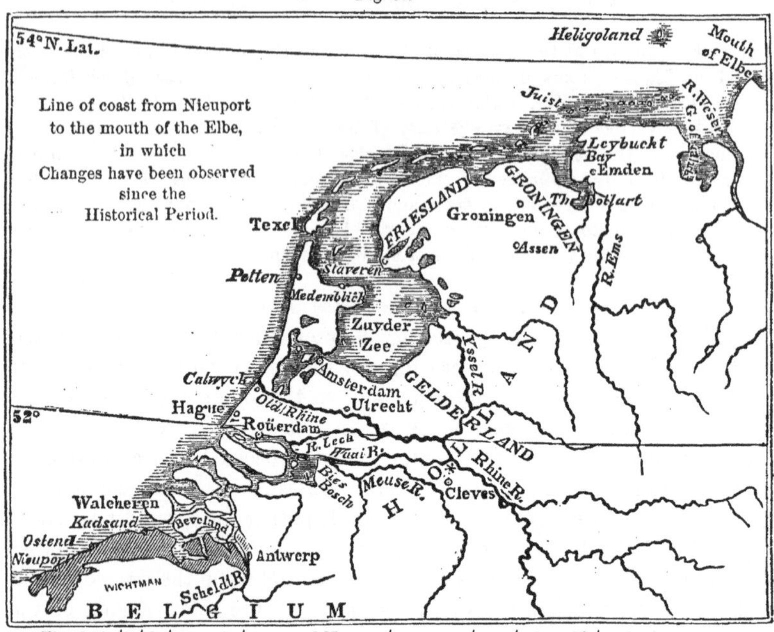

The dark tint between Antwerp and Nieuport, represents part of the Netherlands which was land in the time of the Romans, then overflowed by the sea before and during the 5th century, and afterwards reconverted into land.

sediment of the Rhine was converting a part of the sea into land, or they may have constituted islands previously.

When the river divides north of Cleves, the left arm takes the name of the Waal; and the right, retaining that of the Rhine, is connected, a little farther to the north, by an artificial canal with the river Yssel. The Rhine then flowing westward divides again southeast of Utrecht, and from this point it takes the name of the Leck, a name which was given to distinguish it from the northern arm called the old Rhine, which was sanded up until the year 1825, when a channel was cut for it, by which it now enters the sea at Catwyck. It is common, in all great deltas, that the principal channels of discharge should shift from time to time, but in Holland so many magnificent canals have been constructed, and have so diverted, from time to time, the course of the waters, that the geographical changes in this delta are endless, and their history, since the Roman era, forms a complicated topic of antiquarian research. The present head of the delta is about forty geographical miles from the nearest part of the gulf called the Zuyder Zee, and more than twice that distance from the general coast-line. The present head of the delta of the Nile is about 80 or 90 geographical miles from the sea; that of the Ganges, as before stated, 220; and that of the Mississippi about 180, reckoning from the point where the Atchafalaya branches off to the extremity of the new tongue of land in the Gulf of Mexico. But the comparative distance between the heads of deltas and the sea affords no positive data for estimating the relative magnitude of the alluvial tracts formed by their respective rivers, for the ramifications depend on many varying and temporary circumstances, and the area over which they extend does not hold any constant proportion to the volume of water in the river.

The Rhine therefore has at present three mouths. About two-thirds of its waters flow to the sea by the Waal, and the remainder is carried partly to the Zuyder Zee by the Yssel, and partly to the ocean by the Leck. As the whole coast to the south as far as Ostend, and on the north to the entrance of the Baltic, has, with few exceptions, from time immemorial, yielded to the force of the waves, it is evident that the common delta of the Rhine, Meuse, and Scheldt, for these three rivers may all be considered as discharging their waters into the same part of the sea, would, if its advance had not been checked, have become extremely prominent; and even if it had remained stationary, would long ere this have projected far beyond the rounded outline of the coast, like that strip of land already described at the mouth of the Mississippi. But we find, on the contrary, that the islands which skirt the coast have not only lessened in size, but in number also, while great bays have been formed in the interior by incursions of the sea.

In order to explain the incessant advance of the ocean on the shores and inland country of Holland, M. E. de Beaumont has suggested that there has in all probability been a general depression or sinking of the land below its former level over a wide area. Such a change of level would enable the sea to break through the ancient line of sand-banks and islands which protected the coast,—would lead to the enlargement of bays, the formation of new estuaries, and ultimately to the entire submergence of land. These views appear to be supported by the fact that several peat-mosses of fresh-water origin now occur under the level of the sea, especially on the site of the Zuyder Zee and Lake Elevo, presently to be mentioned. Several excavations also made for wells at Utrecht, Amsterdam, and Rotterdam have proved, that below the level of the ocean, the soil near the coast consists of alternations of sand with marine shells, and beds of peat and clay, which have been traced to the depth of fifty feet and upwards.*

I have said that the coast to the south as far as Ostend has given way. This statement may at first seem opposed to the fact, that the tract between Antwerp and Nieuport, shaded black in the annexed map (fig. 38), although now dry land, and supporting a large population, has, within the historical period, been covered with the sea. This region, however, consisted, in the time of the Romans, of woods, marshes, and peat-mosses, protected from the ocean by a chain of sandy dunes, which were afterwards broken through during storms, especially in the fifth century. The waters of the sea during these irruptions threw down upon the barren peat a horizontal bed of fertile

* E. de Beaumont, Géologie Pratique, vol. i. p. 316, and ibid. p. 260.

clay, which is in some places three yards thick, full of recent shells and works of art. The inhabitants, by the aid of embankments and the sand dunes of the coast, have succeeded, although not without frequent disasters, in defending the soil thus raised by the marine deposit.*

Inroads of the Sea in Holland.—If we pass to the northward of the territory just alluded to, and cross the Scheldt, we find that between the fourteenth and eighteenth centuries parts of the islands Walcheren and Beveland were swept away, and several populous districts of Kadzand, losses which far more than counterbalance the gain of land caused by the sanding up of some pre-existing creeks. In 1658 the Island Orisant was annihilated. One of the most memorable inroads of the sea occurred in 1421, when the tide, pouring into the mouth of the united Meuse and Waal, burst through a dam in the district between Dort and Gertrudenberg, and overflowed seventy-two villages, forming a large sheet of water called the Bies Bosch. (See map, fig. 38.) Thirty-five of the villages were irretrievably lost, and no vestige, even of their ruins, was afterwards seen. The rest were redeemed, and the site of the others, though still very generally represented on maps as an estuary, has in fact been gradually filled up by alluvial deposits, and had become in 1835, as I was informed by Professor Moll, an immense plain, yielding abundant crops of hay, though still uninhabited. To the north of the Meuse is a long line of shore covered with sand dunes, where great encroachments have taken place from time to time, in consequence chiefly of the prevalence of southeasterly winds, which blow down the sands towards the sea. The church of Scheveningen, not far from the Hague, was once in the middle of the village, and now stands on the shore, half the place having been overwhelmed by the waves in 1570. Catwyck, once far from the sea, is now upon the shore; two of its streets having been overflowed, and land torn away to the extent of 200 yards, in 1719. It is only by the aid of embankments that Petten, and several other places farther north, have been defended against the sea.

Formation of the Zuyder Zee and Straits of Staveren.—Still more important are the changes which have taken place on the coast opposite the right arm of the Rhine, or the Yssel, where the ocean has burst through a large isthmus, and entered the inland lake Flevo, which, in ancient times, was, according to Pomponius Mela, formed by the overflowing of the Rhine over certain lowlands. It appears that, in the time of Tacitus, there were several lakes on the present site of the Zuyder Zee, between Friesland and Holland. The successive inroads by which these and a great part of the adjoining territory, were transformed into a great gulf, began about the commencement, and were completed towards the close, of the thirteenth century. Alting gives the following relation of the occurrence, drawn from manuscript documents of contemporary inhabitants of the neighboring prov-

* Belpaire, Mém. de l'Acad. Roy. de Bruxelles, tom. x. 1837. Dumont, Bulletin of the same Soc. tom. v. p. 643.

inces. In the year 1205, the island now called Wieringen, to the south of the Texel, was still a part of the mainland, but during several high floods, of which the dates are given, ending in December, 1251, it was separated from the continent. By subsequent incursions the sea consumed great parts of the rich and populous isthmus, a low tract which stretched on the north of Lake Elevo, between Staveren in Friesland and Medemblick in Holland, till at length a breach was completed about the year 1282, and afterwards widened. Great destruction ot land took place when the sea first broke in, and many towns were swept away; but there was afterwards a reaction to a certain extent, large tracts, at first submerged, having been gradually redeemed. The new straits south of Staveren are more than half the width of those of Dover, but are very shallow, the greatest depth not exceeding two or three fathoms. The new bay is of a somewhat circular form, and between *thirty* and *forty* miles in diameter. How much of this space may formerly have been occupied by Lake Flevo is unknown. (See map, fig. 38.)

Destruction of islands.—A series of islands stretching from the Texel to the mouths of the Weser and Elbe are probably the last relics of a tract once continuous. They have greatly diminished in size, and have lost about a third of their number, since the time of Pliny; for that naturalist counted twenty-three islands between the Texel and Eider, whereas there are now only sixteen, including Heligoland and Neuwerk.* The island of Heligoland, at the mouth of the Elbe, consists of a rock of red marl of the Keuper formation (of the Germans), and is bounded by perpendicular red cliffs, above 200 feet high. Although, according to some accounts, it has been greatly reduced in size since the year 800, M. Wiebel assures us, that the ancient map by Meyer cannot be depended upon, and that the island, according to the description still extant by Adam of Bremen, was not much larger than now, in the time of Charlemagne. On comparing the map made in the year 1793 by the Danish engineer Wessel, the average encroachment of the sea on the cliffs, between that period and the year 1848 (or about half a century), did not amount to more than three feet.† On the other hand, some few islands have extended their bounds in one direction, or become connected with others, by the sanding-up of channels; but even these, like Juist, have generally given way as much on the north towards the sea as they have gained on the south, or land side.

The Dollart formed.—While the delta of the Rhine has suffered so materially from the movements of the ocean, it can hardly be supposed that minor rivers on the same coast should have been permitted to extend their deltas. It appears that in the time of the Romans there was an alluvial plain of great fertility, where the Ems entered the sea

* Von Hoff, vol. i. p. 364.
† Quart. Journ. Geol. Soc. vol. iv. p. 32; Memoirs.

by three arms. This low country stretched between Groningen and Friesland, and sent out a peninsula to the northeast towards Emden. A flood in 1277 first destroyed part of the peninsula. Other inundations followed at different periods throughout the fifteenth century. In 1507, a part only of Torum, a considerable town, remained standing; and in spite of the erection of dams, the remainder of that place, together with market-towns, villages, and monasteries, to the number of fifty, were finally overwhelmed. The new gulf, which was called the Dollart, although small in comparison to the Zuyder Zee, occupied no less than six square miles at first; but part of this space was, in the course of the two following centuries, again redeemed from the sea. The small bay of Leybucht, farther north, was formed in a similar manner in the thirteenth century; and the bay of Harlbucht in the middle of the sixteenth. Both of these have since been partially reconverted into dry land. Another new estuary, called the Gulf of Jahde, near the mouth of the Weser, scarcely inferior in size to the Dollart, has been gradually hollowed out since the year 1016, between which era and 1651 a space of about four square miles has been added to the sea. The rivulet which now enters this inlet is very small; but Arens conjectures that an arm of the Weser had once an outlet in that direction.

Coast of Sleswick.—Farther north we find so many records of waste on the western coast of Sleswick, as to lead us to anticipate that, at no distant period in the history of the physical geography of Europe, Jutland may become an island, and the ocean may obtain a more direct entrance into the Baltic. Indeed, the temporary insulation of the northern extremity of Jutland has been affected no less than four times within the records of history, the ocean having as often made a breach through the bar of sand, which usually excludes it from the Lym Fiord. This long frith is 120 miles in length including its windings, and communicates at its eastern end with the Baltic. The last irruption of salt water happened in 1824, and the fiord was still open in 1837, when some vessels of thirty tons' burden passed through.

The Marsh islands between the rivers Elbe and Eider are mere banks, like the lands formed of the "warp" in the Humber, protected by dikes. Some of them, after having been inhabited with security for more than ten centuries, have been suddenly overwhelmed. In this manner, in 1216, no less than ten thousand of the inhabitants of Eiderstede and Ditmarsch perished; and on the 11th of October, 1634, the islands and the whole coast, as far as Jutland, suffered by a dreadful deluge.

Destruction of Northstrand by the sea.—Northstrand, up to the year 1240, was, with the islands Sylt and Föhr, so nearly connected with the mainland as to appear a peninsula, and was called North Friesland, a highly cultivated and populous district. It measured from nine to eleven geographical miles from north to south, and six to eight from east to west. In the above-mentioned year it was torn asunder from the continent, and in part overwhelmed. The Isle of Northstrand, thus formed, was, towards the end of the sixteenth century, only four geographical miles in circumference, and was still celebrated for its cultivation and numerous population. After many losses, it still contained nine thousand inhabitants. At last, in the year 1634, on the evening of the 11th of October, a flood passed over the whole island, whereby 1300 houses, with many churches, were lost; fifty thousand head of cattle perished, and above six thousand men. Three small islets, one of them still called Northstrand, alone remained, which are now continually wasting.

The redundancy of river water in the Baltic, especially during the melting of ice and snow in spring, causes in general an outward current through the channel called the Cattegat. But after a continuance of northwesterly gales, especially during the height of the spring-tides, the Atlantic rises, and pouring a flood of water into the Baltic, commits dreadful devastations on the isles of the Danish Archipelago. This current even acts, though with diminished force, as far eastward as the vicinity of Dantzic.* Accounts written during the last ten centuries attest the wearing down of promontories on the Danish coast, the deepening of gulfs, the severing of peninsulas from the mainland, and the waste of islands, while in several cases marsh land, defended for centuries by dikes, has at last been overflowed, and thousands of the inhabitants whelmed in the waves. Thus the island Barsoe, on the coast of Sleswick, has lost, year after year, an acre at a time, and the island Alsen suffers in like manner.

Cimbrian deluge.—As we have already seen that during the flood before mentioned, 6000 men and 50,000 head of cattle perished on Northstrand on the western coast of Jutland, we are all well prepared to find that this peninsula, the Cimbrica Chersonesus of the ancients, has from a remote period been the theatre of like catastrophes. Accordingly, Strabo records a story, although he treats it as an incredible fiction, that, during a high tide, the ocean rose upon this coast so rapidly, that men on horseback were scarcely able to escape.† Florus, alluding to the same tradition, says, "Cimbri, Teutoni, atque Tigurini, ab extremis Galliæ profugi, cùm terras eorum inundasset Oceanus, novas sedes toto orbe quærebant."‡ This event, commonly called the "Cimbrian Deluge," is supposed to have happened about three centuries before the Christian era; but it is not improbable that the principal catastrophe was preceded and followed by many devastations like those experienced in modern times on the islands and shores of Jutland, and such calamities may well be conceived to have forced on the migration of some maritime tribes.

Inroads of the sea on the eastern shores of North America.—After so many authentic details respecting the destruction of the coast in parts of Europe best known, it will be unnecessary to multiply examples of

* See examples in Von Hoff, vol. i. p. 73, who cites Pisansky.

† Book vii. Cimbri.

‡ Lib. iii. cap 3.

analogous changes in more distant regions of the world. It must not, however, be imagined that our own seas form any exception to the general rule. Thus, for example, if we pass over to the eastern coast of North America, where the tides rise, in the Bay of Fundy, to a great elevation, we find many facts attesting the incessant demolition of land. Cliffs, often several hundred feet high, composed of sandstone, red marl, and other rocks, which border that bay and its numerous estuaries, are perpetually undermined. The ruins of these cliffs are gradually carried, in the form of mud, sand, and large boulders, into the Atlantic by powerful currents, aided at certain seasons by drift ice, which forms along the coast, and freezes round large stones.

At Cape May, on the north side of Delaware Bay, in the United States, the encroachment of the sea was shown by observations made consecutively for sixteen years, from 1804 to 1820, to average about nine feet a year;* and at Sullivan's Island, which lies on the north side of the entrance of the harbor of Charleston, in South Carolina, the sea carried away a quarter of a mile of land in three years, ending in 1786.†

Tidal wave called "the Bore."—Before concluding my remarks on the action of the tides, I must not omit to mention the wave called "the Bore," which is sometimes produced in a river where a large body of water is made to rise suddenly, in consequence of the contraction of the channel. This wave terminates abruptly on the inland side; because the quantity of water contained in it is so great, and its motion so rapid, that time is not allowed for the surface of the river to be immediately raised by means of transmitted pressure. A tide wave thus rendered abrupt has a close analogy, observes Mr. Whewell, to the waves which curl over and break on a shelving shore.‡

The Bore which enters the Severn, where the phenomenon is of almost daily occurrence, is sometimes nine feet high, and at spring-tides rushes up the estuary with extraordinary rapidity. The finest example which I have seen of this wave was at Nova Scotia,§ where the tide is said to rise in some places seventy feet perpendicular, and to be the highest in the world. In the large estuary of the Shubenacadie, which connects with another estuary called the Basin of Mines, itself an embranchment of the Bay of Fundy, a vast body of water comes rushing up, with a roaring noise, into a long narrow channel, and while it is ascending, has all the appearance of pouring down a slope as steep as that of the celebrated rapids of the St. Lawrence. In picturesque effect, however, it bears no comparison, for instead of the transparent green water and snow-white foam of the St. Lawrence, the whole current of the Shubenacadie is turbid and densely charged with red mud. The same phenomenon is frequently witnessed in the principal branches of the Ganges and in the Megna as before mentioned (p. 279). "In

* New Monthly Mag. vol. vi. p. 69.

† Von Hoff, vol. i. p. 96.

‡ Phil. Trans. 1833, p. 204.

§ See Lyell's Travels in North America, in 1842, vol. ii. p. 166. London, 1845.

the Hoogly," says Rennell, "the Bore commences at Hoogly Point, the place where the river first contracts itself, and is perceptible above Hoogly Town; and so quick is its motion, that it hardly employs four hours in travelling from one to the other, though the distance is nearly seventy miles. At Calcutta it sometimes occasions an instantaneous rise of five feet; and both here, and in every other part of its track, the boats, on its approach, immediately quit the shore, and make for safety to the middle of the river. In the channels, between the islands in the mouth of the Megna, the height of the Bore is said to exceed twelve feet; and is so terrific in its appearance, and dangerous in its consequences, that no boat will venture to pass at spring-tide."* These waves may sometimes cause inundations, undermine cliffs, and still more frequently sweep away trees and land animals from low shores, so that they may be carried down, and ultimately imbedded in fluviatile or submarine deposits.

CURRENTS IN INLAND LAKES AND SEAS.

In such large bodies of water as the North American lakes, the continuance of a strong wind in one direction often causes the elevation of the water, and its accumulation on the leeward side; and while the equilibrium is restoring itself, powerful currents are occasioned. In October, 1833, a strong current in Lake Erie, caused partly by the set of the waters towards the outlet of the lake, and partly by the prevailing wind, burst a passage through the extensive peninsula called Long Point, and soon excavated a channel more than nine feet deep and nine hundred feet wide. Its width and depth have since increased, and a new and costly pier has been erected; for it is hoped that this event will permanently improve the navigation of Lake Erie for steamboats.† On the opposite, or southern coast of this lake, in front of the town of Cleveland, the degradation of the cliffs had been so rapid for several years preceding a survey made in 1837, as to threaten many towns with demolition.‡ In the Black Sea, also, although free from tides, we learn from Pallas that there is a sufficiently strong current to undermine the cliffs in many parts, and particularly in the Crimea.

Straits of Gibraltar.—It is well known that a powerful current sets constantly from the Atlantic into the Mediterranean, and its influence extends along the whole southern borders of that sea, and even to the shores of Asia Minor. Captain Smyth found, during his survey, that the central current ran constantly at the rate of from three to six miles an hour eastward into the Mediterranean, the body of water being three miles and a half wide. But there are also two lateral currents—one on the European, and one on the African side; each of them about two miles and a half broad, and flowing at about the same rate as the central stream. These lateral currents ebb and flow with the tide, setting alter-

* Rennell, Phil. Trans. 1781. † MS. of Capt. Bayfield, R. N.
‡ Silliman's Journ. vol. xxxiv. p. 349.

nately into the Mediterranean and into the Atlantic. The excess of water constantly flowing in is very great, and there is only one cause to which this can be attributed, the loss of water in the Mediterranean by evaporation. That the level of this sea should be considerably depressed by this cause is quite conceivable, since we know that the winds blowing from the shores of Africa are hot and dry; and hygrometrical experiments recently made in Malta and other places, show that the mean quantity of moisture in the air investing the Mediterranean is equal only to one half of that in the atmosphere of England. The temperature also of the great inland sea is upon an average higher, by $3\frac{1}{2}°$ of Fahrenheit, than the eastern part of the Atlantic Ocean in the same latitude, which must greatly promote its evaporation. The Black Sea being situated in a higher latitude, and being the receptacle of rivers flowing from the north, is much colder, and its expenditure far less; accordingly it does not draw any supply from the Mediterranean, but, on the contrary, contributes to it by a current flowing outwards, for the most part of the year, through the Dardanelles. The discharge, however, at the Bosphorus is so small, when compared to the volume of water carried in by rivers, as to imply a great amount of evaporation in the Black Sea.

Whether salt be precipitated in the Mediterranean.—It is, however, objected, that evaporation carries away only fresh water, and that the current from the Atlantic is continually bringing in salt water: why, then, do not the component parts of the waters of the Mediterranean vary? or how can they remain so nearly the same as those of the ocean? Some have imagined that the excess of salt might be carried away by an under-current running in a contrary direction to the superior; and this hypothesis appeared to receive confirmation from a late discovery, that the water taken up about fifty miles within the Straits, from a depth of 670 fathoms, contained a quantity of salt *four times greater* than the water of the surface. Dr. Wollaston,* who analyzed this water obtained by Captain Smyth, truly inferred that an under-current of such denser water flowing outward, if of equal breadth and depth with the current near the surface, would carry out as much salt below as is brought in above, although it moved with less than one-fourth part of the velocity, and would thus prevent a perpetual increase of saltness in the Mediterranean beyond that existing in the Atlantic. It was also remarked by others, that the result would be the same, if the swiftness being equal, the inferior current had only one-fourth of the volume of the superior. At the same time there appeared reason to conclude that this great specific gravity was only acquired by water at immense depths; for two specimens of the water, taken within the Mediterranean, at the distance of some hundred miles from the Straits, and at depths of 400 and even 450 fathoms, were found by Dr. Wollaston not to exceed in density that of many ordinary samples of sea-water. Such being the case, we can now prove that the vast amount of salt brought into the

* Phil. Trans. 1829, part i. p. 29.

Mediterranean *does not* pass out again by the Straits; for it appears by Captain Smyth's soundings, which Dr. Wallaston had not seen, that between the capes of Trafalgar and Spartel, which are twenty-two miles apart, and where the Straits are shallowest, the deepest part, which is on the side of Cape Spartel, is only 220 fathoms. It is therefore evident, that if water sinks in certain parts of the Mediterranean, in consequence of the increase of its specific gravity, to greater depths than 220 fathoms, it can never flow out again into the Atlantic, since it must be stopped by the submarine barrier which crosses the shallowest part of the Straits of Gibraltar.

The idea of the existence of a counter-current, at a certain depth, first originated in the following circumstances:—M. De l'Aigle, commander of a privateer called the Phœnix of Marseilles, gave chase to a Dutch merchant-ship, near Ceuta Point, and coming up with her in the middle of the gut, between Tariffa and Tangier, gave her one broadside, which directly sunk her. A few days after, the sunken ship, with her cargo of brandy and oil, was cast ashore near Tangier, which is at least four leagues to the westward of the place where she went down, and to which she must have floated in a direction contrary to the course of the *central* current.* This fact, however, affords no evidence of an under-current, because the ship, when it approached the coast, would necessarily be within the influence of a lateral current, which running westward twice every twenty-four hours, might have brought back the vessel to Tangier.

What, then, becomes of the excess of salt?—for this is an inquiry of the highest geological interest. The Rhone, the Po, the Nile, and many hundred minor streams and springs, pour annually into the Mediterranean large quantities of carbonate of lime, together with iron, magnesia, silica, alumina, sulphur, and other mineral ingredients in a state of chemical solution. To explain why the influx of this matter does not alter the composition of this sea has never been regarded as a difficulty; for it is known that calcareous rocks are forming in the delta of the Rhone, in the Adriatic, on the coast of Asia Minor, and in other localities. Precipitation is acknowledged to be the means whereby the surplus mineral matter is disposed of, after the consumption of a certain portion in the secretions of testacea, zoophytes, and other marine animals. But before muriate of soda can, in like manner, be precipitated, the whole Mediterranean ought, according to the received principles of chemistry, to become as much saturated with salt as Lake Aral, the Dead Sea, or the brine-springs of Cheshire.

It is undoubtedly true, in regard to small bodies of water, that every particle must be fully saturated with muriate of soda before a single crystal of salt can be formed; such is probably the case in all natural salterns: such, for example, as those described by travellers as occurring on the western borders of the Black Sea, where extensive marshes

* Phil. Trans. 1724.

are said to be covered by thin films of salt after a rapid evaporation of sea-water. The salt *étangs* of the Rhone, where salt has sometimes been precipitated in considerable abundance, have been already mentioned. In regard to the depth of the Mediterranean, it appears that between Gibraltar and Ceuta, Captain Smyth sounded to the enormous depth of 950 fathoms, and found there a gravelly bottom, with fragments of broken shells. Saussure sounded to the depth of two thousand feet, within a few yards of the shore, at Nice; and M. Bérard has lately fathomed to the depth of more than six thousand feet in several places without reaching the bottom.*

The central abysses, therefore, of this sea are, in all likelihood, at least as deep as the Alps are high; and, as at the depth of seven hundred fathoms only, water has been found to contain a proportion of salt four times greater than at the surface, we may presume that the excess of salt may be much greater at the depth of two or three miles. After evaporation, the surface water becomes impregnated with a slight excess of salt, and its specific gravity being thus increased, it instantly falls to the bottom, while lighter water rises to the top, or flows in laterally, being always supplied by rivers and the current from the Atlantic. The heavier fluid, when it arrives at the bottom, cannot stop if it can gain access to any lower part of the bed of the sea, not previously occupied by water of the same density.

How far this accumulation of brine can extend before the inferior strata of water will part with any of their salt, and what difference in such a chemical process the immense pressure of the incumbent ocean, or the escape of heated vapors, thermal springs, or submarine volcanic eruptions, might occasion, are questions which cannot be answered in the present state of science.

The Straits of Gibraltar are said to become gradually wider by the wearing down of the cliffs on each side at many points; and the current sets along the coast of Africa, so as to cause considerable inroads in various parts, particularly near Carthage. Near the Canopic mouth of the Nile, at Aboukir, the coast was greatly devastated in the year 1784, when a small island was nearly consumed. By a series of similar operations, the old site of the cities of Nicropolis, Taposiris, Parva and Canopus, have become a sand-bank.†

* Bull. de la Soc. Géol. de France,—Résumé, p. 72, 1832.

† Clarke's Travels in Europe, Asia, and Africa, vol. iii. pp. 340 and 363, 4th edition.

CHAPTER XXI.

REPRODUCTIVE EFFECTS OF TIDES AND CURRENTS.

Estuaries, how formed—Silting up of estuaries does not compensate the loss of land on the borders of the ocean—Bed of the German Ocean—Composition and extent of its sand-banks—Strata deposited by currents in the English channel—On the shores of the Mediterranean—At the mouths of the Amazon, Orinoco, and Mississippi—Wide area over which strata may be formed by this cause.

From the facts enumerated in the last chapter, it appears that on the borders of the ocean, currents and tides co-operating with the waves of the sea are most powerful instruments in the destruction and transportation of rocks; and as numerous tributaries discharge their alluvial burden into the channel of one great river, so we find that many rivers deliver their earthy contents to one marine current, to be borne by it to a distance, and deposited in some deep receptacle of the ocean. The current, besides receiving this tribute of sedimentary matter from streams draining the land, acts also itself on the coast, as does a river on the cliffs which bound a valley. Yet the waste of cliffs by marine currents constitutes on the whole a very insignificant portion of the denudation annually effected by aqueous causes, as I shall point out in the sequel of this chapter (p. 339).

In inland seas, where the tides are insensible, or on those parts of the borders of the ocean where they are feeble, it is scarcely possible to prevent a harbor at a river's mouth from silting up; for a bar of sand or mud is formed at points where the velocity of the turbid river is checked by the sea, or where the river and a marine current neutralize each other's force. For the current, as we have seen, may, like the river, hold in suspension a large quantity of sediment, or, co-operating with the waves, may cause the progressive motion of a shingle beach in one direction. I have already alluded to the erection of piers and groins at certain places on our southern coast, to arrest the course of the shingle and sand (see p. 318). The immediate effect of these temporary obstacles is to cause a great accumulation of pebbles on one side of the barrier, after which the beach still moves on round the end of the pier at a greater distance from the land. This system, however, is often attended with a serious evil, for during storms the waves throw suddenly into the harbor the vast heap of pebbles which have collected for years behind the groin or pier, as happened during a great gale (Jan. 1839) at Dover.

The formation and keeping open of large estuaries are due to the *combined influence* of tidal currents and rivers; for when the tide rises, a large body of water suddenly enters the mouth of the river, where, becoming confined within narrower bounds, while its momentum is not destroyed, it is urged on, and, having to pass through a contracted channel, rises and runs with increased velocity, just as a stream when it reaches the arch of a bridge scarcely large enough to give passage to its waters, rushes with a steep fall through the arch. During the ascent of the tide, a body of fresh water, flowing down in an opposite direction from the higher country, is arrested in its course for several hours; and thus a large lake of fresh and brackish water is accumulated, which, when the sea ebbs, is let loose, as on the removal of an artificial sluice or dam. By the force of this retiring water, the alluvial sediment both of the river and of the sea is swept away, and transported to such a distance from the mouth of the estuary, that a small part only can return with the next tide.

It sometimes happens, that during a violent storm a large bar of sand is suddenly made to shift its position, so as to prevent the free influx of the tides, or efflux of river water. Thus about the year 1500 the sands at Bayonne were suddenly thrown across the mouth of the Adour. That river, flowing back upon itself, soon forced a passage to the northward along the sandy plain of Capbreton, till at last it reached the sea at Boucau, at the distance of *seven leagues* from the point where it had formerly entered. It was not till the year 1579 that the celebrated architect Louis de Foix undertook, at the desire of Henry III., to reopen the ancient channel, which he at last effected with great difficulty.*

In the estuary of the Thames at London, and in the Gironde, the tide rises only for five hours and ebbs seven, and in all estuaries the water requires a longer time to run down than up; so that the preponderating force is always in the direction which tends to keep open a deep and broad passage. But for reasons already explained, there is naturally a tendency in all estuaries to silt up partially, since eddies, and backwaters, and points where opposing streams meet, are very numerous, and constantly change their position.

Many writers have declared that the gain on our eastern coast, since the earliest periods of history, has more than counterbalanced the loss; but they have been at no pains to calculate the amount of loss, and have often forgotten that, while the new acquisitions are manifest, there are rarely any natural monuments to attest the former existence of the land that has been carried away. They have also taken into their account those tracts artificially recovered, which are often of great agricultural importance, and may remain secure, perhaps, for thousands of years, but which are only a few feet above the mean level of the sea, and are therefore exposed to be overflowed again by a small proportion of the force required to move cliffs of considerable height on our shores. If it were true that the area of land annually abandoned by the sea in estuaries were equal to that invaded by it, there would still be no compensation *in kind*.

The tidal current which flows out from the northwest, and bears against the eastern coast of England, transports, as we have seen, mate-

* Nouvelle Chronique de la Ville de Bayonne, pp. 113, 139: 1827.

rials of various kinds. Aided by the waves, it undermines and sweeps away the granite, gneiss, trap-rocks, and sandstone of Shetland, and removes the gravel and loam of the cliffs of Holderness, Norfolk, and Suffolk, which are between twenty and three hundred feet in height, and which waste at various rates of from one foot to six yards annually. It also bears away, in co-operation with the Thames and the tides, the strata of London clay on the coast of Essex and Sheppey. The sea at the same time consumes the chalk with its flints for many miles continuously on the shores of Kent and Sussex—commits annual ravages on the freshwater beds, capped by a thick covering of chalk-flint gravel, in Hampshire, and continually saps the foundations of the Portland limestone. It receives, besides, during the rainy months, large supplies of pebbles, sand, and mud, which numerous streams from the Grampians, Cheviots, and other chains, send down to the sea. To what regions, then, is all this matter consigned? It is not retained in mechanical suspension by the waters of the ocean, nor does it mix with them in a state of chemical solution—it is deposited *somewhere*, yet certainly not in the immediate neighborhood of our shores; for, in that case, there would soon be a cessation of the encroachment of the sea, and large tracts of low land, like Romney Marsh, would almost everywhere encircle our island.

As there is now a depth of water exceeding thirty feet, in some spots where towns like Dunwich flourished but a few centuries ago, it is clear that the current not only carries far away the materials of the wasted cliffs, but is capable also of excavating the bed of the sea to a certain moderate depth.

So great is the quantity of matter held in suspension by the tidal current on our shores, that the waters are in some places artificially introduced into certain lands below the level of the sea; and by repeating this operation, which is called "warping," for two or three years, considerable tracts have been raised, in the estuary of the Humber, to the height of about six feet. If a current, charged with such materials, meets with deep depressions in the bed of the ocean, it must often fill them up; just as a river, when it meets with a lake in its course, fills it gradually with sediment.

I have said (p. 337) that the action of the waves and currents on sea-cliffs, or their power to remove matter from above to below the sea-level, is insignificant in comparison with the power of rivers to perform the same task. As an illustration we may take the coast of Holderness described in the last chapter (p. 304). It is composed, as we have seen, of very destructible materials, is thirty-six miles long, and its average height may be taken at forty feet. As it has wasted away at the rate of two and a quarter yards annually, for a long period, it will be found on calculation that the quantity of matter thrown down into the sea every year, and removed by the current, amounts to 51,321,600 cubic feet. It has been shown that the united Ganges and Brahmapootra carry down to the Bay of Bengal 40,000,000,000 of cubic feet of solid matter every year, so that their transporting power is no less than 780 times greater than that of the sea on the coast above-mentioned; and in order to produce a result equal to that of the two Indian rivers, we must have a line of wasting coast, like that of Holderness, nearly 28,000 miles in length, or longer than the entire circumference of the globe by above 3000 miles. The reason of so great a difference in the results may be understood when we reflect that the operations of the ocean are limited to a single line of cliff surrounding a large area, whereas great rivers with their tributaries, and the mountain torrents which flow into them, act simultaneously on a length of bank almost indefinite.

Nevertheless we are by no means entitled to infer, that the denuding force of the great ocean is a geological cause of small efficacy, or inferior to that of rivers. Its chief influence is exerted at moderate depths below the surface, on all those areas which are slowly rising, or are attempting, as it were, to rise above the sea. From data hitherto obtained respecting subterranean movements, we can scarcely speculate on an average rate of upheaval of more than two or three feet in a century. An elevation to this amount is taking place in Scandinavia, and probably in many submarine areas as vast as those which we know to be sinking from the proofs derived from circular lagoon islands or coral atolls. (See chap. 50.) Suppose strata as destructible as those of the Wealden, or the lower and upper cretaceous formation, or the tertiary deposits of the British Isles to be thus slowly upheaved, how readily might they all be swept away by waves and currents in an open sea! How entirely might each stratum disappear as it was brought up successively and exposed to the breakers! Shoals of wide extent might be produced, but it is difficult to conceive how any continent could ever be formed under such circumstances. Were it not indeed for the hardness and toughness of the crystalline and volcanic rocks, which are often capable of resisting the action of the waves, few lands might ever emerge from the midst of an open sea.

Supposed filling up of the German Ocean.—The German Ocean is deepest on the Norwegian side, where the soundings give 190 fathoms; but the mean depth of the whole basin may be stated at no more than thirty-one fathoms.* The bed of this sea is traversed by several enormous banks, the greatest of which is the Dogger Bank, extending for upwards of 354 miles from north to south. The whole superficies of these shoals is equal to about one-third of the whole extent of England and Scotland. The average height of the banks measures, according to Mr. Stevenson, about seventy-eight feet; the upper portion of them consisting of fine and coarse siliceous sand, mixed with comminuted corals and shells.† It had been supposed that these vast submarine hills were made up bodily of loose materials supplied from the waste of the English, Dutch, and other coasts; but the survey of the North Sea,

* Stevenson on bed of German Ocean, Ed. Phil. Journ. No. v. p. 44: 1820.
† Stevenson, ibid. p. 47: 1820.

conducted by Captain Hewett, affords ground for suspecting this opinion to be erroneous. If such immense mounds of sand and mud had been accumulated under the influence of currents, the same causes ought nearly to have reduced to one level the entire bottom of the German Ocean; instead of which some long narrow ravines are found to intersect the banks. One of these varies from seventeen to forty-four fathoms in depth, and has very precipitous sides; in one part, called the "Inner Silver Pits," it is fifty-five fathoms deep. The shallowest parts of the Dogger Bank were found to be forty-two feet under water, except in one place, where the wreck of a ship had caused a shoal. Such uniformity in the minimum depth of water seems to imply that the currents, which vary in their velocity from a mile to two miles and a half per hour, have power to prevent the accumulation of drift matter in places of less depth.

Strata deposited by currents.—It appears extraordinary, that in some tracts of the sea, adjoining the coast of England, where we know that currents are not only sweeping along rocky masses, thrown down, from time to time, from the high cliffs, but also occasionally scooping out channels in the regular strata, there should exist fragile shells and tender zoophytes in abundance, which live uninjured by these violent movements. The ocean, however, is in this respect a counterpart of the land; and as, on the continents, rivers may undermine their banks, uproot trees, and roll along sand and gravel, while their waters are inhabited by testacea and fish, and their alluvial plains are adorned with rich vegetation and forests, so the sea may be traversed by rapid currents, and its bed may here and there suffer great local derangement, without any interruption of the general order and tranquillity. It has been ascertained by soundings in all parts of the world, that where new deposits are taking place in the sea, coarse sand and small pebbles commonly occur near the shore, while farther from land, and in deeper water, finer sand and broken shells are spread out over the bottom. Still farther out, the finest mud and ooze are alone met with. Mr. Austen observes that this rule holds good in every part of the English Channel examined by him. He also informs us, that where the tidal current runs rapidly in what are called "races," where surface undulations are perceived in the calmest weather, over deep banks, the discoloration of the water does not arise from the power of such a current to disturb the bottom at a depth of 40 or 80 fathoms, as some have supposed. In these cases, a column of water sometimes 500 feet in height, is moving onwards with the tide clear and transparent above, while the lower portion holds fine sediment in suspension (a fact ascertained by soundings), when suddenly it impinges upon a bank, and its height is reduced to 300 feet. It is thus made to boil up and flow off at the surface, a process which forces up the lower strata of water charged with fine particles of mud, which in their passage from the coast had gradually sunk to a depth of 300 feet or more.*

* Robt. A. C. Austen, Quart. Journ. Geol. Soc. vol. vi. p. 76.

One important character in the formations produced by currents is, the immense extent over which they may be the means of diffusing homogeneous mixtures, for these are often coextensive with a great line of coast; and, by comparison with their deposits, the deltas of rivers must shrink into significance. In the Mediterranean, the same current which is rapidly destroying many parts of the African coast, between the Straits of Gibraltar and the Nile, checks also the growth of the delta of the Nile, and drifts the sediment of that great river to the eastward. To this source may be attributed the rapid accretions of land on parts of the Syrian shores where rivers do not enter.

Among the greatest deposits now in progress, and of which the distribution is chiefly determined by currents, we may class those between the mouths of the Amazon and the southern coast of North America. Captain Sabine found that the equatorial current before mentioned (p. 292) was running with the rapidity of four miles an hour where it crosses the stream of the Amazon, which river preserves part of its original impulse, and has its waters not wholly mingled with those of the ocean at the distance of 300 miles from its mouth.* The sediment of the Amazon is thus constantly carried to the northwest as far as to the mouths of the Orinoco, and an immense tract of swamp is formed along the coast of Guiana, with a long range of muddy shoals bordering the marshes, and becoming converted into land.† The sediment of the Orinoco is partly detained, and settles near its mouth, causing the shores of Trinidad to extend rapidly, and is partly swept away into the Carribean Sea by the Guinea current. According to Humboldt, much sediment is carried again out of the Carribean Sea into the Gulf of Mexico.

It should not be overlooked that marine currents, even on coasts where there are no large rivers, may still be the agents of spreading not only sand and pebbles, but the finest mud, far and wide over the bottom of the ocean. *For several thousand miles* along the western coast of South America, comprising the larger parts of Peru and Chili, there is a perpetual rolling of shingle along the shore, part of which, as Mr. Darwin has shown, are incessantly reduced to the finest mud by the waves, and swept into the depths of the Pacific by the tides and currents. The same author however has remarked that, notwithstanding the great force of the waves on that shore, all rocks 60 feet under water are covered by sea-weed, showing that the bed of the sea is not denuded at that depth, the effects of the winds being comparatively superficial.

In regard to the distribution of sediment by currents it may be observed, that the rate of subsidence of the finer mud carried down by every great river into the ocean, or of that caused by the rolling of the waves upon a shore, must be extremely slow; for the more minute the separate particles of mud, the slower will they sink to the bottom, and

* Experiments to determine the Figure of the Earth, &c. p. 445.
† Lochead on Nat. Hist. of Guiana, Edin. Trans. vol. iv.

the sooner will they acquire what is called their terminal velocity. It is well known that a solid body, descending through a resisting medium, falls by the force of gravity, which is constant, but its motion is resisted by the medium more and more as its velocity increases, until the resistance becomes sufficient to counteract the farther increase of velocity. For example, a leaden ball, one inch diameter, falling through air of density as at the earth's surface, will never acquire greater velocity than 260 feet per second, and, in water, its greatest velocity will be 8 feet 6 inches per second. If the diameter of the ball were $\frac{1}{100}$ of an inch, the terminal velocities in air would be 26 feet, and in water ·86 of a foot per second.

Now, every chemist is familiar with the fact, that minute particles descend with extreme slowness through water, the extent of their surface being very great in proportion to their weight, and the resistance of the fluid depending on the amount of surface. A precipitate of sulphate of baryta, for example, will sometimes require more than five or six hours to subside one inch ;* while oxalate and phosphate of lime require nearly an hour to subside about an inch and a half and two inches respectively,† so exceedingly small are the particles of which these substances consist.

When we recollect that the depth of the ocean is supposed frequently to exceed three miles, and that currents run through different parts of that ocean at the rate of four miles an hour, and when at the same time we consider that some fine mud carried away from the mouths of rivers and from sea-beaches, where there is a heavy surf, as well as the impalpable powder showered down by volcanoes, may subside at the rate of only an inch per hour, we shall be prepared to find examples of the transportation of sediment over areas of indefinite extent.

It is not uncommon for the emery powder used in polishing glass to take more than an hour to sink one foot. Suppose mud composed of coarser particles to fall at the rate of two feet per hour, and these to be discharged into that part of the Gulf Stream which preserves a mean velocity of three miles an hour for a distance of two thousand miles ; in twenty-eight days these particles will be carried 2016 miles, and will have fallen only to a depth of 224 fathoms.

In this example, however, it is assumed that the current retains its superficial velocity at the depth of 224 fathoms, for which we have as yet no data, although we have seen that the motion of a current may continue at the depth of 100 fathoms. (See above, p. 28.) Experiments should be made to ascertain the rate of currents at considerable distances from the surface, and the time taken by the finest sediment to settle in sea-water of a given depth, and then the geologist may determine the area over which homogeneous mixtures may be simultaneously distributed in certain seas.

* On the authority of Mr. Faraday. † On the authority of Mr. R. Phillips.

CHAPTER XXII.

IGNEOUS CAUSES.

Changes of the inorganic world, *continued*—Igneous causes—Division of the subject—Distinct volcanic regions—Region of the Andes—System of volcanoes extending from the Aleutian isles to the Molucca and Sunda islands—Polynesian archipelago—Volcanic region extending from Central Asia to the Azores—Tradition of deluges on the shores of the Bosphorus, Hellespont, and Grecian isles—Periodical alternation of earthquakes in Syria and Southern Italy—Western limits of the European region—Earthquakes rarer and more feeble as we recede from the centres of volcanic action. Extinct volcanoes not to be included in lines of active vents.

We have hitherto considered the changes wrought, since the times of history and tradition, by the continued action of aqueous causes on the earth's surface; and we have next to examine those resulting from igneous agency. As the rivers and springs on the land, and the tides and currents in the sea, have, with some slight modifications, been fixed and constant to certain localities from the earliest periods of which we have any records, so the volcano and the earthquake have, with few exceptions, continued, during the same lapse of time, to disturb the same regions. But as there are signs, on almost every part of our continent, of great power having been exerted by running water on the surface of the land, and by waves, tides, and currents on cliffs bordering the sea, where, in modern times, no rivers have excavated, and no waves or tidal currents undermined—so we find signs of volcanic vents and violent subterranean movements in places where the action of fire or internal heat has long been dormant. We can explain why the intensity of the force of aqueous causes should be developed in succession in different districts. Currents, for example, tides, and the waves of the sea, cannot destroy coasts, shape out or silt up estuaries, break through isthmuses, and annihilate islands, form shoals in one place, and remove them from another, without the direction and position of their destroying and transporting power becoming transferred to new localities. Neither can the relative levels of the earth's crust, above and beneath the waters, vary from time to time, as they are admitted to have varied at former periods, and as it will be demonstrated that they still do, without the continents being, in the course of ages, modified, and even entirely altered, in their external configuration. Such events must clearly be accompanied by a complete change in the volume, velocity, and direction of the streams and land floods to which certain regions give passage. That we should find, therefore, cliffs where the sea once committed ravages, and from which it has now retired—estuaries where high tides once rose, but which are now dried up—valleys hollowed out by water, where no streams now flow, is no more than we should expect; these and similar phenomena are the necessary consequences of physical causes now in operation; and if there be no instability in the laws of nature, similar fluctuations must recur again and again in time to come.

But, however natural it may be that the force of running water in numerous valleys, and of tides and currents in many tracts of the sea, should now be *spent*, it is by no means so easy to explain why the violence of the earthquake and the fire of the volcano should also have become locally extinct at successive periods. We can look back to the time when the marine strata, whereon the great mass of Etna rests, had no existence; and that time is extremely modern in the earth's history. This alone affords ground for anticipating that the eruptions of Etna will one day cease.

> Nec quæ sulfureis ardet fornacibus Ætna
> Ignea semper erit, *neque enim fuit ignea semper*,
> (OVID, *Metam.* lib. 15–340,)

are the memorable words which are put into the mouth of Pythagoras by the Roman poet, and they are followed by speculations as to the cause of volcanic vents shifting their positions. Whatever doubts the philosopher expresses as to the nature of these causes, it is assumed, as incontrovertible, that the points of eruption will hereafter vary, *because they have formerly done so;* a principle of reasoning which, as I have endeavored to show in former chapters, has been too much set at naught by some of the earlier schools of geology, which refused to conclude that great revolutions in the earth's surface are now in progress, or that they will take place hereafter, *because* they have often been repeated in former ages.

Division of the subject.—Volcanic action may be defined to be "the influence exerted by the heated interior of the earth on its external covering." If we adopt this definition, without connecting it, as Humboldt has done, with the theory of secular refrigeration, or the cooling down of an original heated and fluid nucleus, we may then class under a general head all the subterranean phenomena, whether of volcanoes, or earthquakes, and those insensible movements of the land, by which, as will afterwards appear, large districts may be depressed or elevated, without convulsions. According to this view, I shall consider first, the volcano; secondly, the earthquake; thirdly, the rising or sinking of land in countries where there are no volcanoes or earthquakes; fourthly, the probable *causes* of the changes which result from subterranean agency.

It is a very general opinion that earthquakes and volcanoes have a common origin; for both are confined to certain regions, although the subterranean movements are least violent in the immediate proximity of volcanic vents, especially where the discharge of aeriform fluids and melted rock is made constantly from the same crater. But as there are particular regions, to which both the points of eruption and the movements of great earthquakes are confined, I shall begin by tracing out the geographical boundaries of some of these, that the reader may be aware of the magnificent scale on which the agency of subterranean fire is now simultaneously developed. Over the whole of the vast tracts alluded to, active volcanic vents are distributed at intervals, and most commonly arranged in a linear direction. Throughout the intermediate spaces there is often abundant evidence that the subterranean fire is at work continuously, for the ground is convulsed from time to time by earthquakes; gaseous vapors, especially carbonic acid gas, are disengaged plentifully from the soil; springs often issue at a very high temperature, and their waters are usually impregnated with the same mineral matters as are discharged by volcanoes during eruptions.

VOLCANIC REGIONS.

Region of the Andes.—Of these great regions, that of the Andes of South America is one of the best defined, extending from the southward of Chili to the northward of Quito, from about lat. 43° S. to about 2° N. of the equator. In this range, however, comprehending forty-five degrees of latitude, there is an alternation on a grand scale of districts of active with those of extinct volcanoes, or which, if not spent, have at least been dormant for the last three centuries. How long an interval of rest may entitle us to consider a volcano as entirely extinct is not easily determined; but we know that in Ischia there intervened between two consecutive eruptions a pause of seventeen centuries; and the discovery of America is an event of far too recent a date to allow us even to conjecture whether different portions of the Andes, nearly the whole of which are subject to earthquakes, may not experience alternately a cessation and renewal of eruptions.

The first line of active vents which have been seen in eruption in the Andes extends from lat. 43° 28′ S.; or, from Yantales, opposite the isle of Chiloe, to Coquimbo, in lat. 30° S.; to these thirteen degrees of latitude succeed more than eight degrees in which no recent volcanic eruptions have been observed. We then come to the volcanoes of Bolivia and Peru, reaching six degrees from S. to N., or from lat. 21° S. to lat. 15° S. Between the Peruvian volcanoes and those of Quito, another space intervenes of no less than fourteen degrees of latitude, said to be free from volcanic action so far as yet known. The volcanoes of Quito then succeed, beginning about 100 geographical miles south of the equator, and continuing for about 130 miles north of the line, when there occurs another undisturbed interval of more than six degrees of latitude, after which we arrive at the volcanoes of Guatemala or Central America, north of the Isthmus of Panama.*

Having thus traced out the line from south to north, I may first state, in regard to the numerous vents of Chili, that the volcanoes of Yantales

* See Von Buch's Description of Canary Islands (Paris, ed. 1836) for a valuable sketch of the principal volcanoes of the globe.

and Osorno were in eruption during the great earthquake of 1835, at the same moment that the land was shaken in Chiloe, and in some parts of the Chilian coast permanently upheaved; whilst at Juan Fernandez, at the distance of no less than 720 geographical miles from Yantales, an eruption took place beneath the sea. Some of the volcanoes of Chili are of great height, as that of Antuco, in lat. 37° 40′ S., the summit of which is at least 16,000 feet above the sea. From the flanks of this volcano, at a great height, immense currents of lava have issued, one of which flowed in the year 1828. This event is said to be an exception in the general rule; few volcanoes in the Andes, and none of those in Quito, having been seen in modern times to pour out lava, but having merely ejected vapor or scoriæ.

Both the basaltic (or augitic) lavas, and those of the felspathic class, occur in Chili and other parts of the Andes; but the volcanic rocks of the felspathic family are said by Von Buch to be generally not trachyte, but a rock which has been called andesite, or a mixture of augite and albite. The last-mentioned mineral contains soda instead of the potash found in common felspar.

The volcano of Rancagua, lat. 34° 15′ S., is said to be always throwing out ashes and vapors like Stromboli, a proof of the permanently heated state of certain parts of the interior of the earth below. A year rarely passes in Chili without some slight shocks of earthquakes, and in certain districts not a month. Those shocks which come from the side of the ocean are the most violent, and the same is said to be the case in Peru. The town of Copiapo was laid waste by this terrible scourge in the years 1773, 1796, and 1819, or in both cases after regular intervals of twenty-three years. There have, however, been other shocks in that country in the periods intervening between the dates above mentioned, although probably all less severe, at least on the exact site of Copiapo. The evidence against a regular recurrence of volcanic convulsions at stated periods is so strong as a general fact, that we must be on our guard against attaching too much importance to a few striking but probably accidental coincidences. Among these last might be adduced the case of Lima, violently shaken by an earthquake on the 17th of June, 1578, and again on the very same day, 1678; or the eruptions of Coseguina in the year 1709 and 1809, which are the only two recorded of that volcano previous to that of 1835.*

Of the permanent upheaval of land after earthquakes in Chili, I shall have occasion to speak in the next chapter, when it will also be seen that great shocks often coincide with eruptions, either submarine or from the cones of the Andes, showing the identity of the force which elevates continents with that which causes volcanic outbursts.†

The space between Chili and Peru, in which no volcanic action has been observed, is 160 nautical leagues from south to north. It is, however, as Von Buch observes, that part of the Andes which is least

* Darwin, Geol. Trans. 2d series, vol. v. p. 612. † Ibid. p. 606.

known, being thinly peopled, and in some parts entirely desert. The volcanoes of Peru rise from a lofty platform to vast heights above the level of the sea, from 17,000 to 20,000 feet. The lava which has issued from Viejo, lat. 16° 55′ S., accompanied by pumice, is composed of a mixture of crystals of albitic felspar, hornblende, and mica, a rock which has been considered as one of the varieties of andesite. Some tremendous earthquakes which have visited Peru in modern times will be mentioned in a subsequent chapter.

The volcanoes of Quito, occurring between the second degree of south and the third degree of north latitude, rise to vast elevations above the sea, many of them being between 14,000 and 18,000 feet high. The Indians of Lican have a tradition that the mountain called L'Altar, or Capac Urcu, which means "the chief," was once the highest of those near the equator, being higher than Chimborazo; but in the reign of Ouainia Abomatha, before the discovery of America, a prodigious eruption took place, which lasted eight years, and broke it down. The fragments of trachyte, says M. Boussingault, which once formed the conical summit of this celebrated mountain, are at this day spread over the plain.* Cotopaxi is the most lofty of all the South American volcanoes which have been in a state of activity in modern times, its height being 18,858 feet; and its eruptions have been more frequent and destructive than those of any other mountain. It is a perfect cone, usually covered with an enormous bed of snow, which has, however, been sometimes melted suddenly during an eruption; as in January, 1803, for example, when the snows were dissolved in one night.

Deluges are often caused in the Andes by the liquefaction of great masses of snow, and sometimes by the rending open, during earthquakes, of subterranean cavities filled with water. In these inundations fine volcanic sand, loose stones, and other materials which the water meets with in its descent, are swept away, and a vast quantity of mud, called "moya," is thus formed and carried down into the lower regions. Mud derived from this source descended, in 1797, from the sides of Tunguragua in Quito, and filled valleys a thousand feet wide to the depth of six hundred feet, damming up rivers and causing lakes. In these currents and lakes of moya, thousands of small fish are sometimes enveloped, which, according to Humboldt, have lived and multiplied in subterranean cavities. So great a quantity of these fish were ejected from the volcano of Imbaburu in 1691, that fevers, which prevailed at the period, were attributed to the effluvia arising from the putrid animal matter.

In Quito, many important revolutions in the physical features of the country are said to have resulted, within the memory of man, from the earthquakes by which it has been convulsed. M. Boussingault declares his belief, that if a full register had been kept of all the convulsions experienced here and in other populous districts of the Andes, it would

* Bull. de la Soc. Géol. tom. vi. p. 55.

be found that the trembling of the earth had been incessant. The frequency of the movement, he thinks, is not due to volcanic explosions, but to the continual falling in of masses of rock which have been fractured and upheaved in a solid form at a comparatively recent epoch; but a longer series of observations would be requisite to confirm this opinion. According to the same author, the height of several mountains of the Andes has diminished in modern times.*

The great crest or cordillera of the Andes is depressed at the Isthmus of Panama to a height of about 1000 feet, and at the lowest point of separation between the two seas near the Gulf of San Miguel, to 150 feet. What some geographers regard as a continuation of that chain in Central America lies to the east of a series of volcanoes, many of which are active in the provinces of Pasto, Popayan, and Guatemala. Coseguina, on the south side of the Gulf of Fonseca, was in eruption in January, 1835, and some of its ashes fell at Truxillo, on the shores of the Gulf of Mexico. What is still more remarkable, on the same day, at Kingston, in Jamaica, the same shower of ashes fell, having been carried by an upper counter-current against the regular east wind which was then blowing. Kingston is about 700 miles distant from Coseguina, and these ashes must have been more than four days in the air, having travelled 170 miles a day. Eight leagues to the southward of the crater, the ashes covered the ground to the depth of three yards and a half, destroying the woods and dwellings. Thousands of cattle perished, their bodies being in many instances one mass of scorched flesh. Deer and other wild animals sought the towns for protection; many birds and quadrupeds were found suffocated in the ashes, and the neighboring streams were strewed with dead fish.† Such facts throw light on geological monuments, for in the ashes thrown out at remote periods from the volcanoes of Auvergne, now extinct, we find the bones and skeletons of lost species of quadrupeds.

Mexico.—The great volcanic chain, after having thus pursued its course for several thousand miles from south to north, sends off a branch in a new direction in Mexico, in the parallel of the city of that name, and is prolonged in a great platform between the eighteenth and twenty-second degrees of north latitude. Five active volcanoes traverse Mexico from west to east—Tuxtla, Orizaba, Popocatepetl, Jorullo, and Colima. Jorullo, which is in the centre of the great platform, is no less than 120 miles from the nearest ocean—an important circumstance, as showing that the proximity of the sea is not a necessary condition, although certainly a very general characteristic of the position of active volcanoes. The extraordinary eruption of this mountain, in 1759, will be described in the sequel. If the line which connects these five vents be prolonged in a westerly direction, it cuts the volcanic group of islands called the Isles of Revillagigedo.

* Bull. de la Soc. Géol. de France, tom. vi. p. 56.
† Caldcleugh, Phil. Trans. 1836, p. 27.

To the north of Mexico there are said to be three, or according to some, five volcanoes in the peninsula of California; and a volcano is reported to have been in eruption in the N. W. coast of America, near the Colombia river, lat. 45° 37′ N.

West Indies.—To return to the Andes of Quito: Von Buch inclines to the belief that if we were better acquainted with the region to the east of the Madalena, and with New Granada and the Caraccas, we might find the volcanic chain of the Andes to be connected with that of the West Indian or Carribee Islands. The truth of this conjecture has almost been set at rest by the eruption, in 1848, of the volcano of Zamba, in New Grenada, at the mouth of the river Madalena.*

Of the West Indian islands there are two parallel series: the one to the west, which are all volcanic, and which rise to the height of several thousand feet; the others to the east, for the most part composed of calcareous rocks, and very low. In the former or volcanic series, are Granada, St. Vincent, St. Lucia, Martinique, Dominica, Guadaloupe, Montserrat, Nevis, and St. Eustace. In the calcareous chain are Tobago, Barbadoes, Mariegallante, Grandeterre, Desirade, Antigua, Barbuda, St. Bartholomew, and St. Martin. The most considerable eruptions in modern times have been those of St. Vincent. Great earthquakes have agitated St. Domingo, as will be seen in the twenty-ninth chapter.

I have before mentioned (p. 270) the violent earthquake which in 1812 convulsed the valley of the Mississippi at New Madrid, for the space of 300 miles in length, of which more will be said in the twenty-seventh chapter. This happened exactly at the same time as the great earthquake of Caraccas, so that it is possible that these two points are parts of one subterranean volcanic region. The island of Jamaica, with a tract of the contiguous sea, has often experienced tremendous shocks; and these are frequent along a line extending from Jamaica to St. Domingo and Porto Rico.

Thus it will be seen that, without taking account of the West Indian and Mexican branches, a linear train of volcanoes and tracts shaken by earthquakes may be traced from the island of Chiloe and opposite coast to Mexico, or even perhaps to the mouth of the Colombia river—a distance upon the whole as great as from the pole to the equator. In regard to the western limits of the region, they lie deep beneath the waves of the Pacific, and must continue unknown to us. On the east they are not prolonged, except where they include the West Indian Islands, to a great distance; for there seem to be no indications of volcanic disturbances in Buenos Ayres, Brazil, and the United States of North America.

Volcanic region from the Aleutian Isles to the Moluccas and Isles of Sunda.—On a scale which equals or surpasses that of the Andes, is another line of volcanic action, which commences, on the north, with the Aleutian Isles in Russian America, and extends, first in a westerly

* Comptes Rendus, 1849, vol. xxix. p. 531.

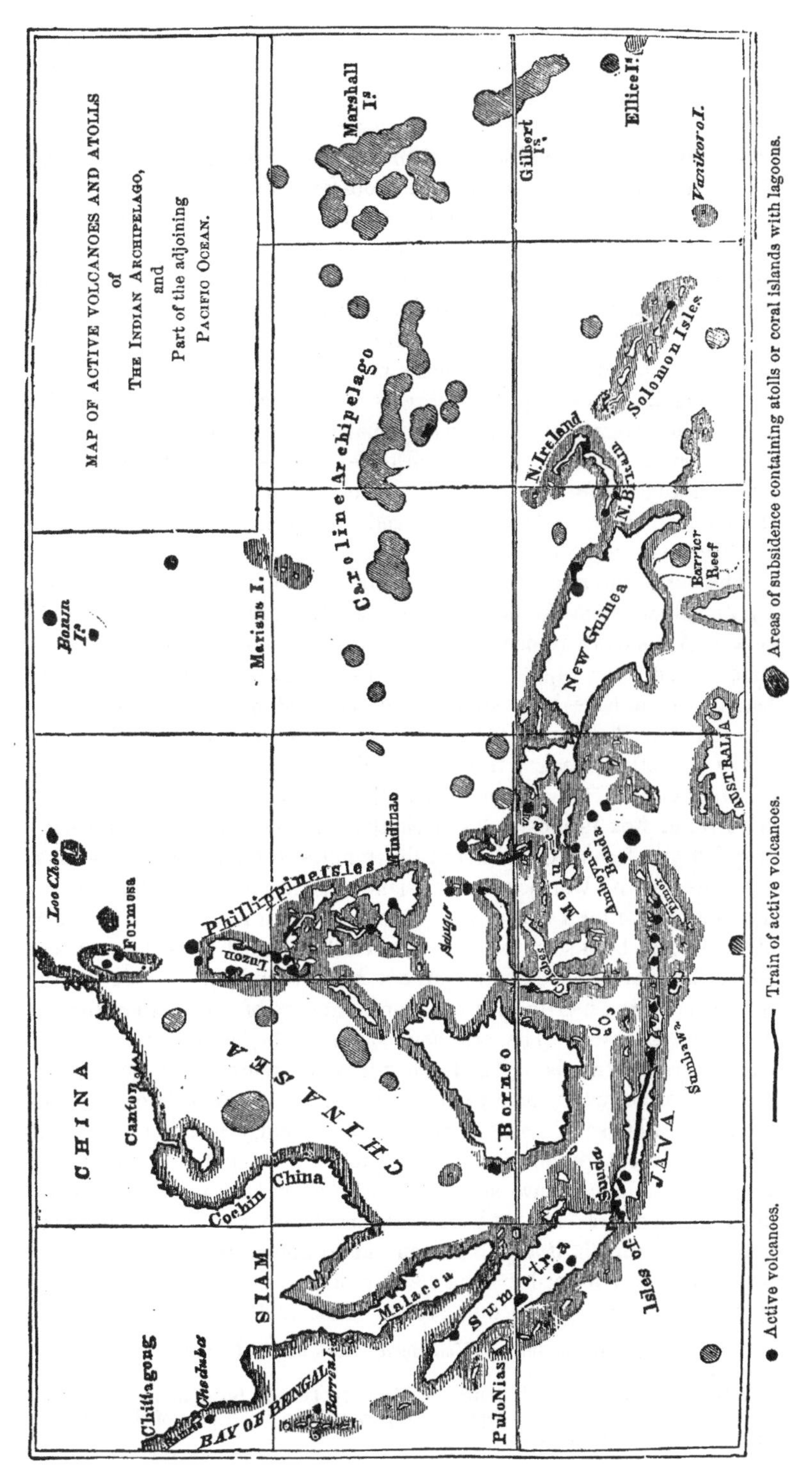
MAP OF ACTIVE VOLCANOES AND ATOLLS
of
THE INDIAN ARCHIPELAGO,
and
Part of the adjoining
PACIFIC OCEAN.
Marshall Is.
Gilbert Is.
Ellice Is.
Vanikoro I.
Caroline Archipelago
Solomon Isles
N. Ireland
N. Britain
Barrier Reef
New Guinea
Marians I.
Bonin Is.
Loo Choo
Formosa
Philippine Isles
Luzon
Mindinao
AUSTRALIA
Timor
Banda
Amboyna
Moluccas
Celebes
CHINA
Canton
CHINA SEA
China
Cochin
Borneo
Sumbawa
JAVA
Sunda
SIAM
Malacca
Sumatra
Isles of
Pulo Nias
Chittagong
Cheduba
BAY OF BENGAL
Barren I.
Active volcanoes.
Train of active volcanoes.
Areas of subsidence containing atolls or coral islands with lagoons.

direction for nearly 200 geographical miles, and then southwards, with few interruptions, throughout a space of between sixty and seventy degrees of latitude to the Moluccas, where it sends off a branch to the southeast while the principal train continues westerly through Sumbawa and Java to Sumatra, and then in a northwesterly direction to the Bay of Bengal.* This volcanic line, observes Von Buch, may be said to follow throughout its course the external border of the continent of Asia; while the branch which has been alluded to as striking southeast from the Moluccas, passes from New Guinea to New Zealand, conforming, though somewhat rudely, to the outline of Australia.†

The connection, however, of the New Guinea volcanoes with the line in Java (as laid down in Von Buch's map) is not clearly made out. By consulting Darwin's map of coral reefs and active volcanoes,‡ the reader will see that we might almost with equal propriety include the Mariana and Bonin volcanoes in a band with New Guinea. Or if we allow so much latitude in framing zones of volcanic action, we must also suppose the New Hebrides, Solomon Isles, and New Ireland to constitute one line (see map, fig. 39, p. 351).

The northern extremity of the volcanic region of Asia, as described by Von Buch, is on the borders of Cook's Inlet, northeast of the Peninsula of Alaska, where one volcano, in about the sixtieth degree of latitude, is said to be 14,000 feet high. In Alaska itself are cones of vast height, which have been seen in eruption, and which are covered for two-thirds of their height downwards with perpetual snow. The summit of the loftiest peak is truncated, and is said to have fallen in during an eruption in 1786. From Alaska the line is continued through the Aleutian or Fox Islands to Kamtschatka. In the Aleutian Archipelago eruptions are frequent, and about thirty miles to the north of Unalaska, near the Isle of Umnack, a new island was formed in 1796. It was first observed after a storm, at a point in the sea from which a column of smoke had been seen to rise. Flames then issued from the new islet which illuminated the country for ten miles round; a frightful earthquake shook the new-formed cone, and showers of stones were thrown as far as Umnack. The eruption continued for several months, and eight years afterwards, in 1804, when it was explored by some hunters, the soil was so hot in some places that they could not walk on it. According to Langsdorf and others, this new island, which is now several thousand feet high, and two or three miles in circumference, has been continually found to have increased in size when successively visited by different travellers; but we have no accurate means of determining how much of its growth, if any, has been due to upheaval, or how far it has been exclusively formed by the ejection of ashes and streams of lava. It seems, however, to be well attested that earthquakes of the most

* See map of volcanic lines in Von Buch's work on the Canaries.

† Von Buch, ibid. p. 409.

‡ Darwin, Structure and Distrib. of Coral reefs, &c., London, 1842. In the subjoined map, fig. 39, I have copied with permission a small part of the valuable map accompanying this work.

terrific description agitate and alter the bed of the sea and surface of the land throughout this tract.

The line is continued in the southern extremity of the Peninsula of Kamtschatka, where there are many active volcanoes, which, in some eruptions, have scattered ashes to immense distances. The largest and most active of these is Klutschew, lat. 56° 3′ N., which rises at once from the sea to the prodigious height of 15,000 feet. Within 700 feet of the summit, Erman saw, in 1829, a current of lava, emitting a vivid light, flow down the northwest side to the foot of the cone. A flow of lava from the summit of Mont Blanc to its base in the valley of Chamouni would afford but an inadequate idea of the declivity down which this current descended. Large quantities of ice and snow opposed for a time a barrier to the lava, until at length the fiery torrent overcame, by its heat and pressure, this obstacle, and poured down the mountain side with a frightful noise, which was heard for a distance of more than fifty miles.*

The Kurile chain of islands constitutes the prolongation of the Kamtschatka range, where a train of volcanic mountains, nine of which are known to have been in eruption, trends in a southerly direction. The line is then continued to the southwest in the great island of Jesso, and again in Nipon, the principal of the Japanese group. It then extends by Loo Choo and Formosa to the Philippine Islands, and thence by Sangir and the northeastern extremity of Celebes to the Moluccas (see map, fig. 39). Afterwards it passes westward through Sumbawa to Java.

There are said to be thirty-eight considerable volcanoes in Java, some of which are more than 10,000 feet high. They are remarkable for the quantity of sulphur and sulphureous vapors which they discharge. They rarely emit lava, but rivers of mud issue from them, like the moya of the Andes of Quito. The memorable eruption of Galongoon, in 1822, will be described in the twenty-fifth chapter. The crater of Taschem, at the eastern extremity of Java, contains a lake strongly impregnated with sulphuric acid, a quarter of a mile long, from which a river of acid water issues, which supports no living creature, nor can fish live in the sea near its confluence. There is an extinct crater near Batur, called Guevo Upas, or the Valley of Poison, about half a mile in circumference, which is justly an object of terror to the inhabitants of the country. Every living being which penetrates into this valley falls down dead, and the soil is covered with the carcasses of tigers, deer, birds, and even the bones of men; all killed by the abundant emanations of carbonic acid gas, by which the bottom of the valley is filled.

In another crater in this land of wonders, near the volcano of Talaga Bodas, we learn from M. Reinwardt, that the sulphureous exhalations have killed tigers, birds, and innumerable insects; and the soft parts of these animals, such as as the fibres, muscles, nails, hair, and skin, are

* Von Buch, Descrip. des Iles Canar. p. 450, who cites Erman and others.

very well preserved, while the bones are corroded, and entirely destroyed.

We learn from observations made in 1844, by Mr. Jukes, that a recent tertiary formation composed of limestone and resembling the coral rock of a fringing reef, clings to the flanks of all the volcanic islands from the east end of Timor to the west end of Java. These modern calcareous strata are often white and chalk-like, sometimes 1000 feet and upwards above the sea, regularly stratified in thick horizontal beds, and they show that there has been a general elevation of these islands at a comparatively modern period.*

The same linear arrangement which is observed in Java holds good in the volcanoes of Sumatra, some of which are of great height, as Berapi, which is more than 12,000 feet above the sea, and is continually smoking. Hot springs are abundant at its base. The volcanic line then inclines slightly to the northwest, and points to Barren Island, lat. 12° 15′ N., in the Bay of Bengal. This volcano was in eruption in 1792, and will be described in the twenty-sixth chapter. The volcanic train then extends, according to Dr. Macclelland, to the island of Narcondam, lat. 13° 22′ N., which is a cone seven or eight hundred feet high, rising from deep water, and said to present signs of lava currents descending from the crater to the base. Afterwards the train stretches in the same direction to the volcanic island of Ramree, about lat. 19° N., and the adjoining island of Cheduba, which is represented in old charts as a burning mountain. Thus we arrive at the Chittagong coast, which in 1762 was convulsed by a tremendous earthquake (see chap. 29).†

To enumerate all the volcanic regions of the Indian and Pacific oceans would lead me far beyond the proper limits of this treatise ; but it will appear in the last chapter of this volume, when coral reefs are treated of, that the islands of the Pacific consist alternately of linear groups of two classes, the one lofty, and containing active volcanoes, and marine strata above the sea-level, and which have been undergoing upheaval in modern times; the other very low, consisting of reefs of coral, usually with lagoons in their centres, and in which there is evidence of a gradual subsidence of the ground. The extent and direction of these parallel volcanic bands have been depicted with great care by Darwin in his map before cited (p. 351).

The most remarkable theatre of volcanic activity in the Northern Pacific—or, perhaps, in the whole world—occurs in the Sandwich Islands, which have been admirably treated of in a recent work by Mr. Dana.‡

Volcanic region from central Asia to the Azores.—Another great region of subterranean disturbance is that which has been imagined to extend through a large part of Central Asia to the Azores, that is to

* Paper read at meeting of Brit. Assoc. Southampton, Sept. 1846.
† Macclelland, Report on Coal and Min. Resources of India. Calcutta, 1838.
‡ Geology of the American Exploring Expedition. See also Lyell's Manual, "Sandwich I. Volcanoes"—Index.

say, from China and Tartary through Lake Aral and the Caspian to the Caucasus, and the countries bordering the Black Sea, then again through part of Asia Minor to Syria, and westward to the Grecian Islands, Greece, Naples, Sicily, the southern part of Spain, Portugal and the Azores. Respecting the eastern extremity of this line in China, we have little information, but many violent earthquakes are known to have occurred there. The volcano said to have been in eruption in the seventh century in Central Tartary is situated on the northern declivity of the Celestial Mountains, not far distant from the large lake called Issikoul; and Humboldt mentions other vents and solfataras in the same quarter, which are all worthy of notice, as being far more distant from the ocean (260 geographical miles) than any other known points of eruption.

We find on the western shores of the Caspian, in the country round Baku, a tract called the Field of Fire, which continually emits inflammable gas, while springs of naphtha and petroleum occur in the same vicinity, as also mud volcanoes. Syria and Palestine abound in volcanic appearances, and very extensive areas have been shaken, at different periods, with great destruction of cities and loss of lives. Continual mention is made in history of the ravages committed by earthquakes in Sidon, Tyre, Berytus, Laodicea, and Antioch, and in the Island of Cyprus. The country around the Dead Sea appears evidently, from the accounts of modern travellers, to be volcanic. A district near Smyrna, in Asia Minor, was termed by the Greeks Catacecaumene, or "the burnt up," where there is a large arid territory, without trees, and with a cindery soil.* This country was visited in 1841 by Mr. W. J. Hamilton, who found in the valley of the Hermus perfect cones of scoriæ, with lava-streams, like those of Auvergne, conforming to the existing river-channels, and with their surface undecomposed.†

Grecian Archipelago.—Proceeding westwards, we reach the Grecian Archipelago, where Santorin, afterwards to be described, is the grand centre of volcanic action.

It was Von Buch's opinion that the volcanoes of Greece were arranged in a line running N. N. W. and S. S. E., and that they afforded the only example in Europe of active volcanoes having a linear direction; but M. Virlet, on the contrary, announces as the result of his investigations, made during the French expedition to the Morea in 1829, that there is no one determinate line of direction for the volcanic phenomena in Greece, whether we follow the points of eruptions, or the earthquakes, or any other signs of igneous agency.‡

Macedonia, Thrace, and Epirus, have always been subject to earthquakes, and the Ionian Isles are continually convulsed.

Respecting Southern Italy, Sicily, and the Lipari Isles, it is unnecessary to enlarge here, as I shall have occasion again to allude to them.

* Strabo, ed Fal., p. 900.
† Researches in Asia Minor, vol. ii. p. 39.
‡ Virlet, Bulletin de la Soc. Géol. de France, tom. iii. p. 109.

I may mention, however, that a band of volcanic action has been traced by Dr. Daubeny across the Italian Peninsula, from Ischia to Mount Vultur, in Apulia, the commencement of the line being found in the hot springs of Ischia, after which it is prolonged through Vesuvius to the Lago d'Ansanto, where gases similar to those of Vesuvius are evolved. Its farther extension strikes Mount Vultur, a lofty cone composed of tuff and lava, from one side of which carbonic acid and sulphuretted hydrogen are emitted.*

Traditions of deluges.—The traditions which have come down to us from remote ages of great inundations said to have happened in Greece and on the confines of the Grecian settlements, had doubtless their origin in a series of local catastrophes, caused principally by earthquakes. The frequent migrations of the earlier inhabitants, and the total want of written annals long after the first settlement of each country, make it impossible for us at this distance of time to fix either the true localities or probable dates of these events. The first philosophical writers of Greece were, therefore, as much at a loss as ourselves to offer a reasonable conjecture on these points, or to decide how many catastrophes might sometimes have become confounded in one tale, or how much this tale may have been amplified, in after times, or obscured by mythological fiction. The floods of Ogyges and Deucalion are commonly said to have happened before the Trojan war; that of Ogyges more than seventeen, and that of Deucalion more than fifteen centuries before our era. As to the Ogygian flood, it is generally described as having laid waste Attica, and was referred by some writers to a great overflowing of rivers, to which cause Aristotle also attributed the deluge of Deucalion, which, he says, affected Hellas only, or the central part of Thessaly. Others imagined the same event to have been due to an earthquake, which drew down masses of rock, and stopped up the course of the Peneus in the narrow defile between mounts Ossa and Olympus.

As to the deluge of Samothrace, which is generally referred to a distinct date, it appears that the shores of that small island and the adjoining mainland of Asia were inundated by the sea. Diodorus Siculus says that the inhabitants had time to take refuge in the mountains, and save themselves by flight; he also relates, that long after the event the fishermen of the island drew up in their nets the capitals of columns, which were the remains of cities submerged by that terrible catastrophe."† These statements scarcely leave any doubt that there occurred, at the period alluded to, a subsidence of the coast, accompanied by earthquakes and inroads of the sea. It is not impossible that the story of the bursting of the Black Sea through the Thracian Bosphorus into the Grecian Archipelago, which accompanied, and, as some say, caused the Samothracian deluge, may have reference to a wave, or succession of waves, raised in the Euxine by the same convulsion.

* Daubeny on Mount Vultur, Ashmolean Memoirs. Oxford, 1835.

† Book v. ch. xlvi.—See letter of M. Virlet, Bulletin de la Soc. Géol. de France, tom. ii. p. 341.

We know that subterranean movements and volcanic eruptions are often attended not only by incursions of the sea, but also by violent rains, and the complete derangement of the river drainage of the inland country, and by the damming up of the outlets of lakes by landslips, or obstructions in the courses of subterranean rivers, such as abound in Thessaly and the Morea. We need not therefore be surprised at the variety of causes assigned for the traditional floods of Greece, by Herodotus, Aristotle, Diodorus, Strabo, and others. As to the area embraced, had all the Grecian deluges occurred simultaneously, instead of being spread over many centuries, and had they, instead of being extremely local, reached at once from the Euxine to the southwestern limit of the Peloponnese, and from Macedonia to Rhodes, the devastation would still have been more limited than that which visited Chili in 1835, when a volcanic eruption broke out in the Andes, opposite Chiloe, and another at Juan Fernandez, distant 720 geographical miles, at the same time that several lofty cones, in the Cordillera, 400 miles to the eastward of that island, threw out vapor and ignited matter. Throughout a great part of the space thus recently shaken in South America, cities were laid in ruins, or the land was permanently upheaved, or mountainous waves rolled inland from the Pacific.

Periodical alternation of Earthquakes in Syria and Southern Italy.— It has been remarked by Von Hoff, that from the commencement of the thirteenth to the latter half of the seventeenth century, there was an almost entire cessation of earthquakes in Syria and Judea; and, during this interval of quiescence, the Archipelago, together with part of the adjacent coast of Lesser Asia, as also Southern Italy and Sicily, suffered greatly from earthquakes; while volcanic eruptions were unusually frequent in the same regions. A more extended comparison, also, of the history of the subterranean convulsions of these tracts seems to confirm the opinion, that a violent crisis of commotion never visits both at the same time. It is impossible for us to declare, as yet, whether this phenomenon is constant in this and other regions, because we can rarely trace back a connected series of events farther than a few centuries; but it is well known that, where numerous vents are clustered together within a small area, as in many archipelagoes for instance, two of them are never in violent eruption at once. If the action of one becomes very great for a century or more, the others assume the appearance of spent volcanoes. It is, therefore, not improbable that separate provinces of the same great range of volcanic fires may hold a relation to one deep-seated focus, analogous to that which the apertures of a small group bear to some more superficial rent or cavity. Thus, for example, we may conjecture that, at a comparatively small distance from the surface, Ischia and Vesuvius mutually communicate with certain fissures, and that each affords relief alternately to elastic fluids and lava there generated. So we may suppose Southern Italy and Syria to be connected, at a much greater depth, with a lower part of the very same system of fissures; in which case any obstruction occurring in one duct may have the effect of causing almost all the vapor and melted matter to be forced up the other, and if they cannot get vent, they may be the cause of violent earthquakes. Some objections advanced against this doctrine that "volcanoes act as safety-valves," will be considered in the sequel.*

The northeastern portion of Africa, including Egypt, which lies six or seven degrees south of the volcanic line already traced, has been almost always exempt from earthquakes; but the northwestern portion, especially Fez and Morocco, which fall within the line, suffer greatly from time to time. The southern part of Spain also, and Portugal, have generally been exposed to the same scourge simultaneously with Northern Africa. The provinces of Malaga, Murcia, and Granada, and in Portugal the country round Lisbon, are recorded at several periods to have been devastated by great earthquakes. It will be seen, from Michell's account of the great Lisbon shock, in 1755, that the first movement proceeded from the bed of the ocean ten or fifteen leagues from the coast. So late as February 2, 1816, when Lisbon was vehemently shaken, two ships felt a shock in the ocean west from Lisbon; one of them at the distance of 120, and the other 262 French leagues from the coast†—a fact which is more interesting, because a line drawn through the Grecian Archipelago, the volcanic region of Southern Italy, Sicily, Southern Spain, and Portugal, will, if prolonged westward through the ocean, strike the volcanic group of the Azores, which may possibly therefore have a submarine connection with the European line.

In regard to the volcanic system of Southern Europe, it may be observed, that there is a central tract where the greatest earthquakes prevail, in which rocks are shattered, mountains rent, the surface elevated or depressed, and cities laid in ruins. On each side of this line of greatest commotion there are parallel bands of country where the shocks are less violent. At a still greater distance (as in Northern Italy, for example, extending to the foot of the Alps), there are spaces where the shocks are much rarer and more feeble, yet possibly of sufficient force to cause, by continued repetition, some appreciable alteration in the external form of the earth's crust. Beyond these limits, again, all countries are liable to slight tremors, at distant intervals of time, when some great crisis of subterranean movement agitates an adjoining volcanic region; but these may be considered as mere vibrations, propagated mechanically through the external covering of the globe, as sounds travel almost to indefinite distances through the air. Shocks of this kind have been felt in England, Scotland, Northern France, and Germany—particularly during the Lisbon earthquake. But these countries cannot, on this account, be supposed to constitute parts of the southern volcanic region, any more than the Shetland and Orkney islands can be considered as belonging to the Icelandic circle, because the sands ejected from Hecla have been wafted thither by the winds.

* See ch. 32, *Cause of Volcanic Eruptions.*

† Verneur, Journal des Voyages, tom. iv. p. 111. Von Hoff, vol. ii. p. 275.

Besides the continuous spaces of subterranean disturbance, of which we have merely sketched the outline, there are other disconnected volcanic groups, of which several will be mentioned hereafter.

Lines of active and extinct Volcanoes not to be confounded.—We must always be careful to distinguish between lines of extinct and active volcanoes, even where they appear to run in the same direction; for ancient and modern systems may interfere with each other. Already, indeed, we have proof that this is the case; so that it is not by geographical position, but by reference to the species of organic beings alone, whether aquatic or terrestrial, whose remains occur in beds interstratified with lavas, that we can clearly distinguish the relative age of volcanoes of which no eruptions are recorded. Had Southern Italy been known to civilized nations for as short a period as America, we should have had no record of eruptions in Ischia; yet we might have assured ourselves that the lavas of that isle had flowed since the Mediterranean was inhabited by the species of testacea now living in the Neapolitan seas. With this assurance, it would not have been rash to include the numerous vents of that island in the modern volcanic group of Campania.

On similar grounds we may infer, without much hesitation, that the eruptions of Etna, and the modern earthquakes of Calabria, are a continuation of that action which, at a somewhat earlier period, produced the submarine lavas of the Val di Noto in Sicily. But on the other hand, the lavas of the Euganean hills and the Vicentin, although not wholly beyond the range of earthquakes in Northern Italy, must not be confounded with any existing volcanic system; for when they flowed, the seas were inhabited by animals almost all of them distinct from those now known to live, whether in the Mediterranean or other parts of the globe.

CHAPTER XXIII.

VOLCANIC DISTRICT OF NAPLES.

History of the volcanic eruptions in the district round Naples—Early convulsions in the island of Ischia—Numerous cones thrown up there—Lake Avernus—The Solfatara—Renewal of the eruptions of Vesuvius, A. D. 79—Pliny's description of the phenomena—His silence respecting the destruction of Herculaneum and Pompeii—Subsequent history of Vesuvius—Lava discharged in Ischia in 1302—Pause in the eruptions of Vesuvius—Monte Nuovo thrown up—Uniformity of the volcanic operations of Vesuvius and Phlegræan Fields in ancient and modern times.

I SHALL next give a sketch of the history of some of the volcanic vents dispersed throughout the great regions before described, and consider the composition and arrangement of their lavas and ejected matter. The only volcanic region known to the ancients was that of the Mediterranean; and even of this they have transmitted to us very imperfect records relating to the eruptions of the three principal districts, namely, that round Naples, that of Sicily and its isles, and that of the Grecian Archipelago. By far the most connected series of records throughout a long period relates to the first of these provinces; and these cannot be too attentively considered, as much historical information is indispensable in order to enable us to obtain a clear view of the connection and alternate mode of action of the different vents in a single volcanic group.

Early convulsions in the Island of Ischia.—The Neapolitan volcanoes extend from Vesuvius, through the Phlegræan Fields, to Procida and Ischia, in a somewhat linear arrangement, ranging from the northeast to the southwest, as will be seen in the annexed map of the volcanic district of Naples (fig. 40). Within the space above limited, the volcanic force is sometimes developed in single eruptions from a considerable number of irregularly scattered points; but a great part of its action has been confined to one principal and habitual vent, Vesuvius or Somma. Before the Christian era, from the remotest periods of which we have any tradition, this principal vent was in a state of inactivity. But terrific convulsions then took place from time to time in Ischia (Pithecusa), and seem to have extended to the neighboring isle of Procida (Prochyta); for Strabo* mentions a story of Procida having been torn asunder from Ischia; and Pliny† derives its name from its having been poured forth by an eruption from Ischia.

The present circumference of Ischia along the water's edge is eighteen miles, its length from west to east about five, and its breadth from north

* Lib. v.

† Nat. Hist. lib. iii. c. 6.

to south three miles. Several Greek colonies which settled there before the Christian era were compelled to abandon it in consequence of the violence of the eruptions. First the Erythræans, and afterwards the Chalcidians, are mentioned as having been driven out by earthquakes and igneous exhalations. A colony was afterwards established by Hiero,

Fig 40.

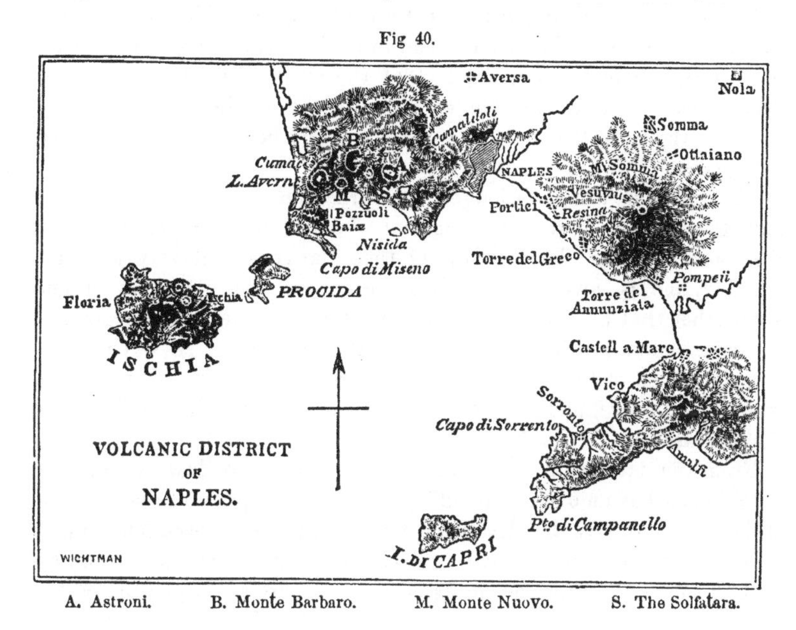

A. Astroni. B. Monte Barbaro. M. Monte Nuovo. S. The Solfatara.

king of Syracuse, about 380 years before the Christian era; but when they had built a fortress, they were compelled by an eruption to fly, and never again returned. Strabo tells us that Timæus recorded a tradition, that, a little before his time, Epomeus, the principal mountain in the centre of the island, vomited fire during great earthquakes; that the land between it and the coast had ejected much fiery matter, which flowed into the sea, and that the sea receded for the distance of three stadia, and then returning, overflowed the island. This eruption is supposed by some to have been that which formed the crater of Monte Corvo on one of the higher flanks of Epomeo, above Foria, the lava-current of which may still be traced, by aid of the scoriæ on its surface, from the crater to the sea.

To one of the subsequent eruptions in the lower parts of the isle, which caused the expulsion of the first Greek colony, Monte Rotaro has been attributed, and it bears every mark of recent origin. The cone, which I examined in 1828, is remarkably perfect, and has a crater on its summit precisely resembling that of Monte Nuovo near Naples; but the hill is larger, and resembles some of the more considerable cones of single eruption near Clermont in Auvergne, and, like some of them, it has given vent to a lava-stream at its base, instead of its summit. A small ravine swept out by a torrent exposes the structure of the cone, which is composed of innumerable inclined and slightly undulating layers of pumice, scoriæ, white lapilli, and enormous angular blocks of trachyte. These last have evidently been thrown out by violent explosions, like those which in 1822 launched from Vesuvius a mass of augitic lava, of many tons' weight, to the distance of three miles, which fell in the garden of Prince Ottajano. The cone of Rotaro is covered with the arbutus, and other beautiful evergreens. Such is the strength of the virgin soil, that the shrubs have become almost arborescent; and the growth of some of the smaller wild plants has been so vigorous, that botanists have scarcely been able to recognize the species.

The eruption which dislodged the Syracusan colony is supposed to have given rise to that mighty current which forms the promontory of Zaro and Caruso. The surface of these lavas is still very arid and bristling, and is covered with black scoriæ; so that it is not without great labor that human industry has redeemed some small spots, and converted them into vineyards. Upon the produce of these vineyards the population of the island is almost entirely supported. It amounted when I was there, in 1828, to about twenty-five thousand, and was on the increase.

From the date of the great eruption last alluded to, down to our own time, Ischia has enjoyed tranquillity, with the exception of one emission of lava hereafter to be described, which, although it occasioned much local damage, does not appear to have devastated the whole country, in the manner of more ancient explosions. There are, upon the whole,

Fig 41.

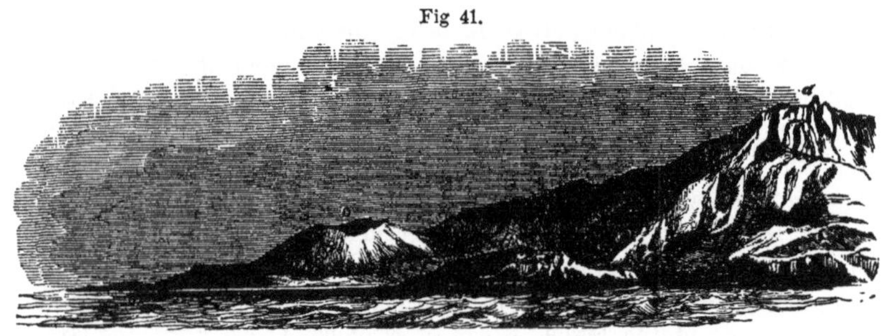

Part of Ischia seen from the West.

a. Monte Epomeo or San Niccola. *b.* Monte Vico.
c. Another of the minor cones with a crater.*

on different parts of Epomeo, or scattered through the lower tracts of Ischia, twelve considerable volcanic cones which have been thrown up since the island was raised above the surface of the deep; and many streams of lava may have flowed, like that of "Arso" in 1302, without cones having been produced; so that this island may, for ages before the period of the remotest traditions, have served as a safety-valve to the whole Terra di Lavoro, while the fires of Vesuvius were dormant.

* See Poulett Scrope, Geol. Trans. 2d series, vol. ii. pl. 34

Lake Avernus.—It seems also clear that Avernus, a circular lake near Puzzuoli, about half a mile in diameter, which is now a salubrious and cheerful spot, once exhaled mephitic vapors, such as are often emitted by craters after eruptions. There is no reason for discrediting the account of Lucretius, that birds could not fly over it without being stifled, although they may now frequent it uninjured.* There must have been a time when this crater was in action; and for many centuries afterwards it may have deserved the appellation of "atri jauna Ditis," emitting, perhaps, gases as destructive of animal life as those suffocating vapors given out by Lake Quilotoa, in Quito, in 1797, by which whole herds of cattle on its shores were killed,† or as those deleterious emanations which annihilated all the cattle in the island of Lancerote, one of the Canaries, in 1730.‡ Bory St. Vincent mentions, that in the same isle birds fell lifeless to the ground; and Sir William Hamilton informs us that he picked up dead birds on Vesuvius during an eruption.

Solfatara.—The Solfatara, near Puzzuoli, which may be considered as a nearly extinguished crater, appears, by the accounts of Strabo and others, to have been before the Christian era in very much the same state as at present, giving vent continually to aqueous vapor, together with sulphureous and muriatic acid gases, like those evolved by Vesuvius.

Ancient history of Vesuvius.—Such, then, were the points where the subterranean fires obtained vent, from the earliest period to which tradition reaches back, down to the first century of the Christian era; but we then arrive at a crisis in the volcanic action of this district—one of the most interesting events witnessed by man during the brief period throughout which he has observed the physical changes on the earth's surface. From the first colonization of Southern Italy by the Greeks, Vesuvius afforded no other indications of its volcanic character than such as the naturalist might infer, from the analogy of its structure to other volcanoes. These were recognized by Strabo, but Pliny did not include the mountain in his list of active vents. The ancient cone was of a very regular form, terminating not as at present in two peaks, but with a summit which presented, when seen from a distance, the even outline of an abruptly truncated cone. On the summit, as we learn from Plutarch, there was a crater with steep cliffs, and having its interior overgrown with wild vines, and with a sterile plain at the bottom. On the exterior, the flanks of the mountain were clothed with fertile fields richly cultivated, and at its base were the populous cities of Herculaneum and Pompeii. But the scene of repose was at length doomed to cease, and the volcanic fire was recalled to the main channel, which at some former unknown period had given passage to repeated streams of melted lava, sand, and scoriæ.

* De Rerum Nat. vi. 740.—Forbes, on Bay of Naples, Edin. Journ. of Sci. No iii. new series, p. 87. Jan. 1830.

† Humboldt, Voy. p. 317.

‡ Von Buch, Ueber einen vulcanischen Ausbruch auf der Insel Lanzerote.

Renewal of its eruptions.—The first symptom of the revival of the energies of this volcano was the occurrence of an earthquake in the year 63 after Christ, which did considerable injury to the cities in its vicinity. From that time to the year 79 slight shocks were frequent; and in the month of August of that year they became more numerous and violent, till they ended at length in an eruption. The elder Pliny, who commanded the Roman fleet, was then stationed at Misenum; and in his anxiety to obtain a near view of the phenomena, he lost his life, being suffocated by sulphureous vapors. His nephew, the younger Pliny, remained at Misenum, and has given us, in his Letters, a lively description of the awful scene. A dense column of vapor was first seen rising vertically from Vesuvius, and then spreading itself out laterally, so that its upper portion resembled the head, and its lower the trunk of the pine, which characterizes the Italian landscape. This black cloud was pierced occasionally by flashes of fire, as vivid as lightning, succeeded by darkness more profound than night. Ashes fell even upon the ships at Misenum, and caused a shoal in one part of the sea—the ground rocked, and the sea receded from the shores, so that many marine animals were seen on the dry sand. The appearances above described agree perfectly with those witnessed in more recent eruptions, especially those of Monte Nuovo, in 1538, and of Vesuvius in 1822.

The younger Pliny, although giving a circumstantial detail of so many physical facts, and describing the eruption and earthquake, and the shower of ashes which fell at Stabiæ, makes no allusion to the sudden overwhelming of two large and populous cities, Herculaneum and Pompeii. In explanation of this omission, it has been suggested that his chief object was simply to give Tacitus a full account of the particulars of his uncle's death. It is worthy, however, of remark, that had the buried cities never been discovered, the accounts transmitted to us of their tragical end might well have been discredited by the majority, so vague and general are the narratives, or so long subsequent to the event. Tacitus, the friend and contemporary of Pliny, when adverting in general terms to the convulsions, says merely that "cities were consumed or buried."*

Suetonius, although he alludes to the eruption incidentally, is silent as to the cities. They are mentioned by Martial, in an epigram, as immersed in cinders; but the first historian who alludes to them by name is Dion Cassius,† who flourished about a century and a half after Pliny. He appears to have derived his information from the traditions of the inhabitants, and to have recorded, without discrimination, all the facts and fables which he could collect. He tells us, "that during the eruption a multitude of men of superhuman stature, resembling giants, appeared, sometimes on the mountain, and sometimes in the environs—that stones and smoke were thrown out, the sun was hidden, and then the giants seemed to rise again, while the sounds of trumpets were heard, &c., &c.; and finally," he relates, "two entire cities, Herculaneum

* Haustæ aut obrutæ urbes.—Hist. lib. i.

† Hist. Rom. lib. lxvi.

and Pompeii, were buried under showers of ashes, while all the people were sitting in the theatre." That many of these circumstances were invented, would have been obvious, even without the aid of Pliny's letters; and the examination of Herculaneum and Pompeii enables us to prove, that none of the people were destroyed in the theatres, and indeed that there were very few of the inhabitants who did not escape from both cities. Yet some lives were lost, and there was ample foundation for the tale in its most essential particulars.

It does not appear that in the year 79 any lava flowed from Vesuvius; the ejected substances, perhaps, consisted entirely of lapilli, sand, and fragments of older lava, as when Monte Nuovo was thrown up in 1538. The first era at which we have authentic accounts of the flowing of a stream of lava, is the year 1036, which is the seventh eruption from the revival of the fires of the volcano. A few years afterwards, in 1049, another eruption is mentioned, and another in 1138 (or 1139), after which a great pause ensued of 168 years. During this long interval of repose, two minor vents opened at distant points. First, it is on tradition that an eruption took place from the Solfatara, in the year 1198, during the reign of Frederick II., Emperor of Germany; and although no circumstantial detail of the event has reached us from those dark ages, we may receive the fact without hesitation.* Nothing more, however, can be attributed to this eruption, as Mr. Scrope observes, than the discharge of a light and scoriform trachytic lava, of recent aspect, resting upon the strata of loose tuff which covers the principal mass of trachyte.†

Volcanic eruption in Ischia, 1302.—The other occurrence is well authenticated—the eruption, in the year 1302, of a lava-stream from a new vent on the southeast end of the Island of Ischia. During part of 1301, earthquakes had succeeded one another with fearful rapidity; and they terminated at last with the discharge of a lava-stream from a point named the Campo del Arso, not far from the town of Ischia. This lava ran quite down to the sea—a distance of about two miles; in color it varies from iron-gray to reddish black, and is remarkable for the glassy felspar which it contains. Its surface is almost as sterile, after a period of five centuries, as if it had cooled down yesterday. A few scantlings of wild thyme, and two or three other dwarfish plants, alone appear in the interstices of the scoriæ, while the Vesuvian lava of 1767 is already covered with a luxuriant vegetation. Pontanus, whose country-house was burnt and overwhelmed, describes the dreadful scene as having lasted two months.‡ Many houses were swallowed up, and a partial emigration of the inhabitants followed. This eruption produced no cone, but only a slight depression, hardly deserving the name of a cra-

* The earliest authority, says Mr. Forbes, given for this fact, appears to be Capaccio, quoted in the Terra Tremante of Bonito.—Edin. Journ. of Sci. &c. No. i. new series, p. 127. July, 1829.

† Geol. Trans. second series, vol. ii. p. 346.

‡ Lib. vi. de Bello Neap in Grævii Thesaur.

ter, where heaps of black and red scoriæ lie scattered around. Until this eruption, Ischia is generally believed to have enjoyed an interval of rest for about seventeen centuries; but Julius Obsequens,* who flourished A. D. 214, refers to some volcanic convulsions in the year 662 after the building of Rome (91 B. C.) As Pliny, who lived a century before Obsequens, does not enumerate this among other volcanic eruptions, the statement of the latter author is supposed to have been erroneous; but it would be more consistent, for reasons before stated, to disregard the silence of Pliny, and to conclude, that some kind of subterranean commotion, probably of no great violence, happened at the period alluded to.

History of Vesuvius after 1138.—To return to Vesuvius:—the next eruption occurred in 1306; between which era and 1631 there was only one other (in 1500), and that a slight one. It has been remarked, that throughout this period Etna was in a state of such unusual activity, as to lend countenance to the idea that the great Sicilian volcano may sometimes serve as a channel of discharge to elastic fluids and lava that would otherwise rise to the vents in Campania.

Formation of Monte Nuovo, 1538.—The great pause was also marked by a memorable event in the Phlegræan Fields—the sudden formation of a new mountain in 1538, of which we have received authentic accounts from contemporary writers.

The height of this mountain, called ever since Monte Nuovo, has been determined by the Italian mineralogist Pini, to be 440 English feet

Fig. 42.

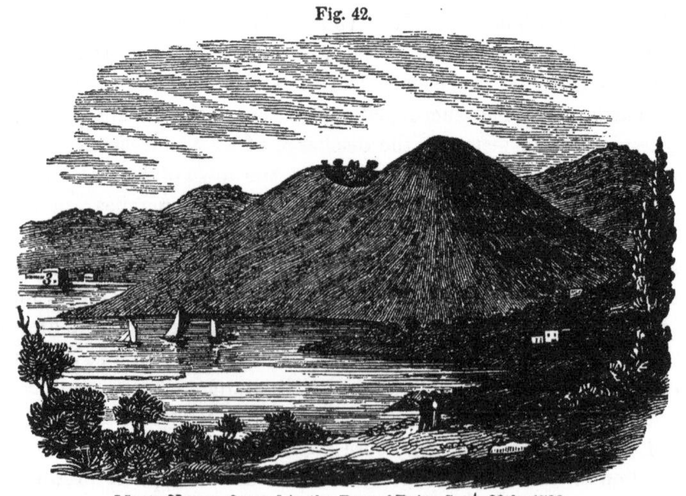

Monte Nuovo, formed in the Bay of Baiæ, Sept. 29th, 1538.
1. Cone of Monte Nuovo. 2. Brim of crater of ditto.
3. Thermal spring, called Baths of Nero, or Stufe di Tritoli.

above the level of the bay; its base is about eight thousand feet, or more than a mile and a half in circumference. According to Pini, the

* Prodig. libel. c. cxiv.

depth of the crater is 421 English feet from the summit of the hill, so that its bottom is only nineteen feet above the level of the sea. The cone is declared, by the best authorities, to stand partly on the site of the Lucrine Lake (4, fig. 43),* which was nothing more than the crater of a pre-existent volcano, and was almost entirely filled during the explosion of 1538. Nothing now remains but a shallow pool, separated from the sea by an elevated beach, raised artificially.

Fig 43.

The Phlegræan Fields.

1. Monte Nuovo.
2. Monte Barbaro.
3. Lake Avernus.
4. Lucrine Lake.
5. The Solfatara.
6. Puzzuoli.
7. Bay of Baiæ.

Sir William Hamilton has given us two original letters describing this eruption. The first, by Falconi, dated 1538, contains the following passages.† "It is now two years since there have been frequent earthquakes at Puzzuoli, Naples, and the neighboring parts. On the day and in the night before the eruption (of Monte Nuovo), above twenty shocks, great and small, were felt. The eruption began on the 29th of September, 1538. It was on a Sunday, about one o'clock in the night, when flames of fire were seen between the hot baths and Tripergola. In a short time the fire increased to such a degree, that it burst open the earth in this place, and threw up so great a quantity of ashes and pumice-stones, mixed with water, as covered the whole country. The next morning (after the formation of Monte Nuovo) the poor inhabitants of Puzzuoli quitted their habitations in terror, covered with the muddy and black shower which continued the whole day in that country—flying from death, but with death painted in their countenances. Some with their children in their arms, some with sacks full of their goods; others leading an ass, loaded with their frightened family, towards Na-

* This representation of the Phlegræan Fields is reduced from part of Plate xxxi. of Sir William Hamilton's great work "Campi Phlegræi." The faithfulness of his colored delineations of the scenery of that country cannot be too highly praised.

† Campi Phlegræi, p. 70.

ples; others carrying quantities of birds, of various sorts, that had fallen dead at the beginning of the eruption; others, again, with fish which they had found, and which were to be met with in plenty on the shore, the sea having left them dry for a considerable time. I accompanied Signor Moramaldo to behold the wonderful effects of the eruption. The sea had retired on the side of Baiæ, abandoning a considerable tract, and the shore appeared almost entirely dry, from the quantity of ashes and broken pumice-stones thrown up by the eruption. I saw two springs in the newly discovered ruins; one before the house that was the queen's, of hot and salt water," &c.

So far Falconi: the other account is by Pietro Giacomo di Toledo, which begins thus:—"It is now two years since this province of Campagna has been afflicted with earthquakes, the country about Puzzuoli much more so than any other parts; but the 27th and the 28th of the month of September last, the earthquakes did not cease day or night in the town of Puzzuoli: that plain which lies between Lake Avernus, the Monte Barbaro, and the sea, was *raised a little,* and many cracks were made in it, from some of which issued water; at the same time the sea, immediately joining the plain, *dried up about two hundred paces,* so that the fish were left on the sand a prey to the inhabitants of Puzzuoli. At last, on the 29th of the same month, about two o'clock in the night, the earth opened near the lake, and discovered a horrid mouth, from which were vomited furiously smoke, fire, stones, and mud, composed of ashes, making at the time of its opening a noise like the loudest thunder. The stones which followed were by the flames converted to pumice, and some of these were *larger than an ox.* The stones went about as high as a cross-bow can carry, and then fell down, sometimes on the edge, and sometimes into the mouth itself. The mud was of the color of ashes, and at first very liquid, then by degrees less so, and in such quantities, that in less than twelve hours, with the help of the above-mentioned stones, a mountain was raised of 1000 paces in height. Not only Puzzuoli and the neighboring country was full of this mud, but the city of Naples also; so that many of its palaces were defaced by it. Now this eruption lasted two nights and two days without intermission, though, it is true, not always with the same force; the third day the eruption ceased, and I went up with many people to the top of the new hill, and saw down into its mouth, which was a round cavity about a quarter of a mile in circumference, in the middle of which, the stones which had fallen were boiling up, just as a caldron of water boils on the fire. The fourth day it began to throw up again, and the seventh much more, but still with less violence than the first night. At this time many persons who were on the hill were knocked down by the stones and killed, or smothered with the smoke. In the day the smoke still continues, and you often see fire in the midst of it in the night-time."*

It will be seen that both these accounts, written immediately after the

* Campi Phlegræi, p. 77.

birth of Monte Nuovo, agree in stating that the sea retired; and one mentions that its bottom was upraised; but they attribute the origin of the new hill exclusively to the jets of mud, showers of scoriæ, and large fragments of rock, cast out from a central orifice, for several days and nights. Baron Von Buch, however, in his excellent work on the Canary Islands, and volcanic phenomena in general, has declared his opinion that the crone and crater of Monte Nuovo were formed, not in the manner above described, but by the upheaval of solid beds of white tuff, which were previously horizontal, but which were pushed up in 1538, so as to dip away in all directions from the centre, with the same inclination as the sloping surface of the cone itself. "It is an error," he says, "to imagine that this hill was formed by eruption, or by the ejection of pumice, scoriæ, and other incoherent matter; for the solid beds of upraised tuff are visible all round the crater, and it is merely the superficial covering of the cone which is made up of ejected scoriæ."*

In confirmation of this view, M. Dufrénoy has cited a passage from the works of Porzio, a celebrated physician of that period, to prove that in 1538 the ground where Monte Nuovo stands was pushed up in the form of a great bubble or blister, which on bursting, gave origin to the present deep crater. Porzio, says, "that after two days and nights of violent earthquakes, the sea retired for nearly 200 yards; so that the inhabitants could collect great numbers of fish on this part of the shore, and see some springs of fresh water which rose up there. At length, on the third day of the calends of October (September 29), they saw a large tract of ground intervening between the foot of Monte Barbaro, and part of the sea, near the Lake Avernus, rise, and suddenly assume the form of an incipient hill; and at two o'clock at night, this heap of earth, opening as it were its mouth, vomited, with a loud noise, flames, pumice-stones, and ashes."†

So late as the year 1846 a fourth manuscript (written immediately after the eruption) was discovered and published in Germany. It was written in 1538 by Francesco del Nero,‡ who mentions the drying up of the bed of the sea near Puzzuoli, which enabled the inhabitants of the town to carry off loads of fish. About eight o'clock in the morning of the 29th September, the earth sunk down about 14 feet in that place where the volcanic orifice now appears, and there issued forth a small stream of water, at first cold, and afterwards tepid. At noon, on the same day, the earth began to swell up in the same spot where it had sunk down 14 feet, so as to form a hill. About this time fire

* P. 347. Paris, 1836.

† "Magnus terræ tractus, qui inter radices montis, quem Barbarum incolæ appellant, et mare juxta Avernum jacet, sese erigere videbatur, et montis subitò nascentis figuram imitari. Eo ipso die horâ noctis II., iste terræ cumulus, aperto veluti ore, magno cum fremitu, magnos ignes evomuit; pumicesque, et lapides, cineresque."—Porzio, Opera Omnis, Medica, Phil., et Mathemat., in unum collecta, 1736, cited by Dufrénoy, Mém. pour servir à une Description Géologique de la France, tom. iv. p. 274.

‡ See Neues Jahr Buch for 1846, and a translation in the Quarterly Journ. of the Geol. Soc. for 1847, vol iii. p. 20, Memoirs.

issued forth, and gave rise to the great gulf, "with such a force, noise, and shining light, that I, who was standing in my garden, was seized with terror. Forty minutes afterwards, although unwell, I got upon a neighboring height, from which I saw all that took place, and by my troth it was a splendid fire, that threw up for a long time much earth and many stones, which fell back again all round the gulf, in a semicircle of from one to three bow-shots in diameter, and, filling up part of the sea, formed a hill nearly of the height of Monte Morello. Masses of earth and stones, as large as an ox, were shot up from the fiery gulf into the air, to a height which I estimate at a mile and a half. When they descended, some were dry, others in a soft muddy state." He concludes by alluding again to the sinking of the ground, and the elevation of it which followed, and says that to him it was inconceivable how such a mass of stones and ashes could have been poured forth from the gulf. He also refers to the account which Porzio was to draw up for the Viceroy.

On comparing these four accounts, recorded by eye-witnesses, there appears to be no real discrepancy between them. It seems clear that the ground first sunk down 14 feet on the site of the future volcano, and after having subsided it was again propelled upwards by the lava mingled with steam and gases, which were about to burst forth. Jets of red-hot lava, fragments of fractured rock, and occasionally mud composed of a mixture of pumice, tuff, and sea-water, were hurled into the air. Some of the blocks of stone were very large, leading us to infer that the ground which sank and rose again was much shattered and torn to pieces by the elastic vapors. The whole hill was not formed at once, but by an intermittent action extending over a week or more. It seems that the chasm opened between Tripergola and the baths in its suburbs, and that the ejected materials fell and buried that small town. A considerable part, however, of the hill was formed in less than twenty-four hours, and in the same manner as on a smaller scale the mud cones of the air volcanoes are produced, with a cavity in the middle. There is no difficulty in conceiving that the pumiceous mud, if so thrown out, may have set into a kind of stone on drying, just as some cements, composed of volcanic ashes, are known to consolidate with facility.

I am informed that Baron Von Buch discovered some marine shells of existing species, such as occur fossil in the tuff of the neighborhood, in beds exposed low down in the walls of the crater of Monte Nuovo. These may have been ejected in the mud mixed with sea-water which was cast out of the boiling gulf; or, as Signor Arcangelo Scacchi has suggested,* they may have been derived from the older tuff, which contains marine shells of recent species. The same observer remarks that Porzio's account upon the whole corroborates the doctrine of the cone having been formed by eruption, in proof of which he cites the

* Mem. Roy. Acad. Nap. 1849.

following passage:—"But what was truly astonishing, a hill of pumice-stones and ashes was heaped up round the gulf to the height of a mile in a single night."* Signor Scacchi also adds that the ancient temple of Apollo, now at the foot of Monte Nuovo, and the walls of which still retain their perfect perpendicularity, could not possibly have maintained that position had the cone of Monte Nuovo really been the result of upheaval.

Tripergola was much frequented as a watering-place, and contained a hospital for those who resorted there for the benefit of the thermal springs; and it appears that there were no fewer than three inns in the principal street. Had Porzio stated that any of these buildings, or the ruins of them, were seen by himself or others raised up above the plain, a short time before the first eruption, so as to stand on the summit or slope of a newly-raised hillock, we might have been compelled, by so circumstantial a narrative, to adopt M. Dufrénoy's interpretation.

But in the absence of such evidence, we must appeal to the crater itself, where we behold a section of the whole mountain, without being able to detect any original nucleus of upheaved rock distinct from the rest; on the contrary, the whole mass is similar throughout in composition, and the cone very symmetrical in form; nor are there any clefts, such as might be looked for, as the effect of the sudden upthrow of stony masses. M. C. Prevost has well remarked, that if beds of solid and non-elastic materials had yielded to a violent pressure directed from below upward, we should find not simply a deep empty cavity, but an irregular opening, where many rents converged; and these rents would be now seen breaking through the walls of the crater, widening as they approach the centre. (See Fig. 44, *a*, *b*.)† Not a single fissure of this kind is observable in the interior of Monte Nuovo, where the walls of the crater are continuous and entire; nor are there any dikes implying that rents had existed, which were afterwards filled with lava or other matter.

Fig. 44.

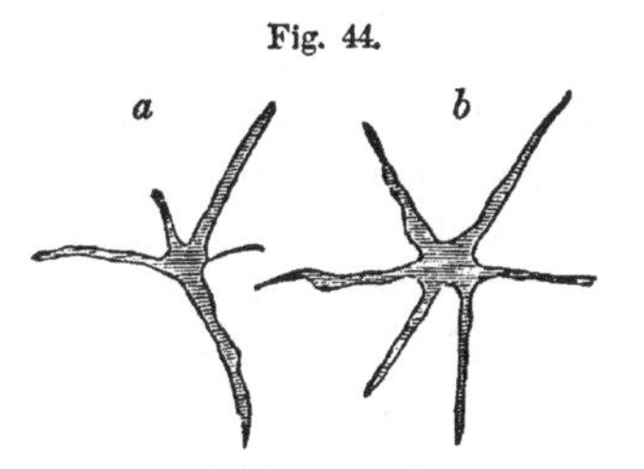

It has moreover been often urged by Von Buch, De Beaumont, and others, who ascribe the conical form of volcanoes chiefly to upheaval from below, that in such mountains there are a great number of deep rents and ravines, which diverge on all sides like the spokes of a wheel, from near the central axis to the circumference or base of the cone, as in the case of Palma, Cantal, and Teneriffe. Yet the entire absence of such divergent fissures or ravines, in such cases as Monte Nuovo, Somma, or Etna, is passed by unnoticed, and appears to have raised in their minds no objection to their favorite theory.

It is, indeed, admitted by M. Dufrénoy that there are some facts

* "Verum quod omnem superat admirationem, mons circum eam voraginem ex pummicibus et cincere plusquàm mille passuum altitudine unâ nocte congestus aspicitur."

† Mém. de la Soc. Géol. de France, tom. ii. p. 91.

which it is very difficult to reconcile with his own view of Porzio's record. Thus, for example, there are certain Roman monuments at the base of Monte Nuovo, and on the borders of Lake Avernus, such as the temples of Apollo (before mentioned) and Pluto, which do not seem to have suffered in the least degree by the supposed upheaval. "The walls which still exist have preserved their vertical position, and the vaults are in the same state as other monuments on the shores of the Bay of Baiæ. The long gallery which led to the Sibyl's Cave, on the other side of Lake Avernus, has in like manner escaped injury, the roof of the gallery remaining perfectly horizontal, the only change being that the soil of the chamber in which the Sibyl gave out her oracles is now covered by a few inches of water, which merely indicates a slight alteration in the level of Lake Avernus."* On the supposition, then, that pre-existing beds of pumiceous tuff were upraised in 1538, so as to form Monte Nuovo, it is acknowledged that the perfectly undisturbed state of the contiguous soil on which these ancient monuments stand, is very different from what might have been expected.

Mr. Darwin, in his "Volcanic Islands," has described several crateriform hills in the Galapagos Archipelago as composed of tuff which has evidently flowed like mud, and yet on consolidating has preserved an inclination of twenty and even thirty degrees. The tuff does not fold in continuous sheets round the hills as would have happened if they had been formed by the upheaval of horizontal layers. The author describes the composition of the tuff as very similar to that of Monte Nuovo, and the high angles at which the beds slope, both those which have flowed and those which have fallen in the form of ashes, entirely removes the difficulty supposed by M. Dufrénoy to exist in regard to the slope of Monte Nuovo, where it exceeds an angle of 18° to 20°.† Mr. Dana, also, in his account of the Sandwich Islands,‡ shows that in the "cinder cones" of that region, the strata have an original inclination of between 35° and 40°, while in the "tufa cones" formed near the sea, the beds slope at about an angle of 30°. The same naturalist also observed in the Samoan or Navigator Islands in Polynesia, that fragments of fresh coral had been thrown up together with volcanic matter to the height of 200 feet above the level of the sea in cones of tufa.§

I shall again revert to the doctrine of the origin of volcanic cones by upheaval, when speaking of Vesuvius, Etna, and Santorin, and shall now merely add, that, in 1538, the whole coast, from Monte Nuovo to beyond Puzzuoli, was upraised to the height of many feet above the bed of the Mediterranean, and has since retained the greater part of the elevation then acquired. The proofs of these remarkable changes of level will be considered at length when the phenomena of the temple of Serapis are described.‖

* Dufrénoy, Mem. pour servir, &c. p. 277.
† Darwin's Volcanic Islands, 106, note.
‡ Geology of the American Exploring Expedition, in 1838–1842, p. 354.
§ Ibid. p. 328.
‖ See chap. 29.

Volcanoes of the Phlegræan Fields.—Immediately adjoining Monte Nuovo is the larger volcanic cone of Monte Barbaro (2, fig. 43, p. 367), the "Gaurus inanis" of Juvenal—an appellation given to it probably from its deep circular crater, which is about a mile in diameter. Large as is this cone, it was probably produced by a single eruption; and it does not, perhaps, exceed in magnitude some of the largest of those formed in Ischia, within the historical era. It is composed chiefly of indurated tufa like Monte Nuovo, stratified conformably to its conical surface. This hill was once very celebrated for its wines, and is still covered with vineyards; but when the vine is not in leaf it has a sterile appearance, and, late in the year, when seen from the beautiful Bay of Baiæ, it often contrasts so strongly in verdure with Monte Nuovo, which is always clothed with arbutus, myrtle, and other wild evergreens, that a stranger might well imagine the cone of older date to be that thrown up in the sixteenth century.*

There is nothing, indeed, so calculated to instruct the geologist as the striking manner in which the recent volcanic hills of Ischia, and that now under consideration, blend with the surrounding landscape. Nothing seems wanting or redundant; every part of the picture is in such perfect harmony with the rest, that the whole has the appearance of having been called into existence by a single effort of creative power. Yet what other result could we have anticipated if nature has ever been governed by the same laws? Each new mountain thrown up—each new tract of land raised or depressed by earthquakes—should be in perfect accordance with those previously formed, if the entire configuration of the surface has been due to a long series of similar disturbances. Were it true that the greater part of the dry land originated simultaneously in its present state, at some era of paroxysmal convulsion, and that additions were afterwards made slowly and successively during a period of comparative repose; then, indeed, there might be reason to expect a strong line of demarcation between the signs of the ancient and modern changes. But the very continuity of the plan, and the perfect identity of the causes, are to many a source of deception; since by producing a unity of effect, they lead them to exaggerate the energy of the agents which operated in the earlier ages. In the absence of all historical information, they are as unable to separate the dates of the origin of different portions of our continents, as the stranger is to determine, by their physical features alone, the distinct ages of Monte Nuovo, Monte Barbara, Astroni, and the Solfatara.

The vast scale and violence of the volcanic operations in Campania, in the olden time, has been a theme of declamation, and has been contrasted with the comparative state of quiescence of this delightful region in the modern era. Instead of inferring, from analogy, that the ancient Vesuvius was always at rest when the craters of the Phlegræan Fields

* Hamilton (writing in 1770) says, "the new mountain produces as yet but a very slender vegetation."—Campi Phlegræi, p. 69. This remark was no longer applicable when I saw it, in 1828.

were burning—that each cone rose in succession,—and that many years, and often centuries, of repose intervened between different eruptions,—geologists seem to have generally conjectured that the whole group sprung up from the ground at once, like the soldiers of Cadmus when he sowed the dragon's teeth. As well might they endeavor to persuade us that on these Phlegræan Fields, as the poets feigned, the giants warred with Jove, ere yet the puny race of mortals were in being.

Modern eruptions of Vesuvius.—For nearly a century after the birth of Monte Nuovo, Vesuvius continued in a state of tranquillity. There had been no violent eruption for 492 years; and it appears that the crater was then exactly in the condition of the present extinct volcano of Astroni, near Naples. Bracini, who visited Vesuvius not long before the eruption of 1631, gives the following interesting description of the interior:—"The crater was five miles in circumference, and about a thousand paces deep: its sides were covered with brushwood, and at the bottom there was a plain on which cattle grazed. In the woody parts wild boars frequently harbored. In one part of the plain, covered with ashes, were three small pools, one filled with hot and bitter water, another salter than the sea, and a third hot, but tasteless."* But at length these forests and grassy plains were consumed, being suddenly blown into the air, and their ashes scattered to the winds. In December, 1631, seven streams of lava poured at once from the crater, and overflowed several villages, on the flanks and at the foot of the mountain. Resina, partly built over the ancient site of Herculaneum, was consumed by the fiery torrent. Great floods of mud were as destructive as the lava itself,—no uncommon occurrence during these catastrophes; for such is the violence of rains produced by the evolutions of aqueous vapor, that torrents of water descend the cone, and becoming charged with impalpable volcanic dust, and rolling along loose ashes, acquire sufficient consistency to deserve their ordinary appellation of "aqueous lavas."

A brief period of repose ensued, which lasted only until the year 1666, from which time to the present there has been a constant series of eruptions, with rarely an interval of rest exceeding ten years. During these three centuries, no irregular volcanic agency has convulsed other points in this district. Brieslak remarked, that such irregular convulsions had occurred in the Bay of Naples in every second century; as, for example, the eruption of the Solfatara, in the twelfth; of the lava of Arso, in Ischia, in the fourteenth; and of Monte Nuovo in the sixteenth; but the eighteenth has formed an exception to this rule, and this seems accounted for by the unprecedented number of eruptions of Vesuvius during that period; whereas, when the new vents opened, there had always been, as we have seen, a long intermittence of activity in the principal volcano.

* Hamilton's Campi Phlegræi, folio, vol. i. p. 62; and Brieslak, Campanie, tome i. p. 186.

CHAPTER XXIV.

VOLCANIC DISTRICT OF NAPLES—*continued.*

Dimensions and structure of the cone of Vesuvius—Fluidity and motion of lava—Dikes—Alluviums called "aqueous lavas"—Origin and composition of the matter enveloping Herculaneum and Pompeii—Condition and contents of the buried cities—Small number of skeletons—State of preservation of animal and vegetable substances—Rolls of papyrus—Stabiæ—Torre del Greco—Concluding remarks on the Campanian volcanoes.

Structure of the cone of Vesuvius.—BETWEEN the end of the eighteenth century and the year 1822, the great crater of Vesuvius had been gradually filled by lava boiling up from below, and by scoriæ falling from the explosions of minor mouths which were formed at intervals on its bottom and sides. In place of a regular cavity, therefore, there was a rough and rocky plain, covered with blocks of lava and scoriæ, and cut by numerous fissures, from which clouds of vapor were evolved. But this state of things was totally changed by the eruption of October, 1822, when violent explosions, during the space of more than twenty days, broke up and threw out all this accumulated mass, so as to leave an immense gulf or chasm, of an irregular, but somewhat elliptical shape, about three miles in circumference when measured along the very sinuous and irregular line of its extreme margin, but somewhat less than three quarters of a mile in its longest diameter, which was directed from N. E. to S. W.* The depth of this tremendous abyss has been variously estimated; for from the hour of its formation it increased daily by the dilapidation of its sides. It measured, at first, according to the account of some authors, two thousand feet in depth from the extreme part of the existing summit;† but Mr. Scrope, when he saw it, soon after the eruption, estimated its depth at less than half that amount. More than eight hundred feet of the cone was carried away by the explosions, so that the mountain was reduced in height from about 4200 to 3400 feet.‡

As we ascend the sloping sides, the volcano appears a mass of loose materials—a mere heap of rubbish, thrown together without the slightest order; but on arriving at the brim of the crater, and obtaining a view of the interior, we are agreeably surprised to discover that the conformation of the whole displays in every part the most perfect symmetry and arrangement. The materials are disposed in regular strata, slightly undulating, appearing, when viewed in front, to be disposed in horizontal planes. But, as we make the circuit of the edge of the cra-

* Account of the Eruption of Vesuvius in October, 1822, by G. P. Scrope, Esq., Journ. of Sci. &c. vol. xv. p. 175.

† Mr. Forbes, Account of Mount Vesuvius, Edin. Journ. of Sci. No. xviii. p. 195. Oct. 1828.

‡ Ibid. p. 194.

ter, and observe the cliffs by which it is encircled projecting or receding in salient or retiring angles, we behold transverse sections of the currents of lava and beds of sand and scoriæ, and recognize their true dip. We then discover that they incline outwards from the axis of the cone, at angles varying from 30° to 40°. The whole cone, in fact, is composed of a number of concentric coatings of alternating lavas, sand, and scoriæ. Every shower of ashes which has fallen from above, and every stream of lava descending from the lips of the crater, have conformed to the outward surface of the hill, so that one conical envelope may be said to have been successively folded round another, until the aggregation of the whole mountain was completed. The marked separation into distinct beds results from the different colors and degrees of coarseness in the sands, scoriæ, and lava, and the alternation of these with each other. The greatest difficulty, on the first view, is to conceive how so much regularity can be produced, notwithstanding the unequal distribution of sand and scoriæ, driven by prevailing winds in particular eruptions, and the small breadth of each sheet of lava as it first flows out from the crater.

But, on a closer examination, we find that the appearance of extreme uniformity is delusive; for when a number of beds thin out gradually, and at different points, the eye does not without difficulty recognize the termination of any one stratum, but usually supposes it continuous with some other, which at a short distance may lie precisely in the same plane. The slight undulations, moreover, produced by inequalities on the sides of the hill on which the successive layers were moulded, assist the deception. As countless beds of sand and scoriæ constitute the greater part of the whole mass, these may sometimes mantle continuously round the whole cone; and even lava streams may be of considerable breadth when first they overflow, and since, in some eruptions, a considerable part of the upper portion of the cone breaks down at once, may form a sheet extending as far as the space which the eye usually takes in, in a single section.

The high inclination of some of the beds, and the firm union of the particles even where there is evidently no cement, is another striking feature in the volcanic tuffs and breccias, which seems at first not very easy of explanation. But the last great eruption afforded ample illustration of the manner in which these strata are formed. Fragments of lava, scoriæ, pumice, and sand, when they fall at slight distances from the summit, are only half cooled down from a state of fusion, and are afterwards acted upon by the heat from within, and by fumeroles or small crevices in the cone through which hot vapors are disengaged. Thus heated, the ejected fragments cohere together strongly; and the whole mass acquires such consistency in a few days, that fragments cannot be detached without a smart blow of the hammer. At the same time sand and scoriæ, ejected to a greater distance, remain incoherent.*

* Monticelli and Covelli, Storia di Fenon. del Vesuv. en 1821–23.

Sir William Hamilton, in his description of the eruption of 1779, says that jets of liquid lava, mixed with stones and scoriæ, were thrown up to the height of at least ten thousand feet, having the appearance of a column of fire.* Some of these were directed by the winds towards Ottajano, and some of them falling almost perpendicularly, still red-hot and liquid, on Vesuvius, covered its whole cone, part of the mountain of Somma, and the valley between them. The falling matter being nearly as vividly inflamed as that which was continually issuing fresh from the crater, formed with it one complete body of fire, which could not be less than two miles and a half in breadth, and of the extraordinary height above mentioned, casting a heat to the distance of at least six miles round it. Dr. Clarke, also, in his account of the eruption of 1793, says that millions of red-hot stones were shot into the air full half the height of the cone itself, and then bending, fell all round in a fine arch. On another occasion he says that, as they fell, they covered nearly half the cone with fire.

The same author has also described the different appearance of the lava at its source, and at some distance from it, when it had descended into the plains below. At the point where it issued, in 1793, from an arched chasm in the side of the mountain, the vivid torrent rushed with the velocity of a flood. It was in perfect fusion, unattended with any scoriæ on its surface, or any gross materials not in a state of complete solution. It flowed with the translucency of honey, "in regular channels, cut finer than art can imitate, and glowing with all the splendor of the sun."—"Sir William Hamilton," he continues, "had conceived that no stones thrown upon a current of lava would make any impression. I was soon convinced of the contrary. Light bodies, indeed, of five, ten, and fifteen pounds' weight, made little or no impression even at the source; but bodies of sixty, seventy, and eighty pounds were seen to form a kind of bed on the surface of the lava, and float away with it. A stone of three hundred weight, that had been thrown out by the crater, lay near the source of the current of lava: I raised it upon one end, and then let it fall in upon the liquid lava; when it gradually sunk beneath the surface, and disappeared. If I wished to describe the manner in which it acted upon the lava, I should say that it was like a loaf of bread thrown into a bowl of very thick honey, which gradually involves itself in the heavy liquid, and then slowly sinks to the bottom.

"The lava, at a small distance from its source, acquires a darker tint upon its surface, is less easily acted upon, and, as the stream widens, the surface, having lost its state of perfect solution, grows harder and harder, and cracks into innumerable fragments of very porous matter, to which they give the name of scoriæ, and the appearance of which has led many to suppose that it proceeded thus from the mountain. There is, however, no truth in this. All lava, at its first exit from its

* Campi Phlegræi.

native volcano, flows out in a liquid state, and all equally in fusion. The appearance of the scoriæ is to be attributed only to the action of the external air, and not to any difference in the materials which compose it, since any lava whatever, separated from its channel, and exposed to the action of the external air, immediately cracks, becomes porous, and alters its form. As we proceeded downwards, this became more and more evident; and the same lava which at its original source flowed in perfect solution, undivided, and free from incumbrances of any kind, a little farther down had its surface loaded with scoriæ in such a manner, that, upon its arrival at the bottom of the mountain, the whole current resembled nothing so much as a heap of unconnected cinders from an iron-foundry." In another place he says that "the rivers of lava in the plain resembled a vast heap of cinders, or the scoriæ of an iron-foundry, rolling slowly along, and falling with a rattling noise over one another."* Von Buch, who was in company with MM. de Humboldt and Gay-Lussac, describes the lava of 1805 (the most fluid on record) as shooting suddenly before their eyes from top to bottom of the cone in one single instant. Professor J. D. Forbes remarks that the length of the slope of the cone proper being about 1300 feet, this motion must correspond to a velocity of many hundred feet in a few seconds, without interpreting Von Buch's expression literally. The same lava, when it reached the level road at Torre del Greco, moved at the rate of only eighteen inches per minute, or three-tenths of an inch per second.† "Although common lava," observes Professor Forbes, "is nearly as liquid as melted iron, when it issues from the orifice of the crater, its fluidity rapidly diminishes, and as it becomes more and more burdened by the consolidated slag through which it has to force its way, its velocity of motion diminishes in an almost inconceivable degree; and at length, when it ceases to present the slightest external trace of fluidity, its movement can only be ascertained by careful and repeated observations, just as in the case of a glacier."‡

It appears that the intensity of the light and heat of the lava varies considerably at different periods of the same eruption, as in that of Vesuvius in 1819 and 1820, when Sir H. Davy remarked different degrees of vividness in the white heat at the point where the lava originated.§

When the expressions "flame" and "smoke" are used in describing volcanic appearances, they must generally be understood in a figurative sense. We are informed, indeed, by M. Abich, that he distinctly saw, in the eruption of Vesuvius in 1834, the flame of burning hydrogen;‖ but what is usually mistaken for flame consists of vapor or scoriæ, and impalpable dust illuminated by that vivid light which is emitted from the crater below, where the lava is said to glow with the splendor of

* Otter's Life of Dr. Clarke.

† Phil. Trans. 1846, p. 154.

‡ Ibid. p. 148.

§ Ibid. p. 241.

‖ Bulletin de la Soc. Géol. de France, tom. vii. p. 43; and Illustrations of Vesuvius and Etna, p. 3.

the sun. The clouds of apparent smoke are formed either of aqueous and other vapor, or of finely comminuted scoriæ.

Dikes in the recent cone, how formed.—The inclined strata before mentioned which dip outwards in all directions from the axis of the cone of Vesuvius, are intersected by veins or dikes of compact lava, for the most part in a vertical position. In 1828 these were seen to be about seven in number, some of them not less than four or five hundred feet in height, and thinning out before they reached the uppermost part of the cone. Being harder than the beds through which they pass, they have decomposed less rapidly, and therefore stand out in relief. When I visited Vesuvius, in November, 1828, I was prevented from descending into the crater by the constant ejections then thrown out; so that I got sight of three only of the dikes; but Signor Monticelli had previously had drawings made of the whole, which he showed me. The dikes which I saw were on that side of the cone which is encircled by Somma. The eruption before mentioned, of 1828, began in March, and in the November following the ejected matter had filled up nearly one-third of the deep abyss formed at the close of the eruption of 1822. In November I found a single black cone at the bottom of the crater continually throwing out scoriæ, while on the exterior of the cone I observed the lava of 1822, which had flowed out six years before, not yet cool, and still evolving much heat and vapor from crevices.

Hoffmann, in 1832, saw on the north side of Vesuvius, near the peak called Palo, a great many parallel bands of lava, some from six to eight feet thick, alternating with scoriæ and conglomerate. These beds, he says, were cut through by many dikes, some of them five feet broad. They resemble those of Somma, the stone being composed of grains of leucite and augite.*

There can be no doubt that the dikes above mentioned have been produced by the filling up of open fissures with liquid lava; but of the date of their formation we know nothing farther than that they are all subsequent to the year 79, and, relatively speaking, that they are more modern than all the lavas and scoriæ which they intersect. A considerable number of the upper strata are not traversed by them. That the earthquakes, which almost invariably precede eruptions, occasion rents in the mass, is well known; and, in 1822, three months before the lava flowed out, open fissures, evolving hot vapors, were numerous. It is clear that such rents must be ejected with melted matter when the column of lava rises, so that the origin of the dikes is easily explained, as also the great solidity and cystalline nature of the rock composing them, which has been formed by lava cooling slowly under great pressure.

It has been suggested that the frequent rending of volcanic cones during eruptions may be connected with the gradual and successive upheaval of the whole mass in such a manner as to increase the inclination of the beds composing the cone; and in accordance with the

* Geognost. Beobachtungen, &c., p. 182. Berlin, 1839.

hypothesis before proposed for the origin of Monte Nuovo, Von Buch supposes that the present cone of Vesuvius was formed in the year 79, not by eruption, but by upheaval. It was not produced by the repeated superposition of scoriæ and lava cast out or flowing from a central source, but by the uplifting of strata previously horizontal. The entire cone rose at once, such as we now see it, from the interior and middle of Somma, and has since received no accession of height, but, on the contrary, has ever since been diminishing in elevation.*

Although I consider this hypothesis of Von Buch to be quite untenable, I may mention some facts which may at first sight seem to favor it. These are recorded by M. Abich in his account of the Vesuvian eruptions of 1833 and 1834, a work illustrated by excellent engravings of the volcanic phenomena which he witnessed.† It appears that, in the year 1834, the great crater of Vesuvius had been filled up nearly to the top with lava, which had consolidated and formed a level and unbroken plain, except that a small cone thrown up by the ejection of scoriæ rose in the middle of it like an island in a lake. At length this plain of lava was broken by a fissure which passed from N. E. to S. W., and along this line a great number of minute cones emitting vapor were formed. The first act of formation of these minor cones is said to have consisted of a partial upheaval of beds of lava previously horizontal, and which had been rendered flexible by the heat and tension of elastic fluids, which, rising from below, escaped from the centre of each new monticule. There would be considerable analogy between this mode of origin and that ascribed by Von Buch to Vesuvius and Somma, if the dimensions of the upraised masses were not on so different a scale, and if it was safe to reason from the inflation of bladders of half-fused lava, from fifteen to twenty-five feet in height, to mountains attaining an altitude of several thousand feet, and having their component strata strengthened by intersecting dikes of solid lava.

At the same time M. Abich mentions, that when, in August, 1834, a great subsidence took place in the platform of lava within the great crater, so that the structure of the central cone was laid open, it was seen to have been evidently formed, *not by upheaval*, but by the fall of cinders and scoriæ which had been thrown out during successive eruptions.‡

Previous to the year 79, Vesuvius appears, from the description of its figure given by Strabo, to have been a truncated cone, having a level and even outline as seen from a distance. That it had a crater on its summit, we may infer from a passage in Plutarch, on which Dr. Daubeny has judiciously commented in his treatise on volcanoes.§ The walls of the crater were evidently entire, except on one side, where there was a single narrow breach. When Spartacus, in the year 72, encamped his

* Von Buch, Descrip. Phys. des Iles Canaries, p. 342. Paris, 1836.
† Vues Illust. de Phénom. Géol. Observ. sur le Vésuve et l'Etna. Berlin, 1837.
‡ Ibid. p. 2.
§ 2d edit. 1848, p. 216.

gladiators in this hollow, Clodius, the prætor, besieged him there, keeping the single outlet carefully guarded, and then let down his soldiers by scaling-ladders over the steep precipices which surrounded the crater, at the bottom of which the insurgents were encamped. On the side towards the sea, the walls of this original cavity, which must have been three miles in diameter, have been destroyed, and Brieslak was the first to announce the opinion that this destruction happened during the tremendous eruption which occurred in 79, when the new cone, now called Vesuvius, was thrown up, which stands encircled on three sides by the ruins of the ancient cone, called Monte Somma.

In the annexed diagram (fig. 45) it will be seen that on the side of Vesuvius opposite to that where a portion of the ancient cone of Somma (*a*) still remains, is a projection (*b*) called the Pedamentina, which some

Fig. 45.

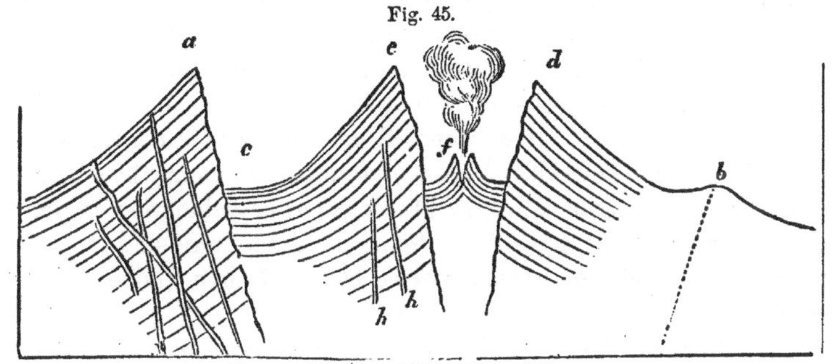

Supposed section of Vesuvius and Somma.

a, Monte Somma, or the remains of the ancient cone of Vesuvius.
b, The Pedamentina, a terrace-like projection, encircling the base of the recent cone of Vesuvius on the south side.
c, Atrio del Cavallo.*
d, *e*, Crater left by eruption of 1822.
f, Small cone thrown up in 1828, at the bottom of the great crater.
g, *g*, Dikes intersecting Somma.
h, *h*, Dikes intersecting the recent cone of Vesuvius.

have supposed to be part of the circumference of the ancient crater broken down towards the sea, and over the edge of which the lavas of the modern Vesuvius have poured; the axis of the present cone of Vesuvius being, according to Visconti, precisely equidistant from the escarpment of Somma and the Pedamentina.

In the same diagram I have represented the slanting beds of the cone of Vesuvius as becoming horizontal in the Atrio del Cavallo (at *c*), where the base of the new cone meets the precipitous escarpment of Somma; for when the lava flows down to this point, as happened in 1822, its descending course is arrested, and it then runs in another direction along this small valley, circling round the base of the cone. Sand and scoriæ, also, blown by the winds, collect at the base of the cone, and are then swept away by torrents; so that there is always here a flattish plain, as represented. In the same manner, the small

* So called from travellers leaving their horses and mules there when they prepare to ascend the cone on foot.

interior cone (f) must be composed of sloping beds, terminating in a horizontal plain; for, while this monticule was gradually gaining height by successive ejections of lava and scoriæ, in 1828, it was always surrounded by a flat pool of semi-fluid lava, into which scoriæ and sand were thrown.

In the steep simicircular escarpment of Somma, which faces the modern Vesuvius, we see a great number of sheets of lava inclined at an angle of about 26°. They alternate with scoriæ, and are intersected by numerous dikes, which, like the sheets of lava, are composed chiefly of augite, with crystals of leucite, but the rock in the dikes is more compact, having cooled and consolidated under greater pressure. Some of the dikes cut through and shift others, so that they have evidently been formed during successive eruptions. While the higher region of Somma is made up of these igneous products, there appear on its flanks, for some depth from the surface, as seen in a ravine called the "Fossa Grande," beds of white pumiceous tuff, resembling the tuff which, at Pausilippo, and other places, near Naples, contain shells of living Mediterranean species. It is supposed by Pilla, Von Buch, and others, that the tufaceous beds, which rise in Somma to more than half the height of that mountain, are, in like manner, of submarine origin, because a few sea-shells have been found in them, here and there, together with serpulæ of recent species attached to included blocks of limestone.*

It is contended, therefore, that as these strata were once accumulated beneath the sea, they may have been subjected as they rose to such an upward movement as may have given rise to a conical hill; and this hypothesis, it is said, acquires confirmation from the fact, that the sheets of lava near the summit of Somma are so compact and crystalline, and of such breadth individually, as would not have been the case had they run down a steep slope. They must, therefore, have consolidated on a nearly level surface, and have been subsequently uplifted into their present inclined position.

Unfortunately there are no sections of sufficient depth and continuity on the flanks of Somma, to reveal to us clearly the relations of the lava, scoriæ, and associated dikes, forming the highest part of the mountain, with the marine tuffs observed on its declivity. Both may, perhaps, have been produced contemporaneously when Somma raised its head, like Stromboli, above the sea, its sides and base being then submerged. Such a state of things may be indicated by a fact noticed by Von Buch, namely, that the pumiceous beds of Naples, when they approach Somma, contain fragments of the peculiar leucitic lava proper to that mountain, which are not found in the same tuff at a greater distance.† Portions, therefore, of this lava were either thrown

* Dufrénoy, Mém. pour servir à une Descrip. Géol. de la France, tom. iv. p. 294.

† Descrip. Phys. des Iles Canaries, p. 344.

out by explosions, or torn off by the waves, during the deposition of the pumiceous strata beneath the sea.

We have as yet but a scanty acquaintance with the laws which regulate the flow of lava beneath water, or the arrangement of scoriæ and volcanic dust on the sides of a submarine cone. There can, however, be little doubt that showers of ejected matter may settle on a steep slope, and may include shells and the remains of aquatic animals, which flourish in the intervals between eruptions. Lava under the pressure of water would be less porous; but, as Dr. Daubeny suggests, it may retain its fluidity longer than in the open air; for the rapidity with which heated bodies are cooled by being plunged into water arises chiefly from the conversion of the lower portions of water into steam, which steam absorbing much heat, immediately ascends, and is reconverted into water. But under the pressure of a deep ocean, the heat of the lava would be carried off more slowly, and only by the circulation of ascending and descending currents of water, those portions nearest the source of heat becoming specifically light, and consequently displacing the water above. This kind of circulation would take place with much less rapidity than in the atmosphere, inasmuch as the expansion of water by equal increments of heat is less considerable than that of air.*

We learn from the valuable observations made by Mr. Dana on the active volcanoes of the Sandwich Islands, that large sheets of compact basaltic lava have been poured out of craters at the top or near the summits of flattened domes higher than Etna, as in the case of Mount Loa for example, where a copious stream two miles broad and twenty-five miles long proceeded from an opening 13,000 feet above the level of the sea. The usual slope of these sheets of lava is between 5° and 10°; but Mr. Dana convinced himself that, owing to the suddenness with which they cool in the air, some lavas may occasionally form on slopes equalling 25°, and still preserve a considerable compactness of texture. It is even proved, he says, from what he saw in the great lateral crater of Kilauea, on the flanks of Mount Loa, that a mass of such melted rock may consolidate at an inclination of 30°, and be continuous for 300 or 400 feet. Such masses are narrow, he admits, "but if the source had been more generous, they would have had a greater breadth, and by a succession of ejections overspreading each cooled layer, a considerable thickness might have been attained."† The same author has also shown, as before mentioned, that in the "cinder cones" of the Sandwich Islands, the strata have an original inclination of between 35° and 40°.‡

Mr. Scrope, writing in 1827, attributed the formation of a volcanic cone chiefly to matter ejected from a central orifice, but partly to the injection of lava into dikes, and "to that force of gaseous expansion, the

* See Daubeny's Volcanoes, p. 400.

† Geol. of American Explor. Exped. p. 359, note. Mr. Dana informed me (Sept. 1852), that an angle of 60° instead of 30°, was given by mistake in his work.

‡ Ibid. p. 354.

intensity of which, in the central parts of the cone, is attested by local earthquakes, which so often accompany eruptions.* It is the opinion of MM. Von Buch, De Beaumont, and Dufrénoy, that the sheets of lava on Somma are so uniform and compact, that their original inclination did not exceed four or five degrees, and that four-fifths, therefore, of their present slope is due to their having been subsequently tilted and upraised. Notwithstanding the light thrown by M. de Beaumont on the laws regulating the flow and consolidation of lava, I do not conceive that these laws are as yet sufficiently determined to warrant us in assigning so much of the inclined position of the beds of Somma to the subsequent rending and dislocation of the cone. Even if this were admitted, it is far more in harmony with the usual mode of development of volcanic forces to suppose the movement which modified the shape of the cone to have been intermittent and gradual, and not to have consisted of a single effort, or one sudden and violent convulsion.†

Vesuvian lavas.—The lavas of Somma are characterized by containing disseminated crystals of leucite (called, by the French, amphigène), a mineral said to be very rare in the modern lavas of Vesuvius, which are in general much more scoriaceous and less crystalline than those of Somma.‡

At the fortress near Torre del Greco a section is exposed, fifteen feet in height, of a current which ran into the sea; and it evinces, especially in the lower part, a decided tendency to divide into rude columns. A still more striking example may be seen to the west of Torre del Annunziata, near Forte Scassato, where the mass is laid open to the depth of twenty feet. In both these cases, however, the rock may rather be said to be divided into numerous perpendicular fissures, than to be prismatic, although the same picturesque effect is produced. In the lava-currents of Central France (those of the Vivarais, in particular), the uppermost portion, often forty feet or more in thickness, is an amorphous mass passing downwards into lava irregularly prismatic; and under this there is a foundation of regular and vertical columns; but these lavas are often one hundred feet or more in thickness. We can scarcely expect to discover the same phenomenon in the shallow currents of Vesuvius, where the lowest part has cooled more rapidly, although it may be looked for in modern streams in Iceland, which exceed even those of ancient France in volume.

Mr. Scrope mentions that, in the cliffs encircling the modern crater of Vesuvius, he saw many currents offering a columnar division, and some almost as regularly prismatic as any ranges of the older basalts; and he adds, that in some the spheroidal concretionary structure, on a large scale, was equally conspicuous.§ Brieslak‖ also informs us that, in the

* Geol. Trans. 2d series, vol. ii. p. 341.

† See a paper by the Author on "Craters of Denudation," Quart. Journ. Geol. Soc. 1850.

‡ Dufrénoy, Mém. pour servir, &c. tom. iv. p. 285.

§ Journal of Science, vol. xv. p. 177.

‖ Voy. dans la Campanie, tome i. p. 201.

siliceous lava of 1737, which contains augite, leucite, and crystals of felspar, he found very regular prisms in a quarry near Torre del Greco; an observation confirmed by modern authorities.*

Effects of decomposition on lavas.—The decomposition of some of the felspathic lavas, either by simple weathering, or by gaseous emanations, converts them from a hard to a soft clayey state, so that they no longer retain the smallest resemblance to rocks cooled down from a state of fusion. The exhalations of sulphuretted hydrogen and muriatic acid, which are disengaged continually from the Solfatara, also produce curious changes on the trachyte of that nearly extinct volcano: the rock is bleached, and becomes porous, fissile, and honey-combed, till at length it crumbles into a white siliceous powder.† Numerous globular concretions, composed of concentric laminæ, are also formed by the same vapors in this decomposed rock.‡

Vesuvian minerals.—A great variety of minerals are found in the lavas of Vesuvius and Somma; augite, leucite, felspar, mica, olivine, and sulphur are most abundant. It is an extraordinary fact, that in an area of three square miles round Vesuvius, a greater number of simple minerals have been found than in any spot of the same dimensions on the surface of the globe. Häuy enumerated only 380 species of simple minerals as known to him; and no less than eighty-two had been found on Vesuvius and in the tuffs on the flanks of Somma before the end of the year 1828.§ Many of these are peculiar to that locality. Some mineralogists have conjectured that the greater part of these were not of Vesuvian origin, but thrown up in fragments from some older formation, through which the gaseous explosions burst. But none of the older rocks in Italy, or elsewhere, contain such an assemblage of mineral products; and the hypothesis seems to have been prompted by a disinclination to admit that, in times so recent in the earth's history, the laboratory of nature could have been so prolific in the creation of new and rare compounds. Had Vesuvius been a volcano of high antiquity, formed when nature

> Wanton'd as in her prime, and play'd at will
> Her virgin fancies,

it would have been readily admitted that these, or a much greater variety of substances, had been sublimed in the crevices of lava, just as several new earthy and metallic compounds are known to have been produced by fumeroles, since the eruption of 1822.

Mass enveloping Herculaneum and Pompeii.—In addition to the ejections which fall on the cone, and that much greater mass which finds its way gradually to the neighboring sea, there is a third portion, often of no inconsiderable thickness, composed of alluviums, spread over the valleys and plains at small distances from the volcano. Aqueous

* Mr. Forbes, Edin. Journ. of Sci. No. xviii. Oct. 1828.
† Daubeny on Volcanoes, p. 169.
‡ Scrope, Geol. Trans. second series, vol. ii. p. 346.
§ Monticelli and Covelli, Prodrom. della Mineral. Vesuv.

vapors are evolved copiously from volcanic craters during eruptions, and often for a long time subsequently to the discharge of scoriæ and lava: these vapors are condensed in the cold atmosphere surrounding the high volcanic peak, and heavy rains are thus caused. The floods thus occasioned, sweep along the impalpable dust and light scoriæ, till a current of mud is produced, which is called in Campania "lava d' acqua," and is often more dreaded than an igneous stream (lava di fuoco), from the greater velocity with which it moves. So late as the 27th of October, 1822, one of these alluviums descended the cone of Vesuvius, and, after overspreading much cultivated soil, flowed suddenly into the villages of St. Sebastian and Massa, where, filling the streets and interior of some of the houses, it suffocated seven persons. It will, therefore, happen very frequently that, towards the base of a volcanic cone, alternations will be found of lava, alluvium, and showers of ashes.

To which of these two latter divisions the mass enveloping Herculaneum and Pompeii should be referred, has been a question of the keenest controversy; but the discussion might have been shortened, if the combatants had reflected that, whether volcanic sand and ashes were conveyed to the towns by running water, or through the air, during an eruption, the interior of buildings, so long as the roofs remain entire, together with all underground vaults and cellars, could be filled only by an *alluvium.* We learn from history, that a heavy shower of sand, pumice, and lapilli, sufficiently great to render Pompeii and Herculaneum uninhabitable, fell for eight successive days and nights in the year 79, accompanied by violent rains.* We ought, therefore, to find a very close resemblance between the strata covering these towns and those composing the minor cones of the Phlegræan Fields, accumulated rapidly, like Monte Nuovo, during a continued shower of ejected matter; with this difference however, that the strata incumbent on the cities would be horizontal, whereas those on the cones are highly inclined; and that large angular fragments of rock, which are thrown out near the vent, would be wanting at a distance where small lapilli only can be found. Accordingly, with these exceptions, no identity can be more perfect than the form and distribution of the matter at the base of Monte Nuovo, as laid open by the encroaching sea, and the appearance of the beds superimposed on Pompeii. That city is covered with numerous alternations of different horizontal beds of tuff and lapilli, for the most part thin, and subdivided into very fine layers. I observed the following section near the amphitheatre, in November, 1828—(descending series):—

* The great eruption, in 1822, caused a covering only a few inches thick on Pompeii. Several feet are mentioned by Prof. J. D. Forbes.—Ed. Journ. of Science, No. xix. p. 131, Jan. 1829. But he must have measured in spots where it had drifted. The dust and ashes were five feet thick at the top of the crater, and decreased gradually to ten inches at Torre del Annunziata. The size and weight of the ejected fragments diminished very regularly in the same continuous stratum, as the distance from the centre of projection was greater.

	Feet.	Inches.
1. Black sparkling sand from the eruption of 1822, containing minute regularly formed crystals of augite and tourmaline -	0	2½
2. Vegetable mould - - - - - - - -	3	0
3. Brown incoherent tuff, full of *pisolitic globules* in layers, from half an inch to three inches in thickness - - - -	1	6
4. Small scoriæ and white lapilli - - - - - - -	0	3
5. Brown earthy tuff, with numerous pisolitic globules - -	0	9
6. Brown earthy tuff, with lapilli divided into layers - -	4	0
7. Layer of whitish lapilli - - - - - - -	0	1
8. Gray solid tuff - - - - - - - -	0	3
9. Pumice and white lapilli - - - - - - -	0	3
	10	3½

Many of the ashes in these beds are vitrified, and harsh to the touch. Crystals of leucite, both fresh and farinaceous, have been found intermixed.* The depth of the bed of ashes above the houses is variable, but seldom exceeds twelve or fourteen feet, and it is said that the higher part of the amphitheatre always projected above the surface; though if this were the case, it seems inexplicable that the city should never have been discovered till the year 1750. It will be observed in the above section that two of the brown, half-consolidated tuffs are filled with small pisolitic globules. This circumstance is not alluded to in the animated controversy which the Royal Academy of Naples maintained with one of their members, Signor Lippi, as to the origin of the strata incumbent on Pompeii. The mode of aggregation of these globules has been fully explained by Mr. Scrope, who saw them formed in great numbers in 1822, by rain falling during the eruption on fine volcanic sand, and sometimes also produced like hail in the air, by the mutual attraction of the minutest particles of fine damp sand. Their occurrence, therefore, agrees remarkably well with the account of heavy rain, and showers of sand and ashes recorded in history.†

Lippi entitled his work, "Fù il fuoco o l' acqua che sotterò Pompei ed Ercolano?"‡ and he contended that neither were the two cities destroyed in the year 79, nor by a volcanic eruption, but purely by the agency of water charged with transported matter. His letters, wherein he endeavored to dispense, as far as possible, with igneous agency, even at the foot of the volcano, were dedicated, with great propriety, to Werner, and afford an amusing illustration of the polemic style in which geological writers of that day indulged themselves. His arguments were partly of an historical nature, derived from the silence of contemporary historians, respecting the fate of the cities, which, as we have already stated, is most remarkable, and partly drawn from physical proofs. He pointed out with great clearness the resemblance of the tufaceous matter in the vaults and cellars at Herculaneum and Pompeii to aqueous alluviums, and its distinctness from ejections which had fallen through the air. Nothing, he observes, but moist pasty matter could have received the impression of a woman's breast, which was found in a vault at Pompeii,

* Forbes, Ed. Journ. of Sci. No. xix. p. 130, Jan. 1829.

† Scrope, Geol. Trans. second series, vol. ii. p. 346.

‡ Napoli, 1816.

or have given the cast of a statue discovered in the theatre at Herculaneum. It was objected to him, that the heat of the tuff in Herculaneum and Pompeii was proved by the carbonization of the timber, corn, papyrus-rolls, and other vegetable substances there discovered; but Lippi replied with truth, that the papyri would have been burnt up if they had come in contact with fire, and that their being only carbonized was a clear demonstration of their having been enveloped, like fossil wood, in a sediment deposited from water. The Academicians, in their report on his pamphlet, assert, that when the amphitheatre was first cleared out, the matter was arranged on the steps in a succession of concave layers, accommodating themselves to the interior form of the building, just as snow would lie if it had fallen there. This observation is highly interesting, and points to the difference between the stratification of ashes in an open building and of mud derived from the same in the interior of edifices and cellars. Nor ought we to call the allegation in question, because it could not be substantiated at the time of the controversy after the matter had been all removed; although Lippi took advantage of this removal, and met the argument of his antagonists by requiring them to prove the fact. There is decisive evidence that no stream of lava has ever reached Pompeii since it was first built, although the foundations of the town stand upon the old leucitic lava of Somma; several streams of which, with tuff interposed, had been cut through in excavations.

Infusorial beds covering Pompeii.—A most singular and unexpected discovery has been recently made (1844–5) by Professor Ehrenberg, respecting the remote origin of many of the layers of ashes and pumice enveloping Pompeii. They are, he says, in great part, of organic and freshwater origin, consisting of the siliceous cases of microscopic infusoria. What is still more surprising, this fact proves to be by no means an isolated or solitary example of an intimate relation between organic life and the results of volcanic activity. On the Rhine, several beds of tuff and pumiceous conglomerate, resembling the mass incumbent upon Pompeii and closely connected with extinct volcanoes, are now ascertained to be made up to a great extent of the siliceous cases of infusoria (or Diatomaceæ), invisible to the naked eye, and often half fused.* No less than 94 distinct species have already been detected in one mass of this kind, more than 150 feet thick, at Hochsimmer, on the left bank of the Rhine, near the Laacher-see. Some of these Rhenish infusorial accumulations appear to have fallen in showers, others to have been poured out of lake-craters in the form of mud, as in the Brohl valley.

In Mexico, Peru, the Isle of France, and several other volcanic regions, analogous phenomena have been observed, and everywhere the species of infusoria belong to freshwater and terrestrial genera, except in the case of the Patagonian pumiceous tuffs, specimens of which, brought

* Not a few of the organic bodies, called by Ehrenberg "infusoria," such as Galionella and Bacillaria, have been recently claimed by many botanists as belonging to the vegetable kingdom, and are referred to the classes called Diatomaceæ and Desmidiæ.

home by Mr. Darwin, are found to contain the remains of marine animalcules. In various kinds of pumice ejected by volcanoes, the microscope has revealed to Professor Ehrenberg the siliceous cases of infusoria often half obliterated by the action of heat, and the fine dust thrown out into the air during eruptions, is sometimes referable to these most minute organic substances, brought up from considerable depths, and sometimes mingled with small particles of vegetable matter.

In what manner did the solid coverings of these most minute plants and animalcules, which can only originate and increase at the surface of the earth, sink down and penetrate into subterranean cavities, so as to be ejected from the volcanic orifices? We have of late years become familiar with the fact, in the process of boring Artesian wells, that the seeds of plants, the remains of insects, and even small fish, with other organic bodies, are carried in an uninjured state by the underground circulation of waters, to the depth of many hundred feet. With still greater facility in a volcanic region we may conjecture, that water and mud full of invisible infusoria may be sucked down, from time to time, into subterranean rents and hollows in cavernous lava which has been permeated by gases, or in rocks dislocated by earthquakes. It often happens that a lake which has endured for centuries in a volcanic crater, disappears suddenly on the approach of a new eruption. Violent shocks agitate the surrounding region, and ponds, rivers, and wells are dried up. Large cavities far below may thus become filled with fen-mud chiefly composed of the more indestructible and siliceous portions of infusoria, destined perhaps to be one day ejected in a fragmentary or half-fused state, yet without the obliteration of all traces of organic structure.*

Herculaneum.—It was remarked that no lava has flowed over the site of Pompeii, since that city was built, but with Herculaneum the case is different. Although the substance which fills the interior of the houses and the vaults must have been introduced in a state of mud, like that found in similar situations in Pompeii; yet the superincumbent mass differs wholly in composition and thickness. Herculaneum was situated several miles nearer to the volcano, and has, therefore, been always more exposed to be covered, not only by showers of ashes, but by alluviums and streams of lava. Accordingly, masses of both have accu-

* See Ehrenberg, Proceedings (Berichte) of the Royal Acad. of Sci. Berlin, 1844, 1845, and an excellent abstract of his papers by Mr. Ansted in the Quart. Journ. of the Geol. Soc. London, No. 7, Aug. 1846. In regard to marine infusoria found in volcanic tuff, it is well known that on the shores of the island of Cephalonia in the Mediterranean (Proceedings, Geol. Soc. vol. ii. p. 220), there is a cavity in the rock, into which the sea has been flowing for ages, and many others doubtless exist in the leaky bottom of the ocean. The marine current has been rushing in for many years, and as the infusoria inhabiting the waters of the Mediterranean are exceedingly abundant, a vast store of their cases may accumulate in submarine caverns (the water, perhaps, being converted into steam, and so escaping upwards), and they may then be cast up again to furnish the materials of volcanic tuff, should an eruption occur like that which produced Graham Island, off the coast of Sicily, in 1831.

mulated on each other above the city, to a depth of nowhere less than 70, and in many places of 112 feet.†

The tuff which envelops the buildings consists of comminuted volcanic ashes, mixed with pumice. A mask imbedded in this matrix has left a cast, the sharpness of which was compared by Hamilton to those in plaster of Paris; nor was the mask in the least degree scorched, as if it had been imbedded in heated matter. This tuff is porous; and, when first excavated, is soft and easily worked, but acquires a considerable degree of induration on exposure to the air. Above this lowest stratum is placed, according to Hamilton, "the matter of six eruptions," each separated from the other by veins of good soil. In these soils Lippi states that he collected a considerable number of land shells—an observation which is no doubt correct; for many snails burrow in soft soils, and some Italian species descend, when they hybernate, to the depth of five feet and more from the surface. Della Torre also informs us that there is in one part of this superimposed mass a bed of true siliceous lava (*lava di pietra dura*); and, as no such current is believed to have flowed till near one thousand years after the destruction of Herculaneum, we must conclude, that the origin of a large part of the covering of Herculaneum was long subsequent to the first inhumation of the place. That city, as well as Pompeii, was a seaport. Herculaneum is still very near the shore, but a tract of land, a mile in length, intervenes between the borders of the Bay of Naples and Pompeii. In both cases the gain of land is due to the filling up of the bed of the sea with volcanic matter, and not to elevation by earthquakes, for there has been no change in the relative level of land and sea. Pompeii stood on a slight eminence composed of the lavas of the ancient Vesuvius, and flights of steps led down to the water's edge. The lowermost of these steps are said to be still on an exact level with the sea.

Condition and contents of the buried cities.—After these observations on the nature of the strata enveloping and surrounding the cities, we may proceed to consider their internal condition and contents, so far at least as they offer facts of geological interest. Notwithstanding the much greater depth at which Herculaneum was buried, it was discovered before Pompeii, by the accidental circumstance of a well being sunk, in 1713, which came right down upon the theatre, where the statues of Hercules and Cleopatra were soon found. Whether this city or Pompeii, both of them founded by Greek colonies, was the more considerable, is not yet determined; but both are mentioned by ancient authors as among the seven most flourishing cities in Campania. The walls of Pompeii were three miles in circumference; but we have, as yet, no certain knowledge of the dimensions of Herculaneum. In the latter place the theatre alone is open for inspection; the Forum, Temple of Jupiter, and other buildings, having been filled up with rubbish as the workmen proceeded, owing to the difficulty of removing it from so

* Hamilton, Observ. on Mount Vesuvius, p. 94. London, 1774.

great a depth below ground. Even the theatre is only seen by torch-light, and the most interesting information, perhaps, which the geologist obtains there, is the continual formation of stalactite in the galleries cut through the tuff; for there is a constant percolation of water charged with carbonate of lime mixed with a small portion of magnesia. Such mineral waters must, in the course of time, create great changes in many rocks; especially in lavas, the pores of which they may fill with calcareous spar, so as to convert them into amygdaloids. Some geologists, therefore, are unreasonable when they expect that volcanic rocks of remote eras should accord precisely with those of modern date; since it is obvious that many of those produced in our own time will not long retain the same aspect and internal composition.

Both at Herculaneum and Pompeii, temples have been found with inscriptions commemorating the rebuilding of the edifices after they had been thrown down by an earthquake.* This earthquake happened in the reign of Nero, sixteen years before the cities were overwhelmed. In Pompeii, one-fourth of which is now laid open to the day, both the public and private buildings bear testimony to the catastrophe. The walls are rent, and in many places traversed by fissures still open. Columns are lying on the ground only half hewn from huge blocks of travertin, and the temple for which they were designed is seen half repaired. In some few places the pavement had sunk in, but in general it was undisturbed, consisting of large irregular flags of lava joined neatly together, in which the carriage wheels have often worn ruts an inch and a half deep. In the wider streets, the ruts are numerous and irregular; in the narrower, there are only two, one on each side, which are very conspicuous. It is impossible not to look with some interest even on these ruts, which were worn by chariot wheels more than seventeen centuries ago; and, independently of their antiquity, it is remarkable to see such deep incisions so continuous in a stone of great hardness.

Small number of skeletons.—A very small number of skeletons have been discovered in either city; and it is clear that most of the inhabitants not only found time to escape, but also to carry with them the principal part of their valuable effects. In the barracks at Pompeii were the skeletons of two soldiers chained to the stocks, and in the vaults of a country-house in the suburbs were the skeletons of seventeen persons, who appear to have fled there to escape from the shower of ashes. They were found inclosed in an indurated tuff, and in this matrix was preserved a perfect cast of a woman, perhaps the mistress of the house, with an infant in her arms. Although her form was imprinted on the rock, nothing but the bones remained. To these a chain of gold was suspended, and on the fingers of the skeletons were rings with jewels. Against the sides of the same vault was ranged a long line of earthen amphoræ.

* Swinburne and Lalande. Paderni, Phil. Trans. 1758, vol. i. p. 619.

The writings scribbled by the soldiers on the walls of their barracks, and the names of the owners of each house written over the doors, are still perfectly legible. The colors of fresco paintings on the stuccoed walls in the interior of buildings are almost as vivid as if they were just finished. There are public fountains decorated with shells laid out in patterns in the same fashion as those now seen in the town of Naples; and in the room of a painter, who was perhaps a naturalist, a large collection of shells was found, comprising a great variety of Mediterranean species, in as good a state of preservation as if they had remained for the same number of years in a museum. A comparison of these remains, with those found so generally in a fossil state would not assist us in obtaining the least insight into the time required to produce a certain degree of decomposition or mineralization; for, although under favorable circumstances much greater alteration might doubtless have been brought about in a shorter period, yet the example before us shows that an inhumation of seventeen centuries may sometimes effect nothing towards the reduction of shells to the state in which fossils are usually found.

The wooden beams in the houses at Herculaneum are black on the exterior, but, when cleft open, they appear to be almost in the state of ordinary wood, and the progress made by the whole mass towards the state of lignite is scarcely appreciable. Some animal and vegetable substances of more perishable kinds have of course suffered much change and decay, yet the state of preservation of these is truly remarkable. Fishing-nets are very abundant in both cities, often quite entire; and their number at Pompeii is the more interesting from the sea being now, as we stated, a mile distant. Linen has been found at Herculaneum, with the texture well defined; and in a fruiterer's shop in that city were discovered vessels full of almonds, chestnuts, walnuts, and fruit of the "carubiere," all distinctly recognizable from their shape. A loaf, also, still retaining its form, was found in a baker's shop, with his name stamped upon it. On the counter of an apothecary was a box of pills converted into a fine earthy substance; and by the side of it a small cylindrical roll evidently prepared to be cut into pills. By the side of these was a jar containing medicinal herbs. In 1827, moist olives were found in a square glass-case, and "caviare," or roe of a fish, in a state of wonderful preservation. An examination of these curious condiments has been published by Covelli of Naples, and they are preserved hermetically sealed in the museum there.*

Papyri.—There is a marked difference in the condition and appearance of the animal and vegetable substances found at Pompeii and Herculaneum; those of Pompeii being penetrated by a gray pulverulent tuff, those in Herculaneum seeming to have been first enveloped by a paste which consolidated round them, and then allowed them to become slowly carbonized. Some of the rolls of papyrus at Pompeii

* Prof. J. D. Forbes, Edin. Journ. of Sci. No. xix. p. 130, Jan. 1829.

still retain their form; but the writing, and indeed almost all the vegetable matter, appear to have vanished, and to have been replaced by volcanic tuff somewhat pulverulent. At Herculaneum the earthy matter has scarcely ever penetrated; and the vegetable substance of the papyrus has become a thin friable black matter, almost resembling in appearance the tinder which remains when stiff paper has been burnt, in which the letters may still be sometimes traced. The small bundles of papyri, composed of five or six rolls tied up together, had sometimes lain horizontally, and were pressed in that direction, but sometimes they had been placed in a vertical position. Small tickets were attached to each bundle, on which the title of the work was inscribed. In one case only have the sheets been found with writing on both sides of the pages. So numerous are the obliterations and corrections, that many must have been original manuscripts. The variety of handwritings is quite extraordinary: nearly all are written in Greek, but there are a few in Latin. They were almost all found in a suburban villa in the library of one private individual; and the titles of four hundred of those least injured, which have been read, are found to be unimportant works, but all entirely new, chiefly relating to music, rhetoric, and cookery. There are two volumes of Epicurus "On Nature," and the others are mostly by writers of the same school, only one fragment having been discovered, by an opponent of the Epicurean system, Chrysippus.*

Probability of future discoveries of MSS.—In the opinion of some antiquaries, not one-hundredth part of the city has yet been explored: and the quarters hitherto cleared out at a great expense, are those where there was the least probability of discovering manuscripts. As Italy could already boast her splendid Roman amphitheatres and Greek temples, it was a matter of secondary interest to add to their number those in the dark and dripping galleries of Herculaneum; and having so many of the masterpieces of ancient art, we could have dispensed with the inferior busts and statues which could alone have been expected to reward our researches in the ruins of a provincial town. But from the moment that it was ascertained that rolls of papyrus preserved in this city could still be deciphered, every exertion ought to have been steadily and exclusively directed towards the discovery of other libraries. Private dwellings should have been searched, before so much labor and expense were consumed in examining public edifices. A small portion of that zeal and enlightened spirit which prompted the late French and Tuscan expedition to Egypt might long ere this, in a country nearer home, have snatched from oblivion some of the lost works of the Augustan age, or of eminent Greek historians and philosophers. A single roll of papyrus might have disclosed more

* In one of the manuscripts which was in the hands of the interpreters when I visited the museum in 1828, the author indulges in the speculation that all the Homeric personages were allegorical—that Agamemnon was the ether, Achilles the sun, Helen the earth, Paris the air, Hector the moon, &c.

matter of intense interest than all that was ever written in hieroglyphics.

Stabiæ.—Besides the cities already mentioned, Stabiæ, a small town about six miles from Vesuvius, and near the site of the modern Castel-a-Mare (see map of volcanic district of Naples), was overwhelmed during the eruption of 79. Pliny mentions that, when his uncle was there, he was obliged to make his escape, so great was the quantity of falling stones and ashes. In the ruins of this place, a few skeletons have been found buried in volcanic ejections, together with some antiquities of no great value, and rolls of papyrus, which, like those of Pompeii, were illegible.

Torre del Greco overflowed by lava.—Of the towns hitherto mentioned, Herculaneum alone has been overflowed by a stream of melted matter; but this did not, as we have seen, enter or injure the buildings, which were previously enveloped or covered over with tuff. But burning torrents have often taken their course through the streets of Torre del Greco, and consumed or inclosed a large portion of the town in solid rock. It seems probable that the destruction of three thousand of its inhabitants in 1631, which some accounts attribute to boiling water, was principally due to one of those alluvial floods which we before mentioned: but, in 1737, the lava itself flowed through the eastern side of the town, and afterwards reached the sea; and, in 1794, another current, rolling over the western side, filled the streets and houses, and killed more than four hundred persons. The main street is now quarried through this lava, which supplied building stones for new houses erected where others had been annihilated. The church was half buried in a rocky mass, but the upper portion served as the foundation of a new edifice.

The number of the population at present is estimated at fifteen thousand; and a satisfactory answer may readily be returned to those who inquire how the inhabitants can be so "inattentive to the voice of time and the warnings of nature,"* as to rebuild their dwellings on a spot so often devastated. No neighboring site unoccupied by a town, or which would not be equally insecure, combines the same advantages of proximity to the capital, to the sea, and to the rich lands on the flanks of Vesuvius. If the present population were exiled, they would immediately be replaced by another, for the same reason that the Maremma of Tuscany and the Campagna di Roma will never be depopulated, although the malaria fever commits more havoc in a few years than the Vesuvian lavas in as many centuries. The district around Naples supplies one amongst innumerable examples, that those regions where the surface is most frequently renewed, and where the renovation is accompanied, at different intervals of time, by partial destruction of animal and vegetable life, may nevertheless be amongst the most habitable and delightful on our globe.

* Sir H. Davy, Consolations in Travel, p. 66.

I have already made a similar remark when speaking of tracts where aqueous causes are now most active; and the observation applies as well to parts of the surface which are the abode of aquatic animals, as to those which support terrestrial species. The sloping sides of Vesuvius give nourishment to a vigorous and healthy population of about eighty thousand souls; and the surrounding hills and plains, together with several of the adjoining isles, owe the fertility of their soil to matter ejected by prior eruptions. Had the fundamental limestone of the Apennines remained uncovered throughout the whole area, the country could not have sustained a twentieth part of its present inhabitants. This will be apparent to every geologist who has marked the change in the agricultural character of the soil the moment he has passed the utmost boundary of the volcanic ejections, as when, for example, at the distance of about seven miles from Vesuvius, he leaves the plain and ascends the declivity of the Sorrentine Hills.

Yet, favored as this region has been by Nature from time immemorial, the signs of the changes imprinted on it during the period that it has served as the habitation of man may appear in after-ages to indicate a series of unparalleled disasters. Let us suppose that at some future time the Mediterranean should form a gulf of the great ocean, and that the waves and tidal current should encroach on the shores of Campania, as it now advances upon the eastern coast of England; the geologist will then behold the towns already buried, and many more which will evidently be entombed hereafter, laid open in the steep cliffs, where he will discover buildings superimposed above each other, with thick intervening strata of tuff or lava—some unscathed by fire, like those of Herculaneum and Pompeii; others half melted down, as in Torre del Greco; and many shattered and thrown about in strange confusion, as in Tripergola, beneath Monte Nuovo. Among the ruins will be seen skeletons of men, and impressions of the human form stamped in solid rocks of tuff. Nor will the signs of earthquakes be wanting. The pavement of part of the Domitian Way, and the temple of the Nymphs, submerged at high tide, will be uncovered at low water, the columns remaining erect and uninjured. Other temples which had once sunk down, like that of Serapis, will be found to have been upraised again by subsequent movements. If they who study these phenomena, and speculate on their causes, assume that there were periods when the laws of Nature or the whole course of natural events differed greatly from those observed in their own time, they will scarcely hesitate to refer the wonderful monuments in question to those primeval ages. When they consider the numerous proofs of reiterated catastrophes to which the region was subject, they may, perhaps, commiserate the unhappy fate of beings condemned to inhabit a planet during its nascent and chaotic state, and feel grateful that their favored race has escaped such scenes of anarchy and misrule.

Yet what was the real condition of Campania during those years of dire convulsion? "A climate where heaven's breath smells sweet and wooingly—a vigorous and luxuriant nature unparalleled in its productions—a coast which was once the fairy-land of poets, and the favorite retreat of great men. Even the tyrants of the creation loved this alluring region, spared it, adorned it, lived in it, died in it."* The inhabitants, indeed, have enjoyed no immunity from the calamities which are the lot of mankind; but the principal evils which they have suffered must be attributed to moral, not to physical, causes—to disastrous events over which man might have exercised a control, rather than to the inevitable catastrophes which result from subterranean agency. When Spartacus encamped his army of ten thousand gladiators in the old extinct crater of Vesuvius, the volcano was more justly a subject of terror to Campania, than it has ever been since the rekindling of its fires.

CHAPTER XXV.

ETNA.

External physiognomy of Etna—Lateral cones—Their successive obliteration—Early eruptions—Monti Rossi in 1669—Towns overflowed by lava—Part of Catania overflowed—Mode of advance of a current of lava—Subterranean caverns—Marine strata at base of Etna—Val del Bove not an ancient crater—Its scenery—Form, composition, and origin of the dikes—Linear direction of cones formed in 1811 and 1819—Lavas and breccias—Flood produced by the melting of snow by lava—Glacier covered by a lava stream—Val del Bove how formed—Structure and origin of the cone of Etna—Whether the inclined sheets of lava were originally horizontal—Antiquity of Etna—Whether signs of diluvial waves are observable on Etna.

External physiognomy of Etna.—AFTER Vesuvius, our most authentic records relate to Etna, which rises near the sea in solitary grandeur to the height of nearly eleven thousand feet.† The base of the cone is almost circular, and eighty-seven English miles in circumference; but if we include the whole district over which its lavas extend, the circuit is probably twice that extent.

Divided into three regions.—The cone is divided by nature into three distinct zones, called the *fertile*, the *woody*, and the *desert* regions. The first of these, comprising the delightful country around the skirts of the mountain, is well cultivated, thickly inhabited, and covered with olives,

* Forsyth's Italy, vol. ii.

† In 1815, Captain Smyth ascertained, trigonometrically, that the height of Etna was 10,874 feet. The Catanians, disappointed that their mountain had lost nearly 2000 feet of the height assigned to it by Recupero, refused to acquiesce in the decision. Afterwards, in 1824, Sir J. Herschel, not being aware of Captain Smyth's conclusions, determined by careful barometrical measurement that the height was 10,872½ feet. This singular agreement of results so differently obtained was spoken of by Herschel as "a happy accident;" but Dr. Wollaston remarked that "it was one of those accidents which would not have happened to two fools."

vines, corn, fruit-trees, and aromatic herbs. Higher up, the woody region encircles the mountain—an extensive forest six or seven miles in width, affording pasturage for numerous flocks. The trees are of various species, the chestnut, oak, and pine being most luxuriant; while in some tracts are groves of cork and beech. Above the forest is the desert region, a waste of black lava and scoriæ; where, on a kind of plain, rises a cone of eruption to the height of about eleven hundred feet, from which sulphureous vapors are continually evolved.

Cones produced by lateral eruption.—The most grand and original feature in the physiognomy of Etna is the multitude of minor cones which are distributed over its flanks, and which are most abundant in the woody region. These, although they appear but trifling irregularities when viewed from a distance as subordinate parts of so imposing and colossal a mountain, would, nevertheless, be deemed hills of considerable altitude in almost any other region. Without enumerating numerous monticules of ashes thrown out at different points, there are about eighty of these secondary volcanoes, of considerable dimensions; fifty-two on the west and north, and twenty-seven on the east side of Etna. One of the largest, called Monte Minardo, near Bronte, is upwards of 700 feet in height, and a double hill near Nicolosi, called Monti Rossi, formed in 1669, is 450 feet high, and the base two miles in circumference; so that it somewhat exceeds in size Monte Nuovo, before described. Yet it ranks only as a cone of the second magnitude amongst those produced by the lateral eruptions of Etna. On looking down from the lower borders of the desert region, these volcanoes present us with one of the most delightful and characteristic scenes in Europe. They afford every variety of height and size, and are arranged in beautiful and picturesque groups. However uniform they may appear when seen from the sea, or the plains below, nothing can be more diversified than their shape when we look from above into their craters, one side of which is generally broken down. There are, indeed, few objects in nature more picturesque than a wooded volcanic crater. The cones situated in the higher parts of the forest zone are chiefly clothed with lofty pines; while those at a lower elevation are adorned with chestnuts, oaks, beech, and holm.

Successive obliteration of these cones.—The history of the eruptions of Etna, imperfect and interrupted as it is, affords us, nevertheless, much insight into the manner in which the whole mountain has successively attained its present magnitude and internal structure. The principal cone has more than once fallen in and been reproduced. In 1444 it was 320 feet high, and fell in after the earthquakes of 1537. In the year 1693, when a violent earthquake shook the whole of Sicily, and killed sixty thousand persons, the cone lost so much of its height, says Boccone, that it could not be seen from several places in Valdemone, from which it was before visible. The greater number of eruptions happen either from the great crater, or from lateral openings in the desert region. When hills are thrown up in the middle zone, and project beyond the general level, they gradually lose their height during subsequent eruptions; for when lava runs down from the upper parts of the mountain, and encounters any of these hills, the stream is divided, and flows round them so as to elevate the gently sloping grounds from which they rise. In this manner a deduction is often made at once of twenty or thirty feet, or even more, from their height. Thus, one of the minor cones, called Monte Peluso, was diminished in altitude by a great lava stream which encircled it in 1444; and another current has recently taken the same course—yet this hill still remains four or five hundred feet high.

There is a cone called Monte Nucilla near Nicolosi, round the base of which several successive currents have flowed, and showers of ashes have fallen, since the time of history, till at last, during an eruption in 1536, the surrounding plain was so raised, that the top of the cone alone was left projecting above the general level. Monte Nero, situated above the Grotta dell' Capre, was in 1766 almost submerged by a current: and Monte Capreolo afforded, in the year 1669, a curious example of one of the last stages of obliteration; for a lava stream, descending on a high ridge which had been built up by the continued superposition of successive lavas, flowed directly into the crater, and nearly filled it. The lava, therefore, of each new lateral cone tends to detract from the relative height of lower cones above their base: so that the flanks of Etna, sloping with a gentle inclination, envelop in succession a great multitude of minor volcanoes, while new ones spring up from time to time.

Early eruptions of Etna.—Etna appears to have been in activity from the earliest times of tradition; for Diodorus Siculus mentions an eruption which caused a district to be deserted by the Sicani before the Trojan war. Thucydides informs us, that in the sixth year of the Peloponnesian war, or in the spring of the year 425 B. C., a lava stream ravaged the environs of Catania, and this he says was the third eruption which had happened in Sicily since the colonization of that island by the Greeks.* The second of the three eruptions alluded to by the historian took place in the year 475 B. C., and was that so poetically described by Pindar, two years afterwards, in his first Pythian ode:—

κιων
Δ' ουρανια συνεχει
Νιφοεσσ' Αιτνα, πανετες
Χιονος οξειας τιθηνα.

In these and the seven verses which follow, a graphic description is given of Etna, such as it appeared five centuries before the Christian era, and such as it has been seen when in eruption in modern times. The poet is only making a passing allusion to the Sicilian volcano, as the mountain under which Typhœus lay buried, yet by a few touches of his master-hand every striking feature of the scene has been faithfully por-

* Book iii. at the end.

trayed. We are told of "the snowy Etna, the pillar of heaven—the nurse of everlasting frost, in whose deep caverns lie concealed the fountains of unapproachable fire—a stream of eddying smoke by day—a bright and ruddy flame by night; and burning rocks rolled down with loud uproar into the sea."

Eruption of 1669—*Monti Rossi formed.*—The great eruption which happened in the year 1669 is the first which claims particular attention. An earthquake had levelled to the ground all the houses in Nicolosi, a town situated near the lower margin of the woody region, about twenty miles from the summit of Etna, and ten from the sea at Catania. Two gulfs then opened near that town,, from whence sand and scoriæ were thrown up in such quantity, that in the course of three or four months a double cone was formed, called Monti Rossi, about 450 feet high. But

Fig. 46.

Minor cones on the flanks of Etna.

1. Monti Rossi, near Nicolosi, formed in 1669. 2. Vampeluso ?*

the most extraordinary phenomenon occurred at the commencement of the convulsion in the plain of S. Lio. A fissure six feet broad, and of unknown depth, opened with a loud crash, and ran in a somewhat tortuous course to within a mile of the summit of Etna. Its direction was from north to south, and its length twelve miles. It emitted a most vivid light. Five other parallel fissures of considerable length afterwards opened, one after the other, and emitted smoke, and gave out bellowing sounds which were heard at the distance of forty miles. This case seems to present the geologist with an illustration of the manner in which those continuous dikes of vertical porphyry were formed, which are seen to traverse some of the older lavas of Etna; for the light emitted from the great rent of S. Lio appears to indicate that the fissure was filled to a certain height with incandescent lava, probably to the height of an orifice not far distant from Monti Rossi, which at that time opened and poured out a lava current. When the melted matter in such a rent

* The hill which I have here introduced was called by my guide Vampolara, but the name given in the text is the nearest to this which I find in Gemmellaro's Catalogue of Minor Cones.

has cooled, it must become a solid wall or dike, intersecting the older rocks of which the mountain is composed; similar rents have been observed during subsequent eruptions, as in 1832, when they ran in all directions from the centre of the volcano. It has been justly remarked by M. Elie de Beaumont, that such star-shaped fractures may indicate a slight upheaval of the whole of Etna. They may be the signs of the stretching of the mass, which may thus be raised gradually by a force from below.*

The lava current of 1669, before alluded to, soon reached in its course a minor cone called Mompiliere, at the base of which it entered a subterranean grotto, communicating with a suite of those caverns which are so common in the lavas of Etna. Here it appears to have melted down some of the vaulted foundations of the hill, so that the whole of that cone became slightly depressed and traversed by numerous open fissures.

Part of Catania destroyed.—The lava, after overflowing fourteen towns and villages, some having a population of between three and four thousand inhabitants, arrived at length at the walls of Catania. These had been purposely raised to protect the city; but the burning flood accumulated till it rose to the top of the rampart, which was sixty feet in height, and then it fell in a fiery cascade and overwhelmed part of the city. The wall, however, was not thrown down, but was discovered long afterwards by excavations made in the rock by the Prince of Biscari; so that the traveller may now see the solid lava curling over the top of the rampart as if still in the very act of falling.

This great current performed the first thirteen miles of its course in twenty days, or at the rate of 162 feet per hour, but required twenty-three days for the last two miles, giving a velocity of only twenty-two feet per hour; and we learn from Dolomieu that the stream moved during part of its course at the rate of 1500 feet an hour, and in others it took several days to cover a few yards.† When it entered the sea it was still six hundred yards broad, and forty feet deep. It covered some territories in the environs of Catania which had never before been visited by the lavas of Etna. While moving on, its surface was in general a mass of solid rock; and its mode of advancing, as is usual with lava streams, was by the occasional fissuring of the solid walls. A gentleman of Catania, named Pappalardo, desiring to secure the city from the approach of the threatening torrent, went out with a party of fifty men whom he had dressed in skins to protect them from the heat, and armed with iron crows and hooks. They broke open one of the solid walls which flanked the current near Belpasso, and immediately forth issued a rivulet of melted matter which took the direction of Paternó; but the inhabitants of that town, being alarmed for their safety, took up arms and put a stop to farther operations.‡

* Mém. pour servir, &c. tom. iv. p. 116.

† See Prof. J. D. Forbes, Phil. Trans. 1846, p. 155, on Velocity of Lava.

‡ Ferrara, Descriz. dell' Etna, p. 108.

As another illustration of the solidity of the walls of an advancing lava stream, I may mention an adventure related by Recupero, who, in 1766, had ascended a small hill formed of ancient volcanic matter, to behold the slow and gradual approach of a fiery current, two miles and a half broad; when suddenly two small threads of liquid matter issuing from a crevice detached themselves from the main stream, and ran rapidly towards the hill. He and his guide had just time to escape, when they saw the hill, which was fifty feet in height, surrounded, and in a quarter of an hour melted down into the burning mass, so as to flow on with it.

But it must not be supposed that this complete fusion of rocky matter coming in contact with lava is of universal, or even common, occurrence. It probably happens when fresh portions of incandescent matter come successively in contact with fusible materials. In many of the dikes which intersect the tuffs and lavas of Etna, there is scarcely any perceptible alteration effected by heat on the edges of the horizontal beds, in contact with the vertical and more crystalline mass. On the side of Mompiliere, one of the towns overflowed in the great eruption above described, an excavation was made in 1704; and by immense labor the workmen reached, at the depth of thirty-five feet, the gate of the principal church, were there were three statues, held in high veneration. One of these, together with a bell, some money, and other articles, were extracted in a good state of preservation from beneath a great arch formed by the lava. It seems very extraordinary that any works of art, not encased with tuff, like those in Herculaneum, should have escaped fusion in hollow spaces left open in this lava-current, which was so hot at Catania eight years after it entered the town, that it was impossible to hold the hand in some of the crevices.

Subterranean caverns on Etna.—Mention was made of the entrance of a lava-stream into a subterranean grotto, whereby the foundations of a hill were partially undermined. Such underground passages are among the most curious features on Etna, and appear to have been produced by the hardening of the lava, during the escape of great volumes of elastic fluids, which are often discharged for many days in succession, after the crisis of the eruption is over. Near Nicolosi, not far from Monti Rossi, one of these great openings may be seen, called the Fossa della Palomba, 625 feet in circumference at its mouth, and seventy-eight deep. After reaching the bottom of this, we enter another dark cavity, and then others in succession, sometimes descending precipices by means of ladders. At length the vaults terminate in a great gallery ninety feet long, and from fifteen to fifty broad, beyond which there is still a passage, never yet explored; so that the extent of these caverns remains unknown.* The walls and roofs of these great vaults are composed of rough and bristling scoriæ, of the most fantastic forms.

Marine strata at base of Etna.—If we skirt the fertile region at the

* Ferrara, Descriz. dell' Etna. Palermo, 1818.

base of Etna on its southern and eastern sides, we behold marine strata of clay sand, and volcanic tuff, cropping out from beneath the modern lavas. The marine fossil shells occurring in these strata are all of them, or nearly all, identical with species now inhabiting the Mediterranean; and as they appear at the height of from 600 to 800 feet above the sea near Catania, they clearly prove that there has been in this region, as in other parts of Sicily farther to the south, an upward movement of the ancient bed of the sea. It is fair, therefore, to infer that the whole mountain, with the exception of those parts which are of very modern origin, has participated in this upheaval.

If we view Etna from the south, we see the marine deposits above alluded to, forming a low line of hills (*e, e*, Fig. 47), or a steep inland slope or cliff (*f*), as in the annexed drawing taken from the limestone platform of Primosole. It should be observed however, in reference to this view, that the height of the volcanic cone is ten times greater than the hills at its base (*e, e*), although it appears less elevated, because the summit of the cone is ten or twelve times more distant from the plain of Catania than is Licodia.

Fig. 47.

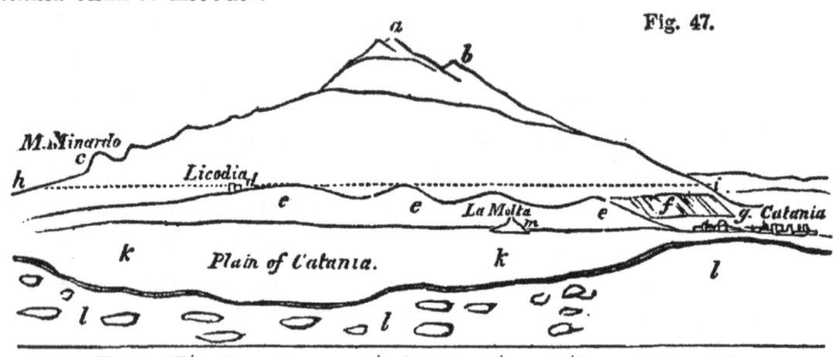

View of Etna from the summit of the limestone platform of Primosole.

a, Highest cone. *b*, Montagnuola.
c, Monte Minardo, with smaller lateral cones above.
d, Town of Licodia dei Monaci.
e, Marine formation called creta, argillaceous and sandy beds with a few shells, and associated volcanic rocks.
f, Escarpment of stratified subaqueous volcanic tuff, &c., northwest of Catania.
g, Town of Catania.
h i, Dotted line expressing the highest boundary along which the marine strata are occasionally seen.
k, Plain of Catania.
l, Limestone platform of Primosole of the Newer Pliocene period.
m, La Motta di Catania.

The mountain is in general of a very symmetrical form, a flattened cone broken on its eastern side, by a deep valley, called the Val del Bove, or in the provincial dialect of the peasants, "Val di Bué," for here the herdsman

—— "in reductâ valle *mugientium*
Prospectat errantes greges."

Dr. Buckland was, I believe, the first English geologist who examined this valley with attention, and I am indebted to him for having described it to me, before I visited Sicily, as more worthy of attention than any single spot in that island, or perhaps in Europe.

The Val del Bove commences near the summit of Etna, and descend-

PLATE III

VIEW LOOKING UP THE VAL DEL BOVE, ETNA

ing into the woody region, is farther continued on one side by a second and narrower valley, called the Val di Calanna. Below this another, named the Val di St. Giacomo, begins,—a long narrow ravine, which is prolonged to the neighborhood of Zaffarana (*e*, fig. 48), on the confines of the fertile region. These natural incisions into the side of the volcano are of such depth that they expose to view a great part of the struc-

Fig. 48.

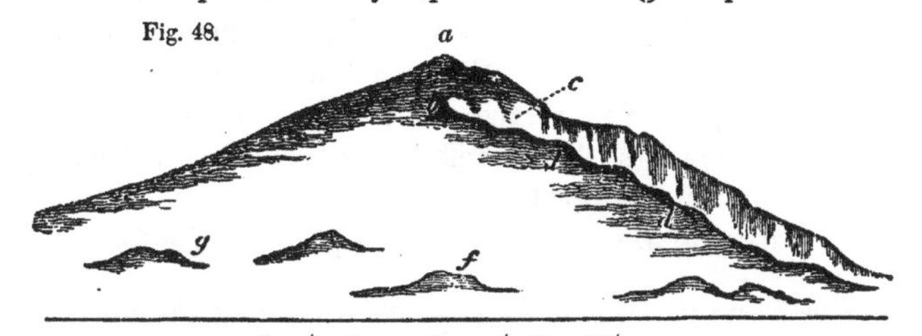

Great valley on the east side of Etna.

a, Highest cone. *b*, Montagnuola.
c, Head of Val del Bove. *d*, *d*, Serre del Solfizio.
e, Village of Zaffarana on the lower border of the woody region.
f, One of the lateral cones. *g*, Monti Rossi.

ture of the entire mass, which, in the Val del Bove, is laid open to the depth of from 3000 to above 4000 feet from the summit of Etna. The geologist thus enjoys an opportunity of ascertaining how far the internal conformation of the cone corresponds with what he might have anticipated as the result of that mode of increase which has been witnessed during the historical era.

Description of Plate III.—The accompanying view (Pl. III.) is part of a panoramic sketch which I made in November, 1828, and may assist the reader in comprehending some topographical details to be alluded to in the sequel, although it can convey no idea of the picturesque grandeur of the scene.

The great lava-currents of 1819 and 1811 are seen pouring down from the higher parts of the valley, overrunning the forests of the great plain, and rising up in the foreground on the left with a rugged surface, on which many hillocks and depressions appear, such as often characterize a lava-current immediately after its consolidation.

The small cone, No. 7, was formed in 1811, and was still smoking when I saw it in 1828. The other small volcano to the left, from which vapor is issuing, was, I believe, one of those formed in 1819.

The following are the names of some of the other points indicated in the sketch:—

1, Montagnuola.
2, Torre del Filosofo.
3, Highest cone.
4, Lepra.
5, Finocchio.
6, Capra.
7, Cone of 1811.
8, Cima del Asino.
9, Musara.
10, Zocolaro.
11, Rocca di Calanna.

Description of Plate IV.—The second view (Pl. IV.) represents the same valley as seen from above, or looking directly down the Val del Bove, from the summit of the principal crater formed in 1819.* I am

* This view is taken from a sketch made by Mr. James Bridges, corrected after comparison with several sketches of my own.

unable to point out the precise spot which this crater would occupy in the view represented in Plate III.; but I conceive that it would appear in the face of the great precipice, near which the smoke issuing from the cone No. 7 is made to terminate. There are many ledges of rock on the face of that precipice where eruptions have occurred.

The circular form of the Val del Bove is well shown in this view. (Pl. IV.) To the right and left are the lofty precipices which form the southern and northern sides of the great valley, and which are intersected by dikes projecting in the manner afterwards to be described. In the distance appears the "fertile region" of Etna, extending like a great plain along the sea-coast.

The spots particularly referred to in the plate are the following:—

a, Cape Spartivento, in Italy, of which the outline is seen in the distance.
b, The promontory of Taormino, on the Sicilian coast.
c, The river Alcantra.
d, The small village of Riposto.
f, The town of Aci Reale.
g, Cyclopian islands, or "Faraglioni," in the Bay of Trezza.
h, The great harbor of Syracuse.
k, The Lake of Lentini.
i, The city of Catania, near which is marked the course of the lava which flowed from the Monti Rossi in 1669, and destroyed part of the city.
l, To the left of the view is the crater of 1811, which is also shown at No. 7 in Plate III.
m, Rock of Musara, also seen at No. 9 in Plate III.
e, Valley of Calanna.

The Val del Bove is of truly magnificent dimensions, a vast amphitheatre four or five miles in diameter, surrounded by nearly vertical precipices, varying from 1000 to above 3000 feet in height, the loftiest being at the upper end, and the height gradually diminishing on both sides. The feature which first strikes the geologist as distinguishing the boundary cliffs of this valley, is the prodigious multitude of verticle dikes which are seen in all directions traversing the volcanic beds. The circular form of this great chasm, and the occurrence of these countless dikes, amounting perhaps to several thousands in number, so forcibly recalled to my mind the phenomena of the Atrio del Cavallo, on Vesuvius, that I at first imagined that I had entered a vast crater, on a scale as far exceeding that of Somma, as Etna surpasses Vesuvius in magnitude.

But I was soon undeceived when I had attentively explored the different sides of the great amphitheatre, in order to satisfy myself whether the semicircular wall of the Val del Bove had ever formed the boundary of a crater, and whether the beds had the same quâquâ-versal dip which is so beautifully exhibited in the escarpment of Somma. Had the supposed analogy between Somma and the Val del Bove held true, the tufts and lavas at the head of the valley would have dipped to the west, those on the north side towards the north, and those on the southern side to the south. But such I did not find to be the inclination of the beds; they all dip towards the sea, or nearly east, as in the valleys of St. Giacomo and Calanna below.

PLATE IV

VIEW OF THE VAL DEL BOVE, ETNA, AS SEEN FROM ABOVE, OR FROM THE CRATER OF 1819

Scenery of the Val del Bove.—Let the reader picture to himself a large amphitheatre, five miles in diameter, and surrounded on three sides by precipices from 2000 to 3000 feet in height. If he has beheld that most picturesque scene in the chain of the Pyrenees, the celebrated "cirque of Gavarnie," he may form some conception of the magnificent circle of precipitous rocks which inclose, on three sides, the great plain of the Val del Bove. This plain has been deluged by repeated streams of lava; and although it appears almost level, when viewed from a distance, it is, in fact, more uneven than the surface of the most tempestuous sea. Besides the minor irregularities of the lava, the valley is in one part interrupted by a ridge of rocks, two of which, Musara and Capra, are very prominent. It can hardly be said that they

> ——— "like giants stand
> To sentinel enchanted land;"

for although, like the Trosachs, in the Highlands of Scotland, they are of gigantic dimensions, and appear almost isolated, as seen from many points, yet the stern and severe grandeur of the scenery which they adorn is not such as would be selected by a poet for a vale of enchantment. The character of the scene would accord far better with Milton's picture of the infernal world; and if we imagine ourselves to behold in motion, in the darkness of the night, one of those fiery currents which have so often traversed the great valley, we may well recall

> ——— "yon dreary plain, forlorn and wild,
> The seat of desolation, void of light,
> Save what the glimmering of these livid flames
> Casts pale and dreadful."

The face of the precipices already mentioned is broken in the most picturesque manner by the vertical walls of lava which traverse them. These masses usually stand out in relief, are exceedingly diversified in form, and of immense altitude. In the autumn, their black outline may often be seen relieved by clouds of fleecy vapor which settle behind them, and do not disperse until mid-day, continuing to fill the valley while the sun is shining on every other part of Sicily, and on the higher regions of Etna.

As soon as the vapors begin to rise, the changes of scene are varied in the highest degree, different rocks being unveiled and hidden by turns, and the summit of Etna often breaking through the clouds for a moment with its dazzling snows, and being then as suddenly withdrawn from the view.

An unusual silence prevails; for there are no torrents dashing from the rocks, nor any movement of running water in this valley such as may almost invariably be heard in mountainous regions. Every drop of water that falls from the heavens, or flows from the melting ice and snow, is instantly absorbed by the porous lava; and such is the dearth of springs, that the herdsman is compelled to supply his flocks, during the hot season, from stores of snow laid up in hollows of the mountain during winter.

The strips of green herbage and forest land, which have here and there escaped the burning lavas, serve, by contrast, to heighten the desolation of the scene. When I visited the valley, nine years after the eruption of 1819, I saw hundreds of trees, or rather the white skeletons of trees, on the borders of the black lava, the trunks and branches being all leafless, and deprived of their bark by the scorching heat emitted from the melted rock; an image recalling those beautiful lines —

> ——— "As when heaven's fire
> Hath scath'd the forest oaks, or mountain pines,
> With singed top their stately growth, though bare,
> Stands on the blasted heath."

Form, composition, and origin of the dikes.—But without indulging the imagination any longer in descriptions of scenery, I may observe that the dikes before mentioned form unquestionably the most interesting geological phenomenon in the Val del Bove. Some of these are composed of trachyte, others of compact blue basalt with olivine. They vary in breadth from two to twenty feet and upwards, and usually project from the face of the cliffs, as represented in the annexed drawing (fig. 49).

Fig. 49.

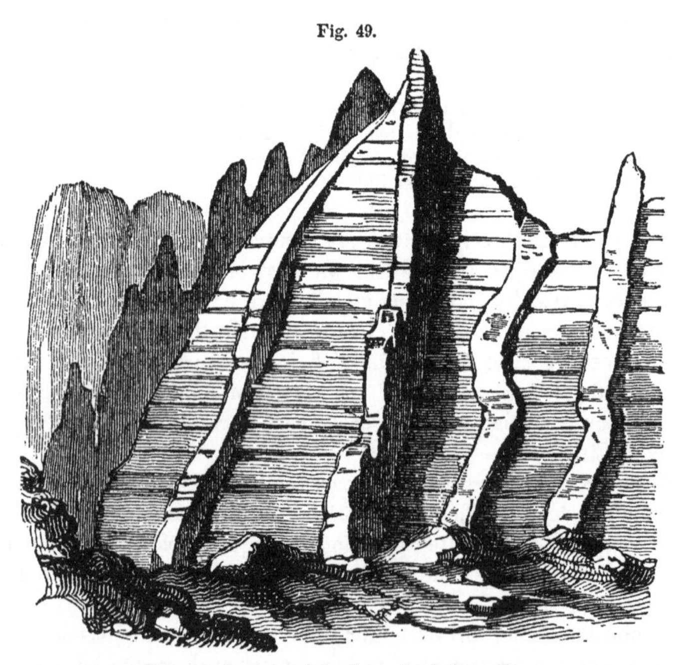

Dikes at the base of the Serre del Solfizio, Etna.

They consist of harder materials than the strata which they traverse, and therefore waste away less rapidly under the influence of that repeated congelation and thawing to which the rocks in this zone of Etna are exposed. The dikes are for the most part vertical, but sometimes they run in a tortuous course through the tuffs and breccias, as represented in fig. 50. In the escarpment of Somma, where similar walls of lava cut through alternating beds of sand and scoriæ, a coating of coal-black rock, approaching in its nature and appearance to pitchstone, is seen at the contact of the dike with the intersected beds. I did not observe such parting layers at the junction of the Etnean dikes which I examined, but they may perhaps be discoverable.

Fig. 50.

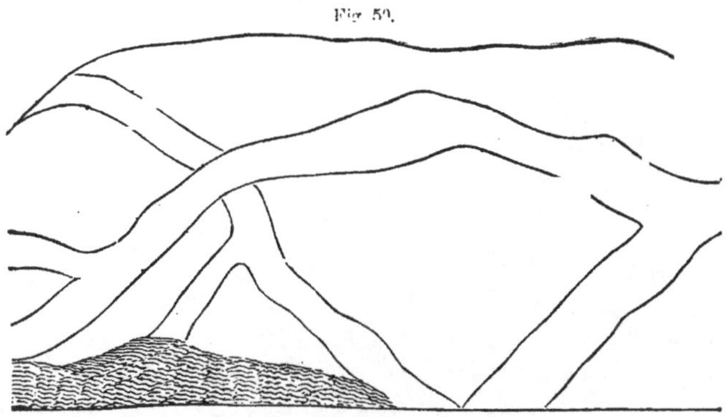

Tortuous veins of lava at Punto di Giumento, Etna.

The geographical position of these dikes is most interesting, as they are very numerous near the head of the Val del Bove, where the cones of 1811 and 1819 were thrown up, as also in that zone of the mountain where lateral eruptions are frequent; whereas in the valley of Calanna, which is below that parallel, and in a region where lateral eruptions are extremely rare, scarcely any dikes are seen, and none whatever still lower in the valley of St. Giacomo. This is precisely what we might have expected, if we consider the vertical fissures now filled with rock to have been the feeders of lateral cones, or, in other words, the channels which gave passage to the lava-currents and scoriæ that have issued from vents in the forest zone. In other parts of Etna there may be numerous dikes at as low a level as the Valley of Calanna, because the line of lateral eruptions is not everywhere at the same height above the sea; but in the section above alluded to, there appeared to me an obvious connection between the frequency of dikes and of lateral eruptions.

Some fissures may have been filled from above, but I did not see any which, by terminating downwards, gave proof of such an origin. Almost all the isolated masses in the Val del Bove, such as Capra, Musara, and others, are traversed by dikes, and may, perhaps, have partly owed their preservation to that circumstance, if at least the action of occasional floods has been one of the destroying causes in the Val del Bove; for there is nothing which affords so much protection to a mass of strata against the undermining action of running water as a perpendicular dike of hard rock.

In the accompanying drawing (fig. 51), the flowing of the lavas of 1811 and 1819, between the rocks Finochio, Capra, and Musara, is represented. The height of the two last-mentioned isolated masses has been much diminished by the elevation of their base, caused by these currents. They may, perhaps, be the remnants of lateral cones which existed before the Val del Bove was formed, and may hereafter be once more buried by the lavas that are now accumulating in the valley.

Fig. 51.

View of the rocks Finochio, Capra, and Musara, Val del Bove.

From no point of view are the dikes more conspicuous than from the summit of the highest cone of Etna; a view of some of them is given in the annexed drawing. (Fig. 52.)

Eruption of 1811.—I have alluded to the streams of lava which were poured forth in 1811 and 1819. Gemmellaro, who witnessed these eruptions, informs us that the great crater in 1811 first testified by its loud detonations that a column of lava had ascended to near the summit of the mountain. A violent shock was then felt, and a stream broke out from the side of the cone, at no great distance from its apex. Shortly after this had ceased to flow, a second stream burst forth at another opening, considerably below the first; then a third still lower, and so on till seven different issues had been thus successively formed, all lying upon the same straight line. It has been supposed that this line was a perpendicular rent in the internal framework of the mountain, which rent was probably not produced at one shock, but prolonged successively downwards, by the lateral pressure and intense heat of the internal column of lava, as it subsided by gradual discharge through each vent.*

Eruption of 1819.—In 1819 three large mouths or caverns opened very near those which were formed in the eruptions of 1811, from which flames, red-hot cinders, and sand were thrown up with loud explosions. A few minutes afterwards another mouth opened below, from which

* Scrope on Volcanoes, p. 153.

Fig. 52.

View from the summit of Etna into the Val del Bove *

flames and smoke issued; and finally a fifth, lower still, whence a torrent of lava flowed, which spread itself with great velocity over the deep and broad valley called "Val del Bove." This stream flowed two miles in the first twenty-four hours, and nearly as far in the succeeding day and night. The three original mouths at length united into one large crater, and sent forth lava, as did the inferior apertures, so that an enormous torrent poured down the "Val del Bove." When it arrived at a vast and almost perpendicular precipice, at the head of the Valley of Calanna, it poured over in a cascade, and, being hardened in its descent, made an inconceivable crash as it was dashed against the bottom. So immense was the column of dust raised by the abrasion of the tufaceous hill over which the hardened mass descended, that the Catanians were in great alarm, supposing a new eruption to have burst out in the woody region, exceeding in violence that near the summit of Etna.

Mode of advance of the lava.—Of the cones thrown up during this eruption, not more than two are of sufficient magnitude to be numbered among those eighty which were before described as adorning the flanks of Etna. The surface of the lava which deluged the "Val del Bove," consists of rocky and *angular blocks,* tossed together in the

* This drawing is part of a panoramic sketch which I made from the summit of the cone, December 1, 1828, when every part of Etna was free from clouds except the Val del Bove. The small cone, and the crater nearest the foreground, were among those formed during the eruptions of 1810 and 1811.

utmost disorder. Nothing can be more rugged, or more unlike the smooth and even superficies, which those who are unacquainted with volcanic countries may have pictured to themselves, in a mass of matter which had consolidated from a liquid state. Mr. Scrope observed this current in the year 1819, slowly advancing down a considerable slope, at the rate of about a yard an hour, nine months after its emission. The lower stratum being arrested by the resistance of the ground, the upper or central part gradually protruded itself, and, being unsupported, fell down. This in its turn was covered by a mass of more liquid lava, which swelled over it from above. The current had all the appearance of a huge heap of rough and large cinders rolling over and over upon itself by the effect of an extremely slow propulsion from behind. The contraction of the crust as it solidified, and the friction of the scoriform cakes against one another, produced a crackling sound. Within the crevices a dull red heat might be seen by night, and vapor issuing in considerable quantity was visible by day.*

It was stated that when the lava of 1819 arrived at the head of the Valley of Calanna, after flowing down the Val del Bove, it descended in a cascade. This stream, in fact, like many previous currents of lava which have flowed down successively from the higher regions of Etna, was turned by a great promontory projecting from the southern side of the Val del Bove. This promontory consists of the hills called Zocolaro and Calanna, and of a ridge of inferior height which connects them. (See fig. 53.)

Fig. 53.

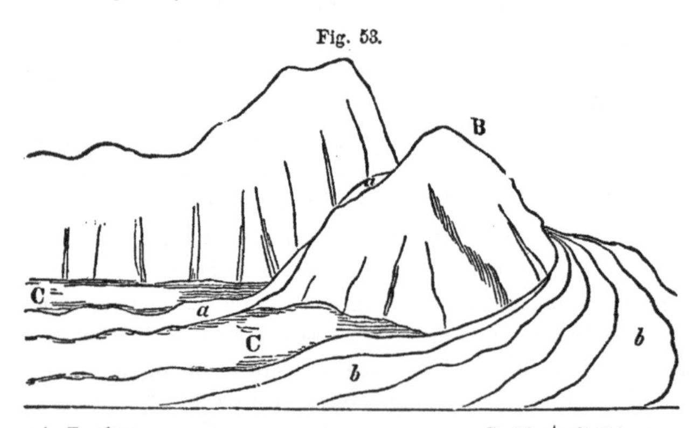

A, Zocolaro. B, Monte di Calanna.
C, Plain at the head of the Valley of Calanna.
a, Lava of 1819 descending the precipice and flowing through the valley.
b, Lavas of 1811 and 1819 flowing round the hill of Calanna.

It happened in 1811 and 1819 that the flows of lava overtopped the ridge intervening between the hills of Zocolaro and Calanna, so that they fell in a cascade over a lofty precipice, and began to fill up the valley of Calanna (*a*, fig. 53). Other portions of the same lava-current (*b*) flowed round the promontory, and they exhibit one of the peculiar

* Scrope on Volcanoes, p. 102.

characteristics of such streams, namely that of becoming solid externally, even while yet in motion. Instead of thinning out gradually at their edges, their sides may often be compared to two rocky walls which are sometimes inclined at an angle of between thirty and forty degrees. When such streams are turned from their course by a projecting rock, they move right onwards in a new direction; and in the Valley of Calanna a considerable space has thus been left between the steep sides of the lavas *b b*, so deflected, and the precipitous escarpment of Zocolaro, A, which bounds the plain C.

Lavas and breccias.—In regard to the volcanic masses which are intersected by dikes in the Val del Bove, they consist in great part of graystone lavas, of an intermediate character between basalt and trachyte, and partly of porphyritic lava resembling trachyte, but to which that name cannot, according to Von Buch and G. Rose, be in strictness applied, because the felspar belongs to the variety called Labradorite. There is great similarity in the composition of the ancient and modern lavas of Etna, both consisting of felspar, augite, olivine, and titaniferous iron. The alternating breccias are made up of scoriæ, sand, and angular blocks of lava. Many of these fragments may have been thrown out by volcanic explosions, which, falling on the hardened surface of moving lava-currents, may have been carried to a considerable distance. It may also happen that when lava advances very slowly, in the manner of the flow of 1819, the angular masses resulting from the frequent breaking of the mass as it rolls over upon itself, may produce these breccias. It is at least certain that the upper portion of the lava-currents of 1811 and 1819 now consist of angular masses to the depth of many yards. D'Aubuisson has compared the surface of one of the ancient lavas of Auvergne to that of a river suddenly frozen over by the stoppage of immense fragments of drift-ice, a description perfectly applicable to these modern Etnean flows. The thickness of the separate beds of conglomerate or breccia which are seen in the same vertical section, is often extremely different, varying from 3 to nearly 50 feet, as I observed in the hill of Calanna.

Flood produced by the melting of snow by lava.—It is possible that some of the breccias or conglomerates may be referred to aqueous causes, as great floods occasionally sweep down the flanks of Etna, when eruptions take place in winter, and when the snows are melted by lava. It is true that running water in general exerts no power on Etna, the rain which falls being immediately imbibed by the porous lavas; so that, vast as is the extent of the mountain, it feeds only a few small rivulets, and these, even, are dry throughout the greater portion of the year. The enormous rounded boulders, therefore, of felspar-porphyry and basalt, a line of which can be traced from the sea, from near Giardini, by Mascali, and Zafarana, to the "Val del Bove," would offer a perplexing problem to the geologist, if history had not preserved the memorials of a tremendous flood which happened in this district in in the year 1755. It appears that two streams of lava flowed in that year, on the 2d of March, from the highest crater; they were immediately precipitated upon an enormous mass of snow which then covered the whole mountain, and was extremely deep near the summit. The sudden melting of this frozen mass, by a fiery torrent three miles in length, produced a frightful inundation, which devastated the sides of the mountain for eight miles in length, and afterwards covered the lower flanks of Etna, where they were less steep, together with the plains near the sea, with great deposits of sand, scoriæ, and blocks of lava.

Many absurd stories circulated in Sicily respecting this event; such as that the water was boiling, and that it was vomited from the highest crater; that it was as salt as the sea, and full of marine shells; but these were mere inventions, to which Recupero, although he relates them as tales of the mountaineers, seems to have attached rather too much importance.

Floods of considerable violence have also been produced on Etna by the fall of heavy rains, aided, probably, by the melting of snow. By this cause alone, in 1761, sixty of the inhabitants of Acicatena were killed, and many of their houses swept away.*

Glacier covered by a lava-stream.—A remarkable discovery was made on Etna in 1828 of a great mass of ice, preserved for many years, perhaps for centuries, from melting, by the singular accident of a current of red-hot lava having flowed over it. The following are the facts in attestation of a phenomenon which must at first sight appear of so paradoxical a character. The extraordinary heat experienced in the South of Europe, during the summer and autumn of 1828, caused the supplies of snow and ice which had been preserved in the spring of that year, for the use of Catania and the adjoining parts of Sicily and the island of Malta, to fail entirely. Great distress was consequently felt for want of a commodity regarded in those countries as one of the necessaries of life rather than an article of luxury, and the abundance of which contributes in some of the larger cities to the salubrity of the water and the general health of the community. The magistrates of Catania applied to Signor M. Gemmellaro, in the hope that his local knowledge of Etna might enable him to point out some crevice or natural grotto on the mountain, where drift-snow was still preserved. Nor were they disappointed; for he had long suspected that a small mass of perennial ice at the foot of the highest cone was part of a large and continuous glacier covered by a lava-current. Having procured a large body of workmen, he quarried into this ice, and proved the superposition of the lava for several hundred yards, so as completely to satisfy himself that nothing but the subsequent flowing of the lava over the ice could account for the position of the glacier. Unfortunately for the geologist, the ice was so extremely hard, and the excavation so expensive, that there is no probability of the operations being renewed.

On the first of December, 1828, I visited this spot, which is on the

* Ferara, Descriz. dell' Etna, p. 116.

southeast side of the cone, and not far above the Casa Inglese; but the fresh snow had already nearly filled up the new opening, so that it had only the appearance of the mouth of a grotto. I do not, however, question the accuracy of the conclusion of Signor Gemmellaro, who, being well acquainted with all the appearances of drift-snow in the fissures and cavities of Etna, had recognized, even before the late excavations, the peculiarity of the position of the ice in this locality. We may suppose that, at the commencement of the eruption, a deep mass of drift-snow had been covered by volcanic sand showered down upon it before the descent of the lava. A dense stratum of this fine dust mixed with scoriæ is well known to be an extremely bad conductor of heat; and the shepherds in the higher regions of Etna are accustomed to provide water for their flocks during summer, by strewing a layer of volcanic sand a few inches thick over the snow, which effectually prevents the heat of the sun from penetrating.

Suppose the mass of snow to have been preserved from liquefaction until the lower part of the lava had consolidated, we may then readily conceive that a glacier thus protected, at the height of ten thousand feet above the level of the sea, would endure as long as the snows of Mont Blanc, unless melted by volcanic heat from below. When I visited the great crater in the beginning of winter (December 1st, 1828), I found the crevices in the interior incrusted with thick ice, and in some cases hot vapors were actually streaming out between masses of ice and the rugged and steep walls of the crater.*

After the discovery of Signor Gemmellaro, it would not be surprising to find in the cones of the Icelandic volcanoes, which are covered for the most part with perpetual snow, repeated alternations of lava-streams and glaciers. We have, indeed, Lieutenant Kendall's authority for the fact that Deception Island, in New South Shetland, lat. 62° 55′ S., is principally composed of alternate layers of volcanic ashes and ice.†

Origin of the Val del Bove.—It is recorded, as will be stated in the history of earthquakes (ch. 29), that in the year 1772 a great subsidence took place on Papandayang, the largest volcano in the island of Java; an extent of ground *fifteen miles in length, and six in breadth,* covered by no less than forty villages, was engulphed, and the cone lost 4000 feet of its height. In like manner the summit of Carguairazo, one of the loftiest of the Andes of Quito, fell in on the 19th July, 1698; and another mountain of still greater altitude in the same chain, called Capac Urcu, a short time before the conquest of America by the Spaniards.

* Mr. Nasmyth, the inventor of the steam-hammer, has lately illustrated, by a very striking experiment, the non-conductibility of a thin layer of dry sand and clay. Into a caldron of iron one-fourth of an inch thick, lined with sand and clay five-eighths of an inch thick, he poured eight tons of melted iron at a white heat. After the fused metal had been twenty minutes in the caldron the palm of the hand could be applied to the outside without inconvenience, and after forty minutes there was not heat enough to singe writing-paper. This fact may help us to explain how strata in contact with dikes, or beds of fused matter, have sometimes escaped without perceptible alteration by heat.

† Journ. of Roy. Geograph. Soc. vol. i. p. 64.

It will also be seen in the next chapter that, so late as the year 1822, during a violent earthquake and volcanic eruption in Java, one side of the mountain called Galongoon, which was covered by a dense forest, became an enormous gulf in the form of a semicircle. The new cavity was about midway between the summit and the plain, and surrounded by steep rocks.

Now we might imagine a similar event, or a series of subsidences to have formerly occurred on the eastern side of Etna, although such catastrophes have not been witnessed in modern times, or only on a very trifling scale. A narrow ravine, about a mile long, twenty feet wide, and from twenty to thirty-six in depth, has been formed, within the historical era, on the flanks of the volcano, near the town of Mascalucia; and a small circular tract, called the Cisterna, near the summit, sank down in the year 1792, to the depth of about forty feet, and left on all sides of the chasm a vertical section of the beds, exactly resembling those which are seen in the precipices of the Val del Bove. At some remote periods, therefore, we might suppose more extensive portions of the mountain to have fallen in during great earthquakes.

But we ought not to exclude entirely from our speculations another possible agency, by which the great cavity may in part at least have been excavated, namely, the denuding action of the sea. Whether its waves may once have had access to the great valley before the ancient portion of Etna was upheaved to its present elevation, is a question which will naturally present itself to every geologist. Marine shells have been traced to a height of 800 feet above the base of Etna, and would doubtless be seen to ascend much higher, were not the structure of the lower region of the mountain concealed by floods of lava. We cannot ascertain to what extent a change in the relative level of land and sea may have been carried in this spot, but we know that some of the tertiary strata in Sicily of no ancient date reach a height of 3000 feet, and the marine deposits on the flanks of Etna, full of recent species of shells, may ascend to equal or greater heights. The narrow Valley of Calanna leading out of the Val del Bove, and that of San Giacomo lower down, have much the appearance of ravines swept out by aqueous action.

Structure and origin of the cone of Etna.—Our data for framing a correct theory of the manner in which the cone of Etna has acquired its present dimensions and internal structure are very imperfect, because it is on its eastern side only, in the Val del Bove above described, that we see a deep section exposed. Even here we obtain no insight into the interior composition of the mountain beyond a depth of between three and four thousand feet below the base of that highest cone, which has been several times destroyed and renewed. The precipices seen at the head of the Val del Bove, in the escarpment called the Serre del Solfizio, exhibit merely the same series of alternating lavas and breccias, which, descending with a general dip towards the sea, form the boundary cliffs of all other parts of the Val del Bove. If then we estimate the height of Etna at about 11,000 feet, we may say that we know from actual observation less than one-half of its component materials, assuming it to extend downwards to the level of the sea; namely, first, the highest cone, which is about 1000 feet above its base; and, secondly, the alternations of lava, tuff, and volcanic breccia, which constitute the rocks between the Cisterna, near the base of the upper cone, and the foot of the precipices at the head of the Val del Bove. At the lowest point to which the vertical section extends, there are no signs of any approach to a termination of the purely volcanic mass, which may perhaps penetrate many thousand feet farther downwards. There is, indeed, a rock called Rocca Gianicola, near the foot of the great escarpment, which consists of a large mass between 150 and 200 feet wide, not divided into beds, and almost resembling granite in its structure, although agreeing very closely in mineral composition with the lavas of Etna in general.* This mass may doubtless be taken as a representative of those crystalline or plutonic formations which would be met with in abundance if we could descend to greater depths in the direction of the central axis of the mountain. For a great body of geological evidence leads us to conclude, that rocks of this class result from the consolidation, under great pressure, of melted matter, which has risen up and filled rents and chasms, such, for example, as may communicate with the principal and minor vents of eruption in a volcano like Etna.

But, if we speculate on the nature of the formation which the lava may have pierced in its way upwards, we may fairly presume that a portion of these consist of marine tertiary rocks, like those of the neighboring Val di Noto, or those which skirt the borders of the Etnean cone, on its southern and eastern sides. Etna may, in fact, have been at first an insular volcano, raising its summit but slightly above the level of the sea; but we have no grounds for concluding that any of the beds exposed in the deep section of the Val del Bove have formed a part of such a marine accumulation. On the contrary, all the usual signs of subaqueous origin are wanting; and even if we believe the foundations of the mountain to have been laid in the sea, we could not expect this portion to be made visible in sections which only proceed downwards from the summit through one-half the thickness of the mountain, especially as the highest points attained by the tertiary strata in other parts of Sicily very rarely exceed 3000 feet above the sea.

On the eastern and southern base of Etna, a marine deposit, already alluded to, is traced up to the height of 800 or 1000 feet, before it becomes concealed beneath that covering of modern lavas which is continually extending its limits during successive eruptions, and prevents us from ascertaining how much higher the marine strata may ascend. As the imbedded shells belong almost entirely to species now inhabiting the Mediterranean, it is evident that there has been here an upheaval of the region at the base of Etna at a very modern period. It is fair, there-

* Hoffman, Geognost. Beobachtungen, p. 701. Berlin, 1839.

fore, to infer that the volcanic nucleus of the mountain, partly perhaps of submarine, and partly of subaerial origin, participated in this movement, and was carried up bodily. Now, in proportion as a cone gains height by such a movement, combined with the cumulative effects of eruptions, throwing out matter successively from one or more central vents, the hydrostatic pressure of the columns of lava augments with their increasing height, until the time arrives when the flanks of the cone can no longer resist the increased pressure; and from that period they give way more readily, lateral outbursts becoming more frequent. Hence, independently of any local expansion of the fractured volcanic mass, those general causes by which the modern tertiary strata of a great part of Sicily have been raised to the height of several thousand feet above their original level, would tend naturally to render the discharge of lava and scoriæ from the summit of Etna less copious, and the lateral discharge greater.

If, then, a conical or dome-shaped mass of volcanic materials was accumulated to the height of 4000, or perhaps 7000 feet, before the up ward movement began, or, what is much more probable, during the continuance of the upward movement, that ancient mass would not be buried under the products of newer eruptions, because these last would then be poured out chiefly at a lower level.

Since I visited Etna in 1828, M. de Beaumont has published a most valuable memoir on the structure and origin of that mountain, which he examined in 1834;* and an excellent description of it has also appeared in the posthumous work of Hoffmann.†

In M. de Beaumont's essay, in which he has explained his views with uncommon perspicuity and talent, he maintains that all the alternating stony and fragmentary beds, more than 3000 feet thick, which are exposed in the Val del Bove, were formed originally on a surface so nearly flat that the slope never exceeded three degrees. From this horizontal position they were at length heaved up suddenly (d'un seul coup) into a great mountain, to which no important additions have since been made. Prior to this upthrow, a platform is supposed to have existed above the level of the sea, in which various fissures opened; and from these melted matter was poured forth again and again, which spread itself around in thin sheets of uniform thickness. From the same rents issued showers of scoriæ and fragmentary matter, which were spread out so as to form equally uniform and horizontal beds, intervening between the sheets of lava. But although, by the continued repetition of these operations, a vast pile of volcanic matter, 4000 feet or more in thickness, was built up precisely in that region where Etna now rises, and to which nothing similar was produced elsewhere in Sicily, still we are told that Etna was not yet a mountain. No hypothetical diagram has been given to help us to conceive how this great mass of materials of supramarine origin

* Mém. pour servir, &c., tom. iv. Paris, 1838.
† Geognost. Beobachtungen, &c. Berlin, 1839.

could have been disposed of in horizontal beds, so as not to constitute an eminence towering far above the rest of Sicily; but it is assumed that a powerful force from below at length burst suddenly through the horizontal formation, uplifted it to a considerable height, and caused the beds to be, in many places, highly inclined. This elevatory force was not all expended on a single central point as Von Buch has imagined in the case of Palma, Teneriffe, or Somma, but rather followed for a short distance a linear direction.*

Among other objections that may be advanced against the theory above proposed, I may mention, first, that the increasing number of dikes as we approach the head of the Val del Bove, or the middle of Etna, and the great thickness of lava, scoriæ, and conglomerates in that region, imply that the great centre of eruption was always where it now is, or nearly at the same point, and there must, therefore, have been a tendency, from the beginning, to a conical or dome-shaped arrangement in the ejected materials. Secondly, were we to admit a great number of separate points of eruption, scattered over a plain or platform, there must have been a great number of cones thrown up over these different vents; and these hills, some of which would probably be as lofty as those now seen on the flanks of Etna, or from 300 to 750 feet in height, would break the continuity of the sheets of lava, while they would become gradually enveloped by them. The ejected materials, moreover, would slope at a high angle on the sides of these cones, and where they fell on the surrounding plain, would form strata thicker near the base of each cone than at a distance.

What then are the facts, it will be asked, to account for which this hypothesis of original horizontality, followed by a single and sudden effort of upheaval, which gave to the beds their present slope, has been invented? M. de Beaumont observes, that in the boundary precipices of the Val del Bove, sheets of lava and intercalated beds of cinders, mixed with pulverulent and fragmentary matter evidently cast out during eruptions, are sometimes inclined at steep angles, varying from 15° to 27°. It is impossible, he says, that the lavas could have flowed originally on planes so steeply inclined, for streams which descend a slope even of 10° from narrow stripes, and never acquire such a compact texture. Their thickness, moreover, always inconsiderable, varies with every variation of steepness, in the declivity down which they flow; whereas, in several parts of the Val del Bove, the sheets of lava are continuous for great distances, in spite of their steep inclination, and are often compact, and perfectly parallel one to the other, even where there are more than 100 beds of interpolated fragmentary matter.

The intersecting dikes also terminate upwards in many instances, at different elevations, and blend (or, as M. de Beaumont terms it, articulate) with sheets of lava, which they meet at right angles. It is therefore assumed that such dikes were the feeders of the streams of lava

* De Beaumont, Mém. pour servir, &c. tom. iv. pp. 187, 188.

with which they unite, and they are supposed to prove that the platform, on the surface of which the melted matter was poured out, was at first so flat, that the fluid mass spread freely and equally in every direction, and not towards one point only of the compass, as would happen if it had descended the sloping sides of a cone. This argument is ingeniously and plainly put in the following terms:—"Had the melted matter poured down an inclined plane, after issuing from a rent, the sheet of lava would, after consolidation, have formed an elbow with the dike, like the upper bar of the letter F, instead of extending itself on both sides like that of a T."* It is also contended that a series of sheets of lava, formed on a conical or dome-shaped mountain, would have been more numerous at points farthest from the central axis, since every dike which had been the source of a lava-stream, must have poured its contents downwards, and never upwards.

In reference to the facts here stated, I may mention that the dikes which I saw in the Val del Bove were either vertical, or made almost all of them a near approach to the perpendicular, which could not have been the case had they been the feeders of horizontal beds of lava, and had they consequently joined them originally at right angles, for then the dikes, as at *a*, *b*, *c*, fig. 54, ought subsequently to have acquired a

Fig. 54.

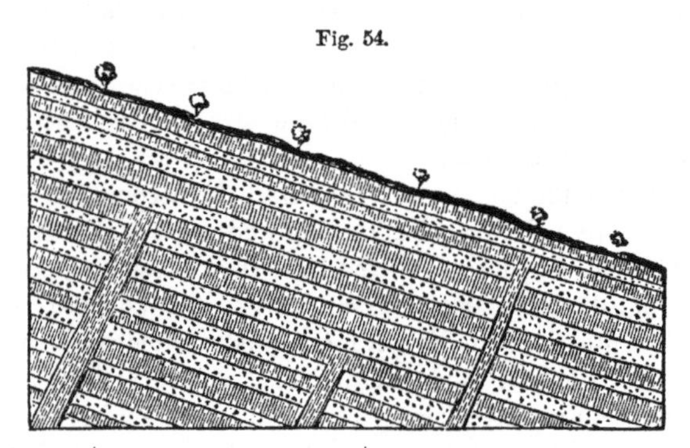

Dikes as they would now appear had they been originally perpendicular.

considerable slope, like the beds which they intersect. I may also urge another objection to the views above set forth, namely, that had the dikes been linear vents, or orifices of eruption, we must suppose the inter-stratified scoriæ and lapilli, as well as the lavas, to have come out of them, and in that case the irregular heaping up of fragmentary matter around the vents would, as before hinted, have disturbed that uniform thickness and parallelism of the beds which M. de Beaumont describes.

If, however, some of the sheets of lava join the dikes in such a manner, as to imply that they were in a melted state simultaneously with the contents of the fissures,—a point not easily ascertained, where the

* Mém. pour servir, tom. iv. p. 149.

precipices are for the most part inaccessible,—the fact may admit of a different interpretation from that proposed by the French geologists. Rents like those before alluded to (p. 399), which opened in the plain of S. Lio in 1669, filled below with incandescent lava, may have lain in the way of currents of melted matter descending from higher openings. In that case, the matter of the current would have flowed into the fissure and mixed with the lava at its bottom. Numerous open rents of this kind are described by Mr. Dana as having been caused, during a late eruption, in one of the volcanic domes of the Sandwich Islands. They remained open at various heights on the slopes of the great cone, running in different directions, and demonstrate the possibility of future junctions of slightly inclined lava-streams with perpendicular walls of lava.

To me, therefore, it appears far more easy to explain the uniform thickness and parallelism of so many lavas and beds of fragmentary matter seen in the Val del Bove, by supposing them to have issued successively out of one or more higher vents near the summit of a great dome, than to imagine them to have proceeded from lateral dikes or rents opening in a level plain. In the Sandwich Islands, we have examples of volcanic domes 15,000 feet high, produced by successive outpourings from vents at or near the summit. One of these, Mount Loa, has a slope in all directions of 6° 30′; another, Mount Kea, a mean inclination of 7° 46′. That their lavas may occasionally consolidate on slopes of 25°, and even more, and still preserve considerable solidity of texture, has been already stated; see above, p. 383.

We know not how large a quantity of modern lava may have been poured into the bottom of the Val del Bove, yet we perceive that eruptions breaking forth near the centre of Etna have already made some progress in filling up this great hollow. Even within the memory of persons now living, the rocks of Musara and Capra have, as before stated, lost much of their height and picturesque grandeur by the piling up of recent lavas round their base (see fig. 51, p. 408), and the great chasm has intercepted many streams which would otherwise have deluged the fertile region below, as has happened on the side of Catania. The volcanic forces are now laboring, therefore, to repair the breach which subsidence has caused on one side of the great cone; and unless their energy should decline, or a new sinking take place, they may in time efface this inequality. In that event, the restored portion will always be unconformable to the more ancient part, yet it will consist, like it, of alternating beds of lava, scoriæ, and conglomerates, which, with all their irregularities, will have a general slope from the centre and summit of Etna towards the sea.

I shall conclude, then, by remarking that I conceive the general inclination of the alternating stony and fragmentary beds of the Val del Bove, from the axis of Etna towards its circumference or base, and the greater thickness of the volcanic pile as we approach the central parts of the mountain, to be due to the preponderance of eruptions from that centre. These gave rise, from the first, to a dome-shaped mass, which has ever since been increasing in height and area, being fractured again and again by the expansive force of vapors, and the several parts made to cohere together more firmly after the solidification of the lava with which every open fissure and chasm has been filled. At the same time the cone may have gained a portion of its height by the elevatory effect of such dislocating movements, and the sheets of lava may have acquired in some places a greater, in others a less, inclination than that which at first belonged to them.

But had the mountain been due solely, or even principally, to upheaval, its structure would have resembled that which geologists have so often recognized in dome-shaped hills, or certain elevated regions, which all consider as having been thrust up by a force from below. In this case there is often an elliptical cavity at the summit, due partly to the fracture of the upraised rocks, but still more to aqueous denudation, as they rose out of the sea. The central cavity, or valley, exposes to view the subjacent formation *c*, fig. 55, and the incumbent mass dips away on all

Fig. 55.

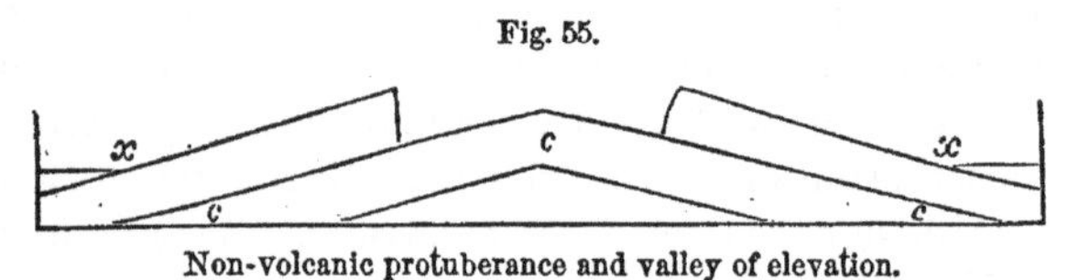

Non-volcanic protuberance and valley of elevation.

sides from the axis, but has no tendency to thin out near the base of the dome, or at *x*, *x*; whereas at this point the volcanic mass terminates (see fig. 56) and allows the fundamental rock *c* to appear at the surface. In the last diagram, the more ordinary case is represented of a great hollow or crater at the summit of the volcanic cone; but instead of

Fig. 56.

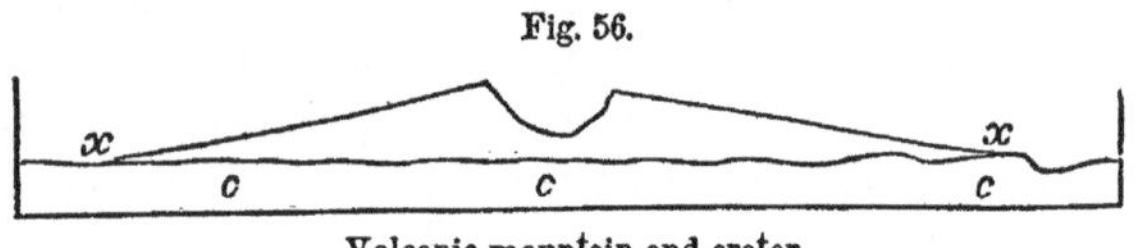

Volcanic mountain and crater.

this, we have seen that in the case of Etna there is a deep lateral depression, called the Val del Bove, the upper part of which approaches near to the central axis, and the origin of which we have attributed to subsidence.

Antiquity of the cone of Etna.—It was before remarked that confined notions in regard to the quantity of past time have tended, more than any other prepossessions, to retard the progress of sound theoretical views in geology;* the inadequacy of our conceptions of the earth's antiquity having cramped the freedom of our speculations in this science, very much in the same way as a belief in the existence of a vaulted firmament once retarded the progress of astronomy. It was not until Descartes assumed the indefinite extent of the celestial spaces, and removed the supposed boundaries of the universe, that just opinions began

* P. 62, *supra*.

to be entertained of the relative distances of the heavenly bodies; and until we habituate ourselves to contemplate the possibility of an indefinite lapse of ages having been comprised within each of the modern periods of the earth's history, we shall be in danger of forming most erroneous and partial views in geology.

If history had bequeathed to us a faithful record of the eruptions of Etna, and a hundred other of the principal active volcanoes of the globe, during the last three thousand years,—if we had an exact account of the volume of lava and matter ejected during that period, and the times of their production,—we might, perhaps, be able to form a correct estimate of the average rate of the growth of a volcanic cone. For we might obtain a mean result from the comparison of the eruptions of so great a number of vents, however irregular might be the development of the igneous action in any one of them, if contemplated singly during a brief period.

It would be necessary to balance protracted periods of inaction against the occasional outburst of paroxysmal explosions. Sometimes we should have evidence of a repose of seventeen centuries, like that which was interposed in Ischia, between the end of the fourth century B. C., and the beginning of the fourteenth century of our era.* Occasionally a tremendous eruption, like that of Jorullo, would be recorded, giving rise, at once, to a considerable mountain.

If we desire to approximate to the age of a cone such as Etna, we ought first to obtain some data in regard to the thickness of matter which has been added during the historical era, and then endeavor to estimate the time required for the accumulation of such alternating lavas and beds of sand and scoriæ as are superimposed upon each other in the Val del Bove; afterwards we should try to deduce, from observations on other volcanoes, the more or less rapid increase of burning mountains in all the different stages of their growth.

There is a considerable analogy between the mode of increase of a volcanic cone and that of trees of *exogenous* growth. These trees augment, both in height and diameter, by the successive application externally of cone upon cone of new ligneous matter; so that if we make a transverse section near the base of the trunk, we intersect a much greater number of layers than nearer to the summit. When branches occasionally shoot out from the trunk, they first pierce the bark, and then, after growing to a certain size, if they chance to be broken off, they may become inclosed in the body of the tree, as it augments in size, forming knots in the wood, which are themselves composed of layers of ligneous matter, cone within cone.

In like manner, a volcanic mountain, as we have seen, consists of a succession of conical masses enveloping others, while lateral cones, having a similar internal structure, often project, in the first instance, like branches from the surface of the main cone, and then becoming buried again, are hidden like the knots of a tree.

* See p. 366

We can ascertain the age of an oak or pine by counting the number of concentric rings of annual growth seen in a transverse section near the base, so that we may know the date at which the seedling began to vegetate. The Baobab-tree of Senegal (*Adansonia digitata*) is supposed to exceed almost any other in longevity. Adanson inferred that one which he measured, and found to be thirty feet in diameter, had attained the age of 5150 years. Having made an incision to a certain depth, he first counted three hundred rings of annual growth, and observed what thickness the tree had gained in that period. The average rate of growth of younger trees, of the same species, was then ascertained, and the calculation made according to a supposed mean rate of increase. De Candolle considers it not improbable that the celebrated Taxodium of Chapultepec, in Mexico (*Cupressus disticha*, Linn.), which is 117 feet in circumference, may be still more aged.*

It is, however, impossible, until more data are collected respecting the average intensity of the volcanic action, to make any thing like an approximation to the age of a cone like Etna; because, in this case, the successive envelopes of lava and scoriæ are not continuous, like the layers of wood in a tree, and afford us no definite measure of time. Each conical envelope is made up of a great number of distinct lava-currents and showers of sand and scoriæ, differing in quantity, and which may have been accumulated in unequal periods of time. Yet we cannot fail to form the most exalted conception of the antiquity of this mountain, when we consider that its base is about ninety miles in circumference; so that it would require ninety flows of lava, each a mile in breadth at their termination, to raise the present foot of the volcano as much as the average height of one lava-current.

There are no records within the historical era which lead to the opinion that the altitude of Etna has materially varied within the last two thousand years. Of the eighty most conspicuous minor cones which adorn its flanks, only one of the largest, Monti Rossi, has been produced within the times of authentic history. Even this hill, thrown up in the year 1669, although 450 feet in height, only ranks as a cone of second magnitude. Monte Minardo, near Bronte, rises, even now, to the height of 750 feet, although its base has been elevated by more modern lavas and ejections. The dimensions of these larger cones appear to bear testimony to *paroxysms* of volcanic activity, after which we may conclude, from analogy, that the fires of Etna remained dormant for many years—since nearly a century of rest has sometimes followed a violent eruption in the historical era. It must also be remembered, that of the small number of eruptions which occur in a century, one only is estimated to issue from the summit of Etna for every two that proceed from the sides. Nor do all the lateral eruptions give rise to such cones as would be reckoned amongst the smallest of the eighty hills above enumerated; some of them produce merely insignificant monticules, which are soon afterwards buried by showers of ashes.

* On the Longevity of Trees, Bibliot. Univ., May, 1831.

How many years then must we not suppose to have been expended in the formation of the eighty cones? It is difficult to imagine that a fourth part of them have originated during the last thirty centuries. But if we conjecture the whole of them to have been formed in twelve thousand years, how inconsiderable an era would this portion of time constitute in the history of the volcano! If we could strip off from Etna all the lateral monticules now visible, together with the lavas and scoriæ that have been poured out from them, and from the highest crater, during the period of their growth, the diminution of the entire mass would be extremely slight: Etna might lose, perhaps, several miles in diameter at its base, and some hundreds of feet in elevation; but it would still be the loftiest of Sicilian mountains, studded with other cones, which would be recalled, as it were, into existence by the removal of the rocks under which they are now buried.

There seems nothing in the deep sections of the Val del Bove to indicate that the lava-currents of remote periods were greater in volume than those of modern times; and there are abundant proofs that the countless beds of solid rock and scoriæ were accumulated, as now, in succession. On the grounds, therefore, already explained, we must infer that a mass so many thousand feet in thickness must have required an immense series of ages anterior to our historical periods for its growth; yet the whole must be regarded as the product of a modern portion of the tertiary epoch. Such, at least, is the conclusion that seems to follow from geological data, which show that the oldest parts of the mountain, if not of posterior date to the marine strata around its base, were at least of coeval origin.

Some geologists contend, that the sudden elevation of large continents from beneath the waters of the sea have again and again produced waves which have swept over vast regions of the earth.* But it is clear that no devastating wave has passed over the forest zone of Etna since any of the lateral cones before mentioned were thrown up; for none of these heaps of loose sand and scoriæ could have resisted for a moment the denuding action of a violent flood. To some, perhaps, it may appear that hills of such incoherent materials cannot be of very great antiquity, because the mere action of the atmosphere must, in the course of several thousand years, have obliterated their original forms. But there is no weight in this objection; for the older hills are covered with trees and herbage, which protect them from waste; and, in regard to the newer ones, such is the porosity of their component materials, that the rain which falls upon them is instantly absorbed; and for the same reason that the rivers on Etna have a subterranean course, there are none descending the sides of the minor cones.

No sensible alteration has been observed in the form of these cones since the earliest periods of which there are memorials; and there seems no reason for anticipating that in the course of the next ten

* Sedgwick, Anniv. Address to Geol. Soc. p. 35. Feb. 1831.

thousand or twenty thousand years they will undergo any great alteration in their appearance, unless they should be shattered by earthquakes or covered by volcanic ejections.

In other parts of Europe, as in Auvergne and Velay, in France, similar loose cones of scoriæ, probably of as high antiquity as the whole mass of Etna, stand uninjured at inferior elevations above the level of the sea.

CHAPTER XXVI.

Volcanic eruption in Iceland in 1783—New island thrown up—Lava currents of Skaptár Jokul, in same year—their immense volume—Eruption of Jorullo in Mexico—Humboldt's theory of the convexity of the plain of Malpais—Eruption of Galongoon in Java—Submarine volcanoes—Graham island, formed in 1831—Volcanic archipelagoes—Submarine eruptions in mid-Atlantic—The Canaries—Teneriffe—Cones thrown up in Lancerote, 1730–36—Santorin and its contiguous isles—Barren island in the Bay of Bengal—Mud volcanoes—Mineral composition of volcanic products.

Volcanic eruptions in Iceland.—WITH the exception of Etna and Vesuvius, the most complete chronological records of a series of eruptions are those of Iceland, for their history reaches as far back as the ninth century of our era; and, from the beginning of the twelfth century, there is clear evidence that, during the whole period, there has never been an interval of more than forty, and very rarely one of twenty years, without either an eruption or a great earthquake. So intense is the energy of the volcanic action in this region, that some eruptions of Hecla have lasted six years without ceasing. Earthquakes have often shaken the whole island at once, causing great changes in the interior, such as the sinking down of hills, the rending of mountains, the desertion by rivers of their channels, and the appearance of new lakes.* New islands have often been thrown up near the coast, some of which still exist; while others have disappeared, either by subsidence or the action of the waves.

In the interval between eruptions, innumerable hot springs afford vent to subterranean heat, and solfataras discharge copious streams of inflammable matter. The volcanoes in different parts of this island are observed, like those of the Phlegræan Fields, to be in activity by turns, one vent often serving for a time as a safety-valve to the rest. Many cones are often thrown up in one eruption, and in this case they take a linear direction, running generally from northeast to southwest, from the northeastern part of the island, where the volcano Krabla lies, to the promontory Reykianas.

* Von Hoff, vol. ii. p. 393.

New island thrown up in 1783.—The convulsions of the year 1783 appear to have been more tremendous than any recorded in the modern annals of Iceland; and the original Danish narrative of the catastrophe, drawn up in great detail, has since been substantiated by several English travellers, particularly in regard to the prodigious extent of country laid waste, and the volume of lava produced.* About a month previous to the eruption on the mainland, a submarine volcano burst forth in the sea in lat. 63° 25′ N., long. 23° 44′ W., at a distance of thirty miles in a southwest direction from Cape Reykianas, and ejected so much pumice, that the ocean was covered with that substance to the distance of 150 miles, and ships were considerably impeded in their course. A new island was thrown up, consisting of high cliffs, within which fire, smoke, and pumice were emitted from two or three different points. This island was claimed by his Danish Majesty, who denominated it Nyöe, or the New Island; but before a year had elapsed, the sea resumed its ancient domain, and nothing was left but a reef of rocks from five to thirty fathoms under water.

Great eruption of Skaptár Jokul.—Earthquakes which had long been felt in Iceland, became violent on the 11th of June, 1783, when Skaptár Jokul, distant nearly 200 miles from Nyöe, threw out a torrent of lava which flowed down into the river Skaptâ, and completely dried it up. The channel of the river was between high rocks, in many places from four hundred to six hundred feet in depth, and near two hundred in breadth. Not only did the lava fill up this great defile to the brink, but it overflowed the adjacent fields to a considerable extent. The burning flood, on issuing from the confined rocky gorge, was then arrested for some time by a deep lake, which formerly existed in the course of the river, between Skaptardal and Aa, which it entirely filled. The current then advanced again, and reaching some ancient lava full of subterraneous caverns, penetrated and melted down part of it; and in some places, where the steam could not gain vent, it blew up the rock, throwing fragments to the height of more than 150 feet. On the 18th of June another ejection of liquid lava rushed from the volcano, which flowed down with amazing velocity over the surface of the first stream. By the damming up of the mouths of some of the tributaries of the Skaptâ, many villages were completely overflowed with water, and thus great destruction of property was caused. The lava, after flowing for several days, was precipitated down a tremendous cataract called Stapafoss, where it filled a profound abyss, which that great waterfall had

* The first narrative of the eruption was drawn up by Stephenson, then Chief Justice in Iceland, appointed Commissioner by the King of Denmark for estimating the damage done to the country, that relief might be afforded to the sufferers. Henderson was enabled to correct some of the measurements given by Stephenson, of the depth, width, and length of the lava currents, by reference to the MS. of Mr. Paulson, who visited the tract in 1794, and examined the lava with attention. (Journal of a Residence in Iceland, &c. p. 229.) Some of the principal facts are also corroborated by Sir William Hooker, in his "Tour in Iceland," vol. ii. p. 128.

been hollowing out for ages, and after this, the fiery current again continued its course.

On the third of August, fresh floods of lava still pouring from the volcano, a new branch was sent off in a different direction; for the channel of the Skaptâ was now so entirely choked up, and every opening to the west and north so obstructed, that the melted matter was forced to take a new course, so that it ran in a southeast direction, and discharged itself into the bed of the river Hverfisfliot, where a scene of destruction scarcely inferior to the former was occasioned. These Icelaudic lavas (like the ancient streams which are met with in Auvergne, and other provinces of Central France), are stated by Stephenson to have accumulated to a prodigious depth in narrow rocky gorges; but when they came to wide alluvial plains, they spread themselves out into broad burning lakes, sometimes from twelve to fifteen miles wide, and one hundred feet deep. When the "fiery lake" which filled up the lower portion of the valley of the Skaptâ had been augmented by new supplies, the lava flowed up the course of the river to the foot of the hills from whence the Skaptâ takes its rise. This affords a parallel case to one which can be shown to have happened at a remote era in the volcanic region of the Vivarais in France, where lava issued from the cone of Thueyts, and while one branch ran down, another more powerful stream flowed up the channel of the river Ardêche.

The sides of the valley of the Skaptâ present superb ranges of basaltic columns of older lava, resembling those which are laid open in the valleys descending from Mont Dor, in Auvergne, where more modern lava-currents, on a scale very inferior in magnitude to those of Iceland, have also usurped the beds of the existing rivers. The eruption of Skaptár Jokul did not entirely cease till the end of two years; and when Mr. Paulson visited the tract eleven years afterwards, in 1794, he found columns of smoke still rising from parts of the lava, and several rents filled with hot water.*

Although the population of Iceland was very much scattered, and did not exceed fifty thousand, no less than twenty villages were destroyed, besides those inundated by water; and more than nine thousand human beings perished, together with an immense number of cattle, partly by the depredations of the lava, partly by the noxious vapors which impregnated the air, and, in part, by the famine caused by showers of ashes throughout the island, and the desertion of the coasts by the fish.

Immense volume of the lava.—But the extraordinary volume of melted matter produced in this eruption deserves the particular attention of the geologist. Of the two branches, which flowed in nearly opposite directions, the greatest was fifty, and the lesser forty miles in length. The extreme breadth which the Skaptâ branch attained in the low countries was from twelve to fifteen miles, that of the other about seven. The ordinary height of both currents was one hundred feet, but in narrow

* Henderson's Journal, &c. p. 228.

defiles it sometimes amounted to six hundred. Professor Bischoff has calculated that the mass of lava brought up from the subterranean regions by this single eruption "surpassed in magnitude the bulk of Mont Blanc."* But a more distinct idea will be formed of the dimensions of the two streams, if we consider how striking a feature they would now form in the geology of England, had they been poured out on the bottom of the sea after the deposition and before the elevation of our secondary and tertiary rocks. The same causes which have excavated valleys through parts of our marine strata, once continuous, might have acted with equal force on the igneous rocks, leaving, at the same time, a sufficient portion undestroyed to enable us to discover their former extent. Let us, then, imagine the termination of the Skaptá branch of lava to rest on the escarpment of the inferior and middle oolite, where it commands the vale of Gloucester. The great platform might be one hundred feet thick, and from ten to fifteen miles broad, exceeding any which can be found in Central France. We may also suppose great tabular masses to occur at intervals, capping the summit of the Cotswold Hills between Gloucester and Oxford, by Northleach, Burford, and other towns. The wide valley of the Oxford clay would then occasion an interruption for many miles; but the same rocks might recur on the summit of Cumnor and Shotover Hills, and all the other oolitic eminences of that district. On the chalk of Berkshire, extensive plateaus, six or seven miles wide, would again be formed; and lastly, crowning the highest sands of Highgate and Hampstead, we might behold some remnants of the current five or six hundred feet in thickness, causing those hills to rival, or even to surpass, in height, Salisbury Craigs and Arthur's Seat.

The distance between the extreme points here indicated would not exceed ninety miles in a direct line; and we might then add, at the distance of nearly two hundred miles from London, along the coast of Dorsetshire and Devonshire, for example, a great mass of igneous rocks, to represent those of contemporary origin, which were produced beneath the level of the sea, where the island of Nyöe rose up.

Volume of ancient and modern flows of lava compared.—Yet, gigantic as must appear the scale of these modern volcanic operations, we must be content to regard them as perfectly insignificant in comparison to currents of the primeval ages, if we embrace the theoretical views of many geologists, which were not inaccurately expressed by the late Professor Alexander Brongniart, when he declared that "aux époques géognostiques anciennes, tous les phénomènes géologiques se passoient dans des dimensions *centuples* de celles qu'ils présentent aujourd'hui."† Had Skaptár Jokul, therefore, been a volcano of the olden time, it would have poured forth lavas at a single eruption a hundred times more voluminous than those which were witnessed by the present generation in

* Jameson's Phil. Journ. vol. xxvi. p. 291.
† Tableau des Terrains qui composent l'Ecorce du Globe, p. 52. Paris, 1829.

1783. But it may, on the contrary, be affirmed that, among the older formations, no igneous rock of such colossal magnitude has yet been met with; nay, it would be most difficult to point out a mass of ancient date (distinctly referable to a single eruption) which would even rival in volume the matter poured out from Skaptár Jokul in 1783.

Eruption of Jorullo in 1759.—As another example of the stupendous scale of modern volcanic eruptions, I may mention that of Jorullo in Mexico, in 1759. The great region to which this mountain belongs has already been described. The plain of Malpais forms part of an elevated platform, between two and three thousand feet above the level of the sea, and is bounded by hills composed of basalt, trachyte, and volcanic tuff, clearly indicating that the country had previously, though probably at a remote period, been the theatre of igneous action. From the era of the discovery of the New World to the middle of the last century, the district had remained undisturbed, and the space, now the site of the volcano, which is thirty-six leagues distant from the nearest sea, was occupied by fertile fields of sugar-cane and indigo, and watered by the two brooks Cuitimba and San Pedro. In the month of June, 1759, hollow sounds of an alarming nature were heard, and earthquakes succeeded each other for two months, until, at the end of September, flames issued from the ground, and fragments of burning rocks were thrown to prodigious heights. Six volcanic cones, composed of scoriæ and fragmentary lava, were formed on the line of a chasm which ran in the direction from N. N. E. to S. S. W. The least of these cones was 300 feet in height; and Jorullo, the central volcano, was elevated 1600 feet above the level of the plain. It sent forth great streams of basaltic lava, containing included fragments of granitic rocks, and its ejections did not cease till the month of February, 1760.*

Humboldt visited the country more than forty years after this occurrence, and was informed by the Indians, that when they returned, long after the catastrophe, to the plain, they found the ground uninhabitable from the excessive heat. When he himself visited the place, there appeared, around the base of the cones, and spreading from them, as from a centre, over an extent of four square miles, a mass of matter of a convex form, about 550 feet high at its junction with the cones, and gradually sloping from them in all directions towards the plain. This mass was still in a heated state, the temperature in the fissures being on

Fig. 57.

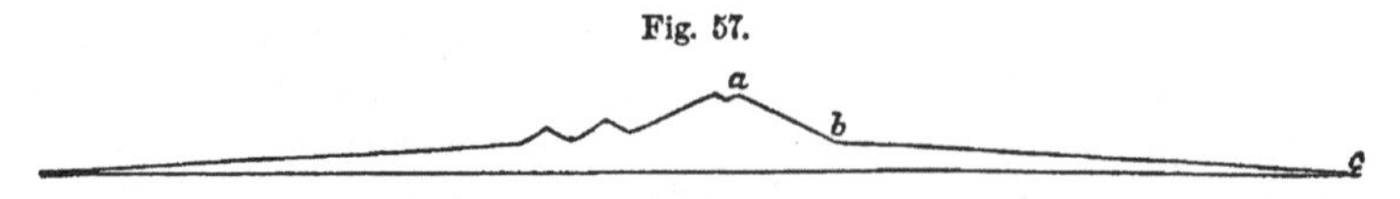

a, Summit of Jorullo. *b*, *c*, Inclined plane sloping at an angle of 6° from the base of the cones.

the decrease from year to year, but in 1780 it was still sufficient to light a cigar at the depth of a few inches. On this slightly convex protuber-

* Daubeny on Volcanoes, p. 337.

ance, the slope of which must form an angle of about 6° with the horizon, were thousands of flattish conical mounds, from six to nine feet high, which, as well as large fissures traversing the plain, acted as fumeroles, giving out clouds of sulphurous acid and hot aqueous vapor. The two small rivers before mentioned disappeared during the eruption, losing themselves below the eastern extremity of the plain, and reappearing as hot springs at its western limit.

Cause of the convexity of the plain of Malpais.—Humboldt attributed the convexity of the plain to inflation from below; supposing the ground, for four square miles in extent, to have risen up in the shape of a bladder to the elevation of 550 feet above the plain in the highest part. But Mr. Scrope has suggested that the phenomena may be accounted for far more naturally, by supposing that lava flowing simultaneously from the different orifices, and principally from Jorullo, united into a sort of pool or lake. As they were poured forth on a surface previously flat, they would, if their liquidity was not very great, remain thickest and deepest near their source, and diminish in bulk from thence towards the limits of the space which they covered. Fresh supplies were probably emitted successively during the course of an eruption which lasted more than half a year; and some of these, resting on those first emitted, might only spread to a small distance from the foot of the cone, where they would necessarily accumulate to a great height. The average slope of the great dome-shaped volcanoes of the Sandwich Islands, formed almost exclusively of lava, with scarce any scoriæ, is between 6° 30′ and 7° 46′, so that the inclination of the convex mass around Jorullo, if we adopt Mr. Scrope's explanation (see fig. 57), is quite in accordance with the known laws which govern the flow of lava.

The showers, also, of loose and pulverulent matter from the six craters, and principally from Jorullo, would be composed of heavier and more bulky particles near the cones, and would raise the ground at their base, where, mixing with rain, they might have given rise to the stratum of black clay, which is described as covering the lava. The small conical mounds (called "hornitos," or little ovens) may resemble those five or six small hillocks which existed in 1823 on the Vesuvian lava, and sent forth columns of vapor, having been produced by the disengagement of elastic fluids heaping up small dome-shaped masses of lava. The fissures mentioned by Humboldt as of frequent occurrence, are such as might naturally accompany the consolidation of a thick bed of lava, contracting as it congeals; and the disappearance of rivers is the usual result of the occupation of the lower part of a valley or plain by lava, of which there are many beautiful examples in the old lava-currents of Auvergne. The heat of the "hornitos" is stated to have diminished from the first; and Mr. Bullock, who visited the spot many years after Humboldt, found the temperature of the hot spring very low,—a fact which seems clearly to indicate the gradual congelation of a subjacent bed of lava, which from its immense thickness may have been enabled to retain its heat for half a century. The reader may be reminded, that when we thus suppose the lava near the volcano to have been, together with the ejected ashes, more than five hundred feet in depth, we merely assign a thickness which the current of Skaptár Jokul attained in some places in 1783.

Hollow sound of the plain when struck.—Another argument adduced in support of the theory of inflation from below, was, the hollow sound made by the steps of a horse upon the plain; which, however, proves nothing more than that the materials of which the convex mass is composed are light and porous. The sound called "rimbombo" by the Italians is very commonly returned by *made ground* when struck sharply; and has been observed not only on the sides of Vesuvius and other volcanic cones where there is a cavity below, but in such regions as the Campagna di Roma, composed in a great measure of tuff and porous volcanic rocks. The reverberation, however, may perhaps be assisted by grottoes and caverns, for these may be as numerous in the lavas of Jorullo as in many of those of Etna; but their existence would lend no countenance to the hypothesis of a great arched cavity, four square miles in extent, and in the centre 550 feet high.*

No recent eruptions of Jorullo.—In a former edition I stated that I had been informed by Captain Vetch, that in 1819 a tower at Guadalaxara was thrown down by an earthquake, and that ashes, supposed to have come from Jorullo, fell at the same time at Guanaxuato, a town situated 140 English miles from the volcano. But Mr. Burkhardt, a German director of mines, who examined Jorullo in 1827, ascertained that there had been no eruption there since Humboldt's visit in 1803. He went to the bottom of the crater, and observed a slight evolution of sulphurous acid vapors, but the "hornitos" had entirely ceased to send forth steam. During the twenty-four years intervening between his visit and that of Humboldt, vegetation had made great progress on the flanks of the new hills; the rich soil of the surrounding country was once more covered with luxuriant crops of sugar-cane and indigo, and there was an abundant growth of natural underwood on all the uncultivated tracts.†

Galongoon, Java, 1822.—The mountain of Galongoon (or Galung Gung) was in 1822 covered by a dense forest, and situated in a fruitful and thickly-peopled part of Java. There was a circular hollow at its summit, but no tradition existed of any former eruption. In July, 1822, the waters of the river Kunir, one of those which flowed from its flanks, became for a time hot and turbid. On the 8th of October following a loud explosion was heard, the earth shook, and immense columns of hot water and boiling mud, mixed with burning brimstone, ashes, and lapilli, of the size of nuts, were projected from the mountain like a waterspout, with such prodigious violence that large quantities fell beyond the river Tandoi, which is forty miles distant. Every valley within the range of

* See Scrope on Volcanoes, p. 267.
† Leonhard and Bronn's Neues Jahrbuch, 1835, p. 36.

this eruption became filled with a burning torrent, and the rivers, swollen with hot water and mud, overflowed their banks, and carried away great numbers of the people, who were endeavoring to escape, and the bodies of cattle, wild beasts, and birds. A space of twenty-four miles between the mountain and the river Tandoi was covered to such a depth with bluish mud that people were buried in their houses, and not a trace of the numerous villages and plantations throughout that extent was visible. Within this space the bodies of those who perished were buried in mud and concealed, but near the limits of the volcanic action they were exposed, and strewed over the ground in great numbers, partly boiled and partly burnt.

It was remarked, that the boiling mud and cinders were projected with such violence from the mountain, that while many remote villages were utterly destroyed and buried, others much nearer the volcano were scarcely injured.

The first eruption lasted nearly five hours, and on the following days the rain fell in torrents, and the rivers, densely charged with mud, deluged the country far and wide. At the end of four days (October 12th) a second eruption occurred more violent than the first, in which hot water and mud were again vomited, and great blocks of basalt were thrown to the distance of seven miles from the volcano. There was at the same time a violent earthquake, and in one account it is stated that the face of the mountain was utterly changed, its summits broken down, and one side, which had been covered with trees, became an enormous gulf in the form of a semicircle. This cavity was about midway between the summit and the plain, and surrounded by steep rocks, said to be newly heaped up during the eruption. New hills and valleys are said to have been formed, and the rivers Banjarang and Wulan changed their course, and in one night (October 12th) 2000 persons were killed.

The first intimation which the inhabitants of Bandong received of this calamity on the 8th of October, was the news that the river Wulna was bearing down into the sea the dead bodies of men, and the carcasses of stags, rhinoceroses, tigers, and other animals. The Dutch painter Payen determined to travel from thence to the volcano, and he found that the quantity of the ashes diminished as he approached the base of the mountain. He alludes to the altered form of the mountain after the 12th, but does not describe the new semicircular gulf on its side.

The official accounts state that 114 villages were destroyed, and above 4000 persons killed.*

Submarine volcanoes.—Although we have every reason to believe that volcanic eruptions as well as earthquakes are common in the bed of the sea, it was not to be expected that many opportunities would occur to scientific observers of witnessing the phenomena. The crews of vessels have sometimes reported that they have seen in different places sulphur-

* Van der Boon Mesch, de Incendiis Montium Javæ, &c. Lugd. Bat. 1826; and Official Report of the President, Baron Van der Capellen; also, Von Buch, Iles Canar. p. 424.

ous smoke, flame, jets of water, and steam, rising up from the sea, or they have observed the waters greatly discolored, and in a state of violent agitation as if boiling. New shoals have also been encountered, or a reef of rocks just emerging above the surface, where previously there was always supposed to have been deep water. On some few occasions the gradual formation of an island by a submarine eruption has been observed, as that of Sabrina, in the year 1811, off St. Michael's in the Azores. The throwing up of ashes in that case, and the formation of a cone about three hundred feet in height, with a crater in the centre, closely resembled the phenomena usually accompanying a volcanic eruption on land. Sabrina was soon washed away by the waves. Previous eruptions in the same part of the sea were recorded to have happened in 1691 and 1720. The rise of Nyöe, also, a small island off the coast of Iceland, in 1783, has already been alluded to; and another volcanic isle was produced by an eruption near Reikiavig, on the same coast, in June, 1830.*

Graham Island†, 1831.—We have still more recent and minute information respecting the appearance, in 1831, of a new volcanic island in the Mediterranean, between the S. W. coast of Sicily and that projecting part of the African coast where ancient Carthage stood. The site of the island was not any part of the great shoal, or bank, called "Nerita," as was first asserted, but a spot where Captain W. H. Smyth had found, in his survey a few years before, a depth of more than one hundred fathoms water.‡

The position of the island (lat. 37° 8′ 30″ N., long. 12° 42′ 15″ E.) was about thirty miles S. W. of Sciacca, in Sicily, and thirty-three miles N. E. of Pantellaria.§ On the 28th of June, about a fortnight before the eruption was visible, Sir Pulteney Malcolm, in passing over the spot in his ship, felt the shocks of an earthquake, as if he had struck on a sand-bank; and the same shocks were felt on the west coast of Sicily, in a direction from S. W. to N. E. About the 10th of July, John Corrao, the captain of a Sicilian vessel, reported that, as he passed near the place, he saw a column of water like a water-spout, sixty feet high, and 800 yards in circumference, rising from the sea, and soon afterwards a dense steam in its place, which ascended to the height of 1800 feet. The same Corrao, on his return from Girgenti, on the 18th of July, found a small island, twelve feet high with a crater in its centre, ejecting volcanic matter, and immense columns of vapor; the sea around being

* Journ. de Géol. tome i.

† In a former edition, I selected the name of Sciacca out of seven which had been proposed; but the Royal and Geographical Societies have now adopted Graham Island; a name given by Capt. Senhouse, R. N., the first who succeeded in landing on it. The seven rival names are Nerita, Ferdinanda, Hotham, Graham, Corrao, Sciacca, Julia. As the isle was visible for only about three months, this is an instance of a wanton multiplication of synonyms which has scarcely ever been outdone even in the annals of zoology and botany.

‡ Phil. Trans. 1832, p. 255.

§ Journ. of Roy. Geograph. Soc. 1830–31.

Fig. 58.

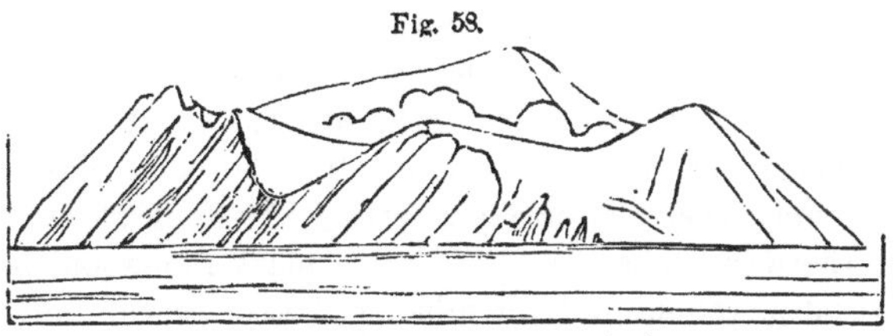

Form of the cliffs of Graham Island, as seen from S. S. E., distant one mile, 7th August, 1831.*

covered with floating cinders and dead fish. The scoriæ were of a chocolate color, and the water which boiled in the circular basin was of

Fig. 59.

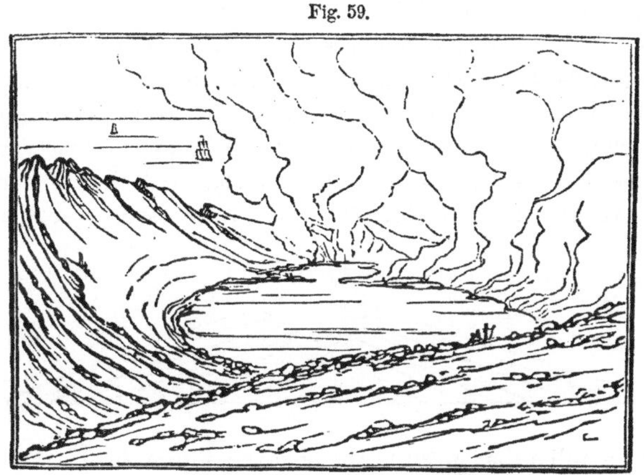

View of the interior of Graham Island, 29th Sept., 1831.

a dingy red. The eruption continued with great violence to the end of the same month; at which time the island was visited by several per-

Fig. 60.

Graham Island, 29th Sept., 1831.†

* Phil. Trans. part. ii. 1832, reduced from drawings by Capt. Wodehouse, R. N.

† In the annexed sketch (fig. 60), drawn by M. Joinville, who accompanied M. C. Prevost, the beds seem to slope towards the centre of the crater; but I am informed by M. Prevost that these lines were not intended by the artist to represent the dip of the beds.

sons, and among others by Capt. Swinburne, R. N., and M. Hoffmann, the Prussian geologist. It was then from fifty to ninety feet in height, and three-quarters of a mile in circumference. By the 4th of August it became, according to some accounts, above 200 feet high, and three miles in circumference; after which it began to diminish in size by the action of the waves, and it was only two miles round on the 25th of August; and on the 3d of September, when it was carefully examined by Captain Wodehouse, only three-fifths of a mile in circumference; its greatest height being then 107 feet. At this time the crater was about 780 feet in circumference. On the 29th of September, when it was visited by Mons. C. Prevost, its circumference was reduced to about 700 yards. It was composed entirely of incoherent ejected matter, scoriæ, pumice, and lapilli, forming regular strata, some of which are described as having been parallel to the steep inward slope of the crater, while the rest were inclined outwards, like those of Vesuvius.* When the arrangement of the ejected materials has been determined by their falling continually on two steep slopes, that of the external cone and that of the crater, which is always a hollow inverted cone, a transverse section would

Fig. 61.

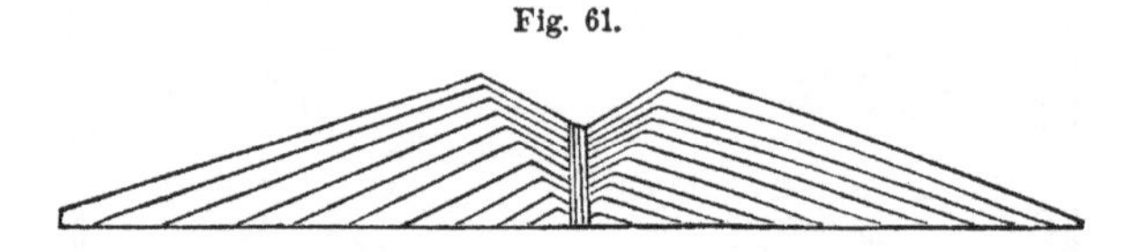

probably resemble that given in the annexed figure (61). But when I visited Vesuvius, in 1828, I saw no beds of scoriæ inclined towards the axis of the cone. (See fig. 45, p. 381.) Such may have once existed; but the explosions or subsidences, or whatever causes produced the great crater of 1822, had possibly destroyed them.

Few of the pieces of stone thrown out from Graham Island exceeded a foot in diameter. Some fragments of dolomitic limestone were intermixed; but these were the only non-volcanic substances. During the month of August, there occurred on the S. W. side of the new island a violent ebullition and agitation of the sea, accompanied by the constant ascension of a column of dense white steam, indicating the existence of a second vent at no great depth from the surface. Towards the close of October, no vestige of the crater remained, and the island was nearly levelled with the surface of the ocean, with the exception, at one point, of a small monticule of sand and scoriæ. It was reported that, at the commencement of the year following (1832), there was a depth of 150 feet where the island had been: but this account was quite erroneous; for in the early part of that year Captain Swinburne found a shoal and discolored water there, and towards the end of 1833 a dangerous reef existed of an oval figure, about three-fifths of a mile in extent. In the centre was a black rock, of the diameter of about

* See Memoir by M. C. Prevost, Ann. des Sci. Nat. tom. xxiv.

twenty-six fathoms, from nine to eleven feet under water; and round this rock are banks of black volcanic stones and loose sand. At the distance of sixty fathoms from this central mass, the depth increased rapidly. There was also a second shoal at the distance of 450 feet S. W. of the great reef, with fifteen feet water over it, also composed of rock, surrounded by deep sea. We can scarcely doubt that the rock in the middle of the larger reef is solid lava, which rose up in the principal crater, and that the second shoal marks the site of the submarine eruption observed in August, 1831, to the S. W. of the island.

From the whole of the facts above detailed, it appears that a hill eight hundred feet or more in height was formed by a submarine volcanic vent, of which the upper part (only about two hundred feet high) emerged above the waters, so as to form an island. This cone must have been equal in size to one of the largest of the lateral volcanoes on the flanks of Etna, and about half the height of the mountain Jorullo in Mexico, which was formed in the course of nine months, in 1759. In the centre of the new volcano a large cavity was kept open by gaseous discharges, which threw out scoriæ; and fluid lava probably rose up in this cavity. It is not uncommon for small subsidiary craters to open near the summit of a cone, and one of these may have been formed in the case of Graham Island; a vent, perhaps, connected with the main channel of discharge which gave passage in that direction to elastic fluids, scoriæ, and melted lava. It does not appear that, either from this duct, or from the principal vent, there was any overflowing of lava; but melted rock may have flowed from the flanks or base of the cone (a common occurrence on land), and may have spread in a broad sheet over the bottom of the sea.

The dotted lines in the annexed figure are an imaginary restoration of the upper part of the cone, now removed by the waves: the strong lines represent the part of the volcano which is still under water: in the centre is a great column, or dike, of solid lava, two hundred feet in diameter, supposed to fill the space by which the gaseous fluids rose; and on each side of the dike is a stratified mass of scoriæ and fragmen-

Fig. 62.

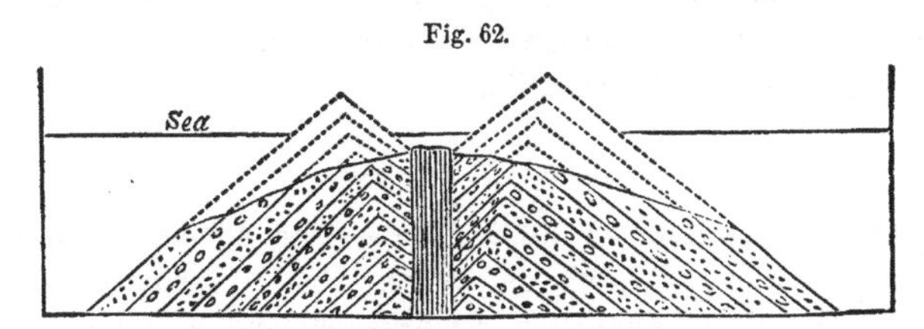

Supposed section of Graham Island. (C. Maclaren.*)

tary lava. The solid nucleus of the reef, where the black rock is now found, withstands the movements of the sea; while the surrounding loose tuffs are cut away to a somewhat lower level. In this manner the

* Geol. of Fife and the Lothians, p. 41. Edin. 1839.

lava, which was the lowest part of the island, or, to speak more correctly, which scarcely ever rose above the level of the sea when the island existed, has now become the highest point in the reef.

No appearances observed, either during the eruption or since the island disappeared, gave the least support to the opinion promulgated by some writers, that part of the ancient bed of the sea had been lifted up bodily.

The solid products, says Dr. John Davy, whether they consisted of sand, light cinders, or vesicular lava, differed more in form than in composition. The lava contained augite; and the specific gravity was 2·07 and 2·70. When the light spongy cinder, which floated on the sea, was reduced to fine powder by trituration, and the greater part of the entangled air got rid of, it was found to be of the specific gravity 2·64; and that of some of the sand which fell in the eruption was 2·75;* so that the materials equalled ordinary granites in weight and solidity. The only gas evolved in any considerable quantity was carbonic acid.†

Submarine eruptions in mid-Atlantic.—In the Nautical Magazine for 1835, p. 642, and for 1838, p. 361, and in the Comptes Rendus, April, 1838, accounts are given of a series of volcanic phenomena, earthquakes, troubled water, floating scoriæ and columns of smoke, which have been observed at intervals since the middle of the last century, in a space of open sea between longitudes 20° and 22° west, about half a degree south of the equator. These facts, says Mr. Darwin, seem to show, that an island or an archipelago is in process of formation in the middle of the Atlantic; a line joining St. Helena and Ascension would, if prolonged, intersect this slowly nascent focus of volcanic action.‡ Should land be eventually formed here, it will not be the first that has been produced by igneous action in this ocean since it was inhabited by the existing species of testacea. At Porto Praya in St. Jago, one of the Azores, a horizontal, calcareous stratum occurs, containing shells of *recent* marine species, covered by a great sheet of basalt eighty feet thick.§ It would be difficult to estimate too highly the commercial and political importance which a group of islands might acquire, if in the next two or three thousand years they should rise in mid-ocean between St. Helena and Ascension.

CANARY ISLANDS.

Eruption in Lancerote, 1730 *to* 1736.—The effects of an eruption which happened in Lancerote, one of the Canary Islands, between the years 1730 and 1736, were very remarkable; and a detailed description has been published by Von Buch, who had an opportunity, when he visited that island in 1815, of comparing the accounts transmitted to us of the event, with the present state and geological appearances of the

* Phil. Trans. 1832, p. 243.

† Ibid. p. 249

‡ Darwin's Volcanic Islands, p. 92.

§ Ibid. p. 6.

country.* On the 1st of September, 1730, the earth split open on a sudden two leagues from Yaira. In one night a considerable hill of ejected matter was thrown up; and, a few days later, another vent opened, and gave out a lava-stream, which overran Chinanfaya and other villages. It flowed first rapidly, like water, but became afterwards heavy and slow, like honey. On the 7th of September an immense rock was protruded from the bottom of the lava with a noise like thunder, and the stream was forced to change its course from N. to N. W., so that St. Catalina and other villages were overflowed.

Whether this mass was protruded by an earthquake, or was a mass of ancient lava, blown up like that before mentioned in 1783 in Iceland, is not explained.

On the 11th of September more lava flowed out, and covered the village of Maso entirely, and for the space of eight days precipitated itself with a horrible roar into the sea. Dead fish floated on the waters in indescribable multitudes, or were thrown dying on the shore. After a brief interval of repose, three new openings broke forth immediately from the site of the consumed St. Catalina, and sent out an enormous quantity of lapilli, sand, and ashes. On the 28th of October the cattle throughout the whole country dropped lifeless to the ground, suffocated by putrid vapors, which condensed and fell down in drops. On the 1st of December a lava-stream reached the sea, and formed an island, round which dead fish were strewed.

Number of cones thrown up.—It is unnecessary here to give the details of the overwhelming of other places by fiery torrents, or of a storm which was equally new and terrifying to the inhabitants, as they had never known one in their country before. On the 10th of January, 1731, a high hill was thrown up, which, on the same day, precipitated itself back again into its own crater; fiery brooks of lava flowed from it to the sea. On the 3d of February a new cone arose. Others were thrown up in March, and poured forth lava-streams. Numerous other volcanic cones were subsequently formed in succession, till at last their number amounted to about thirty. In June, 1731, during a renewal of the eruptions, all the banks and shores in the western part of the island were covered with dying fish, of different species, some of which had never before been seen. Smoke and flame arose from the sea, with loud detonations. These dreadful commotions lasted without interruption for *five successive years*, so that a great emigration of the inhabitants became necessary.

Their linear direction.—As to the height of the new cones, Von Buch was assured that the formerly great and flourishing St. Catalina lay buried under hills 400 feet in height; and he observes that the most elevated cone of the series rose 600 feet above its base, and 1378 feet above the sea, and that several others were nearly as high. The new

* This account was principally derived by Von Buch from the MS. of Don Andrea Lorenzo Curbeto, curate of Yaira, the point where the eruption began. —Ueber einen vulcanischen Ausbruch auf der Insel Lanzerote.

vents were all arranged *in one line*, about two geographical miles long, and in a direction nearly east and west. If we admit the probability of Von Buch's conjecture, that these vents opened along the line of a cleft, it seems necessary to suppose that this subterranean fissure was only prolonged upwards to the surface by degrees, and that the rent was narrow at first, as is usually the case with fissures caused by earthquakes. Lava and elastic fluids might escape from some point on the rent where there was least resistance, till, the first aperture becoming obstructed by ejections and the consolidation of lava, other orifices burst open in succession along the line of the original fissure. Von Buch found that each crater was lowest on that side on which lava had issued; but some craters were not breached, and were without any lava streams. In one of these were open fissures, out of which hot vapors rose, which in 1815 raised the thermometer to 145° Fahrenheit, and was probably at the boiling point lower down. The exhalations seemed to consist of aqueous vapor; yet they could not be pure steam, for the crevices were incrusted on either side by siliceous sinter (an opal-like hydrate of silica of a white color), which extended almost to the middle. This important fact attests the length of time during which chemical processes continue after eruptions, and how open fissures may be filled up laterally by mineral matter, sublimed from volcanic exhalations. The lavas of this eruption covered nearly a third of the whole island, often forming on slightly inclined planes great horizontal sheets several square leagues in area, resembling very much the basaltic platforms of Auvergne.

Pretended distinction between ancient and modern lavas.—One of the new lavas was observed to contain masses of olivine of an olive-green color, resembling those which occur in one of the lavas of the Vivarais. Von Buch supposes the great crystals of olivine to have been derived from a previously existing basalt melted up by the new volcanoes; but we have scarcely sufficient data to bear out such a conjecture. The older rocks of the island consist, in a great measure, of that kind of basaltic lava called dolerite, sometimes columnar, and partly of common basalt and amygdaloid. Some recent lavas assumed, on entering the sea, a prismatic form, and so much resembled the older lavas of the Canaries, that the only geological distinction which Von Buch appears to have been able to draw between them was, that they did not alternate with conglomerates, like the ancient basalts. Some modern writers have endeavored to discover, in the abundance of these conglomerates, a proof of the dissimilarity of the volcanic action in ancient and modern times; but this character is more probably attributable to the difference between submarine operations and those on the land. All the blocks and imperfectly rounded fragments of lava, transported during the intervals of eruption, by rivers and torrents, into the adjoining sea, or torn by the continued action of the waves from cliffs which are undermined, must accumulate in stratified breccias and conglomerates, and be covered again and again by other lavas. This is now taking place on the shores of Sicily, between Catania and Trezza, where the sea breaks down and covers the shore with blocks and pebbles of the modern lavas of Etna; and on parts of the coast of Ischia, where numerous currents of trachyte are in like manner undermined in lofty precipices. So often, then, as an island is raised in a volcanic archipelago by earthquakes from the deep, the fundamental and (relatively to all above) the oldest lava will often be distinguishable from those formed by subsequent eruptions on dry land, by their alternation with beds of sandstone and fragmentary rocks.

The supposed want of identity, then, between the volcanic phenomena of different epochs resolves itself partly at least into the marked difference between the operations simultaneously in progress, above and below the waters. Such, indeed, is the source, as was before stated in the First Book (Chap. V.), of many of our strongest theoretical prejudices in geology. No sooner do we study and endeavor to explain submarine appearances, than we feel, to use a common expression, out of our element; and unwilling to concede that our extreme ignorance of processes now continually going on can be the cause of our perplexity, we take refuge in a "pre-existent order of nature."

Recent formation of oolitic travertin in Lancerote.—Throughout a considerable part of Lancerote, the old lavas are covered by a thin stratum of limestone, from an inch to two feet in thickness. It is of a hard stalactitic nature, sometimes oolitic, like the Jura limestone, and contains fragments of lava and terrestrial shells, chiefly helices and spiral bulimi. It sometimes rises to the height of 800 feet above the level of the sea. Von Buch imagines that this remarkable superstratum has been produced by the furious northwest storms, which in winter drive the spray of the sea in clouds over the whole island; from whence calcareous particles may be deposited stalactitically. Mr. Darwin informs me that he found a limestone in St. Helena, the harder parts of which correspond precisely to the stone of Lancerote. He attributes the origin of this rock in St. Helena not to the spray of the sea, but to drifting by violent winds of the finer particles of shells from the sea-beach. Some parts of this drift are subsequently dissolved by atmospheric moisture, and redeposited, so as to convert calcareous sand into oolite.

Recent eruption in Lancerote.—From the year 1736 to 1815, when Von Buch visited Lancerote, there had been no eruption; but, in August, 1824, a crater opened near the port of Rescif, and formed by its ejections, in the space of twenty-four hours, a considerable hill. Violent earthquakes preceded and accompanied this eruption.*

Teneriffe.—The Peak of Teneriffe is about 12,000 feet high, and stands, says Von Buch, like a tower encircled by its fosse and bastion. The bastion consists, like the semicircular escarpment of Somma turned towards Vesuvius, of precipitous cliffs, composed of trachyte, basalt, coarse conglomerates, and tuffs, traversed by volcanic dikes, mostly vertical, and of

* Férussac, Bulletin des Sci. Nat. tome v. p. 45: 1825.

basalt. These cliffs vary in height from 1000 to 1800 feet, and are supposed by Von Buch to have been heaved up into their present position by a force exerted from below, in accordance with the theory proposed by the same author for the origin of the cones of Vesuvius and Etna. According to the observations of M. Deville in 1839*, the trachytes are often granitoid in their aspect, and contain instead of glassy felspar the allied mineral called oligoclase, which had been previously considered as characteristic of more ancient igneous rocks. The same traveller supposes, although he found no limestone or trace of fossils in any of the rocks of Teneriffe, that the alternating trachytes and trachytic conglomerates originated beneath the sea. If this opinion be correct, and it is at least very probable, geologists may still speculate on two modes in which the mass of the island acquired its present form and elevation above the sea. 1st, The advocates of Von Buch's crater-of-elevation hypothesis may imagine that a succession of horizontally superimposed beds were upheaved by a sudden movement, and tilted so as to dip in all directions outwards from the centre of a new dome-shaped eminence, in the middle of which a large opening or bowl-shaped cavity was produced. 2dly, Or according to the theory which to me appears preferable, a submarine hill in the form of a flattened dome may have gradually accumulated, partly below the waters and partly above by the continued outpourings of sheets of lava and the ejection of ashes from a central orifice. In this case the dikes would represent the fissures, which were filled during successive eruptions, and the original inclination of the beds may have been increased by the distension and upheaval of the mass during reiterated convulsions, acting most forcibly at or near the channel of discharge, which would become partially sealed up with lava from time to time, and then be burst open again during eruptions. At length the whole island may have been raised bodily out of the sea by a gradual upward movement.

Whatever theory we adopt, we must always explain the abrupt termination of the dikes and layers of trachyte and basalt in the steep walls of the escarpments surrounding the great crater by supposing the removal of part of the materials once prolonged farther inward towards the centre. If, according to the elevation-crater hypothesis, a series of sheets of lava and ashes originally spread over a level and even surface have been violently broken and uplifted, why do not the opposite walls of the chasm correspond in such a manner as to imply by their present outline that they were formerly united? It is evident that the precipices on opposite sides of the crateriform hollow would not fit if brought together, there being no projecting masses in one wall to enter into indentations in the other, as would happen with the sides of many mineral veins, trap-dikes, and faults, could we extract the intrusive matter now separating them, and reunite the rocks which have been fractured and disjoined.

The highest crater of the peak has merely disengaged sulphureous

* Comptes Rendus Acad. Sci. Paris, Juin, 1846.

vapors ever since it has been known to Europeans; but an eruption happened in June, 1798, not far from the summit, and others are recorded, which poured out streams of lava from great heights, besides many which have broken out nearer the level of the sea. All these, however, seem to be dependent on one great centre of eruption, or on that open channel communicating between the interior of the earth and the atmosphere, which terminates in the highest crater of the peak.

We may consider Teneriffe, then, as having been from a remote period the principal and habitual vent of the volcanic archipelago of the Canaries. The discharges which have taken place in the contiguous isles of Palma, Lancerote, and the rest, may be of a subsidiary kind, and have probably been most frequent and violent when the greater crater has been partially sealed up, just as the violent eruptions of Ischia or that of Monte Nuovo coincided with the dormant state of Vesuvius.

SANTORIN.

The Gulf of Santorin, in the Grecian Archipelago, has been for two thousand years a scene of active volcanic operations. The largest of the three outer islands of the group (to which the general name of Santorin is given) is called Thera (or sometimes Santorin), and forms more than two-thirds of the circuit of the gulf (see Map, fig. 63, p. 442). The length of the exterior coast-line of this and the other two islands named Therasia and Aspronisi, taken together, amounts to about thirty miles, and that of the inner coast-line of the same islands to about eighteen miles. In the middle of the gulf are three other islands, called the Little, the New, and the Old "Kaimenis," or "Burnt Islands." The accompanying map has been reduced from a recent survey executed in 1848 by Captain Graves, R. N., and shortly to be published by the Admiralty.

Pliny informs us that the year 186, B. C., gave birth to the Old Kaimeni, also called Hiera, or the "Sacred Isle," and in the year 19 of our era "Thia" (the Divine) made its appearance above water, and was soon joined by subsequent eruptions to the older island, from which it was only 250 paces distant. The Old Kaimeni also increased successively in size in 726 and in 1427. A century and a half later, in 1573, another eruption produced the cone and crater called Micra-Kaimeni, or "the Small Burnt Island." The next great event which we find recorded occurred in 1650, when a submarine outbreak violently agitated the sea, at a point three and a half miles to the N. E. of Thera, and which gave rise to a shoal (see A in the map) carefully examined during the late survey in 1848 by Captain Graves, and found to have ten fathoms water over it, the sea deepening around it in all directions. This eruption lasted three months, covering the sea with floating pumice. At the same time an earthquake destroyed many houses in Thera, while the sea broke upon the coast and overthrew two churches, exposing to view two villages, one on each side of the mountain of St. Stephen, both of which

Fig. 63.

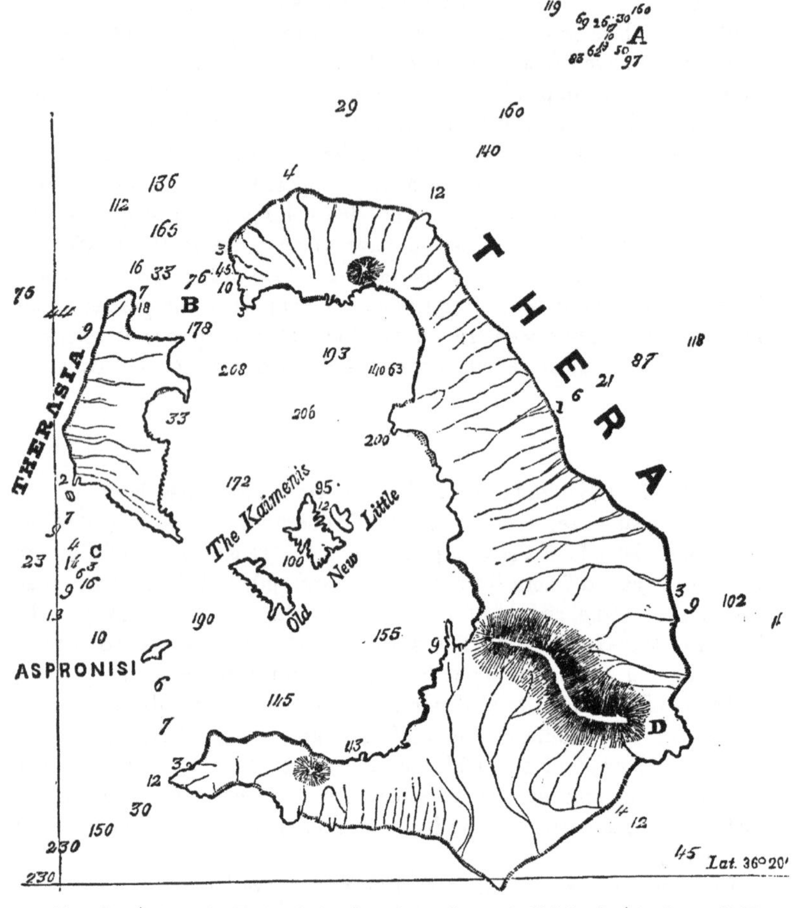

Map of Santorin in the Grecian Archipelago, from a Survey in 1848, by Captain Graves, R. N.

The soundings are given in fathoms.

A, Shoal formed by submarine volcanic eruption in 1650.
B, Northern entrance. C, Mansell's Rock.
D, Mount St. Elias, 1887 feet high.

Fig. 64.

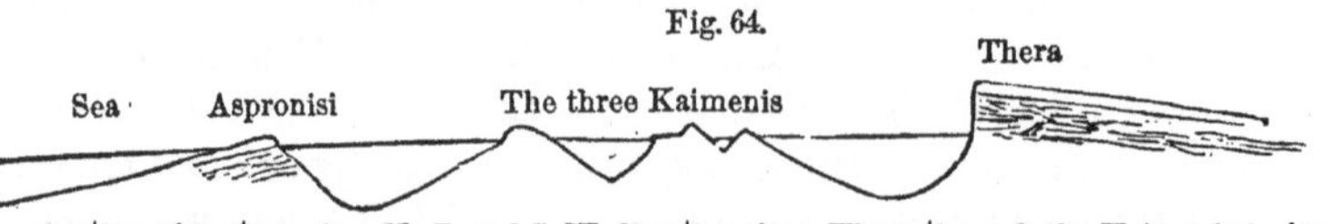

Section of Santorin, in a N. E. and S. W. direction, from Thera through the Kaimenis to Aspronisi

Fig. 65.

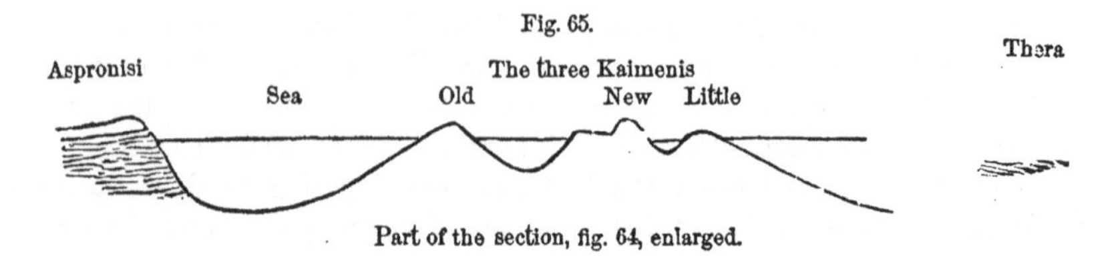

Part of the section, fig. 64, enlarged.

must have been overwhelmed by showers of volcanic matter during some previous eruptions of unknown date.* The accompanying evolution of sulphur and hydrogen issuing from the sea killed more than fifty persons, and above 1000 domestic animals. A wave, also, 50 feet high, broke upon the rocks of the Isle of Nia, about four leagues distant, and advanced 450 yards into the interior of the Island of Sikino. Lastly, in 1707 and 1709, Nea-Kaimeni, or the New Burnt Island, was formed between the two others, Palaia and Micra, the Old and Little isles. This isle was composed originally of two distinct parts; the first which rose was called the White Island, composed of a mass of pumice, extremely porous. Goree, the Jesuit, who was then in Santorin, says that the rock "cut like bread," and that, when the inhabitants landed on it, they found a multitude of full-grown fresh oysters adhering to it, which they ate.† This mass was afterwards covered, in great part, by the matter ejected from the crater of a twin-island formed simultaneously, and called Black Island, consisting of brown trachyte. The trachytic lava which rose on this spot appears to have been a long time in an intumescent state, for the New Kaimeni was sometimes lowered on one side while it gained height on the other, and rocks rose up in the sea at different distances from the shore and then disappeared again. The eruption was renewed at intervals during the years 1711 and 1712, and at length a cone was piled up to the height of 330 feet above the level of the sea, its exterior slope forming an angle of 33° with the horizon, and the crater on its summit being 80 yards in diameter. In addition to the two points of subaerial eruption on the New and Little Kaimenis, two other cones, indicating the sites of submarine outbursts of unknown date, were discovered under water near the Kaimenis during the late survey.

In regard to the "White Island," which was described and visited by Goree in 1707, we are indebted to Mr. Edward Forbes for having, in 1842, carefully investigated the layer of pumiceous ash of which it is constituted. He obtained from it many shells of marine genera, Pectunculus, Arca, Cardita, Trochus, and others, both univalve and bivalve, all of recent Mediterranean species. They were in a fine state of preservation, the bivalves with the epidermis remaining, and valves closed, showing that they had been suddenly destroyed. Mr. Forbes, from his study of the habits of the mollusca living at different depths in the Mediterranean, was able to decide that such an assemblage of species could not have lived at a less depth than 220 feet, so that a bodily upheaval of the mass to that amount must have taken place in order to bring up this bed of ashes and shells to the level of the sea, and they now rise five or six feet above that level.‡

We may compare this partial elevation of solid matter to the rise of a hardened crust of scoriæ, such as is usually formed on the surface of lava-currents, even while they are in motion, and which, although stony and capable of supporting heavy weights, may be upraised without

* Virlet, Bull. de la Soc. Géol. de France, tom. iii. p. 103.

† Phil. Trans. No. 332.

‡ E. Forbes, Brit. Association, Report for 1843

bursting by the intumescence of the melted matter below. That the upheaval was merely local is proved by the fact that the neighboring Kaimenis did not participate in the movement, still less the three more distant or outer islands before mentioned. The history, therefore, of the Kaimenis shows that they have been the result of intermittent action, and it lends no support to the hypothesis of the sudden distension of horizontal beds blown up like a bladder by a single paroxysmal effort of expansive gases.

It will be seen by the accompanying map and sections, that the Kaimenis are arranged in a linear direction, running N. E. and S. W., in a manner different from that represented in the older charts. In their longest diameter they form at their base a ridge nearly bisecting the gulf or crater (see sections, figs. 64, 65).

On considering these facts we are naturally led to compare the smaller and newer islands in the centre of the gulf to the modern cone of Vesuvius, surrounded by the older semicircular escarpment of Somma, or to liken them to the Peak of Teneriffe before described, as surrounded by its "fosse and bastion." This idea will appear to be still more fully confirmed when we study the soundings taken during the late hydrographical survey. Thera, which constitutes alone more than two-thirds of the outer circuit, presents everywhere towards the gulf, high and steep precipices composed of rocks of volcanic origin. In all places near the base of its cliffs, a depth of from 800 to 1000 feet of water was found, and Lieut. Leycester informs us* that if the gulf, which is six miles in diameter, could be drained, a bowl-shaped cavity would appear with walls 2449 feet high in some places, and even on the southwest side, where it is lowest, nowhere less than 1200 feet high; while the Kaimenis would be seen to form in the centre a huge mountain five and a half miles in circumference at its base, with three principal summits (the Old, the New, and the Little Burnt Islands) rising severally to the heights of 1251, 1629, and 1158 feet above the bottom of the abyss. The rim of the great caldron thus exposed would be observed to be in all parts perfect and unbroken, except at one point where there is a deep and long chasm or channel, known by mariners as "the northern entrance" (B, fig. 63) between Thera and Therasia, and called by Lieut. Leycester "the door into the crater." It is no less than 1170 feet deep, and constitutes, as will appear by the soundings (see map), a remarkable feature in the bed of the sea. There is no corresponding channel passing out from the gulf into the Mediterranean at any other point in the circuit between the outer islands, the greatest depth there ranging from 7 to 66 feet.

We may conceive, therefore, if at some former time the whole mass of Santorin stood at a higher level by 1200 feet, that this single ravine or narrow valley now forming "the northern entrance," was the passage by which the sea entered a circular bay and swept out in the form of

* See a paper read to the Geographical Society in 1849.

mud and pebbles, the materials derived by denudation from wasting cliffs. In this manner the original crater may have been slowly widened and deepened, after which the whole archipelago may have been partially submerged to its present depth.

That such oscillations of level may in the course of ages have taken place, will be the more readily admitted when we state that part of Thera has actually sunk down in modern times, as, for example, during the great earthquake before alluded to, which happened in 1650. The subsidence alluded to is proved not only by tradition, but by the fact that a road which formerly led between two places on the east coast of Thera is now twelve fathoms under water.

MM. Boblaye and Virlet mention,* that the waves are constantly undermining and encroaching on the cliffs of Therasia and Aspronisi, and shoals or submarine ledges were found, during the late survey, to occur round a great part of these islands, attesting the recent progress of denudation. M. Virlet also remarks, in regard to the separation of the three islands forming the walls of the crater, that the channels between them are all to the W. and N. W., the quarter most exposed to the waves and currents.

Mr. Darwin, in his work on volcanic islands, has shown that in the Mauritius and in Santiago, there is an external circle of basaltic rocks of vast diameter, in the interior of which more modern eruptions have taken place, the older rocks dipping away from the central space in every direction, as in the outer islands of Santorin. He refers the numerous breaches, some of them very wide in the external ramparts of those islands, to the denuding action of the sea. Every geologist, therefore, will be prepared to call in the aid of the same powerful cause, to account for the removal of a large part of the rocks which must once have occupied the interior space, in the same manner as they attribute the abstraction of matter from elliptical "valleys of elevation," such as those of Woolhope and the Wealden in England, to the waves and currents of the sea.

Thera, Therasia, and Aspronisi are all composed of volcanic matter, except the southern part of Thera, where Mount St. Elias rises to three times the height of the loftiest of the igneous rocks, reaching an elevation of 1887 feet above the sea.† This mountain is formed of granular limestone and argillaceous schist, and must have been originally a submarine eminence in the bed of the Mediterranean, before the volcanic cone, one side of the base of which now abuts against it, was formed. The inclination, strike, and fractures of the calcareous and argillaceous strata of St. Elias have no relation to the great cone, but, according to M. Bory St. Vincent, have the same direction as those of the other isles of the Grecian Archipelago, namely, from N. N. W. to S. S. E. Each of the three islands, Thera, Therasia, and Aspronisi,

* Bull. de la Soc. Géol. de France, tome iii.
† Virlet, Bull. de la Soc. Géol. de France, tome iii. p. 103.

is capped by an enormous mass of white tufaceous conglomerate, from forty to fifty feet thick, beneath which are beds of trachytic lava and tuff, having a gentle inclination of only 3° or 4°. Each bed is usually very narrow and discontinuous, the successive layers being moulded or dove-tailed, as M. Virlet expresses it, into the inequalities of the previously existing surface, on which showers of cinders or streams of melted matter have been poured. Nothing, therefore, seems more evident than that we have in Santorin the basal remains of a great ruined cone, or flattened dome; and the absence of dikes in the cliffs surrounding the gulf would indicate that the eruptions took place originally, as they have done in the last two thousand years, not near the margin but in the centre of the space now occupied by the gulf. The central portions of the dome have since been removed by engulfment, or denudation, or by both these causes.

An important fact is adduced by M. Virlet, to show that the gentle dip of the lava-streams in the three outer islands towards all points of the compass, away from the centre of the gulf, has not been due to the upheaval of horizontal beds, as conjectured by Von Buch, who had not visited Santorin.* The French geologist found that the vesicles or pores of the trachytic masses were lengthened out in the several directions in which they would have flowed if they had descended from the axis of a cone once occupying the centre of the crater. For it is well known that the bubbles of confined gas in a fluid in motion assume an oval form, and the direction of their longer axis coincides always with that of the stream.

On a review, therefore, of all the facts now brought to light respecting Santorin, I attribute the moderate slope of the beds in Thera and the other external islands to their having originally descended the inclined flanks of a large volcanic cone, the principal orifice or vents of eruption having been always situated where they are now, in or near the centre of the space occupied by the gulf or crater—in other words, where the outburst of the Kaimenis has been witnessed in historical times. The single long and deep opening into the crater is a feature common to all those remnants of ancient volcanoes, the central portions of which have been removed, and is probably connected with aqueous denudation. This denuding process has been the work of ages when the sea was admitted into an original crater, and has taken place during the gradual emergence of the island from the sea, or during various oscillations in its level.

The volcanic island of St. Paul in the midst of the Indian Ocean, lat. 38° 44′ S., long. 77° 37′ E., surveyed by Capt. Blackwood in 1842, seems to exemplify the first stage in the formation of such an archipelago as that of Santorin. We have there a crater one mile in diameter, surrounded by steep and lofty cliffs on every side save one, where the sea enters by a single passage nearly dry at low water. In the interior of

* Poggendorf's Annalen, 1836, p. 183.

the small circular bay or crater there is a depth of 30 fathoms, or 180 feet. The surface of the island slopes away on all sides from the crest of the rocks encircling the crater.*

Barren Island.—There is great analogy between the structure of Barren Island in the Bay of Bengal, lat. 12° 15′, and that of Santorin

Fig. 66.

Cone and crater of Barren Island, in the Bay of Bengal. Height of the central cone (according to Capt. Miller, in 1834), 500 feet.

last described. When seen from the ocean, this island presents, on almost all sides, a surface of bare rocks, rising, with a moderate acclivity, towards the interior; but at one point there is a cleft by which we can penetrate into the centre, and there discover that it is occupied by a great circular basin, filled by the waters of the sea, and bordered all around by steep rocks, in the midst of which rises a volcanic cone, very frequently in eruption. The summit of this cone is about 500 feet in height, corresponding to that of the circular border which incloses the basin; so that it can be seen from the sea only through the ravine. It is most probable that the exterior inclosure of Barren Island (*c*, *d*, fig. 67) is nothing more than the remains of a truncated cone *c*, *a*, *b*, *d*,

Fig. 67.

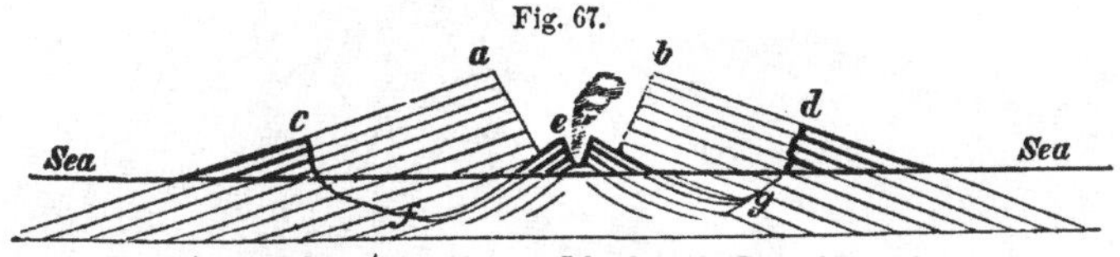

Supposed section of Barren Island, in the Bay of Bengal.

a great portion of which has been removed by engulfment, explosion, or denudation, which may have preceded the formation of the new interior cone *f*, *e*, *g*.†

MUD VOLCANOES.

Iceland.—Mr. R. Bunsen, in his account of the pseudo-volcanic phenomena of Iceland, describes many valleys where sulphurous and aqueous vapors burst forth with a hissing sound, from the hot soil

* See Admiralty Chart, with views and sections, 1842.

† For height of cone and references, see Buist, Volcanoes of India, Trans. Bombay Geol. Soc. vol. x. p. 143.

formed of volcanic tuff. In such spots a pool of boiling water is seen, in which a bluish-black argillaceous paste rises in huge bubbles. These bubbles on bursting throw the boiling mud to a height of fifteen feet and upwards, accumulating it in ledges round the crater or basin of the spring.

Baku on the Caspian.—The formation of a new mud volcano was witnessed on the 27th of November, 1827, at Tokmali, on the peninsula of Abscheron, east of Baku. Flames blazed up to an extraordinary height for a space of three hours, and continued for twenty hours to rise about three feet above a crater, from which mud was ejected. At another point in the same district where flames issued, fragments of rock of large size were hurled up into the air, and scattered around.*

Sicily.—At a place called Macaluba, near Girgenti in Sicily, are several conical mounds from ten to thirty feet in height, with small craters at their summits, from which cold water, mixed with mud and bitumen, is cast out. Bubbles of carbonic acid and carburetted hydrogen gas are also disengaged from these springs, and at certain periods with such violence, as to throw the mud to the height of 200 feet. These "air volcanoes," as they are sometimes termed, are known to have been in the same state of activity for the last fifteen centuries; and Dr. Daubeny imagines that the gases which escape may be generated

Fig. 68.

Mud cones and craters of Hinglaj near Beila, district of Lus, 120 miles northwest of mouth of Indus. From original drawing by Capt. Robertson. (See Map, p. 460.)

* Humboldt's Cosmos.

by the slow combustion of beds of sulphur, which is actually in progress in the blue clay, out of which the springs rise.* But as the gases are similar to those disengaged in volcanic eruptions, and as they have continued to stream out for so long a period, they may perhaps be derived from a more deep-seated source.

Beila in India.—In the district of Luss or Lus, south of Beila, about 120 miles N. W. of Cutch and the mouths of the Indus (see Map, fig. 71, p. 460), numerous mud volcanoes are scattered over an area of probably not less than 1000 square miles. Some of these have been well described by Captain Hart, and subsequently by Captain Robertson, who has paid a visit to that region, and made sketches of them, which he has kindly placed at my disposal. From one of these the annexed view has been selected. These conical hills occur to the westward of the Hara mountains and the river Hubb. (See Map, p. 460.) One of the cones is 400 feet high, composed of light-colored earth, and having at its summit a crater thirty yards in diameter. The liquid mud which fills the crater is continually disturbed by air-bubbles, and here and there is cast up in small jets.†

Mineral composition of volcanic products.—The mineral called felspar forms in general more than half of the mass of modern lavas. When it is in great excess, lavas are called trachytic: they consist generally of a base of compact felspar, in which crystals of glassy felspar are disseminated.‡ When augite (or pyroxene) predominates, lavas are termed basaltic. They contain about 50 per cent. of silica, or much less than the trachytes, in which there is usually about 75 per cent. of that mineral. They also contain about 11 per cent. of protoxide of iron, and as much of lime, both of which are wanting, or only in insignificant quantities in the trachytic rocks.‡ But lavas occur of an intermediate composition between the trachytic and basaltic, which from their color have been called graystones. The abundance of quartz, forming distinct crystals or concretions, characterizes the granitic and other ancient rocks, now generally considered by geologists as of igneous origin; whereas that mineral is rarely exhibited in a separate form in recent lavas, although silica enters so largely into their composition. Hornblende, so common in hypogene rocks, or those commonly called "primary," is rare in modern lava; nor does it enter largely into rocks of any age in which augite abounds. It should, however, be stated, that the experiments of Mr. Gustav Rose have made it very questionable, whether the minerals called hornblende and augite can be separated as distinct species, as their different varieties seem to pass into each other, whether we consider the characters derived from their angles of crystallization, their chemical composition, or their specific gravity. The difference in form of the two substances may be explained by the different circumstances under which

* Daubeny, Volcanoes, p. 267.

† See Buist, Volcanoes of India, Trans. Bombay Geol. Soc. vol. x. p. 154, and Captain Robertson, Journ. of Roy. Asiat. Soc. 1850.

‡ See Glossary. § Bunsen, Volcanic Rocks of Iceland.

they have been produced, the form of hornblende being the result of slower cooling. Crystals of augite have been met with in the scoriæ of furnaces, but never those of hornblende; and crystals of augite have been obtained by melting hornblende in a platina crucible; but hornblende itself has not been formed artificially.* Mica occurs plentifully in some recent trachytes, but is rarely present where augite is in excess.

Frequency of eruptions, and nature of subterranean igneous rocks.—When we speak of the igneous rocks of our own times, we mean that small portion which, in violent eruptions, is forced up by elastic fluids to the surface of the earth,—the sand, scoriæ, and lava, which cool in the open air. But we cannot obtain access to that which is congealed far beneath the surface under great pressure, equal to that of many hundred, or many thousand atmospheres.

During the last century, about fifty eruptions are recorded of the five European volcanic districts, of Vesuvius, Etna, Volcano, Santorin, and Iceland; but many beneath the sea in the Grecian archipelago and near Iceland may doubtless have passed unnoticed. If some of them produced no lava, others, on the contrary, like that of Skaptár Jokul, in 1783, poured out melted matter for five or six years consecutively; which cases, being reckoned as single eruptions, will compensate for those of inferior strength. Now, if we consider the active volcanoes of Europe to constitute about a fortieth part of those already known on the globe, and calculate that, one with another, they are about equal in activity to the burning mountains in other districts, we may then compute that there happen on the earth about 2000 eruptions in the course of a century, or about twenty every year.

However inconsiderable, therefore, may be the superficial rocks which the operations of fire produce on the surface, we must suppose the subterranean changes now constantly in progress to be on the grandest scale. The loftiest volcanic cones must be as insignificant, when contrasted to the products of fire in the nether regions, as are the deposits formed in shallow estuaries when compared to submarine formations accumulating in the abysses of the ocean. In regard to the characters of these volcanic rocks, formed in our own times in the bowels of the earth, whether in rents and caverns, or by the cooling of lakes of melted lava, we may safely infer that the rocks are heavier and less porous than ordinary lavas, and more crystalline, although composed of the same mineral ingredients. As the hardest crystals produced artificially in the laboratory require the longest time for their formation, so we must suppose that where the cooling down of melted matter takes place by insensible degrees, in the course of ages, a variety of minerals will be produced far harder than any formed by natural processes within the short period of human observation.

These subterranean volcanic rocks, moreover, cannot be stratified in the same manner as sedimentary deposits from water, although it is evident that when great masses consolidate from a state of fusion, they may

* Bulletin de la Soc. Géol. de France, tom. ii. p. 206.

separate into natural divisions; for this is seen to be the case in many lava-currents. We may also expect that the rocks in question will often be rent by earthquakes, since these are common in volcanic regions; and the fissures will be often injected with similar matter, so that dikes of crystalline rock will traverse masses of similar composition. It is also clear, that no organic remains can be included in such masses, as also that these deep-seated igneous formations considered in mass must underlie all the strata containing organic remains, because the heat proceeds from below upwards, and the intensity required to reduce the mineral ingredients to a fluid state must destroy all organic bodies in rocks included in the midst of them.

If by a continued series of elevatory movements, such masses shall hereafter be brought up to the surface, in the same manner as sedimentary marine strata have, in the course of ages, been upheaved to the summit of the loftiest mountains, it is not difficult to foresee what perplexing problems may be presented to the geologist. He may then, perhaps, study in some mountain-chain the very rocks produced at the depth of several miles beneath the Andes, Iceland, or Java, in the time of Leibnitz, and draw from them the same conclusion which that philosopher derived from certain igneous products of high antiquity; for he conceived our globe to have been, for an indefinite period, in the state of a comet, without an ocean, and uninhabitable alike by aquatic or terrestrial animals.

CHAPTER XXVII.

EARTHQUAKES AND THEIR EFFECTS.

Earthquakes and their effects—Deficiency of ancient accounts—Ordinary atmospheric phenomena—Changes produced by earthquakes in modern times considered in chronological order—Earthquake in Syria, 1837—Earthquakes in Chili in 1837 and 1835—Isle of Santa Maria raised ten feet—Chili, 1822—Extent of country elevated—Aleppo and Ionian Isles—Earthquake of Cutch in 1819—Subsidence in the Delta of the Indus—Island of Sumbawa in 1815—Earthquake of Caraccas in 1812—Shocks at New Madrid in 1811 in the valley of the Mississippi—Aleutian Islands in 1806—Reflections on the earthquakes of the nineteenth century—Earthquake in Quito, Quebec, &c.—Java, 1786—Sinking down of large tracts.

IN the sketch before given of the geographical boundaries of volcanic regions, I stated, that although the points of eruption are but thinly scattered, constituting mere spots on the surface of those vast districts, yet the subterranean movements extend simultaneously over immense areas. We may now proceed to consider the changes which these movements produce on the surface, and in the internal structure of the earth's crust.

Deficiency of ancient accounts.—It is only within the last century and a half, since Hooke first promulgated, in 1688, his views respecting the connection between geological phenomena and earthquakes, that the permanent changes affected by these convulsions have excited attention. Before that time, the narrative of the historian was almost exclusively confined to the number of human beings who perished, the number of cities laid in ruins, the value of property destroyed, or certain atmospheric appearances which dazzled or terrified the observers. The creation of a new lake, the engulfing of a new city, or the raising of a new island, are sometimes, it is true, adverted to, as being too obvious, or of too much geographical or political interest to be passed over in silence. But no researches were made expressly with a view of ascertaining the amount of depression or elevation of the ground, or any particular alterations in the relative position of sea and land; and very little distinction was made between the raising of soil by volcanic ejections, and the upheaving of it by forces acting from below. The same remark applies to a very large proportion of modern accounts: and how much reason we have to regret this deficiency of information appears from this, that in every instance where a spirit of scientific inquiry has animated the eye-witnesses of these events, facts calculated to throw light on former modifications of the earth's structure are recorded.

Phenomena attending earthquakes.—As I shall confine myself almost entirely, in the following notice of earthquakes, to the changes brought about by them in the configuration of the earth's crust, I may mention, generally, some accompaniments of these terrible events which are almost uniformly commemorated in history, that it may be unnecessary to advert to them again. Irregularities in the seasons preceding or following the shocks; sudden gusts of wind, interrupted by dead calms; violent rains at unusual seasons, or in countries where such phenomena are almost unknown; a reddening of the sun's disk, and a haziness in the air, often continued for months; an evolution of electric matter, or of inflammable gas from the soil, with sulphurous and mephitic vapors; noises underground, like the running of carriages, or the discharge of artillery, or distant thunder; animals uttering cries of distress, and evincing extraordinary alarm, being more sensitive than men of the slightest movement; a sensation like sea-sickness, and a dizziness in the head, experienced by men:—these, and other phenomena, less connected with our present subject as geologists, have recurred again and again at distant ages, and in all parts of the globe.

I shall now begin the enumeration of earthquakes with the latest authentic narratives, and so carry back the survey retrospectively, that I may bring before the reader, in the first place, the minute and circumstantial details of modern times, and thus enable him, by observing the extraordinary amount of change within the last 150 years, to perceive how great must be the deficiency in the meager annals of earlier eras.

EARTHQUAKES OF THE NINETEENTH CENTURY.*

Syria, January, 1837.—It has been remarked that earthquakes affect elongated areas. The violent shock which devastated Syria in 1837 was felt on a line 500 miles in length by 90 in breadth:† more than 6000 persons perished; deep rents were caused in solid rocks, and new hot springs burst out at Tabereah.

Chili—Valdivia, 1837.—One of the latest earthquakes by which the position of solid land is known to have been permanently altered is that which occurred in Chili, on November 7th, 1837. On that day Valdivia was destroyed by an earthquake, and a whaler, commanded by Captain Coste, was violently shaken at sea, and lost her masts, in lat. 43° 38′ S. in sight of the land. The captain went on the 11th of December following to a spot near the island of Lemus, one of the Chonos archipelago, where he had anchored two years before, and found that the bottom of the sea had been raised more than eight feet. Some rocks formerly covered at all times by the sea were now constantly exposed, and an enormous quantity of shells and fish in a decaying state, which had been thrown there by the waves, or suddenly laid dry during the earthquake, attested the recent date of the occurrence. The whole coast was strewed with uprooted trees.‡

Chili—Conception, 1835.—Fortunately we have a still more detailed account of the geographical changes produced in the same country on the 20th of February, 1835. An earthquake was then felt at all places between Copiapo and Chiloe, from north to south, and from Mendoza to Juan Fernandez, from east to west. "Vessels," says Mr. Caldcleugh, "navigating the Pacific, within 100 miles of the coast, experienced the shock with considerable force."§ Conception, Talcahuano, Chillan, and other towns were thrown down. From the account of Captain Fitz Roy, R. N., who was then employed in surveying the coast, we learn that after the shock the sea retired in the Bay of Conception, and the vessels grounded, even those which had been lying in seven fathoms water: all the shoals were visible, and soon afterwards a wave rushed in and then retreated, and was followed by two other waves. The vertical height of these waves does not appear to have been much greater than from sixteen to twenty feet, although they rose to much greater heights when they broke upon a sloping beach.

According to Mr. Caldcleugh and Mr. Darwin, the whole volcanic

* Since the publication of the first edition of this work, numerous accounts of recent earthquakes have been published; but as they do not illustrate any new principle, I cannot insert them, as they would enlarge too much the size of my work. The late Von Hoff published from time to time, in Poggendorf's Annalen, lists of earthquakes which happened between 1821 and 1836; and, by consulting these, the reader will perceive that every month is signalized by one or many convulsions in some part of the globe. See also Mallet's Dynamics of Earthquakes, Trans. Roy. Irish Acad. 1846; and "Earthquakes," Admiralty Manual, 1849; also Hopkins' Report, Brit. Assoc. 1847–8.

† Darwin, Geol. Proceedings, vol. ii. p. 658.

‡ Dumoulin, Comptes Rendus de l'Acad. des Sci. Oct. 1838, p. 706.

§ Phil. Trans. 1836, p. 21.

Fig. 69.

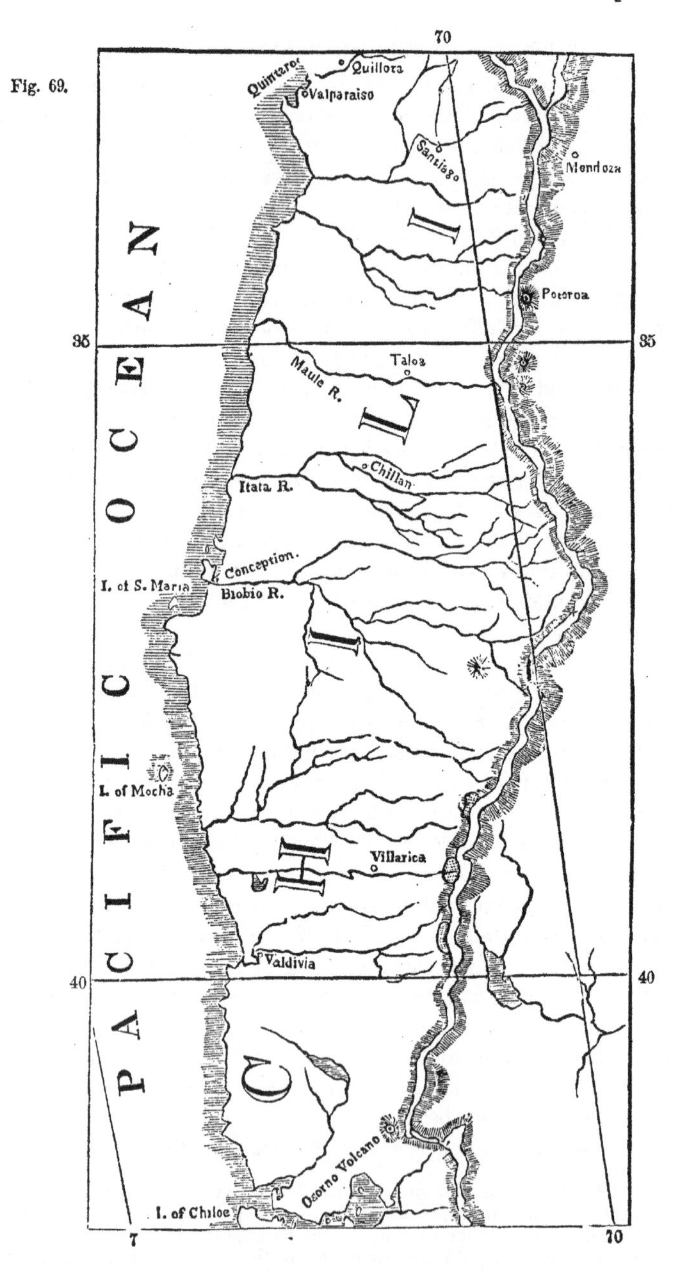

chain of the Chilian Andes, a range 150 miles in length, was in a state of unusual activity, both during the shocks and for some time preceding and after the convulsion, and lava was seen to flow from the crater of Osorno. (See Map, fig. 69.) The island of Juan Fernandez, distant 365 geographical miles from Chili, was violently shaken at the same time, and devastated by a great wave. A submarine volcano broke out there near Bacalao Head, about a mile from the shore, in sixty-nine fathoms water, and illumined the whole island during the night.*

"At Conception," says Captain Fitz Roy, "the earth opened and closed rapidly in numerous places. The direction of the cracks was not uniform, though generally from southeast to northwest. The earth was not quiet for three days after the great shock, and more than 300 shocks were counted between the 20th February and the 4th of March. The loose earth of the valley of the Biobio was everywhere parted from the solid rocks which bound the plain, there being an opening between them from an inch to a foot in width.

Fig. 70.

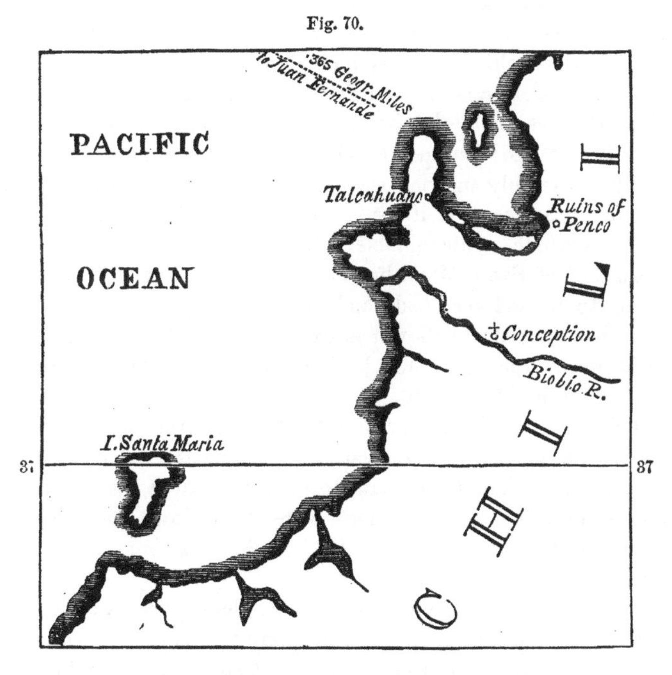

"For some days after the 20th of February, the sea at Talcahuano," says Captain Fitz Roy, "did not rise to the usual marks by four or five feet vertically. When walking on the shore, even at high water, beds of dead mussels, numerous chitons, and limpets, and withered sea-weed, still adhering, though lifeless, to the rocks on which they had lived, everywhere met the eye." But this difference in the relative level of the land and sea gradually diminished, till in the middle of April the water rose again to within two feet of the former high-water mark. It might be supposed that these changes of level merely indicated a temporary disturbance in the set of the currents or in the height of the tides at Talcahuano; but, on considering what occurred in the neighboring island of Santa Maria, Captain Fitz Roy concluded that the land had

* Phil. Trans. 1826.

been raised four or five feet in February, and that it had returned in April to within two or three feet of its former level.

Santa Maria, the island just alluded to, is about seven miles long and two broad, and about twenty-five miles southwest of Conception. (See Map, fig. 70.) The phenomena observed there are most important. "It appeared," says Captain Fitz Roy, who visited Santa Maria twice, the first time at the end of March, and afterwards in the beginning of April, "that the southern extremity of the island had been raised eight feet, the middle nine, and the northern end upwards of ten feet. On steep rocks, where vertical measures could be correctly taken, beds of dead mussels were found ten feet above high-water mark. One foot lower than the highest bed of mussels, a few limpets and chitons were seen adhering to the rock where they had grown. Two feet lower than the same, dead mussels, chitons, and limpets were abundant.

"An extensive rocky flat lies around the northern parts of Santa Maria. Before the earthquake this flat was covered by the sea, some projecting rocks only showing themselves. Now, the whole flat is exposed, and square acres of it are covered with dead shell-fish, the stench arising from which is abominable. By this elevation of the land the southern port of Santa Maria has been almost destroyed; little shelter remaining there, and very bad landing." The surrounding sea is also stated to have become shallower in exactly the same proportion as the land had risen; the soundings having diminished a fathom and a half everywhere around the island.

At Tubal, also, to the southeast of Santa Maria, the land was raised six feet, at Mocha two feet, but no elevation could be ascertained at Valdivia.

Among other effects of the catastrophe, it is stated that cattle standing on a steep slope, near the shore, were rolled down into the sea, and many others were washed off by the great wave from low land and drowned.*

In November of the same year (1835), Conception was shaken by a severe earthquake, and on the same day Osorno, at the distance of 400 miles, renewed its activity. These facts prove not only the connection of earthquakes with volcanic eruptions in this region, but also the vast extent of the subterranean areas over which the disturbing cause acts simultaneously.

Ischia, 1828.—On the 2d of February the whole island of Ischia was shaken by an earthquake, and in the October following I found all the houses in Casamicciol still without their roofs. On the sides of a ravine between that town and Forio, I saw masses of greenish tuff which had been thrown down. The hot-spring of Rita, which was nearest the centre of the movement, was ascertained by M. Covelli to have increased in temperature, showing, as he observes, that the explosion took place below the reservoirs which heat the thermal waters.†

* Darwin's Journ. of Travels in South America, Voyage of Beagle, p. 372.
† Biblioth. Univ. Oct. 1828, p. 157.

Bogota, 1827.—On the 16th of November, 1827, the plain of Bogota, in New Granada, or Colombia, was convulsed by an earthquake, and a great number of towns were thrown down. Torrents of rain swelled the Magdalena, sweeping along vast quantities of mud and other substances, which emitted a sulphurous vapor and destroyed the fish. Popayan, which is distant 200 geographical miles S. S. W. of Bogota, suffered greatly. Wide crevices appeared in the road of Guanacas, leaving no doubt that the whole of the Cordilleras sustained a powerful shock. Other fissures opened near Costa, in the plains of Bogota, into which the river Tunza immediately began to flow.* It is worthy of remark, that in all such cases the ancient gravel bed of a river is deserted and a new one formed at a lower level; so that a want of relation in the position of alluvial beds of the existing water-courses may be no test of the high antiquity of such deposits, at least in countries habitually convulsed by earthquakes. Extraordinary rains accompanied the shocks before mentioned; and two volcanoes are said to have been in eruption in the mountain-chain nearest to Bogota.

Chili, 1822.—On the 19th of November, 1822, the coast of Chili was visited by a most destructive earthquake. The shock was felt simultaneously throughout a space of 1200 miles from north to south. St. Jago, Valparaiso, and some other places, were greatly injured. When the district round Valparaiso was examined on the morning after the shock, it was found that the coast for a considerable distance was raised above its former level.† At Valparaiso the elevation was three feet, and at Quintero about four feet. Part of the bed of the sea, says Mrs. Graham, remained bare and dry at high water, "with beds of oysters, mussels, and other shells adhering to the rocks on which they grew, the fish being all dead, and exhaling most offensive effluvia.‡

An old wreck of a ship, which before could not be approached, became accessible from the land, although its distance from the original sea-shore had not altered. It was observed that the water-course of a mill, at the distance of about a mile from the sea, gained a fall of fourteen inches, in little more than one hundred yards; and from this fact it is inferred that the rise in some parts of the inland country was far more considerable than on the borders of the ocean.§ Part of the coast thus elevated consisted of granite, in which parallel fissures were caused, some of which were traced for a mile and a half inland. Cones of earth about four feet high were thrown up in several districts, by the forcing up of water mixed with sand through funnel-shaped hollows,—a phenomenon very common in Calabria, and the explanation of which will hereafter be considered. Those houses in Chili of which the foundations were on rock were less damaged than such as were built on alluvial soil.

Mr. Cruickshanks, an English botanist, who resided in the country

* Phil. Mag. July 1828, p. 37

† Geol. Trans. vol. i. 2d ser., and Journ. of Sci. 1824, vol. xvii. p. 40.

‡ Geol. Trans. vol. i. 2d ser. p. 415. § Journ. of Sci. vol. xvii. p. 42.

during the earthquake, has informed me that some rocks of greenstone at Quintero, a few hundred yards from the beach, which had always been under water till the shock of 1822, have since been uncovered when the tide is at half-ebb: and he states that, after the earthquake, it was the general belief of the fishermen and inhabitants of the Chilian coast, *not* that the land had risen, but that the ocean had permanently retreated.

Dr. Meyen, a Prussian traveller, who visited Valparaiso in 1831, says that on examining the rocks both north and south of the town, nine years after the event, he found, in corroboration of Mrs. Graham's account, that remains of animals and sea-weed, the *Lessonia* of Bory de St. Vincent, which has a firm ligneous stem, still adhered to those rocks which in 1822 had been elevated above high-water mark.* According to the same author, the whole coast of Central Chili was raised about four feet, and banks of marine shells were laid dry on many parts of the coast. He observed similar banks, elevated at unknown periods, in several places, especially at Copiapo, where the species all agree with those now living in the ocean. Mr. Freyer also, who resided some years in South America, has confirmed these statements;† and Mr. Darwin obtained evidence that the remains of an ancient wall, formerly washed by the sea, and now 11½ feet above high-water mark, acquired several feet of this additional elevation during the earthquake of 1822.‡

The shocks continued up to the end of September, 1823; even then, forty-eight hours seldom passed without one, and sometimes two or three were felt during twenty-four hours. Mrs. Graham observed, after the earthquake of 1822, that besides a beach newly raised above high-water mark, there were several older elevated lines of beach, one above the other, consisting of shingle mixed with shells extending in a parallel direction to the shore, to the height of fifty feet above the sea.§

Extent of country elevated.—By some observers it has been supposed that the whole country from the foot of the Andes to a great distance under the sea was upraised in 1822, the greatest rise being at the distance of about two miles from the shore. "The rise upon the coast was from two to four feet:—at the distance of a mile inland it must have been from five to six or seven feet."‖ It has also been conjectured by the same eye-witnesses to the convulsion, that the area over which this permanent alteration of level extended may have been equal to 100,000 square miles. Although the increased fall of certain water-courses may have afforded some ground for this conjecture, it must be considered as very hypothetical, and the estimate may have exceeded or greatly fallen short of the truth. It may nevertheless be useful to reflect on the enormous amount of change which this single convulsion occasioned, if the

* Reise um die Erde; and see Dr. Meyen's letter cited Foreign Quart. Rev No. 33, p. 13, 1836.

† Geol. Soc. Proceedings, No. xl. p. 179, Feb. 1835.

‡ Proceed. Geol. Soc. vol. ii. p. 447.

§ Geol. Trans. vol. i. 2d ser. p. 415.

‖ Journal of Science, vol. xvii. pp. 40, 45.

extent of country moved upward really amounted to 100,000 square miles,—an extent just equal to half the area of France, or about five-sixths of the area of Great Britain and Ireland. If we suppose the elevation to have been only three feet on an average, it will be seen that the mass of rock added to the continent of America by the movement, or, in other words, the mass previously below the level of the sea, and after the shocks permanently above it, must have contained fifty-seven cubic miles in bulk; which would be sufficient to form a conical mountain two miles high (or about as high as Etna), with a circumference at the base of nearly thirty-three miles. We may take the mean specific gravity of the rock at 2·655,—a fair average, and a convenient one in such computations, because at such a rate a cubic yard weighs two tons. Then, assuming the great pyramid of Egypt, if solid, to weigh, in accordance with an estimate before given, six million tons, we may state the rock added to the continent by the Chilian earthquake to have more than equalled 100,000 pyramids.

But it must always be borne in mind that the weight of rock here alluded to constituted but an insignificant part of the whole amount which the volcanic forces had to overcome. The whole thickness of rock between the surface of Chili and the subterranean foci of volcanic action may be many miles or leagues deep. Say that the thickness was only two miles, even then the mass which changed place and rose three feet being 200,000 cubic miles in volume, must have exceeded in weight 363 million pyramids.

It may be instructing to consider these results in connection with others already obtained from a different source, and to compare the working of two antagonistic forces—the levelling power of running water, and the expansive energy of subterranean heat. How long, it may be asked, would the Ganges require, according to data before explained (p. 283), to transport to the sea a quantity of solid matter equal to that which may have been added to the land by the Chilian earthquake? The discharge of mud in one year by the Ganges was estimated at 20,000 million cubic feet. According to that estimate it would require about four centuries (or 418 years) before the river could bear down from the continent into the sea a mass equal to that gained by the Chilian earthquake. In about half that time, perhaps, the united waters of the Ganges and Burrampooter might accomplish the operation.

Cutch, 1819.—A violent earthquake occurred at Cutch, in the delta of the Indus, on the 16th of June, 1819. (See Map, fig. 71.) The principal town, Bhooj, was converted into a heap of ruins, and its stone buildings were thrown down. The movement was felt over an area having a radius of 1000 miles from Bhooj, and extending to Khatmandoo, Calcutta, and Pondicherry.* The vibrations were felt in North-west India, at a distance of 800 miles, after an interval of about fifteen minutes after the earthquake at Bhooj. At Ahmedabad the great

* See Asiatic Journal, vol. i.

Fig. 71.

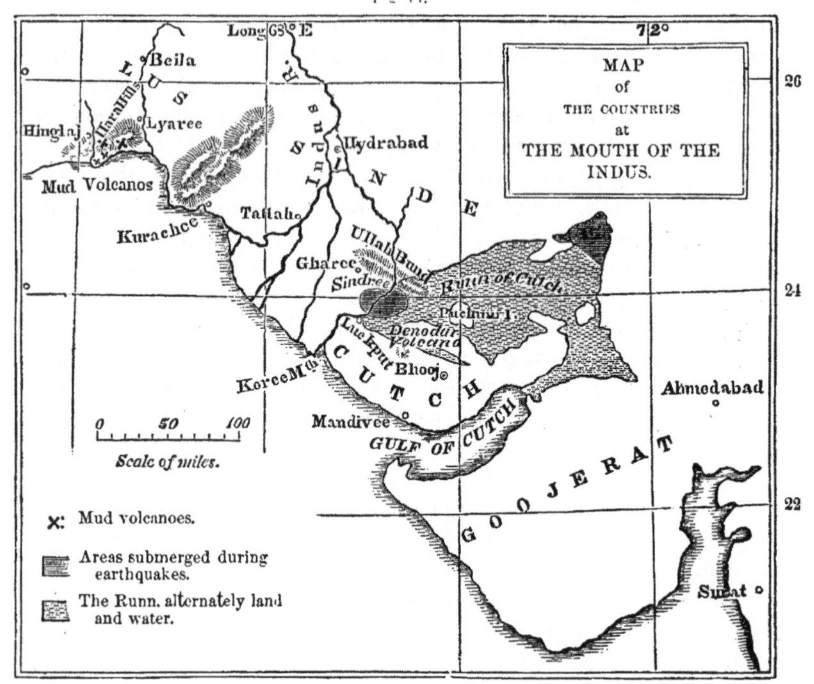

mosque, erected by Sultan Ahmed nearly 450 years before, fell to the ground, attesting how long a period had elapsed since a shock of similar violence had visited that point. At Anjar, the fort, with its tower and guns, was hurled to the ground in one common mass of ruin. The shocks continued until the 20th; when, thirty miles northwest from Bhooj, the volcano called Denodur is said by some to have sent forth flames, but Capt. Grant was unable to authenticate this statement.

Subsidence in the delta of the Indus.—Although the ruin of towns was great, the face of nature in the inland country, says Captain Macmurdo, was not visibly altered. In the hills some large masses only of rock and soil were detached from the precipices; but the eastern and almost deserted channel of the Indus, which bounds the province of Cutch, was greatly changed. This estuary, or inlet of the sea, was, before the earthquake, fordable at Luckput, being only about a foot deep when the tide was at ebb, and at flood-tide never more than six feet; but it was deepened at the fort of Luckput, after the shock, to more than *eighteen feet at low water.** On sounding other parts of the channel, it was found, that where previously the depth of the water at flood never exceeded one or two feet, it had become from four to ten feet deep. By these and other remarkable changes of level, a part of the inland navigation of that country, which had been closed for centuries, became again practicable.

* Macmurdo Ed. Phil. Journ. iv. 106.

Fig. 72.

Fort of Sindree, on the eastern branch of the Indus, before it was submerged by the earthquake of 1819, from a sketch of Capt. Grindlay, made in 1808.

*Fort and village submerged.**—The fort and village of Sindree, on the eastern arm of the Indus, above Luckput, are stated by the same writer to have been overflowed; and, after the shock, the tops of the houses and wall were alone to be seen above the water, for the houses, although submerged, were not cast down. Had they been situated, therefore, in the interior, where so many forts were levelled to the ground, their site would, perhaps, have been regarded as having remained comparatively unmoved. Hence we may suspect that great permanent upheavings and depressions of soil may be the result of earthquakes, without the inhabitants being in the least degree conscious of any change of level.

A more recent survey of Cutch, by Sir A. Burnes, who was not in communication with Capt. Macmurdo, confirms the facts above enumerated, and adds many important details.† That officer examined the delta of the Indus in 1826 and 1828, and from his account it appears that, when Sindree subsided in June, 1819, the sea flowed in by the eastern mouth of the Indus, and in a few hours converted a tract of land, 2000 square miles in area, into an inland sea, or lagoon. Neither the rush of the sea into this new depression, nor the movement of the earthquake, threw down entirely the small fort of Sindree, one of the four towers, the northwestern, still continuing to stand; and, the day after the earthquake, the inhabitants who had ascended to the top of this tower, saved themselves in boats.‡

* I was indebted to my friend the late Sir Alexander Burnes for the accompanying sketch (fig. 72) of the fort of Sindree, as it appeared eleven years before the earthquake.

† This Memoir is now in the Library of the Royal Asiatic Society of London.

‡ Several particulars not given in the earlier edition were afterwards obtained by me from personal communication with Sir A. Burnes in London.

Elevation of the Ullah Bund.—Immediately after the shock, the inhabitants of Sindree saw, at the distance of five miles and a half from their village, a long elevated mound, where previously there had been a low and perfectly level plain. (See Map, fig. 71.) To this uplifted tract they gave the name of "Ullah Bund," or the "Mound of God," to distinguish it from several artificial dams previously thrown across the eastern arm of the Indus.

Extent of country raised.—It has been ascertained that this new-raised country is *upwards of fifty miles* in length from east to west, running parallel to that line of subsidence before mentioned, which caused the grounds around Sindree to be flooded. The range of this elevation extends from Puchum Island towards Gharee; its breadth from north to south is conjectured to be in some parts *sixteen miles*, and its greatest ascertained height above the original level of the delta is ten feet,—an elevation which appears to the eye to be very uniform throughout.

For several years after the convulsion of 1819, the course of the Indus was very unsettled, and at length, in 1826, the river threw a vast body of water into its eastern arm, that called the Phurraun, above Sindree; and forcing its way in a more direct course to the sea, burst through all the artificial dams which had been thrown across its channel, and at length cut right through the "Ullah Bund," whereby a natural section was obtained. In the perpendicular cliffs thus laid open Sir A. Burnes found that the upraised lands consisted of clay filled with shells. The new channel of the river where it intersected the "bund" was eighteen feet deep, and forty yards in width; but in 1828 the channel was still farther enlarged. The Indus, when it first opened this new passage, threw such a body of water into the new mere, or salt lagoon, of Sindree, that it became fresh for many months; but it had recovered its saltness in 1828, when the supply of river-water was less copious, and finally it became more salt than the sea, in consequence, as the natives suggested to Sir A. Burnes, of the saline particles with which the "Runn of Cutch" is impregnated.

In 1828 Sir A. Burnes went in a boat to the ruins of Sindree, where a single remaining tower was seen in the midst of a wide expanse of sea. The tops of the ruined walls still rose two or three feet above the level of the water; and standing on one of these, he could behold nothing in the horizon but water, except in one direction, where a blue streak of land to the north indicated the Ullah Bund. This scene presents to the imagination a lively picture of the revolutions now in progress on the earth—a waste of waters where a few years before all was land, and the only land visible consisting of ground uplifted by a recent earthquake.

Ten years after the visit of Sir A. Burnes above alluded to, my friend, Captain Grant, F. G. S., of the Bombay Engineers, had the kindness to send at my request a native surveyor to make a plan of Sindree and Ullah Bund, in March, 1838. From his description it appears that, at that season, the driest of the whole year, he found the channel traversing the Bund to be 100 yards wide, without water, and incrusted with salt. He was told that it has now only four or five feet of water in it after rains. The sides or banks were nearly perpendicular, and nine feet in height. The lagoon has diminished both in area and depth, and part near the fort was dry land. The annexed drawing, made by Captain Grant from the surveyor's plan, shows the appearance of the fort in the midst of the lake, as seen in 1838 from the west, or from the same point as that from which Captain Grindlay's sketch (see fig. 72) was taken in 1808, before the earthquake.

Fig. 73.

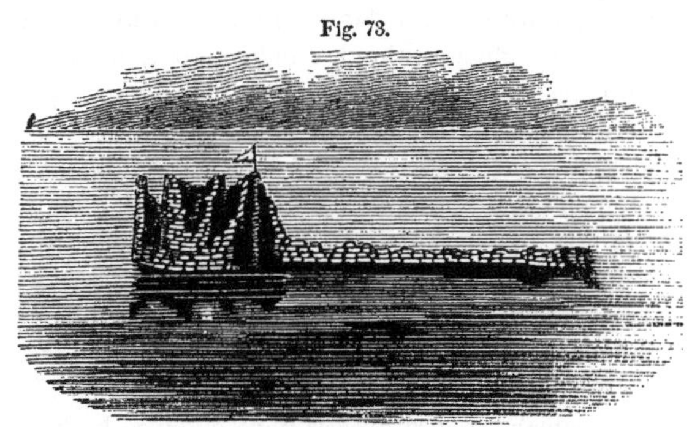

View of the Fort of Sindree, from the west, in March, 1838.

The Runn of Cutch is a flat region of a very peculiar character, and no less than 7000 square miles in area: a greater superficial extent than Yorkshire, or about one-fourth the area of Ireland. It is not a desert of moving sand, nor a marsh, but evidently the dried-up bed of an inland sea, which for a great part of every year has a hard and dry bottom uncovered by weeds or grass, and only supporting here and there a few tamarisks. But during the monsoons, when the sea runs high, the salt-water driven up from the Gulf of Cutch and the creeks at Luckput overflows a large part of the Runn, especially after rains, when the soaked ground permits the sea-water to spread rapidly. The Runn is also liable to be overflowed occasionally in some parts by river-water: and it is remarkable that the only portion which was ever highly cultivated (that anciently called Sayra) is now permanently submerged. The surface of the Runn is sometimes incrusted with salt about an inch in depth, in consequence of the evaporation of the sea-water. Islands rise up in some parts of the waste, and the boundary lands form bays and promontories. The natives have various traditions respecting the former separation of Cutch and Sinde by a bay of the sea, and the drying up of the district called the Runn. But these tales, besides the usual uncertainty of oral tradition, are farther obscured by mythological fictions. The conversion of the Runn into land is chiefly ascribed to the miraculous powers of a Hindoo saint, by name Damorath (or Dhoorunnath), who had previously done penance for twelve years on the summit of Denodur hill. Captain Grant infers, on various grounds, that this saint flourished about the eleventh or twelfth century of our era. In proof of the drying up of the Runn, some towns far inland are still pointed out as having once been ancient ports. It has, moreover, been always said that ships were wrecked and engulphed by the great catastrophe; and in the jets of black muddy water thrown out of fissures in that region, in 1819, there were cast up numerous pieces of wrought-iron and ship nails.* Cones of sand six or eight feet in height were at the same time thrown up on these lands.†

We must not conclude without alluding to a *moral* phenomenon connected with this tremendous catastrophe, which we regard as highly deserving the attention of geologists. It is stated by Sir A. Burnes, that "these wonderful events passed *unheeded* by the inhabitants of Cutch;" for the region convulsed, though once fertile, had for a long period been reduced to sterility by want of irrigation, so that the natives were indifferent as to its fate. Now it is to this profound apathy which all but highly civilized nations feel, in regard to physical events not having an immediate influence on their worldly fortunes, that we must ascribe the extraordinary dearth of historical information concerning changes of the earth's surface, which modern observations show to be by no means of rare occurrence in the ordinary course of nature.

Since the above account was written, a description has been published of more recent geographical changes in the district of Cutch, near the mouth of the Koree, or eastern branch of the Indus, which happened in June, 1845. A large area seems to have subsided, and the Sindree lake had become a salt marsh.‡

Island of Sumbawa, 1815.—In April, 1815, one of the most frightful eruptions recorded in history occurred in the province of Tomboro, in the island of Sumbawa (see Map, fig. 39, p. 351), about 200 miles from the eastern extremity of Java. In April of the year preceding the volcano had been observed in a state of considerable activity, ashes having fallen upon the decks of vessels which sailed past the coast.§ The eruption of 1815 began on the 5th of April, but was most violent on the 11th and 12th, and did not entirely cease till July. The sound of the explosions was heard in Sumatra, at the distance of 970 geographical miles in a direct line; and at Ternate, in an opposite direction, at the distance of 720 miles. Out of a population of 12,000, in the province of Tomboro, only twenty-six individuals survived. Violent whirlwinds carried up men, horses, cattle, and whatever else came within their influence into the air; tore up the largest trees by the roots, and covered the whole sea with floating timber.‖ Great tracts of land were covered by lava, several streams of which, issuing from the crater of the Tomboro mountain, reached the sea. So heavy was the fall of ashes, that they broke into the Resident's house at Bima, forty miles east of the volcano, and rendered it as well as many other dwellings in the town uninhabitable. On the side of Java the ashes were carried to the distance of 300

* Capt. Burnes' Account. vol. iv. p. 106.

† Capt. Macmurdo's Memoir, Ed. Phil. Journ.

‡ Quart. Geol. Journ. vol. ii. p. 103.

§ MS. of J. Crawfurd, Esq.

‖ Raffles' Java, vol. i. p. 28.

miles, and 217 towards Celebes, in sufficient quantity to darken the air. The floating cinders to the westward of Sumatra formed, on the 12th of April, a mass two feet thick, and several miles in extent, through which ships with difficulty forced their way.

The darkness occasioned in the daytime by the ashes in Java was so profound, that nothing equal to it was ever witnessed in the darkest night. Although this volcanic dust when it fell was an impalpable powder, it was of considerable weight when compressed, a pint of it weighing twelve ounces and three quarters. "Some of the finest particles," says Mr. Crawfurd, "were transported to the islands of Amboyna and Banda, which last is about 800 miles east from the site of the volcano, although the southeast monsoon was then at its height." They must have been projected, therefore, into the upper regions of the atmosphere, where a counter-current prevailed.

Along the sea-coast of Sumbawa and the adjacent isles, the sea rose suddenly to the height of from two to twelve feet, a great wave rushing up the estuaries, and then suddenly subsiding. Although the wind at Bima was still during the whole time, the sea rolled in upon the shore, and filled the lower parts of the houses with water a foot deep. Every prow and boat was forced from the anchorage, and driven on shore.

The town called Tomboro, on the west side of Sumbawa, was overflowed by the sea, which encroached upon the shore so that the water remained permanently eighteen feet deep in places where there was land before. Here we may observe, that the amount of subsidence of land was apparent, in spite of the ashes, which would naturally have caused the limits of the coast to be extended.

The area over which tremulous noises and other volcanic effects extended, was 1000 English miles in circumference, including the whole of the Molucca Islands, Java, a considerable portion of Celebes, Sumatra, and Borneo. In the island of Amboyna, in the same month and year, the ground opened, threw out water, and then closed again.*

In conclusion, I may remind the reader, that but for the accidental presence of Sir Stamford Raffles, then Governor of Java, we should scarcely have heard in Europe of this tremendous catastrophe. He required all the residents in the various districts under his authority to send in a statement of the circumstances which occurred within their own knowledge; but, valuable as were their communications, they are often calculated to excite rather than to satisfy the curiosity of the geologist. They mention that similar effects, though in a less degree, had, about seven years before, accompanied an eruption of Carang Assam, a volcano in the island of Bali, west of Sumatra; but no particulars of that great catastrophe are recorded.†

Caraccas, 1812.—On the 26th of March, 1812, several violent shocks of an earthquake were felt in Caraccas. The surface undulated like a boiling liquid, and terrific sounds were heard underground. The whole

* Raffles' Hist. of Java, vol. i. p. 25. Ed. Phil. Journ. vol. iii. p. 389.

† Life and Services of Sir Stamford Raffles, p. 241. London, 1830.

city with its splendid churches was in an instant a heap of ruins, under which 10,000 of the inhabitants were buried. On the 5th of April, enormous rocks were detached from the mountains. It was believed that the mountain Silla lost from 300 to 360 feet of its height by subsidence; but this was an opinion not founded on any measurement. On the 27th of April, a volcano in St. Vincent's threw out ashes; and, on the 30th, lava flowed from its crater into the sea, while its explosions were heard at a distance equal to that between Vesuvius and Switzerland, the sound being transmitted, as Humboldt supposes, through the ground. During the earthquake which destroyed Caraccas, an immense quantity of water was thrown out at Valecillo, near Valencia, as also at Porto Cabello, through openings in the earth; and in the Lake Maracaybo the water sank. Humboldt observed that the Cordilleras, composed of gneiss and mica slate, and the country immediately at their feet, were more violently shaken than the plains.*

South Carolina and New Madrid, Missouri, 1811–12.—Previous to the destruction of La Guayra and Caraccas, in 1812, earthquakes were felt in South Carolina; and the shocks continued till those cities were destroyed. The valley also of the Mississippi, from the village of New Madrid to the mouth of the Ohio in one direction, and to the St. Francis in another, was convulsed in such a degree as to create new lakes and islands. It has been remarkéd by Humboldt in his Cosmos, that the earthquake of New Madrid presents one of the few examples on record of the incessant quaking of the ground for several successive months *far from any volcano*. Flint, the geographer, who visited the country seven years after the event, informs us, that a tract of many miles in extent, near the Little Prairie, became covered with water three or four feet deep; and when the water disappeared a stratum of sand was left in its place. Large lakes of twenty miles in extent were formed in the course of an hour, and others were drained. The grave-yard at New Madrid was precipitated into the bed of the Mississippi; and it is stated that the ground whereon the town is built, and the river-bank for fifteen miles above, sank eight feet below their former level.† The neighboring forest presented for some years afterwards "a singular scene of confusion; the trees standing inclined in every direction, and many having their trunks and branches broken."‡

The inhabitants relate that the earth rose in great undulations; and when these reached a certain fearful height, the soil burst, and vast volumes of water, sand, and pit-coal were discharged as high as the tops of the trees. Flint saw hundreds of these deep chasms remaining in an alluvial soil, seven years after. The people in the country, although inexperienced in such convulsions, had remarked that the chasms in the earth were in a direction from S. W. to N. E.; and they accordingly felled the tallest trees, and laying them at right angles to the chasms,

* Humboldt's Pers. Nar. vol. iv. p. 12; and Ed. Phil. Journ. vol. i. p. 272: 1819
† Cramer's Navigator, p. 243. Pittsburgh, 1821.
‡ Long's Exped. to the Rocky Mountains, vol. iii. p. 184.

stationed themselves upon them. By this invention, when chasms opened more than once under these trees, several persons were prevented from being swallowed up.* At one period during this earthquake, the ground not far below New Madrid swelled up so as to arrest the Mississippi in its course, and to cause a temporary reflux of its waves. The motion of some of the shocks is described as having been horizontal, and of others perpendicular; and the vertical movement is said to have been much less desolating than the horizontal.

The above account has been reprinted exactly as it appeared in former editions of this work, compiled from the authorities which I have cited; but having more recently (March, 1846) had an opportunity myself of visiting the disturbed region of the Mississippi, and conversing with many eye-witnesses of the catastrophe, I am able to confirm the truth of those statements, and to add some remarks on the present face and features of the country. I skirted, as was before related (p. 270), part of the territory immediately west of New Madrid, called "the sunk country," which was for the first time permanently submerged during the earthquake of 1811–12. It is said to extend along the course of the White Water and its tributaries for a distance of between 70 and 80 miles north and south, and 30 miles east and west. I saw on its borders many full-grown trees still standing leafless, the bottoms of their trunks several feet under water, and a still greater number lying prostrate. An active vegetation of aquatic plants is already beginning to fill up some of the shallows, and the sediment washed in by occasional floods when the Mississippi rises to an extraordinary height contributes to convert the sunk region into marsh and forest land. Even on the dry ground along the confines of the submerged area, I observed in some places that all the trees of prior date to 1811 were dead and leafless, though standing erect and entire. They are supposed to have been killed by the loosening of their roots during the repeated undulations which passed through the ground for three months in succession.

Mr. Bringier, an experienced engineer of New Orleans, who was on horseback near New Madrid when some of the severest shocks were experienced, related to me (in 1846), that "as the waves advanced the trees bent down, and the instant afterwards, while recovering their position, they often met those of other trees similarly inclined, so that their branches becoming interlocked, they were prevented from righting themselves again. The transit of the wave through the woods was marked by the crashing noise of countless boughs, first heard on one side and then on the other. At the same time powerful jets of water, mixed with sand, mud, and fragments of coaly matter, were cast up, endangering the lives of both horse and rider."

I was curious to ascertain whether any vestiges still remained of these fountains of mud and water, and carefully examined between New

* Silliman's Journ. Jan. 1829.

Madrid and the Little Prairie several "sink holes," as they are termed. They consist of cavities from 10 to 30 yards in width, and 20 feet or more in depth, and are very conspicuous, interrupting the level surface of a flat alluvial plain. I saw abundance of sand, which some of the present inhabitants saw spouting from these deep holes, also fragments of decayed wood and black bituminous shale, probably drifted down at some former period in the main channel of the Mississippi, from the coal-fields farther north. I also found numerous rents in the soil left by the earthquake, some of them still several feet wide, and a yard or two in depth, although the action of rains, frost, and occasional inundations, and especially the leaves of trees blown into them in countless numbers every autumn, have done much to fill them up. I measured the direction of some of the fissures, which usually varied from 10 to 45 degrees W. of north, and were often parallel to each other; I found, however, a considerable diversity in their direction. Many of them are traceable for half a mile and upwards, but they might easily be mistaken for artificial trenches if resident settlers were not there to assure us that within their recollection they were "as deep as wells." Fragments of coaly shale were strewed along the edges of some of these open fissures, together with white sand, in the same manner as round the "sink holes."*

Among other monuments of the changes wrought in 1811–12, I explored the bed of the lake called Eulalie, near New Madrid, 300 yards long by 100 yards in width, which was suddenly drained during the earthquake. The parallel fissures by which the waters escaped are not yet entirely closed, and all the trees growing on its bottom were at the time of my visit less than 34 years old. They consisted of cotton-wood, willows, and honey-locust, and other species, differing from those clothing the surrounding higher grounds, which are more elevated by 12 or 15 feet. On them the hickory, the black and white oak, the gum and other trees, many of them of ancient date, were flourishing.

Aleutian Islands, 1806.—In the year 1806, a new island, in the form of a peak, with some low conical hills upon it, is said to have risen from the sea among the Aleutian Islands, east of Kamtschatka. According to Langsdorf,† it was four geographical miles in circumference; and Von Buch infers from its magnitude, and from its not having again subsided below the level of the sea, that it did not consist merely of ejected matter, but of a solid rock of trachyte upheaved.‡ Another extraordinary eruption happened in the spring of the year 1814, in the sea near Unalaschka, in the same archipelago. A new isle was then produced of considerable size, and with a peak three thousand feet high, which remained standing for a year afterwards, though with somewhat diminished height.

Although it is not improbable that earthquakes accompanying these

* See Lyell's Second Visit to the United States, ch. xxxiii.
† Bemerkungen auf einer Reise um die Welt. bd. ii. s. 209.
‡ Neue Allgem. Geogr. Ephemer. bd. iii. s. 348.

tremendous eruptions may have heaved up part of the bed of the sea, yet the circumstance of the islands not having disappeared like Sabrina (see p. 416), may have arisen from the emission of lava. If Jorullo, for example, in 1759, had risen from a shallow sea to the height of 1600 feet, instead of attaining that elevation above the Mexican plateau, the massive current of basaltic lava which poured out from its crater would have enabled it to withstand, for a long period, the action of a turbulent sea.

Reflections on the earthquakes of the nineteenth century.—We are now about to pass on to the events of the eighteenth century; but before we leave the consideration of those already enumerated, let us pause for a moment, and reflect how many remarkable facts of geological interest are afforded by the earthquakes above described, though they constitute but a small part of the convulsions even of the last forty years. New rocks have risen from the waters; new hot springs have burst out, and the temperature of others has been raised; the coast of Chili has been thrice permanently elevated; a considerable tract in the delta of the Indus has sunk down, and some of its shallow channels have become navigable; an adjoining part of the same district, upwards of fifty miles in length and sixteen in breadth, has been raised about ten feet above its former level; part of the great plain of the Mississippi, for a distance of eighty miles in length by thirty in breadth, has sunk down several feet; the town of Tomboro has been submerged, and twelve thousand of the inhabitants of Sumbawa have been destroyed. Yet, with a knowledge of these terrific catastrophes, witnessed during so brief a period by the present generation, will the geologist declare with perfect composure that the earth has at length settled into a state of repose? Will he continue to assert that the changes of relative level of land and sea, so common in former ages of the world, have now ceased? If, in the face of so many striking facts, he persists in maintaining this favorite dogma, it is in vain to hope that, by accumulating the proofs of similar convulsions during a series of antecedent ages, we shall shake his tenacity of purpose:—

Si fractus illabatur orbis
Impavidum ferient ruinæ.

EARTHQUAKES OF THE EIGHTEENTH CENTURY.

Quito, 1797.—On the morning of February 4th, 1797, the volcano of Tunguragua in Quito, and the surrounding district, for forty leagues from south to north, and twenty leagues from west to east, experienced an undulating movement, which lasted four minutes. The same shock was felt over a tract of 170 leagues from south to north, from Piura to Popayan; and 140 from west to east, from the sea to the river Napo. In the smaller district first mentioned, where the movement was more intense, every town was levelled to the ground; and Riobamba, Quero, and other places, were buried under masses detached from the mountains. At the foot of Tunguragua the earth was rent open in several places; and streams of water and fetid mud, called "moya," poured out, overflowing and wasting every thing. In valleys 1000 feet broad, the water of these floods reached to the height of 600 feet; and the mud deposit barred up the course of the river, so as to form lakes, which in some places continued for more than eighty days. Flames and suffocating vapors escaped from the lake Quilotoa, and killed all the cattle on its shores. The shocks continued all February and March; and on the 5th of April they recurred with almost as much violence as at first. We are told that the form of the surface in the district most shaken was entirely altered, but no exact measurements are given whereby we may estimate the degree of elevation or subsidence.* Indeed it would be difficult, except in the immediate neighborhood of the sea, to obtain any certain standard of comparison if the levels were really as much altered as the narrations imply.

Cumana, 1797.—In the same year, on the 14th of December, the small Antilles experienced subterranean movements, and four-fifths of the town of Cumana was shaken down by a vertical shock. The form of the shoal of Mornerouge, at the mouth of the river Bourdones, was changed by an upheaving of the ground.†

Canada—Quebec, 1791.—We learn from Captain Bayfield's memoirs, that earthquakes are very frequent on the shore of the estuary of the St. Lawrence, of force sufficient at times to split walls and throw down chimneys. Such were the effects experienced in December, 1721, in St. Paul's Bay, about fifty miles N. E. from Quebec; and the inhabitants say, that about every twenty-five years a violent earthquake returns, which lasts forty days. In the History of Canada, it is stated that, in 1663, a tremendous convulsion lasted six months, extending from Quebec to Tadeausac,—a distance of about 130 miles. The ice on the river was broken up, and many landslips caused.‡

Caraccas, 1790.—In the Caraccas, near where the Caura joins the Orinoco, between the towns San Pedro de Alcantara and San Francisco de Aripao, an earthquake, on St. Matthew's day 1790, caused a sinking in of the granitic soil, and left a lake 800 yards in diameter, and from eighty to one hundred in depth. It was a portion of the forest of Aripao which subsided, and the trees remained green for several months under water.§

Sicily, 1790.—On the 18th of March in the same year, at S. Maria di Niscemi, some miles from Terranuova, near the south coast of Sicily, the ground gradually sunk down for a circumference of three Italian miles, during seven shocks; and, in one place, to the depth of thirty feet. It continued to subside to the end of the month. Several fissures sent forth sulphur, petroleum, steam, and hot water, and a stream of

* Cavanilles, Journ. de Phys. tome xlix. p. 230. Gilbert's Annalen, bd. vi. Humboldt's Voy. p. 317.

† Humboldt's Voy., Relat. Hist., part. i. p. 309.

‡ Macgregor's Travels in America.

§ Humboldt's Voy., Relat. Hist., part. ii. p. 632.

mud, which flowed for two hours, and covered a space sixty feet long and thirty broad. This happened far from both the ancient and modern volcanic district, in a group of strata consisting chiefly of blue clay.*

Java, 1786.—About the year 1786, an earthquake was felt at intervals, for the period of four months, in the neighborhood of Batur, in Java, and an eruption followed. Various rents were formed, which emitted a sulphurous vapor; separate tracts sunk away, and were swallowed by the earth. Into one of these the rivulet Dotog entered, and afterwards continued to follow a subterraneous course. The village of Jampang was buried in the ground, with thirty-eight of its inhabitants, who had not time to escape. We are indebted to Dr. Horsfield for having verified the above-mentioned facts.†

CHAPTER XXVIII.

EARTHQUAKE IN CALABRIA, 1783.

Earthquake in Calabria, February 5, 1783—Shocks continued to the end of the year 1786—Authorities—Area convulsed—Geological structure of the district—Difficulty of ascertaining changes of level—Subsidence of the quay at Messina—Movement in the stones of two obelisks—Shift or fault in the Round Tower of Terranuova—Opening and closing of fissures—Large edifices engulfed—Dimensions of new caverns and fissures—Gradual closing in of rents—Bounding of detached masses into the air—Landslips—Buildings transported entire to great distances—New lakes—Funnel-shaped hollows in alluvial plains—Currents of mud—Fall of cliffs, and shore near Scilla inundated—State of Stromboli and Etna during the shocks—How earthquakes contribute to the formation of valleys—Concluding remarks.

Calabria, 1783.—OF the numerous earthquakes which have occurred in different parts of the globe, during the last 100 years, that of Calabria, in 1783, is almost the only one of which the geologist can be said to have such a circumstantial account as to enable him fully to appreciate the changes which this cause is capable of producing in the lapse of ages. The shocks began in February, 1783, and lasted for nearly four years, to the end of 1786. Neither in duration, nor in violence, nor in the extent of territory moved, was this convulsion remarkable, when contrasted with many experienced in other countries, both during the last and present century; nor were the alterations which it occasioned in the relative level of hill and valley, land and sea, so great as those effected by some subterranean movements in South America, in later times. The importance of the earthquake in question arises from the circumstance, that Calabria is the only spot hitherto visited, both

* Ferrara, Camp. fl., p. 51.

† Batav. Trans. vol. viii. p. 141.

during and after the convulsions, by men possessing sufficient leisure, zeal, and scientific information, to enable them to collect and describe with accuracy the physical facts which throw light on geological questions.

Authorities.—Among the numerous authorities, Vivenzio, physician to the king of Naples, transmitted to the court a regular statement of his observations during the continuance of the shocks; and his narrative is drawn up with care and clearness.* Francesco Antonio Grimaldi, then secretary of war, visited the different provinces at the king's command, and published a most detailed description of the permanent changes in the surface.† He measured the length, breadth, and depth

Fig. 42.

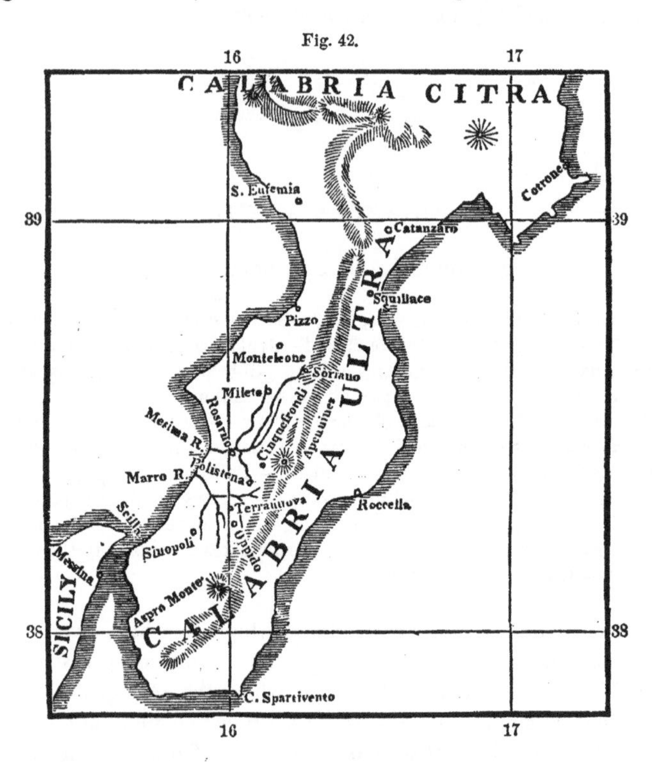

of the different fissures and gulfs which opened, and ascertained their number in many provinces. His comments, moreover, on the reports of the inhabitants, and his explanations of their relations, are judicious and instructive. Pignataro, a physician residing at Monteleone, a town placed in the very centre of the convulsions, kept a register of the shocks, distinguishing them into four classes, according to their degree of violence. From his work, it appears that, in the year 1783, the number was 949, of which 501 were shocks of the first degree of force;

* Istoria de' Tremuoti della Calabria del 1783.

† Descriz. de' Tremuoti Accad. nelle Calabria nel 1783. Napoli, 1784.

and in the following year there were 151, of which 98 were of the first magnitude.

Count Ippolito, also, and many others, wrote descriptions of the earthquake; and the Royal Academy of Naples, not satisfied with these and other observations, sent a deputation from their own body into Calabria, before the shocks had ceased, who were accompanied by artists instructed to illustrate by drawings the physical changes of the district, and the state of ruined towns and edifices. Unfortunately these artists were not very successful in their representations of the condition of the country, particularly when they attempted to express, on a large scale, the extraordinary revolutions which many of the great and minor river-courses underwent. But many of the plates published by the Academy are valuable; and as they are little known, I shall frequently avail myself of them to illustrate the facts about to be described.*

In addition to these Neapolitan sources of information, our countryman, Sir William Hamilton, surveyed the district, not without some personal risk, before the shocks had ceased; and his sketch, published in the Philosophical Transactions, supplies many facts that would otherwise have been lost. He has explained, in a rational manner, many events which, as related in the language of some eye-witnesses, appeared marvellous and incredible. Dolomieu also examined Calabria during the catastrophe, and wrote an account of the earthquake, correcting a mistake into which Hamilton had fallen, who supposed that a part of the tract shaken had consisted of volcanic tuff. It is, indeed, a circumstance which enhances the geological interest of the commotions which so often modify the surface of Calabria, that they are confined to a country where there are neither ancient nor modern rocks of volcanic or trappean origin; so that at some future time, when the era of disturbance shall have passed by, the cause of former revolutions will be as latent as in parts of Great Britain now occupied exclusively by ancient marine formations.

Extent of the area convulsed.—The convulsion of the earth, sea, and air extended over the whole of Calabria Ultra, the southeast part of Calabria Citra, and across the sea to Messina and its environs; a district lying between the 38th and 39th degrees of latitude. The concussion was perceptible over a great part of Sicily, and as far north as Naples; but the surface over which the shocks acted so forcibly as to excite intense alarm did not generally exceed 500 square miles in area. The soil of that part of Calabria is composed chiefly, like the southern part of Sicily, of calcareo-argillaceous strata of great thickness, containing marine shells. This clay is sometimes associated with beds of sand and limestone. For the most part these formations resemble in appearance and consistency the Subapennine marls, with their accompanying sands and sandstones; and the whole group bears considerable resemblance,

* Istoria de' Fenomeni del Tremoto, &c., nell' An. 1783, posta in luce dalla Real. Accad., &c. di Nap. Napoli, 1783, fol.

in the yielding nature of its materials, to most of our tertiary deposits in France and England. Chronologically considered, however, the Calabrian formations are comparatively of modern date, often abounding in fossil shells referable to species now living in the Mediterranean.

We learn from Vivencio, that on the 20th and 26th of March, 1783, earthquakes occurred in the islands of Zante, Cephalonia, and St. Maura; and in the last-mentioned island several public edifices and private houses were overthrown, and many people destroyed.

If the city of Oppido, in Calabria Ultra, be taken as a centre, and round that centre a circle be described, with a radius of twenty-two miles, this space will comprehend the surface of the country which suffered the greatest alteration, and where all the towns and villages were destroyed. The first shock, of February 5th, 1783, threw down, in two minutes, the greater part of the houses in all the cities, towns, and villages, from the western flanks of the Apennines in Calabria Ultra to Messina in Sicily, and convulsed the whole surface of the country. Another occurred on the 28th of March, with almost equal violence. The granitic chain which passes through Calabria from north to south, and attains the height of many thousand feet, was shaken but slightly by the first shock, but more rudely by some which followed.

Some writers have asserted that the wave-like movements which were propagated through the recent strata, from west to east, became very violent when they reached the point of junction with the granite, as if a reaction was produced where the undulatory movement of the soft strata was suddenly arrested by the more solid rocks. But the statement of Dolomieu on this subject is most interesting, and perhaps, in a geological point of view, the most important of all the observations which are recorded.* The Apennines, he says, which consist in great part of hard and solid granite, with some micaceous and argillaceous schists, form bare mountains with steep sides, and exhibit marks of great degradation. At their base newer strata are seen of sand and clay, mingled with shells; a marine deposit containing such ingredients as would result from the decomposition of granite. The surface of this newer (*tertiary*) formation constitutes what is called the plain of Calabria—a platform which is flat and level, except where intersected by narrow valleys or ravines, which rivers and torrents have excavated sometimes to the depth of six hundred feet. The sides of these ravines are almost perpendicular; for the superior stratum, being bound together by the roots of trees, prevents the formation of a sloping bank. The usual effect of the earthquake, he continues, was to disconnect all those masses which either had not sufficient bases for their bulk, or which was supported only by lateral adherence. Hence it follows that throughout almost the whole length of the chain, the soil which adhered to the granite at the base of the mountains Caulone, Esope, Sagra, and

* Dissertation on the Calabrian Earthquake, &c., translated in Pinkerton's Voyages and Travels, vol. v.

Aspramonte, slid over the solid and steeply inclined nucleus, and descended somewhat lower, leaving almost uninterruptedly from St. George to beyond St. Christina, a distance of from nine to ten miles, a chasm between the solid granitic nucleus and the sandy soil. Many lands slipping thus were carried to a considerable distance from their former position, so as entirely to cover others; and disputes arose as to whom the property which had thus shifted its place should belong.

From this account of Dolomieu we might anticipate, as the result of a continuance of such earthquakes, first, a longitudinal valley following the line of junction of the older and newer rocks; secondly, greater disturbance in the newer strata near the point of contact than at a greater distance from the mountains; phenomena very common in other parts of Italy at the junction of the Apennine and Subapennine formations.

Mr. Mallet, in his valuable essay on the Dynamics of Earthquakes,* offers the following explanation of the fact to which Dolomieu has called attention. When a wave of elastic compression, of which he considers the earth-wave to consist, passes abruptly from a body having an extremely low elasticity, such as clay and gravel, into another like granite, whose elasticity is remarkably high, it changes not only its velocity but in part also its course, a portion being reflected and a portion refracted. The wave being thus sent back again produces a shock in the opposite direction, doing great damage to buildings on the surface by thus returning upon itself. At the same time, the shocks are at once eased when they get into the more elastic materials of the granitic mountains.

The surface of the country during the Calabrian earthquakes often heaved like the billows of a swelling sea, which produced a swimming in the head, like sea-sickness. It is particularly stated, in almost all the accounts, that just before each shock the clouds appeared motionless; and, although no explanation is offered of this phenomenon, it is obviously the same as that observed in a ship at sea when it pitches violently. The clouds seem arrested in their career as often as the vessel rises in a direction contrary to their course; so that the Calabrians must have experienced precisely the same motion on the land.

Trees, supported by their trunks, sometimes bent during the shocks to the earth, and touched it with their tops. This is mentioned as a well-known fact by Dolomieu; and he assures us that he was always on his guard against the spirit of exaggeration in which the vulgar are ever ready to indulge when relating these wonderful occurrences.

It is impossible to suppose that these waves, which are described in Italy and other regions of earthquakes as passing along the solid surface of the earth in a given direction like a billow on the sea, have any strict analogy with the undulations of a fluid. They are doubtless the effects of vibrations, radiating from some deep-seated point, each of which on reaching the surface lifts up the ground, and then allows it again to subside. As the distance between the source of the subterranean move-

* Proceed. Roy. Irish Acad. 1846, p. 26.

ment and the surface must vary according to the outline of the country, so the vibratory jar will reach different points in succession.

The Academicians relate that in some of the cities of Calabria effects were produced seeming to indicate a whirling or vorticose movement. Thus, for example, two obelisks (fig. 75) placed at the extremities of a

Fig. 75.

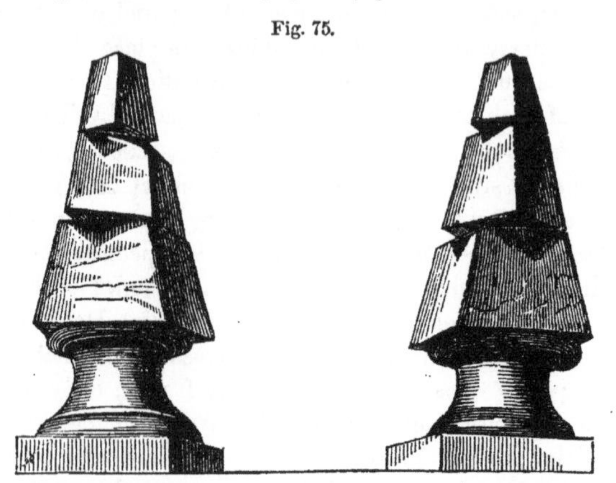

Shifts in the stones of two obelisks in the Convent of St. Bruno.

magnificent façade in the convent of S. Bruno, in a small town called Stefano del Bosco, were observed to have undergone a movement of a singular kind. The shock which agitated the building is described as having been horizontal and vorticose. The pedestal of each obelisk remained in its original place; but the separate stones above were turned partially round, and removed sometimes nine inches from their position without falling.

It has been suggested by Mr. Darwin that this kind of displacement may be due to a vibratory rather than a whirling motion;* and more lately Mr. Mallet, in the paper already cited, has offered a very ingenious solution of the problem. He refers the twisting simply to an elastic wave, which has moved the pedestal forwards and back again, by an alternate horizontal motion within narrow limits, and he has succeeded in showing that a rectilinear movement in the ground may have sufficed to cause an incumbent body to turn partially round upon its bed, provided a certain relation exist between the position of the centre of gravity of the body and its centre of adherence.†

I shall now consider, in the first place, that class of physical changes produced by the earthquake which are connected with alterations in the relative level of the different parts of the land; and afterwards describe those which are more immediately connected with the derangement of the regular drainage of the country, and where the force of running water co-operated with that of the earthquake.

Difficulty of ascertaining changes of level.—In regard to alterations of

* Journal of a Naturalist, p. 376, and ii. ib. 308.
† Proceedings Roy. Irish Acad. 1846, pp. 14–16.

relative level, none of the accounts establish that they were on a considerable scale; but it must always be remembered that, in proportion to the area moved is the difficulty of proving that the general level has undergone any change, unless the sea-coast happens to have participated in the principal movement. Even then it is often impossible to determine whether an elevation or depression even of several feet has occurred, because there is nothing to attract notice in a band of shingle and sand of unequal breadth above the level of the sea running parallel to a coast; such bands generally marking the point reached by the waves during spring tides, or the most violent tempests. The scientific investigator has not sufficient topographical knowledge to discover whether the extent of beach has diminished or increased; and he who has the necessary local information, scarcely ever feels any interest in ascertaining the amount of the rise or fall of the ground. Add to this the great difficulty of making correct observations, in consequence of the enormous waves which roll in upon a coast during an earthquake, and efface every landmark near the shore.

Subsidence of the quay at Messina.—It is evidently in seaports alone that we can look for very accurate indications of slight changes of level; and when we find them, we may presume that they would not be rare at other points, if equal facilities of comparing relative altitudes were afforded. Grimaldi states (and his account is confirmed by Hamilton and others), that at Messina, in Sicily, the shore was rent; and the soil along the port, which before the shock was perfectly level, was found afterwards to be inclined towards the sea,—the sea itself near the "Banchina" becoming deeper, and its bottom in several places disordered. The quay also sunk down about fourteen inches below the level of the sea, and the houses in its vicinity were much fissured. (*Phil. Trans.* 1783.)

Among various proofs of partial elevation and depression in the interior, the Academicians mention, in their Survey, that the ground was sometimes on the same level on both sides of new ravines and fissures, but sometimes there had been a considerable shifting, either by the upheaving of one side, or the subsidence of the other. Thus, on the sides of long rents in the territory of Soriano, the stratified masses had altered their relative position to the extent of from eight to fourteen palms (six to ten and a half feet).

Polistena.—Similar shifts in the strata are alluded to in the territory of Polistena, where there appeared innumerable fissures in the earth. One of these was of great length and depth; and in parts the level of the corresponding sides was greatly changed. (See fig. 76.)

Terranuova.—In the town of Terranuova some houses were seen uplifted above the common level, and others adjoining sunk down into the earth. In several streets the soil appeared thrust up, and abutted against the walls of houses: a large circular tower of solid masonry, part of which had withstood the general destruction, was divided by a vertical rent, and one side was upraised, and the foundations heaved out of the ground. It was compared by the Academicians to a great tooth

Fig. 76.

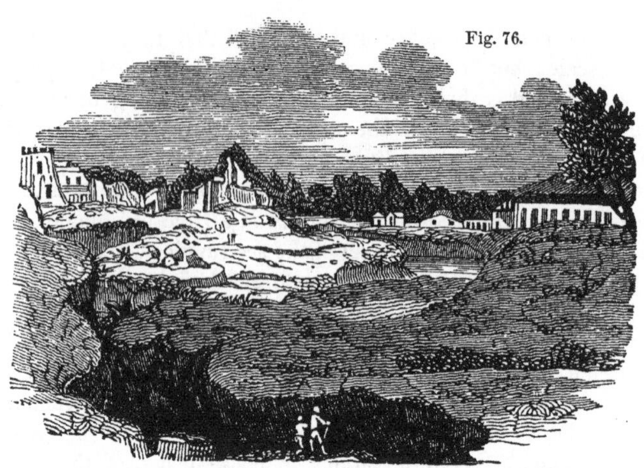

Deep fissure, near Polistena, caused by the earthquake of 1783.

half extracted from the alveolus, with the upper part of the fangs exposed. (See fig. 77.)

Along the line of this shift, or "fault," as it would be termed technically by miners, the walls were found to adhere firmly to each other, and to fit so well, that the only signs of their having been disunited was the want of correspondence in the courses of stone on either side of the rent.

Fig. 77.

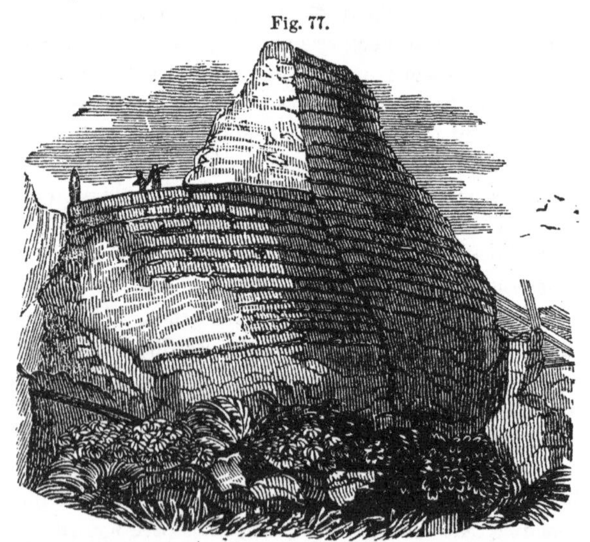

Shift or "fault" in the Round Tower of Terranuova in Calabria, occasioned by the earthquake of 1783.

Dolomieu saw a stone well in the convent of the Augustins at Terranuova, which had the appearance of having been driven out of the earth. It resembled a small tower eight or nine feet in height, and a little inclined. This effect, he says, was produced by the consolidation and consequent sinking of the sandy soil in which the well was dug.

In some walls which had been thrown down, or violently shaken, in Monteleone, the separate stones were parted from the mortar, so as to leave an exact mould where they had rested; whereas in other cases the mortar was ground to dust between the stones.

It appears that the wave-like motions often produced effects of the most capricious kind. Thus, in some streets of Monteleone, every house was thrown down but one; in others, all but two; and the buildings which were spared were often scarcely in the least degree injured. In many cities of Calabria, all the most solid buildings were thrown down, while those which were slightly built escaped; but at Rosarno, as also at Messina in Sicily, it was precisely the reverse, the massive edifices being the only ones that stood.

Fissures.—It appears evident that a great part of the rending and fissuring of the ground was the effect of a violent motion from below upwards; and in a multitude of cases where the rents and chasms opened and closed alternately, we must suppose that the earth was by turns heaved up, and then let fall again.* We may conceive the same effect to be produced on a small scale, if, by some mechanical force, a pavement composed of large flags of stone should be raised up, and then allowed to fall suddenly, so as to resume its original position. If any small pebbles happened to be lying on the line of contact of two flags, they would fall into the opening when the pavement rose, and be swallowed up, so that no trace of them would appear after the subsidence of the stones. In the same manner, when the earth was upheaved, large houses, trees, cattle, and men were engulfed in an instant in chasms and fissures; and when the ground sank down again, the earth closed upon them, so that no vestige of them was discoverable on the surface. In many instances, individuals were swallowed up by one shock, and then thrown out again alive, together with large jets of water, by the shock which immediately succeeded.

Fig. 78.

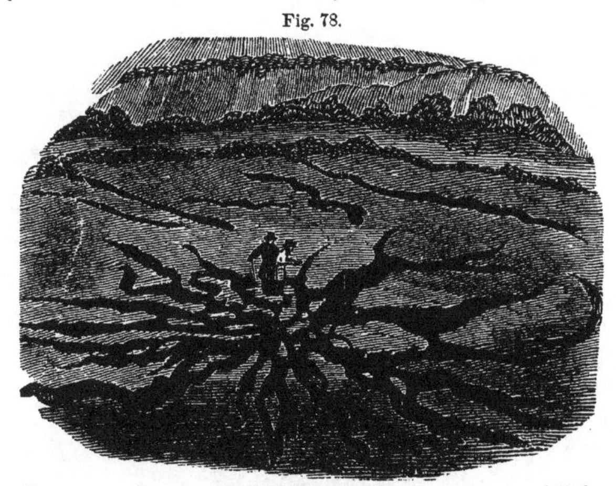

Fissures near Jerocarne, in Calabria, caused by the earthquake of 1783.

* See Mr. Mallet's attempt to controvert this view, p. 32 ibid.

At Jerocarne, a country which, according to the Academicians, was *lacerated* in a most extraordinary manner, the fissures ran in every direction, "like cracks on a broken pane of glass" (see fig. 78); and as a great portion of them remained open after the shocks, it is very possible that this country was permanently upraised. It was usual, as we learn from Dolomieu, for the chasms and fissures throughout Calabria, to run parallel to the course of some pre-existing gorges in their neighborhood.

Houses engulfed.—In the vicinity of Oppido, the central point from which the earthquake diffused its violent movements, many houses were swallowed up by the yawning earth, which closed immediately over them. In the adjacent district, also, of Cannamaria four farm-houses, several oil-stores, and some spacious dwelling-houses were so completely engulfed in one chasm, that not a vestige of them was afterwards discernible. The same phenomena occurred at Terranuova, S. Christina, and Sinopoli. The Academicians state particularly, that when deep abysses had opened in the argillaceous strata of Terranuova, and houses had sunk into them, the sides of the chasms closed with such violence, that, on excavating afterwards to recover articles of value, the workmen found the contents and detached parts of the buildings jammed together so as to become one compact mass. It is unnecessary to accumulate examples of similar occurrences; but so many are well authenticated during this earthquake in Calabria, that we may, without hesitation, yield assent to the accounts of catastrophes of the same kind repeated again and again in history, where whole towns are declared to have been engulfed, and nothing but a pool of water or tract of sand left in their place.

Chasm formed near Oppido.—On the sloping side of a hill near Oppido a great chasm opened; and, although a large quantity of soil was precipitated into the abyss, together with a considerable number of olive-trees and part of a vineyard, a great gulf remained after the shock, in the form of an amphitheatre, 500 feet long and 200 feet deep. (See fig. 79.)

Fig. 79.

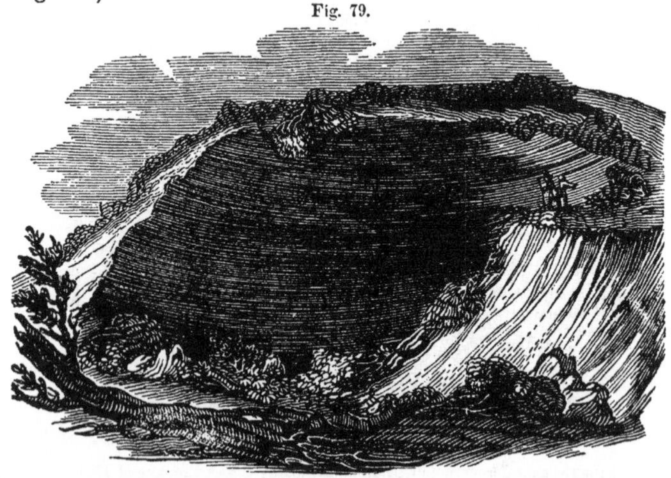

Chasm formed by the earthquake of 1783, near Oppido in Calabria.

Dimensions of new fissures and chasms.—According to Grimaldi, many fissures and chasms, formed by the first shock of February 5th, were greatly widened, lengthened, and deepened by the violent convulsions of March 28th. In the territory of San Fili this observer found a new ravine, half a mile in length, two feet and a half broad, and twenty-five feet deep; and another of similar dimensions in the territory of Rosarno. A ravine *nearly a mile long*, 105 feet broad and thirty feet deep, opened in the district of Plaisano, where, also, two gulfs were caused—one in a place called Cerzulle, three-quarters of a mile long, 150 feet broad, and above *one hundred feet deep;* and another at La Fortuna, nearly a quarter of a mile long, above thirty feet in breadth, and no less than 225 feet deep.

In the district of Fosolano three gulfs opened: one of these measured 300 feet square, and above thirty feet deep; another was nearly half a mile long, fifteen feet broad, and above thirty-feet deep; the third was 750 feet square. Lastly, a calcareous mountain, called Zefirio, at the southern extremity of the Italian peninsula, was cleft in two for the length of nearly half a mile, and an irregular breadth of many feet. Some of these chasms were in the form of a crescent. The annexed cut (fig. 80) represents one by no means remarkable for

Fig. 80.

Chasm in the hill of St. Angelo, near Soriano, in Calabria, caused by the earthquake of 1783.

its dimensions, which remained open by the side of a small pass over the hill of St. Angelo, near Soriano. The small river Mesima is seen in the foreground.

Formation of circular hollows and new lakes.—In the report of the Academy, we find that some plains were covered with circular hollows, for the most part about the size of carriage-wheels, but often somewhat larger or smaller. When filled with water to within a foot or two of the surface, they appeared like wells; but, in general, they were filled with dry sand, sometimes with a concave surface, and at other times convex. (See fig. 81.) On digging down, they found them to be funnel-

Fig. 81.

Circular hollows in the plain of Rosarno, formed by the earthquake of 1783.

shaped, and the moist loose sand in the centre marked the tube up which the water spouted. The annexed cut (fig. 82) represents a section of one of these inverted cones when the water had disappeared, and nothing but dry micaceous sand remained.

Fig. 82.

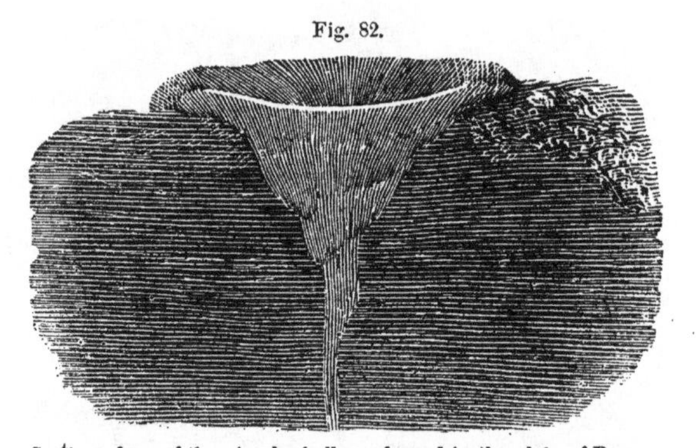

Section of one of the circular hollows formed in the plain of Rosarno.

A small circular pond of similar character was formed not far from Polistena (see fig. 83); and in the vicinity of Seminara, a lake was suddenly caused by the opening of a great chasm, from the bottom of which water issued. This lake was called Lago del Tolfilo. It extended 1785 feet in length, by 937 in breadth, and 52 in depth. The inhabitants, dreading the miasma of this stagnant pool, endeavored, at great cost, to drain it by canals, but without success, as it was fed by springs issuing from the bottom of the deep chasm.

Vivenzio states, that near Sitizzano a valley was nearly filled up to a level with the high grounds on each side, by the enormous masses de-

Fig. 83.

Circular pond near Polistena, in Calabria, caused by the earthquake in 1783.

tached from the boundary hills, and cast down into the course of two streams. By this barrier a lake was formed of great depth, about two miles long and a mile broad. The same author mentions that, upon the whole, there were fifty lakes occasioned during the convulsions: and he assigns localities to all of these. The government surveyors enumerated 215 lakes; but they included in this number many small ponds.

Cones of sand thrown up.—Many of the appearances exhibited in the alluvial plains, such as springs spouting up their water like fountains at the moment of the shock, have been supposed to indicate the alternate rising and sinking of the ground. The first effect of the more violent shocks was usually to dry up the rivers, but they immediately afterwards overflowed their banks. In marshy places, an immense number of cones of sand were thrown up. These appearances Hamilton explains, by supposing that the first movement raised the fissured plain from below upwards, so that the rivers and stagnant waters in bogs sank down, or at least were not upraised with the soil. But when the ground returned with violence to its former position, the water was thrown up in jets through fissures.*

The phenomenon, according to Mr. Mallet, may be simply an accident contingent on the principal cause of disturbance, the rapid transit of the earth-wave. "The sources," he says, "of copious springs usually lie in flat plates or fissures filled with water, whether issuing from solid rock, or from loose materials; now, if a vein, or thin flat cavity filled with water, be in such a position that the plane of the plate of water or fissure be transverse to the line of transit of the earth-wave, the effect of the arrival of the earth-wave at the watery fissure will be, at the instant, to compress its walls more or less together, and so squeeze out the water, which will, for a moment, gush up at the spring-head like a fountain, and again remain in repose after the transit of the wave."

Gradual closing in of fissures.—Sir W. Hamilton was shown several

* Phil. Trans. vol. lxxiii. p. 180.

deep fissures in the vicinity of Mileto, which, although not one of them was above a foot in breadth, had opened so wide during the earthquake as to swallow an ox and nearly one hundred goats. The Academicians also found, on their return through districts which they had passed at the commencement of their tour, that many rents had, in that short interval, gradually closed in, so that their width had diminished several feet, and the opposite walls had sometimes nearly met. It is natural that this should happen in argillaceous strata, while, in more solid rocks, we may expect that fissures will remain open for ages. Should this be ascertained to be a general fact in countries convulsed by earthquakes, it may afford a satisfactory explanation of a common phenomenon in mineral veins. Such veins often retain their full size so long as the rocks consist of limestone, granite, or other indurated materials; but they contract their dimensions, become mere threads, or are even entirely cut off, where masses of an argillaceous nature are interposed. If we suppose the filling up of fissures with metallic and other ingredients to be a process requiring ages for its completion, it is obvious that the opposite walls of rents, where strata consist of yielding materials, must collapse or approach very near to each other before sufficient time is allowed for the accretion of a large quantity of veinstone.

Thermal waters augmented.—It is stated by Grimaldi, that the thermal waters of St. Eufemia, in Terra di Amato, which first burst out during the earthquake of 1638, acquired, in February, 1783, an augmentation both in quantity and degree of heat. This fact appears to indicate a connection between the heat of the interior and the fissures caused by the Calabrian earthquakes, notwithstanding the absence of volcanic rocks, either ancient or modern, in that district.

Bounding of detached masses into the air.—The violence of the movement of the ground upwards was singularly illustrated by what the Academicians call the "sbalzo," or bounding into the air, to the height of several yards, of masses slightly adhering to the surface. In some towns a great part of the pavement stones were thrown up, and found lying with their lower sides uppermost. In these cases, we must suppose that they were propelled upwards by the momentum which they had acquired; and that the adhesion of one end of the mass being greater than that of the other, a rotatory motion had been communicated to them. When the stone was projected to a sufficient height to perform somewhat more than a quarter of a revolution in the air, it pitched down on its edge, and fell with its lower side uppermost.

Effects of earthquakes on the excavations of valleys.—The next class of effects to be considered, are those more immediately connected with the formation of valleys, in which the action of water was often combined with that of the earthquake. The country agitated was composed, as before stated, chiefly of argillaceous strata, intersected by deep narrow valleys, sometimes from 500 to 600 feet deep. As the boundary cliffs were in great part vertical, it will readily be conceived that, amidst the various movements of the earth, the precipices overhanging rivers, being without support on one side, were often thrown down. We find, indeed, that inundations produced by obstructions in river-courses are among the most disastrous consequences of great earthquakes in all parts of the world, for the alluvial plains in the bottoms of valleys are usually the most fertile and well-peopled parts of the whole country; and whether the site of a town is above or below a temporary barrier in the channel of a river, it is exposed to injury by the waters either of a lake or flood.

Landslips.—From each side of the deep valley or ravine of Terranuova enormous masses of the adjoining flat country were detached, and cast down into the course of the river, so as to give rise to great lakes. Oaks, olive-trees, vineyards, and corn, were often seen growing at the bottom of the ravine, as little injured as their former companions, which still continued to flourish in the plain above, at least 500 feet higher, and at the distance of about three-quarters of a mile. In one part of this ravine was an enormous mass, 200 feet high and about 400 feet at its base, which had been detached by some former earthquake. It is well attested, that this mass travelled down the ravine nearly four miles, having been put in motion by the earthquake of the 5th of February. Hamilton, after examining the spot, declared that this phenomenon might be accounted for by the declivity of the valley, the great abundance of rain which fell, and the great weight of the alluvial matter which pressed behind it. Dolomieu also alludes to the fresh impulse derived from other masses falling, and pressing upon the rear of those first set in motion.

The first account sent to Naples of the two great slides or landslips above alluded to, which caused a great lake near Terranuova, was couched in these words:—"Two mountains on the opposite sides of a valley walked from their original position until they met in the middle of the plain, and there joining together, they intercepted the course of a river," &c. The expressions here used resemble singularly those applied to phenomena, probably very analogous, which are said to have occurred at Fez, during the great Lisbon earthquake, as also in Jamaica and Java at other periods.

Not far from Soriano, which was levelled to the ground by the great shock of February, a small valley, containing a beautiful olive-grove, called Fra Ramondo, underwent a most extraordinary revolution. Innumerable fissures first traversed the river-plain in all directions, and absorbed the water until the argillaceous substratum became soaked, so that a great part of it was reduced to a state of fluid paste. Strange alterations in the outline of the ground were the consequence, as the soil to a great depth was easily moulded into any form. In addition to this change, the ruins of the neighboring hills were precipitated into the hollow; and while many olives were uprooted, others remained growing on the fallen masses, and inclined at various angles (see fig. 84). The small river Caridi was entirely concealed for many days; and when at length it reappeared, it had shaped for itself an entirely new channel.

Buildings transported entire to great distances.—Near Seminara an

Fig 84.

Changes of the surface at Fra Ramondo, near Soriano, in Calabria.

1, Portion of a hill covered with olives thrown down.
2, New bed of the river Caridi. 3, Town of Soriano.

extensive olive-ground and orchard were hurled to a distance of two hundred feet, into a valley sixty feet in depth. At the same time a deep chasm was riven in another part of the high platform from which the orchard had been detached, and the river immediately entered the fissure, leaving its former bed completely dry. A small inhabited house, standing on the mass of earth carried down into the valley, went along with it entire, and without injury to the inhabitants. The olive-trees, also, continued to grow on the land which had slid into the valley, and bore the same year an abundant crop of fruit.

Two tracts of land on which a great part of the town of Polistena stood, consisting of some hundreds of houses, were detached into a contiguous ravine, and nearly across it, about half a mile from their original site; and what is most extraordinary, several of the inhabitants were dug out from the ruins alive and unhurt.

Two tenements, near Mileto, called the Macini and Vaticano, occupying an extent of ground about a mile long and half a mile broad, were carried for a mile down a valley. A thatched cottage, together with large olive and mulberry trees, most of which remained erect, were carried uninjured to this extraordinary distance. According to Hamilton, the surface removed had been long undermined by rivulets, which were afterwards in full view on the bare spot deserted by the tenements. The earthquake seems to have opened a passage in the adjoining argillaceous hills, which admitted water charged with loose soil into the subterranean channels of the rivulets immediately under the tenements, so that the foundations of the ground set in motion by the earthquake were loosened. Another example of subsidence, where the edifices were not destroyed, is mentioned by Grimaldi, as having taken place in the city of Catanzaro, the capital of the province of that name. The houses in the quarter called San Giuseppe subsided with the ground to various depths from two to four feet, but the buildings remained uninjured.

It would be tedious, and our space would not permit us, to follow the different authors through their local details of landslips produced in minor valleys; but they are highly interesting, as showing to how great an extent the power of rivers to widen valleys, and to carry away large portions of soil towards the sea, is increased where earthquakes are of periodical occurrence. Among other territories, that of Cinquefrondi, was greatly convulsed, various portions of soil being raised or sunk, and innumerable fissures traversing the country in all directions (see fig. 85).

Fig. 85.

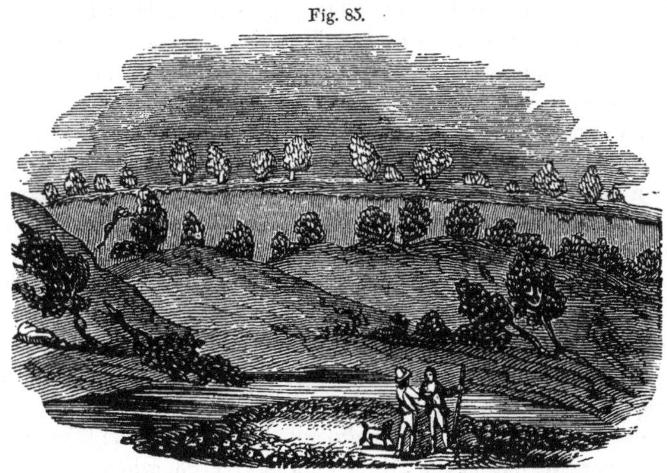

Landslips near Cinquefrondi, caused by the earthquake of 1783.

Along the flanks of a small valley in this district there appears to have been an almost uninterrupted line of landslips.

Currents of mud.—Near S. Lucido, among other places, the soil is described as having been "dissolved," so that large torrents of mud inundated all the low grounds, like lava. Just emerging from this mud, the tops only of trees and of the ruins of farm-houses were seen. Two miles from Laureana, the swampy soil in two ravines became filled with calcareous matter, which oozed out from the ground immediately before the first great shock. This mud, rapidly accumulating, began, ere long, to roll onward, like a flood of lava, into the valley, where the two streams uniting, moved forward with increased impetus from east to west. It now presented a breadth of 225 feet by 15 in depth, and, before it ceased to move, covered a surface equal in length to an Italian mile. In its progress it overwhelmed a flock of thirty goats, and tore up by the roots many olive and mulberry trees, which floated like ships upon its surface. When this calcareous lava had ceased to move, it gradually became dry and hard, during which process the mass was lowered seven feet and a half. It contained fragments of earth of a ferruginous color, and emitting a sulphureous smell.

Fall of the sea-cliffs.—Along the sea-coast of the Straits of Messina, near the celebrated rock of Scilla, the fall of huge masses detached from the bold and lofty cliffs overwhelmed many villas and gardens. At Gian Greco, a continuous line of cliff, for a mile in length, was thrown down. Great agitation was frequently observed in the bed of the sea during the shocks, and, on those parts of the coast where the movement was most violent, all kinds of fish were taken in abundance, and with unusual facility. Some rare species, as that called Cicirelli, which usually lie buried in the sand, were taken on the surface of the waters in great quantity. The sea is said to have boiled up near Messina, and to have been agitated as if by a copious discharge of vapors from its bottom.

Shore near Scilla inundated.—The prince of Scilla had persuaded a great part of his vassals to betake themselves to their fishing-boats for safety, and he himself had gone on board. On the night of the 5th of February, when some of the people were sleeping in the boats, and others on a level plain slightly elevated above the sea, the earth rocked, and suddenly a great mass was torn from the contiguous Mount Jaci, and thrown down with a dreadful crash upon the plain. Immediately afterwards, the sea, rising more than twenty feet above the level of this low tract, rolled foaming over it, and swept away the multitude. It then retreated, but soon rushed back again with greater violence, bringing with it some of the people and animals it had carried away. At the same time every boat was sunk or dashed against the beach, and some of them were swept far inland. The aged prince, with 1430 of his people, was destroyed.

State of Stromboli and Etna during the shocks.—The inhabitants of Pizzo remarked that on the 5th of February, 1783, when the first great shock afflicted Calabria, the volcano of Stromboli, which is in full view of that town, and at the distance of about fifty miles, smoked less, and threw up a less quantity of inflamed matter than it had done for some years previously. On the other hand, the great crater of Etna is said to have given out a considerable quantity of vapor towards the beginning, and Stromboli towards the close, of the commotions. But as no eruption happened from either of these great vents during the whole earthquake, the sources of the Calabrian convulsions, and of the volcanic fires of Etna and Stromboli, appear to be very independent of each other; unless, indeed, they have the same mutual relation as Vesuvius and the volcanoes of the Phlegræan Fields and Ischia, a violent disturbance in one district serving as a safety-valve to the other, and both never being in full activity at once.

Excavation of valleys.—It is impossible for the geologist to consider attentively the effect of this single earthquake of 1783, and to look forward to the alterations in the physical condition of the country to which a continued series of such movements will hereafter give rise, without perceiving that the formation of valleys by running water can never be understood, if we consider the question independently of the agency of earthquakes. It must not be imagined that rivers only begin to act when a country is already elevated far above the level of the sea, for their action must of necessity be most powerful while land is *rising* and *sinking* by successive movements. Whether Calabria is now undergoing any considerable change of relative level, in regard to the sea, or is, upon the whole, nearly stationary, is a question which our observations, confined almost entirely to the last half century, cannot possibly enable us to determine. But we know that strata, containing species of shells identical with those now living in the contiguous parts of the Mediterranean, have been raised in that country, as they have in Sicily, to the height of several thousand feet.

Now, those geologists who grant that the present course of Nature in the inanimate world has continued the same since the existing species of animals were in being, will not feel surprised that the Calabrian streams and rivers have cut out of such comparatively modern strata a great system of valleys, varying in depth from fifty to six hundred feet, and often several miles wide, if they consider how numerous may have been the shocks which accompanied the uplifting of those recent marine strata to so prodigious a height. Some speculators, indeed, who disregard the analogy of existing nature, and who are always ready to assume that her forces were more energetic in by-gone ages, may dispense with a long series of movements, and suppose that Calabria " rose like an exhalation" from the deep, after the manner of Milton's Pandemonium. But such an hypothesis would deprive them of that peculiar removing force required to form a regular system of deep and wide valleys; for *time*, which they are so unwilling to assume, is essential to the operation. Time must be allowed in the intervals between distinct convulsions, for running water to clear away the ruins caused by landslips, otherwise the fallen masses will serve as buttresses, and prevent the succeeding earthquake from exerting its full power. The sides of the valley must be again cut away by the stream, and made to form precipices and overhanging cliffs, before the next shock can take effect in the same manner.

Possibly the direction of the succeeding shock may not coincide with that of the valley; a great extent of adjacent country being equally shaken. Still it will usually happen that no permanent geographical change will be produced except in valleys. In them alone will occur landslips from the boundary cliffs, and these will frequently divert the stream from its accustomed course, causing the original ravine to become both wider and more tortuous in its direction.

If a single convulsion of extreme violence should agitate at once an entire hydrographical basin, or if the shocks should follow each other too rapidly, the previously existing valleys would be annihilated, instead of being modified and enlarged. Every stream might in that case be compelled to begin its operations anew, and to shape out new channels, instead of continuing to deepen and widen those already excavated. But if the subterranean movements have been intermittent, and if sufficient periods have always intervened between the severer shocks to allow the drainage of the country to be nearly restored to its original state, then are both the kind and degree of force supplied by which running water may hollow out valleys of any depth or size consistent with the elevation above the sea which the districts drained by them may have attained.

When we read of the drying up and desertion of the channels of rivers, the accounts most frequently refer to their deflection into some other part of the same alluvial plain, perhaps several miles distant. Under certain circumstances a change of level may undoubtedly force the water to flow over into some distinct hydrographical basin; but even then it will fall immediately into some other system of valleys already formed.

We learn from history that, ever since the first Greek colonists settled in Calabria, that region has been subject to devastation by earthquakes; and, for the last century and a half, ten years have seldom elapsed without a shock; but the severer convulsions have not only been separated by intervals of twenty, fifty, or one hundred years, but have not affected precisely the same points when they recurred. Thus the earthquake of 1783, although confined within the same geographical limits as that of 1638, and not very inferior in violence, visited, according to Grimaldi, very different districts. The points where the local intensity of the force is developed being thus perpetually varied, more time is allowed for the removal of separate mountain masses thrown into river-channels by each shock.

Number of persons who perished during the earthquake.—The number of persons who perished during the earthquake in the two Calabrias and Sicily, is estimated by Hamilton at about forty thousand; and about twenty thousand more died by epidemics, which were caused by insufficient nourishment, exposure to the atmosphere, and malaria, arising from the new stagnant lakes and pools.

By far the greater number were buried under the ruins of their houses; but many were burnt to death in the conflagrations which almost invariably followed the shocks. These fires raged the more violently in some cities, such as Oppido, from the immense magazines of oil which were consumed.

Many persons were engulfed in deep fissures, especially the peasants when flying across the open country, and their skeletons may perhaps be buried in the earth to this day, at the depth of several hundred feet.

When Dolomieu visited Messina after the shock of Feb. 5th, he describes the city as still presenting, at least at a distance, an imperfect image of its ancient splendor. Every house was injured, but the walls were standing; the whole population had taken refuge in wooden huts in the neighborhood, and all was solitude and silence in the streets: it seemed as if the city had been desolated by the plague, and the impression made upon his feelings was that of melancholy and sadness. "But when I passed over to Calabria, and first beheld Polistena, the scene of horror almost deprived me of my faculties; my mind was filled with mingled compassion and terror; nothing had escaped; all was levelled with the dust; not a single house or piece of wall remained; on all sides were heaps of stone so destitute of form, that they gave no conception of there ever having been a town on the spot. The stench of the dead bodies still rose from the ruins. I conversed with many persons who had been buried for three, four, and even for five days; I questioned them respecting their sensations in so dreadful a situation, and they agreed that of all the physical evils they endured, thirst was the most intolerable; and that their mental agony was increased by the idea that they were abandoned by their friends, who might have rendered them assistance."*

It is supposed that about a fourth part of the inhabitants of Polistena, and of some other towns, were buried alive, and might have been saved had there been no want of hands; but in so general a calamity, where each was occupied with his own misfortunes or those of his family, aid could rarely be obtained. Neither tears, nor supplications, nor promises of high rewards were listened to. Many acts of self-devotion, prompted by parental and conjugal tenderness, or by friendship, or the gratitude of faithful servants, are recorded; but individual exertions were, for the most part, ineffectual. It frequently happened, that persons in search of those most dear to them could hear their moans,—could recognize their voices—were certain of the exact spot where they lay buried beneath their feet, yet could afford them no succor. The piled mass resisted all their strength, and rendered their efforts of no avail.

At Terranuova, four Augustin monks, who had taken refuge in a vaulted sacristy, the arch of which continued to support an immense pile of ruins, made their cries heard for the space of four days. One only of the brethren of the whole convent was saved, and "of what avail was his strength to remove the enormous weight of rubbish which had overwhelmed his companions?" He heard their voices die away gradually; and when afterwards their four corpses were disinterred, they were found clasped in each other's arms. Affecting narratives are preserved of mothers saved after the fifth, sixth, and even seventh day of their interment, when their infants or children had perished with hunger.

It might have been imagined that the sight of sufferings such as these would have been sufficient to awaken sentiments of humanity and pity in the most savage breasts; but while some acts of heroism are related, nothing could exceed the general atrocity of conduct displayed by the Calabrian peasants: they abandoned the farms, and flocked in great numbers into the towns—not to rescue their countrymen from a lingering death, but to plunder. They dashed through the streets, fearless of danger, amid tottering walls and clouds of dust, trampling beneath their feet the bodies of the wounded and half-buried, and often stripping them, while yet living, of their clothes.†

Concluding remarks.—But to enter more fully into these details would be foreign to the purpose of the present work, and several volumes

* Pinkerton's Voyages and Travels, vol. v. as cited above, p. 455, note.
† Dolomieu, ibid.

would be required to give the reader a just idea of the sufferings which the inhabitants of many populous districts have undergone during the earthquakes of the last 150 years. A bare mention of the loss of life—as that fifty or a hundred thousand souls perished in one catastrophe—conveys to the reader no idea of the extent of misery inflicted: we must learn, from the narratives of eye-witnesses, the various forms in which death was encountered, the numbers who escaped with loss of limbs or serious bodily injuries, and the multitude who were suddenly reduced to penury and want. It has been often remarked, that the dread of earthquakes is strongest in the minds of those who have experienced them most frequently; whereas, in the case of almost every other danger, familiarity with peril renders men intrepid. The reason is obvious—scarcely any part of the mischief apprehended in this instance is imaginary; the first shock is often the most destructive; and, as it may occur in the dead of the night, or if by day, without giving the least warning of its approach, no forethought can guard against it; and when the convulsion has begun, no skill, or courage, or presence of mind, can point out the path of safety. During the intervals, of uncertain duration, between the more fatal shocks, slight tremors of the soil are not unfrequent; and as these sometimes precede more violent convulsions, they become a source of anxiety and alarm. The terror arising from this cause alone is of itself no inconsiderable evil.

Although sentiments of pure religion are frequently awakened by these awful visitations, yet we more commonly find that an habitual state of fear, a sense of helplessness, and a belief in the futility of all human exertions, prepare the minds of the vulgar for the influence of a demoralizing superstition.

Where earthquakes are frequent, there can never be perfect security of property under the best government; industry cannot be assured of reaping the fruits of its labor; and the most daring acts of outrage may occasionally be perpetrated with impunity, when the arm of the law is paralyzed by the general consternation. It is hardly necessary to add, that the progress of civilization and national wealth must be retarded by convulsions which level cities to the ground, destroy harbors, render roads impassable, and cause the most cultivated valley-plains to be covered with lakes, or the ruins of adjoining hills.

Those geologists who imagine that, at remote periods ere man became a sojourner on earth, the volcanic agency was more energetic than now, should be careful to found their opinion on strict geological evidence, and not permit themselves to be biased, as they have often been, by a notion, that the disturbing force would probably be mitigated for the sake of man.

I shall endeavor to point out in the sequel, that the general tendency of subterranean movements, when their effects are considered for a sufficient lapse of ages, is eminently beneficial, and that they constitute an essential part of that mechanism by which the integrity of the habitable surface is preserved, and the very existence and perpetuation of dry land secured. Why the working of this same machinery should be attended with so much evil, is a mystery far beyond the reach of our philosophy, and must probably remain so until we are permitted to investigate, not our planet alone and its inhabitants, but other parts of the moral and material universe with which they may be connected. Could our survey embrace other worlds, and the events, not of a few centuries only, but of periods as indefinite as those with which geology renders us familiar, some apparent contradictions might be reconciled, and some difficulties would doubtless be cleared up. But even then, as our capacities are finite, while the scheme of the universe may be infinite, both in time and space, it is presumptuous to suppose that all sources of doubt and perplexity would ever be removed. On the contrary, they might, perhaps, go on augmenting in number, although our confidence in the wisdom of the plan of Nature should increase at the same time; for it has been justly said, that the greater the circle of light, the greater the boundary of darkness by which it is surrounded.*

CHAPTER XXIX.

EARTHQUAKES—*continued.*

Earthquake of Java, 1772—Truncation of a lofty cone—St. Domingo, 1770—Lisbon, 1755—Great area over which the shocks extended—Retreat of the sea—Proposed explanations—Conception Bay, 1750—Permanent elevation—Peru, 1746—Java, 1699—Rivers obstructed by landslips—Subsidence in Sicily, 1693—Moluccas, 1693—Jamaica, 1692—Large tracts engulfed—Portion of Port Royal sunk—Amount of change in the last 150 years—Elevation and subsidence of land in Bay of Baiæ—Evidence of the same afforded by the Temple of Serapis.

In the preceding chapters we have considered a small part only of those earthquakes which have occurred during the last seventy years, of which accurate and authentic descriptions happen to have been recorded. In examining those of earlier date, we find their number so great that allusion can be made to a few only respecting which information of peculiar geological interest has been obtained.

Java, 1772.—*Truncation of a lofty cone.*—In the year 1772, Papandayang, formerly one of the loftiest volcanoes in the island of Java, was in eruption. Before all the inhabitants on the declivities of the mountain could save themselves by flight, the ground began to give way, and a great part of the volcano fell in and disappeared. It is estimated that an extent of ground of the mountain itself and its immediate envi-

* Sir H. Davy's Consolations in Travel, p. 246.

rons, fifteen miles long and full six broad, was by this commotion swallowed up in the bowels of the earth. Forty villages were destroyed, some being engulfed and some covered by the substances thrown out on this occasion, and 2957 of the inhabitants perished. A proportionate number of cattle were also killed, and most of the plantations of cotton, indigo, and coffee in the adjacent districts were buried under the volcanic matter. This catastrophe appears to have resembled, although on a grander scale, that of the ancient Vesuvius in the year 79. The cone was reduced in height from 9000 to about 5000 feet; and, as vapors still escape from the crater on its summit, a new cone may one day rise out of the ruins of the ancient mountain, as the modern Vesuvius has risen from the remains of Somma.*

St. Domingo, 1770.—During a tremendous earthquake which destroyed a great part of St. Domingo, innumerable fissures were caused throughout the island, from which mephitic vapors emanated and produced an epidemic. *Hot springs* burst forth in many places where there had been no water before; but after a time they ceased to flow.†

In a previous earthquake, in November, 1751, a violent shock destroyed the capital, Port au Prince, and part of the coast, twenty leagues in length, sank down, and has ever since formed a bay of the sea.‡

Hindostan, 1762.—The town of Chittagong, in Bengal, was violently shaken by an earthquake, on the 2d of April, 1762, the earth opening in many places, and throwing up water and mud of a sulphureous smell. At a place called Bardavan, a large river was dried up; and at Bar Charra, near the sea, a tract of ground sunk down, and 200 people, with all their cattle, were lost. It is said, that sixty square miles of the Chittagong coast suddenly and permanently subsided during this earthquake, and that Ces-lung-Toom, one of the Mug mountains, entirely disappeared, and another sank so low, that its summit only remained visible. Four hills are also described as having been variously rent asunder, leaving open chasms from thirty to sixty feet in width. Towns which subsided several cubits, were overflowed with water; among others, Deep Gong, which was submerged to the depth of seven cubits. Two volcanoes are said to have opened in the Secta Cunda hills. The shock was also felt at Calcutta.§ While the Chittagong coast was sinking, a corresponding rise of the ground took place at the island of Ramree, and at Cheduba (see Map, fig. 39, p. 351).‖

Lisbon, 1755.—In no part of the volcanic region of southern Europe

* Dr. Horsfield, Batav. Trans. vol. viii. p. 26. Dr. H. informs me that he has seen this truncated mountain; and, though he did not ascend it, he has conversed with those who have examined it. Raffles' account (History of Java, vol. i.) is derived from Horsfield.

† Essai sur l'Hist. Nat. de l'Isle de St. Domingue. Paris, 1776.

‡ Hist. de l'Acad. des Sciences. 1752, Paris.

§ M'Clelland's Report on Min. Resources of India: 1838, Calcutta. For other particulars, see Phil. Trans. vol. liii.

‖ Journ. Asiat. Soc. Bengal, vol. x. pp. 351, 433.

has so tremendous an earthquake occurred in modern times, as that which began on the 1st of November, 1755, at Lisbon. A sound of thunder was heard underground, and immediately afterwards a violent shock threw down the greater part of that city. In the course of about six minutes, sixty thousand persons perished. The sea first retired and laid the bar dry; it then rolled in, rising fifty feet or more above its ordinary level. The mountains of Arrabida, Estrella, Julio, Marvan, and Cintra, being some of the largest in Portugal, were impetuously shaken, as it were, from their very foundations; and some of them opened at their summits, which were split and rent in a wonderful manner, huge masses of them being thrown down into the subjacent valleys.* Flames are related to have issued from these mountains, which are supposed to have been electric; they are also said to have smoked; but vast clouds of dust may have given rise to this appearance.

The area over which this convulsion extended is very remarkable. It has been computed, says Humboldt,† that on the 1st November, 1755, a portion of the earth's surface four times greater than the extent of Europe was simultaneously shaken. The shock was felt in the Alps, and on the coast of Sweden, in small inland lakes on the shores of the Baltic, in Thuringia, and in the flat country of northern Germany. The thermal springs of Toplitz dried up, and again returned, inundating every thing with water discolored by ochre. In the islands of Antigua, Barbadoes, and Martinique in the West Indies, where the tide usually rises little more than two feet, it suddenly rose above twenty feet, the water being discolored and of an inky blackness. The movement was also sensible in the great lakes of Canada. At Algiers and Fez, in the north of Africa, the agitation of the earth was as violent as in Spain and Portugal; and at the distance of eight leagues from Morocco, a village with the inhabitants, to the number of about 8000 or 10,000 persons, are said to have been swallowed up; the earth soon afterwards closing over them.

Subsidence of the quay.—Among other extraordinary events related to have occurred at Lisbon during the catastrophe was the subsidence of a new quay, built entirely of marble at an immense expense. A great concourse of people had collected there for safety, as a spot where they might be beyond the reach of falling ruins; but suddenly the quay sank down with all the people on it, and not one of the dead bodies ever floated to the surface. A great number of boats and small vessels anchored near it, all full of people, were swallowed up, as in a whirlpool.‡ No fragments of these wrecks ever rose again to the surface, and the water in the place where the quay had stood is stated, in many accounts, to be unfathomable; but Whitehurst says he ascertained it to be one hundred fathoms.§

* Hist. and Philos. of Earthquakes, p. 317.

† Cosmos, vol. i.

‡ Rev. C. Davy's Letters, vol. ii. Letter ii. p. 12, who was at Lisbon at the time, and ascertained that the boats and vessels said to have been swallowed were missing.

§ On the Formation of the Earth, p. 55.

Circumstantial as are the contemporary narratives, I learn from a correspondent, Mr. F. Freeman, in 1841, that no part of the Tagus was then more than thirty feet deep at high tide, and an examination of the position of the new quay, and the memorials preserved of the time and manner in which it was built, rendered the statement of so great a subsidence in 1755 quite unintelligible. Perhaps a deep narrow chasm, such as was before described in Calabria (p. 481), opened and closed again in the bed of the Tagus, after swallowing up some incumbent buildings and vessels. We have already seen that such openings may collapse after the shock suddenly, or, in places where the strata are of soft and yielding materials, very gradually. According to the observations made at Lisbon, in 1837, by Mr. Sharpe, the destroying effects of this earthquake were confined to the tertiary strata, and were most violent on the blue clay, on which the lower part of the city is constructed. Not a building, he says, on the secondary limestone or the basalt was injured.*

Shocks felt at sea.—The shock was felt at sea, on the deck of a ship to the west of Lisbon, and produced very much the same sensation as on dry land. Off St. Lucar, the captain of the ship Nancy felt his vessel so violently shaken, that he thought she had struck the ground; but, on heaving the lead, found a great depth of water. Captain Clark, from Denia, in latitude 36° 24′ N., between nine and ten in the morning, had his ship shaken and strained as if she had struck upon a rock, so that the seams of the deck opened, and the compass was overturned in the binnacle. Another ship, forty leagues west of St. Vincent, experienced so violent a concussion, that the men were thrown a foot and a half perpendicularly up from the deck.

Rate at which the movement travelled.—The agitation of lakes, rivers, and springs, in Great Britain, was remarkable. At Loch Lomond, in Scotland, for example, the water, without the least apparent cause, rose against its banks, and then subsided below its usual level. The greatest perpendicular height of this swell was two feet four inches. It is said that the movement of this earthquake was undulatory, and that it travelled at the rate of twenty miles a minute, its velocity being calculated by the intervals between the time when the first shock was felt at Lisbon, and its time of occurrence at other distant places.†

Great wave and retreat of the sea.—A great wave swept over the coast of Spain, and is said to have been sixty feet high at Cadiz. At Tangier, in Africa, it rose and fell eighteen times on the coast. At Funchal, in Madeira, it rose full fifteen feet perpendicular above high-water mark, although the tide, which ebbs and flows there seven feet, was then at half-ebb. Besides entering the city, and committing great havoc, it overflowed other seaports in the island. At Kinsale, in Ireland, a body of water rushed into the harbor, whirled round several vessels,‡ and poured into the market-place.

† Geol. Soc. Proceedings, No. 60, p. 36. 1838.
‡ Michell on Earthquakes, Phil. Trans. vol. li. p. 566. 1760.

It was before stated that the sea first retired at Lisbon; and this retreat of the ocean from the shore, at the commencement of an earthquake, and its subsequent return in a violent wave, is a common occurrence. In order to account for the phenomenon, Michell imagined a subsidence at the bottom of the sea, from the giving way of the roof of some cavity in consequence of a vacuum produced by the condensation of steam. Such condensation, he observes, might be the first effect of the introduction of a large body of water into fissures and cavities already filled with steam, before there has been sufficient time for the heat of the incandescent lava to turn so large a supply of water into steam, which being soon accomplished causes a greater explosion.

Another proposed explanation is, the sudden rise of the land, which would cause the sea to abandon immediately the ancient line of coast; and if the shore, after being thus heaved up, should fall again to its original level, the ocean would return. This theory, however, will not account for the facts observed during the Lisbon earthquake; for the retreat preceded the wave, not only on the coast of Portugal, but also at the island of Madeira, and several other places. If the upheaving of the coast of Portugal had caused the retreat, the motion of the waters, when propagated to Madeira, would have produced a wave previous to the retreat. Nor could the motion of the waters at Madeira have been caused by a different local earthquake; for the shock travelled from Lisbon to Madeira in two hours, which agrees with the time which it required to reach other places equally distant.*

The following is another solution of the problem, which has been offered:—Suppose a portion of the bed of the sea to be suddenly upheaved; the first effect will be to raise over the elevated part a body of water, the momentum of which will carry it much above the level it will afterwards assume, causing a draught or receding of the water from the neighboring coasts, followed immediately by the return of the displaced water, which will also be impelled by its momentum much farther and higher on the coast than its former level.†

Mr. Darwin, when alluding to similar waves on the coast of Chili, states his opinion, that "the whole phenomenon is due to a common undulation in the water, proceeding from a line or point of disturbance some little way distant. If the waves," he says, "sent off from the paddles of a steam-vessel be watched breaking on the sloping shore of a still river, the water will be seen first to retire two or three feet, and then to return in little breakers, precisely analogous to those consequent on an earthquake." He also adds, that "the earthquake-wave occurs some time after the shock, the water at first retiring both from the shores of the mainland and of outlying islands, and then returning in mountainous breakers. Their size is modified by the form of the neighboring coast; for it is ascertained in South America, that places situa-

* Michell, Phil. Trans. vol. li. p. 614.
† Quarterly Review, No. lxxxvi. p. 459.

ted at the head of shoaling bays have suffered most, whereas towns like Valparaiso, seated close on the border of a profound ocean, have never been inundated, though severely shaken by earthquakes."*

More recently (February, 1846), Mr. Mallet, in his memoir above cited (p. 475), has endeavored to bring to bear on this difficult subject the more advanced knowledge obtained of late years respecting the true theory of waves. He conceives that when the origin of the shock is beneath the deep ocean, one wave is propagated through the land, and another moving with inferior velocity is formed on the surface of the ocean. This last rolls in upon the land long after the earth-wave has arrived and spent itself. However irreconcilable it may be to our common notions of solid bodies, to imagine them capable of transmitting, with such extreme velocity, motions analogous to tidal waves, it seems nevertheless certain that such undulations are produced, and it is supposed that when the shock passes a given point, each particle of the solid earth describes an ellipse in space. The facility with which all the particles of a solid mass can be made to vibrate may be illustrated, says Gay Lussac, by many familiar examples. If we apply the ear to one end of a long wooden beam, and listen attentively when the other end is struck by a pin's head, we hear the shock distinctly; which shows that every fibre throughout the whole length has been made to vibrate. The rattling of carriages on the pavement shakes the largest edifices; and in the quarries underneath some quarters in Paris, it is found that the movement is communicated through a considerable thickness of rock.†

The great sea-wave originating directly over the centre of disturbance is propagated, as Michell correctly stated, in every direction, like the circle upon a pond when a pebble is dropped into it, the different rates at which it moves depending (as he also suggested) on variations in the depth of the water. This wave of the sea, says Mr. Mallet, is raised by the impulse of the shock immediately below it, which in great earthquakes lifts up the ground two or three feet perpendicularly. The velocity of the shock, or earth-wave, is greater because it "depends upon a function of the elasticity of the crust of the earth, whereas the velocity of the sea-wave depends upon a function of the depth of the sea."

"Although the shock in its passage under the deep ocean gives no trace of its progress, it no sooner gets into soundings or shallow water, than it gives rise to another and smaller wave of the sea. It carries, as it were, upon its back, this lesser aqueous undulation; a long narrow ridge of water which corresponds in form and velocity to itself, being pushed up by the partial elevation of the bottom. It is this small wave, called technically the 'forced sea-wave,' which communicates the earthquake-shock to ships at sea, as if they had struck upon a rock. It breaks upon a coast at the same moment that the shock reaches it, and some-

* Darwin's Travels in South America, &c., 1832 to 1836. Voyage of H. M. S. Beagle, vol. iii. p. 377.

† Ann. de Ch. et de Ph., tom. xxii. p. 428.

times it may cause an apparent slight recession from the shore, followed by its flowing up somewhat higher than the usual tide mark: this will happen where the beach is very sloping, as is usual where the sea is shallow, for then the velocity of the low flat earth-wave is such, that it slips as it were, from under the undulation in the fluid above. It does this at the moment of reaching the beach, which it elevates by a vertical height equal to its own, and as instantly lets drop again to its former level."

"While the shock propagated through the solid earth has thus travelled with extra rapidity to the land, the great sea-wave has been following at a slower pace, though advancing at the rate of several miles in a minute. It consists, in the deep ocean, of a long low swell of enormous volume, having an equal slope before and behind, and that so gentle that it might pass under a ship without being noticed. But when it reaches the edge of soundings, its front slope, like that of a tidal wave under similar circumstances, becomes short and steep, while its rear slope is long and gentle. If there be water of some depth close into shore, this great wave may roll in long after the shock, and do little damage; but if the shore be shelving, there will be first a retreat of the water, and then the wave will break upon the beach and roll in far upon the land."*

The various opinions which have been offered by Michell and later writers, respecting the remote causes of earthquake shocks in the interior of the earth, will more properly be discussed in the thirty-second chapter.

Chili, 1751.—On the 24th of May, 1751, the ancient town of Conception, otherwise called Penco, was totally destroyed by an earthquake, and the sea rolled over it. (See plan of the bay, fig. 70, p. 455.) The ancient port was rendered entirely useless, and the inhabitants built another town about ten miles from the sea-coast, in order to be beyond the reach of similar inundations. At the same time, a colony recently settled on the sea-shore of Juan Fernandez was almost entirely overwhelmed by a wave which broke upon the shore.

It has been already stated, that in 1835, or eighty-four years after the destruction of Penco, the same coast was overwhelmed by a similar flood from the sea during an earthquake; and it is also known that twenty-one years before (or in 1730), a like wave rolled over these fated shores, in which many of the inhabitants perished. A series of similar catastrophes has also been tracked back as far as the year 1590,† beyond which we have no memorials save those of oral tradition. Molina, who has recorded the customs and legends of the aborigines, tells us, that the Araucanian Indians, a tribe inhabiting the country between the Andes and the Pacific, including the part now called Chili, "had among them a tradition of a great deluge, in which only a few persons were

* Mallet, Proceed. Roy. Irish Acad. 1846.

† See Father Acosta's work; and Sir Woodbine Parish, Geol. Soc. Proceedings, vol. ii. p. 215.

saved, who took refuge upon a high mountain called Thegtheg, "the thundering," which had three points. Whenever a violent earthquake occurs, these people fly for safety to the mountains, assigning as a reason, that they are fearful, after the shock, that the sea will again return and deluge the world.*

Notwithstanding the tendency of writers in his day to refer all traditionary inundations to one remote period, Molina remarks that this flood of the Araucanians "was probably very different from that of Noah." We have, indeed, no means of conjecturing how long this same tribe had flourished in Chili, but we can scarcely doubt, that if its experience reached back even for three or four centuries, several inroads of the ocean must have occurred within that period. But the memory of a succession of physical events, similar in kind, though distinct in time, can never be preserved by a people destitute of written annals. Before two or three generations have passed away all dates are forgotten, and even the events themselves, unless they have given origin to some customs, or religious rites and ceremonies. Oftentimes the incidents of many different earthquakes and floods become blended together in the same narrative; and in such cases the single catastrophe is described in terms so exaggerated, or is so disguised by mythological fictions, as to be utterly valueless to the antiquary or philosopher.

Proofs of elevation of twenty-four feet.—During a late survey of Conception Bay, Captain Beechey and Sir E. Belcher discovered that the ancient harbor, which formerly admitted all large merchant vessels which went round the Cape, is now occupied by a reef of sandstone, certain points of which project above the sea at low water, the greater part being very shallow. A tract of a mile and a half in length, where, according to the report of the inhabitants, the water was formerly four or five fathoms deep, is now a shoal; consisting, as our hydrographers found, of hard sandstone, so that it cannot be supposed to have been formed by recent deposits of the river Biobio, an arm of which carries down loose micaceous sand into the same bay.

It is impossible at this distance of time to affirm that the bed of the sea was uplifted at once to the height of twenty-four feet, during the single earthquake of 1751, because other movements may have occurred subsequently; but it is said, that ever since the shock of 1751, no vessels have been able to approach within a mile and a half of the ancient port of Penco. (See Map, p. 455.) In proof of the former elevation of the coast near Penco our surveyors found above high-water mark an enormous bed of shells of the same species as those now living in the bay, filled with micaceous sand like that which the Biobio now conveys to the bay. These shells, as well as others, which cover the adjoining hills of mica-schist to the height of several hundred feet, have lately been examined by experienced conchologists in London, and identified

* Molina, Hist. of Chili, vol. ii.

with those taken at the same time in a living state from the bay and its neighborhood.*

Ulloa, therefore, was perfectly correct in his statement that, at various heights above the sea between Talcahuano and Conception, "mines were found of various sorts of shells used for lime of the very same kinds as those found in the adjoining sea." Among them he mentions the great mussel called Choros, and two others which he describes. Some of these, he says, are entire, and others broken; they occur at the bottom of the sea, in four, six, ten, or twelve fathom water, where they adhere to a sea-plant called Cochayuyo. They are taken in dredges, and have no resemblance to those found on the shore or in shallow water; yet beds of them occur at various heights on the hills. "I was the more pleased with the sight," he adds, "as it appeared to me a convincing proof of the universality of the deluge, although I am not ignorant that some have attributed their position to other causes."† It has, however, been ascertained that the foundation of the Castle of Penco was so low in 1835, or at so inconsiderable an elevation above the highest spring tides, as to discountenance the idea of any permanent upheaval in modern times, on the site of that ancient port; but no exact measurements or levellings appear as yet to have been made to determine this point, which is the more worthy of investigation, because it may throw some light on an opinion often promulgated of late years, that there is a tendency in the Chilian coast, after each upheaval, to sink gradually and return towards its former position.

Peru, 1746.—Peru was visited, on the 28th of October, 1746, by a tremendous earthquake. In the first twenty-four hours, two hundred shocks were experienced. The ocean twice retired and returned impetuously upon the land: Lima was destroyed, and part of the coast near Callao was converted into a bay: four other harbors, among which were Cavalla and Guanape, shared the same fate. There were twenty-three ships and vessels, great and small, in the harbor of Callao, of which nineteen were sunk; and the other four, among which was a frigate called St. Fermin, were carried by the force of the waves to a great distance up the country, and left on dry ground at a considerable height above the sea. The number of inhabitants in this city amounted to four thousand. Two hundred only escaped, twenty-two of whom were saved on a small fragment of the fort of Vera Cruz, which remained as the only memorial of the town after this dreadful inundation. Other portions of its site were completely covered with heaps of sand and gravel.

A volcano in Lucanas burst forth the same night, and such quantities of water descended from the cone that the whole country was overflowed; and in the mountain near Pataz, called Conversiones de Caxamarquilla, three other volcanoes burst out, and frightful torrents of water swept down their sides.‡

* Captain Belcher has shown me these shells, and the collection has been examined by Mr. Broderip.

† Ulloa's Voyage to South America, vol. ii. book viii. ch. vi

‡ Ibid. vol. ii. book vii. ch. vii

There are several records of prior convulsions in Peru, accompanied by similar inroads in the sea, one of which happened fifty-nine years before (in 1687), when the ocean, according to Ulloa, first retired and then returned in a mountainous wave, overwhelming Callao and its environs, with the miserable inhabitants.* This same wave, according to Lionel Wafer, carried ships a league into the country, and drowned man and beast for fifty leagues along the shore.† Inundations of still earlier dates are carefully recorded by Ulloa, Wafer, Acosta, and various writers, who describe them as having expended their chief fury, some on one part of the coast and some on another.

But all authentic accounts cease when we ascend to the era of the conquest of Peru by the Spaniards. The ancient Peruvians, although far removed from barbarism, were without written annals, and therefore unable to preserve a distinct recollection of a long series of natural events. They had, however, according to Antonio de Herrera, who, in the beginning of the seventeenth century, investigated their antiquities, a tradition, "that many years before the reign of the Incas, at a time when the country was very populous, there happened a great flood; the sea breaking out beyond its bounds, so that the land was covered with water and all the people perished. To this the Guacas, inhabiting the vale of Xausca, and the natives of Chiquito, in the province of Callao, add that some persons remained in the hollows and caves of the highest mountains, who again peopled the land. Others of the mountain people affirm that all perished in the deluge, only six persons being saved on a float, from whom descended all the inhabitants of that country."‡

On the mainland near Lima, and on the neighboring island of San Lorenzo, Mr. Darwin found proofs that the ancient bed of the sea had been raised to the height of more than eighty feet above water within the human epoch, strata having been discovered at that altitude, containing pieces of cotton thread and plaited rush, together with sea-weed and marine shells.§ The same author learnt from Mr. Gill, a civil engineer, that he discovered in the interior near Lima, between Casma and Huaraz, the dried-up channel of a large river, sometimes worn through solid rock, which, instead of continually ascending towards its source, has, in one place, a steep downward slope in that direction, for a ridge or line of hills has been uplifted directly across the bed of the stream, which is now arched. By these changes the water has been turned into some other course; and a district, once fertile, and still covered with ruins, and bearing the marks of ancient cultivation, has been converted into a desert.‖

Java, 1699.—On the 5th of January, 1699, a terrible earthquake visited Java, and no less than 208 considerable shocks were reckoned.

* Ulloa's Voyage, vol. ii. p. 82.
† Wafer, cited by Sir W. Parish, Geol. Soc. Proceedings, vol. ii. p. 215.
‡ Hist. of America, decad. iii. book xi. ch. i.
§ Darwin's Journal, p. 451. ‖ Ibid. p. 418.

Many houses in Batavia were overturned, and the flame and noise of a volcanic eruption were seen and heard in that city, which were afterwards found to proceed from Mount Salek,* a volcano six days' journey distant. Next morning the Batavian river, which has its rise from that mountain, became very high and muddy, and brought down abundance of bushes and trees, half burnt. The channel of the river being stopped up, the water overflowed the country round the gardens about the town, and some of the streets, so that fishes lay dead in them. All the fish in the river, except the carps, were killed by the mud and turbid water. A great number of drowned buffaloes, tigers, rhinoceroses, deer, apes, and other wild beasts, were brought down by the current; and, "notwithstanding," observes one of the writers, "that a crocodile is amphibious, several of them were found dead among the rest."†

It is stated that seven hills bounding the river sank down; by which is merely meant, as by similar expressions in the description of the Calabrian earthquakes, seven great landslips. These hills, descending some from one side of the valley and some from the other, filled the channel, and the waters then finding their way under the mass, flowed out thick and muddy. The Tangaran river was also dammed up by nine hills, and in its channel were large quantities of drift trees. Seven of its tributaries also are said to have been "covered up with earth." A high tract of forest land, between the two great rivers before mentioned, is described as having been changed into an open country, destitute of trees, the surface being spread over with fine red clay. This part of the account may, perhaps, merely refer to the sliding down of woody tracts into the valleys, as happened to so many extensive vineyards and olive-grounds in Calabria, in 1783. The close packing of large trees in the Batavian river is represented as very remarkable, and it attests in a striking manner the destruction of soil bordering the valleys which had been caused by floods and landslips.‡

Quito, 1698.—In Quito, on the 19th of July, 1698, during an earthquake, a great part of the crater and summit of the volcano Carguairazo fell in, and a stream of water and mud issued from the broken sides of the hill.§

Sicily, 1693.—Shocks of earthquakes spread over all Sicily in 1693, and on the 11th of January the city of Catania and forty-nine other places were levelled to the ground, and about one hundred thousand people killed. The bottom of the sea, says Vicentino Bonajutus, sank down considerably, both in ports, inclosed bays, and open parts of the coast, and water bubbled up along the shores. Numerous long fissures of various breadths were caused, which threw out sulphurous water; and one of them, in the plain of Catania (the delta of the Simeto), at the distance of four miles from the sea, sent forth water as salt as the sea. The stone buildings of a street in the city of Noto, for the length

* Misspelt "Sales" in Hooke's Account.

† Hooke's Posthumous Works, p. 437. 1705.

‡ Phil. Trans. 1700.

§ Humboldt, Atl. Pit. p. 106.

of half a mile, sank into the ground, and remained hanging on one side. In another street, an opening large enough to swallow a man and horse appeared.*

Moluccas, 1693.—The small Isle of Sorea, which consists of one great volcano, was in eruption in the year 1693. Different parts of the cone fell, one after the other, into a deep crater, until almost half the space of the island was converted into a fiery lake. Most of the inhabitants fled to Banda; but great pieces of the mountain continued to fall down, so that the lake of lava became wider; and finally the whole population was compelled to emigrate. It is stated that, in proportion as the burning lake increased in size, the earthquakes were less vehement.†

Jamaica, 1692.—In the year 1692, the island of Jamaica was visited by a violent earthquake; the ground swelled and heaved like a rolling sea, and was traversed by numerous cracks, two or three hundred of which were often seen at a time, opening and then closing rapidly again. Many people were swallowed up in these rents; some the earth caught by the middle, and squeezed to death; the heads of others only appeared above ground; and some were first engulfed, and then cast up again with great quantities of water. Such was the devastation, that even in Port Royal, then the capital, where more houses are said to have been left standing than in the whole island besides, three-quarters of the buildings, together with the ground they stood on, sank down with their inhabitants entirely under water.

Subsidence in the harbor.—The large storehouses on the harbor side subsided, so as to be twenty-four, thirty-six, and forty-eight feet under water; yet many of them appear to have remained standing, for it is stated that, after the earthquake, the mast-heads of several ships wrecked in the harbor, together with the chimney-tops of houses, were just seen projecting above the waves. A tract of land round the town, about a thousand acres in extent, sank down in less than one minute, during the first shock, and the sea immediately rolled in. The Swan frigate, which was repairing in the wharf, was driven over the tops of many buildings, and then thrown upon one of the roofs, through which it broke. The breadth of one of the streets is said to have been doubled by the earthquake.

According to Sir H. De la Beche, the part of Port Royal described as having sunk was built upon newly formed land, consisting of sand, in which piles had been driven; and the *settlement* of this loose sand, charged with the weight of heavy houses, may, he suggests, have given rise to the subsidence alluded to.‡

There have undoubtedly been instances in Calabria and elsewhere of slides of land on which the houses have still remained standing; and it is possible that such may have been the case at Port Royal. The

* Phil. Trans. 1693–4. † Phil. Trans. 1693.
‡ Manual of Geol. p. 133, second edition.

fact at least of submergence is unquestionable, for I was informed by the late Admiral Sir Charles Hamilton that he frequently saw the submerged houses of Port Royal in the year 1780, in that part of the harbor which lies between the town and the usual anchorage of men-of-war. Bryan Edwards also says, in his history of the West Indies, that in 1793 the *ruins* were visible in clear weather from the boats which sailed over them.* Lastly, Lieutenant B. Jeffery, R. N., tells me that, being engaged in a survey between the years 1824 and 1835, he repeatedly visited the site in question, where the depth of the water is from four to six fathoms, and whenever there was but little wind perceived distinct traces of houses. He saw these more clearly when he used the instrument called the "diver's eye," which is let down below the ripple of the wave.†

At several thousand places in Jamaica the earth is related to have opened. On the north of the island several plantations, with their inhabitants, were swallowed up, and a lake appeared in their place, covering above a thousand acres, which afterwards dried up, leaving nothing but sand and gravel, without the least sign that there had ever been a house or a tree there. Several tenements at Yallows were buried under land-slips; and one plantation was removed half a mile from its place, the crops continuing to grow upon it uninjured. Between Spanish Town and Sixteen-mile Walk, the high and perpendicular cliffs bounding the river fell in, stopped the passage of the river and flooded the latter place for nine days, so that the people "concluded it had been sunk as Port Royal was." But the flood at length subsided, for the river had found some new passage at a great distance.

Mountains shattered.—The Blue and other of the highest mountains are declared to have been strangely torn and rent. They appeared shattered and half-naked, no longer affording a fine green prospect, as before, but stripped of their woods and natural verdure. The rivers on these mountains first ceased to flow for about twenty-four hours, and then brought down into the sea, at Port Royal and other places, several hundred thousand tons of timber, which looked like floating islands on the ocean. The trees were in general barked, most of their branches having been torn off in the descent. It is particularly remarked in this, as in the narratives of so many earthquakes, that fish were taken in great numbers on the coast during the shocks. The correspondents of Sir Hans Sloane, who collected with care the accounts of eye-witnesses of the catastrophe, refer constantly to *subsidences*, and some supposed the whole of Jamaica to have sunk down.‡

Reflections on the amount of change in the last one hundred and sixty years.—I have now only enumerated some few of the earthquakes of the last 160 years, respecting which facts illustrative of geological inquiries are on record. Even if my limits permitted, it would be an unprofit-

* Vol. i. p. 235, 8vo ed. 3 vols. 1801. † Letter to the Author, May, 1838.
‡ Phil. Trans. 1694.

able task to examine all the obscure and ambiguous narratives of similar events of earlier epochs; although, if the places were now examined by geologists well practised in the art of interpreting the monuments of physical changes, many events which have happened within the historical era might doubtless be still determined with precision. It must not be imagined that, in the above sketch of the occurrences of a short period, I have given an account of all, or even the greater part, of the mutations which the earth has undergone by the agency of subterranean movements. Thus, for example, the earthquake of Aleppo, in the present century, and of Syria, in the middle of the eighteenth, would doubtless have afforded numerous phenomena, of great geological importance, had those catastrophes been described by scientific observers. The shocks in Syria in 1759, were protracted for three months, throughout a space of ten thousand square leagues: an area compared to which that of the Calabrian earthquake in 1783 was insignificant. Accon, Saphat, Balbeck, Damascus, Sidon, Tripoli, and many other places, were almost entirely levelled to the ground. Many thousands of the inhabitants perished in each; and, in the valley of Balbeck alone, 20,000 men are said to have been victims to the convulsion. In the absence of scientific accounts, it would be as irrelevant to our present purpose to enter into a detailed account of such calamities, as to follow the tract of an invading army, to enumerate the cities burnt or rased to the ground, and reckon the number of individuals who perished by famine or the sword.

Deficiency of historical records.—If such, then, be the amount of ascertained changes in the last 160 years, notwithstanding the extreme deficiency of our records during that brief period, how important must we presume the physical revolutions to have been in the course of thirty or forty centuries, during which some countries habitually convulsed by earthquakes have been peopled by civilized nations! Towns engulfed during one earthquake may, by repeated shocks, have sunk to great depths beneath the surface, while the ruins remain as imperishable as the hardest rocks in which they are inclosed. Buildings and cities, submerged, for a time, beneath seas or lakes, and covered with sedimentary deposits, must, in some places, have been re-elevated to considerable heights above the level of the ocean. The signs of these events have, probably, been rendered visible by subsequent mutations, as by the encroachments of the sea upon the coast, by deep excavations made by torrents and rivers, by the opening of new ravines, and chasms, and other effects of natural agents, so active in districts agitated by subterranean movements.

If it be asked why, if such wonderful monuments exist, so few have hitherto been brought to light, we reply—because they have not been searched for. In order to rescue from oblivion the memorials of former occurrences, the inquirer must know what he may reasonably expect to discover, and under what peculiar local circumstances. He must be acquainted with the action and effect of physical causes, in order to recognize, explain, and describe correctly the phenomena when they present themselves.

The best known of the great volcanic regions, of which the boundaries were sketched in the twenty-second chapter, is that which includes Southern Europe, Northern Africa, and Central Asia; yet nearly the whole, even of this region, must be laid down, in a geological map, as "Terra Incognita." Even Calabria may be regarded as unexplored, as also Spain, Portugal, the Barbary States, the Ionian Isles, Asia Minor, Cyprus, Syria, and the countries between the Caspian and Black seas. We are, in truth, beginning to obtain some insight into one small spot of that great zone of volcanic disturbance, the district around Naples; a tract by no means remarkable for the violence of the earthquakes which have convulsed it.

If, in this part of Campania, we are enabled to establish that considerable changes in the relative level of land and sea have taken place since the Christian era, it is all that we could have expected; and it is to recent antiquarian and geological research, not to history, that we are principally indebted for the information. I shall now proceed to lay before the reader some of the results of modern investigations in the Bay of Baiæ and the adjoining coast.

PROOFS OF ELEVATION AND SUBSIDENCE IN THE BAY OF BAIÆ.

Temple of Jupiter Serapis.—This celebrated monument of antiquity, a representation of which is given in the frontispiece,* affords in itself

Fig. 86.

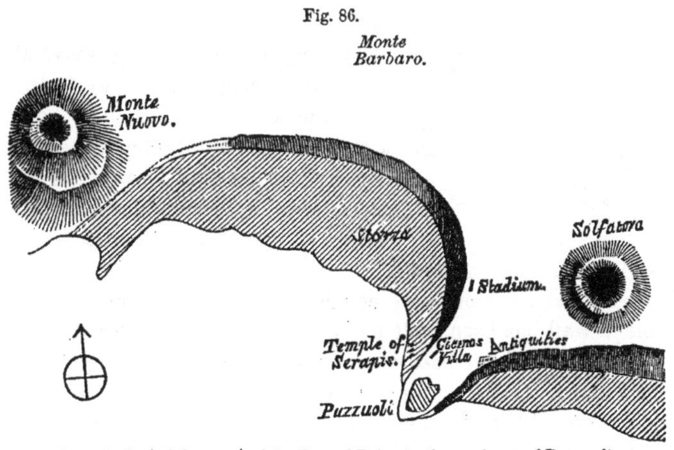

Ground plan of the coast of the Bay of Baiæ, in the environs of Puzzuoli.

alone, unequivocal evidence that the relative level of land and sea has changed twice at Puzzuoli since the Christian era; and each movement,

* This view of the temple (substituted for one by A. de Jorio, given in the earlier editions) has been reduced from part of a beautiful colored drawing taken in 1836, with the aid of the camera lucida, by Mr. I'Anson, to illustrate a paper by Mr. Babbage on the temple, read March, 1834, and published in the Quart. Journ. of the Geol. Soc. of London, vol. iii. 1847.

both of elevation and subsidence, has exceeded twenty feet. Before examining these proofs, I may observe, that a geological examination of the coast of Baiæ, both on the north and south of Puzzuoli, establishes, in the most satisfactory manner, an elevation, at no remote period, of more than twenty feet, and, at one point, of more than thirty feet; and the evidence of this change would have been complete, if even the temple had, to this day, remained undiscovered.

Coast south of Puzzuoli.—If we coast along the shore from Naples to Puzzuoli, we find, on approaching the latter place, that the lofty and precipitous cliffs of indurated tuff, resembling that of which Naples is built, retire slightly from the sea; and that a low level tract of fertile land, of a very different aspect, intervenes between the present sea-beach and what was evidently the ancient line of coast.

The inland cliff may be seen opposite the small island of Nisida, about two miles and a half southeast of Puzzuoli (see Map, fig. 40, p. 361), where, at the height of thirty-two feet above the level of the sea, Mr. Babbage observed an ancient mark, such as might have been worn by the waves; and, upon farther examination, discovered that, along that line, the face of the perpendicular rock, consisting of very hard tuff, was covered with barnacles (*Balanus sulcatus*, Lamk.), and drilled by boring testacea. Some of the hollows of the lithodomi contained the shells; while others were filled with the valves of a species of Arca.* Nearer to Puzzuoli, the inland cliff is eighty feet high, and as perpendicular as if it was still undermined by the waves. At its base, a new deposit, constituting the fertile tract above alluded to, attains a height of about twenty feet above the sea; and, since it is composed of regular sedimentary deposits, containing marine shells, its position proves that, subsequently to its formation, there has been a change of more than twenty feet in the relative level of land and sea.

Fig. 87.

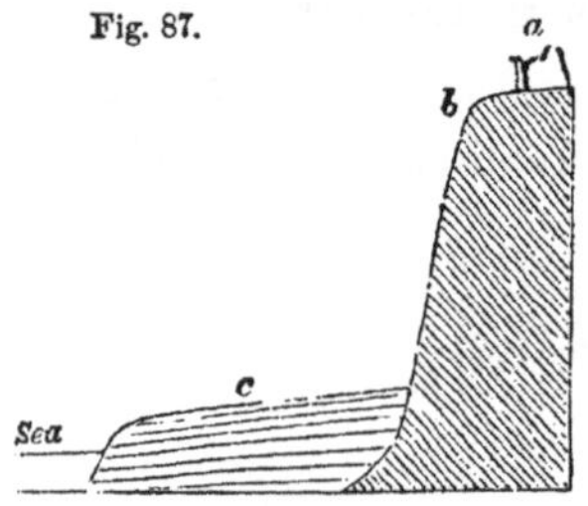

a, Antiquities on hill S. E. of Puzzuoli (see ground plan, fig. 86).
b, Ancient cliff now inland.
c, Terrace composed of recent submarine deposit.

The sea encroaches on these new incoherent strata; and as the soil is valuable, a wall has been built for its protection; but when I visited the spot in 1828, the waves had swept away part of this rampart, and exposed to view a regular series of strata of tuff, more or less argillaceous, alternating with beds of pumice and lapilli, and containing great abundance of marine shells, of species now common on this coast, and amongst them *Cardium rusticum, Ostrea edulis, Donax trunculus*, Lamk., and others. The strata vary from about a foot to a foot and a half in thickness, and one of them contains abundantly remains of works of art, tiles, squares of mosaic pavement of different colors, and small sculptured or-

* Mr. Babbage examined this spot in company with Sir Edmund Head in June, 1828, and has shown me numerous specimens of the shells collected there, and in the Temple of Serapis.

Fig. 88.

1. Puzzuoli. 2. Temple of Serapis. 3. Caligula's Bridge. 4. Monte Barbaro. 5. Monte Nuovo. 6. Baths of Nero. 7. Baiæ.
8. Castle of Baiæ. 9. Bauli. 10. Cape Misenum. 11. Mount Epomeo in Ischia. 12. South part of Ischia.

naments, perfectly uninjured. Intermixed with these I collected some teeth of the pig and ox. These fragments of building occur below as well as above strata containing marine shells. Puzzuoli itself stands chiefly on a promontory of the older tufaceous formation, which cuts off the new deposit, although I detected a small patch of the latter in a garden under the town.

From the town the ruins of a mole, called Caligula's Bridge, run out into the sea (see fig. 88, p. 509).* This mole, which is believed to be eighteen centuries old, consists of a number of piers and arches, thirteen of which are now standing, and two others appear to have been overthrown. Mr. Babbage found, on the sixth pier, perforations of lithodomi four feet above the level of the sea; and, near the termination of the mole on the last pier but one, marks of the same, ten feet above the level of the sea, together with great numbers of balani and flustra. The depth of the sea, at a very small distance from most of the piers, is from thirty to fifty feet.

Coast north of Puzzuoli.—If we then pass to the north of Puzzuoli, and examine the coast between that town and Monte Nuovo, we find a repetition of analogous phenomena. The sloping sides of Monte Barbaro slant down within a short distance of the coast, and terminate in an inland cliff of moderate elevation, to which the geologist perceives at

Fig. 89.

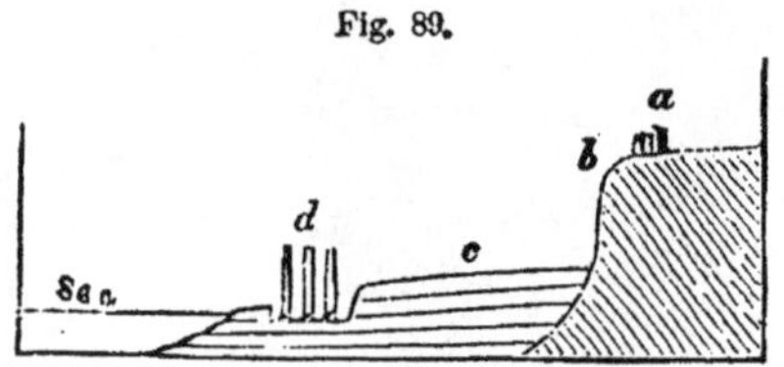

a, Remains of Cicero's villa, N. side of Puzzuoli.† *b*, Ancient cliff now inland.
c, Terrace (called La Starza) composed of recent submarine deposits.
d, Temple of Serapis.

once that the sea must, at some former period, have extended. Between this cliff and the sea is a low plain or terrace, called La Starza (*c*, fig. 89), corresponding to that before described on the southeast of the town; and as the sea encroaches rapidly, fresh sections of the strata may readily be obtained, of which the annexed is an example.

Section on the shore north of the town of Puzzuoli:—

	Ft.	In.
1. Vegetable soil	1	0
2. Horizontal beds of pumice and scoriæ, with broken fragments of unrolled bricks, bones of animals, and marine shells	1	6
3. Beds of lapilli, containing abundance of marine shells, principally *Cardium rusticum*, *Donax trunculus*, Lam., *Ostrea edulis*, *Triton cutaceum*, Lam., and *Buccinum serratum*, Brocchi, the beds varying in thickness from one to eighteen inches	10	0
4. Argillaceous tuff, containing bricks and fragments of buildings not rounded by attrition	1	6

* This view is taken from Sir W. Hamilton, Campi Phlegræi, plate 26.

† This spot here indicated on the summit of the cliff is that from which Hamilton's view, plate 26, Campi Phlegræi (reduced in fig. 88, p. 509) is taken, and on which, he says, Cicero's villa, called the Academia, anciently stood.

The thickness of many of these beds varies greatly as we trace them along the shore, and sometimes the whole group rises to a greater height than at the point above described. The surface of the tract which they compose appears to slope gently upwards towards the base of the old cliffs.

Now, if such appearances presented themselves on the coast of England, a geologist might endeavor to seek an explanation in some local change in the set of the tides and currents: but there are scarce any tides in the Mediterranean; and, to suppose the sea to have sunk generally from twenty to twenty-five feet since the shores of Campania were covered with sumptuous buildings is an hypothesis obviously untenable. The observations, indeed, made during modern surveys on the moles and cothons (docks) constructed by the ancients in various ports of the Mediterranean, have proved that there has been no sensible variation of level in that sea during the last two thousand years.*

Thus we arrive, without the aid of the celebrated temple, at the conclusion, that the recent marine deposit at Puzzuoli was upraised in modern times above the level of the sea, and that not only this change of position, but the accumulation of the modern strata, was posterior to the destruction of many edifices, of which they contain the imbedded remains. If we next examine the evidence afforded by the temple itself, it appears, from the most authentic accounts, that the three pillars now standing erect continued, down to the middle of the last century, almost buried in the new marine strata (*c*, fig. 89). The upper part of each protruding several feet above the surface was concealed by bushes, and had not attracted, until the year 1749, the notice of antiquaries; but, when the soil was removed in 1750, they were seen to form part of the remains of a splendid edifice, the pavement of which was still preserved, and upon it lay a number of columns of African breccia and of granite. The original plan of the building could be traced distinctly: it was of a quadrangular form, seventy feet in diameter, and the roof had been supported by forty-six noble columns, twenty-four of granite and the rest of marble. The large court was surrounded by apartments, supposed to have been used as bathing-rooms; for a thermal spring, still used for medicinal purposes, issues just behind the building, and the water of this spring appears to have been originally conveyed by a marble duct, still extant, into the chambers, and then across the pavement by a groove an inch or two deep, to a conduit made of Roman brickwork, by which it gained the sea.

Many antiquaries have entered into elaborate discussions as to the deity to which this edifice was consecrated. It is admitted that, among other images found in excavating the ruins, there was one of the god Serapis; and at Puzzuoli a marble column was dug up, on which was carved an ancient inscription, of the date of the building of Rome 648 (or B. C. 105), entitled "Lex parieti faciundo." This inscription, written

* On the authority of Captain W. H. Smyth, R. N.

in very obscure Latin, sets forth a contract, between the municipality of the town, and a company of builders who undertook to keep in repair certain public edifices, the Temple of Serapis being mentioned amongst the rest, and described as being near or towards the sea, "mare vorsum." Sir Edmund Head, after studying, in 1828, the topography and antiquities of this district, and the Greek, Roman, and Italian writers on the subject, informed me, that at Alexandria, on the Nile, the chief seat of the worship of Serapis, there was a Serapeum of the same form as this temple at Puzzuoli, and surrounded in like manner by chambers, in which the devotees were accustomed to pass the night, in the hope of receiving during sleep a revelation from the god, as to the nature and cure of their diseases. Hence it was very natural that the priests of Serapis, a pantheistic divinity, who, among other usurpations, had appropriated to himself the attributes of Esculapius, should regard the hot spring as a suitable appendage to the temple, although the original Serapeum of Alexandria could boast no such medicinal waters. Signor Carelli* and others, in objecting to these views, have insisted on the fact, that the worship of Serapis, which we know prevailed at Rome in the days of Catullus (in the first century before Christ), was prohibited by the Roman Senate, during the reign of the Emperor Tiberius. But there is little doubt that, during the reigns of that emperor's successors, the shrines of the Egyptian god were again thronged by zealous votaries; and in no place more so than at Puteoli (now Puzzuoli), one of the principal marts for the produce of Alexandria.

Without entering farther into an inquiry which is not strictly geological, I shall designate this valuable relic of antiquity by its generally received name, and proceed to consider the memorials of physical changes inscribed on the three standing columns in most legible characters by the hand of Nature. (See Frontispiece.) These pillars, which have been carved each out of a single block of marble, are forty feet three inches and a half in height. A horizontal fissure nearly intersects one of the columns; the other two are entire. They are all slightly out of the perpendicular, inclining somewhat to the southwest, that is, towards the sea.† Their surface is smooth and uninjured to the height of about twelve feet above their pedestals. Above this is a zone, about nine feet in height, where the marble has been pierced by a species of marine perforating bivalve—*Lithodomus*, Cuv.‡ The holes of these animals are pear-shaped, the external opening being minute, and gradually increasing downwards. At the bottom of the cavities, many shells are still found, notwithstanding the great numbers that have been

* Dissertazione sulla Sagra Archittetura degli Antichi.

† This appears from the measurement of Captain Basil Hall, R. N., Proceedings of Geol. Soc., No. 38, p. 114; see also Patchwork, by the same author, vol. iii. p. 158. The fact of the three standing columns having been each formed out of a single stone was first pointed out to me by Mr. James Hall, and is important, as helping to explain why they were not shaken down.

‡ *Modiola lithophaga*, Lam. *Mytilus lithophagus*, Linn.

taken out by visitors; in many the valves of a species of arca, an animal which conceals itself in small hollows, occur. The perforations are so considerable in depth and size, that they manifest a long-continued abode of the lithodomi in the columns, for, as the inhabitant grows older and increases in size, it bores a larger cavity, to correspond with the increased magnitude of its shell. We must, consequently, infer a long-continued immersion of the pillars in sea-water, at a time when the lower part was covered up and protected by marine, fresh-water, and volcanic strata, afterwards to be described, and by the rubbish of buildings; the highest part, at the same time, projecting above the waters, and being consequently weathered, but not materially injured. (See fig. 90, p. 514.)

On the pavement of the temple lie some columns of marble, which are also perforated in certain parts; one, for example, to the length of eight feet, while, for the length of four feet, it is uninjured. Several of these broken columns are eaten into, not only on the exterior, but on the cross fracture, and, on some of them, other marine animals (serpulæ, &c.) have fixed themselves.* All the granite pillars are untouched by lithodomi. The platform of the temple, which is not perfectly even, was, when I visited it in 1828, about one foot below high-water mark (for there are small tides in the bay of Naples); and the sea, which was only one hundred feet distant, soaked through the intervening soil. The upper part of the perforations, therefore, were at least twenty-three feet above high-water mark; and it is clear that the columns must have continued for a long time in an erect position, immersed in salt water, and then the submerged portion must have been upraised to the height of about twenty-three feet above the level of the sea.

By excavations carried on in 1828, below the marble pavement on which the columns stand, another costly pavement of mosaic was found, at the depth of about five feet below the upper one (*a*, *b*, fig. 90). The existence of these two pavements, at different levels, clearly implies some subsidence previously to the building of the more modern temple which had rendered it necessary to construct the new floor at a higher level.

We have already seen (p. 512) that a temple of Serapis existed long before the Christian era. The change of level just mentioned must have taken place some time before the end of the second century, for inscriptions have been found in the temple, from which we learn that Septimius Severus adorned its walls with precious marbles, between the years 194 and 211 of our era, and the emperor Alexander Severus displayed the like munificence between the years 222 and 235.† From that era there is an entire dearth of historical information for a period of more than twelve centuries, except the significant fact that Alaric and his Goths sacked Puzzuoli in 456, and that Genseric did the like in 545, A. D. Yet we have fortunately a series of natural archives self-registered during the

* *Serpula contortuplicata*, Linn., and *Vermilia triquetra*, Lam. These species, as well as the *Lithodomus*, are now inhabitants of the neighboring sea.

† Brieslak, Voy. dans la Campanie, tom. ii. p. 167.

dark ages, by which many events which occurred in and about the temple are revealed to us. These natural records consist partly of deposits, which envelop the pillars below the zone of lithodomous perforations, and partly of those which surround the outer walls of the temple. Mr. Babbage, after a minute examination of these, has shown (see p. 507, note) that incrustations on the walls of the exterior chambers and on the floor of the building demonstrate that the pavement did not sink down suddenly, but was depressed by a gradual movement. The sea first entered the court or atrium and mingled its waters partially with those of the hot spring. From this brackish medium a dark calcareous precipitate (*c c*, fig. 90) was thrown down, which became, in the course of time,

Fig. 90.

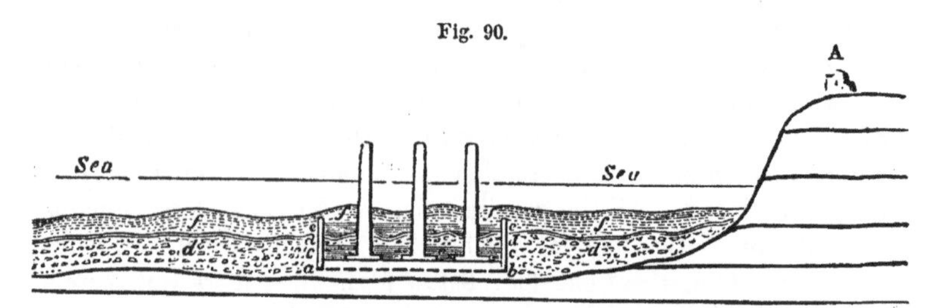

Temple of Serapis at its period of greatest depression.

a b, Ancient mosaic pavement.
c c, Dark marine incrustation.
d d, First filling up, shower of ashes.
e e, Freshwater calcareous deposit.
f f, Second filling up.
A, Stadium.

more than two feet thick, including some serpulæ in it. The presence of these annelids teaches us that the water was salt or brackish. After this period the temple was filled up with an irregular mass of volcanic tuff (*d d*, fig. 90), probably derived from an eruption of the neighboring crater of the Solfatara, to the height of from five to nine feet above the pavement. Over this again a purely freshwater deposit of carbonate of lime (*e e*, fig. 90) accumulated with an *uneven* bottom since it necessarily accommodated itself to the irregular outline of the upper surface of the volcanic shower before thrown down. The top of the same deposit (a freshwater limestone) was perfectly even and flat, bespeaking an ancient water level. It is suggested by Mr. Babbage that this freshwater lake may have been caused by the fall of ashes which choked up the channel previously communicating with the sea, so that the hot spring threw down calcareous matter in the atrium, without any marine intermixture. To the freshwater limestone succeeded another irregular mass of volcanic ashes and rubbish (*f f*, fig. 90), some of it perhaps washed in by the waves of the sea during a storm, its surface rising to ten or eleven feet above the pavement. And thus we arrive at the period of greatest depression expressed in the accompanying diagram, when the lower half of the pillars were enveloped in the deposits above enumerated, and the uppermost twenty feet were exposed in the atmosphere, the remaining or middle portion, about nine feet long, being for years immersed in salt water and drilled by perforating bivalves. After this period other strata, consisting of showers of volcanic ashes and materials washed in during storms, covered up the pillars to the height in some places of thirty-five feet above the pavement. The exact time when these enveloping masses were heaped up, and how much of them were formed during submergence, and how much after the re-elevation of the temple, cannot be made out with certainty.

The period of deep submergence was certainly antecedent to the close of the fifteenth century. Professor James Forbes* has reminded us of a passage in an old Italian writer Loffredo, who says that in 1530, or fifty years before he wrote, which was in 1580, the sea washed the base of the hills which rise from the flat land called La Starza, as represented in fig. 90, so that, to quote his words, "a person might then have fished from the site of those ruins which are now called the stadium" (A, fig. 90).

But we know from other evidence that the upward movement had begun before 1530, for the Canonico Andrea di Jorio cites two authentic documents in illustration of this point. The first, dated Oct. 1503, is a deed written in Italian, by which Ferdinand and Isabella grant to the University of Puzzuoli a portion of land, "where the sea is drying up" (che va seccando el mare); the second, a document in Latin, dated May 23, 1511, or nearly eight years after, by which Ferdinand grants to the city a certain territory around Puzzuoli, where the ground *is dried up* from the sea (desiccatum).†

The principal elevation, however, of the low tract unquestionably took place at the time of the great eruption of Monte Nuovo in 1538. That event and the earthquakes which preceded it have been already described (p. 368); and we have seen that two of the eye-witnesses of the convulsion, Falconi and Giacomo di Toledo, agree in declaring that the sea abandoned a considerable tract of the shore, so that fish were taken by the inhabitants; and, among other things, Falconi mentions that he saw two springs *in the newly discovered ruins.*

The flat land, when first upraised, must have been more extensive than now, for the sea encroaches somewhat rapidly, both to the north and southeast of Puzzuoli. The coast had, when I examined it in 1828, given way more than a foot in a twelvemonth; and I was assured, by fishermen in the bay, that it has lost ground near Puzzuoli, to the extent of thirty feet, within their memory.

It is, moreover, very probable that the land rose to a greater height at first before it ceased to move upwards, than the level at which it was observed to stand when the temple was rediscovered in 1749, for we learn from a memoir of Niccolini, published in 1838, that since the beginning of the nineteenth century, the temple of Serapis has subsided more than two feet. That learned architect visited the ruins frequently, for the sake of making drawings, in the beginning of the year 1807, and

* Ed. Journ. of Science, new series, No. II. p. 281.
† Sul Tempio di Serap. ch. viii.

was in the habit of remaining there throughout the day, yet never saw the pavement overflowed by the sea, except occasionally when the south wind blew violently. On his return, sixteen years after, to superintend some excavations ordered by the king of Naples, he found the pavement covered by sea-water twice every day at high tide, so that he was obliged to place there a line of stones to stand upon. This induced him to make a series of observations from Oct. 1822 to July 1838, by which means he ascertained that the ground had been and was sinking, at the average rate of about seven millimetres a year, or about one inch in four years; so that, in 1838, fish were caught every day on that part of the pavement where, in 1807, there was never a drop of water in calm weather.*

On inquiring still more recently as to the condition of the temple and the continuance of the sinking of the ground, I learn from Signor Scacchi in a letter, dated June 1852, that the downward movement has ceased for several years, or has at least become almost inappreciable. During an examination undertaken by him at my request in the summer of that year (1852), he observed that the rising tide spread first over the seaward side of the flat surface of the pedestals of each column (confirming the fact previously noticed by others, that they are out of the perpendicular); and he also remarked that the water gained unequally on the base of each pillar, in such a manner as to prove that they have neither the same amount of inclination, nor lean precisely in the same direction.

From what was said before (p. 510), we saw that the marine shells in the strata forming the plain called La Starza, considered separately, establish the fact of an upheaval of the ground to the height of twenty-three feet and upwards. The temple proves much more, because it could not have been built originally under water, and must therefore first have sunk down twenty feet at least below the waves, to be afterwards restored to its original position. Yet if such was the order of events we ought to meet with other independent signs of a like subsidence round the margin of a bay once so studded with buildings as the Bay of Baiæ. Accordingly memorials of such submergence are not wanting. About a mile northwest of the temple of Serapis, and about 500 feet from the shore, are the ruins of a temple of Neptune and others of a temple of the Nymphs, now under water. The columns of the former edifice stand erect in five feet water, their upper portions just rising to the surface of the sea. The pedestals are doubtless buried in the sand or mud; so that, if this part of the bottom of the bay should hereafter be elevated, the exhumation of these temples might take place after the manner of that of Serapis. Both these buildings probably participated in the movement which raised the Starza; but either they were deeper under water than the temple of Serapis, or they were not raised

* Tavola Metrica Chronologica, &c. Napoli, 1838. Mr. Smith, of Jordan Hill, writing in 1847, estimated the rate of subsidence, at that period, at *one inch* annually. Quart. Journ. Geol. Soc. vol. iii. p. 237.

up again to so great a height. There are also two Roman roads under water in the bay, one reaching from Puzzuoli to the Lucrine Lake, which may still be seen, and the other near the castle of Baiæ (No. 8, fig. 88, p. 509). The ancient mole, too, of Puzzuoli (No. 4, ibid.) before alluded to, has the water up to a considerable height of the arches; whereas Brieslak justly observes, it is next to certain that the piers must formerly have reached the surface before the springing of the arches;* so that, although the phenomena before described prove that this mole has been uplifted ten feet above the level at which it once stood, it is still evident that it has not yet been restored to its original position.

A modern writer also reminds us, that these effects are not so local as some would have us to believe; for on the opposite side of the Bay of Naples, on the Sorrentine coast, which, as well as Puzzuoli, is subject to earthquakes, a road, with some fragments of Roman buildings, is covered to some depth by the sea. In the island of Capri, also, which is situated some way out at sea, in the opening of the Bay of Naples, one of the palaces of Tiberius is now covered with water.†

That buildings should have been submerged, and afterwards upheaved, without being entirely reduced to a heap of ruins, will appear no anomaly, when we recollect that, in the year 1819, when the delta of the Indus sank down, the houses within the fort of Sindree subsided beneath the waves without being overthrown. In like manner, in the year 1692, the buildings round the harbor of Port Royal, in Jamaica, descended suddenly to the depth of between thirty and fifty feet under the sea without falling. Even on small portions of land transported to a distance of a mile down a declivity, tenements, like those near Mileto, in Calabria, were carried entire. At Valparaiso buildings were left standing in 1822, when their foundations, together with a long tract of the Chilian coast, were permanently upraised to the height of several feet. It is still more easy to conceive that an edifice may escape falling during the upheaval or subsidence of land, if the walls are supported on the exterior and interior with a deposit like that which surrounded and filled to the height of ten or eleven feet the temple of Serapis all the time it was sinking, and which enveloped it to more than twice that height when it was rising again to its original level.

We can scarcely avoid the conclusion, as Mr. Babbage has hinted, "that the action of heat is in some way or other the cause of the phenomena of the change of level of the temple. Its own hot spring, its immediate contiguity to the Solfatara, its nearness to the Monte Nuovo, the hot spring at the baths of Nero (No. 6, fig. 88), on the opposite side of the Bay of Baiæ; the boiling springs and ancient volcanoes of Ischia on one side and Vesuvius on the other, are the most prominent

* Voy. dans la Campanie, tome ii. p. 162.

† Mr. Forbes, Physical Notices of the Bay of Naples. Ed. Journ. of Sci., No. II., new series, p. 280. October, 1829. When I visited Puzzuoli, and arrived at the above conclusions, I knew nothing of Mr. Forbes's observations, which I first saw on my return to England the year following.

of a multitude of facts which point to that conclusion."* And when we reflect on the dates of the principal oscillations of level, and the volcanic history of the country before described (chap. 23), we seem to discover a connection between each era of upheaval and a local development of volcanic heat, and again between each era of depression and the local quiescence or dormant condition of the subterranean igneous causes. Thus for example, before the Christian era, when so many vents were in frequent eruption in Ischia, and when Avernus and other points in the Phlegræan Fields were celebrated for their volcanic aspect and character, the ground on which the temple stood was several feet above water. Vesuvius was then regarded as a spent volcano; but when, after the Christian era, the fires of that mountain were rekindled, scarcely a single outburst was ever witnessed in Ischia, or around the Bay of Baiæ. Then the temple was sinking. Vesuvius, at a subsequent period, became nearly dormant for five centuries preceding the great outbreak of 1631 (see p. 374), and in that interval the Solfatara was in eruption A. D. 1198, Ischia in 1302, and Monte Nuovo was formed in 1538. Then the foundations on which the temple stood were rising again. Lastly, Vesuvius once more became a most active vent, and has been so ever since, and during the same lapse of time the area of the temple, so far as we know any thing of its history, has been subsiding.

These phenomena would agree well with the hypothesis, that when the subterranean heat is on the increase, and when lava is forming without obtaining an easy vent, like that afforded by a great habitual chimney, such as Vesuvius, the incumbent surface is uplifted; but when the heated rocks below are cooling and contracting, and sheets of lava are slowly consolidating and diminishing in volume, then the incumbent land subsides.

Signor Niccolini, when he ascertained in 1838 that the relative levels of the floor of the temple and of the sea were slowly changing from year to year, embraced the opinion that it was the sea which was rising. But Signor Capocci successfully controverted this view, appealing to many appearances which attest the local character of the movements of the adjoining country, besides the historical fact that in 1538, when the sea retired permanently 200 yards from the ancient shore at Puzzuoli, there was no simultaneous retreat of the waters from Naples, Castelamare, and Ischia.†

Permanence of the ocean's level.—In concluding this subject I may observe, that the interminable controversies to which the phenomena of the Bay of Baiæ gave rise, have sprung from an extreme reluctance to admit that the land, rather than the sea, is subject alternately to rise and fall. Had it been assumed that the level of the ocean was invariable, on the ground that no fluctuations have as yet been clearly established, and that, on the other hand, the continents are inconstant in

* Quart. Journ. Geol. Soc. 1847, vol. iii. p. 203.
† Nuove Ricerche sul Temp. di Serap.

their level, as has been demonstrated by the most unequivocal proofs again and again, from the time of Strabo to our own times, the appearances of the temple at Puzzuoli could never have been regarded as enigmatical. Even if contemporary accounts had not distinctly attested the upraising of the coast, this explanation should have been proposed in the first instance as the most natural, instead of being now adopted unwillingly when all others have failed.

To the strong prejudices still existing in regard to the mobility of the land, we may attribute the rarity of such discoveries as have been recently brought to light in the Bay of Baiæ and the Bay of Conception. A false theory, it is well known, may render us blind to facts which are opposed to our prepossessions, or may conceal from us their true import when we behold them. But it is time that the geologist should, in some degree, overcome those first and natural impressions, which induced the poets of old to select the rock as the emblem of firmness—the sea as the image of inconstancy. Our modern poet, in a more philosophical spirit, saw in the sea "The image of eternity," and has finely contrasted the fleeting existence of the successive empires which have flourished and fallen on the borders of the ocean with its own unchanged stability.

> ——— Their decay
> Has dried up realms to deserts:—not so thou,
> Unchangeable, save to thy wild wave's play:
> Time writes no wrinkle on thine azure brow;
> Such as creation's dawn beheld, thou rollest now.
>
> CHILDE HAROLD, Canto iv.

CHAPTER XXX.

ELEVATION AND SUBSIDENCE OF LAND WITHOUT EARTHQUAKES.

Changes in the relative level of land and sea in regions not volcanic—Opinion of Celsius that the waters of the Baltic Sea and Northern Ocean were sinking—Objections raised to his opinion—Proofs of the stability of the sea level in the Baltic—Playfair's hypothesis that the land was rising in Sweden—Opinion of Von Buch—Marks cut on the rocks—Survey of these in 1820—Facility of detecting slight alterations of level on coast of Sweden—Shores of the ocean also rising—Area upheaved—Shelly deposits of Uddevalla—Of Stockholm, containing fossil shells characteristic of the Baltic—Subsidence in south of Sweden—Fishing hut buried under marine strata—Upheaval in Sweden not always in horizontal planes—Sinking of land in Greenland—Bearing of these facts on geology.

WE have now considered the phenomena of volcanoes and earthquakes according to the division of the subject before proposed (p. 345), and have next to turn our attention to those slow and insensible changes in the relative level of land and sea which take place in countries remote from volcanoes, and where no violent earthquakes have occurred within the period of human observation. Early in the last century the Swedish naturalist, Celsius, expressed his opinion that, the waters, both of the Baltic and Northern Ocean, were gradually subsiding. From numerous observations, he inferred that the rate of depression was about fifty Swedish inches in a century.* In support of this position, he alleged that there were many rocks both on the shores of the Baltic and the ocean known to have been once sunken reefs, and dangerous to navigators, but which were in his time above water—that the waters of the Gulf of Bothnia had been gradually converted into land; several ancient ports having been changed into inland cities, small islands joined to the continent, and old fishing-grounds deserted as being too shallow, or entirely dried up. Celsius also maintained, that the evidence of the change rested not only on modern observations, but on the authority of the ancient geographers, who had stated that Scandinavia was formerly an island. This island, he argued, must in the course of centuries, by the gradual retreat of the sea, have become connected with the continent; an event which he supposed to have happened after the time of Pliny, and before the ninth century of our era.

To this argument it was objected that the ancients were so ignorant of the geography of the most northern parts of Europe, that their authority was entitled to no weight; and that their representation of Scandinavia as an island, might with more propriety be adduced to prove the scantiness of their information, than to confirm so bold an hypothesis. It was also remarked that if the land which connected Scandinavia with the main continent was laid dry between the time of Pliny and the ninth century, to the extent to which it is known to have risen above the sea at the latter period, the rate of depression could not have been uniform, as was pretended; for it ought to have fallen much more rapidly between the ninth and eighteenth centuries.

Many of the proofs relied on by Celsius and his followers were immediately controverted by several philosophers, who saw clearly that a fall of the sea in any one region could not take place without a general sinking of the waters over the whole globe: they denied that this was the fact, or that the depression was universal, even in the Baltic. In proof of the stability of the level of that sea, they appealed to the position of the island of Saltholm, not far from Copenhagen. This island is so low, that in autumn and winter it is permanently overflowed; and it is only dry in summer, when it serves for pasturing cattle. It appears, from the documents of the year 1280, that Saltholm was then also in the same state, and exactly on a level with the mean height of the sea, instead of having been about twenty feet under water, as it ought to have been, according to the computation of Celsius. Several towns, also, on the shores of the Baltic, as Lubeck, Wismar, Rostock, Stralsund, and others, after six and even eight hundred years, are as little elevated above the sea as at the era of their foundation, being now close to the water's edge. The lowest part of Dantzic was no higher than the mean

* The Swedish measure scarcely differs from ours; the foot being divided into twelve inches, and being less than ours by three-eighths of an inch only.

level of the sea in the year 1000; and after eight centuries its relative position remains exactly the same.*

Several of the examples of the gain of land and shallowing of the sea pointed out by Celsius, and afterwards by Linnæus, who embraced the same opinions, were ascribed by others to the deposition of sediment at points where rivers entered; and, undoubtedly, Celsius had not sufficiently distinguished between changes due to these causes and such as would arise if the waters of the ocean itself were diminishing. Many large rivers descending from a mountainous country, at the head of the Gulf of Bothnia, enter the sea charged with sand, mud, and pebbles; and it was said that in these places the low land had advanced rapidly, especially near Torneo. At Piteo also, half a mile had been gained in forty-five years; at Luleo,† no less than a mile in twenty-eight years; facts which might all be admitted consistently with the assumption that the level of the Baltic has remained unchanged, like that of the Adriatic, during a period when the plains of the Po and the Adige have greatly extended their area.

It was also alleged that certain insular rocks, once entirely covered with water, had at length protruded themselves above the waves, and grown, in the course of a century and a half, to be eight feet high. The following attempt was made to explain away this phenomenon:— In the Baltic, large erratic blocks, as well as sand and smaller stones which lie on shoals, are liable every year to be frozen into the ice, where the sea freezes to the depth of five or six feet. On the melting of the snow in spring, when the sea rises about half a fathom, numerous ice-islands float away, bearing up these rocky fragments so as to convey them to a distance; and if they are driven by the waves upon shoals, they may convert them into islands by depositing the blocks; if stranded upon low islands, they may considerably augment their height.

Browallius, also, and some other Swedish naturalists, affirmed that some islands were lower than formerly; and that, by reference to this kind of evidence, there was equally good reason for contending that the level of the Baltic was gradually rising. They also added another curious proof of the permanency of the water level, at some points at least, for many centuries. On the Finland coast were some large pines, growing close to the water's edge; these were cut down, and, by counting the concentric rings of annual growth, as seen in a transverse section of the trunk, it was demonstrated that they had stood there for four hundred years. Now, according to the Celsian hypothesis, the sea had sunk about fifteen feet during that period, in which case the germination and early growth of these pines must have been, for many seasons, below the level of the water. In like manner it was asserted, that the lower walls of many ancient castles, such as those of Sonder-

* For a full account of the Celsian controversy, we may refer our readers to Von Hoff, Geschichte, &c. vol. i. p. 439.

† Piteo, Luleo, and Obo are spelt, in many English maps, Pitea, Lulea, Abo; the *a* is not sounded in the Swedish diphthong *ao* or *å*.

Fig. 91.

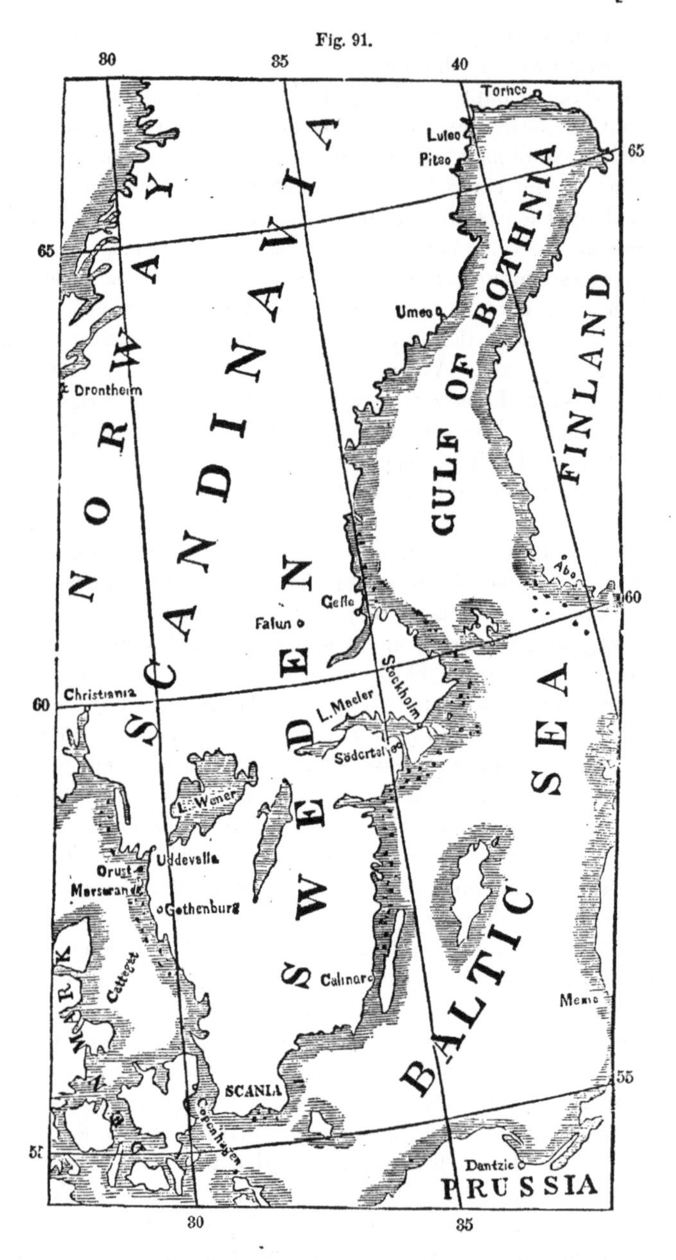

burg and Åbo, reached then to the water's edge, and must, therefore, according to the theory of Celsius, have been originally constructed below the level of the sea.

In reply to this last argument, Colonel Hällstrom, a Swedish engineer, well acquainted with the Finland coast, assured me, that the base of the walls of the castle of Åbo is now ten feet above the water, so that there may have been a considerable rise of the land at that point since the building was erected.

Playfair, in his "Illustrations of the Huttonian Theory," in 1802, admitted the sufficiency of the proofs adduced by Celsius, but attributed the change of level to the movement of the land, rather than to a diminution of the waters. He observed, "that in order to depress or elevate the absolute level of the sea, by a given quantity, in any one place, we must depress or elevate it by the same quantity over the whole surface of the earth; whereas no such necessity exists with respect to the elevation or depression of the land."* The hypothesis of the rising of the land he adds, "agrees well with the Huttonian theory, which holds, that our continents are subject to be acted upon by the expansive forces of the mineral regions; that by these forces they have been actually raised up, and are sustained by them in their present situation.†

In the year 1807, Von Buch, after returning from a tour in Scandinavia, announced his conviction, "that the whole country, from Frederickshall in Norway to Åbo in Finland, and perhaps as far as St. Petersburgh, was slowly and insensibly rising." He also suggested "that Sweden may rise more than Norway, and the northern more than the southern part."‡ He was led to these conclusions principally by information obtained from the inhabitants and pilots, and in part by the occurrence of marine shells of recent species, which he had found at several points on the coast of Norway above the level of the sea. He also mentions the marks set on the rocks. Von Buch, therefore, has the merit of being the first geologist who, after a personal examination of the evidence, declared in favor of the rise of land in Scandinavia.

The attention excited by this subject in the early part of the last century, induced many philosophers in Sweden to endeavor to determine, by accurate observations, whether the standard level of the Baltic was really subject to periodical variations; and under their direction, lines or grooves, indicating the ordinary level of the water on a calm day, together with the date of the year, were chiselled out upon the rocks. In 1820–21, all the marks made before those years were examined by the officers of the pilotage establishment of Sweden; and in their report to the Royal Academy of Stockholm they declared, that on comparing the level of the sea at the time of their observations with that indicated by the ancient marks, they found that the Baltic was lower relatively to the land in certain places, but the amount of change during equal periods of time had not been everywhere the same. During their survey, they cut new marks for the guidance of future observers, several of which I had an opportunity of examining fourteen years after (in the summer of 1834), and in that interval the land appeared to me to have risen at certain places north of Stockholm four or five inches. I also convinced myself, during my visit to Sweden, after conversing with many civil engineers, pilots, and fishermen, and

* Sect. 393. † Sect. 398. ‡ Transl. of his Travels, p. 387.

after examining some of the ancient marks, that the evidence formerly adduced in favor of the change of level, both on the coasts of Sweden and Finland, was full and satisfactory.* The alteration of level evidently diminishes as we proceed from the northern parts of the Gulf of Bothnia towards the south, being very slight around Stockholm. Some writers have indeed represented the rate of depression of the waters at Stockholm as very considerable, because certain houses in that city which are built on piles have sunk down within the memory of persons still living, so as to be out of the perpendicular; and this in consequence of the tops of the piles giving way and decaying, owing to a fall of the waters which has exposed them to be alternately wet and dry. The houses alluded to are situated on the borders of Lake Maeler, a large lake, the outlet of which joins the Baltic, in the middle of Stockholm. This lake is certainly lower than formerly; but the principal cause of the change is not the elevation of the land, but the removal of two old bridges built on piles, which formerly obstructed the discharge of the fresh water into the sea. Another cause is the opening, in the year 1819, of a new canal at Södertelje, a place south of Stockholm, by means of which a new line of communication was formed between Lake Maeler and the Baltic.†

It will naturally be asked, whether the mean level of a sea like the Baltic can ever be determined so exactly as to permit us to appreciate a variation of level, amounting only to one or two feet. In reply, I may observe, that, except near the Cattegat, there are no tides in the Baltic; and it is only when particular winds have prevailed for several days in succession, or at certain seasons when there has been an unusually abundant influx of river water, or when these causes have combined, that this sea is made to rise two or three feet above its standard level. The fluctuations due to these causes are nearly the same from year to year; so that the pilots and fishermen believe and apparently with reason, that they can mark a deviation, even of a few inches, from the ordinary or mean height of the waters.

There are, moreover, peculiarities in the configuration of the shores of Norway and Sweden, which facilitate in a remarkable degree the appreciation of slight changes in the relative level of land and water. It has often been said, that there are two coasts, an inner and an outer one; the inner being the shore of the main land; the outer one, a fringe of countless rocky islands of all dimensions, called the skär (*shair*). Boats and small vessels make their coasting voyages within this skär: for here they may sail in smooth water, even when the sea without is strongly agitated. But the navigation is very intri-

* In the earlier editions I expressed many doubts as to the validity of the proofs of a gradual rise of land in Sweden. A detailed statement of the observations which I made in 1834, and which led me to change my opinion, will be found in the Philosophical Transactions for 1835, part i.

† See Professor Johnston's Paper, Ed. New Phil. Journ. No. 29, July 1833; and my remarks, Phil. Trans. 1835, p. 12.

cate, and the pilot must possess a perfect acquaintance with the breadth and depth of every narrow channel, and the position of innumerable sunken rocks. If on such a coast the land rises one or two feet in the course of half a century, the minute topography of the skär is entirely altered. To a stranger, indeed, who revisits it after an interval of many years, its general aspect remains the same; but the inhabitant finds that he can no longer penetrate with his boat through channels where he formerly passed, and he can tell of countless other changes in the height and breadth of isolated rocks, now exposed, but once only seen through the clear water.

The rocks of gneiss, mica-schist, and quartz are usually very hard on this coast, slow to decompose, and, when protected from the breakers, remaining for ages unaltered in their form. Hence it is easy to mark the stages of their progressive emergence by the aid of natural and artificial marks imprinted on them. Besides the summits of *fixed* rocks, there are numerous erratic blocks of vast size strewed over the shoals and islands in the skär, which have been probably drifted by ice in the manner before suggested.* All these are observed to have increased in height and dimension with the last half century. Some, which were formerly known as dangerous sunken rocks, are now only hidden when the water is highest. On their first appearance, they usually present a smooth, bare, rounded protuberance, a few feet or yards in diameter; and a single sea-gull often appropriates to itself this resting-place, resorting there to devour its prey. Similar points, in the mean time, have grown to long reefs, and are constantly whitened by a multitude of sea-fowl; while others have been changed from a reef, annually submerged, to a small islet, on which a few lichens, a fir-seedling, and a few blades of grass, attest that the shoal has at length been fairly changed into dry land. Thousands of wooded islands around show the great alterations which time can work. In the course of centuries also, the spaces intervening between the existing islands may be laid dry, and become grassy plains encircled by heights well clothed with lofty firs. This last step of the process, by which long fiords and narrow channels, once separating wooded islands, are deserted by the sea, has been exemplified within the memory of living witnesses on several parts of the coast.

Had the apparent fall of the waters been observed in the Baltic only, we might have endeavored to explain the phenomenon by local causes affecting that sea alone. For instance, the channel by which the Baltic discharges its surplus waters into the Atlantic, might be supposed to have been gradually widened and deepened by the waves and currents, in which case a fall of the water like that before alluded to in Lake Mæler, might have occurred. But the lowering of level would, in that case have been uniform and universal, and the waters could not have sunk at Torneo, while they retained their former level at Copenhagen. Such an explanation is also untenable on other grounds; for it is a fact,

* See p. 522; also chap. 15, *supra*.

as Celsius long ago affirmed, that the alteration of level extends to the western shores of Sweden, bordering the ocean. The signs of elevation observed between Uddevalla and Gothenburg are as well established as those on the shores of the Bothnian Gulf. Among the places where they may be studied, are the islands of Marstrand and Gulholmen, the last-mentioned locality being one of those particularly pointed out by Celsius.

The inhabitants there and elsewhere affirm, that the rate of the sinking of the sea (or elevation of land) varies in different and adjoining districts, being greatest at points where the land is low. But in this they are deceived; for they measure the amount of rise by the area gained, which is most considerable where the land descends with a gentle slope into the sea. In the same manner, some advocates of the Celsian theory formerly appealed to the increase of lands near the mouths of rivers, not sufficiently adverting to the fact, that if the bed of the sea is rising, the change will always be most sensible where the bottom has been previously rendered shallow; whereas, at a distance from these points where the scarped granitic cliffs plunge at once into deep water, a much greater amount of elevation is necessary to produce an equally conspicuous change.

As to the area in northern Europe which is subject to this slow upheaving movement, we have not as yet sufficient data for estimating it correctly. It seems probable, however, that it reaches from Gothenburg to Torneo, and from thence to the North Cape, the rate of elevation increasing always as we proceed farther northwards. The two extremities of this line are more than a thousand geographical miles distant from each other; and as both terminate in the ocean, we know not how much farther the motion may be prolonged under water. As to the breadth of the tract, its limits are equally uncertain, though it evidently extends across the widest parts of the Gulf of Bothnia, and may probably stretch far into the interior, both of Sweden and Finland. Now if the elevation continue, a larger part of the Gulf of Bothnia will be turned into land, as also more of the ocean off the west coast of Sweden between Gothenburg and Uddevalla; and on the other hand, if the change has been going on for thousands of years at the rate of several feet in a century, large tracts of what is now land must have been submarine at periods comparatively modern. It is natural therefore to inquire whether there are any signs of the recent sojourn of the sea on districts now inland? The answer is most satisfactory.—Near Uddevalla and the neighboring coastland, we find upraised deposits of shells belonging to species such as now live in the ocean; while on the opposite or eastern side of Sweden, near Stockholm, Gefle, and other places bordering the Bothnian Gulf, there are analogous beds containing shells of species characteristic of the Baltic.

Von Buch announced in 1807, that he had discovered in Norway and at Uddevalla in Sweden, beds of shells of existing species, at considerable heights above the sea. Since that time, other naturalists have confirmed his observation; and, according to Ström, deposits occur at an elevation of more than 400 feet above the sea in the northern part of Norway. M. Alex. Brongniart, when he visited Uddevalla, ascertained that one of the principal masses of shells, that of Capellbacken, is raised more than 200 feet above the sea, resting on rocks of gneiss, all the species being identical with those now inhabiting the contiguous ocean. The same naturalist also stated, that on examining with care the surface of the gneiss, immediately above the ancient shelly deposit, he found barnacles (*balani*) adhering to the rocks, showing that the sea had remained there for a long time. I was fortunate enough to be able to verify this observation by finding in the summer of 1834, at Kured, about two miles north of Uddevalla, and at the height of more than 100 feet above the sea, a surface of gneiss newly laid open by the partial removal of a mass of shells used largely in the district for making lime and repairing the roads. So firmly did these barnacles adhere to the gneiss, that I broke off portions of the rock with the shells attached. The face of the gneiss was also incrusted with small zoophytes (*Cellepora?* Lam.); but had these or the barnacles been exposed in the atmosphere ever since the elevation of the rocks above the sea, they would doubtless have decomposed and been obliterated.

The town of Uddevalla (see Map, p. 523) stands at the head of a narrow creek overhung by steep and barren rocks of gneiss, of which all the adjacent country is composed, except in the low grounds and bottoms of valleys, where strata of sand, clay, and marl frequently hide the fundamental rocks. To these newer and horizontal deposits the fossil shells above mentioned belong, and similar marine remains are found at various heights above the sea on the opposite island of Orust. The extreme distance from the sea to which such fossils extend is as yet unknown; but they have been already found at Trollhättan in digging the canal there, and still farther inland on the northern borders of Lake Wener, fifty miles from the sea, at an elevation of 200 feet near Lake Rogvarpen.

To pass to the Baltic: I observed near its shores at Södertelje, sixteen miles S. W. of Stockholm, strata of sand, clay, and marl, more than 100 feet high, and containing shells of species now inhabiting the Bothnian Gulf. These consist partly of marine and partly of freshwater species; but they are few in number, the brackishness of the water appearing to be very unfavorable to the development of testacea. The most abundant species are the common cockle and the common mussel and periwinkle of our shores (*Cardium edule, Mytilus edulis,* and *Littorina littorea*), together with a small tellina (*T. Baltica*) and a few minute univalves allied to *Paludina ulva.* These live in the same water as a *Lymneus*, a *Neritina* (*N. fluviatilis*), and some other freshwater shells.

But the marine mollusks of the Baltic above mentioned, although very numerous in individuals, are dwarfish in size, scarcely ever attaining a third of the average dimensions which they acquire in the salter waters of the ocean. By this character alone a geologist would generally be able to recognize an assemblage of Baltic fossils as distinguished from those derived from a deposit in the ocean. The absence also of oysters, barnacles, whelks, scallops, limpets (*ostrea, balanus, buccinum, pecten, patella*), and many other forms abounding alike in the sea near Uddevalla, and in the fossiliferous deposits of modern date on that coast, supplies an additional negative character of the greatest value; distinguishing assemblages of Baltic from those of oceanic shells. Now the strata containing Baltic shells are found in many localities near Stockholm, Upsala, and Gefle, and will probably be discovered everywhere around the borders of the Bothnian Gulf; for I have seen similar remains brought from Finland, in marl resembling that found near Stockholm. The utmost distance to which these deposits have yet been traced inland, is on the southern shores of Lake Macler, at a place seventy miles from the sea.* Hence it appears from the distinct assemblage of fossil shells found on the eastern and western coasts of Sweden, that the Baltic has been for a long period separated as now from the ocean, although the intervening tract of land was once much narrower, even after both seas had become inhabited by all the existing species of testacea.

As no accurate observations on the rise of the Swedish coast refer to periods more remote than a century and a half from the present time, and as traditional information, and that derived from ancient buildings on the coast, do not enable the antiquary to trace back any monuments of change for more than five or six centuries, we cannot declare whether the rate of the upheaving force is uniform during very long periods. In those districts where the fossil shells are found at the height of more than 200 feet above the ocean, as at Uddevalla, Orust, and Lake Rogvarpen, the present rate of rise seems less than four feet in a century. Even at that rate it would have required five thousand years to lift up those deposits. But as the movement is now very different in different places, it may also have varied much in intensity at different eras.

We have, moreover, yet to learn not only whether the motion proceeds always at the same rate, but also whether it has been uniformly *in one direction*. The level of the land may oscillate; and for centuries there may be a depression, and afterwards a re-elevation, of the same district. Some phenomena in the neighborhood of Stockholm appear to me only explicable on the supposition of the alternate rising and sinking of the ground since the country was inhabited by man. In digging a canal, in 1819, at Södertelje, about sixteen miles to the south of Stockholm, to unite Lake Macler with the Baltic, marine strata, containing fossil shells of Baltic species, were passed through. At a depth of about sixty feet, they came down upon what seems to have been a buried fishing-hut, constructed of wood in a state of decomposition, which soon crumbled away on exposure to the air. The lowest part, however, which had stood on a level with the sea, was in a more perfect state of preservation. On

* See a paper by the Author, Phil. Trans. 1835, part i.

the floor of this hut was a rude fireplace, consisting of a ring of stones, and within this were cinders and charred wood. On the outside lay boughs of the fir, cut as with an axe, with the leaves or needles still attached. It seems very difficult to explain the position of this buried hut, without imagining, as in the case of the temple of Serapis (see p. 486), first a subsidence to the depth of more than sixty feet, then a re-elevation. During the period of submergence, the hut must have become covered over with gravel and shelly marl, under which not only the hut, but several vessels also were found, of a very antique form, and having their timbers fastened together by wooden pegs instead of nails.*

Whether any of the land in Norway is now rising, must be determined by future investigations. Marine fossil shells, of recent species, have been collected from inland places near Drontheim; but Mr. Everest, in his "Travels through Norway," informs us that the small island of Munkholm, which is an insulated rock in the harbor of Drontheim, affords conclusive evidence of the land having in that region remained stationary for the last eight centuries. The area of this isle does not exceed that of a small village, and by an official survey, its highest point has been determined to be twenty-three feet above the mean high-water mark, that is, the mean between neap and spring tides. Now, a monastery was founded there by Canute the Great, A. D. 1028, and thirty-three years before that time it was in use as a common place of execution. According to the assumed average rate of rise in Sweden (about forty inches in a century), we should be obliged to suppose that this island had been three feet eight inches below high-water mark when it was originally chosen as the site of the monastery.

Professor Keilhau of Christiania, after collecting the observations of his predecessors respecting former changes of level in Norway, and combining them with his own, has made the fact of a general change of level at a modern period, that is to say, within the period of the actual testaceous fauna, very evident. He infers that the whole country from Cape Lindesnæs to Cape North, and beyond that as far as the fortress of Vardhuus, has been gradually upraised, and on the southeast coast the elevation has amounted to more than 600 feet. The marks which denote the ancient coast-line are so nearly horizontal that the deviation from horizontality, although the measurements have been made at a great number of points, is too small to be appreciated.

More recently (1844), however, it appears from the researches of M. Bravais, member of the French scientific commission of the North, that in the Gulf of Alten in Finmark, the most northerly part of Norway,

* See my paper before referred to, Phil. Trans. 1835, part i. p. 8, 9. Attempts have been since made to explain away the position of this hut, by conjecturing that a more recent trench had been previously dug here, which had become filled up in time by sand drifted by the wind. The engineers who superintended the works in 1819, and with whom I conversed, had considered every hypothesis of the kind, but could not so explain the facts.

there are two distinct lines of upraised ancient sea-coast, one above the other, which are not parallel, and both of them imply that within a distance of fifty miles a considerable slope can be detected in such a direction as to show that the ancient shores have undergone a greater amount of upheaval in proportion as we advance inland.*

It has been already stated, that, in proceeding from the North Cape to Stockholm, the rate of upheaval diminishes from several feet to a few inches in a century. To the south of Stockholm, the upward movement ceases, and at length in Scania, or the southernmost part of Sweden, it appears to give place to a movement in an opposite direction. In proof of this fact, Professor Nilsson observes, in the first place, that there are no elevated beds of recent marine shells in Scania like those farther to the north. Secondly, Linnæus, with a view of ascertaining whether the waters of the Baltic were retiring from the Scanian shore, measured, in 1749, the distance between the sea and a large stone near Trelleborg. This same stone was, in 1836, a hundred feet nearer the water's edge than in Linnæus's time, or eighty-seven years before. Thirdly, there is also a submerged peat moss, consisting of land and freshwater plants, beneath the sea at a point to which no peat could have been drifted down by any river. Fourthly, and what is still more conclusive, it is found that in seaport towns, all along the coast of Scania, there are streets below the high-water level of the Baltic, and in some cases below the level of the lowest tide. Thus, when the wind is high at Malmo, the water overflows one of the present streets, and some years ago some excavations showed an ancient street in the same place eight feet lower, and it was then seen that there had been an artificial raising of the ground, doubtless in consequence of that subsidence. There is also a street at Trelleborg, and another at Skanör, a few inches below high-water mark, and a street at Ystad is exactly on a level with the sea, at which it could not have been originally built.

The inferences deduced from the foregoing facts are in perfect harmony with the proofs brought to light by two Danish investigators, Dr. Pingel and Captain Graah, of the sinking down of part of the west coast of Greenland, for a space of more than 600 miles from north to south. The observations of Captain Graah were made during a survey of Greenland in 1823–24; and afterwards in 1828–29; those by Dr. Pingel were made in 1830–32. It appears from various signs and traditions, that the coast has been subsiding for the last four centuries from the firth called Igaliko, in lat. 60° 43′ N. to Disco Bay, extending to nearly the 69th degree of north latitude. Ancient buildings on low rocky islands and on the shore of the main land have been gradually submerged, and experience has taught the aboriginal Greenlander never to build his hut near the water's edge. In one case the Moravian settlers have been obliged more than once to move inland the poles upon

* Quart. Journ. of Geol. Soc. No. 4, p. 534. M. Bravais' observations were verified in 1849 by Mr. R. Chambers in his "Tracings of N. of Europe," p. 208.

which their large boats were set, and the old poles still remain beneath the water as silent witnesses of the change.*

The probable cause of the movements above alluded to, whether of elevation or depression, will be more appropriately discussed in the following chapters, when the origin of subterranean heat is considered. But I may remark here, that the rise of Scandinavia has naturally been regarded as a very singular and scarcely credible phenomenon, because no region on the globe has been more free within the times of authentic history from violent earthquakes. In common, indeed, with our own island and with almost every spot on the globe, some movements have been, at different periods, experienced, both in Norway and Sweden. But some of these, as for example during the Lisbon earthquake in 1755, may have been mere vibrations or undulatory movements of the earth's crust prolonged from a great distance. Others, however, have been sufficiently local to indicate a source of disturbance immediately under the country itself. Notwithstanding these shocks, Scandinavia has, upon the whole, been as tranquil in modern times, and as free from subterrauean convulsions, as any region of equal extent on the globe. There is also another circumstance which has made the change of level in Sweden appear anomalous, and has for a long time caused the proofs of the fact to be received with reluctance. Volcanic action, as we have seen, is usually intermittent: and the variations of level to which it has given rise have taken place by starts, not by a prolonged and insensible movement similar to that experienced in Sweden. Yet, as we enlarge our experience of modern changes, we discover instances in which the volcanic eruption, the earthquake, and the permanent rise or fall of land, whether slow or sudden, are all connected. The union of these various circumstances was exemplified in the case of the temple of Serapis, described in the last chapter, and we might derive other illustrations from the events of the present century in South America.

Some writers, indeed, have imagined that there is geological evidence in Norway, of the sudden upheaval of land to a considerable height at successive periods, since the era when the sea was inhabited by the living species of testacea. They point in proof to certain horizontal lines of inland cliffs and sea-beaches containing recent shells at various heights above the level of the sea.† But these appearances, when truly interpreted, simply prove that there have been long pauses in the process of upheaval or subsidence. They mark eras at which the level of the sea has remained stationary for ages, and during which new strata were deposited near the shore in some places, while in others the waves and currents had time to hollow out rocks, undermine cliffs, and throw up long ranges of shingle. They undoubtedly show that the movement has not been always uniform or continuous, but they do not establish the fact of any sudden alterations of level.

* See Proceedings of Geol. Soc. No. 42, p. 208. I also conversed with Dr. Pingel on the subject at Copenhagen in 1834.

† Keilhau, Bulletin de la Soc. Géol. de France, tom. vii. p. 18.

When we are once assured of the reality of the gradual rise of a large region, it enables us to account for many geological appearances otherwise of very difficult explanation. There are large continental tracts and high table-lands where the strata are nearly horizontal, bearing no marks of having been thrown up by violent convulsions, nor by a series of movements, such as those which occur in the Andes, and cause the earth to be rent open, and raised or depressed from time to time, while large masses are engulfed in subterranean cavities. The result of a series of such earthquakes might be to produce in a great lapse of ages a country of shattered, inclined, and perhaps vertical strata. But a movement like that of Scandinavia would cause the bed of the sea, and all the strata recently formed in it, to be upheaved so gradually, that it would merely seem as if the ocean had formerly stood at a higher level, and had slowly and tranquilly sunk down into its present bed.

The fact also of a very gradual and insensible elevation of land may explain many geological movements of denudation, on a grand scale. If, for example, instead of the hard granitic rocks of Norway and Sweden, a large part of the bed of the Atlantic, consisting chiefly of soft strata, should rise up century after century, at the rate of about half an inch, or an inch, in a year, how easily might oceanic currents sweep away the thin film of matter thus brought up annually within the sphere of aqueous denudation! The tract, when it finally emerged, might present table-lands and ridges of horizontal strata, with intervening valleys and vast plains, where originally, and during its period of submergence, the surface was level and nearly uniform.

These speculations relate to superficial changes; but others must be continually in progress in the subterranean regions. The foundations of the country, thus gradually uplifted in Sweden, must be undergoing important modifications. Whether we ascribe these to the expansion of solid matter by continually increasing heat, or to the liquefaction of rock, or to the crystallization of a dense fluid, or the accumulation of pent-up gases, in whatever conjectures we indulge, we can never doubt for a moment, that at some unknown depth beneath Sweden and the Baltic, the structure of the globe is in our own times becoming changed from day to day, throughout a space probably more than a thousand miles in length, and several hundred in breadth.

CHAPTER XXXI.

CAUSES OF EARTHQUAKES AND VOLCANOES.

Intimate connection between the causes of volcanoes and earthquakes—Supposed original state of fusion of the planet—Universal fluidity not proved by spheroidal figure of the earth—Attempt to calculate the thickness of the solid crust of the earth by precessional motion—Heat in mines increasing with the depth—Objections to the supposed intense heat of a central fluid—Whether chemical changes may produce volcanic heat—Currents of electricity circulating in the earth's crust.

It will hardly be questioned, after the description before given of the phenomena of earthquakes and volcanoes, that both of these agents have, to a certain extent, a common origin; and I may now, therefore, proceed to inquire into their probable causes. But first, it may be well to recapitulate some of those points of relation and analogy which lead naturally to the conclusion that they spring from a common source.

The regions convulsed by violent earthquakes include within them the site of all the active volcanoes. Earthquakes, sometimes local, sometimes extending over vast areas, often precede volcanic eruptions. The subterranean movement and the eruption return again and again, at irregular intervals of time, and with unequal degrees of force, to the same spots. The action of either may continue for a few hours, or for several consecutive years. Paroxysmal convulsions are usually followed, in both cases, by long periods of tranquillity. Thermal and mineral springs are abundant in countries of earthquakes and active volcanoes. Lastly, hot springs situated in districts considerably distant from volcanic vents have been observed to have their temperature suddenly raised, and the volume of their water augmented, by subterranean movements.

All these appearances are evidently more or less connected with the passage of heat from the interior of the earth to the surface; and where there are active volcanoes, there must exist, at some unknown depth below, enormous masses of matter intensely heated, and, in many instances, in a constant state of fusion. We have first, then, to inquire, whence is this heat derived?

It has long been a favorite conjecture, that the whole of our planet was originally in a state of igneous fusion, and that the central parts still retain a great portion of their primitive heat. Some have imagined, with the late Sir W. Herschel, that the elementary matter of the earth may have been first in a gaseous state, resembling those nebulæ which we behold in the heavens, and which are of dimensions so vast, that some of them would fill the orbits of the remotest planets of our system. The increased power of the telescope has of late years resolved the greater number of these nebulous appearances into clusters of stars, but so long as they were confidently supposed to consist of aeriform matter it was a favorite conjecture that they might, if concentrated, form solid spheres; and it was also imagined that the evolution of heat, attendant on condensation, might retain the materials of the new globes in a state of igneous fusion.

Without dwelling on such speculations, which can only have a distant bearing on geology, we may consider how far the spheroidal form of the earth affords sufficient ground for presuming that its primitive condition was one of universal fluidity. The discussion of this question would be superfluous, were the doctrine of original fluidity less popular; for it may well be asked, why the globe should be supposed to have had a pristine shape different from the present one?—why the terrestrial materials, when first called into existence, or assembled together in one place, should not have been subject to rotation, so as to assume at once that form which alone could retain their several parts in a state of equilibrium?

Let us, however, concede that the statical figure may be a modification of some other pre-existing form, and suppose the globe to have been at first a perfect and quiescent sphere, covered with a uniform ocean—what would happen when it was made to turn round on its axis with its present velocity? This problem has been considered by Playfair in his Illustrations, and he has decided, that if the surface of the earth, as laid down in Hutton's theory, has been repeatedly changed by the transportation of the detritus of the land to the bottom of the sea, the figure of the planet must in that case, whatever it may have been originally, be brought at length to coincide with the spheroid of equilibrium.* Sir John Herschel also, in reference to the same hypothesis, observes, "a centrifugal force would in that case be generated, whose general tendency would be to urge the water at every point of the surface to *recede* from the *axis*. A rotation might indeed be conceived so swift as to flirt the whole ocean from the surface, like water from a mop. But this would require a far greater velocity than what we now speak of. In the case supposed, the *weight* of the water would still keep it *on* the earth; and the tendency to recede from the axis *could* only be satisfied therefore by the water leaving the poles, and flowing towards the equator; there heaping itself up in a ridge, and being retained in opposition to its weight or natural tendency towards the centre by the pressure thus caused. This, however, could not take place without laying dry the polar regions, so that protuberant land would appear at the poles, and a zone of ocean be disposed around the equator. This would be the first or immediate effect. Let us now see what would afterwards happen if things were allowed to take their natural course.

"The sea is constantly beating on the land, grinding it down, and scattering its worn-off particles and fragments, in the state of sand and pebbles, over its bed. Geological facts afford abundant proof that the

* Illust. of Hutt. Theory, § 435–443.

existing continents have all of them undergone this process even more than once, and been entirely torn in fragments, or reduced to powder, and submerged and reconstructed. Land, in this view of the subject, loses its attribute of fixity. As a mass it might hold together in opposition to forces which the water freely obeys; but in its state of successive or simultaneous degradation, when disseminated through the water, in the state of sand or mud, it is subject to all the impulses of that fluid. In the lapse of time, then, the protuberant land would be destroyed, and spread over the bottom of the ocean, filling up the lower parts, and tending continually to remodel the surface of the solid nucleus, in correspondence with the *form of equilibrium.* Thus after a sufficient lapse of time, in the case of an earth in rotation, the polar protuberances would gradually be cut down and disappear, being transferred to the equator (as being *then* the *deepest sea*), till the earth would assume by degrees the form we observe it to have—that of a flattened or *oblate* ellipsoid.

"We are far from meaning here to trace the process *by which* the earth really assumed its actual form; all we intend is to show that this is the form to which, under a condition of a rotation on its axis, it must *tend,* and which it would attain even if originally and (so to speak) perversely constituted otherwise."*

In this passage, the author has contemplated the superficial effects of aqueous causes only; but neither he nor Playfair seem to have followed out the same inquiry with reference to another part of Hutton's system; namely, that which assumes the successive fusion by heat of different parts of the solid earth. Yet the progress of geology has continually strengthened the evidence in favor of the doctrine that local variations of temperature have melted one part after another of the earth's crust, and this influence has perhaps extended downwards to the very centre. If, therefore, before the globe had assumed its present form, it was made to revolve on its axis, all matter to which freedom of motion was given by fusion, must before consolidating have been impelled towards the equatorial regions in obedience to the centrifugal force. Thus lava flowing out in superficial streams would have its motion retarded when its direction was towards the pole, accelerated when towards the equator; or if lakes and seas of lava existed beneath the earth's crust in equatorial regions, as probably now beneath the Peruvian Andes, the imprisoned fluid would force outwards and permanently upheave the overlying rocks. The statical figure, therefore, of the terrestrial spheroid (of which the longest diameter exceeds the shortest by about twenty-five miles), may have been the result of gradual and even of existing causes, and not of a primitive, universal, and simultaneous fluidity.†

Experiments made with the pendulum, and observations on the manner in which the earth attracts the moon, have shown that our planet is

* Herschel's Astronomy, chap. iii.

† See Hennessy, On Changes in Earth's Figure, &c. Journ. Geol. Soc. Dublin, 1849; and Proc. Roy. Irish Acad. vol. iv. p. 337.

not an empty sphere, but, on the contrary, that its interior, whether solid or fluid, has a higher specific gravity than the exterior. It has also been inferred, that there is a regular increase in density from the surface towards the centre, and that the equatorial protuberance is continued inwards; that is to say, that layers of equal density are arranged elliptically, and symmetrically, from the exterior to the centre. These conclusions, however, have been deduced rather as a consequence of the hypothesis of primitive and simultaneous fluidity than proved by experiment. The inequalities in the moon's motion, by which some have endeavored to confirm them, are so extremely slight, that the opinion can be regarded as little more than a probable conjecture.

The mean density of the earth has been computed by Laplace to be about 5½, or more than five times that of water. Now the specific gravity of many of our rocks is from 2½ to 3, and the greater part of the metals range between that density and 21. Hence some have imagined that the terrestrial nucleus may be metallic—that it may correspond, for example, with the specific gravity of iron, which is about 7. But here a curious question arises in regard to the form which materials, whether fluid or solid, might assume, if subjected to the enormous pressure which must obtain at the earth's centre. Water, if it continued to decrease in volume according to the rate of compressibility deduced from experiment, would have its density doubled at the depth of ninety-three miles, and be as heavy as mercury at the depth of 362 miles. Dr. Young computed that, at the earth's centre, steel would be compressed into one-fourth, and stone into one-eighth of its bulk.* It is more than probable, however, that after a certain degree of condensation, the compressibility of bodies may be governed by laws altogether different from those which we can put to the test of experiment; but the limit is still undetermined, and the subject is involved in such obscurity, that we cannot wonder at the variety of notions which have been entertained respecting the nature and conditions of the central nucleus. Some have conceived it to be fluid, others solid; some have imagined it to have a cavernous structure, and have even endeavored to confirm this opinion by appealing to observed irregularities in the vibrations of the pendulum in certain countries.

An attempt has recently been made by Mr. Hopkins to determine the least thickness which can be assigned to the solid crust of the globe, if we assume the whole to have been once perfectly fluid, and a certain portion of the exterior to have acquired solidity by gradual refrigeration. This result he has endeavored to obtain by a new solution of the delicate problem of the precessional motion of the pole of the earth. It is well known that while the earth revolves round the sun the direction of its axis remains very nearly the same, *i. e.* its different positions in space are all nearly parallel to each other. This parallelism, however, is not

* Young's Lectures, and Mrs. Somerville's Connection of the Physical Sciences, p. 90.

accurately preserved, so that the axis, instead of coming exactly into the position which it occupied a year before, becomes inclined to it at a very small angle, but always retaining very nearly the same inclination to the plane of the earth's orbit. This motion of the pole changes the position of the equinoxes by about fifty seconds annually, and always in the same direction. Thus the pole-star, after a certain time, will entirely lose its claim to that appellation, until in the course of somewhat more than 25,000 years the earth's axis shall again occupy its present angular position, and again point very nearly as now to the pole-star. This motion of the axis is called *precession.* It is caused by the attraction of the sun and moon, and principally the moon, on the protuberant parts of the earth's equator; and if these parts were solid to a great depth, the motion thus produced would differ considerably from that which would exist if they were perfectly fluid, and incrusted over with a thin shell only a few miles thick. In other words, the disturbing action of the moon will not be the same upon a globe all solid and upon one nearly all fluid, or it will not be the same upon a globe in which the solid shell forms one-half of the mass, and another in which it forms only one-tenth.

Mr. Hopkins has, therefore, calculated the amount of precessional motion which would result if we assume the earth to be constituted as above stated; *i. e.* fluid internally, and enveloped by a solid shell; and he finds that the amount will not agree with the observed motion, unless the crust of the earth be of a certain thickness. In calculating the exact amount some ambiguity arises in consequence of our ignorance of the effect of pressure in promoting the solidification of matter at high temperatures. The hypothesis least favorable for a great thickness is found to be that which assumes the pressure to produce no effect on the process of solidification. Even on this extreme assumption the thickness of the solid crust must be nearly *four hundred miles*, and this would lead to the remarkable result that the proportion of the solid to the fluid part would be as 49 to 51, or, to speak in round numbers, there would be nearly as much solid as fluid matter in the globe. The conclusion, however, which Mr. Hopkins announces as that to which his researches have finally conducted him, is thus expressed: "Upon the whole, then, we may venture to assert that the minimum thickness of the crust of the globe, which can be deemed consistent with the observed amount of precession, cannot be less than one-fourth or one-fifth of the earth's radius." That is from 800 to 1000 miles.*

It will be remarked, that this is a *minimum*, and any still *greater* amount would be quite consistent with the actual phenomena; the calculations not being opposed to the supposition of the general solidity of the entire globe. Nor do they preclude us from imagining that great lakes or seas of melted matter may be distributed through a shell 400 or 800

* Phil. Trans. 1839, and Researches in Physical Geology, 1st, 2d, and 3d series, London, 1839–1842; also on Phenomena and Theory of Volcanoes, Report Brit. Assoc. 1847.

miles thick, provided they be so inclosed as to move with it, whatever motion of rotation may be communicated by the disturbing forces of the sun and moon.

Central heat.—The hypothesis of internal fluidity calls for the more attentive consideration, as it has been found that the heat in mines augments in proportion as we descend. Observations have been made, not only on the temperature of the air in mines, but on that of the rocks, and on the water issuing from them. The mean rate of increase, calculated from results obtained in six of the deepest coal mines in Durham and Northumberland, is 1° Fahr. for a descent of forty-four English feet.* A series of observations, made in several of the principal lead and silver mines in Saxony, gave 1° Fahr. for every sixty-five feet. In this case, the bulb of the thermometer was introduced into cavities purposely cut in the solid rock at depths varying from 200 to above 900 feet. But in other mines of the same country, it was necessary to descend thrice as far for each degree of temperature.†

A thermometer was fixed in the rock of the Dolcoath mine, in Cornwall, by Mr. Fox, at the great depth of 1380 feet, and frequently observed during eighteen months; the mean temperature was 68° Fahr., that of the surface being 50°, which gives 1° for every seventy-five feet.

Kupffer, after an extensive comparison of the results in different countries, makes the increase 1° F. for about every thirty-seven English feet.‡ M. Cordier announces, as the result of his experiments and observations on the temperature of the interior of the earth, that the heat increases rapidly with the depth; but the increase does not follow the same law over the whole earth, being twice or three times as much in one country as in another, and these differences are not in constant relation either with the latitudes or longitudes of places.§ He is of opinion, however, that the increase would not be overstated at 1° Cent. for every twenty-five metres, or about 1° F. for every forty-five feet.‖ The experimental well bored at Grenelle, near Paris, gave about 1° F. for every sixty English feet, when they had reached a depth of 1312 feet.

Some writers have endeavored to refer these phenomena (which, however discordant as to the ratio of increasing heat, appear all to point one way) to the condensation of air constantly descending from the surface into the mines. For the air under pressure would give out latent heat, on the same principle as it becomes colder when rarefied in the higher regions of the atmosphere. But, besides that the quantity of heat is greater than could be supposed to flow from this source, the argument has been answered in a satisfactory manner by Mr. Fox, who has shown, that in the mines of Cornwall the ascending have generally a higher

* Ed. Journ. of Sci. April, 1832.

† Cordier, Mém. de l'Instit. tom. vii.

‡ Pog. Ann. tom. xv. p. 159.

§ See M. Cordier's Memoir on the Temperature of the Interior of the Earth, read to the Academy of Sciences, 4th June, 1827.—Edin. New Phil. Journal, No viii. p. 273.

‖ Cordier, Mém. de l'Instit. tom. vii.

temperature than the descending aerial currents. The difference between them was found to vary from 9° to 17° F.; a proof that, instead of imparting heat, these currents actually carry off a large quantity from the mines.*

If we adopt M. Cordier's estimate of 1° F. for every 45 feet of depth as the mean result, and assume, with the advocates of central fluidity, that the increasing temperature is continued downwards, we should reach the ordinary boiling point of water at about two miles below the surface, and at the depth of about twenty-four miles should arrive at the melting point of iron, a heat sufficient to fuse almost every known

Fig. 92.

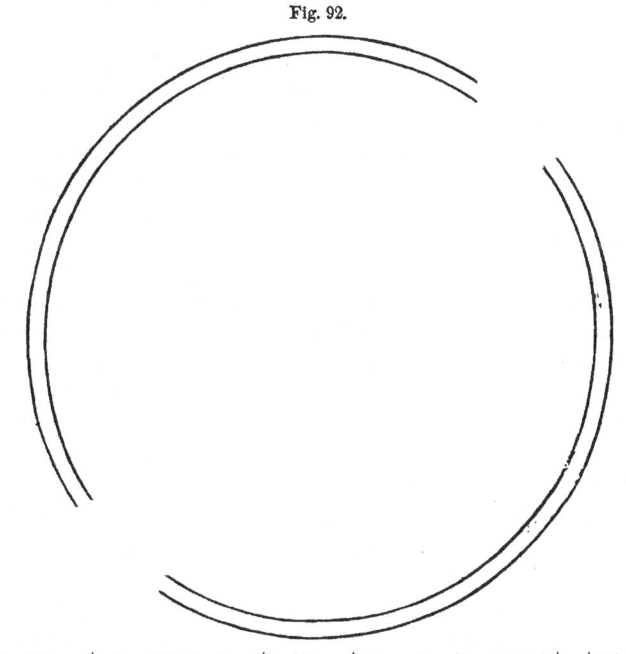

Section of the earth, in which the breadth of the outer boundary line represents a thickness of 25 miles; the space between the circles, including the breadth of the lines, 200 miles.

substance. The temperature of melted iron was estimated at 21,000° F., by Wedgwood; but his pyrometer gives, as is now demonstrated, very erroneous results. Professor Daniell ascertained that the point of fusion is 2786° F.†

According to Mr. Daniell's scale, we ought to encounter the internal

* Phil. Mag. and Ann. Feb. 1830.

† The heat was measured in Wedgwood's pyrometer by the contraction of pure clay, which is reduced in volume when heated, first by the loss of its water of combination, and afterwards, on the application of more intense heat, by incipient vitrification. The expansion of platina is the test employed by Mr. Daniell in his pyrometer, and this has been found to yield uniform and constant results, such as are in perfect harmony with conclusions drawn from various other independent sources. The instrument for which the author received the Rumford Medal from the Royal Society, in 1833, is described in the Phil. Trans. 1830, part ii., and 1831, part ii.

melted matter before penetrating through a thickness represented by that of the outer circular line in the annexed diagram (fig. 92); whereas, if the other or less correct scale be adopted, we should meet with it at some point between the two circles; the space between them, together with the lines themselves, representing a crust of 200 miles in depth. In either case, we must be prepared to maintain that a temperature many times greater than that sufficient to melt the most refractory substances known to us, is sustained at the centre of the globe; while a comparatively thin crust, resting upon the fluid, remains unmelted; or is even, according to M. Cordier, increasing in thickness, by the continual addition of new internal layers solidified during the process of refrigeration.

The mathematical calculations of Fourier, on the passage of heat through conducting bodies, have been since appealed to in support of these views; for he has shown that it is compatible with theory that the present temperature of the surface might coexist with an intense heat at a certain depth below. But his reasoning seems to be confined to the conduction of heat through solid bodies; and the conditions of the problem are wholly altered when we reason about a fluid nucleus, as we must do if it be assumed that the heat augments from the surface to the interior, according to the rate observed in mines. For when the heat of the lower portion of a fluid is increased, a circulation begins throughout the mass, by the ascent of hotter, and the descent of colder currents. And this circulation, which is quite distinct from the mode in which heat is propagated through solid bodies, must evidently occur in the supposed central ocean, if the laws of fluids and of heat are the same there as upon the surface.

In Mr. Daniell's experiments for obtaining a measure of the heat of bodies at their point of fusion, he invariably found that it was impossible to raise the heat of a large crucible of melted iron, gold, or silver, a single degree beyond the melting point, so long as a bar of the respective metals was kept immersed in the fluid portions. So in regard to other substances, however great the quantities fused, their temperature could not be raised while any solid pieces immersed in them remained unmelted; every accession of heat being instantly absorbed during their liquefaction. These results are, in fact, no more than the extension of a principle previously established, that so long as a fragment of ice remains in water, we cannot raise the temperature of the water above 32° F.

If, then, the heat of the earth's centre amount to 450,000° F., as M. Cordier deems highly probable, that is to say, about twenty times the heat of melted iron, even according to Wedgwood's scale, and upwards of 160 times according to the improved pyrometer, it is clear that the upper parts of the fluid mass could not long have a temperature only just sufficient to melt rocks. There must be a continual tendency towards a uniform heat; and until this were accomplished, by the interchange of portions of fluid of different densities, the surface could not begin to consolidate. Nor, on the hypothesis of primitive fluidity, can we conceive any crust to have been formed until *the whole* planet had cooled down to about the temperature of incipient fusion.

It cannot be objected that hydrostatic pressure would prevent a tendency to equalization of temperature; for, as far as observations have yet been made, it is found that the waters of deep lakes and seas are governed by the same laws as a shallow pool; and no experiments indicate that solids resist fusion under high pressure. The arguments, indeed, now controverted, always proceed on the admission that the internal nucleus is in a state of fusion.

It may be said that we may stand upon the hardened surface of a lava-current while it is still in motion,—nay, may descend into the crater of Vesuvius after an eruption, and stand on the scoriæ while every crevice shows that the rock is red-hot two or three feet below us; and at a somewhat greater depth, all is, perhaps, in a state of fusion. May not, then, a much more intense heat be expected at the depth of several hundred yards, or miles? The answer is,—that until a great quantity of heat has been given off, either by the emission of lava, or in a latent form by the evolution of steam and gas, the melted matter continues to boil in the crater of a volcano. But ebullition ceases when there is no longer a sufficient supply of heat from below, and then a crust of lava may form on the top, and showers of scoriæ may then descend upon the surface, and remain unmelted. If the internal heat be raised again, ebullition will recommence, and soon fuse the superficial crust. So in the case of the moving current, we may safely assume that no part of the liquid beneath the hardened surface is much above the temperature sufficient to retain it in a state of fluidity.

It may assist us in forming a clearer view of the doctrine now controverted, if we consider what would happen were a globe of homogeneous composition placed under circumstances analogous, in regard to the distribution of heat, to those above stated. If the whole planet, for example, were composed of water covered with a spheroidal crust of ice fifty miles thick, and with an interior ocean having a central heat about two hundred times that of the melting point of ice, or 6400° F.; and if, between the surface and the centre, there was every intermediate degree of temperature between that of melting ice and that of the central nucleus;—could such a state of things last for a moment? If it must be conceded, in this case, that the whole spheroid would be instantly in a state of violent ebullition, that the ice (instead of being strengthened annually by new internal layers) would soon melt, and form part of an atmosphere of steam—on what principle can it be maintained that analogous effects would not follow, in regard to the earth, under the conditions assumed in the theory of central heat?*

* The above remarks are reprinted verbatim from my third edition, May, 1834. A memoir was afterwards communicated by M. Poisson to the Academy of Sciences, January, 1837, on the solid parts of the globe, containing an epitome of a work entitled "Théorie Mathématique de la Chaleur," published in 1835. In this memoir he controverts the doctrine of the high temperature of a central fluid on

M. Cordier admits that there must be tides in the internal melted ocean; but their effect, he says, has become feeble, although originally, when the fluidity of the globe was perfect, "the rise and fall of these ancient land tides could not have been less than from thirteen to sixteen feet." Now, granting for a moment, that these tides have become so feeble as to be incapable of causing the fissured shell of the earth to be first uplifted and then depressed every six hours, still may we not ask whether, during eruptions, the lava, which is supposed to communicate with a great central ocean, would not rise and fall sensibly in a crater such as Stromboli, where there is always melted matter in a state of ebullition?

Whether chemical changes may produce volcanic heat.—Having now explained the reasons which have induced me to question the hypothesis of central heat as the primary source of volcanic action, it remains to consider what has been termed the chemical theory of volcanoes. It is well known that many, perhaps all, of the substances of which the earth is composed are continually undergoing chemical changes. To what depth these processes may be continued downwards must, in a great degree, be matter of conjecture; but there is no reason to suspect that, if we could descend to a great distance from the surface, we should find elementary substances differing essentially from those with which we are acquainted.

All the solid, fluid, and gaseous bodies known to us consist of a very small number of these elementary substances variously combined: the total number of elements at present known is less than sixty; and not half of these enter into the composition of the more abundant inorganic productions. Some portions of such compounds are daily undergoing decomposition, and their constituent parts being set free are passing into new combinations. These processes are by no means confined to minerals at the earth's surface, and are very often accompanied by the evolution of heat, which is intense in proportion to the rapidity of the combinations. At the same time there is a development of electricity.

The spontaneous combustion of beds of bituminous shale, and of refuse coal thrown out of mines, is generally due to the decomposition of pyrites; and it is the contact of air and water which brings about the change. Heat results from the oxidation of the sulphur and iron, though on what principle heat is generated, when two or more bodies having a strong affinity for each other unite suddenly, is wholly unexplained.

Electricity a source of volcanic heat.—It has already been stated, that chemical changes develop electricity; which, in its turn, becomes a powerful disturbing cause. As a chemical agent, says Davy, its silent and slow operation in the economy of nature is much more important than its grand and impressive operation in lightning and thunder. It may be considered, not only as directly producing an infinite variety of

similar grounds to those above stated. He imagines, that if the globe ever passed from a liquid to a solid state by radiation of heat, the central nucleus must have begun to cool and consolidate first.

changes, but as influencing almost all which take place; it would seem, indeed, that chemical attraction itself is only a peculiar form of the exhibition of electrical attraction.*

Now that it has been demonstrated that magnetism and electricity are always associated, and are perhaps only different conditions of the same power, the phenomena of terrestrial magnetism have become of no ordinary interest to the geologist. Soon after the first great discoveries of Oersted in electro-magnetism, Ampère suggested that all the phenomena of the magnetic needle might be explained by supposing currents of electricity to circulate constantly in the shell of the globe in directions parallel to the magnetic equator. This theory has acquired additional consistency the farther we have advanced in science; and according to the experiments of Mr. Fox, on the electro-magnetic properties of metalliferous veins, some trace of electric currents seems to have been detected in the interior of the earth.†

Some philosophers ascribe these currents to the chemical action going on in the superficial parts of the globe to which air and water have the readiest access; while others refer them, in part at least, to thermo-electricity excited by the solar rays on the surface of the earth during its rotation; successive parts of the atmosphere, land, and sea being exposed to the influence of the sun, and then cooled again in the night. That this idea is not a mere speculation, is proved by the correspondence of the diurnal variations of the magnet with the apparent motion of the sun; and by the greater amount of variation in summer than in winter, and during the day than in the night. M. de la Rive, although conceding that such minor variations of the needle may be due to thermo-electricity, contends that the general phenomena of terrestrial magnetism must be attributed to currents far more intense; which, though liable to secular fluctuations, act with much greater constancy and regularity than the causes which produce the diurnal variations.‡ The remark seems just; yet it is difficult to assign limits to the accumulated influence even of a very feeble force constantly acting on the whole surface of the earth. This subject, however, must evidently remain obscure, until we become acquainted with the causes which give a determinate direction to the supposed electric currents. Already the experiments of Faraday on the rotation of magnets have led him to speculate on the manner in which the earth, when once it had become magnetic, might produce electric currents within itself, in consequence of its diurnal rotation.§ We have seen also in a former chapter (p. 129) that the recent observations of Schwabe, 1852, have led Col. Sabine to the discovery of a connection between certain periodical changes, which take place in the spots on the sun, and a certain cycle of variations in terrestrial magnetism. These seem to point to the existence of a solar

* Consolations in Travel, p. 271. † Phil. Trans. 1830, p. 399.
‡ Biblioth. Univers. 1833, Electricité.
§ Phil. Trans. 1832, p. 176; also pp. 172, 173, &c.

magnetic period, and suggest the idea of the sun's magnetism exerting an influence on the mass of our planet.

In regard to thermo-electricity, I may remark, that it may be generated by great inequalities of temperature, arising from a partial distribution of volcanic heat. Wherever, for example, masses of rock occur of great horizontal extent, and of considerable depth, which are at one point in a state of fusion (as beneath some active volcano); at another, red-hot; and at a third, comparatively cold—strong thermo-electric action may be excited.

Some, perhaps, may object, that this is reasoning in a circle; first to introduce electricity as one of the primary causes of volcanic heat, and then to derive the same heat from thermo-electric currents. But there must, in truth, be much reciprocal action between the agents now under consideration; and it is very difficult to decide which should be regarded as the prime mover, or to see where the train of changes, once begun, would terminate. Whether subterranean electric currents if once excited might sometimes possess the decomposing power of the voltaic pile, is a question not perhaps easily answered in the present state of science; but such a power, if developed, would at once supply us with a never-failing source of chemical action from which volcanic heat might be derived.

Recapitulation.—Before entering, in the next chapter, still farther into the inquiry, how far the phenomena of volcanoes and earthquakes accord with the hypothesis of a continued generation of heat by chemical action, it may be desirable to recapitulate, in a few words, the conclusions already obtained.

1st. The primary causes of the volcano and the earthquake are, to a great extent, the same, and must be connected with the passage of heat from the interior to the surface.

2dly. This heat has been referred, by many, to a supposed state of igneous fusion of the central parts of the planet when it was first created, of which a part still remains in the interior, but is always diminishing in intensity.

3dly. The spheroidal figure of the earth, adduced in support of this theory, does not of necessity imply a universal and simultaneous fluidity, in the beginning; for supposing the original figure of our planet had been strictly spherical—which, however, is a gratuitous assumption, resting on no established analogy—still the statical figure must have been assumed, if sufficient time be allowed, by the gradual operation of the centrifugal force, acting on the materials brought successively within its action by aqueous and igneous causes.

4thly. It appears, from experiment, that the heat in mines increases progressively with their depth; and if the ratio of increase be continued uniformly from the surface to the interior, the whole globe, with the exception of a small external shell, must be fluid, and the central parts must have a temperature many times higher than that of melted iron.

5thly. But the theory adopted by M. Cordier and others, which maintains the actual existence of such a state of things, seems wholly inconsistent with the laws which regulate the circulation of heat through fluid bodies. For, if the central heat were as intense as is represented, there must be a circulation of currents, tending to equalize the temperature of the resulting fluids, and the solid crust itself would be melted.

6thly. Instead of an original central heat, we may, perhaps, refer the heat of the interior to chemical changes constantly going on in the earth's crust; for the general effect of chemical combination is the evolution of heat and electricity, which in their turn become sources of new chemical changes.

CHAPTER XXXII.

CAUSES OF EARTHQUAKES AND VOLCANOES—*continued.*

Review of the proofs of internal heat—Theory of an unoxidated metallic nucleus—Whether the decomposition of water may be a source of volcanic heat—Geysers of Iceland—Causes of earthquakes—Wavelike motion—Expansive power of liquid gases—Connection between the state of the atmosphere and earthquakes—Permanent upheaval and subsidence of land—Expansion of rocks by heat—The balance of dry land how preserved—Subsidence in excess—Conclusion.

When we reflect that the largest mountains are but insignificant protuberances upon the surface of the earth, and that these mountains are nevertheless composed of different parts which have been formed in succession, we may well feel surprise that the central fluidity of the planet should have been called in to account for volcanic phenomena. To suppose the entire globe to be in a state of igneous fusion, with the exception of a solid shell, not more than from thirty to one hundred miles thick, and to imagine that the central heat of this fluid spheroid exceeds by more than two hundred times that of liquid lava, is to introduce a force altogether disproportionate to the effects which it is required to explain.

The ordinary repose of the surface implies, on the contrary, an inertness in the internal mass which is truly wonderful. When we consider the combustible nature of the elements of the earth, so far as they are known to us,—the facility with which their compounds may be decomposed and made to enter into new combinations,—the quantity of heat which they evolve during these processes; when we recollect the expansive power of steam, and that water itself is composed of two gases which, by their union, produce intense heat; when we call to mind the number of explosive and detonating compounds which have been already discovered, we may be allowed to share the astonishment of Pliny, that a single day should pass without a general conflagration:—"Excedit profectò omnia miracula, ullum diem fuisse quo non cuncta conflagrarent."*

The signs of internal heat observable on the surface of the earth do not necessarily indicate the permanent existence of subterranean heated masses, whether fluid or solid, by any means so vast as our continents and seas; yet how insignificant would these appear if distributed through an external shell of the globe one or two hundred miles in depth! The principal facts in proof of the accumulation of heat below the surface may be summed up in a few words. Several volcanoes are constantly in eruption, as Stromboli and Nicaragua; others are known to have been active for periods of 60, or even 150 years, as those of Sangay in Quito, Popocatepetl in Mexico, and the volcano of the Isle of Bourbon. Many craters emit hot vapors in the intervals between eruptions, and solfataras evolve incessantly the same gases as volcanoes. Steam of high temperature has continued for more than twenty centuries to issue from the "stufas," as the Italians call them; thermal springs abound not only in regions of earthquakes, but are found in almost all countries, however distant from active vents; and, lastly, the temperature in the mines of various parts of the world is found to increase in proportion as we descend.

The diagram (fig. 93) in the next page, may convey some idea of the proportion which our continents and the ocean bear to the radius of the earth.† If all the land were about as high as the Himalaya mountains, and the ocean everywhere as deep as the Pacific, the whole of both might be contained within a space expressed by the thickness of the line *a b*; and masses of nearly equal volume might be placed in the space marked by the line *c d*, in the interior. Seas of lava, therefore, of the size of the Mediterranean, or even of the Atlantic, would be as nothing if distributed through such an outer shell of the globe as is represented by the shaded portion of the figure *a b c d*. If throughout that space we imagine electro-chemical causes to be continually in operation, even of very feeble power, they might give rise to heat which, if accumulated at certain points, might melt or render red-hot entire mountains, or sustain the temperature of stufas and hot springs for ages.

Theory of an unoxidated metallic nucleus.—When Sir H. Davy first discovered the metallic basis of the earths and alkalies, he threw out the idea that those metals might abound in an unoxidized state in the subterranean regions to which water must occasionally penetrate. Whenever this happened, gaseous matter would be set free, the metals would combine with the oxygen of the water, and sufficient heat might be evolved to melt the surrounding rocks. This hypothesis, although afterwards abandoned by its author, was at first very favorably received both by the chemist and the geologist: for silica, alumina, lime, soda, and oxide of iron,—substances of which lavas are principally composed,—

* Hist. Mundi, lib. ii. c. 107.

† Reduced, by permission, from a figure in plate 40 of Sir H. De la Beche's Geological Sections and Views.

would all result from the contact of the inflammable metals alluded to with water. But whence this abundant store of unsaturated metals in the interior? It was assumed that, in the beginning of things, the

Fig. 93.

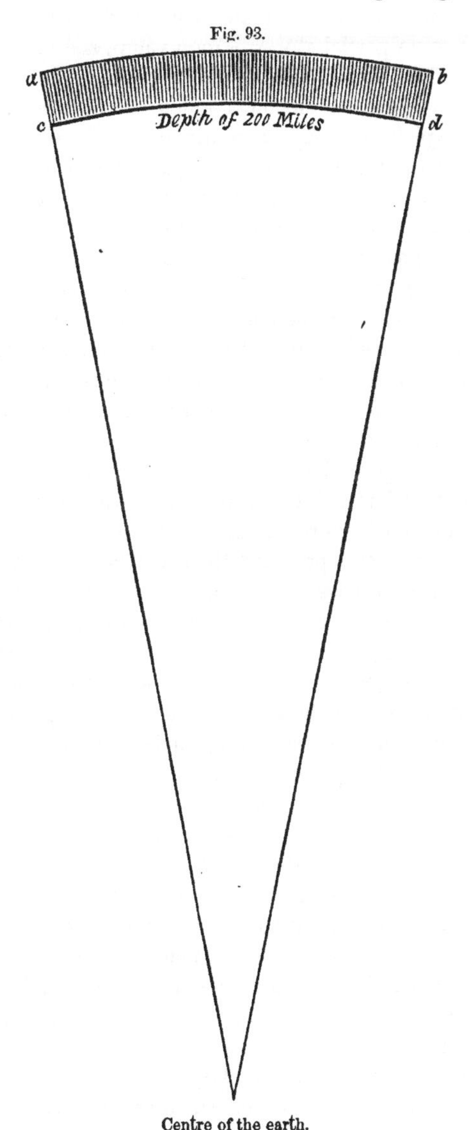

Centre of the earth.

nucleus of the earth was mainly composed of inflammable metals, and that oxidation went on with intense energy at first; till at length, when a superficial crust of oxides had been formed, the chemical action became more and more languid.

This speculation, like all others respecting the primitive state of the earth's nucleus, rests unavoidably on arbitrary assumptions. But we may fairly inquire whether any existing causes may have the power of deoxidating the earthy and alkaline compounds formed from time to time by the action of water upon the metallic bases. If so, and if the original crust or nucleus of the planet contained distributed through it here and there some partial stores of potassium, sodium, and other metallic bases, these might be oxidated and again deoxidated, so as to sustain for ages a permanent chemical action. Yet even then we should be unable to explain why such a continuous circle of operations, after having been kept up for thousands of years in one district, should entirely cease, and why another region, which had enjoyed a respite from volcanic action for one or many geological periods, should become a theatre for the development of subterranean heat.

It is well known to chemists, that the metallization of oxides, the most difficult to reduce, may be effected by hydrogen brought into contact with them at a red heat; and it is more than probable that the production of potassium itself, in the common gun-barrel process, is due to the power of nascent hydrogen derived from the water which the hydrated oxide contains. According to the recent experiments, also, of Faraday, it would appear that every case of metallic reduction by voltaic agency, from saline solutions, in which water is present, is due to the secondary action of hydrogen upon the oxide; both of these being determined to the negative pole and then reacting upon one another.

It is admitted that intense heat would be produced by the occasional contact of water with the metallic bases; and it is certain that, during the process of saturation, vast volumes of hydrogen must be evolved. The hydrogen, thus generated, might permeate the crust of the earth in different directions, and become stored up for ages in fissures and caverns, sometimes in a liquid form, under the necessary pressure. Whenever, at any subsequent period, in consequence of the changes effected by earthquakes in the shell of the earth, this gas happened to come in contact with metallic oxides at a high temperature, the reduction of these oxides might be the result.

No theory seems at first more startling than that which represents water as affording an inexhaustible supply of fuel to the volcanic fires; yet is it by no means visionary. It is a fact that must not be overlooked, that while a great number of volcanoes are entirely submarine, the remainder occur for the most part in islands or maritime tracts. There are a few exceptions; but some of these, observes Dr. Daubeny, are near inland salt lakes, as in Central Tartary; while others form part of a train of volcanoes, the extremities of which are near the sea.

Sir H. Davy suggested that, when the sea is distant, as in the case of some of the South American volcanoes, they may still be supplied with water from subterranean lakes; since, according to Humboldt, large quantities of fish are often thrown out during eruptions.* Mr. Dana also, in his valuable and original observations on the volcanoes of the

* Phil. Trans. 1828, p. 250.

Sandwich Islands, reminds us of the prodigious volume of atmospheric water which must be absorbed into the interior of such large and lofty domes, composed as they are entirely of porous lava. To this source alone he refers the production of the steam by which the melted matter is propelled upwards, even to the summit of cones three miles in height.*

When treating of springs and overflowing wells, I have stated that porous rocks are percolated by fresh water to great depths, and that sea-water probably penetrates in the same manner through the rocks which form the bed of the ocean. But, besides this universal circulation in regions not far from the surface, it must be supposed that, wherever earthquakes prevail, much larger bodies of water will be forced by the pressure of the ocean into fissures at great depths, or swallowed up in chasms; in the same manner as on the land, towns, houses, cattle, and trees are sometimes engulfed. It will be remembered, that these chasms often close again after houses have fallen into them; and for the same reason, when water has penetrated to a mass of melted lava, the steam into which it is converted may often rush out at a different aperture from that by which the water entered.

The gases, it is said, exhaled from volcanoes, together with steam, are such as would result from the decomposition of salt water, and the fumes which escape from the Vesuvian lava have been observed to deposit common salt.† The emission of free muriatic acid gas in great quantities is also thought by many to favor the theory of the decomposition of the salt contained in sea-water. It has been objected, however, that M. Boussingault did not meet with this gas in his examination of the elastic fluids evolved from the volcanoes of equatorial America; which only give out aqueous vapor (in very large quantity), carbonic acid gas, sulphurous acid gas, and sometimes fumes of sulphur.‡ In reply, Dr. Daubeny has remarked, that muriatic acid may have ceased to be disengaged, because the volcanic action has become languid in equatorial America, and sea-water may no longer obtain admission.

M. Gay Lussac, while he avows his opinion that the decomposition of water contributes largely to volcanic action, called attention, nevertheless, to the supposed fact, that hydrogen had not been detected in a separate form among the gaseous products of volcanoes; nor can it, he says, be present; for, in that case, it would be inflamed in the air by the red-hot stones thrown out during an eruption. Dr. Davy, in his account of Graham Island, says, "I watched when the lightning was most vivid, and the eruption of the greatest degree of violence, to see if there was any inflammation occasioned by this natural electric spark—any indication of the presence of inflammable gas; but in vain."§

May not the hydrogen, Gay Lussac inquires, be combined with chlorine, and produce muriatic acid? for this gas has been observed to be evolved from Vesuvius—and the chlorine may have been derived from

* Geology of American Exploring Expedition, p. 369.
† Davy, Phil. Trans. 1828, p. 244.
‡ Ann. de Chim. et de Phys. tom. iii. p. 181.
§ Phil. Trans. 1832, p. 240.

sea salt; which was, in fact, extracted by simple washing from the Vesuvian lava of 1822, in the proportion of nine per cent.* But it was answered, that Sir H. Davy's experiments had shown, that hydrogen is not combustible when mixed with muriatic acid gas; so that if muriatic gas was evolved in large quantities, the hydrogen might be present without inflammation.† M. Abich, on the other hand, assures us, "that although it be true that vapor illuminated by incandescent lava has often been mistaken for flame," yet he clearly detected in the eruption of Vesuvius in 1834 the flame of hydrogen.‡

M. Gay Lussac, in the memoir just alluded to, expressed doubt as to the presence of sulphurous acid; but the abundant disengagement of this gas during eruptions has been since ascertained: and thus all difficulty in regard to the general absence of hydrogen in an inflammable state is removed; for, as Dr. Daubeny suggests, the hydrogen of decomposed water may unite with sulphur to form sulphuretted hydrogen gas, and this gas will then be mingled with the sulphurous acid as it rises to the crater. It is shown by experiment, that these gases mutually decompose each other when mixed where steam is present; the hydrogen of the one immediately uniting with the oxygen of the other to form water, while the excess of sulphurous acid alone escapes into the atmosphere. Sulphur is at the same time precipitated.

This explanation is sufficient; but it may also be observed that the flame of hydrogen would rarely be visible during an eruption; as that gas, when inflamed in a pure state, burns with a very faint blue flame, which even in the night could hardly be perceptible by the side of red-hot and incandescent cinders. Its immediate conversion into water when inflamed in the atmosphere, might also account for its not appearing in a separate form.

Dr. Daubeny is of opinion that water containing atmospheric air may descend from the surface of the earth to the volcanic foci, and that the same process of combustion by which water is decomposed may deprive such subterranean air of its oxygen. In this manner he explains the great quantities of nitrogen evolved from volcanic vents and thermal waters, and the fact that air disengaged from the earth in volcanic regions is either wholly or in part deprived of its oxygen.

Sir H. Davy, in his memoir on the "Phenomena of Volcanoes," remarks, that there was every reason to suppose in Vesuvius the existence of a descending current of air; and he imagined that subterranean cavities which threw out large volumes of steam during the eruption, might afterwards, in the quiet state of the volcano, become filled with atmospheric air.§ The presence of ammoniacal salts in volcanic emanations, and of ammonia (which is in part composed of nitrogen) in lava, favors greatly the notion of air as well as water being deoxidated in the interior of the earth.‖

* Ann. de Chim. et de Phys. tom. xxii.
† Quart. Journ. of Sci. 1823, p. 132, note by editor.
‡ Phénom. Géol. &c. p. 3.
§ Phil. Trans. 1828.
‖ See Daubeny, Encyc. Metrop. part 40.

It has been alleged by Professor Bischoff that the slight specific gravity of the metals of the alkalies is fatal to Davy's hypothesis, for if the mean density of the earth, as determined by astronomers, surpass that of all kinds of rocks, these metals cannot exist, at least not in great quantities in the interior of the earth.* But Dr. Daubeny has shown, that if we take the united specific gravity of potassium, sodium, silicon, iron, and all the materials which, when united with oxygen, constitute ordinary lava, and then compare their weight with lava of equal bulk, the difference is not very material, the specific gravity of the lava only exceeding by about one-fourth that of the unoxidized metals. Besides, at great depths, the metallic bases of the earths and alkalies may very probably be rendered heavier by pressure.† Nor is it fair to embarrass the chemical theory of volcanoes with a doctrine so purely gratuitous, as that which supposes the entire nucleus of the planet to have been at first composed of unoxidated metals.

Professor Bunsen of Marburg, after analyzing the gases which escape from the volcanic fumeroles and solfataras of Iceland, and after calculating the quantity of hydrogen evolved between two eruptions, affirms, in contradiction of opinions previously entertained, that the hydrogen bears a perfect relation in quantity to the magnitude of the streams of lava, assuming the fusion of these last to have been the result of the heat evolved during the oxidation of alkaline and earthy metals, and this to have been brought about by the decomposition of water. Yet after having thus succeeded in removing the principal objection once so triumphantly urged against Davy's hypothesis, Bunsen concludes by declaring that the hydrogen evolved in volcanic regions cannot have been generated by the decomposition of water coming in contact with alkaline and earthy metallic bases. For, says the Professor, this process presupposes the prevalence of a temperature in which carbonic acid cannot exist in contact with hydrogen without suffering a partial reduction to carbonic oxide; "and not a trace of carbonic oxide is ever found in volcanic exhalations."‡ At the same time it will be seen, by consulting the able memoirs of the Marburg chemist, that he supposes many energetic kinds of chemical action to be continually going on in the interior of the earth, capable of causing the disengagement of hydrogen; and there can be no doubt that this gas may be a source of innumerable new changes, capable of producing the local development of internal heat.

Cause of volcanic eruptions.—The most probable causes of a volcanic outburst at the surface have been in a great degree anticipated in the preceding speculations on the liquefaction of rocks and the generation of gases. When a minute hole is bored in a tube filled with gas condensed into a liquid, the whole becomes instantly aeriform, or, as some writers have expressed it, "flashes into vapor," and often bursts the tube. Such

* Jam. Ed. New Phil. Journ. No. li. p. 31.

† See Daubeny's Reply to Bischoff, Jam. Ed. New Phil. Journ. No. lii. p. 291; and note in No. liii. p. 158.

‡ Poggend. Ann. 1851 translated, Sci. Mem. 1852.

an experiment may represent the mode in which gaseous matter may rush through a rent in the rocks, and continue to escape for days or weeks through a small orifice, with an explosive power sufficient to reduce every substance which opposes its passage into small fragments or even dust. Lava may be propelled upwards at the same time, and ejected in the form of scoriæ. In some places, where the fluid lava lies at the bottom of a deep fissure, communicating on the one hand with the surface, and on the other with a cavern in which a considerable body of vapor has been formed, there may be an efflux of lava, followed by the escape of gas. Eruptions often commence and close with the discharge of vapor; and, when this is the case, the next outburst may be expected to take place by the same vent, for the concluding evolution of elastic fluids will keep open the duct, and leave it unobstructed.

The breaking out of lava from the side or base of a lofty cone, rather than from the summit, may be attributed to the hydrostatic pressure to which the flanks of the mountain are exposed, when the column of lava has risen to a great height. Or if, before it has reached the top, there should happen to be any stoppage in the main duct, the upward pressure of the ascending column of gas and lava may burst a lateral opening.

In the case however of Mount Loa, in the Sandwich Islands, there appears to be a singular want of connection or sympathy between the eruptions of the central and the great lateral vent. The great volcanic cone alluded to rises to the height of 13,760 feet above the level of the sea, having a crater at its summit, from which powerful streams of lava have flowed in recent times, and having another still larger crater, called Kilauea, on its southeastern slope, about 4000 feet above the sea. This lateral cavity resembles a huge quarry cut in the mountain's side, being about 1000 feet deep when in its ordinary state. It is seven miles and a half in circuit, and scattered over its bottom, at different levels, are lakes and pools of lava, always in a state of ebullition. The liquid in one of these will sometimes sink 100 or 150 feet, while it is overflowing in another at a higher elevation, there being, it should seem, no communication between them. In like manner, lava overflows in the summit crater of Mount Loa, nearly 14,000 feet high, while the great lateral cauldron just alluded to (of Kilauea) continues as tranquil as usual, affording no relief to any part of the gases or melted matter which are forcing their way upwards in the centre of the mountain. "How," asks Mr. Dana, "if there were any subterranean channel connecting the two great vents, could this want of sympathy exist? How, according to the laws of hydrostatic pressure, can a column of fluid stand 10,000 feet higher in one leg of the siphon than in the other?" The eruptions, he observes, are not paroxysmal; on the contrary, the lava rises slowly and gradually to the summit of the lofty cone, and then escapes there without any commotion manifesting itself in Kilauea, a gulf always open on the flanks of the same mountain. One conclusion, he says, is certain, namely, *that volcanoes are no safety valves* as they have been called; for here two independent and apparently isolated centres of volcanic activity, only sixteen miles distant from each other, are sustained in one and the same cone.*

Without pretending to solve this enigma, I cannot refrain from remarking, that the supposed independence of several orifices of eruption in one crater like Kilauea, when adduced in confirmation of the doctrine of two distinct sources of volcanic action underneath one mountain, proves too much. No one can doubt, that the pools of lava in Kilauea have been derived from some common reservoir, and have resulted from a combination of causes commonly called volcanic, which are at work in the interior at some unknown distance below. These causes have given rise in Mount Loa to eruptions from many points, but principally from one centre, so that a vast dome of ejected matter has been piled up. The subsidiary crater has evidently never given much relief to the imprisoned, heated, and liquefied matter, for Kilauea does not form a lateral protuberance interfering with the general shape or uniform outline of Mount Loa.

Geysers of Iceland.—As aqueous vapor constitutes the most abundant of the aeriform products of volcanoes in eruption, it may be well to consider attentively a case in which steam is exclusively the moving power —that of the Geysers of Iceland. These intermittent hot springs occur in a district situated in the southwestern division of Iceland, where nearly one hundred of them are said to break out within a circle of two miles. That the water is of atmospheric origin, derived from rain and melted snow, is proved, says Professor Bunsen, by the nitrogen which rises from them either pure or mixed with other gases. The springs rise through a thick current of lava, which may perhaps have flowed from Mount Hecla, the summit of that volcano being seen from the spot at the distance of more than thirty miles. In this district the rushing of water is sometimes heard in chasms beneath the surface; for here, as on Etna, rivers flow in subterranean channels through the porous and cavernous lavas. It has more than once happened, after earthquakes, that some of the boiling fountains have increased or diminished in violence and volume, or entirely ceased, or that new ones have made their appearance—changes which may be explained by the opening of new rents and the closing of pre-existing fissures.

Few of the Geysers play longer than five or six minutes at a time, although sometimes half an hour. The intervals between their eruptions are for the most part very irregular. The Great Geyser rises out of a spacious basin at the summit of a circular mound composed of siliceous incrustations deposited from the spray of its waters. The diameter of this basin, in one direction, is fifty-six feet, and forty-six in another. (See fig. 94.) In the centre is a pipe seventy-eight feet in perpendicular depth, and from eight to ten feet in diameter, but gradually widening, as it rises into the basin. The inside of the basin is whitish, consisting of a siliceous crust, and perfectly smooth, as are likewise two small

* Proceed. Americ. Assoc. 1849.

channels on the sides of the mound, down which the water escapes when the bowl is filled to the margin. The circular basin is sometimes empty, as represented in the following sketch; but is usually filled with beautifully transparent water in a state of ebullition. During the rise of the boiling water in the pipe, especially when the ebullition is most violent, and when the water is thrown up in jets, subterranean noises are heard, like the distant firing of cannon, and the earth is slightly shaken. The sound then increases and the motion becomes more violent,

Fig. 94.

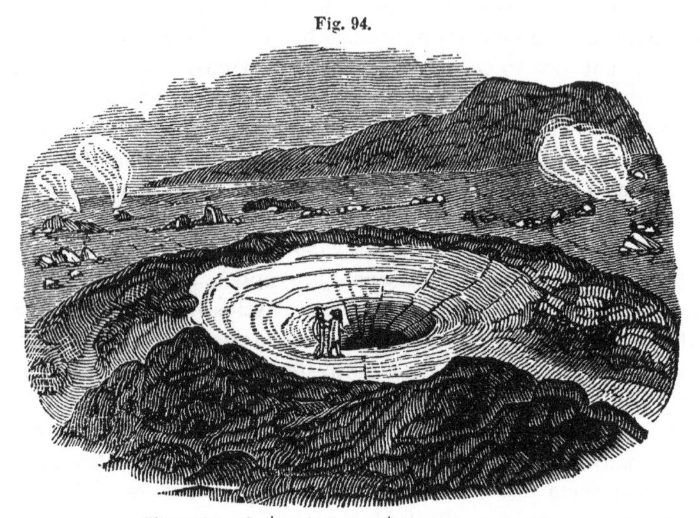

View of the Crater of the Great Geyser in Iceland.*

till at length a column of water is thrown up, with loud explosions, to the height of one or two hundred feet. After playing for a time like an artificial fountain, and giving off great clouds of vapor, the pipe or tube is emptied; and a column of steam, rushing up with amazing force and a thundering noise, terminates the eruption.

If stones are thrown into the crater, they are instantly ejected; and such is the explosive force, that very hard rocks are sometimes shivered by it into small pieces. Henderson found that by throwing a great quantity of large stones into the pipe of Strockr, one of the Geysers, he could bring on an eruption in a few minutes.† The fragments of stone, as well as the boiling water, were thrown in that case to a much greater height than usual. After the water had been ejected, a column of steam continued to rush up with a deafening roar for nearly an hour; but the Geyser, as if exhausted by this effort, did not send out a fresh eruption when its usual interval of rest had elapsed. The account given by Sir George Mackenzie of a Geyser which he saw in eruption in 1810 (see fig. 95), agrees perfectly with the above description by Henderson. The

* Reduced from a sketch given by Sir W. J. Hooker, in his Tour in Iceland, vol. i. p. 149.

† Journal of a Residence in Iceland, p. 74.

steam and water rose for half an hour to the height of 70 feet, and the white column remained perpendicular notwithstanding a brisk gale of wind which was blowing against it. Stones thrown into the pipe were projected to a greater height than the water. To leeward of the vapor a heavy shower of rain was seen to fall.*

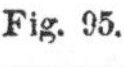
Fig. 95.

Eruption of the New Geyser in 1810. (Mackenzie.)

Among the different theories proposed to account for these phenomena, I shall first mention one suggested by Sir. J. Herschel. An imitation of these jets, he says, may be produced on a small scale, by heating red hot the stem of a tobacco pipe, filling the bowl with water, and so inclining the pipe as to let the water run through the stem. Its escape, instead of taking place in a continued stream, is then performed by a succession of violent explosions, at first of steam alone, then of water mixed with steam; and, as the pipe cools, almost wholly of water. At every such paroxysmal escape of the water, a portion is driven back, accompanied with steam, into the bowl. The intervals between the explosions depend on the heat, length, and inclination of the pipe; their continuance, on its thickness and conducting power.† The application of

* Mackenzie's Iceland.

† MS. read to Geol. Soc. of London, Feb. 29, 1832.

this experiment to the Geysers merely requires that a subterranean stream, flowing through the pores and crevices of lava, should suddenly reach a fissure in which the rock is red hot or nearly so. Steam would immediately be formed, which, rushing up the fissure, might force up water along with it to the surface, while, at the same time, part of the steam might drive back the water of the supply for a certain distance towards its source. And when, after the space of some minutes, the steam was all condensed, the water would return, and a repetition of the phenomena take place.

There is, however, another mode of explaining the action of the Geyser, perhaps more probable than that above described. Suppose water percolating from the surface of the earth to penetrate into the subterranean cavity A D (fig. 96) by the fissures F F, while, at the same time,

Fig. 96.

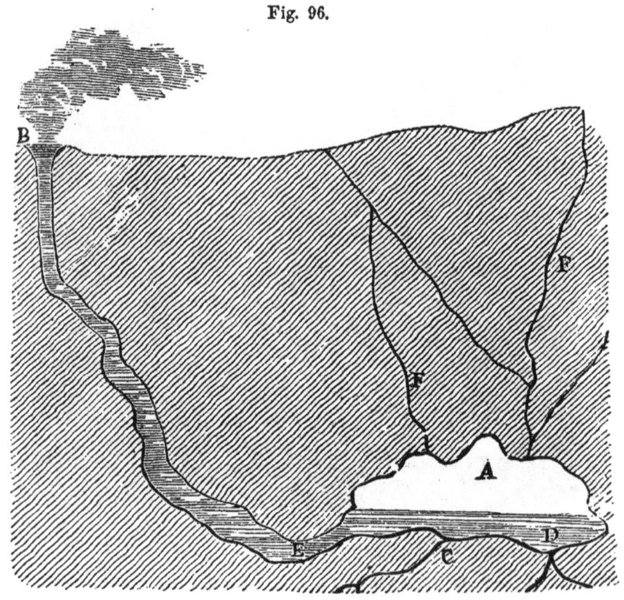

Supposed reservoir and pipe of a Geyser in Iceland.*

steam at an extremely high temperature, such as is commonly given out from the rents of lava currents during congelation, emanates from the fissures C. A portion of the steam is at first condensed into water, while the temperature of the water is raised by the latent heat thus evolved, till, at last, the lower part of the cavity is filled with boiling water and the upper with steam under high pressure. The expansive force of the steam becomes, at length, so great, that the water is forced up the fissure or pipe E B, and runs over the rim of the basin. When the pressure is thus diminished, the steam in the upper part of the cavity A expands, until all the water D is driven into the pipe; and when this

* From Sir George Mackenzie's Iceland.

happens, the steam, being the lighter of the two fluids, rushes up through the water with great velocity. If the pipe be choked up artificially, even for a few minutes, a great increase of heat must take place; for it is prevented from escaping in a latent form in steam; so that the water is made to boil more violently, and this brings on an eruption.

Professor Bunsen, before cited, adopts this theory to account for the play of the "Little Geyser," but says it will not explain the phenomena of the Great one. He considers this, like the others, to be a thermal spring, having a narrow funnel-shaped tube in the upper part of its course, where the walls of the channel have become coated over with siliceous incrustations. At the mouth of this tube the water has a temperature, corresponding to the pressure of the atmosphere, of about 212° Fahr., but at a certain depth below it is much hotter. This the professor succeeded in proving by experiment; a thermometer suspended by a string in the pipe rising to 266° Fahr., or no less than 48 degrees above the boiling point. After the column of water has been expelled, what remains in the basin and pipe is found to be much cooled.

Previously to these experiments of Bunsen and Descloizeaux, made in Iceland in 1846, it would scarcely have been supposed possible that the lower part of a free and open column of water could be raised so much in temperature without causing a circulation of ascending and descending currents, followed by an almost immediate equalization of heat. Such circulation is no doubt impeded greatly by the sides of the well not being vertical, and by numerous contractions of its diameter, but the phenomenon may be chiefly due to another cause. According to recent experiments on the cohesion of liquids by Mr. Donny of Ghent, it appears that when water is freed from all admixture of air, its temperature can be raised, even under ordinary atmospheric pressure, to 275° Fahr., so much does the cohesion of its molecules increase* when they are not separated by particles of air. As water long boiled becomes more and more deprived of air, it is probably very free from such intermixture at the bottom of the Geysers.

Among other results of the experiments of Bunsen and his companion, they convinced themselves that the column of fluid filling the tube is constantly receiving accessions of hot water from below, while it becomes cooler above by evaporation on the broad surface of the basin. They also came to a conclusion of no small interest, as bearing on the probable mechanism of ordinary volcanic eruptions, namely that the tube itself is the main seat or focus of mechanical force. This was proved by letting down stones suspended by strings to various depths. Those which were sunk to considerable distances from the surface were not cast up again, whereas those nearer the mouth of the tube were ejected to great heights. Other experiments also were made tending to demonstrate the singular fact, that there is often scarce any motion below, when a violent rush of steam and water is taking place above. It seems

* See Mr. Horner's Anniversary Address, Quart. Journ. Geol. Soc. 1847, liii.

that when a lofty column of water possesses a temperature increasing with the depth, any slight ebullition or disturbance of equilibrium in the upper portion may first force up water into the basin, and then cause it to flow over the edge. A lower portion, thus suddenly relieved of part of its pressure, expands and is converted into vapor more rapidly than the first, owing to its greater heat. This allows the next subjacent stratum, which is much hotter, to rise and flash into a gaseous form; and this process goes on till the ebullition has descended from the middle to near the bottom of the funnel.*

In speculating, therefore, on the mechanism of an ordinary volcanic eruption, we may suppose that large subterranean cavities exist at the depth of some miles below the surface of the earth, in which melted lava accumulates; and when water containing the usual mixture of air penetrates into these, the steam thus generated may press upon the lava and force it up the duct of a volcano, in the same manner as a column of water is driven up the pipe of a Geyser. In other cases we may suppose a continuous column of liquid lava mixed with *red-hot water* (for water may exist in that state, as Professor Bunsen reminds us, under pressure), and this column may have a temperature regularly increasing downwards. A disturbance of equilibrium may first bring on an eruption near the surface, by the expansion and conversion into gas of entangled water and other constituents of what we call lava, so as to occasion a diminution of pressure. More steam would then be liberated, carrying up with it jets of melted rock, which being hurled up into the air may fall in showers of ashes on the surrounding country, and at length, by the arrival of lava and water more and more heated at the orifice of the duct or the crater of the volcano, expansive power may be acquired sufficient to expel a massive current of lava. After the eruption has ceased, a period of tranquillity succeeds, during which fresh accessions of heat are communicated from below, and additional masses of rock fused by degrees, while at the same time atmospheric or sea water is descending from the surface. At length the conditions required for a new outburst are obtained, and another cycle of similar changes is renewed.

Causes of earthquakes—wave-like motion.—I shall now proceed to examine the manner in which the heat of the interior may give rise to earthquakes. One of the most common phenomena attending subterranean movements, is the undulatory motion of the ground. And this, says Michell, will seem less extraordinary, if we call to mind the extreme elasticity of the earth and the compressibility of even the most solid materials. Large districts, he suggests, may rest on fluid lava; and, when this is disturbed, its motions may be propagated through the incumbent rocks. He also adds the following ingenious speculation:—"As a small quantity of vapor almost instantly generated at some con-

* Liebig's Annalen der Chimie und Pharmacie, translated in "Reports and Memoirs" of Cavendish Soc. London, 1848.

siderable depth below the surface of the earth will produce a vibratory motion, so a very large quantity (whether it be generated almost instantly, or in any small portion of time) will produce a wave-like motion. The manner in which this wave-like motion will be propagated may, in some measure, be represented by the following experiment:— Suppose a large cloth, or carpet (spread upon a floor), to be raised at one edge, and then suddenly brought down again to the floor; the air under it, being by this means propelled, will pass along till it escapes at the opposite side, raising the cloth in a wave all the way as it goes. In like manner, a large quantity of vapor may be conceived to raise the earth in a wave, as it passes along between the strata, which it may easily separate in a horizontal direction, there being little or no cohesion between one stratum and another. The part of the earth that is first raised being bent from its natural form, will endeavor to restore itself by its elasticity; and the parts next to it being to have their weight supported by the vapor, which will insinuate itself under them, will be raised in their turn, till it either finds some vent, or is again condensed by the cold into water, and by that means prevented from proceeding any farther."* In a memoir published in 1843, on the structure of the Appalachian chain, by the Professors Rogers,† the following hypothesis is proposed as "simpler and more in accordance with dynamical considerations, and the recorded observations on earthquakes."—"In place," say they, "of supposing it possible for a body of vapor or gaseous matter to pass horizontally between the strata, or even between the crust and the fluid lava upon which it floats, and with which it must be closely entangled, we are inclined to attribute the movement to an *actual pulsation*, engendered in the *molten matter itself*, by a linear disruption under enormous tension, giving vent explosively to elastic vapors, escaping either to the surface, or into cavernous spaces beneath. According to this supposition, the movement of the subterranean vapors would be *towards*, and not from, the disrupted belt, and the oscillation of the crust would originate in the tremendous and sudden disturbance of the previous pressure on the surface of the lava mass below, brought about by the instantaneous and violent rending of the overlying strata."

This theory requires us to admit that the crust of the earth is so flexible, that it can assume the form, and follow the motion of an undulation in the fluid below. Even if we grant this, says Mr. Mallet, another more serious objection presents itself, viz. the great velocity attributed to the transit of the wave in the subterranean sea of lava. We are called upon to admit that the speed of the wave below equals that of the true earthquake shock at the surface, which is so immense, that it is not inferior to the velocity of sound in the same solids. But the undulation in the fluid below must follow the laws of a tidal wave, or of the great sea-wave already spoken of. "Its velocity, like that of the tidal wave

* On the Cause and Phenomena of Earthquakes, Phil. Trans. vol. li. sec. 58, 1760.

† Trans. of Assoc. of American Geol. 1840–1842, p. 520.

of our seas, will be a function of its length and of the depth of the fluid, diminished in this case by certain considerations as to the density and degree of viscidity of the liquid; and although it would be at present impossible, for want of data, to calculate the exact velocity with which this subterraneous lava-wave could move, it may be certainly affirmed that its velocity would be immeasurably short of the observed or theoretic velocity of the great earth-wave, or true shock in earthquakes."*

Liquid gases.—The rending and upheaving of continental masses are operations which are not difficult to explain, when we are once convinced that heat, of sufficient power, not only to melt but to reduce to a gaseous form a great variety of substances, is accumulated in certain parts of the interior. We see that elastic fluids are capable of projecting solid masses to immense heights in the air; and the volcano of Cotopaxi has been known to throw out, to the distance of eight or nine miles, a mass of rock about one hundred cubic yards in volume. When we observe these aeriform fluids rushing out from particular vents for months, or even years, continuously, what power may we not expect them to exert in other places, where they happen to be confined under an enormous weight of rock?

The experiments of Faraday and others have shown, within the last twelve years, that many of the gases, including all those which are most copiously disengaged from volcanic vents, as the carbonic, sulphurous, and muriatic acids, may be condensed into liquids by pressure. At temperatures of from 30° to 50° F., the pressure required for this purpose varies from fifteen to fifty atmospheres; and this amount of pressure we may regard as very insignificant in the operations of nature. A column of Vesuvian lava that would reach from the lip of the crater to the level of the sea, must be equal to about three hundred atmospheres; so that, at depths which may be termed moderate in the interior of the crust of the earth, the gases may be condensed into liquids, even at very high temperatures. The method employed to reduce some of these gases to a liquid state is, to confine the materials, from the mutual action of which they are evolved, in tubes hermetically sealed, so that the accumulated pressure of the vapor, as it rises and expands, may force some part of it to assume the liquid state. A similar process may, and indeed must, frequently take place in subterranean caverns and fissures, or even in the pores and cells of many rocks; by which means, a much greater store of expansive power may be *packed* into a small space than could happen if these vapors had not the property of becoming liquid. For, although the gas occupies much less room in a liquid state, yet it exerts exactly the same pressure upon the sides of the containing cavity as if it remained in the form of vapor.

If a tube, whether of glass or other materials, filled with condensed gas, have its temperature slightly raised, it will often burst; for a slight increment of heat causes the elasticity of the gas to increase in a very

* Mallet, p. 39.

high ratio. We have only to suppose certain rocks, permeated by these liquid gases (as porous strata are sometimes filled with water), to have their temperature raised some hundred degrees, and we obtain a power capable of lifting superincumbent masses of almost any conceivable thickness; while, if the depth at which the gas is confined be great, there is no reason to suppose that any other appearances would be witnessed by the inhabitants of the surface than vibratory movements and rents, from which no vapor might escape. In making their way through fissures a very few miles only in length, or in forcing a passage through soft yielding strata, the vapors may be cooled and absorbed by water. For water has a strong affinity to several of the gases, and will absorb large quantities, with a very slight increase of volume. In this manner, the heat or the volume of springs may be augmented, and their mineral properties made to vary.

Connection between the state of the atmosphere and earthquakes.—The inhabitants of Stromboli, who are mostly fishermen, are said to make use of that volcano as a weather-glass, the eruptions being comparatively feeble when the sky is serene, but increasing in turbulence during tempestuous weather, so that in winter the island often seems to shake from its foundations. Mr. P. Scrope, after calling attention to these and other analogous facts, first started the idea (as long ago as the year 1825) that the diminished pressure of the atmosphere, the concomitant of stormy weather, may modify the intensity of the volcanic action. He suggests that where liquid lava communicates with the surface, as in the crater of Stromboli, it may rise or fall in the vent on the same principle as mercury in a barometer; because the ebullition or expansive power of the steam contained in the lava would be checked by every increase, and augmented by every diminution of weight. In like manner, if a bed of liquid lava be confined at an immense depth below the surface, its expansive force may be counteracted partly by the weight of the incumbent rocks, and also in part by atmospheric pressure acting contemporaneously on a vast superficial area. In that case, if the upheaving force increase gradually in energy, it will at length be restrained by only the slightest degree of superiority in the antagonist or repressive power, and then the equilibrium may be suddenly destroyed by any cause, such as an ascending draught of air, which is capable of depressing the barometer. In this manner we may account for the remarkable coincidence so frequently observed between the state of the weather and subterranean commotions, although it must be admitted that earthquakes and volcanic eruptions react in their turn upon the atmosphere, so that disturbances of the latter are generally the cousequences rather than the forerunners of volcanic disturbances.*

From an elaborate catalogue of the earthquakes experienced in Europe and Syria during the last fifteen centuries, M. Alexis Perrey has deduced the conclusion that the number which happen in the winter

* Scrope on Volcanoes, pp. 58–60.

season preponderates over those which occur in any one of the other seasons of the year, there being, however, some exceptions to this rule, as in the Pyrenees. Curious and valuable as are these data, M. d'Archiac justly remarks, in commenting upon them, that they are not as yet sufficiently extensive or accordant in different regions, to entitle us to deduce any general conclusions from them respecting the laws of subterranean movements throughout the globe.*

Permanent elevation and subsidence.—It is easy to conceive that the shattered rocks may assume an arched form during a convulsion, so that the country above may remain permanently upheaved. In other cases gas may drive before it masses of liquid lava, which may thus be injected into newly opened fissures. The gas having then obtained more room, by the forcing up of the incumbent rocks, may remain at rest; while the lava congealing in the rents may afford a solid foundation for the newly raised district.

Experiments have recently been made in America, by Colonel Totten, to ascertain the ratio according to which some of the stones commonly used in architecture expand with given increments of heat.† It was found impossible, in a country where the annual variation of temperature was more than 90° F., to make a coping of stones, five feet in length, in which the joints should fit so tightly as not to admit water between the stone and the cement; the annual contraction and expansion of the stones causing, at the junctions, small crevices, the width of which varied with the nature of the rock. It was ascertained that fine-grained granite expanded with 1° F. at the rate of ·000004825; while crystalline marble ·000005668; and red sandstone ·000009532, or about twice as much as granite.

Now, according to this law of expansion, a mass of sandstone a mile in thickness, which should have its temperature raised 200 F., would lift a superimposed layer of rock to the height of ten feet above its former level. But, suppose a part of the earth's crust, one hundred miles in thickness and equally expansive, to have its temperature raised 600° or 800°, this might produce an elevation of between two and three thousand feet. The cooling of the same mass might afterwards cause the overlying rocks to sink down again and resume their original position. By such agency we might explain the gradual rise of Scandinavia or the subsidence of Greenland, if this last phenomenon should also be established as a fact on farther inquiry.

It is also possible that as the clay in Wedgwood's pyrometer contracts, by giving off its water, and then, by incipient vitrification; so, large masses of argillaceous strata on the earth's interior may shrink, when subjected to heat and chemical changes, and allow the incumbent rocks to subside gradually.

Moreover, if we suppose that lava cooling slowly at great depths

* Archiac, Hist. des Progrés de la Géol, 1847, vol. i. pp. 605–610.

† Silliman's American Journ. vol. xxii. p. 136. The application of these results to the theory of earthquakes was first suggested to me by Mr. Babbage.

may be converted into various granitic rocks, we obtain another source of depression; for, according to the experiments of Deville and the calculations of Bischoff, the contraction of granite when passing from a melted or plastic to a solid and crystalline state must be more than ten per cent.* The sudden subsidence of land may also be occasioned by subterranean caverns giving way, when gases are condensed, or when they escape through newly-formed crevices. The subtraction, moreover, of matter from certain parts of the interior, by the flowing of lava and of mineral springs, must, in the course of ages, cause vacuities below, so that the undermined surface may at length fall in.

The balance of dry land, how preserved.—In the present state of our knowledge, we cannot pretend to estimate the average number of earthquakes which may happen in the course of a single year. As the area of the ocean is nearly three times that of the land, it is probable that about three submarine earthquakes may occur for one exclusively continental; and when we consider the great frequency of slight movements in certain districts, we can hardly suppose that a day, if, indeed, an hour, ever passes without one or more shocks being experienced in some part of the globe. We have also seen that in Sweden, and other countries, changes in the relative level of sea and land may take place without commotion, and these perhaps produce the most important geographical and geological changes; for the position of land may be altered to a greater amount by an elevation or depression of one inch over a vast area, than by the sinking of a more limited tract, such as the forest of Aripao, to the depth of many fathoms at once.†

It must be evident, from the historical details above given, that the force of subterranean movement, whether intermittent or continuous, whether with or without disturbance, does not operate at random, but is developed in certain regions only; and although the alterations produced during the time required for the occurrence of a few volcanic eruptions may be inconsiderable, we can hardly doubt that, during the ages necessary for the formation of large volcanic cones, composed of thousands of lava currents, shoals might be converted into lofty mountains, and low lands into deep seas.

In a former chapter (p. 198), I have stated that aqueous and igneous agents may be regarded as antagonist forces; the aqueous laboring incessantly to reduce the inequalities of the earth's surface to a level, while the igneous are equally active in renewing the unevenness of the surface. By some geologists it has been thought that the levelling power of running water was opposed rather to the *elevating* force of earthquakes than to their action generally. This opinion is, however, untenable; for the sinking down of the bed of the ocean is one of the means by which the gradual submersion of land is prevented. The depth of the sea cannot be increased at any one point without a universal fall of the waters, nor can any partial deposition of sediment occur without

* Bulletin de la Soc. Géol. 2d series, vol. iv. p. 1312. † See p. 468.

the displacement of a quantity of water of equal volume, which will raise the sea, though in an imperceptible degree, even to the antipodes. The preservation, therefore, of the dry land may sometimes be effected by the subsidence of part of the earth's crust (that part, namely, which is covered by the ocean), and in like manner an upheaving movement must often tend to destroy land; for if it render the bed of the sea more shallow, it will displace a certain quantity of water, and thus tend to submerge low tracts.

Astronomers having proved (see above, p. 129) that there has been no change in the diameter of the earth during the last two thousand years, we may assume it as probable, that the dimensions of the planet remain uniform. If, then, we inquire in what manner the force of earthquakes must be regulated, in order to restore perpetually the inequalities of the surface which the levelling power of water tends to efface, it will be found, that the amount of depression must exceed that of elevation. It would be otherwise if the action of volcanoes and mineral springs were suspended; for then the forcing outwards of the earth's envelope ought to be no more than equal to its sinking in.

To understand this proposition more clearly, it must be borne in mind, that the deposits of rivers and currents probably add as much to the height of lands which are rising, as they take from those which have risen. Suppose a large river to bring down sediment to a part of the ocean two thousand feet deep, and that the depth of this part is gradually reduced by the accumulation of sediment till only a shoal remains, covered by water at high tides; if now an upheaving force should uplift this shoal to the height of 2000 feet, the result would be a mountain 2000 feet high. But had the movement raised the same part of the bottom of the sea before the sediment of the river had filled it up; then, instead of changing a shoal into a mountain 2000 feet high, it would only have converted a deep sea into a shoal.

It appears, then, that the operations of the earthquake are often such as to cause the levelling power of water to counteract itself; and, although the idea may appear paradoxical, we may be sure, wherever we find hills and mountains composed of stratified deposits, that such inequalities of the surface would have had no existence if water, at some former period, had not been laboring to reduce the earth's surface to one level.

But, besides the transfer of matter by running water from the continents to the ocean, there is a constant transportation from below upwards, by mineral springs and volcanic vents. As mountain masses are, in the course of ages, created by the pouring forth of successive streams of lava, so stratified rocks, of great extent, originate from the deposition of carbonate of lime, and other mineral ingredients, with which springs are impregnated. The surface of the land, and portions of the bottom of the sea, being thus raised, the external accessions due to these operations would cause the dimensions of the planet to enlarge continually, if the amount of depression of the earth's crust were no more than equal to the elevation. In order, therefore, that the mean diameter of the earth should remain uniform, and the unevenness of the surface be preserved, it is necessary that the amount of subsidence should be in excess. And such a predominance of depression is far from improbable, on mechanical principles, since every upheaving movement must be expected either to produce caverns in the mass below, or to cause some diminution of its density. Vacuities must, also, arise from the subtraction of the matter poured out from volcanoes and mineral springs, or from the contraction of argillaceous masses by subterranean heat; and the foundations having been thus weakened, the earth's crust, shaken and rent by reiterated convulsions, must, in the course of time, fall in.

If we embrace these views, important geological consequences will follow; since, if there be, upon the whole, more subsidence than elevation, the average depth to which former surfaces have sunk beneath their original level must exceed the height which ancient marine strata have attained above the sea. If, for example, marine strata, about the age of our chalk and greensand, have been lifted up in Europe to an extreme height of more than eleven thousand feet, and a mean elevation of some hundreds, we may conclude that certain parts of the surface, which existed when those strata were deposited, have sunk to an extreme depth of *more than* eleven thousand feet below their original level, and to a mean depth of *more than* a few hundreds.

In regard to faults, also, we must infer, according to the hypothesis now proposed, that a greater number have arisen from the sinking down than from the elevation of rocks.

To conclude: it seems to be rendered probable, by the views above explained, that the constant repair of the land, and the subserviency of our planet to the support of terrestrial as well as aquatic species, are secured by the elevating and depressing power of causes acting in the interior of the earth; which, although so often the source of death and terror to the inhabitants of the globe—visiting in succession every zone, and filling the earth with monuments of ruin and disorder—are nevertheless the agents of a conservative principle above all others essential to the stability of the system.

BOOK III.

CHAPTER XXXIII.

CHANGES OF THE ORGANIC WORLD NOW IN PROGRESS.

Division of the subject—Examination of the question, Whether species have a real existence in nature?—Importance of this question in geology—Sketch of Lamarck's arguments in favor of the transmutation of species, and his conjectures respecting the origin of existing animals and plants—His theory of the transformation of the orang-outang into the human species.

THE last book, from chapters fourteen to thirty-three inclusive, was occupied with the consideration of the changes brought about on the earth's surface, within the period of human observation, by inorganic agents; such, for example, as rivers, marine currents, volcanoes, and earthquakes. But there is another class of phenomena relating to the organic world, which have an equal claim on our attention, if we desire to obtain possession of all the preparatory knowledge respecting the existing course of nature, which may be available in the interpretation of geological monuments. It appeared from our preliminary sketch of the progress of the science, that the most lively interest was excited among its earlier cultivators, by the discovery of the remains of animals and plants in the interior of mountains frequently remote from the sea. Much controversy arose respecting the nature of these remains, the causes which may have brought them into so singular a position, and the want of a specific agreement between them and known animals and plants. To qualify ourselves to form just views on these curious questions, we must first study the present condition of the animate creation on the globe.

This branch of our inquiry naturally divides itself into two parts: first, we may examine the vicissitudes to which species are subject; secondly, the processes by which certain individuals of these species occasionally become fossil. The first of these divisions will lead us, among other topics, to inquire, first, whether species have a real and permanent existence in nature? or whether they are capable, as some naturalists pretend, of being indefinitely modified in the course of a long series of generations? Secondly, whether, if species have a real existence, the individuals composing them have been derived originally from many similar stocks, or each from one only, the descendants of which have spread themselves gradually from a particular point over the habitable lands and waters? Thirdly, how far the duration of each species of animal and plant is limited by its dependence on certain fluctuating and temporary conditions in the state of the animate and inanimate world? Fourthly, whether there be proofs of the successive extermination of species in the ordinary course of nature, and whether there be any reason for conjecturing that new animals and plants are created from time to time, to supply their place?

Whether species have a real existence in nature.—Before we can advance a step in our proposed inquiry, we must be able to define precisely the meaning which we attach to the term species. This is even more necessary in geology than in the ordinary studies of the naturalist; for they who deny that such a thing as a species exists, concede nevertheless that a botanist or zoologist may reason as if the specific character were constant, because they confine their observations to a brief period of time. Just as the geographer, in constructing his maps from century to century, may proceed as if the apparent places of the fixed stars remained absolutely the same, and as if no alteration were brought about by the precession of the equinoxes; so, it is said, in the organic world, the stability of a species may be taken as absolute, if we do not extend our views beyond the narrow period of human history; but let a sufficient number of centuries elapse, to allow of important revolutions in climate, physical geography, and other circumstances, and the characters, say they, of the descendants of common parents may deviate indefinitely from their original type.

Now, if these doctrines be tenable, we are at once presented with a principle of incessant change in the organic world; and no degree of dissimilarity in the plants and animals which may formerly have existed, and are found fossil, would entitle us to conclude that they may not have been the prototypes and progenitors of the species now living. Accordingly M. Geoffroy St. Hilaire has declared his opinion, that there has been an uninterrupted succession in the animal kingdom, effected by means of generation, from the earliest ages of the world up to the present day, and that the ancient animals whose remains have been preserved in the strata, however different, may nevertheless have been the ancestors of those now in being. This notion is not very generally received, but we are not warranted in assuming the contrary, without fully explaining the data and reasoning by which it may be refuted.

I shall begin by stating as concisely as possible all the facts and ingenious arguments by which the theory has been supported; and for this purpose I cannot do better than offer the reader a rapid sketch of Lamarck's statement of the proofs which he regards as confirmatory of the doctrine, and which he has derived partly from the works of his predecessors and in part from original investigations.

His proofs and inferences will be best considered in the order in which they appear to have influenced his mind, and I shall then point out some of the results to which he was led while boldly following out his principles to their legitimate consequences.

Lamarck's arguments in favor of the transmutation of species.—The name of species, observes Lamarck, has been usually applied to "every collection of similar individuals produced by other individuals like themselves."* This definition, he admits, is correct; because every living individual bears a very close resemblance to those from which it springs. But this is not all which is usually implied by the term species; for the majority of naturalists agree with Linnæus in supposing that all the individuals propagated from one stock have certain distinguishing characters in common, which will never vary, and which have remained the same since the creation of each species.

In order to shake this opinion, Lamarck enters upon the following line of argument:—The more we advance in the knowledge of the different organized bodies which cover the surface of the globe, the more our embarrassment increases, to determine what ought to be regarded as a species, and still more how to limit and distinguish genera. In proportion as our collections are enriched, we see almost every void filled up, and all our lines of separation effaced! we are reduced to arbitrary determinations, and are sometimes fain to seize upon the slight differences of mere varieties, in order to form characters for what we choose to call a species; and sometimes we are induced to pronounce individuals but slightly differing, and which others regard as true species, to be varieties.

The greater the abundance of natural objects assembled together, the more do we discover proofs that every thing passes by insensible shades into something else; that even the more remarkable differences are evanescent, and that nature has, for the most part, left us nothing at our disposal for establishing distinctions, save trifling, and, in some respects, puerile particularities.

We find that many genera amongst animals and plants are of such an extent, in consequence of the number of species referred to them, that the study and determination of these last has become almost impracticable. When the species are arranged in a series, and placed near to each other, with due regard to their natural affinities, they each differ in so minute a degree from those next adjoining, that they almost melt into each other, and are in a manner confounded together. If we see isolated species, we may presume the absence of some more closely connected, and which have not yet been discovered. Already are there genera, and even entire orders—nay, whole classes, which present an approximation to the state of things here indicated.

If, when species have been thus placed in a regular series, we select one, and then, making a leap over several intermediate ones, we take a second, at some distance from the first, these two will, on comparison, be seen to be very dissimilar; and it is in this manner that every naturalist begins to study the objects which are at his own door. He then finds it an easy task to establish generic and specific distinctions; and it is only when his experience is enlarged, and when he has made himself master of the intermediate links, that his difficulties and ambiguities begin. But while we are thus compelled to resort to trifling and minute characters in our attempt to separate the species, we find a striking dis-

* Phil. Zool. tom. i. p. 54.

parity between individuals which we know to have descended from a common stock; and these newly acquired peculiarities are regularly transmitted from one generation to another, constituting what are called *races*.

From a great number of facts, continues the author, we learn that in proportion as the individuals of one of our species change their situation, climate, and manner of living, they change also, by little and little, the consistence and proportions of their parts, their form, their faculties, and even their organization, in such a manner that every thing in them comes at last to participate in the mutations to which they have been exposed. Even in the same climate, a great difference of situation and exposure causes individuals to vary; but if these individuals continue to live and to be reproduced under the same difference of circumstances, distinctions are brought about in them which become in some degree essential to their existence. In a word, at the end of many successive generations, these individuals, which originally belonged to another species, are transformed into a new and distinct species.*

Thus, for example, if the seeds of a grass, or any other plant which grows naturally in a moist meadow, be accidentally transported, first to the slope of some neighboring hill, where the soil, although at a greater elevation, is damp enough to allow the plant to live; and if, after having lived there, and having been several times regenerated, it reaches by degrees the drier and almost arid soil of a mountain declivity, it will then, if it succeeds in growing, and perpetuates itself for a series of generations, be so changed that botanists who meet with it will regard it as a particular species.† The unfavorable climate in this case, deficiency of nourishment, exposure to the winds, and other causes, give rise to a stunted and dwarfish race, with some organ more developed than others, and having proportions often quite peculiar.

What nature brings about in a great lapse of time, we occasion suddenly by changing the circumstances in which a species has been accustomed to live. All are aware that vegetables taken from their birthplace, and cultivated in gardens, undergo changes which render them no longer recognizable as the same plants. Many which were naturally hairy become smooth, or nearly so; a great number of such as were creepers and trailed along the ground, rear their stalks and grow erect. Others lose their thorns or asperities; others, again, from the ligneous state which their stem possessed in hot climates, where they were indigenous, pass to the herbaceous; and, among them, some which were perennials become mere annuals. So well do botanists know the effects of such changes of circumstances, that they are averse to describe species from garden specimens, unless they are sure that they have been cultivated for a very short period.

"Is not the cultivated wheat" (*Triticum sativum*), asks Lamarck, "a vegetable brought by man into the state in which we now see it? Let

* Phil. Zool. tom. i. p. 62. † Ibid.

any one tell me in what country a similar plant grows wild, unless where it has escaped from cultivated fields? Where do we find in nature our cabbages, lettuces, and other culinary vegetables, in the state in which they appear in our gardens? Is it not the same in regard to a great quantity of animals which domesticity has changed or considerably modified?"* Our domestic fowls and pigeons are unlike any wild birds. Our domestic ducks and geese have lost the faculty of raising themselves into the higher regions of the air, and crossing extensive countries in their flight, like the wild ducks and wild geese from which they were originally derived. A bird which we breed in a cage cannot, when restored to liberty, fly like others of the same species which have been always free. This small alteration of circumstances, however, has only diminished the power of flight, without modifying the form of any part of the wings. But when individuals of the same race are retained in captivity during a considerable length of time, the form even of their parts is gradually made to differ, especially if climate, nourishment, and other circumstances be also altered.

The numerous races of dogs which we have produced by domesticity are nowhere to be found in a wild state. In nature we should seek in vain for mastiffs, harriers, spaniels, greyhounds, and other races, between which the differences are sometimes so great that they would be readily admitted as specific between wild animals; "yet all these have sprung originally from a single race, at first approaching very near to a wolf, if, indeed, the wolf be not the true type which at some period or other was domesticated by man."

Although important changes in the nature of the places which they inhabit modify the organization of animals as well as vegetables; yet the former, says Lamarck, require more time to complete a considerable degree of transmutation; and, consequently, we are less sensible of such occurrences. Next to a diversity of the medium in which animals or plants may live, the circumstances which have most influence in modifying their organs are differences in exposure, climate, the nature of the soil, and other local particulars. These circumstances are as varied as are the characters of the species, and, like them, pass by insensible shades into each other, there being every intermediate gradation between the opposite extremes. But each locality remains for a very long time the same, and is altered so slowly that we can only become conscious of the reality of the change by consulting geological monuments, by which we learn that the order of things which now reigns in each place has not always prevailed, and by inference anticipate that it will not always continue the same.†

Every considerable alteration in the local circumstances in which each race of animals exists causes a change in their wants, and these new wants excite them to new actions and habits. These actions require the more frequent employment of some parts before but slightly exercised,

* Phil. Zool. tom. i. p. 227. † Ibid. p. 232.

and then greater development follows as a consequence of their more frequent use. Other organs no longer in use are impoverished and diminished in size, nay, are sometimes entirely annihilated, while in their place new parts are insensibly produced for the discharge of new functions.*

I must here interrupt the author's argument, by observing, that no positive fact is cited to exemplify the substitution of some *entirely new* sense, faculty, or organ, in the room of some other suppressed as useless. All the instances adduced go only to prove that the dimensions and strength of members and the perfection of certain attributes may, in a long succession of generations, be lessened and enfeebled by disuse; or, on the contrary, be matured and augmented by active exertion; just as we know that the power of scent is feeble in the greyhound, while its swiftness of pace and its acuteness of sight are remarkable—that the harrier and stag-hound, on the contrary, are comparatively slow in their movements, but excel in the sense of smelling.

It was necessary to point out to the reader this important chasm in the chain of evidence, because he might otherwise imagine that I had merely omitted the illustrations for the sake of brevity; but the plain truth is, that there were no examples to be found; and when Lamarck talks "of the efforts of internal sentiment," "the influence of subtle fluids," and "acts of organization," as causes whereby animals and plants may acquire *new organs*, he substitutes names for things; and, with a disregard to the strict rules of induction, resorts to fictions, as ideal as the "plastic virtue," and other phantoms of the geologists of the middle ages.

It is evident that, if some well-authenticated facts could have been adduced to establish one complete step in the process of transformation, such as the appearance, in individuals descending from a common stock, of a sense or organ entirely new, and a complete disappearance of some other enjoyed by their progenitors, time alone might then be supposed sufficient to bring about any amount of metamorphosis. The gratuitous assumption, therefore, of a point so vital to the theory of transmutation, was unpardonable on the part of its advocate.

But to proceed with the system: it being assumed as an undoubted fact, that a change of external circumstances may cause one organ to become entirely obsolete, and a new one to be developed, such as never before belonged to the species, the following proposition is announced, which, however staggering and absurd it may seem, is logically deduced from the assumed premises. It is not the organs, or, in other words, the nature and form of the parts of the body of an animal, which have given rise to its habits, and its particular faculties; but, on the contrary, its habits, its manner of living, and those of its progenitors, have in the course of time determined the form of its body, the number and condition of its organs—in short, the faculties which it enjoys. Thus otters,

* Phil. Zool. tom. i. p. 234.

beavers, waterfowl, turtles, and frogs, were not made web-footed in order that they might swim; but their wants having attracted them to the water in search of prey, they stretched out the toes of their feet to strike the water and move rapidly along its surface. By the repeated stretching of their toes, the skin which united them at the base acquired a habit of extension, until, in the course of time, the broad membranes which now connect their extremities were formed.

In like manner, the antelope and the gazelle were not endowed with light agile forms, in order that they might escape by flight from carnivorous animals; but, having been exposed to the danger of being devoured by lions, tigers, and other beasts of prey, they were compelled to exert themselves in running with great celerity; a habit which, in the course of many generations, gave rise to the peculiar slenderness of their legs, and the agility and elegance of their forms.

The camelopard was not gifted with a long flexible neck because it was destined to live in the interior of Africa, where the soil was arid and devoid of herbage; but, being reduced by the nature of that country to support itself on the foliage of lofty trees, it contracted a habit of stretching itself up to reach the high boughs, until its neck became so elongated that it could raise its head to the height of twenty feet above the ground.

Another line of argument is then entered upon, in farther corroboration of the instability of species. In order, it is said, that individuals should perpetuate themselves unaltered by generation, those belonging to one species ought never to ally themselves to those of another; but such sexual unions do take place, both among plants and animals; and although the offspring of such irregular connections are usually sterile, yet such is not always the case. Hybrids have sometimes proved prolific, where the disparity between the species was not too great; and by this means alone, says Lamarck, varieties may gradually be created by near alliances, which would become races, and in the course of time would constitute what we term species.*

But if the soundness of all these arguments and inferences be admitted, we are next to inquire, what were the original types of form, organization, and instinct, from which the diversities of character, as now exhibited by animals and plants, have been derived? We know that individuals which are mere varieties of the same species would, if their pedigree could be traced back far enough, terminate in a single stock; so, according to the train of reasoning before described, the species of a genus, and even the genera of a great family, must have had a common point of departure. What, then, was the single stem from which so many varieties of form have ramified? Were there many of these, or are we to refer the origin of the whole animate creation, as the Egyptian priests did that of the universe, to a single egg?

* Phil. Zool. p. 64.

In the absence of any positive data for framing a theory on so obscure a subject, the following considerations were deemed of importance to guide conjecture.

In the first place, if we examine the whole series of known animals, from one extremity to the other, when they are arranged in the order of their natural relations, we find that we may pass progressively, or, at least, with very few interruptions, from beings of more simple to those of a more compound structure; and, in proportion as the complexity of their organization increases, the number and dignity of their faculties increase also. Among plants, a similar approximation to a graduated scale of being is apparent. Secondly, it appears, from geological observations, that plants and animals of more simple organization existed on the globe before the appearance of those of more compound structure, and the latter were successively formed at more modern periods; each new race being more fully developed than the most perfect of the preceding era.

Of the truth of the last-mentioned geological theory, Lamarck seems to have been fully persuaded; and he also shows that he was deeply impressed with a belief prevalent amongst the older naturalists, that the primeval ocean invested the whole planet long after it became the habitation of living beings; and thus he was inclined to assert the priority of the types of marine animals to those of the terrestrial, so as to fancy, for example, that the testacea of the ocean existed first, until some of them, by gradual evolution, were *improved* into those inhabiting the land.

These speculative views had already been, in a great degree, anticipated by Demaillet in his Telliamed, and by several modern writers; so that the tables were completely turned on the philosophers of antiquity, with whom it was a received maxim, that created things were always most perfect when they came first from the hands of their Maker; and that there was a tendency to progressive deterioration in sublunary things when left to themselves—

——— omnia fatis
In pejus ruere, ac retrò sublapsa referri.

So deeply was the faith of the ancient schools of philosophy imbued with this doctrine, that, to check this universal proneness to degeneracy, nothing less than the reintervention of the Deity was thought adequate; and it was held, that thereby the order, excellence, and pristine energy of the moral and physical world had been repeatedly restored.

But when the possibility of the indefinite modification of individuals descending from common parents was once assumed, as also the geological inference respecting the progressive development of organic life, it was natural that the ancient dogma should be rejected, or rather reversed, and that the most simple and imperfect forms and faculties should be conceived to have been the originals whence all others were developed. Accordingly, in conformity to these views, inert matter was supposed to have been first endowed with life; until, in the course of ages, sensation was superadded to mere vitality: sight, hearing, and the other senses were afterwards acquired; then instinct and the mental faculties; until, finally, by virtue of the tendency of things to *progressive improvement*, the irrational was developed in the rational.

The reader, however, will immediately perceive that when all the higher orders of plants and animals were thus supposed to be comparatively modern, and to have been derived in a long series of generations from those of more simple conformation, some farther hypothesis became indispensable, in order to explain why, after an indefinite lapse of ages, there were still so many beings of the simplest structure. Why have the majority of existing creatures remained stationary throughout this long succession of epochs, while others have made such prodigious advances? Why are there such multitudes of infusoria and polyps, or of confervæ and other cryptogamic plants? Why, moreover, has the process of development acted with such unequal and irregular force on those classes of beings which have been greatly perfected, so that there are wide chasms in the series; gaps so enormous, that Lamarck fairly admits we can never expect to fill them up by future discoveries?

The following hypothesis was provided to meet these objections. Nature, we are told, is not an intelligence, nor the Deity; but a delegated power—a mere instrument—a piece of mechanism acting by necessity—an order of things constituted by the Supreme Being, and subject to laws which are the expressions of his will. This Nature is *obliged* to proceed gradually in all her operations; she cannot produce animals and plants of all classes at once, but must always begin by the formation of the most simple kinds, and out of them elaborate the more compound, adding to them, successively, different systems of organs, and multiplying more and more their number and energy.

This nature is daily engaged in the formation of the elementary rudiments of animal and vegetable existence, which correspond to what the ancients termed *spontaneous generation*. She is always beginning anew, day by day, the work of creation, by forming monads, or "rough draughts" (ébauches), which are the only living things she gives birth to *directly*.

There are distinct primary rudiments of plants and animals, and *probably* of each of the great divisions of the animal and vegetable kingdoms.* These are gradually developed into the higher and more perfect classes by the slow but unceasing agency of two influential principles: first, *the tendency to progressive advancement* in organization, accompanied by greater dignity in instinct, intelligence, &c.; secondly, *the force of external circumstances*, or of variations in the physical condition of the earth, or the mutual relations of plants and animals. For, as species spread themselves gradually over the globe, they are exposed from time to time to variations in climate, and to changes in the quantity and quality of their food; they meet with new plants and animals which

* Animaux sans Vert. tom. i. p. 56, Introduction.

assist or retard their development, by supplying them with nutriment, or destroying their foes. The nature, also, of each locality, is in itself fluctuating; so that, even if the relation of other animals and plants were invariable, the habits and organization of species would be modified by the influence of local revolutions.

Now, if the first of these principles, *the tendency to progressive development*, were left to exert itself with perfect freedom, it would give rise, says Lamarck, in the course of ages, to a graduated scale of being, where the most insensible transition might be traced from the simplest to the most compound structure, from the humblest to the most exalted degree of intelligence. But, in consequence of the perpetual interference of the *external causes* before mentioned, this regular order is greatly interfered with, and an approximation only to such a state of things is exhibited by the animate creation, the progress of some races being retarded by unfavorable, and that of others accelerated by favorable, combinations of circumstances. Hence, all kinds of anomalies interrupt the continuity of the plan; and chasms, into which whole genera or families might be inserted, are seen to separate the nearest existing portions of the series.

Lamarck's theory of the transformation of the orang-outang into the human species.—Such is the machinery of the Lamarckian system; but the reader will hardly, perhaps, be able to form a perfect conception of so complicated a piece of mechanism, unless it is exhibited in motion, so that we may see in what manner it can work out, under the author's guidance, all the extraordinary effects which we behold in the present state of the animate creation. I have only space for exhibiting a small part of the entire process by which a complete metamorphosis is achieved, and shall therefore omit the mode by which, after a countless succession of generations, a small gelatinous body is transformed into an oak or an ape; passing on at once to the last grand step in the progressive scheme, by which the orang-outang, having been already evolved out of a monad, is made slowly to attain the attributes and dignity of man.

One of the races of quadrumanous animals which had reached the highest state of perfection, lost, by constraint of circumstances (concerning the exact nature of which tradition is unfortunately silent), the habit of climbing trees, and of hanging on by grasping the boughs with their feet as with hands. The individuals of this race being obliged, for a long series of generations, to use their feet exclusively for walking, and ceasing to employ their hands as feet, were transformed into bimanous animals, and what before were thumbs became mere toes, no separation being required when their feet were used solely for walking. Having acquired a habit of holding themselves upright, their legs and feet assumed, insensibly, a conformation fitted to support them in an erect attitude, till at last these animals could no longer go on all-fours without much inconvenience.

The Angola orang (*Simia troglodytes*, Linn.) is the most perfect of animals; much more so than the Indian orang (*Simia Satyrus*), which has been called the orang-outang, although *both* are *very inferior* to man in corporeal powers and intelligence. These animals frequently hold themselves upright; but their organization has *not yet* been sufficiently modified to sustain them habitually in this attitude, so that the standing posture is very uneasy to them. When the Indian orang is compelled to take flight from pressing danger, he immediately falls down upon all-fours, showing clearly that this was the original position of the animal. Even in man, whose organization, in the course of a long series of generations, has advanced so much farther, the upright posture is fatiguing, and can be supported only for a limited time, and by aid of the contraction of many muscles. If the vertebral column formed the axis of the human body, and supported the head and all the other parts in equilibrium, then might the upright position be a state of repose: but, as the human head does not articulate in the centre of gravity, as the chest, belly, and other parts press almost entirely forward with their whole weight, and as the vertebral column reposes upon an oblique base, a watchful activity is required to prevent the body from falling. Children who have large heads and prominent bellies can hardly walk at the end even of two years; and their frequent tumbles indicate the natural tendency in man to resume the quadrupedal state.

Now, when so much progress had been made by the quadrumanous animals before mentioned, that they could hold themselves habitually in an erect attitude, and were accustomed to a wide range of vision, and ceased to use their jaws for fighting and tearing, or for clipping herbs for food, their snout became gradually shorter, their incisor teeth became vertical, and the facial angle grew more open.

Among other ideas which the natural *tendency to perfection* engendered, the desire of ruling suggested itself, and this race succeeded at length in getting the better of the other animals, and made themselves masters of all those spots on the surface of the globe which best suited them. They drove out the animals which approached nearest them in organization and intelligence, and which were in a condition to dispute with them the good things of this world, forcing them to take refuge in deserts, woods, and wildernesses, where their multiplication was checked, and the progressive development of their faculties retarded; while, in the mean time, the dominant race spread itself in every direction, and lived in large companies, where new wants were successively created, exciting them to industry, and gradually perfecting their means and faculties.

In the supremacy and increased intelligence acquired by the ruling race, we see an illustration of the natural tendency of the organic world to grow more perfect; and, in their influence in repressing the advance of others, an example of one of those disturbing causes before enumerated, that *force of external circumstances* which causes such wide chasms in the regular series of animated being.

When the individuals of the dominant race became very numerous, their ideas greatly increased in number, and they felt the necessity of communicating them to each other, and of augmenting and varying the signs proper for the communication of ideas. Meanwhile the inferior quadrumanous animals, although most of them were gregarious, acquired no new ideas, being persecuted and restless in the deserts, and obliged to fly and conceal themselves, so that they conceived no new wants. Such ideas as they already had remained unaltered, and they could dispense with the communication of the greater part of these. To make themselves, therefore, understood by their fellows, required merely a few movements of the body or limbs—whistling, and the uttering of certain cries varied by the inflexions of the voice.

On the contrary, the individuals of the ascendant race, animated with a desire of interchanging their ideas, which became more and more numerous, were prompted to multiply the means of communication, and were no longer satisfied with mere pantomimic signs, nor even with all the possible inflexions of the voice, but made continual efforts to acquire the power of uttering articulate sounds, employing a few at first, but afterwards varying and perfecting them according to the increase of their wants. The habitual exercise of their throat, tongue, and lips, insensibly modified the conformation of these organs, until they became fitted for the faculty of speech.*

In effecting this mighty change, "the exigencies of the individuals were the sole agents; they gave rise to efforts, and the organs proper for articulating sounds were developed by their habitual employment." Hence, in this peculiar race, the origin of the admirable faculty of speech; hence also the diversity of languages, since the distance of places where the individuals composing the race established themselves soon favored the corruption of conventional signs.†

In conclusion, it may be proper to observe that the above sketch of the Lamarckian theory is no exaggerated picture, and those passages which have probably excited the greatest surprise in the mind of the reader are literal translations from the original.

* Lamarck's Phil. Zool. tom. i. p. 356. † Ibid. p. 357.

CHAPTER XXXIV.

TRANSMUTATION OF SPECIES—*continued.*

Recapitulation of the arguments in favor of the theory of transmutation of species—Their insufficiency—Causes of difficulty in discriminating species—Some varieties possibly more distinct than certain individuals of distinct species—Variability in a species consistent with a belief that the limits of deviation are fixed—No facts of transmutation authenticated—Varieties of the Dog—the Dog and Wolf distinct species—Mummies of various animals from Egypt identical in character with living individuals—Seeds and plants from the Egyptian tombs—Modifications produced in plants by agriculture and gardening.

THE theory of the transmutation of species, considered in the last chapter, has met with some degree of favor from many naturalists, from their desire to dispense, as far as possible, with the repeated intervention of a First Cause, as often as geological monuments attest the successive appearance of new races of animals and plants, and the extinction of those pre-existing. But, independently of a predisposition to account, if possible, for a series of changes in the organic world by the regular action of secondary causes, we have seen that in truth many perplexing difficulties present themselves to one who attempts to establish the nature and reality of the specific character. And if once there appears ground of reasonable doubt, in regard to the constancy of species, the amount of transformation which they are capable of undergoing may seem to resolve itself into a mere question of the quantity of time assigned to the past duration of animate existence.

Before entering upon the reasons which may be adduced for rejecting Lamarck's hypothesis, I shall recapitulate, in a few words, the phenomena, and the whole train of thought, by which I conceive it to have been suggested, and which have gained for this and analogous theories, both in ancient and modern times, a considerable number of votaries.

In the first place, the various groups into which plants and animals may be thrown seem almost invariably, to a beginner, to be so natural, that he is usually convinced at first, as was Linnæus to the last, "that genera are as much founded in nature as the species which compose them." * When by examining the numerous intermediate gradations the student finds all lines of demarcation to be in most instances obliterated, even where they at first appeared most distinct, he grows more and more sceptical as to the real existence of genera, and finally regards them as mere arbitrary and artificial signs, invented, like those

* Genus omne est naturale, in primordio tale creatum, &c. Phil. Bot. § 159. See also ibid. § 162.

which serve to distinguish the heavenly constellations, for the convenience of classification, and having as little pretensions to reality.

Doubts are then engendered in his mind as to whether species may not also be equally unreal. The student is probably first struck with the phenomenon, that some individuals are made to deviate widely from the ordinary type by the force of peculiar circumstances, and with the still more extraordinary fact, that the newly acquired peculiarities are faithfully transmitted to the offspring. How far, he asks, may such variations extend in the course of indefinite periods of time, and during great vicissitudes in the physical condition of the globe? His growing incertitude is at first checked by the reflection that nature has forbidden the intermixture of the descendants of distinct original stocks, or has, at least, entailed sterility on their offspring, thereby preventing their being confounded together, and pointing out that a multitude of distinct types must have been created in the beginning, and must have remained pure and uncorrupted to this day.

Relying on this general law, he endeavors to solve each difficult problem by direct experiment, until he is again astounded by the phenomenon of a prolific hybrid, and still more by an example of a hybrid perpetuating itself throughout several generations in the vegetable world. He then feels himself reduced to the dilemma of choosing between two alternatives; either to reject the test, or to declare that the two species, from the union of which the fruitful progeny has sprung, were mere varieties. If he prefer the latter, he is compelled to question the reality of the distinctness of all other supposed species which differ no more than the parents of such prolific hybrids; for although he may not be enabled immediately to procure, in all such instances, a fruitful offspring; yet experiments show, that after repeated failures, the union of two recognized species may at last, under very favorable circumstances, give birth to a fertile progeny. Such circumstances, therefore, the naturalist may conceive to have occurred again and again, in the course of a great lapse of ages.

His first opinions are now fairly unsettled, and every stay at which he has caught has given way one after another; he is in danger of falling into any new and visionary doctrine which may be presented to him; for he now regards every part of the animate creation as void of stability, and in a state of continual flux. In this mood he encounters the Geologist, who relates to him how there have been endless vicissitudes in the shape and structure of organic beings in former ages —how the approach to the present system of things has been gradual —that there has been a progressive development of organization subservient to the purposes of life, from the most simple to the most complex state—that the appearance of man is the last phenomenon in a long succession of events—and, finally, that a series of physical revolutions can be traced in the inorganic world, coeval and co-extensive with those of organic nature.

These views seem immediately to confirm all his preconceived doubts as to the stability of the specific character, and he begins to think there may exist an inseparable connection between a series of changes in the inanimate world, and the capability of the species to be indefinitely modified by the influence of external circumstances. Henceforth his speculations know no definite bounds; he gives the rein to conjecture, and fancies that the outward form, internal structure, instinctive faculties, nay, that reason itself may have been gradually developed from some of the simplest states of existence—that all animals, that man himself, and the irrational beings, may have had one common origin; that all may be parts of one continuous and progressive scheme of development, from the most imperfect to the more complex; in fine, he renounces his belief in the high genealogy of his species, and looks forward, as if in compensation, to the future perfectibility of man in his physical, intellectual, and moral attributes.

Let us now proceed to consider what is defective in evidence, and what fallacious in reasoning, in the grounds of these strange conclusions. Blumenbach judiciously observes, that "no general rule can be laid down for determining the distinctness of species, as there is no particular class of characters which can serve as a criterion. In each case we must be guided by *analogy* and *probability*." The multitude, in fact, and complexity of the proofs to be weighed is so great, that we can only hope to obtain presumptive evidence, and we must, therefore, be the more careful to derive our general views as much as possible from those observations where the chances of deception are least. We must be on our guard not to tread in the footsteps of the naturalists of the middle ages, who believed the doctrine of spontaneous generation to be applicable to all those parts of the animal and vegetable kingdoms which they least understood, in direct contradiction to the analogy of all the parts best known to them; and who, when at length they found that insects and cryptogamous plants were also propagated from eggs or seeds, still persisted in retaining their old prejudices respecting the infusory animalcules and other minute beings, the generation of which had not then been demonstrated by the microscope to be governed by the same laws.

Lamarck has, indeed, attempted to raise an argument in favor of his system, out of the very confusion which has arisen in the study of some orders of animals and plants, in consequence of the slight shades of difference which separate the new species discovered within the last half century. That the embarrassment of those who attempt to classify and distinguish the new acquisitions, poured in such multitudes into our museums, should increase with the augmentation of their number, is quite natural; since to obviate this, it is not enough that our powers of discrimination should keep pace with the increase of the objects, but we ought to possess greater opportunities of studying each animal and plant in all stages of its growth, and to know profoundly their history, their habits, and physiological characters, throughout several generations;

for, in proportion as the series of known animals grows more complete none can doubt there is a nearer approximation to a graduated scale of being; and thus the most closely allied species will be found to possess a greater number of characters in common.

Causes of the difficulty of discriminating species.—But, in point of fact, our new acquisitions consist, more and more as we advance, of specimens brought from foreign and often very distant and barbarous countries. A large proportion have never even been seen alive by scientific inquirers. Instead of having specimens of the young, the adult, and the aged individuals of each sex, and possessing means of investigating the anatomical structure, the peculiar habits, and instincts of each, what is usually the state of our information? A single specimen, perhaps, of a dried plant, or a stuffed bird or quadruped; a shell, without the soft parts of the animal; an insect in one stage of its numerous transformations;—these are the scanty and imperfect data which the naturalist possesses. Such information may enable us to separate species which stand at a considerable distance from each other; but we have no right to expect any thing but difficulty and ambiguity, if we attempt, from such imperfect opportunities, to obtain distinctive marks for defining the characters of species which are closely related.

If Lamarck could introduce so much certainty and precision into the classification of several thousand species of recent and fossil shells, notwithstanding the extreme remoteness of the organization of these animals from the type of those vertebrated species which are best known, and in the absence of so many of the living inhabitants of shells, we are led to form an exalted conception of the degree of exactness to which specific distinctions are capable of being carried, rather than to call in question their reality.

When our data are so defective, the most acute naturalist must expect to be sometimes at fault, and, like the novice, to overlook essential points of difference, passing unconsciously from one species to another, until, like one who is borne along in a current, he is astonished on looking back, at observing that he has reached a point so remote from that whence he set out.

It is by no means improbable, that, when the series of species of certain genera is very full, they may be found to differ less widely from each other than do the mere varieties or races of certain species. If such a fact could be established, it would, undoubtedly, diminish the chance of our obtaining certainty in our results; but it would by no means overthrow our confidence in the reality of species.

Some mere varieties possibly more distinct than certain individuals of distinct species.—It is almost necessary, indeed, to suppose that varieties will differ in some cases more decidedly than some species, if we admit that there is a graduated scale of being, and assume that the following laws prevail in the economy of the animate creation:—first, that the organization of individuals is capable of being modified to a limited extent, by the force of external causes; secondly, that these modifications are, to a certain extent, transmissible to their offspring; thirdly, that there are fixed limits, beyond which the descendants from common parents can never deviate from a certain type; fourthly, that each species springs from one original stock, and can never be permanently confounded by intermixing with the progeny of any other stock; fifthly, that each species shall endure for a considerable period of time. Now, let us assume, for the present, these rules hypothetically, and see what consequences may naturally be expected to result from them.

We must suppose that when the Author of Nature creates an animal or plant, all the possible circumstances in which its descendants are destined to live are foreseen, and that an organization is conferred upon it which will enable the species to perpetuate itself and survive under all the varying circumstances to which it must be inevitably exposed. Now, the range of variation of circumstances will differ essentially in almost every case. Let us take, for example, any one of the most influential conditions of existence, such as temperature. In some extensive districts near the equator, the thermometer might never vary, throughout several thousand centuries, for more than 20° Fahrenheit; so that if a plant or animal be provided with an organization fitting it to endure such a range, it may continue on the globe for that immense period, although every individual might be liable at once to be cut off by the least possible excess of heat or cold beyond the determinate degree. But if a species be placed in one of the temperate zones, and have a constitution conferred on it capable of supporting a similar range of temperature only, it will inevitably perish before a single year has passed away.

Humboldt has shown that, at Cumana, within the tropics, there is a difference of only 4° Fahr. between the temperature of the warmest and coldest months; whereas, in the temperate zones, the annual variation amounts to about 60°, and the extreme range of the thermometer in Canada is not less than 90°.

The same remark might be applied to any other condition, as food, for example; it may be foreseen that the supply will be regular throughout indefinite periods in one part of the world, and in another very precarious and fluctuating both in kind and quantity. Different qualifications may be required for enabling species to live for a considerable time under circumstances so changeable. If, then, temperature and food be among those external causes which, according to certain laws of animal and vegetable physiology, modify the organization, form, or faculties, of individuals, we instantly perceive that the degrees of variability from a common standard must differ widely in the two cases above supposed; since there is a necessity of accommodating a species in one case to a much greater latitude of circumstances than in the other.

If it be a law, for instance, that scanty sustenance should check those individuals in their growth which are enabled to accommodate themselves to privations of this kind, and that a parent, prevented in this manner from attaining the size proper to its species, should produce a dwarfish offspring, a stunted race will arise, as is remarkably exemplified in some varieties of the horse and dog. The difference of stature in some races of dogs, when compared to others, is as one to five in linear dimensions, making a difference of a hundred-fold in volume.* Now, there is a good reason to believe that species in general are by no means susceptible of existing under a diversity of circumstances, which may give rise to such a disparity in size, and, consequently, there will be a multitude of distinct species, of which no two adult individuals can ever depart so widely from a certain standard of dimensions as the mere varieties of certain other species—the dog, for instance. Now, we have only to suppose that what is true of size, may also hold in regard to color and many other attributes; and it will at once follow, that the degree of possible discordance between varieties of the same species may, in certain cases, exceed the utmost disparity which can arise between two individuals of many distinct species.

The same remarks may hold true in regard to instincts; for, if it be foreseen that one species will have to encounter a great variety of foes, it may be necessary to arm it with great cunning and circumspection, or with courage or other qualities capable of developing themselves on certain occasions; such, for example, as those migratory instincts which are so remarkably exhibited at particular periods, after they have remained dormant for many generations. The history and habits of one variety of such a species may often differ more considerably from some other than those of many distinct species which have no such latitude of accommodation to circumstances.

Extent of known variability in species.—Lamarck has somewhat mis-stated the idea commonly entertained of a species; for it is not true that naturalists in general assume that the organisation of an animal or plant remains absolutely constant, and that it can never vary in any of its parts.† All must be aware that circumstances influence the habits, and that the habits may alter the state of the parts and organs; but the difference of opinion relates to the extent to which these modifications of the habits and organs of a particular species may be carried.

Now, let us first inquire what positive facts can be adduced in the history of known species, to establish a great and permanent amount of change in the form, structure, or instinct of individuals descending from some common stock. The best authenticated examples of the extent to which species can be made to vary may be looked for in the history of domesticated animals and cultivated plants. It usually happens, that those species, both of the animal and vegetable kingdom, which have the greatest pliability of organisation, those which are most capable of accommodating themselves to a great variety of new circumstances, are most serviceable to man. These only can be carried by him into different climates, and can have their properties or instincts variously diversified by differences of nourishment

* Cuvier, Discours Prélimin. p. 128.

† Phil. Zool. tom. i. p. 266.

and habits. If the resources of a species be so limited, and its habits and faculties be of such a confined and local character, that it can only flourish in a few particular spots, it can rarely be of great utility.

We may consider, therefore, that in the domestication of animals and the cultivation of plants, mankind have first selected those species which have the most flexible frames and constitutions, and have then been engaged for ages in conducting a series of experiments, with much patience and at great cost, to ascertain what may be the greatest possible deviation from a common type which can be elicited in these extreme cases.

Varieties of the dog—no transmutation.—The modifications produced in the different races of dogs exhibit the influence of man in the most striking point of view. These animals have been transported into every climate and placed in every variety of circumstances; they have been made, as a modern naturalist observes, the servant, the companion, the guardian, and the intimate friend of man, and the power of a superior genius has had a wonderful influence not only on their forms, but on their manners and intelligence.* Different races have undergone remarkable changes in the quantity and color of their clothing; the dogs of Guinea are almost naked, while those of the arctic circle are covered with a warm coat both of hair and wool, which enables them to bear the most intense cold without inconvenience. There are differences also of another kind no less remarkable, as in size, the length of their muzzles, and the convexity of their foreheads.

But, if we look for some of those essential changes which would be required to lend even the semblance of a foundation for the theory of Lamarck, respecting the growth of new organs and the gradual obliteration of others, we find nothing of the kind. For, in all these varieties of the dog, says Cuvier, the relation of the bones with each other remains essentially the same; the form of the teeth never changes in any perceptible degree, except that, in some individuals, one additional false grinder occasionally appears, sometimes on the one side, and sometimes on the other.† The greatest departure from a common type—and it constitutes the maximum of variation as yet known in the animal kingdom—is exemplified in those races of dogs which have a supernumerary toe on the hind foot with the corresponding tarsal bones; a variety analogous to one presented by six-fingered families of the human race.‡

Lamarck has thrown out as a conjecture, that the wolf may have been the original of the dog; and eminent naturalists are still divided in opinion on this subject. It seems now admitted that both species agree in the period of gestation, and Mr. Owen has been unable

* Dureau de la Malle, An. des Sci. Nat. tom. xxi. p. 53. Sept. 1830.
† Disc. Prél. p. 139. sixth edition. ‡ Ibid.

to confirm the alleged difference in the structure of a part of the intestinal canal.* Mr. Bell inclines to the opinion that all the various races of dogs have descended from one common stock, of which the wolf is the original source.

It is well known that the horse, the ox, the boar, and other domestic animals which have been introduced into South America, and have run wild in many parts, have entirely lost all marks of domesticity, and have reverted to the original characters of their species. But dogs have also become wild in Cuba, Hayti, and in all the Caribbean islands. In the course of the seventeenth century, they hunted in packs from twelve to fifty, or more, in number, and fearlessly attacked herds of wild boars and other animals. It is natural, therefore, to inquire to what form they reverted? Now, they are said by many travellers to have resembled very nearly the shepherd's dog; but it is certain that they were never turned into wolves. They were extremely savage, and their ravages appear to have been as much dreaded as those of wolves; but when any of their whelps were caught, and brought from the woods to the towns, they grew up in the most perfect submission to man.

Many examples might be adduced to prove that the extent to which the alteration of species can be pushed in the domestic state depends on the original capacity of the species to admit of variation. The horse has been as long domesticated as the dog, yet its different races depart much less widely from a common type; the ass has been still less changed, the camel scarcely at all; yet these species have probably been subjected to the influence of domestication as long as the horse.

Mummies of animals in Egyptian tombs identical with species still living.—As the advocates of the theory of transmutation trust much to the slow and insensible changes which time may work, they are accustomed to lament the absence of accurate descriptions, and figures of particular animals and plants, handed down from the earliest periods of history, such as might have afforded data for comparing the condition of species, at two periods considerably remote. But, fortunately, we are in some measure independent of such evidence: for, by a singular accident, the priests of Egypt have bequeathed to us, in their cemeteries, that information which the museums and works of the Greek philosophers have failed to transmit.

For the careful investigation of these documents, we are greatly indebted to the skill and diligence of those naturalists who accompanied the French armies during their brief occupation of Egypt: that conquest of four years, from which we may date the improvement of the modern Egyptians in the arts and sciences, and the rapid progress which has been made of late in our knowledge of the arts and sciences of their remote predecessors. Instead of wasting their whole time, as so many preceding travellers had done, in exclusively collecting human mum-

* Güldenstädt, cited by Pritchard, Phys. Hist. of Mankind, vol. i. p. 96.
† History of British Quadrupeds, p. 200. 1837.

mies, M. Geoffroy and his associates examined diligently, and sent home great numbers of embalmed bodies of consecrated animals, such as the bull, the dog, the cat, the ape, the ichneumon, the crocodile, and the ibis.

To those who have never been accustomed to connect the facts of Natural History with philosophical speculations, who have never raised their conceptions of the end and import of such studies beyond the mere admiration of isolated and beautiful objects, or the exertion of skill in detecting specific differences, it will seem incredible that amidst the din of arms, and the stirring excitement of political movements, so much enthusiasm could have been felt in regard to these precious remains.

In the official report drawn up by the Professors of the Museum at Paris, on the value of these objects, there are some eloquent passages, which may appear extravagant, unless we reflect how fully these naturalists could appreciate the bearing of the facts thus brought to light on the past history of the globe.

"It seems," say they, "as if the superstition of the ancient Egyptians had been inspired by Nature, with a view of transmitting to after ages a monument of her history. That extraordinary and eccentric people, by embalming with so much care the brutes which were the objects of their stupid adoration, have left us in their secret grottoes, cabinets of zoology almost complete. The climate has conspired with the art of embalming to preserve the bodies from corruption, and we can now assure ourselves by our own eyes what was the state of a great number of species three thousand years ago. We can scarcely restrain the transports of our imagination, on beholding thus preserved, with their minutest bones, with the smallest portions of their skin, and in every particular most perfectly recognizable, many an animal, which at Thebes or Memphis, two or three thousand years ago, had its own priests and altars."*

Among the Egyptian mummies thus procured were not only those of numerous wild quadrupeds, birds, and reptiles; but what was perhaps of still higher importance in deciding the great question under discussion, there were the mummies of domestic animals, among which those above mentioned, the bull, the dog, and the cat, were frequent. Now, such was the conformity of the whole of these species to those now living, that there was no more difference, says Cuvier, between them than between the human mummies and the embalmed bodies of men of the present day. Yet some of these animals have since that period been transported by man to almost every climate, and forced to accommodate their habits to the greatest variety of circumstances. The cat, for example, has been carried over the whole earth, and within the last three centuries, has been naturalized in every part of the new world, —from the cold regions of Canada to the tropical plains of Guiana;

* Ann. du Muséum d'Hist. Nat. tom. i. p. 234. 1802. The reporters were MM. Cuvier, Lacépède, and Lamarck.

yet it has scarcely undergone any perceptible mutation, and is still the same animal which was held sacred by the Egyptians.

Of the ox, undoubtedly, there are many very distinct races; but the bull Apis, which was led in solemn processions by the Egyptian priests, did not differ from some of those now living. The black cattle that have run wild in America, where there were many peculiarities in the climate not to be found, perhaps, in any part of the old world, and where scarcely a single plant on which they fed was of precisely the same species, instead of altering their form and habits, have actually reverted to the exact likeness of the aboriginal wild cattle of Europe.

In answer to the arguments drawn from the Egyptian mummies, Lamarck said they were identical with their living descendants in the same country, because the climate and physical geography of the banks of the Nile have remained unaltered for the last thirty centuries. But why, it may be asked, have other individuals of these species retained the same characters in many different quarters of the globe, where the climate and many other conditions are so varied?

Seeds and plants from the Egyptian tombs.—The evidence derived from the Egyptian monuments was not confined to the animal kingdom; the fruits, seeds, and other portions of twenty different plants, were faithfully preserved in the same manner; and among these the common wheat was procured by Delille, from closed vessels in the sepulchres of the kings, the grain of which retained not only their form but even their color; so effectual has proved the process of embalming with bitumen in a dry and equable climate. No difference could be detected between this wheat and that which now grows in the East and elsewhere; and in regard to the barley, I am informed by Mr. Brown, the celebrated botanist, that its identity with the grain of our own times can be tested by the closest comparison. On examining, for example, one of the seeds from Mr. Sam's Egyptian collection in the British Museum, it is found that "the structure of the husks or that part of the flower which is persistent, agrees precisely with the barley of the present day, in having one perfect flower and the filiform rudiments of a second." Some naturalists believe that the perfect identification of the ancient Egyptian cerealia with the varieties now cultivated has been carried still further, by sowing the seeds taken out of the catacombs, and raising plants from them; but we want more evidence of this fact. Certain it is, that when the experiment was recently made in the botanic garden at Kew, with 100 seeds of wheat, barley, and lentils, from the Egyptian collection before mentioned of the British Museum, not one of them would germinate.*

* I by no means wish to express an opinion that seeds cannot retain their vitality after an entombment of 3,000 years; but one of my botanical friends who entertained a philosophical doubt on this subject, being desirous of ascertaining the truth of three or four alleged instances of the germination of "mummy wheat," discovered, on communicating with several Egyptian travellers, that they had procured the grains in question, not directly from the catacombs, but from the Arabs, who are always ready to supply strangers with an

Native country of the common wheat.—And here I may observe that there is an obvious answer to Lamarck's objection, that the botanist cannot point out a country where the common wheat grows wild, unless in places where it may have been derived from neighboring cultivation.* All naturalists are well aware that the geographical distribution of a great number of species is extremely limited; that it was to be expected that every useful plant should first be cultivated successfully in the country where it was indigenous; and that, probably, every station which it partially occupied, when growing wild, would be selected by the agriculturist as best suited to it when artificially increased. Palestine has been conjectured, by a late writer on the cerealia, to have been the original habitation of wheat and barley; a supposition which is rendered the more plausible by Hebrew and Egyptian traditions, and by tracing the migrations of the worship of Ceres, as indicative of the migrations of the plant.†

If we are to infer that some one of the wild grasses has been transformed into the common wheat, and that some animal of the genus *Canis*, still unreclaimed, has been metamorphosed into the dog, merely because we cannot find the domestic dog, or the cultivated wheat, in a state of nature, we may be next called upon to make similar admissions in regard to the camel; for it seems very doubtful whether any race of this species of quadruped is now wild.

Changes in plants produced by cultivation.—But if agriculture, it will be said, does not supply examples of extraordinary changes of form and organization, the horticulturist can, at least, appeal to facts which may confound the preceding train of reasoning. The crab has been transformed into the apple; the sloe into the plum; flowers have changed their color, and become double; and these new characters can be perpetuated by seed; a bitter plant, with wavy sea-green leaves, has been taken from the sea-side, where it grew like wild charlock; has been transplanted into the garden, lost its saltness, and has been metamorphosed into two distinct vegetables, as unlike each other as is each to the parent plant—the red cabbage and the cauliflower. These, and a multitude of analogous facts, are undoubtedly among the wonders of nature, and attest more strongly, perhaps, the extent to which species may be modified, than any examples derived from the animal kingdom. But in these cases we find that we soon reach certain limits, beyond which we are unable to cause the individuals descending from the same stock to vary; while, on the other hand, it is easy to show that these extraordinary varieties could seldom arise, and could never be perpetuated in a wild state for many generations, under any imaginable combination of accidents. They may be regarded as ex-

article now very frequently in demand. The presence of an occasional grain of Indian corn or maize in several of the parcels of grain shown to my friend as coming from the catacombs confirmed his scepticism.

* Phil. Zool., tom. i. p. 227.

† L'Origine et la Patrie des Céréales, &c., Annales des Sciences Natur., tom. ix. p. 61.

treme cases, brought about by human interference, and not as phenomena which indicate a capability of indefinite modification in the natural world.

The propagation of a plant by buds or grafts, and by cuttings, is obviously a mode which nature does not employ; and this multiplication, as well as that produced by roots and layers, seems merely to operate as an extension of the life of an individual, and not as a reproduction of the species such as happens by seed. All plants increased by grafts or layers retain precisely the peculiar qualities of the individual to which they owe their origin, and, like an individual, they have only a determinate existence; in some cases longer, and in others shorter.* It seems now admitted by horticulturists, that none of our garden varieties of fruit are entitled to be considered strictly permanent, but that they wear out after a time;† and we are thus compelled to resort again to seeds; in which case there is so decided a tendency in the seedlings to revert to the original type, that our utmost skill is sometimes baffled in attempting to recover the desired variety.

Varieties of the cabbage.—The different races of cabbages afford, as was admitted, an astonishing example of deviation from a common type; but we can scarcely conceive them to have originated, much less to have lasted for several generations, without the intervention of man. It is only by strong manures that these varieties have been obtained, and in poorer soils they instantly degenerate. If, therefore, we suppose in a state of nature the seed of the wild *Brassica oleracea* to have been wafted from the sea-side to some spot enriched by the dung of animals, and to have there become a cauliflower, it would soon diffuse its seed to some comparatively sterile soils around, and the offspring would relapse to the likeness of the parent stock.

But if we go so far as to imagine the soil, in the spot first occupied, to be constantly manured by herds of wild animals, so as to continue as rich as that of a garden, still the variety could not be maintained; because we know that each of these races is prone to fecundate others, and gardeners are compelled to exert the utmost diligence to prevent cross-breeds. The intermixture of the pollen of varieties growing in the poorer soil around would soon destroy the peculiar characters of the race which occupied the highly manured tract; for, if these accidents so continually happen, in spite of our care, among the culinary varieties, it is easy to see how soon this cause might obliterate every marked singularity in a wild state.

Besides, it is well known that, although the pampered races which we rear in our gardens for use or ornament may often be perpetuated by seed, yet they rarely produce seed in such abundance, or so prolific in quality, as wild individuals; so that if the care of man were withdrawn, the most fertile variety would always, in the end, prevail over the more sterile.

* Smith's Introduction to Botany, p. 138, edit. 1807.
† See Mr. Knight's Observations, Hort. Trans., vol. ii. p. 160.

Similar remarks may be applied to the double flowers, which present such strange anomalies to the botanist. The ovarium, in such cases, is frequently abortive; and the seeds, when prolific, are generally much fewer than where the flowers are single.

Changes caused by soil.—Some curious experiments, recently made on the production of blue instead of red flowers in the *Hydrangea hortensis*, illustrate the immediate effect of certain soils on the colors of the calyx and petals. In garden-mould or compost, the flowers are invariably red; in some kinds of bog-earth they are blue; and the same change is always produced by a particular sort of yellow loam.

Varieties of the primrose.—Linnæus was of opinion that the primrose, oxlip, cowslip, and polyanthus, were only varieties of the same species. The majority of the modern botanists, on the contrary, consider them to be distinct, although some conceived that the oxlip might be a cross between the cowslip and the primrose. Mr. Herbert has lately recorded the following experiment:—"I raised from the natural seed of one umbel of a highly manured red cowslip a primrose, a cowslip, oxlips of the usual and other colors, a black polyanthus, a hose-in-hose cowslip, and a natural primrose bearing its flower on a polyanthus stalk. From the seed of that very hose-in-hose cowslip I have since raised a hose-in-hose primrose. I therefore consider all these to be only local varieties, depending upon soil and situation."* Professor Henslow, of Cambridge, has since confirmed this experiment of Mr. Herbert; so that we have an example, not only of the remarkable varieties which the florist can obtain from a common stock, but of the distinctness of analogous races found in a wild state.†

On what particular ingredient, or quality in the earth, these changes depend, has not yet been ascertained.‡ But gardeners are well aware that particular plants, when placed under the influence of certain circumstances, are changed in various ways, according to the species; and as often as the experiments are repeated, similar results are obtained. The nature of these results, however, depends upon the species, and they are, therefore, part of the specific character; they exhibit the same phenomena, again and again, and indicate certain fixed and invariable relations between the physiological peculiarities of the plant, and the influence of certain external agents. They afford no ground for questioning the instability of species, but rather the contrary; they present us with a class of phenomena, which, when they are more thoroughly understood, may afford some of the best tests for identifying species, and proving that the attributes originally conferred endure so long as any issue of the original stock remains upon the earth.

* Hort. Trans. vol. iv. p. 19.
† Loudon's Mag. of Nat. Hist., Sept. 1830, vol. iii. p. 408.
‡ Hort. Trans. vol. iii. p. 173.

CHAPTER XXXV.

WHETHER SPECIES HAVE A REAL EXISTENCE IN NATURE—*continued.*

Limits of the variability of species—Species susceptible of modification may be altered greatly in a short time, and in a few generations; after which they remain stationary—The animals now subject to man had originally an aptitude to domesticity—Acquired peculiarities which become hereditary have a close connexion with the habits or instincts of the species in a wild state—Some qualities in certain animals have been conferred with a view of their relation to man—Wild elephant domesticated in a few years, but its faculties incapable of further development.]

Variability of a species compared to that of an individual.—I endeavored, in the last chapter, to show, that a belief in the reality of species is not inconsistent with the idea of a considerable degree of variability in the specific character. This opinion, indeed, is little more than an extension of the idea which we must entertain of the identity of an individual, throughout the changes which it is capable of undergoing.

If a quadruped, inhabiting a cold northern latitude, and covered with a warm coat of hair or wool, be transported to a southern climate, it will often, in the course of a few years, shed a considerable portion of its coat, which it gradually recovers on being again restored to its native country. Even there the same changes are, perhaps, superinduced to a certain extent by the return of winter and summer. We know that the Alpine hare (*Lepus variabilis*, Pal.) and the ermine, or stoat, (*Mustela erminea*, Linn.) become white during winter, and again obtain their full color during the warmer season; that the plumage of the ptarmigan undergoes a like metamorphosis in color and quantity, and that the change is equally temporary. We are aware that, if we reclaim some wild animal, and modify its habits and instincts by domestication, it may, if it escapes, become in a few years nearly as wild and untractable as ever; and if the same individual be again retaken, it may be reduced to its former tame state. A plant is sown in a prepared soil, in order that the petals of its flowers may multiply, and their color be heightened or changed: if we then withhold our care, the flowers of this same species become again single. In these, and innumerable other instances, we must suppose that the species was produced with a certain number of qualities; and, in the case of animals, with a variety of instincts, some of which may or may not be developed according to circumstances, or which, after having been called forth, may again become latent when the exciting causes are removed.

Now, the formation of races seems the necessary consequence of such a capability in species to vary, if it be a general law that the offspring should very closely resemble the parent. But, before we can infer that there are no limits to the deviation from an original type which may be brought about in the course of an indefinite number of generations, we ought to have some proof that, in each successive generation, individuals may go on acquiring an equal amount of new peculiarities, under the influence of equal changes of circumstances. The balance of evidence, however, inclines most decidedly on the opposite side; for in all cases we find that the quantity of divergence diminishes after a few generations in a very rapid ratio.

Species susceptible of modification may be greatly altered in a few generations.—It cannot be objected, that it is out of our power to go on varying the circumstances in the same manner as might happen in the natural course of events during some great geological cycle. For in the first place, where a capacity is given to individuals to adapt themselves to new circumstances, it does not generally require a very long period for its development: if, indeed, such were the case, it is not easy to see how the modification would answer the ends proposed, for all the individuals would die before new qualities, habits, or instincts were conferred.

When we have succeeded in naturalizing some tropical plant in a temperate climate, nothing prevents us from attempting gradually to extend its distribution to higher latitudes, or to greater elevations above the level of the sea, allowing equal quantities of time, or an equal number of generations, for habituating the species to successive increments of cold. But every husbandman and gardener is aware that such experiments will fail; and we are more likely to succeed in making some plants, in the course of the first two generations, support a considerable degree of difference of temperature, than a very small difference afterwards, though we persevere for many centuries.

It is the same if we take any other cause instead of temperature; such as the quality of the food, or the kind of dangers to which an animal is exposed, or the soil in which a plant lives. The alteration in habits, form, or organization, is often rapid during a short period; but when the circumstances are made to vary farther, though in ever so slight a degree, all modification ceases, and the individual perishes. Thus some herbivorous quadrupeds may be made to feed partially on fish or flesh; but even these can never be taught to live on some herbs which they reject, and which would even poison them, although the same may be very nutritious to other species of the same natural order. So when man uses force or stratagem against wild animals, the persecuted race soon becomes more cautious, watchful, and cunning; new instincts seem often to be developed, and to become hereditary in the first two or three generations: but let the skill and address of man increase, however gradually, no farther variation can take place, no new qualities are elicited by the increasing dangers. The alteration of the habits of the species has reached a point beyond which no ulterior modification is possible, however indefinite the lapse of ages during which the new circumstances operate. Extirpation then follows, rather than such a transformation as could alone enable the species to perpetuate itself under the new state of things.

Animals now subject to man had originally an aptitude to domesticity.—It has been well observed by M. F. Cuvier and M. Dureau de la Malle, that unless some animals had manifested in a wild state an aptitude to second the efforts of man, their domestication would never have been attempted. If they had all resembled the wolf, the fox, and the hyæna, the patience of the experimentalist would have been exhausted by innumerable failures before he at last succeeded in obtaining some imperfect results; so if the first advantages derived from the cultivation of plants had been elicited by as tedious and costly a process as that by which we now make some slight additional improvements in certain races, we should have remained to this day in ignorance of the greater number of their useful qualities.

Acquired instincts of some animals become hereditary.—It is undoubtedly true, that many new habits and qualities have not only been acquired in recent times by certain races of dogs, but have been transmitted to their offspring. But in these cases it will be observed, that the new peculiarities have an intimate relation to the habits of the animal in a wild state, and therefore do not attest any tendency to a departure to an indefinite extent from the original type of the species. A race of dogs employed for hunting deer in the platform of Sante Fé, in Mexico, affords a beautiful illustration of a new hereditary instinct. The mode of attack, observes M. Roulin, which they employ consists in seizing the animal by the belly and overturning it by a sudden effort, taking advantage of the moment when the body of the deer rests only upon the fore-legs. The weight of the animal thus thrown over is often six times that of its antagonist. The dog of pure breed inherits a disposition to this kind of chase, and never attacks a deer from before while running. Even should the deer, not perceiving him, come directly upon him, the dog steps aside and makes his assault on the flank; whereas other hunting dogs, though of superior strength, and general sagacity, which are brought from Europe, are destitute of this instinct. For want of similar precautions, they are often killed by the deer on the spot, the vertebræ of their neck being dislocated by the violence of the shock.*

A new instinct has also become hereditary in a mongrel race of dogs employed by the inhabitants of the banks of the Magdalena almost exclusively in hunting the white-lipped pecari. The address of these dogs consists in restraining their ardor, and attaching themselves to no animal in particular, but keeping the whole herd in check. Now,

* M. Roulin, Ann. des Sci. Nat. tom. xvi. p. 16. 1829.

among these dogs some are found, which the very first time they are taken to the woods, are acquainted with this mode of attack; whereas, a dog of another breed starts forward at once, is surrounded by the pecari, and, whatever may be his strength, is destroyed in a moment.

Some of our countrymen, engaged of late in conducting one of the principal mining associations in Mexico, that of Real del Monte, carried out with them some English greyhounds of the best breed, to hunt the hares which abound in that country. The great platform which is the scene of sport is at an elevation of about nine thousand feet above the level of the sea, and the mercury in the barometer stands habitually at the height of about nineteen inches. It was found that the greyhounds could not support the fatigues of a long chase in this attenuated atmosphere, and before they could come up with their prey, they lay down gasping for breath; but these same animals have produced whelps which have grown up, and are not in the least degree incommoded by the want of density in the air, but run down the hares with as much ease as the fleetest of their race in this country.

The fixed and deliberate stand of the pointer has with propriety been regarded as a mere modification of a habit, which may have been useful to a wild race accustomed to wind game, and steal upon it by surprise, first pausing for an instant, in order to spring with unerring aim. The faculty of the retriever, however, may justly be regarded as more inexplicable and less easily referable to the instinctive passions of the species. M. Majendie, says a French writer in a recently published memoir, having learnt that there was a race of dogs in England which stopped and brought back game of their own accord, procured a pair, and having obtained a whelp from them, kept it constantly under his eyes, until he had an opportunity of assuring himself that, without having received any instruction, and on the very first day that it was carried to the chase, it brought back game with as much steadiness as dogs which had been schooled into the same manœuvre by means of the whip and collar.

Attributes of animals in their relation to man.—Such attainments, as well as the habits and dispositions which the shepherd's dog and many others inherit, seem to be of a nature and extent which we can hardly explain by supposing them to be modifications of instincts necessary for the preservation of the species in a wild state. When such remarkable habits appear in races of this species we may reasonably conjecture that they were given with no other view than for the use of man and the preservation of the dog, which thus obtains protection.

As a general rule, I fully agree with M. F. Cuvier, that, in studying the habits of animals, we must attempt, as far as possible, to refer their domestic qualities to modifications of instincts which are implanted in them in a state of nature; and that writer has successfully pointed out, in an admirable essay on the domestication of the mammalia*, the true

* Mem. du Mus. d'Hist. Nat.—Jameson, Ed. New Phil. Journ. Nos. 6, 7, 8.

origin of many dispositions which are vulgarly attributed to the influence of education alone. But we should go too far if we did not admit that some of the qualities of particular animals and plants may have been given solely with a view to the connection which it was foreseen would exist between them and man—especially when we see that connexion to be in many cases so intimate, that the greater number, and sometimes, as in the case of the camel, all the individuals of the species which exist on the earth are in subjection to the human race.

We can perceive in a multitude of animals, especially in some of the parasitic tribes, that certain instincts and organs are conferred for the purpose of defence or attack against some other species. Now if we are reluctant to suppose the existence of similar relations between man and the instincts of many of the inferior animals, we adopt an hypothesis no less violent, though in the opposite extreme to that which has led some to imagine the whole animate and inanimate creation to have been made solely for the support, gratification, and instruction of mankind.

Many species, most hostile to our persons or property, multiply, in spite of our efforts to repress them; others, on the contrary, are intentionally augmented many hundred fold in number by our exertions. In such instances, we must imagine the relative resources of man, and of species friendly or inimical to him, to have been prospectively calculated and adjusted. To withhold assent to this supposition, would be to refuse what we must grant in respect to the economy of nature in every other part of the organic creation; for the various species of contemporary plants and animals have obviously their relative forces, nicely balanced, and their respective tastes. passions, and instincts so contrived, that they are all in perfect harmony with each other. In no other manner could it happen that each species, surrounded, as it is, by countless dangers, should be enabled to maintain its ground for periods of considerable duration.

The docility of the individuals of some of our domestic species, extending, as it does, to attainments foreign to their natural habits and faculties, may, perhaps, have been conferred with a view to their association with man. But, lest species should be thereby made to vary indefinitely, we find that such habits are never transmissible by generation.

A pig has been trained to hunt and point game with great activity and steadiness *; and other learned individuals, of the same species, have been taught to spell; but such fortuitous acquirements never become hereditary, for they have no relation whatever to the exigencies of the animal in a wild state, and cannot, therefore, be developments of any instinctive propensities.

Influence of domestication.—An animal in domesticity, says M. F.

* In the New Forest, near Ringwood, Hants, by Mr. Toomer, keeper of Broomy Lodge. I have conversed with witnesses of the fact.

Cuvier, is not essentially in a different situation, in regard to the feeling of restraint, from one left to itself. It lives in society without constraint, because, without doubt, it was a social animal; and it conforms itself to the will of man, because it had a chief, to which, in a wild state, it would have yielded obedience. There is nothing in its new situation that is not conformable to its propensities; it is satisfying its wants by submission to a master, and makes no sacrifice of its natural inclinations. All the social animals, when left to themselves, form herds more or less numerous; and all the individuals of the same herd know each other, are mutually attached, and will not allow a strange individual to join them. In a wild state, moreover, they obey some individual, which, by its superiority, has become the chief of the herd. Our domestic species had, originally, this sociability of disposition; and no solitary species, however easy it may be *to tame it*, has yet afforded true domestic races. We merely, therefore, develope, to our own advantage, propensities which propel the individuals of certain species to draw near to their fellows.

The sheep which we have reared is induced to follow us, as it would be led to follow the flock among which it was brought up; and, when individuals of gregarious species have been accustomed to one master, it is he alone whom they acknowledge as their chief—he only whom they obey. "The elephant allows himself to be directed only by the carnac whom he has adopted; the dog itself, reared in solitude with its master, manifests a hostile disposition towards all others; and every body knows how dangerous it is to be in the midst of a herd of cows, in pasturages that are little frequented, when they have not at their head the keeper who takes care of them.

"Every thing, therefore, tends to convince us, that formerly men were only with regard to the domestic animals, what those who are particularly charged with the care of them still are—namely, members of the society which these animals form among themselves; and, that they are only distinguished, in the general mass, by the authority which they have been enabled to assume from their superiority of intellect. Thus, every social animal which recognizes man as a member, and as the chief of its herd, is a domestic animal. It might even be said, that, from the moment when such an animal admits man as a member of its society, it is domesticated, as man could not enter into such society without becoming the chief of it." *

But the ingenious author whose observations I have here cited, admits that the obedience which the individuals of many domestic species yield indifferently to every person, is without analogy in any state of things which could exist previously to their subjugation by man. Each troop of wild horses, it is true, has some stallion for its chief, who draws after him all the individuals of which the herd is composed; but when a domesticated horse has passed from hand to hand,

* Mém. du Mus. d'Hist. Nat.

and has served several masters, he becomes equally docile towards *any person*, and is subjected to the whole human race. It seems fair to presume that the capability in the instinct of the horse to be thus modified, was given to enable the species to render greater services to man; and, perhaps, the facility with which many other acquired characters become hereditary in various races of the horse, may be explicable only on a like supposition. The amble, for example, a pace to which the domestic races in some parts of Spanish America are exclusively trained, has, in the course of several generations, become hereditary, and is assumed by all the young colts before they are broken in.*

It seems, also, reasonable to conclude, that the power bestowed on the horse, the dog, the ox, the sheep, the cat, and many species of domestic fowls, of supporting almost every climate, was given expressly to enable them to follow man throughout all parts of the globe, in order that we might obtain their services, and they our protection. If it be objected that the elephant which, by the union of strength, intelligence, and docility, can render the greatest services to mankind, is incapable of living in any but the warmest latitudes, we may observe that the quantity of vegetable food required by this quadruped would render its maintenance in the temperate zones too costly, and in the arctic impossible.

Among the changes superinduced by man, none appear, at first sight, more remarkable than the perfect tameness of certain domestic races. It is well known that, at however early an age we obtain possession of the young of many unreclaimed races, they will retain, throughout life, a considerable timidity and apprehensiveness of danger; whereas, after one or two generations, the descendants of the same stock will habitually place the most implicit confidence in man. There is good reason, however, to suspect that such changes are not without analogy in a state of nature; or, to speak more correctly, in situations where man has not interfered.

We learn from Mr. Darwin, that in the Galapagos archipelago, placed directly under the equator, and nearly 600 miles west of the American continent, all the terrestrial birds, as the finches, doves, hawks, and others, are so tame, that they may be killed with a switch. One day, says this author, "a mocking bird alighted on the edge of a pitcher which I held in my hand, and began quietly to sip the water, and allowed me to lift it with the vessel from the ground." Yet formerly, when the first Europeans landed, and found no inhabitants in these islands, the birds were even tamer than now: already they are beginning to acquire that salutary dread of man which in countries long settled is natural even to young birds which have never received any injury. So in the Falkland Islands, both the birds and foxes are entirely without fear of man; whereas, in the adjoining mainland of South

* Dureau de la Malle. Ann. des Sci. Nat., tom. xxi. p. 58.

America, many of the same species of birds are extremely wild; for there they have for ages been persecuted by the natives.*

Dr. Richardson informs us, in his able history of the habits of the North American animals, that, "in the retired parts of the mountains where the hunters had seldom penetrated, there is no difficulty in approaching the Rocky Mountain sheep, which there exhibit *the simplicity of character so remarkable in the domestic species;* but where they have been often fired at, they are exceedingly wild, alarm their companions, on the approach of danger, by a hissing noise, and scale the rocks with a speed and agility that baffle pursuit."†

It is probable, therefore, that as man, in diffusing himself over the globe, has tamed many wild races, so, also, he has made many tame races wild. Had some of the larger carnivorous beasts, capable of scaling the rocks, made their way into the North American mountains before our hunters, a similar alteration in the instincts of the sheep would doubtless have been brought about.

Wild elephants domesticated in a few years.—No animal affords a more striking illustration of the principal points which I have been endeavouring to establish than the elephant; for, in the first place, the wonderful sagacity with which he accommodates himself to the society of man, and the new habits which he contracts, are not the result of time, nor of modifications produced in the course of many generations. These animals will breed in captivity, as is now ascertained, in opposition to the vulgar opinion of many modern naturalists, and in conformity to that of the ancients Ælian and Columella‡: yet it has always been the custom, as the least expensive mode of obtaining them, to capture wild individuals in the forests, usually when full grown; and, in a few years after they are taken—sometimes, it is said, in the space of a few months—their education is completed.

Had the whole species been domesticated from an early period in the history of man, like the camel, their superior intelligence would, doubtless, have been attributed to their long and familiar intercourse with the lord of the creation; but we know that a few years is sufficient to bring about this wonderful change of habits; and although the same individual may continue to receive tuition for a century afterwards, yet it makes no farther progress in the general development of its faculties. Were it otherwise, indeed, the animal would soon deserve more than the poet's epithet of "half-reasoning."

From the authority of our countrymen employed in the late Burmese war, it appears, in corroboration of older accounts, that when elephants are required to execute extraordinary tasks, they may be made to understand that they will receive unusual rewards. Some favourite dainty is shown to them, in the hope of acquiring which the work is done; and so perfectly does the nature of the contract appear to be understood,

* Darwin's Journ. in Voyage of H.M.S. Beagle, p. 475.

† Fauna Boreali-Americana, p. 273.

‡ Mr. Corse on the Habits, &c. of the Elephant, Phil. Trans., 1799.

that the breach of it, on the part of the master, is often attended with danger. In this case, a power has been given to the species to adapt their social instincts to new circumstances with surprising rapidity; but the extent of this change is defined by strict and arbitrary limits. There is no indication of a tendency to continued divergence from certain attributes with which the elephant was originally endued—no ground whatever for anticipating that, in thousands of centuries, any material alteration could ever be effected. All that we can infer from analogy is, that some more useful and peculiar races might probably be formed, if the experiment were fairly tried; and that some individual characteristic, now only casual and temporary, might be perpetuated by generation.

In all cases, therefore, where the domestic qualities exist in animals, they seem to require no lengthened process for their developement; and they appear to have been wholly denied to some classes, which, from their strength and social disposition, might have rendered great services to man; as, for example, the greater part of the quadrumana. The orang-outang, indeed, which, for its resemblance in form to man, and apparently for no other good reason, has been assumed by Lamarck to be the most perfect of the inferior animals, has been tamed by the savages of Borneo, and made to climb lofty trees, and to bring down the fruit. But he is said to yield to his masters an unwilling obedience, and to be held in subjection only by severe discipline. We know nothing of the faculties of this animal which can suggest the idea that it rivals the elephant in intelligence; much less anything which can countenance the dreams of those who have fancied that it might have been transmuted into the "dominant race." One of the baboons of Sumatra (*Simia carpolegus*) appears to be more docile, and is frequently trained by the inhabitants to ascend trees, for the purpose of gathering cocoa-nuts; a service in which the animal is very expert. He selects, says Sir Stamford Raffles, the ripe nuts, with great judgment, and pulls no more than he is ordered.* The capuchin and cacajao monkeys are, according to Humboldt, taught to ascend trees in the same manner, and to throw down fruit on the banks of the lower Orinoco.†

It is for the Lamarckians to explain how it happens that those same savages of Borneo have not themselves acquired, by dint of longing, for many generations, for the power of climbing trees, the elongated arms of the ourang, or even the prehensile tails of some American monkeys: Instead of being reduced to the necessity of subjugating stubborn and untractable brutes, we should naturally have anticipated "that their wants would have excited them to efforts, and that continued efforts would have given rise to new organs;" or rather to the re-acquisition of organs which, in a manner irreconcileable with the principle of the

* Linn. Trans. vol. xiii. p. 244.

† Pers. Narr. of Travels to the Equinoctial Regions of the New Continent in the years 1779—1804.

progressive system, have grown obsolete in tribes of men which have such constant need of them.

Recapitulation.—It follows, then, from the different facts which have been considered in this chapter, that a short period of time is generally sufficient to effect nearly the whole change which an alteration of external circumstances can bring about in the habits of a species, and that such capacity of accommodation to new circumstances is enjoyed in very different degrees, by different species.

Certain qualities appear to be bestowed exclusively with a view to the relations which are destined to exist between different species, and, among others, between certain species and man; but these latter are always so nearly connected with the original habits and propensities of each species in a wild state, that they imply no indefinite capacity of varying from the original type. The acquired habits derived from human tuition are rarely transmitted to the offspring; and when this happens, it is almost universally the case with those merely which have some obvious connexion with the attributes of the species when in a state of independence.

CHAPTER XXXVI.

WHETHER SPECIES HAVE A REAL EXISTENCE IN NATURE—*continued.*

Phenomena of hybrids—Hunter's opinions—Mules not strictly intermediate between parent species—Hybrid plants—Experiments of Kölreuter and Wiegmann—Vegetable hybrids prolific throughout several generations—Why rare in a wild state—Decandolle on hybrid plants—The phenomena of hybrids confirm the distinctness of species—Theory of the gradation in the intelligence of animals as indicated by the facial angle—Doctrine that certain organs of the fœtus in mammalia assume successively the forms of fish, reptile, and bird—Recapitulation.

Phenomena of hybrids.—We have yet to consider another class of phenomena, those relating to the production of hybrids, which have been regarded in a very different light with reference to their bearing on the question of the permanent distinctness of species; some naturalists considering them as affording the strongest of all proofs in favor of the reality of species; others, on the contrary, appealing to them as countenancing the opposite doctrine, that all the varieties of organization and instinct now exhibited in the animal and vegetable kingdoms may have been propagated from a small number of original types.

In regard to the mammifers and birds it is found that no sexual union will take place between races which are remote from each other in their habits and organization; and it is only in species that are very nearly allied that such unions produce offspring. It may be laid down as a general rule, admitting of very few exceptions among quadrupeds, that the hybrid progeny is sterile; and there seem to be no well authenticated examples of the continuance of the mule race beyond one generation. The principal number of observations and experiments relate to the mixed offspring of the horse and the ass; and in this case it is well established that the he-mule can generate, and the she-mule produce. Such cases occur in Spain and Italy, and much more frequently in the West Indies and New Holland; but these mules have never bred in cold climates, seldom in warm regions, and still more rarely in temperate countries.

The hybrid offspring of the she-ass and the stallion, the γιννος of Aristotle, and the hinnus of Pliny, differs from the mule, or the offspring of the ass and mare. In both cases, says Buffon, these animals retain more of the dam than of the sire, not only in the magnitude, but in the figure of the body: whereas, in the form of the head, limbs, and tail, they bear a greater resemblance to the sire. The same naturalist infers, from various experiments respecting cross-breeds between the he-goat and ewe, the dog and she-wolf, the goldfinch and canary-bird, that the male transmits his sex to the greatest number, and that the preponderance of males over females exceeds that which prevails where the parents are of the same species.

Hunter's opinion.—The celebrated John Hunter has observed, that the true distinction of species must ultimately be gathered from their incapacity of propagating with each other, and producing offspring capable of again continuing itself. He was unwilling, however, to admit that the horse and the ass were of the same species, because some rare instances had been adduced of the breeding of mules, although he maintained that the wolf, the dog, and the jackal were all of one species; because he had found, by two experiments, that the dog would breed both with the wolf and the jackal; and that the mule, in each case, would breed again with the dog. In these cases, however, it may be observed, that there was always one parent at least of pure breed, and no proof was obtained that a true hybrid race could be perpetuated; a fact of which I believe no examples are yet recorded, either in regard to mixtures of the horse and ass, or any other of the mammalia.

Should the fact be hereafter ascertained, that two mules can propagate their kind, we must still inquire whether the offspring may not be regarded in the light of a monstrous birth, proceeding from some accidental cause, or, rather, to speak more philosophically, from some general law not yet understood, but which may not be permitted permanently to interfere with those laws of generation by which species may, in general, be prevented from becoming blended. If, for example, we discovered that the progeny of a mule race degenerated greatly, in the first generation, in force, sagacity, or any attribute necessary for its preservation in a state of nature, we might infer that, like a monster, it is a mere temporary and fortuitous variety. Nor does it seem probable that the greater number of such monsters could ever occur unless obtained by art; for, in Hunter's experiments, stratagem or force was, in most instances, employed to bring about the irregular connexion.*

Mules not strictly intermediate between the parent species.—It seems rarely to happen that the mule offspring is truly intermediate in character between the two parents. Thus Hunter mentions that, in his experiments, one of the hybrid pups resembled the wolf much more than the rest of the litter; and we are informed by Wiegmann, that, in a litter lately obtained in the Royal Menagerie at Berlin, from a white pointer and a she-wolf, two of the cubs resembled the common wolf-dog, but the third was like a pointer with hanging ears.

There is undoubtedly a very close analogy between these phenomena and those presented by the intermixture of distinct races of the same species, both in the inferior animals and in man. Dr. Prichard, in his "Physical History of Mankind," cites examples where the peculiarities of the parents have been transmitted very unequally to the offspring; as where children, entirely white, or perfectly black, have sprung from the union of the European and the negro. Sometimes the colour or other peculiarities of one parent, after having failed to show themselves in the immediate progeny, reappear in a subsequent generation; as where a white child is born of two black parents, the grandfather having been a white.†

The same author judiciously observes that, if different species mixed their breed, and hybrid races were often propagated, the animal world would soon present a scene of confusion; its tribes would be every where blended together, and we should perhaps find more hybrid creatures than genuine and uncorrupted races.‡

Hybrid plants.—Kölreuter's experiments.—The history of the vegetable kingdom has been thought to afford more decisive evidence in favour of the theory of the formation of new and permanent species from hybrid stocks. The first accurate experiments in illustration of this curious subject appear to have been made by Kölreuter, who obtained a hybrid from two species of tobacco, *Nicotiana rustica* and *N. paniculata*, which differ greatly in the shape of their leaves, the colour of the corolla, and the height of the stem. The stigma of a plant of *N. rustica* was impregnated with the pollen of a plant of *N. paniculata*. The seed ripened, and produced a hybrid which was intermediate between the two parents, and which, like all the hybrids which this botanist brought up, had imperfect stamens. He afterwards impregnated this hybrid with the pollen of *N. paniculata*, and obtained plants which much more resembled the last. This he continued through several generations, until, by due perseverance, he actually changed the *Nicotiana rustica* into the *Nicotiana paniculata*.

The plan of impregnation adopted, was the cutting off of the anthers

* Phil. Trans. 1787. Additional Remarks, Phil. Trans. 1789. See also Essays by the late Dr. Samuel G. Morton, on Prolific Hybrids, &c.; and on Hybridity as a Test of Species.—American Journ. of Science, vol. iii. 1847.

† Prichard, vol. i. p. 217. ‡ Ibid. p. 97.

of the plant intended for fructification before they had shed pollen, and then laying on foreign pollen upon the stigma.

Wiegmann's experiments.—The same experiment has since been repeated with success by Wiegmann, who found that he could bring back the hybrids to the exact likeness of either parent, by crossing them a sufficient number of times.

The blending of the characters of the parent stocks, in many other of Wiegmann's experiments, was complete; the colour and shape of the leaves and flowers, and even the scent, being intermediate, as in the offspring of the two species of verbascum. An intermarriage, also, between the common onion and the leek (*Allium cepa* and *A. porrum*) gave a mule plant, which, in the character of its leaves and flowers, approached most nearly to the garden onion, but had the elongated bulbous root and smell of the leek.

The same botanist remarks, that vegetable hybrids, when not strictly intermediate, more frequently approach the female than the male parent species; *but they never exhibit characters foreign to both.* A re-cross with one of the original stocks generally causes the mule plant to revert towards that stock; but this is not always the case, the offspring sometimes continuing to exhibit the character of a full hybrid.

In general, the success attending the production and perpetuity of hybrids among plants depends, as in the animal kingdom, on the degree of proximity between the species intermarried. If their organization be very remote, impregnation never takes place; if somewhat less distant, seeds are formed, but always imperfect and sterile. The next degree of relationship yields hybrid seedlings, but these are barren; and it is only when the parent species are very nearly allied that the hybrid race may be perpetuated for several generations. Even in this case the best authenticated examples seem confined to the crossing of hybrids with individuals of pure breed. In none of the experiments most accurately detailed does it appear that both the parents were mules.

Wiegmann diversified as much as possible his mode of bringing about these irregular unions among plants. He often sowed parallel rows, near to each other, of the species from which he desired to breed; and, instead of mutilating, after Kölreuter's fashion, the plants of one of the parent stocks, he merely washed the pollen off their anthers. The branches of the plants in each row were then gently bent towards each other and intertwined; so that the wind, and numerous insects, as they passed from the flowers of one to those of the other species, carried the pollen and produced fecundation.

Vegetable hybrids why rare in a wild state.—The same observer saw a good exemplification of the manner in which hybrids may be formed in a state of nature. Some wallflowers and pinks had been growing in a garden, in a dry sunny situation, and their stigmas had been ripened so as to be moist, and to absorb pollen with avidity, although their anthers were not yet developed. These stigmas became impregnated by pollen blown from some other adjacent plants of the same species; but had they been of different species, and not too remote in their organization, mule races must have resulted.

When, indeed, we consider how busily some insects have been shown to be engaged in conveying anther-dust from flower to flower, especially bees, flower-eating beetles, and the like, it seems a most enigmatical problem how it can happen that promiscuous alliances between distinct species are not perpetually occurring.

How continually do we observe the bees diligently employed in collecting the red and yellow powder by which the stamens of flowers are covered, loading it on their hind legs, and carrying it to their hive for the purpose of feeding their young! In thus providing for their own progeny, these insects assist materially the process of fructification.* Few persons need be reminded that the stamens in certain plants grow on different blossoms from the pistils; and unless the summit of the pistil be touched with the fertilizing dust, the fruit does not swell, nor the seed arrive at maturity. It is by the help of bees chiefly, that the development of the fruit of many such species is secured, the powder which they have collected from the stamens being unconsciously left by them in visiting the pistils.

How often, during the heat of a summer's day, do we see the males of diœcious plants, such as the yew-tree, standing separate from the females, and sending off into the air, upon the slightest breath of wind, clouds of buoyant pollen! That the zephyr should so rarely intervene to fecundate the plants of one species with the anther-dust of others, seems almost to realize the converse of the miracle believed by the credulous herdsmen of the Lusitanian mares—

> Ore omnes versæ in Zephyrum, stant rupibus altis
> Exceptantque leves auras: et sæpe sine ullis
> Conjugiis, vento gravidæ, mirabile dictu.†

But, in the first place, it appears that there is a natural aversion in plants, as well as in animals, to irregular sexual unions; and in most of the successful experiments in the animal and vegetable world, some violence has been used in order to procure impregnation. The stigma imbibes, slowly and reluctantly, the granules of the pollen of another species, even when it is abundantly covered with it; and if it happen that, during this period, ever so slight a quantity of the anther-dust of its own species alight upon it, this is instantly absorbed, and the effect of the foreign pollen destroyed. Besides, it does not often happen that the male and female organs of fructification, in different species, arrive at a state of maturity at precisely the same time. Even where such synchronism does prevail, so that a cross impregnation is effected, the chances are very numerous against the establishment of a hybrid race.

If we consider the vegetable kingdom generally, it must be recollected that even of the seeds which are well ripened, a great part are either

* See Barton on the Geography of Plants, p. 67. † Georg. lib. iii. 273.

eaten by insects, birds, and other animals, or decay for want of room and opportunity to germinate. Unhealthy plants are the first which are cut off by causes prejudicial to the species, being usually stifled by more vigorous individuals of their own kind. If, therefore, the relative fecundity or hardiness of hybrids be in the least degree inferior, they cannot maintain their footing for many generations, even if they were ever produced beyond one generation in a wild state. In the universal struggle for existence, the right of the strongest eventually prevails; and the strength and durability of a race depend mainly on its prolificness, in which hybrids are acknowledged to be deficient.

Centaurea hybrida, a plant which never bears seed, and is supposed to be produced by the frequent intermixture of two well-known species of Centaurea, grows wild upon a hill near Turin. *Ranunculus lacerus*, also sterile, has been produced accidentally at Grenoble, and near Paris, by the union of two Ranunculi; but this occurred in gardens.*

Mr. Herbert's experiments.—Mr. Herbert, in one of his ingenious papers on mule plants, endeavors to account for their non-occurrence in a state of nature, from the circumstance that all the combinations that were likely to occur have already been made many centuries ago, and have formed the various species of botanists; but in our gardens, he says, whenever species, having a certain degree of affinity to each other, are transported from different countries, and brought for the first time into contact, they give rise to hybrid species.† But we have no data, as yet, to warrant the conclusion, that a single permanent hybrid race has ever been formed, even in gardens, by the intermarriage of two allied species brought from distant habitations. Until some fact of this kind is fairly established, and a new species, capable of perpetuating itself in a state of perfect independence of man, can be pointed out, it seems reasonable to call in question entirely this hypothetical source of new species. That varieties do sometimes spring up from cross-breeds, in a natural way, can hardly be doubted; but they probably die out even more rapidly than races propagated by grafts or layers.

Opinion of De Candolle.—De Candolle, whose opinion on a philosophical question of this kind deserves the greatest attention, has observed, in his Essay on Botanical Geography, that the *varieties* of plants range themselves under two general heads: those produced by external circumstances, and those formed by hybridity. After adducing various arguments to show that neither of these causes can explain the permanent diversity of plants indigenous in different regions, he says, in regard to the crossing of races, "I can perfectly comprehend without altogether sharing the opinion, that, where many species of the same genera occur near together, hybrid species may be formed, and I am aware that the great number of species of certain genera which are found in particular regions may be explained in this manner; but I am unable to conceive how any one can regard the same explanation as applicable to species

* Hon. and Rev. W. Herbert, Hort. Trans., vol. iv. p. 41. † Ibid.

which live naturally at great distances. If the three larches, for example, now known in the world, lived in the same localities, I might then believe that one of them was the produce of the crossing of the two others; but I never could admit that the Siberian species has been produced by the crossing of those of Europe and America. I see, then, that there exist in organized beings, permanent differences which cannot be referred to any one of the actual causes of variation, and these differences are what constitute *species*."*

Reality of species confirmed by the phenomena of hybrids.—The most decisive arguments perhaps, amongst many others, against the probability of the derivation of permanent species from cross-breeds, are to be drawn from the fact alluded to by De Candolle, of species having a close affinity to each other occurring in distinct botanical provinces, or countries inhabited by groups of distinct species of indigenous plants; for in this case naturalists, who are not prepared to go the whole length of the transmutationists, are under the necessity of admitting that, in some cases, species which approach very near to each other in their characters, were so created from their origin; an admission fatal to the idea of its being a general law of nature that a few original types only should be formed, and that all intermediate races should spring from the intermixture of those stocks.

This notion, indeed, is wholly at variance with all that we know of hybrid generation; for the phenomena entitle us to affirm, that had the types been at first somewhat distinct, *no cross-breeds would ever have been produced*, much less those prolific races which we now recognize as distinct species.

In regard, moreover, to the permanent propagation of hybrid races among animals, insuperable difficulties present themselves, when we endeavor to conceive the blending together of the different instincts and propensities of two species, so as to insure the preservation of the intermediate race. The common mule, when obtained by human art, may be protected by the power of man; but, in a wild state, it would not have precisely the same wants either as the horse or the ass; and if in consequence of some difference of this kind, it strayed from the herd, it would soon be hunted down by beasts of prey, and destroyed.

If we take some genus of insects, such as the bee, we find that each of the numerous species has some difference in its habits, its mode of collecting honey, or constructing its dwelling, or providing for its young, and other particulars. In the case of the common hive bee, the workers are described, by Kirby and Spence, as being endowed with no less than thirty distinct instincts.† So also we find that, amongst a most numerous class of spiders, there are nearly as many different modes of spinning their webs as there are species. When we recollect how complicated are the relations of these instincts with co-existing species, both

* Essai Elémentaire, &c., 3me partie.
† Intr. to Entom. vol. ii. p. 504. ed. 1817

of the animal and vegetable kingdoms, it is scarcely possible to imagine that a bastard race could spring from the union of two of these species, and retain just so much of the qualities of each parent stock as to preserve its ground in spite of the dangers which surround it.

We might also ask, if a few generic types alone have been *created* among insects, and the intermediate species have proceeded from hybridity, where are those original types, combining, as they ought to do, the elements of all the instincts which have made their appearance in the numerous derivative races? So also in regard to animals of all classes, and of plants; if species are in general of hybrid origin, where are the stocks which combine in themselves the habits, properties, and organs, of which all the intervening species ought to afford us mere modifications?

Recapitulation of the arguments from hybrids.—I shall now conclude this subject by summing up, in a few words, the results to which I have been led by the consideration of the phenomena of hybrids. It appears that the aversion of individuals of distinct species to the sexual union is common to animals and plants; and that it is only when the species approach near to each other in their organization and habits, that any offspring are produced from their connexion. Mules are of extremely rare occurrence in a state of nature, and no examples are yet known of their having procreated in a wild state. But it has been proved, that hybrids are not universally sterile, provided the parent stocks have a near affinity to each other, although the continuation of the mixed race, for several generations, appears hitherto to have been obtained only by crossing the hybrids with individuals of pure species; an experiment which by no means bears out the hypothesis that a true hybrid race could ever be permanently established.

Hence we may infer, that aversion to sexual intercourse is, in general, a good test of the distinctness of original stocks, or of *species;* and the procreation of hybrids is a proof of the near affinity of species. Perhaps, hereafter, the number of generations for which hybrids may be continued, before the race dies out (for it seems usually to degenerate rapidly), may afford the zoologist and botanist an experimental test of the difference in the degree of affinity of allied species.

I may also remark, that if it could have been shown that a single permanent species had ever been produced by hybridity (of which there is no satisfactory proof), it might certainly have lent some countenance to the notions of the ancients respecting the gradual deterioration of created things, but none whatever to Lamarck's theory of their progressive perfectibility, for observations have hitherto shown that there is a tendency in mule animals and plants to degenerate in organization.

It was before remarked, that the theory of progressive development arose partly from an attempt to ingraft the doctrines of the transmutationists upon one of the most popular generalizations in geology. But we have seen in the ninth chapter, that the modern researches of geologists have broken at many points the chain of evidence once supposed to exist in favor of the doctrine, that, at each successive period in the earth's history, animals and plants of a higher grade, or more complex organization, have been created. The recent origin of man, and the absence of all signs of any rational being holding an analogous relation to former states of the animate world, affords one, and perhaps in the present state of science the only argument of much weight in support of the hypothesis of a progressive scheme; but none whatever in favor of the fancied evolution of one species out of another.

Theory of the gradation of intellect as shown by the facial angle.— When the celebrated anatomist, Camper, first attempted to estimate the degrees of sagacity of different animals, and of the races of man, by the measurement of the facial angle, some speculators were bold enough to affirm that certain Simiæ, or apes, differed as little from the more savage races of men, as those do from the human race in general; and that a scale might be traced from "apes with foreheads villanous low" to the African variety of the human species, and from that to the European. The facial angle was measured by drawing a line from the prominent centre of the forehead to the most advanced part of the lower jaw-bone, and observing the angle which it made with the horizontal line; and it was affirmed, that there was a regular series of such angles from birds to the mammalia.

The gradation from the dog to the monkey was said to be perfect, and from that again to man. One of the ape tribe has a facial angle of 42°; and another, which approximated nearest to man in figure, an angle of 50°. To this succeeds (longo sed proximus intervallo) the head of the African negro, which, as well as that of the Calmuck, forms an angle of 70°; while that of the European contains 80°. The Roman painters preferred the angle of 95°; and the character of beauty and sublimity so striking in some works of Grecian sculpture, as in the head of the Apollo, and in the Medusa of Sisocles, is given by an angle which amounts to 100°.*

A great number of valuable facts and curious analogies in comparative anatomy were brought to light during the investigations which were made by Camper, John Hunter, and others, to illustrate this scale of organization; and their facts and generalizations must not be confounded with the fanciful systems which White and others deduced from them.†

That there is some connexion between an elevated and capacious forehead, in certain races of men, and a large developement of the intellectual faculties, seems highly probable; and that a low facial angle is frequently accompanied with inferiority of mental powers, is certain; but the attempt to trace a gradual scale of intelligence through the different species of animals accompanying the modifications of the form of the scull, is a mere visionary speculation. It has been found necessary to

* Prichard's Phys. Hist. of Mankind, vol. i. p. 159.
† Ch. White on the Regular Gradation in Man, &c. 1799.

exaggerate the sagacity of the ape tribe at the expense of the dog; and strange contradictions have arisen in the conclusions deduced from the structure of the elephant; some anatomists being disposed to deny the quadruped the intelligence which he really possesses, because they found that the volume of his brain was small in comparison to that of the other mammalia; while others were inclined to magnify extravagantly the superiority of his intellect, because the vertical height of his skull is so great when compared to its horizontal length

Different races of men are all of one species.—It would be irrelevant to our subject if we were to enter into a farther discussion on these topics; because, even if a graduated scale of organization and intelligence could have been established, it would prove nothing in favor of a tendency, in each species, to attain a higher state of perfection. I may refer the reader to the writings of Blumenbach, Prichard, Lawrence, and more recently Latham*, for convincing proofs that the varieties of form, color, and organization of different races of men, are perfectly consistent with the generally received opinion, that all the individuals of the species have originated from a single pair; and, while they exhibit in man as many diversities of a physiological nature as appear in any other species, they confirm also the opinion of the slight deviation from a common standard of which species are capable.

The power of existing and multiplying in every latitude, and in every variety of situation and climate, which has enabled the great human family to extend itself over the habitable globe, is partly, says Lawrence, the result of physical constitution, and partly of the mental prerogative of man. If he did not possess the most enduring and flexible corporeal frame, his arts would not enable him to be the inhabitant of all climates, and to brave the extremes of heat and cold, and the other destructive influences of local situation.† Yet, notwithstanding this flexibility of bodily frame, we find no signs of indefinite departure from a common standard, and the intermarriages of individuals of the most remote varieties are not less fruitful than between those of the same tribe.

Tiedemann on the brain of the fœtus in vertebrated animals.—There is yet another department of anatomical discovery to which I must allude, because it has appeared to some persons to afford a distant analogy, at least, to that progressive development by which some of the inferior species may have been gradually perfected into those of more complex organization. Tiedemann found, and his discoveries have been most fully confirmed and elucidated by M. Serres, that the brain of the fœtus, in the highest class of vertebrated animals, assumes, in succession, forms, bearing a certain degree of resemblance to those which belong to fishes, reptiles, and birds, before it acquires the additions and modifications which are peculiar to the mammiferous tribe; so that, in the passage from the embryo to the perfect mammifer, there is a typical representa-

* R. G. Latham, The Nat. Hist. of the Varieties of Man, 8vo. London, 1850.
† Lawrence, Lectures on Phys. Zool. and Nat. Hist. of Man, p. 190. Ed. 1823.

tion, it is said, of all those transformations which the primitive species are supposed to have undergone, during a long series of generations, between the present period and the remotest geological era.

"If you examine the brain of the mammalia," says M. Serres, "at an early stage of uterine life, you perceive the cerebral hemispheres consolidated, as in fish, in two vesicles, isolated one from the other; at a later period, you see them affect the configuration of the cerebral hemispheres of reptiles; still later again, they present you with the forms of those of birds; finally they acquire, at the era of birth, and sometimes later, the permanent forms which the adult mammalia present.

"The cerebral hemispheres, then, arrive at the state which we observe in the higher animals only by a series of successive metamorphoses. If we reduce the whole of these evolutions to four periods, we shall see, that in the first are born the cerebral lobes of fishes; and this takes place homogeneously in all classes. The second period will give us the organization of reptiles; the third, the brain of birds; and the fourth, the complex hemispheres of mammalia.

"If we could develope the different parts of the brain of the inferior classes, we should make, in succession, a reptile out of a fish, a bird out of a reptile, and a mammiferous quadruped out of a bird. If, on the contrary, we could starve this organ in the mammalia, we might reduce it successively to the condition of the brain of the three inferior classes.

"Nature often presents us with this last phenomenon in monsters, but never exhibits the first. Among the various deformities which organized beings may experience, they never pass the limits of their own classes to put on the forms of the class above them. Never does a fish elevate itself so as to assume the form of the brain of a reptile; nor does the latter ever attain that of birds; nor the bird that of the mammifer. It may happen that a monster may have two heads; but the conformation of the brain always remains circumscribed narrowly within the limits of its class."*

Dr. Clark of Cambridge, in a memoir on "Fœtal Development" (1845), has shown that the concurrent labours of Valentin, Ratké, and Bischoff disprove the reality of the supposed anatomical analogy between the embryo condition of certain organs in the higher orders, and the perfect structure of the same organs in animals of an inferior class. The hearts and brains, for example, of birds and mammals do not pass through forms which are permanent in fishes and reptiles; there is only just so much resemblance as may point to a unity of plan running through the organization of the whole series of vertebrated animals; but which lends no support whatever to the notion of a gradual transmutation of one species into another; least of all of the passage, in the course of many generations, from an animal of a more simple to one of a more complex structure.

* E. R. A. Serres, Anatomie comparée du Cerveau, illustrated by numerous plates, tome i. 1824.

Recapitulation.—For the reasons, therefore, detailed in this and the two preceding chapters, we may draw the following inferences in regard to the reality of *species* in nature:—

1st. That there is a capacity in all species to accommodate themselves, to a certain extent, to a change of external circumstances, this extent varying greatly, according to the species.

2ndly. When the change of situation which they can endure is great, it is usually attended by some modifications of the form, colour, size, structure, or other particulars; but the mutations thus superinduced are governed by constant laws, and the capability of so varying, forms part of the permanent specific character.

3dly. Some acquired peculiarities, of form, structure, and instinct, are transmissible to the offspring; but these consist of such qualities and attributes only as are intimately related to the natural wants and propensities of the species.

4thly. The entire variation from the original type, which any given kind of change can produce, may usually be effected in a brief period of time, after which no farther deviation can be obtained by continuing to alter the circumstances, though ever so gradually; indefinite divergence, either in the way of improvement or deterioration, being prevented, and the least possible excess beyond the defined limits being fatal to the existence of the individual.

5thly. The intermixture of distinct species is guarded against by the aversion of the individuals composing them to sexual union, or by the sterility of the mule offspring. It does not appear that true hybrid races have ever been perpetuated for several generations, even by the assistance of man; for the cases usually cited relate to the crossing of mules with individuals of pure species, and not to the intermixture of hybrid with hybrid.

6thly. From the above considerations, it appears that species have a real existence in nature; and that each was endowed, at the time of its creation, with the attributes and organization by which it is now distinguished.

CHAPTER XXXVII.

LAWS WHICH REGULATE THE GEOGRAPHICAL DISTRIBUTION OF SPECIES.

Analogy of climate not attended with identity of species—Botanical geography—Stations—Habitations—Distinct provinces of indigenous plants—Vegetation of islands—Marine vegetation—In what manner plants become diffused—Effects of wind, rivers, marine currents—Agency of animals—Many seeds pass through the stomachs of animals and birds undigested—Agency of man in the dispersion of plants, both voluntary and involuntary—Its analogy to that of the inferior animals.

NEXT to determining the question whether species have a real existence, the consideration of the laws which regulate their geographical distribution is a subject of primary importance to the geologist. It is only by studying these laws with attention, by observing the positions which groups of species occupy at present, and inquiring how these may be varied in the course of time by migrations, by changes in physical geography, and other causes, that we can hope to learn whether the duration of species be limited, or in what manner the state of the animate world is affected by the endless vicissitudes of the inanimate.

Different regions inhabited by distinct species.—That different regions of the globe are inhabited by entirely distinct animals and plants, is a fact which has been familiar to all naturalists since Buffon first pointed out the want of *specific* identity between the land quadrupeds of America and those of the Old World. The same phenomenon has, in later times, been forced in a striking manner upon our attention, by the examination of New Holland, where the indigenous species of animals and plants were found to be, almost without exception, distinct from those known in other parts of the world.

But the extent of this parcelling out of the globe amongst different *nations*, as they have been termed, of plants and animals—the universality of a phenomenon so extraordinary and unexpected, may be considered as one of the most interesting facts clearly established by the advance of modern science.

Scarcely fourteen hundred species of plants appear to have been known and described by the Greeks, Romans, and Arabians. At present, more than three thousand species are enumerated, as natives of our own island.* In other parts of the world there have been now collected (1846) upwards of 100,000 species, specimens of which are preserved in European herbariums. It was not to be supposed, therefore, that the ancients should have acquired any correct notions respecting what may be called the geography of plants, although the influence of climate on the character of the vegetation could hardly have escaped their observation.

* Barton's Lectures on the Geography of Plants, p. 2. 1827.

Antecedently to investigation, there was no reason for presuming that the vegetable productions, growing wild in the eastern hemisphere, should be unlike those of the western, in the same latitude; nor that the plants of the Cape of Good Hope should be unlike those of the south of Europe; situations where the climate is little dissimilar. The contrary supposition would have seemed more probable, and we might have anticipated an almost perfect identity in the animals and plants which inhabit corresponding parallels of latitude. The discovery, therefore, that each separate region of the globe, both of the land and water, is occupied by distinct groups of species, and that most of the exceptions to this general rule may be referred to disseminating causes now in operation, is eminently calculated to excite curiosity, and to stimulate us to seek some hypothesis respecting the first introduction of species which may be reconcileable with such phenomena.

Botanical geography.—A comparison of the *plants* of different regions of the globe affords results more to be depended upon in the present state of our knowledge than those relating to the animal kingdom, because the science of botany is more advanced, and probably comprehends a great proportion of the total number of the vegetable productions of the whole earth. Humboldt, in several eloquent passages of his Personal Narrative, was among the first to promulgate philosophical views on this subject. Every hemisphere, says this traveller, produces plants of different species; and it is not by the diversity of climates that we can attempt to explain why equinoctial Africa has no Laurinæ, and the New World no Heaths; why the Calceolariæ are found only in thes outhern hemisphere; why the birds of the continent of India glow with colors less splendid than the birds of the hot parts of America: finally, why the tiger is peculiar to Asia, and the ornithorhynchus to New Holland.*

"We can conceive," he adds, "that a small number of the families of plants, for instance, the Musaceæ and the Palms, cannot belong to very cold regions, on account of their internal structure and the importance of certain organs; but we cannot explain why no one of the family of Melastomas vegetates north of the parallel of thirty degrees; or why no rose-tree belongs to the southern hemisphere. Analogy of climates is often found in the two continents without identity of productions."†

The luminous essay of De Candolle on "Botanical Geography" presents us with the fruits of his own researches and those of Humboldt, Brown, and other eminent botanists, so arranged, that the principal phenomena of the distribution of plants are exhibited in connexion with the causes to which they are chiefly referrible.‡ "It might not, perhaps, be difficult," observes this writer, "to find two points, in the United States and in Europe, or in Equinoctial America and Africa, which present all the same circumstances: as, for example, the same temperature,

* Pers. Nar., vol. v. p. 180.

† Ibid.

‡ Essai Elémentaire de Géographie Botanique. Extrait du 18me vol. du Dict. des Sci. Nat.

the same height above the sea, a similar soil, an equal dose of humidity; yet nearly all, *perhaps all*, the plants in these two similar localities shall be distinct. A certain degree of analogy, indeed, of aspect, and even of structure, might very possibly be discoverable between the plants of the two localities in question; but the *species* would in general be different. Circumstances, therefore, different from those which now determine the *stations*, have had an influence on the *habitations* of plants."

Stations and habitations of plants.—As I shall frequently have occasion to speak of the *stations* and *habitations* of plants in the technical sense in which the terms are used in the above passage, I may remind the geologist that station indicates the peculiar nature of the locality where each species is accustomed to grow, and has reference to climate, soil, humidity, light, elevation above the sea, and other analogous circumstances; whereas, by habitation is meant a general indication of the country where a plant grows wild. Thus the *station* of a plant may be a salt-marsh, a hill-side, the bed of the sea, or a stagnant pool. Its *habitation* may be Europe, North America, or New Holland, between the tropics. The study of stations has been styled the topography, that of habitations the geography, of botany. The terms thus defined, express each a distinct class of ideas, which have been often confounded together, and which are equally applicable in zoology.

In farther illustration of the principle above alluded to, that difference of longitude, independently of any influence of temperature, is accompanied by a great, and sometimes a complete, diversity in the species of plants, De Candolle observes, that, out of 2891 species of phænogamous plants described by Pursh, in the United States, there are only 385 which are found in northern or temperate Europe. MM. Humboldt and Bonpland, in all their travels through equinoctial America, found only twenty-four species (these being all Cyperaceæ and Gramineæ) common to America and any part of the Old World. They collected, it is true, chiefly on the mountains, or the proportion would have been larger; for Dr. J. Hooker informs me that many tropical plants of the New World are identical with African species. Nevertheless, the general discordance of these Floras is very striking. On comparing New Holland with Europe, Mr. Brown ascertained that, out of 4100 species, discovered in Australia, there were only 166 common to Europe, and of this small number there were some few which may have been transported thither by man. Almost all of the 166 species were cryptogamic, and the rest consist, in nearly every case, of phænogamous plants which also inhabit intervening regions.

But what is still more remarkable, in the more widely separated parts of the ancient continent, notwithstanding the existence of an uninterrupted land-communication, the diversity in the specific character of the respective vegetations is almost as striking. Thus there is found one assemblage of species in China, another in the countries bordering the Black Sea and the Caspian, a third in those surrounding the Mediterranean, a fourth in the great platforms of Siberia and Tartary, and so forth.

The distinctness of the groups of indigenous plants, in the same parallel of latitude, is greatest where continents are disjoined by a wide expanse of ocean. In the northern hemisphere, near the pole, where the extremities of Europe, Asia, and America unite or approach near to one another, a considerable number of the same species of plants are found, common to the three continents. But it has been remarked, that these plants, which are thus so widely diffused in the arctic regions, are also found in the chain of the Aleutian islands, which stretch almost across from America to Asia, and which may probably have served as the channel of communication for the partial blending of the Floras of the adjoining regions. It has, indeed, been observed to be a general rule, that plants found at two points very remote from each other occur also in places intermediate.

Dr. J. Hooker informs me that in high latitudes in the southern ocean, in spite of the great extent of the sea, Floras of widely disconnected islands contain many species in common. Perhaps icebergs, transporting to vast distances not only stones, but soil with the seeds of plants, may explain this unusually wide diffusion of insular plants.

In islands very distant from continents the total number of plants is comparatively small; but a large proportion of the species are such as occur nowhere else. In so far as the Flora of such islands is not peculiar to them, it contains, in general, species common to the nearest main lands.* The islands of the great southern ocean exemplify these rules; the easternmost containing more American, and the western more Indian plants.† Madeira and Teneriffe contain many species, and even entire genera, peculiar to them; but they have also plants in common with Portugal, Spain, the Azores, and the north-west coast of Africa.‡

In the Canaries, out of 533 species of phænogamous plants, it is said that 310 are peculiar to these islands, and the rest identical with those of the African continent; but in the Flora of St. Helena, which is so far distant even from the western shores of Africa, there have been found, out of thirty native species of the phænogamous class, only *one or two* which are to be found in any other part of the globe. On the other hand, of sixty cryptogamic plants, collected by Dr. J. Hooker in the same island, twelve only were peculiar.

The natural history of the Galapagos archipelago, described by Mr. Darwin, affords another very instructive illustration of the laws governing the geographical distribution of plants and animals in islands. This group consists of ten principal islands, situated in the Pacific Ocean, under the equator, about 600 miles westward of the coast of South America. As they are all formed of volcanic rocks, many of the craters, of which there are about 2000 in number, having a very fresh aspect, we may

* Prichard, vol. i. p. 36. Brown, Appendix to Flinders.
† Foster, Observations, &c.
‡ Humboldt, Pers. Nar., vol. i. p. 270 of the translation. Prichard, Phys. Hist. of Mankind, vol. i. p. 37.

regard the whole as much more modern in origin than the mass of the adjoining continent; yet neither has the Flora nor Fauna been derived from South America, but consist of species for the most part indigenous, yet stamped with a character decidedly South American.

What is still more singular, there is a difference between the species inhabiting the different islands. Of flowering plants, for example, there are 185 species at present known, and forty cryptogamic, making together 225. One hundred of the former class are new species, probably confined to this archipelago; and of the rest, ten at least have been introduced by man. Of twenty-one species of *Compositæ*, all but one are peculiar, and they belong to twelve genera, no less than ten of which genera are confined to the Galapagos. Dr. Hooker observes, that the type of this Flora has an undoubted relation to that of the western side of South America, and he detects in it no affinity with that of the numerous islands scattered over other parts of the Pacific. So in regard to the birds, reptiles, land-shells, and insects, this archipelago, standing as it does in the Pacific Ocean, is zoologically part of America. Although each small island is not more than fifty or sixty miles apart, and most of them are in sight of each other, formed of precisely the same rocks, rising nearly to an equal height, and placed under a similar climate, they are tenanted each by a different set of beings, the tortoises, mocking-thrushes, finches, beetles, scarcely any of them ever ranging over the whole, and often not even common to any two of the islands.

"The archipelago," says Mr. Darwin, "is a little world within itself, or rather a satellite attached to America; whence it has derived a few stray colonists, and has received the general character of its indigenous productions. One is astonished," he adds, "at the amount of creative force displayed on so many small, barren, and rocky islands, and still more so, at its diverse, yet analogous action on points so near each other. I have said that the Galapagos archipelago might be called a satellite attached to America, but it should rather be called a group of satellites physically similar, organically distinct, yet intimately related to each other, and all related in a marked, though much lesser degree, to the great American continent."*

Number of botanical provinces.—De Candolle has enumerated twenty great botanical provinces inhabited by indigenous or aboriginal plants; and although many of these contain a variety of species which are common to several others, and sometimes to places very remote, yet the lines of demarcation are, upon the whole, astonishingly well defined.† Nor is it likely that the bearing of the evidence on which these general views are founded will ever be materially affected, since they are already confirmed by the examination of nearly one hundred thousand species of plants.

* Voyage of the Beagle, 2d edition, 1845, p. 377.

† See a farther subdivision, by which twenty-seven provinces are made, by M. Alph. De Candolle, son of De Candolle. Monogr. des Campanulées. Paris, 1830.

The entire change of opinion which the contemplation of these phenomena has brought about is worthy of remark. The first travellers were persuaded that they should find, in distant regions, the plants of their own country, and they took a pleasure in giving them the same names. It was some time before this illusion was dissipated; but so fully sensible did botanists at last become of the extreme smallness of the number of phænogamous plants common to different continents, that the ancient Floras fell into disrepute. All grew diffident of the pretended identifications; and we now find that every naturalist is inclined to examine each supposed exception with scrupulous severity.* If they admit the fact, they begin to speculate on the mode whereby the seeds may have been transported from one country into the other, or enquire on which of two continents the plant was indigenous, assuming that a species, like an individual, cannot have two birthplaces.

Marine vegetation.—The marine vegetation is divisible into different systems, like those prevailing on the land; but they are much fewer, as we might have expected, the temperature of the ocean being more uniform than that of the atmosphere, and consequently the dispersion of species from one zone to another being less frequently checked by the intervention of uncongenial climates. The proportion also of land to sea throughout the globe being small, the migration of marine plants is not so often stopped by barriers of land, as is that of the terrestrial species by the ocean. The number of hydrophytes, as they are termed, is very considerable, and their stations are found to be infinitely more varied than could have been anticipated; for while some plants are covered and uncovered daily by the tide, others live at the depth of several hundred feet. Among the known provinces of Algæ, we may mention, 1st, The north circumpolar, from lat. 60° N. to the pole; 2dly, The North Atlantic or the region of Fucus proper and Delesseriæ, extending from lat. 40° N. to lat. 60° N.; 3dly, That of the Mediterranean, which may be regarded as a sub-region of the *fourth* or warmer temperate zone of the Atlantic, between lat. 23° N. and lat. 40° N.; 5thly, The Tropical Atlantic, in which Sargassum, Rhodomelia, Corallinea, and Siphonia abound; 6thly, The South Atlantic, where the Fucus reappears; 7thly, The Antarctic American, comprehending from Chili to Cape Horn, the Falkland Islands, and thence round the world south of latitude 50° S.; 8thly, The Australian and New Zealand, which is very peculiar, being characterized, among other generic forms, by Cystoseiriæ and Fuceæ; 9thly, The Indian Ocean and Red Sea; and, 10thly, The Chinese and Japanese seas.† In addition to the above provinces, there are several others not yet well determined in the Pacific Ocean and elsewhere. There are, however, many species which range through several of these geographical regions of subaqueous vegetation, being common to very remote countries; as, for example, to the coasts of

* De Candolle, Essai Elémen. de Géog. Botan., p. 45.

† I am indebted for the above sketch of distinct regions of algæ to my friend Dr. Joseph Hooker, who refers the botanical student to the labors of Dr. Harvey, of Trinity College, Dublin.

Europe and the United States, and others, to Cape Horn and Van Diemen's Land, the same plants extending also for the most part to the New Zealand sea. Of the *species* strictly antarctic (excluding the New Zealand and Tasmanian groups) Dr. Hooker has identified not less than a fifth part of the whole with British Algæ! Yet is there a much smaller proportion of cosmopolite species among the Algæ than among the terrestrial cellular plants, such as lichens, mosses, and Hepaticæ.

It must always be borne in mind, that the distinctness alluded to between the provinces, whether of subaqueous or terrestrial plants, relates strictly to *species*, and not to forms. In regard to the numerical preponderance of certain forms, and many peculiarities of internal structure, there is usually a marked agreement in the vegetable productions of districts placed in corresponding latitudes, and under similar physical circumstances, however remote their position. Thus there are innumerable points of analogy between the vegetation of the Brazils, equinoctial Africa, and India; and there are also points of difference wherein the plants of these regions are distinguishable from all extra-tropical groups. But there is a very small proportion of the entire number of species common to the three continents. The same may be said, if we compare the plants of the United States with that of the middle of Europe; the species are distinct, but the forms are often so analogous, as to have been styled "geographical representatives." There are very few *species* of phænogamous plants, says Dr. J. Hooker, common to Van Diemen's Land, New Zealand, and Fuegia, but a great many *genera*, and some of them are confined to those three distant regions of the southern hemisphere, being in many instances each severally represented by a single species. The same naturalist also observes that the southern temperate as well as the antarctic regions, possess each of them representatives of some of the genera of the analogous climates of the opposite hemisphere; but very few of the species are identical unless they be such as are equally diffused over other countries, or which inhabit the Andes, by the aid of which they have evidently effected their passage southwards.

Manner in which plants become diffused.— Winds.—Let us now consider what means of diffusion, independently of the agency of man, are possessed by plants, whereby, in the course of ages, they may be enabled to stray from one of the botanical provinces above mentioned to another, and to establish new colonies at a great distance from their birthplace.

The principal of the inanimate agents provided by nature for scattering the seeds of plants over the globe, are the movements of the atmosphere and of the ocean, and the constant flow of water from the mountains to the sea. To begin with the winds: a great number of seeds are furnished with downy and feathery appendages, enabling them, when ripe, to float in the air, and to be wafted easily to great distances by the most gentle breeze. Other plants are fitted for dispersion by means of an attached wing, as in the case of the fir tree, so that they are caught up by the wind as they fall from the cone, and are carried to a distance. Amongst the comparatively small number of plants known to Linnæus, no less than 138 genera are enumerated as having winged seeds.

As winds often prevail for days, weeks, or even months together, in the same direction, these means of transportation may sometimes be without limits; and even the heavier grains may be borne through considerable spaces, in a very short time, during ordinary tempests; for strong gales, which can sweep along grains of sand, often move at the rate of about forty miles an hour, and if the storm be very violent, at the rate of fifty-six miles.* The hurricanes of tropical regions, which root up trees and throw down buildings, sweep along at the rate of ninety miles an hour; so that, for however short a time they prevail, they may carry even the heavier fruits and seeds over friths and seas of considerable width, and doubtless are often the means of introducing into islands the vegetation of adjoining continents. Whirlwinds are also instrumental in bearing along heavy vegetable substances to considerable distances. Slight ones may frequently be observed in our fields, in summer carrying up haycocks into the air, and then letting fall small tufts of hay far and wide over the country; but they are sometimes so powerful as to dry up lakes and ponds, and to break off the boughs of trees, and carry them up in a whirling column of air.

Franklin tells us, in one of his letters, that he saw, in Maryland, a whirlwind which began by taking up the dust which lay in the road, in the form of a sugar loaf with the pointed end downwards, and soon after grew to the height of forty or fifty feet, being twenty or thirty in diameter. It advanced in a direction contrary to the wind; and although the rotary motion of the column was surprisingly rapid, its onward progress was sufficiently slow to allow a man to keep pace with it on foot. Franklin followed it on horseback, accompanied by his son, for three quarters of a mile, and saw it enter a wood, where it twisted and turned round large trees with surprising force. These were carried up in a spiral line, and were seen flying in the air, together with boughs and innumerable leaves, which, from their height, appeared reduced to the apparent size of flies. As this cause operates at different intervals of time throughout a great portion of the earth's surface, it may be the means of bearing not only plants but insects, land testacea and their eggs, with many other species of animals, to points which they could never otherwise have reached, and from which they may then begin to propagate themselves again as from a new centre.

Distribution of cryptogamous plants.—It has been found that a great numerical proportion of the exceptions to the limitation of species to certain quarters of the globe occur in the various tribes of cryptogamic plants. Linnæus observed that, as the germs of plants of this class, such as mosses, fungi, and lichens, consist of an impalpable powder, the particles of which are scarcely visible to the naked eye, there is no difficulty to account for their being dispersed throughout the atmosphere, and carried to every point of the globe, where there is a station fitted for them.

* Annuaire du Bureau des Longitudes.

Lichens in particular ascend to great elevations, sometimes growing two thousand feet above the the line of perpetual snow, at the utmost limits of vegetation, and where the mean temperature is nearly at the freezing point. This elevated position must contribute greatly to facilitate the dispersion of those buoyant particles of which their fructification consists.*

Some have inferred, from the springing up of mushrooms whenever particular soils and decomposed organic matter are mixed together, that the production of fungi is accidental, and not analogous to that of perfect plants. But Fries, whose authority on these questions is entitled to the highest respect, has shown the fallacy of this argument in favor of the old doctrine of equivocal generation. "The sporules of fungi," says this naturalist, "are so infinite, that in a single individual of *Reticularia maxima*, I have counted above ten millions, and so subtile as to be scarcely visible, often resembling thin smoke; so light that they may be raised perhaps by evaporation into the atmosphere, and dispersed in so many ways by the attraction of the sun, by insects, wind, elasticity, adhesion, &c., that it is difficult to conceive a place from which they may be excluded."†

The club-moss called *Lycopodium cernuum* affords a striking example of a cryptogamous plant universally distributed over all equinoctial countries. It scarcely ever passes beyond the northern tropic, except in one instance, where it appears around the hot-springs in the Azores, although it is neither an inhabitant of the Canaries nor Madeira. Doubtless its microscopic sporules are everywhere present, ready to germinate on any spot where they can enjoy throughout the year the proper quantity of warmth, moisture, light, and other conditions essential to the species.

Almost every lichen brought home from the southern hemisphere by the antarctic expedition under Sir James Ross, amounting to no less than 200 species, was ascertained to be also an inhabitant of the northern hemisphere, and almost all of them European.

Agency of rivers and currents.—In considering, in the next place, the instrumentality of the aqueous agents of dispersion, I cannot do better than cite the words of one of our ablest botanical writers. "The mountain stream or torrent," observes Keith, "washes down to the valley the seeds which may accidentally fall into it, or which it may happen to sweep from its banks when it suddenly overflows them. The broad and majestic river, winding along the extensive plain, and traversing the continents of the world, conveys to the distance of many hundreds of miles the seeds that may have vegetated at its source. Thus the southern shores of the Baltic are visited by seeds which grew in the interior of Germany, and the western shores of the Atlantic by seeds that have been generated in the interior of America."‡ Fruits, moreover, indigenous to America and the West Indies, such as that of the *Mimosa scandens*, the cashewnut and others, have been known to be drifted across the Atlantic

* Linn., Tour in Lapland, vol. ii. p. 282.
† Fries, cited by Lindley, Introd. to Nat. Syst. of Botany.
‡ System of Physiological Botany, vol. ii. p. 405.

by the Gulf stream, on the western coasts of Europe, in such a state that they might have vegetated had the climate and soil been favourable. Among these the *Guilandina Bonduc*, a leguminous plant, is particularly mentioned, as having been raised from a seed found on the west coast of Ireland.*

Sir Hans Sloane states, that several kinds of beans cast ashore on the Orkney Isles, and Ireland, but none of which appear to have naturalized themselves, are derived from trees which grow in the West Indies, and many of them in Jamaica. He conjectures that they might have been conveyed by rivers into the sea, and then by the Gulf stream to greater distances, in the same manner as the sea-weed called *Lenticula marina*, or Sargasso, which grows on the rocks about Jamaica, is known to be "carried by the winds and current towards the coast of Florida, and thence into the North American ocean, where it lies very thick on the surface of the sea."†

The absence of liquid matter in the composition of seeds renders them comparatively insensible to heat and cold, so that they may be carried without detriment through climates where the plants themselves would instantly perish. Such is their power of resisting the effects of heat, that Spallanzani mentions some seeds that germinated after having been boiled in water.‡ Sir John Herschel informs me that he has sown at the Cape of Good Hope the seeds of the *Acacia lophanta* after they had remained for twelve hours in water of 140° Fahrenheit, and they germinated far more rapidly than unboiled seeds. He also states that an eminent botanist, Baron Ludwig, could not get the seeds of a species of cedar to grow at the Cape till they were thoroughly boiled.

When therefore, a strong gale, after blowing violently off the land for a time, dies away, and the seeds alight upon the surface of the waters, or wherever the ocean, by eating away the sea-cliffs, throws down into its waves plants which would never otherwise reach the shores, the tides and currents become active instruments in assisting the dissemination of almost all classes of the vegetable kingdom. The pandanus and many other plants have been distributed in this way over the islands of the Pacific. I have before called attention (p. 618.) to the interesting fact that one-fifth of all the algæ found in the antarctic regions in 1841–3, by Dr. J. Hooker, were of species common to the British seas. He has suggested that cold currents which prevail from Cape Horn to the equator, and are there met by other cold water, may by their direct influence, as well as by their temperature, facilitate the passage of antarctic species to the Arctic Ocean. In like manner the migration of certain marine animals from the southern to the northern hemisphere may have been brought about by the same cause.

In a collection of six hundred plants from the neighborhood of the river Zaire, in Africa, Mr. Brown found that thirteen species were also

* Brown, Append. to Tuckey, No. v. p. 481.
† Phil. Trans. 1696.
‡ System of Physiological Botany, vol. ii. p. 403.

met with on the opposite shores of Guiana and Brazil. He remarked that most of these plants were found only on the lower parts of the river Zaire, and were chiefly such as produced seeds capable of retaining their vitality a long time in the currents of the ocean. Dr. J. Hooker informs me that after an examination of a great many insular floras, he has found that no one of the large natural orders is so rich in species common to other countries, as the Leguminosæ. The seeds in this order, which comprises the largest proportion of widely diffused littoral species, are better adapted than those of any other plants for water-carriage.

The migration of plants aided by islands.—Islands, moreover, and even the smallest rocks, play an important part in aiding such migrations; for when seeds alight upon them from the atmosphere, or are thrown up by the surf, they often vegetate, and supply the winds and waves with a repetition of new and uninjured crops of fruit and seeds. These may afterwards pursue their course through the atmosphere, or along the surface of the sea, in the same direction. The number of plants found at any given time on an islet affords us no test whatever of the extent to which it may have co-operated towards this end, since a variety of species may first thrive there and then perish, and be followed by other chance-comers like themselves. If neither St. Helena nor Ascension have promoted the botanical intercourse between the Old and New Worlds, we may easily account for the fact by remembering that they are not only extremely minute and isolated spots, but are also bounded by lofty and precipitous shores without beaches, where the seeds of foreign species could readily establish themselves.

Currents and winds in the arctic regions drift along icebergs covered with an alluvial soil, on which herbs and pine-saplings are seen growing, which may often continue to vegetate on some distant shore where the ice-island is stranded.

Dispersion of marine plants.—With respect to marine vegetation, the seeds, being in their native element, may remain immersed in water without injury for indefinite periods, so that there is no difficulty in conceiving the diffusion of species wherever uncongenial climates, contrary currents, and other causes do not interfere. All are familiar with the sight of the floating sea-weed,

> "Flung from the rock on ocean's foam to sail,
> Where'er the surge may sweep, the tempest's breath prevail."

Remarkable accumulations of that species of sea-weed generally known as gulf-weed, or sargasso, occur on each side of the equator in the Atlantic, Pacific, and Indian Oceans. Columbus and other navigators, who first encountered these banks of algæ in the Northern Atlantic, compared them to vast inundated meadows, and state that they retarded the progress of their vessels. The most extensive bank is a little west of the meridian of Fayal, one of the Azores, between latitudes 35° and 36°: violent north-winds sometimes prevail in this space, and drive the sea-weed to low latitudes, as far as the 24th or even the 20th degree.* Along the northern edge of the Gulf stream Dr. Hooker found *Fucus nodosus*, and *F. serratus*, which he traced all the way from lat. 36° N. to England.

The hollow pod-like receptacle in which the seeds of many algæ are lodged, and the filaments attached to the seed-vessels of others, seem intended to give buoyancy; and I may observe that these hydrophytes are in general *proliferous*, so that the smallest fragment of a branch can be developed into a perfect plant. The seeds, moreover, of the greater number of species are enveloped with a mucous matter like that which surrounds the eggs of some fish, and which not only protects them from injury, but serves to attach them to floating bodies or to rocks.

Agency of animals in the distribution of plants.—But we have as yet considered part only of the fertile resources of nature for conveying seeds to a distance from their place of growth. The various tribes of animals are busily engaged in furthering an object whence they derive such important advantages. Sometimes an express provision is found in the structure of seeds to enable them to adhere firmly by prickles, hooks, and hairs, to the coats of animals, or feathers of the winged tribe, to which they remain attached for weeks, or even months, and are borne along into every region whither birds or quadrupeds may migrate. Linnæus enumerates fifty genera of plants, and the number now known to botanists is much greater, which are armed with hooks, by which, when ripe, they adhere to the coats of animals. Most of these vegetables, he remarks, require a soil enriched with dung. Few have failed to mark the locks of wool hanging on the thorn-bushes, wherever the sheep pass, and it is probable that the wolf or lion never give chase to herbivorous animals without being unconsciously subservient to this part of the vegetable economy.

A deer has strayed from the herd when browsing on some rich pasture, when he is suddenly alarmed by the approach of his foe. He instantly takes to flight, dashing through many a thicket, and swimming across many a river and lake. The seeds of the herbs and shrubs which have adhered to his smoking flanks are washed off again by the waters. The thorny spray is torn off, and fixes itself in its hairy coat, until brushed off again in other thickets and copses. Even on the spot where the victim is devoured many of the seeds which he had swallowed immediately before the chase may be left on the ground uninjured, and ready to spring up in a new soil.

The passage, indeed, of undigested seeds through the stomachs of animals is one of the most efficient causes of the dissemination of plants, and is of all others, perhaps, the most likely to be overlooked. Few are ignorant that a portion of the oats eaten by a horse preserve their germinating faculty in the dung. The fact of their being still nutritious is not lost on the sagacious rook. To many, says Linnæus, it seems extraordinary, and something of a prodigy, that when a field is well tilled and sown with the best wheat, it frequently produces darnel or the wild oat,

* Greville, Introduction to Algæ Britannicæ, p. 12.

especially if it be manured with new dung; they do not consider that the fertility of the smaller seeds is not destroyed in the stomachs of animals.*

Agency of birds.—Some birds of the order Passeres devour the seeds of plants in great quantities, which they eject again in very distant places, without destroying its faculty of vegetation: thus a flight of larks will fill the cleanest field with a great quantity of various kinds of plants, as the melilot trefoil (*Medicago lupulina*), and others whose seeds are so heavy that the wind is not able to scatter them to any distance.† In like manner, the blackbird and misselthrush, when they devour berries in too great quantities, are known to consign them to the earth undigested in their excrement.‡

Pulpy fruits serve quadrupeds and birds as food, while their seeds, often hard and indigestible, pass uninjured through the intestines, and are deposited far from their original place of growth in a condition peculiarly fit for vegetation.§ So well are the farmers, in some parts of England, aware of this fact, that when they desire to raise a quickset hedge in the shortest possible time, they feed turkeys with the haws of the common white-thorn (*Cratægus Oxyacantha*), and then sow the stones which are ejected in their excrement, whereby they gain an entire year in the growth of the plant.‖ Birds, when they pluck cherries, sloes, and haws, fly away with them to some convenient place; and when they have devoured the fruit, drop the stone into the ground. Captain Cook, in his account of the volcanic island of Tanna, one of the New Hebrides, which he visited in his second voyage, makes the following interesting observation:—"Mr. Forster, in his botanical excursion this day, shot a pigeon, in the craw of which was a wild nutmeg."¶ It is easy, therefore, to perceive, that birds in their migrations to great distances, and even across seas, may transport seeds to new isles and continents.

The sudden deaths to which great numbers of frugivorous birds are annually exposed must not be omitted as auxiliary to the transportation of seeds to new habitations. When the sea retires from the shore, and leaves fruits and seeds on the beach, or in the mud of estuaries, it might, by the returning tide, wash them away again, or destroy them by long immersion; but when they are gathered by land birds which frequent the sea side, or by waders and water-fowl, they are often borne inland; and if the bird to whose crop they have been consigned is killed, they may be left to grow up far from the sea. Let such an accident happen but once in a century, or a thousand years, it will be sufficient to spread many of the plants from one continent to another; for in estimating the activity of these causes, we must not consider whether they act slowly in relation

* Linnæus, Amœn. Acad., vol. ii. p. 409.
† Amœn. Acad., vol. iv. Essay 75. § 8.
‡ Ibid., vol. vi. § 22.
§ Smith's Introd. to Phys. and Syst. Botany, p. 304. 1807.
‖ This information was communicated to me by Professor Henslow, of Cambridge.
¶ Book iii. ch. iv.

to the period of our observation, but in reference to the duration of species in general.

Let us trace the operation of this cause in connection with others. A tempestuous wind bears the seeds of a plant many miles through the air, and then delivers them to the ocean; the oceanic current drifts them to a distant continent; by the fall of the tide they become the food of numerous birds, and one of these is seized by a hawk or eagle, which, soaring across hill and dale to a place of retreat, leaves, after devouring its prey, the unpalatable seeds to spring up and flourish in a new soil.

The machinery before adverted to, is so capable of disseminating seeds over almost unbounded spaces, that were we more intimately acquainted with the economy of nature, we might probably explain all the instances which occur of the aberration of plants to great distances from their native countries. The real difficulty which must present itself to every one who contemplates the present geographical distribution of species, is the small number of exceptions to the rule of the non-intermixture of different groups of plants. Why have they not, supposing them to have been ever so distinct originally, become more blended and confounded together in the lapse of ages?

Agency of man in the dispersion of plants.—But in addition to all the agents already enumerated as instrumental in diffusing plants over the globe, we have still to consider man—one of the most important of all. He transports with him, into every region, the vegetables which he cultivates for his wants, and is the involuntary means of spreading a still greater number which are useless to him, or even noxious. "When the introduction of cultivated plants," says De Candolle, "is of recent date, there is no difficulty in tracing their origin; but when it is of high antiquity, we are often ignorant of the true country of the plants on which we feed. No one contests the American origin of the maize or the potatoe; nor the origin, in the Old World, of the coffee-tree, and of wheat. But there are certain objects of culture, of very ancient date, between the tropics, such for example as the banana, of which the origin cannot be verified. Armies, in modern times, have been known to carry, in all directions, grain and cultivated vegetables from one extremity of Europe to the other; and thus have shown us how, in more ancient times, the conquests of Alexander, the distant expeditions of the Romans, and afterwards the crusades, may have transported many plants from one part of the world to the other."*

But, besides the plants used in agriculture, the numbers which have been naturalized by accident, or which man has spread unintentionally, is considerable. One of our old authors, Josselyn, gives a catalogue of such plants as had, in his time, sprung up in the colony since the English planted and kept cattle in New England. They were two-and-twenty in number. The common nettle was the first which the settlers noticed; and the plantain was called by the Indians "Englishman's foot," as if it sprung from their footsteps.†

* De Candolle, Essai Elémen. &c., p. 50.

† Quarterly Review, vol. xxx. p. 8.

"We have introduced every where," observes De Candolle, "some weeds which grow among our various kinds of wheat, and which have been received, perhaps, originally from Asia along with them. Thus, together with the Barbary wheat, the inhabitants of the south of Europe have sown, for many ages, the plants of Algiers and Tunis. With the wools and cottons of the East, or of Barbary, there are often brought into France the grains of exotic plants, some of which naturalize themselves. Of this I will cite a striking example. There is, at the gate of Montpellier, a meadow set apart for drying foreign wool, *after it has been washed.* There hardly passes a year without foreign plants being found naturalized in this drying-ground. I have gathered there *Centaurea parviflora*, *Psoralea palæstina*, and *Hypericum crispum*." This fact is not only illustrative of the aid which man lends inadvertently to the propagation of plants, but it also demonstrates the multiplicity of seeds which are borne about in the woolly and hairy coats of wild animals.

The same botanist mentions instances of plants naturalized in seaports by the ballast of ships; and several examples of others which have spread through Europe from botanical gardens, so as to have become more common than many indigenous species.

It is scarcely a century, says Linnæus, since the Canadian erigeron, or flea-bane, was brought from America to the botanical garden at Paris; and already the seeds have been carried by the winds so that it is diffused over France, the British islands, Italy, Sicily, Holland, and Germany.* Several others are mentioned by the Swedish naturalist, as having been dispersed by similar means. The common thorn-apple (*Datura Stramonium*), observes Willdenow, now grows as a noxious weed throughout all Europe, with the exception of Sweden, Lapland and Russia. It came from the East Indies and Abyssinia to us, and was thus universally spread by certain quacks, who used its seeds as an emetic.† The same plant is now abundant throughout the greater part of the United States, along road-sides and about farm-yards. The yellow monkey-flower, *Mimulus luteus*, a plant from the north-west region of America, has now established itself in various parts of England, and is spreading rapidly.

In hot and ill-cultivated countries, such naturalization takes place more easily. Thus the *Chenopodium ambrosioides*, sown by Mr. Burchell on a point of St. Helena, multiplied so fast in four years as to become one of the commonest weeds in the island, and it has maintained its ground ever since 1845.‡

The most remarkable proof, says De Candolle, of the extent to which man is unconsciously the instrument of dispersing and naturalizing species, is found in the fact, that in New Holland, America, and the Cape of Good Hope, the aboriginal European species exceed in number all the others which have come from any distant regions; so that, in this instance, the influence of man has surpassed that of all the other causes which tend

* Essay on the Habitable Earth, Amœn. Acad., vol. ii. p. 409.
† Principles of Botany, p. 389.
‡ Ibid.

to disseminate plants to remote districts. Of nearly 1600 British flowering plants, it is supposed that about 300 species are naturalized; but a large proportion of these would perish with the discontinuance of agriculture.

Although we are but slightly acquainted, as yet, with the extent of our instrumentality in naturalizing species, yet the facts ascertained afford no small reason to suspect that the number which we introduce unintentionally exceeds all those transported by design. Nor is it unnatural to suppose that the functions, which the inferior beings, extirpated by man, once discharged in the economy of nature, should devolve upon the human race. If we drive many birds of passage from different countries, we are probably required to fulfil their office of carrying seeds, eggs of fish, insects, mollusks, and other creatures, to distant regions: if we extirpate quadrupeds, we must replace them not merely as consumers of the animal and vegetable substances which they devour, but as disseminators of plants, and of the inferior classes of the animal kingdom. I do not mean to insinuate that the very same changes which man brings about, would have taken place by means of the agency of other species, but merely that he supersedes a certain number of agents; and so far as he disperses plants unintentionally, or against his will, his intervention is strictly analogous to that of the species so extirpated.

I may observe, moreover, that if, at former periods, the animals inhabiting any given district have been partially altered by the extinction of some species, and the introduction of others, whether by new creations or by immigration, a change must have taken place in regard to the particular plants conveyed about with them to foreign countries. As, for example, when one set of migratory birds is substituted for another, the countries from and to which seeds are transported are immediately changed. Vicissitudes, therefore, analogous to those which man has occasioned, may have previously attended the springing up of new relations between species in the vegetable and animal worlds.

It may also be remarked, that if man is the most active agent in enlarging, so also is he in circumscribing the geographical boundaries of particular plants. He promotes the migration of some, he retards that of other species; so that, while in many respects he appears to be exerting his power to blend and confound the various provinces of indigenous species, he is, in other ways, instrumental in obstructing the fusion into one group of the inhabitants of contiguous provinces.

Thus, for example, when two botanical regions exist in the same great continent, such as *the European region*, comprehending the central parts of Europe, and those surrounding the Mediterranean, and *the Oriental region*, as it has been termed, embracing the countries adjoining the Black Sea and the Caspian, the interposition between these of thousands of spuare miles of cultivated lands, opposes a new and powerful barrier against the mutual interchange of indigenous plants. Botanists are well aware that garden plants naturalize and diffuse themselves with great facility in comparatively unreclaimed countries, but spread themselves slowly and with difficulty in districts highly cultivated. There are many obvious causes for this difference ; by drainage and culture the natural variety of stations is diminished, and those stray individuals by which the passage of a species from one fit station to another is effected, are no sooner detected by the agriculturist, than they are uprooted as weeds. The larger shrubs and trees, in particular, can scarcely ever escape observation, when they have attained a certain size, and will rarely fail to be cut down if unprofitable.

The same observations are applicable to the interchange of the insects, birds, and quadrupeds of two regions situated like those above alluded to. No beasts of prey are permitted to make their way across the intervening arable tracts. Many birds, and hundreds of insects, which would have found some palatable food amongst the various herbs and trees of the primeval wilderness, are unable to subsist on the olive, the vine, the wheat, and a few trees and grasses favored by man. In addition, therefore, to his direct intervention, man, in this case, operates indirectly to impede the dissemination of plants, by intercepting the migration of animals, many of which would otherwise have been active in transporting seeds from one province to another.

Whether, in the vegetable kingdom, the influence of man will tend, after a considerable lapse of ages, to render the geographical range of *species in general* more extended, as De Candolle seems to anticipate, or whether the compensating agency above alluded to will not counterbalance the exceptions caused by our naturalizations, admits at least of some doubt. In the attempt to form an estimate on this subject, we must be careful not to underrate, or almost overlook, as some appear to have done, the influence of man in checking the diffusion of plants, and restricting their distribution to narrower limits.

CHAPTER XXXVIII.

LAWS WHICH REGULATE THE GEOGRAPHICAL DISTRIBUTION OF SPECIES—*continued.*

Geographical distribution of animals—Buffon on specific distinctness of quadrupeds of Old and New World—Doctrine of "natural barriers"—Different regions of indigenous mammalia—Europe—Africa—India, and Indian Archipelago—Australia—North and South America—Quadrupeds in islands—Range of the Cetacea—Dispersion of quadrupeds—Their powers of swimming—Migratory instincts—Drifting of animals on ice-floes—On floating islands of drift-timber—Migrations of Cetacea—Habitations of birds—Their migrations and facilities of diffusion—Distribution of reptiles, and their power of dissemination.

Geographical distribution of animals.—Although in speculating on "philosophical possibilities," said Buffon, "the same temperature might have been expected, all other circumstances being equal, to produce the same beings in different parts of the globe, both in the animal and vegetable kingdoms, yet it is an undoubted fact, that when America was discovered, its indigenous quadrupeds were all dissimilar to those previously known in the Old World. The elephant, the rhinoceros, the hippopotamus, the camelopard, the camel, the dromedary, the buffalo, the horse, the ass, the lion, the tiger, the apes, the baboons, and a number of other mammalia, were nowhere to be met with on the new continent; while in the old, the American species, of the same great class, were nowhere to be seen—the tapir, the lama, the pecari, the jaguar, the couguar, the agouti, the paca, the coati, and the sloth."

These phenomena, although few in number relatively to the whole animate creation, were so striking and so positive in their nature, that the great French naturalist caught sight at once of a general law in the geographical distribution of organic beings, namely, the limitation of groups of distinct species to regions separated from the rest of the globe by certain natural barriers. It was, therefore, in a truly philosophical spirit that, relying on the clearness of the evidence obtained respecting the larger quadrupeds, he ventured to call in question the identifications announced by some contemporary naturalists of species of animals said to be common to the southern extremities of America and Africa.*

The migration of quadrupeds from one part of the globe to another, observes Dr. Prichard, is prevented by uncongenial climates and the branches of the ocean which intersect continents. "Hence, by a

* Buffon, vol. v.—On the Virginian Opossum.

reference to the geographical site of countries, we may divide the earth into a certain number of regions fitted to become the abodes of particular groups of animals, and we shall find, on inquiry, that each of these provinces, thus conjecturally marked out, is actually inhabited by a distinct nation of quadrupeds." * It will be observed that the language of Buffon respecting "natural barriers," which has since been so popular, would be wholly without meaning if the geographical distribution of organic beings had not led naturalists to adopt very generally *the doctrine of specific centres*, or, in other words, to believe that each species, whether of plant or animal, originated in a single birth-place. Reject this view, and the fact that not a single native quadruped is common to Australia, the Cape of Good Hope, and South America, can in no ways be explained by adverting to the wide extent of intervening ocean, or to the sterile deserts, or the great heat or cold of the climates, through which each species must have passed, before it could migrate from one of those distant regions to another. It might fairly be asked of one who talked of impassable barriers, why the same kangaroos, rhinoceroses, or lamas, should not have been created simultaneously in Australia, Africa, and South America! The horse, the ox, and the dog, although foreign to these countries until introduced by man, are now able to support themselves there in a wild state, and we can scarcely doubt that many of the quadrupeds at present peculiar to Australia, Africa, and South America, might have continued in like manner to inhabit each of the three continents had they been indigenous or could they once have got a footing there as new colonists.

At the same time every zoologist will be willing to concede, that even if the departure of each species from a single centre had not appeared to be part of the plan of Nature, the range of species in general must have become limited, under the influence of a variety of causes, especially in the class of terrestrial mammalia. Scarcely any one of these could be expected to retain as fair a claim to the title of cosmopolite as man, although even the human race, fitted as it is by its bodily constitution and intellectual resources to spread very widely over the earth, is far from being strictly cosmopolite. It is excluded both from the arctic and antarctic circles, from many a wide desert and the summits of many mountain-chains; and lastly, from three-fourths of the globe covered by water, where there are large areas very prolific in animal life, even in the highest order of the vertebrate class. But the *habitations* of species are, as before stated, in reference to plants (see above, p. 614), circumscribed by causes different from those which determine their *stations*, and these causes are clearly connected with the time and place of the original creation of each species.

As the names and characters of land quadrupeds are much better

* Prichard's Phys. Hist. of Mankind, vol. i. p. 54.

known to the general reader than those of other great families of the animal kingdom, I shall select this class to exemplify the zoological provinces into which species are divisible, confining myself, however, to those facts which may help to elucidate some principle, or rule apparently followed by the Author of Nature, in regard to that "mystery of mysteries," the first peopling of the earth with living beings.* First, then, the *European region* comprehends, besides Europe, the borders of the Mediterranean, and even the north of Africa, and extends into Asia, beyond the Oural mountains and the Caspian. Although the species are almost all peculiar, the number of characteristic *genera* is remarkably small. The bear, the fox, the hare, the rabbit, the deer, and almost every European form is found equally in several of the other large provinces of mammalia, where the species are distinct. Even the mole (*Talpa*), although confined to the northern parts of the old world, ranges eastwards, as far as the Himalaya mountains.

2dly. The *African* Fauna, on the other hand, is singularly rich in generic forms, not met with in a living state in any other region. The hippopotamus, for example, of which two very distinct species are known, the giraffe, the Chimpanzee, the blue-faced baboon, the four-fingered monkeys (*Colubus*), many carnivora, such as *Proteles*, allied to the hyæna, and a multitude of other forms, are exclusively African. A few of the species inhabiting the northern confines of this continent, such as the dromedary, lion, and jackall, are also common to Asia; and a much larger number of *forms* belong equally to the great Asiatic province, the species being distinct. The elephant, for example, of Africa is smaller, has a rounder head, and larger ears than the Indian one, and has only three instead of four nails on each hind foot. In like manner, not one of three African species of Rhinoceros agrees with one of the three Indian kinds.

3dly. The *Southern* region of *Africa*, where that continent extends into the temperate zone, constitutes another separate zoological province, surrounded as it is on three sides by the ocean, and cut off from the countries of milder climate in the northern hemisphere, by the intervening torrid zone. In many instances, this region contains the same genera which are found in temperate climates to the northward of the line: but then the southern are different from the northern species. Thus, in the south we find the quagga and the zebra; in the north, the horse, the ass, and the jiggetai of Asia.

The south of Africa is spread out into fine level plains from the tropic to the Cape. In this region, says Pennant, besides the horse genus, of which five species have been found, there are also peculiar species of

* In the above enumeration of the leading zoological provinces of land quadrupeds I have been most kindly assisted by Mr. Waterhouse of the British Museum, author of a most able and comprehensive work on the "Natural History of the Mammalia," now in the course of publication. London, Bailliere, 1846.

rhinoceros, the hog, and the hyrax, among pachydermatous races; and amongst the ruminating, the Cape buffalo, and a variety of remarkable antelopes, as the springbok, the oryx, the gnou, the leucophoë, the pygarga, and several others.*

4thly. The assemblage of quadrupeds in *Madagascar* affords a striking illustration of the laws before alluded to, as governing the distribution of species in islands. Separated from Africa by the Mozambique channel, which is 300 miles wide, Madagascar forms, with two or three small islands in its immediate vicinity, a zoological province by itself, all the species except one, and nearly all the genera, being peculiar. The only exception consists of a small insectivorous quadruped (*Centetes*), found also in the Mauritius, to which place it is supposed to have been taken in ships. The most characteristic feature of this remarkable fauna consists in the number of quadrumana of the Lemur family, no less than six genera of these monkeys being exclusively met with in this island, and a seventh genus of the same, called *Galago*, which alone has any foreign representative, being found, as we might from analogy have anticipated, in the nearest main land. Had the species of quadrupeds in Madagascar agreed with those of the contiguous parts of Africa, as do those of England with the rest of Europe, the naturalist would have inferred that there had been a land communication since the period of the coming in of the existing quadrupeds, whereas we may now conclude that the Mozambique channel has constituted an insuperable barrier to the fusion of the continental fauna with that of the great island during the whole period that has elapsed since the living species were created.

5thly. Another of the great nations of terrestrial mammalia is that of *India*, containing a great variety of peculiar forms, such as the sloth-bear (*Prochilus*), the musk-deer (*Moscus*), the nylghau, the gibbon or long-armed ape, and many others. 6thly. A portion of the islands of the *Indian archipelago* might, perhaps, be considered by some geologists as an appendage of the same province. In fact, we find in the large islands of Java, Sumatra, and Borneo, the same genera, for the most part, as on the continent of India, and some of the same species, *e. g.* the tapir (*Tapirus Malayanus*), the rhinoceros of Sumatra, and some others. Most of the species, however, are distinct, and each island has many, and even a few genera, peculiar to itself. Between eighty and ninety species are known to inhabit Java, and nearly the same number occur in Sumatra. Of these, more than half are common to the two islands. Borneo, which is much less explored, has yielded already upwards of sixty species, more than half of which are met with either in Java or Sumatra. Of the species inhabiting Sumatra and not found in Java, Borneo contains the greater portion. Upon the whole, if these three large islands were united, and a fusion of their respective indigenous mammalia should take place, they would present a fauna related to that of continental India, and comprising about as many species as

* Pennant's Hist. of Quadrupeds, cited by Prichard, Phys. Hist. of Mankind, vol. i. p. 66.

we might expect from analogy to discover in an area of equal extent. The Philippine Islands are peopled with another assemblage of species generically related to the great Indian type.

7thly. But the islands of Celebes, Amboina, Timor, and *New Guinea*, constitute a different region of mammalia more allied to the Australian type, as having an intermixture of marsupial quadrupeds, yet showing an affinity also to the Indian in such forms as the deer (*Cervus*), the weasel (*Viverra*), the pig (*Sus*), the Macaque monkey (*Cercopithecus*), and others. As we proceed in a south-westerly direction, from Celebes to Amboina and thence to New Guinea, we find the Indian types diminishing in number, and the Australian (*i. e.* marsupial forms) increasing. Thus in New Guinea seven species of pouched quadrupeds have been detected, and among them two singular tree-kangaroos; yet only one species of the whole seven, viz. the flying opossum (*Petauris ariel*), is common to the Indian archipelago and the main land of Australia. The greater the zoological affinity, therefore, between the latter and the New Guinea fauna, although it seems in some way connected with geographical proximity, is not to be explained simply by the mutual migration of species from the one to the other.

8thly. When *Australia* was discovered, its land quadrupeds, belonging almost exclusively to the marsupial or pouched tribe, such as the kangaroos, wombats, flying opossums, kangaroo-rats, and others. some feeding on herbs and fruits, others carnivorous, were so novel in their structure and aspect, that they appeared to the naturalist almost as strange as if they were the inhabitants of some other planet. We learn from the recent investigations of Mr. Waterhouse,* that no less than 170 species of marsupial quadrupeds have now been determined, and of the whole number all but thirty-two are exclusively restricted to Australia. Of these thirty-two, nine belong to the islands in the Indian archipelago before mentioned, and the other twenty-three are all species of opossum inhabiting the tropical parts of South America, or a few of them extending into Mexico and California, and one, the Virginian opossum, into the United States.

9thly. It only remains for me to say something of the mammiferous fauna of *North* and *South America*. It has often been said that, where the three continents of Asia, Europe, and North America, approach very near to each other towards the pole, the whole arctic region forms one zoological and botanical province. The narrow straits which separate the old and new world are frozen over in winter, and the distance is farther lessened by intervening islands. Many plants and animals of various classes have accordingly spread over all the arctic lands, being sometimes carried in the same manner as the polar bear, when it is drifted on floating ice from Greenland to Iceland. But on a close inspection of the arctic mammalia, it has been found of late years that a very small number of the American species are identical with those of

* Natural History of the Mammalia, vol. i., on the Marsupials. London, Bailliere, 1846.

Europe or Asia. The genera are, in great part, the same or nearly allied; but the species are rarely identical, and are often very unlike, as in the case of the American badger and that of Europe. Some of the *genera* of arctic America, such as the musk ox (*Ovibos*), are quite peculiar, and the distinctness of the fauna of the great continents goes on increasing in proportion as we trace them southwards, or as they recede farther from each other, and become more and more separated by the ocean. At length we find that the three groups of tropical mammalia, belonging severally to America, Africa, and India, have not a single species in common.

The predominant influence of climate over all the other causes which limit the range of species in the mammalia is perhaps nowhere so conspicuously displayed as in North America. The arctic fauna, so admirably described by Sir John Richardson, has scarcely any species in common with the fauna of the state of New York, which is 600 miles farther south, and comprises about forty distinct mammifers. If again we travel farther south about 600 miles, and enter another zone, running east and west, in South Carolina, Georgia, Alabama, and the contiguous states, we again meet with a new assemblage of land quadrupeds, and this again differs from the fauna of Texas, where frosts are unknown. It will be observed that on this continent there are no great geographical barriers running east and west, such as high snow-clad mountains, barren deserts, or wide arms of the sea, capable of checking the free migration of species from north to south. But notwithstanding the distinctness of those zones of indigenous mammalia, there are some species, such as the buffalo (*Bison Americanus*), the racoon (*Procyon lotor*), and the Virginian opossum (*Didelphis Virginiana*), which have a wider habitation, ranging almost from Canada to the Gulf of Mexico; but they form exceptions to the general rule. The opossum of Texas (*Didelphis carnivora*) is different from that of Virginia, and other species of the same genus inhabit westward of the Rocky Mountains, in California, for example, where almost all the mammalia differ from those of the United States.

10thly. The *West Indian* land quadrupeds are not numerous, but several of them are peculiar; and 11thly, *South America* is the most distinct, with the exception of Australia, of all the provinces into which the mammalia can be classed geographically. The various genera of monkeys, for example, belong to the family Platyrrhini, a large natural division of the quadrumana, so named from their widely separated nostrils. They have a peculiar dentition, and many of them prehensile tails, and are entirely unknown in other quarters of the globe. The sloths and armadillos, the true blood-sucking bats or vampyres (*Phyllostomidæ*), the capybara, the largest of the rodents, the carnivorous coatimondi (*Nasua*), and a great many other forms, are also exclusively characteristic of South America.

"In Peru and Chili," says Humboldt, "the region of the grasses, which is at an elevation of from 12,300 to 15,400 feet, is inhabited by crowds of lama, guanaco, and alpaca. These quadrupeds, which here represent the genus camel of the ancient continent, have not extended themselves either to Brazil or Mexico; because, during their journey, they must necessarily have descended into regions that were too hot for them."* In this passage it will be seen that the doctrine of "specific centres" is tacitly assumed.

Quadrupeds in Islands.—Islands remote from continents, especially those of small size, are either destitute of quadrupeds, except such as have been conveyed to them by man, or contain species peculiar to them. In the Galapagos archipelago no indigenous quadrupeds were found except one mouse, which is supposed to be distinct from any hitherto found elsewhere. A peculiar species of fox is indigenous in the Falkland Islands, and a rat in New Zealand, which last country, notwithstanding its magnitude, is destitute of other mammalia, except bats, and these, says Dr. Prichard, may have made their way along the chain of islands which extend from the shores of New Guinea far into the Southern Pacific. The same author remarks, that among the various groups of fertile islands in the Pacific, no quadrupeds have been met with except the rat and a few bats as above mentioned, and the dog and hog, which appear to have been conveyed thither by the natives from New Guinea. "Rats are to be found even on some desert islands, whither they may have been conveyed by canoes which have occasionally approached the shore. It is known, also, that rats occasionally swim in large numbers to considerable distances."†

Geographical range of the Cetacea.—It is natural to suppose that the geographical range of the different species of Cetacea should be less correctly ascertained than that of the terrestrial mammifers. It is, however, well known that the whales which are obtained by our fishers in the South Seas are distinct from those of the North; and the same dissimilarity has been found in all the other marine animals, of the same class, so far as they have yet been studied by naturalists.

Dispersion of quadrupeds.—Let us now inquire what facilities the various land quadrupeds enjoy of spreading themselves over the surface of the earth. In the first place, as their numbers multiply, all of them, whether they feed on plants, or prey on other animals, are disposed to scatter themselves gradually over as wide an area as is accessible to them. But before they have extended their migrations over a large space, they are usually arrested either by the sea, or a zone of uncongenial climate, or some lofty and unbroken chain of mountains, or a tract already occupied by a hostile and more powerful species.

Their powers of swimming.—Rivers and narrow friths can seldom interfere with their progress; for the greater part of them swim well, and few are without this power when urged by danger and pressing want. Thus, amongst beasts of prey, the tiger is seen swimming about

* Description of the Equatorial Regions.
† Prichard, Phys. Hist. of Mankind, vol. i. p. 75.

among the islands and creeks in the delta of the Ganges, and the jaguar traverses with ease the largest streams in South America.* The bear, also, and the bison, cross the current of the Mississippi. The popular error, that the common swine cannot escape by swimming when thrown into the water, has been contradicted by several curious and well-authenticated instances during the floods in Scotland of 1829. One pig, only six months old, after having been carried down from Garmouth to the bar at the mouth of the Spey, a distance of a quarter of a mile, swam four miles eastward to Port Gordon, and landed safe. Three others, of the same age and litter, swam, at the same time, five miles to the west, and landed at Blackhill.†

In an adult and wild state, these animals would doubtless have been more strong and active, and might, when hard pressed, have performed a much longer voyage. Hence islands remote from the continent may obtain inhabitants by casualties which, like the late storms in Morayshire, may only occur once in many centuries, or thousands of years, under all the same circumstances. It is obvious that powerful tides, winds, and currents may sometimes carry along quadrupeds capable, in like manner, of preserving themselves for hours in the sea, to very considerable distances; and in this way, perhaps, the tapir (*Tapir Indicus*) may have become common to Sumatra and the Malayan peninsula.

To the elephant, in particular, the power of crossing rivers is essential in a wild state, for the quantity of food which a herd of these animals consumes renders it necessary that they should be constantly moving from place to place. The elephant crosses the stream in two ways. If the bed of the river be hard, and the water not of too great a depth, he fords it. But when he crosses great rivers, such as the Ganges and the Niger, the elephant swims deep, so deep, that the end of his trunk only is out of the water; for it is a matter of indifference to him whether his body be completely immersed, provided he can bring the tip of his trunk to the surface, so as to breathe the external air.

Animals of the deer kind frequently take to the water, especially in the rutting season, when the stags are seen, swimming for several leagues at a time, from island to island, in search of the does, especially in the Canadian lakes; and in some countries where there are islands near the sea-shore, they fearlessly enter the sea and swim to them. In hunting excursions, in North America, the elk of that country is frequently pursued for great distances through the water.

The large herbivorous animals, which are gregarious, can never remain long in a confined region, as they consume so much vegetable food. The immense herds of bisons (*Bos Americanus*) which often, in the great valleys of the Mississippi and its tributaries blacken the

* Buffon, vol. v. p. 204.

† Sir T. D. Lauder, Bart., on the Floods in Morayshire, Aug. 1829, p. 302, second edition.

surface of the prairie lands, are continually shifting their quarters, followed by wolves, which prowl about in their rear. "It is no exaggeration," says Mr. James, "to assert, that in one place, on the banks of the Platte, at least ten thousand bisons burst on our sight in an instant. In the morning, we again sought the living picture; but upon all the plain, which last evening was so teeming with noble animals, not one remained."*

Migratory instincts.—Besides the disposition common to the individuals of every species slowly to extend their range in search of food, in proportion as their numbers augment, a migratory instinct often developes itself in an extraordinary manner, when, after an unusually prolific season, or upon a sudden scarcity of provisions, great multitudes are threatened by famine. It may be useful to enumerate some examples of these migrations, because they may put us upon our guard against attributing a high antiquity to a particular species merely because it is diffused over a great space; they show clearly how soon, in a state of nature, a newly created species might spread itself, in every direction, from a single point.

In very severe winters, great numbers of the black bears of America migrate from Canada into the United States; but in milder seasons, when they have been well fed, they remain and hybernate in the north.† The rein-deer, which, in Scandinavia, can scarcely exist to the south of the sixty-fifth parallel, descends, in consequence of the greater coldness of the climate, to the fiftieth degree in Chinese Tartary, and often roves into a country of more southern latitude than any part of England.

In Lapland, and other high latitudes, the common squirrels, whenever they are compelled, by want of provisions, to quit their usual abodes, migrate in amazing numbers, and travel directly forwards, allowing neither rocks nor forests, nor the broadest waters, to turn them from their course. Great numbers are often drowned in attempting to pass friths and rivers. In like manner the small Norway rat sometimes pursues its migrations in a straight line across rivers and lakes; and Pennant informs us, that when the rats, in Kamtschatka, become too numerous, they gather together in the spring, and proceed in great bodies westward, swimming over rivers, lakes, and arms of the sea. Many are drowned or destroyed by water-fowl or fish. As soon as they have crossed the river Penginsk, at the head of the gulf of the same name, they turn southward, and reach the rivers Judoma and Okotsk by the middle of July; a district more than 800 miles distant from their point of departure.

The lemings, also, a small kind of rat, are described as natives of the mountains of Kolen, in Lapland; and once or twice in a quarter of a century they appear in vast numbers, advancing along the ground, and "devouring every green thing." Innumerable bands march from the

* Expedition from Pittsburg to the Rocky Mountains, vol. ii. p. 153.
† Richardson's Fauna Boreali-Americana, p. 16.

Kolen, through Nordland and Finmark, to the Western Ocean, which they immediately enter; and after swimming about for some time, perish.

Fig 97

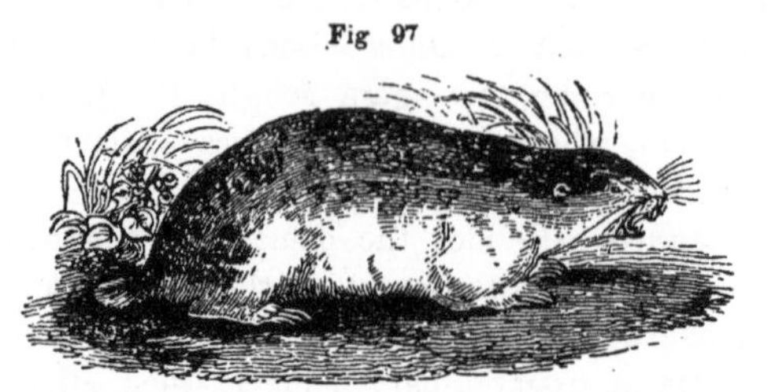

The Leming, or Lapland Marmot (Mus Lemmus, Linn.)

Other bands take their route through Swedish Lapland, to the Bothnian Gulf, where they are drowned in the same manner. They are followed in their journeys by bears, wolves, and foxes, which prey upon them incessantly. They generally move in lines, which are about three feet from each other, and exactly parallel, going directly forward through rivers and lakes; and when they meet with stacks of hay or corn, gnawing their way through them instead of passing round.* These excursions usually precede a rigorous winter, of which the lemings seem in some way forewarned.

Vast troops of the wild ass, or *onager* of the ancients, which inhabit the mountainous deserts of Great Tartary, feed, during the summer, in the tracts east and north of Lake Aral. In the autumn they collect in herds of hundreds, and even thousands, and direct their course towards Persia, to enjoy a warm retreat during winter.† Bands of two or three hundred quaggas, a species of wild ass, are sometimes seen to migrate from the tropical plains of southern Africa to the vicinity of the Malaleveen River. During their migrations they are followed by lions, who slaughter them night by night.‡

The migratory swarms of the springbok, or Cape antelope, afford another illustration of the rapidity with which a species under certain circumstances may be diffused over a continent. When the stagnant pools of the immense deserts south of the Orange River dry up, which often happens after intervals of three or four years, myriads of these animals desert the parched soil, and pour down like a deluge on the cultivated regions near the Cape. The havoc committed by them resembles that of the African locusts; and so crowded are the herds, that "the lion has been seen to walk in the midst of the compressed phalanx with only as much room between him and his victims as the fears of those immediately around could procure by pressing outwards."§

* Phil. Trans., vol. ii. p. 872.

† Wood's Zoography, vol. i. p. 11.

‡ On the authority of Mr. Campbell. Library of Entert. Know., Menageries, vol. i. p. 152.

§ Cuvier's Animal Kingdom by Griffiths, vol. ii. p. 109. Library of Entertaining Knowledge, Menageries, vol. i. p. 366.

Dr. Horsfield mentions a singular fact in regard to the geographical distribution of the *Mydaus meliceps*, an animal intermediate between the polecat and badger. It inhabits Java, and is "confined exclusively to those mountains which have an elevation of more than seven thousand feet above the level of the ocean; on these it occurs with the same regularity as many plants. The long extended surface of Java, abounding with conical points which exceed this elevation, affords many places favorable for its resort. On ascending these mountains, the traveller scarcely fails to meet with this animal, which, from its peculiarities, is universally known to the inhabitants of these elevated tracts, while to those of the plains it is as strange to an animal from a foreign county.

Fig. 98.

Mydaus meliceps, or badger-headed Mydaus. Length, including the tail, 16 inches.

In my visits to the mountainous districts, I uniformly met with it; and, as far as the information of the natives can be relied on, it is found on all the mountains."*

Now, if asked to conjecture how the Mydaus arrived at the elevated regions of each of these isolated mountains, we might say that, before the island was peopled by man, by whom their numbers are now thinned, they may occasionally have multiplied so as to be forced to collect together and migrate: in which case notwithstanding the slowness of their motions, some few would succeed in reaching another mountain, some twenty, or even, perhaps, fifty miles distant; for although the climate of the hot intervening plains would be unfavourable to them, they might support it for a time, and would find there abundance of insects on which they feed. Volcanic eruptions, which, at different times have covered the summits of some of those lofty cones with sterile sand and ashes, may have occasionally contributed to force on these migrations.

Drifting of animals on ice-floes.—The power of the terrestrial mammalia to cross the sea is very limited, and it was before stated that the same species is scarcely ever common to districts widely separated by the ocean. If there be some exceptions to this rule, they generally admit of explanation; for there are natural means whereby some animals may be floated across the water, and the sea may in the course of ages

* Horsfield, Zoological Researches in Java, No. ii., from which the figure is taken.

wear a wide passage through a neck of land, leaving individuals of a species on each side of the new channel. Polar bears are known to have been frequently drifted on the ice from Greenland to Iceland; they can also swim to considerable distances, for Captain Parry, on the return of his ships through Barrow's Straits, met with a bear swimming in the water about midway between the shores, which were about forty miles apart, and where no ice was in sight.* "Near the east coast of Greenland," observes Scoresby, "they have been seen on the ice in such quantities, that they were compared to flocks of sheep on a common; and they are often found on field-ice, above two hundred miles from the shore."† Wolves, in the arctic regions, often venture upon the ice near the shore, for the purpose of preying upon young seals which they surprise when asleep. When these ice-floes get detached, the wolves are often carried out to sea; and though some may be drifted to islands or continents, the greater part of them perish, and have been often heard in this situation howling dreadfully, as they die by famine.‡

During the short summer which visits Melville Island, various plants push forth their leaves and flowers the moment the snow is off the ground, and form a carpet spangled with the most lively colours. These secluded spots are reached annually by herds of musk-oxen and rein-deer, which travel immense distances over dreary and desolate regions, to graze undisturbed on these luxuriant pastures.§ The rein-deer often pass along in the same manner, by the chain of the Aleutian Islands, from Behring's Straits to Kamtschatka, subsisting on the moss found in these islands during their passage.‖ But the musk-ox, notwithstanding its migratory habits, and its long journeys over the ice, does not exist either in Asia or Greenland.¶

On floating islands of drift-wood.—Within the tropics there are no ice-floes; but, as if to compensate for that mode of transportation, there are floating islets of matted trees, which are often borne along through considerable spaces. These are sometimes seen sailing at the distance of fifty or one hundred miles from the mouth of the Ganges, with living trees standing erect upon them. The Amazon, the Congo, and the Orinoco, also produce these verdant rafts, which are formed in the manner already described when speaking of the great raft of the Atchafalaya, an arm of the Mississippi, where a natural bridge of timber, ten miles long, and more than two hundred yards wide, existed for more than forty years, supporting a luxuriant vegetation, and rising and sinking with the water which flowed beneath it.

On these green islets of the Mississippi, observes Malte-Brun, young trees take root, and the pistia and nenuphar display their yellow

* Append. to Parry's Second Voyage, years 1819-20.
† Account of the Arctic Regions, vol. i. p. 518.
‡ Turton in a note to Goldsmith's Nat. Hist., vol. iii. p. 43.
§ Supplement to Parry's First Voyage of Discovery, p. 189
‖ Goldman's American Nat. Hist. vol. i. p. 22.
¶ Dr. Richardson, Brit. Assoc. Report, vol. v. p. 161.

flowers: serpents, birds, and the cayman alligator, come to repose there, and all are sometimes carried to the sea and engulphed in its waters.*

Spix and Martius relate that, during their travels in Brazil, they were exposed to great danger while ascending the Amazon in a canoe, from the vast quantity of drift-wood constantly propelled against them by the current; so much so, that their safety depended on the crew being always on the alert to turn aside the trunks of trees with long poles. The tops alone of some trees appeared above water, others had their roots attached to them with so much soil that they might be compared to floating islets. On these, say the travellers, we saw some very singular assemblages of animals, pursuing peacefully their uncertain way in strange companionship. On one raft were several grave-looking storks, perched by the side of a party of monkeys, who made comical gestures, and burst into loud cries, on seeing the canoe. On another was seen a number of ducks and divers, sitting by a group of squirrels. Next came down upon the stem of a large rotten cedar tree, an enormous crocodile, by the side of a tiger-cat, both animals regarding each other with hostility and mistrust, but the saurian being evidently most at his ease, as conscious of his superior strength.†

Similar green rafts, principally composed of canes and brushwood, are called "camelotes" on the Parana in South America; and they are occasionally carried down by inundations, bearing on them the tiger, cayman, squirrels, and other quadrupeds, which are said to be always terror-stricken on their floating habitation. No less than four tigers (pumas) were landed in this mannner in one night at Monte Video, lat. 35° S., to the great alarm of the inhabitants, who found them prowling about the streets in the morning.‡

In a memoir lately published, a naval officer relates that, as he returned from China by the eastern passage, he fell in, among the Moluccas, with several small floating islands of this kind, covered with mangrove trees interwoven with underwood. The trees and shrubs retained their verdure, receiving nourishment from a stratum of soil which formed a white beach round the margin of each raft, where it was exposed to the washing of the waves and the rays of the sun.§ The occurrence of soil in such situations may easily be explained; for all the natural bridges of timber which occasionally connect the islands of the Ganges, Mississippi, and other rivers, with their banks, are exposed to floods of water, densely charged with sediment.

Captain W. H. Smyth informs me, that, when cruising in the Cornwallis amidst the Philippine Islands, he has more than once seen, after those dreadful hurricanes called typhoons, floating masses of wood, with

* System of Geography, vol. v. p. 157.
† Spix and Martius, Reise, &c., vol. iii. pp. 1011. 1013.
‡ Sir W. Parish's Buenos Ayres, p. 187., and Robertson's Letters on Paraguay, p. 220.
§ United Service Journal, No. xxiv. p. 697.

trees growing upon them, and ships have sometimes been in imminent peril, as often as these islands were mistaken for terra firma, when, in fact, they were in rapid motion.

It is highly interesting to trace, in imagination, the effects of the passage of these rafts from the mouth of a large river to some archipelago, such as those in the South Pacific, raised from the deep, in comparatively modern times, by the operations of the volcano and the earthquake, and the joint labours of coral animals and testacea. If a storm arise, and the frail vessel be wrecked, still many a bird and insect may succeed in gaining, by flight, some island of the newly formed group, while the seeds and berries of herbs and shrubs, which fall into the waves, may be thrown upon the strand. But if the surface of the deep be calm, and the rafts are carried along by a current, or wafted by some slight breath of air fanning the foliage of the green trees, it may arrive, after a passage of several weeks, at the bay of an island, into which its plants and animals may be poured out as from an ark, and thus a colony of several hundred new species may at once be naturalized.

The reader should be reminded, that I merely advert to the transportation of these rafts as of extremely rare and accidental occurrence; but it may account, in tropical countries, for some of the rare exceptions to the general law of the confined range of mammiferous species.

Migrations of the Cetacea.—Many of the Cetacea, the whales of the northern seas for example, are found to desert one tract of the sea, and to visit another very distant, when they are urged by want of food, or danger. The seals also retire from the coast of Greenland in July, return again in September, and depart again in March, to return in June. They proceed in great droves northwards, directing their course where the sea is most free from ice, and are observed to be extremely fat when they set out on this expedition, and very lean when they come home again.*

Species of the Mediterranean, Black Sea, and Caspian identical.—Some naturalists have wondered that the sea-calves, dolphins, and other marine mammalia of the Mediterranean and Black Sea, should be identical with those found in the Caspian: and among other fanciful theories, they have suggested that they may dive through subterranean conduits, and thus pass from one sea into the other. But as the occurrence of wolves and other noxious animals, on both sides of the British Channel, was adduced, by Verstegan and Desmarest, as one of many arguments to prove that England and France were once united; so the correspondence of the aquatic species of the inland seas of Asia with those of the Black Sea tend to confirm the hypothesis, for which there are abundance of independent geological data, that those seas were connected together by straits at no remote period of the earth's history.

Geographical Distribution and Migrations of Birds.

I shall now offer a few observations on some of the other divisions of

* Krantz, vol. i. p. 129., cited by Goldsmith, Nat. Hist., vol. iii. p. 260.

the animal kingdom. Birds, notwithstanding their great locomotive powers, form no exception to the general rules already laid down; but, in this class, as in plants and terrestrial quadrupeds, different groups of species are circumscribed within definite limits. We find, for example, one assemblage in the Brazils, another in the same latitudes in Central Africa, another in India, and a fourth in New Holland. Of twenty-six different species of land birds found in the Galapagos archipelago, all, with the exception of one, are distinct from those inhabiting other parts of the globe;* and in other archipelagos a single island sometimes contains a species found in no other spot on the whole earth; as is exemplified in some of the parrot tribes. In this extensive family, which are, with few exceptions, inhabitants of tropical regions, the American group has not one in common with the African, nor either of these with the parrots of India.†

Another illustration is afforded by that minute and beautiful tribe, the humming-birds. The whole of them are, in the first place, peculiar to the new world; but some species are confined to Mexico, while others exist only in some of the West India Islands, and have not been found elsewhere in the western hemisphere. Yet there are species of this family which have a vast range, as the *Trochilus flammifrons* (or *Mellisuga Kingii*), which is found over a space of 2500 miles on the west coast of South America, from the hot dry country of Lima to the humid forests of Tierra del Fuego. Captain King, during his survey in the years 1826–30, found this bird at the Straits of Magellan, in the month of May—the depth of winter—sucking the flowers of a large species of fuchsia, then in bloom, in the midst of a shower of snow.

The ornithology of our own country affords one well-known and striking exemplification of the law of a limited specific range; for the common grouse (*Tetra scoticus*) occurs nowhere in the known world except in the British isles.

Some species of the vulture tribe are said to be cosmopolites; and the common wild goose (*Anas anser*, Linn.), if we may believe some ornithologists, is a general inhabitant of the globe, being met with from Lapland to the Cape of Good Hope, frequent in Arabia, Persia, China, and Japan, and in the American continent from Hudson's Bay to South Carolina.‡ An extraordinary range has also been attributed to the nightingale, which extends from western Europe to Persia, and still farther. In a work entitled Specchio Comparativo,§ by Charles Bonaparte, many species of birds are enumerated as common to Rome and Philadelphia: the greater part of these are migratory, but some of them, such as the long-eared owl (*Strix otus*), are permanent in both countries. The correspondence of the ornithological fauna of the eastern and

* Darwin's Journal, &c., p. 461.
† Prichard, vol. i. p. 47.
‡ Bewick's Birds, vol. ii. p. 294., who cites Latham.
§ Pisa, 1827 (not sold).

western hemispheres increases considerably, as might have been anticipated, in high northern latitudes.*

Their facilities of diffusion.—In parallel zones of the northern and southern hemispheres, a great general correspondence of form is observable, both in the aquatic and terrestrial birds; but there is rarely any specific identity; and this phenomenon is truly remarkable, when we recollect the readiness with which some birds, not gifted with great powers of flight, shift their quarters to different regions, and the facility with which others, possessing great strength of wing, perform their aërial voyage. Some migrate periodically from high latitudes, to avoid the cold of winter, and the accompaniments of cold,—scarcity of insects and vegetable food; others, it is said, for some particular kinds of nutriment required for rearing their young: for this purpose they often traverse the ocean for thousands of miles, and recross it at other periods, with equal security.

Periodical migrations, no less regular, are mentioned by Humboldt, of many American water-fowl, from one part of the tropics to another, in a zone where there is the same temperature throughout the year. Immense flights of ducks leave the valley of the Orinoco, when the increasing depth of its waters and the flooding of its shores prevent them from catching fish, insects, and aquatic worms. They then betake themselves to the Rio Negro and Amazon, having passed from the eighth and third degrees of north latitude to the first and fourth of south latitude, directing their course south-south-east. In September, when the Orinoco decreases and re-enters its channel, these birds return northwards.†

The insectivorous swallows which visit our island would perish during winter, if they did not annually repair to warmer climes. It is supposed that in these aerial excursions the average rapidity of their flight is not less than fifty miles an hour; so that, when aided by the wind, they soon reach warmer latitudes. Spallanzani calculated that the swallow can fly at the rate of ninety-two miles an hour, and conceived that the rapidity of the swift might be three times greater.‡ The rate of flight of the eider duck (*Anas mollissima*) is said to be ninety miles an hour; and Bachman says that the hawk, wild pigeon (*Columba migratoria*), and several species of wild ducks, in North America, fly at the rate of forty miles an hour, or nearly a thousand miles in twenty-four hours.§

When we reflect how easily different species, in a great lapse of ages, may be each overtaken by gales and hurricanes, and, abandoning themselves to the tempest, be scattered at random through various regions of the earth's surface, where the temperature of the atmosphere, the vegetation, and the animal productions, might be suited to their wants, we shall be prepared to find some species capriciously distributed, and

* Bachman, Silliman's Amer. Journ., No. 61, p. 92.
† Voyage aux Régions Equinoxiales, tome vii. p. 429.
‡ Fleming, Phil. Zool., vol. ii. p. 43.
§ Silliman's Amer. Journ., No. 61. p. 83.

to be sometimes unable to determine the native countries of each. Captain Smyth informs me, that, when engaged in his survey of the Mediterranean, he encountered a gale in the Gulf of Lyons, at the distance of between twenty and thirty leagues from the coast of France, which bore along many land birds of various species, some of which alighted on the ship, while others were thrown with violence against the sails. In this manner islands become tenanted by species of birds inhabiting the nearest mainland.

Geographical Distribution and Dissemination of Reptiles.

A few facts respecting the third great class of vertebrated animals will suffice to show that the plan of nature in regard to their location on the globe is perfectly analogous to that already exemplified in other parts of the organic creation, and has probably been determined by similar causes.

Habitations of reptiles.—Of the great saurians, the gavials which inhabit the Ganges differ from the cayman of America, or the crocodile of the Nile. The monitor of New Holland is specifically distinct from the Indian species; these latter, again, from the African, and all from their congeners in the new world. So in regard to snakes; we find the boa of America represented by the python, a different though nearly allied genus in India. America is the country of the rattlesnake; Africa, of the cerastes; and Asia, of the hooded snake, or cobra di capello. The amphibious genera Siren and Menopoma belong to North America, possessing both lungs and gills, and respiring at pleasure either air or water. The only analogous animal of the old world is the *Proteus anguinus* of the lakes of Lower Carniola, and the grotto of Adelsberg between Trieste and Vienna.*

There is a legend that St. Patrick expelled all reptiles from Ireland; and certain it is that none of the three species of snakes common in England, nor the toad, have been observed there by naturalists. They have our common frog, and our water-newt, and according to Ray (Quad. 264.), the green lizard (*Lacerta viridis*).

Migrations of the larger reptiles.—The range of the large reptiles is, in general, quite as limited as that of some orders of the terrestrial mammalia. The great saurians sometimes cross a considerable tract in order to pass from one river to another; but their motions by land are generally slower than those of quadrupeds. By water, however, they may transport themselves to distant situations more easily. The larger alligator of the Ganges sometimes descends beyond the brackish water of the delta into the sea; and in such cases it might chance to be drifted away by a current, and survive till it reached a shore at some distance; but such casualties are probably very rare.

Turtles migrate in large droves from one part of the ocean to another

* Richardson, Brit. Assoc. Rep., vol. v. p. 202.

during the ovipositing season; and they find their way annually to the island of Ascension, from which the nearest land is about 800 miles distant. Dr. Fleming mentions, that an individual of the hawk's bill turtle (*Chelonia imbricata*), so common in the American seas, has been taken at Papa Stour, one of the West Zetland Islands;* and, according to Sibbald, "the same animal came into Orkney." Another was taken, in 1774, in the Severn, according to Turton. Two instances, also, of the occurrence of the leathern tortoise (*C. coriacea*), on the coast of Cornwall, in 1756, are mentioned by Borlase. These animals of more southern seas can be considered only as stragglers, attracted to our shores during uncommonly warm seasons by an abundant supply of food, or carried by the Gulf stream, or driven by storms to high latitudes.

Some of the smaller reptiles lay their eggs on aquatic plants; and these must often be borne rapidly by rivers, and conveyed to distant regions in a manner similar to the dispersion of seeds before adverted to. But that the larger ophidians may be themselves transported across the seas, is evident from the following most interesting account of the arrival of one at the island of St. Vincent. It is worthy of being recorded, says Mr. Guilding, "that a noble specimen of the *Boa constrictor* was lately conveyed to us by the currents, twisted round the trunk of a large sound cedar tree, which had probably been washed out of the bank by the floods of some great South American river, while its huge folds hung on the branches, as it waited for its prey. The monster was fortunately destroyed after killing a few sheep, and his skeleton now hangs before me in my study, putting me in mind how much reason I might have had to fear in my future rambles through the forests of St. Vincent, had this formidable reptile been a pregnant female, and escaped to a safe retreat."†

CHAPTER XXXIX.

LAWS WHICH REGULATE THE GEOGRAPHICAL DISTRIBUTION OF SPECIES—*continued.*

Geographical distribution and migration of Fish—of Testacea—of Zoophytes—Distribution of Insects—Migratory instincts of some species—Certain types characterize particular countries—Their means of dissemination—Geographical distribution and diffusion of man—Speculations as to the birth-place of the human species—Progress of human population—Drifting of canoes to vast distances—On the involuntary influence of man in extending the range of many other species.

Geographical Distribution and Migrations of Fish.

ALTHOUGH we are less acquainted with the habitations of marine animals than with the grouping of the terrestrial species before described, yet it

* Brit. Animals, p. 149., who cites Sibbald.
† Zool. Journ. vol. iii. p. 406. Dec. 1827.

is well ascertained that their distribution is governed by the same general laws. The testimony borne by MM. Péron and Lesueur to this important fact is remarkably strong. These eminent naturalists, after collecting and describing many thousand species of marine animals which they brought to Europe from the southern hemisphere, insist most emphatically on their distinctness from those north of the equator; and this remark they extend to animals of all classes, from those of a more simple to those of a more complex organization—from the sponges and Medusæ to the Cetacea. "Among all those which we have been able to examine," say they, "with our own eyes, or with regard to which it has appeared to us possible to pronounce with certainty, there is not a single animal of the southern regions which is not distinguished by essential characters from the analogous species in the northern seas."*

On comparing the freshwater fish of Europe and North America, Sir John Richardson remarks, that the only species which is unequivocally common to the two continents is the pike (*Esox lucius*); and it is curious that this fish is unknown to the westward of the Rocky Mountains, the very coast which approaches nearest to the old continent.† According to the same author the genera of freshwater fish in China agree closely with those of the peninsula of India, but the species are not the same. "As in the distribution," he adds, "of marine fish, the interposition of a continent stretching from the tropics far into the temperate or colder parts of the ocean, separate different ichthyological groups; so with respect to the freshwater species, the intrusion of arms of the sea running far to the northwards, or the interposition of a lofty mountain-chain, effects the same thing. The freshwater fish of the Cape of Good Hope and the South American ones, are different from those of India and China, &c."‡

Cuvier and Valenciennes, in their "Histoire des Poissons," observe, that very few species of fish cross the Atlantic. Although their statement is correct, it is found that a great many species are common to the opposite sides of the Indian Ocean, inhabiting alike the Red Sea, the eastern coast of Africa, Madagascar, the Mauritius, the Indian Ocean, the southern seas of China, the Malay archipelago, the northern coasts of Australia, and the whole of Polynesia!§ This very wide diffusion, says Sir J. Richardson, may have been promoted by chains of islands running east and west, which are wanting in the deep Atlantic. An archipelago extending far in longitude, favours the migration of fish by multiplying the places of deposit for spawn along the shores of islands, and on intervening coral banks; and in such places, also, fish find their appropriate food.

The flying fish are found (some stragglers excepted) only between the tropics: in receding from the line, they never approach a higher latitude

* Sur les Habitations des Animaux Marins.—Ann. du Mus., tome. xv., cited by Prichard, Phys. Hist. of Mankind, vol. i. p. 51.

† Brit. Assoc. Reports, vol. v. p. 203.

‡ Report to the Brit. Assoc., 1845, p. 192.

§ Richardson, ibid. p. 190.

than the fortieth parallel. The course of the Gulf stream, however, and the warmth of its water, enable some tropical fish to extend their habitations far into the temperate zone; thus the chætodons which abound in the seas of hot climates, are found among the Bermudas on the thirty-second parallel, where they are preserved in basins inclosed from the sea, as an important article of food for the garrison and inhabitants. Other fish, following the direction of the same great current, range from the coast of Brazil to the banks of Newfoundland.*

All are aware that there are certain fish of passage which have their periodical migrations, like some tribes of birds. The salmon, towards the season of spawning, ascends the rivers for hundreds of miles, leaping up the cataracts which it meets in its course, and then retreats again into the depths of the ocean. The herring and the haddock, after frequenting certain shores, in vast shoals, for a series of years, desert them again, and resort to other stations, followed by the species which prey on them. Eels are said to descend into the sea for the purpose of producing their young, which are seen returning into the fresh water by myriads, extremely small in size, but possessing the power of surmounting every obstacle which occurs in the course of a river, by applying their slimy and glutinous bodies to the surface of rocks, or the gates of a lock, even when dry, and so climbing over it.† Before the year 1800 there were no eels in Lake Wener, the largest inland lake in Sweden, which discharges its waters by the celebrated cataracts of Trolhättan. But I am informed by Professor Nilsson, that since the canal was opened uniting the river Gotha with the lake by a series of nine locks, each of great height, eels have been observed in abundance in the lake. It appears, therefore, that though they were unable to ascend the falls, they have made their way by the locks, by which in a very short space a difference of level of 114 feet is overcome.

Gmelin says, that the Anseres (wild geese, ducks, and others) subsist, in their migrations, on the spawn of fish; and that oftentimes, when they void the spawn, two or three days afterwards, the eggs retain their vitality unimpaired.‡ When there are many disconnected freshwater lakes in a mountainous region, at various elevations, each remote from the other, it has often been deemed inconceivable how they could all become stocked with fish from one common source; but it has been suggested, that the minute eggs of these animals may sometimes be entangled in the feathers of water-fowl. These, when they alight to wash and plume themselves in the water, may often unconsciously contribute to propagate swarms of fish, which, in due season, will supply them with food. Some of the water-beetles, also, as the Dyticidæ, are amphibious, and in the evening quit their lakes and pools, and, flying in the air, transport the minute ova of fishes to distant waters. In this manner some naturalists account for the fry of fish appearing occasion-

* Sir J. Richardson, ibid. p. 190.

† Phil. Trans. 1747, p. 395.

‡ Amœn. Acad., Essay 75.

ally in small pools caused by heavy rains; but the showers of small fish, stated in so many accounts to have fallen from the atmosphere, require farther investigation.

Geographical Distribution and Migrations of Testacea.

The Testacea, of which so great a variety of species occurs in the sea, are a class of animals of peculiar importance to the geologist; because their remains are found in strata of all ages, and generally in a higher state of preservation than those of other organic beings. Climate has a decided influence on the geographical distribution of species in this class; but as there is much greater uniformity of temperature in the waters of the ocean, than in the atmosphere which invests the land, the diffusion of marine mollusks is on the whole more extensive.

Some forms attain their fullest development in warm latitudes; and are often exclusively confined to the torrid zone, as *Nautilus*, *Harpa*, *Terebellum*, *Pyramidella*, *Delphinula*, *Aspergillum*, *Tridacna*, *Cucullœa*, *Crassatella*, *Corbis*, *Perna*, and *Plicatula*. Other forms are limited to one region of the sea, as the *Trigonia* to parts of Australia, and the *Concholepas* to the western coast of South America. The marine species inhabiting the ocean on the opposite sides of the narrow isthmus of Panama, are found to differ almost entirely, as we might have anticipated, since a West Indian mollusk cannot enter the Pacific without coasting round South America, and passing through the inclement climate of Cape Horn. The continuity of the existing lines of continent from north to south, prevents any one species from belting the globe, or from following the direction of the isothermal lines.

Currents also flowing permanently in certain directions, and the influx at certain points of great bodies of fresh water, limit the extension of many species. Those which love deep water are arrested by shoals; others, fitted for shallow seas, cannot migrate across unfathomable abysses. The nature also of the ground has an important influence on the testaceous fauna, both on the land and beneath the waters. Certain species prefer a sandy, others a gravelly, and some a muddy sea-bottom. On the land, limestone is of all rocks the most favourable to the number and propagation of species of the genera Helix, Clausilia, Bulimus, and others. Professor E. Forbes has shown as the result of his labours in dredging in the Ægean Sea, that there are eight well-marked regions of depth, each characterized by its peculiar testaceous fauna. The first of these, called the littoral zone, extends to a depth of two fathoms only; but this narrow belt is inhabited by more than one hundred species. The second region, of which ten fathoms is the inferior limit, is almost equally populous; and a copious list of species is given as characteristic of each region down to the seventh, which lies between the depths of 80 and 105 fathoms, all the inhabited space below this being included in the eighth province, where no less than 65 species of Testacea have been taken. The majority of the shells in this lowest zone are white or transparent. Only two species of Mollusca are common to all the eight regions, namely, *Arca lactea* and *Cerithium lima*.*

Great range of some provinces and species.—In Europe conchologists distinguish between the arctic fauna, the southern boundary of which corresponds with the isothermal line of 32° F., and the Celtic, which, commencing with that limit as its northern frontier, extends southwards to the mouth of the English Channel and Cape Finisterre, in France. From that point begins the Lusitanian fauna, which, according to the recent observations of Mr. M'Andrew (1852), ranges to the Canary Islands. The Mediterranean province is distinct from all those above enumerated, although it has some species in common with each.

The Indo-Pacific region is by far the most extensive of all. It reaches from the Red Sea and the eastern coast of Africa, to the Indian Archipelago, and adjoining parts of the Pacific Ocean. To the geologist it furnishes a fact of no small interest, by teaching us that one group of living species of mollusca may prevail throughout an area exceeding in magnitude the utmost limits we can as yet assign to any assemblage of contemporaneous fossil species. Mr. Cuming obtained more than a hundred species of shells from the eastern coast of Africa identical with those collected by himself at the Philippines and in the eastern coral islands of the Pacific Ocean, a distance equal to that from pole to pole.†

Certain species of the genus *Ianthina* have a very wide range, being common to seas north and south of the equator. They are all provided with a beautifully contrived float, which renders them buoyant, facilitating their dispersion, and enabling them to become active agents in disseminating other species. Captain King took a specimen of *Ianthina fragilis*, alive, a little north of the equator, so loaded with barnacles (*Pentelasmis*) and their ova that the upper part of its shell was invisible. The "Rock Whelk" (*Purpura lapillus*), a well-known British univalve, inhabits both the North Atlantic and North Pacific.

Helix putris (*Succinea putris*, Lam.), so common in Europe, where it reaches from Norway to Italy, is also said to occur in the United States and in Newfoundland. As this animal inhabits constantly the borders of pools and streams where there is much moisture, it is not impossible that different water-fowl have been the agents of spreading some of its minute eggs, which may have been entangled in their feathers. The freshwater snail, *Lymneus palustris*, so abundant in English ponds, ranges uninterruptedly from Europe to Cashmere, and thence to the eastern parts of Asia. *Helix aspersa*, one of the commonest of our larger land-shells, is found in St. Helena and other distant countries. Some conchologists have conjectured that it was accidentally imported into St. Helena in some ship; for it is an eatable species, and these animals are capable of retaining life during long voyages, without air or nourishment.‡

* Report to the Brit. Assoc. 1843, p. 130.
† Quart. Journ., Geol. Soc., 1846, vol. ii. p. 268.
‡ Four individuals of a large species of land shell (*Bulimus*), from Valparaiso, were brought to England by Lieutenant Graves, who accompanied Captain

Perhaps no species has a better claim to be called cosmopolite than one of our British bivalves, *Saxicava rugosa*. It is spread over all the north-polar seas, and ranges in one direction through Europe to Senegal, occurring on both sides of the Atlantic; while in another it finds its way into the North Pacific, and thence to the Indian Ocean. Nor do its migrations cease till it reaches the Australian seas.

A British brachiopod, named *Terebratula caput-serpentis*, is common, according to Professor E. Forbes, to both sides of the North Atlantic, and to the South African and Chinese seas.

Confined range of other species.—Mr. Lowe, in a memoir published in the Cambridge Transactions in 1834, enumerates seventy-one species of land Mollusca, collected by him in the islands of Madeira and Porto Santo, sixty of which belonged to the genus Helix alone, including as sub-genera Bulimus and Achatina, and excluding Vitrina and Clausilia; forty-four of these are new. It is remarkable that very few of the above-mentioned species are common to the neighbouring archipelago of the Canaries; but it is a still more striking fact, that of the sixty species of the three genera above mentioned, thirty-one are natives of Porto Santo; whereas, in Madeira, which contains ten times the superficies, were found but twenty-nine. Of these only four were common to the two islands, which are separated by a distance of only twelve leagues; and two even of these four (namely *Helix rhodostoma* and *H. ventrosa*) are species of general diffusion, common to Madeira, the Canaries, and the south of Europe.*

The confined range of these mollusks may easily be explained, if we admit that species have only one birth-place; and the only problem to be solved would relate to the exceptions—to account for the dissemination of some species throughout several islands, and the European continent. May not the eggs, when washed into the sea by the undermining of cliffs, or blown by a storm from the land, float uninjured to a distant shore?

Their mode of diffusion.—Notwithstanding the proverbially slow motion of snails and mollusks in general, and although many aquatic species adhere constantly to the same rock for their whole lives, they are by no means destitute of provision for disseminating themselves rapidly over a wide area. "Some Mollusca," says Professor E. Forbes, "migrate in their larva state, for all of them undergo a metamorphosis either in the egg or out of the egg. The gasteropoda commence life under the form of a small spiral shell, and an animal furnished with ciliated wings, or lobes, like a pteropod, by means of which it can swim freely, and in this form can migrate with ease through the sea."†

King in his expedition to the Straits of Magellan. They had been a ke up in a box, and enveloped in cotton: two for a space of thirteen, one for seventeen, and a fourth for upwards of twenty months: but, on being exposed by Mr. Broderip to the warmth of a fire in London, and provided with tepid water and leaves, they revived, and lived for several months in Mr. Loddiges' palm-house, till accidentally drowned.

* Camb. Phil. Trans., vol. iv. 1831.

† Edin. New Phil. Journ., April 1844.

We are accustomed to associate in our minds the idea of the greatest locomotive powers with the most mature and perfect state of each species of invertebrate animal, especially when they undergo a series of transformations; but in all the Mollusca the reverse is true. The young fry of the cockle, for example (*Cardium*), possess, when young or in the larva state, an apparatus which enables them both to swim and to be carried along easily by a marine current. (See fig. 99.)

These small bodies here represented, which bear a considerable resemblance to the fry of the univalve, or gasteropodous shells above mentioned, are so minute at first as to be just visible to the naked eye. They begin to move about from the moment they are hatched, by means

Fig. 99.

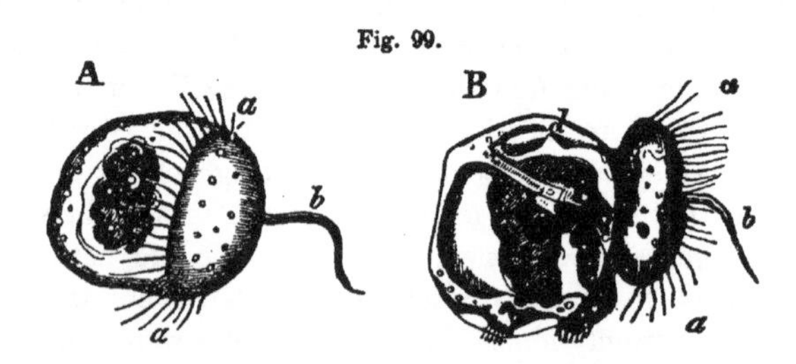

The young fry of a cockle (Cardium pygmæum,) from Loven's Kongl. Vetenskaps. Akadem Handling, 1848.

A, The young just hatched, magnified 100 diameters. B, the same farther advanced.
a, The ciliated organ of locomotion with its filamentous appendage *b*.
c, The rudimentary intestine.
d, The rudimentary shell.

of the long cilia, *a*, *a*, placed on the edges of the locomotive disk or velum. This disk shrinks up as they increase in size, and gradually disappears, no trace of it being visible in the perfect animal.

Some species of shell-bearing Mollusca lay their eggs in a sponge-like nidus, wherein the young remain enveloped for a time after their birth; and this buoyant substance floats far and wide as readily as sea-weed. The young of other viviparous tribes are often borne along entangled in sea-weed. Sometimes they are so light, that, like grains of sand, they can be easily moved by currents. Balani and Serpulæ are sometimes found adhering to floating cocoa-nuts, and even to fragments of pumice. In rivers and lakes, on the other hand, aquatic univalves usually attach their eggs to leaves and sticks which have fallen into the water, and which are liable to be swept away during floods, from tributaries to the main streams, and from thence to all parts of the same basins. Particular species may thus migrate during one season from the head waters of the Mississippi, or any other great river, to countries bordering the sea, at the distance of many thousand miles.

An illustration of the mode of attachment of these eggs will be seen in the annexed cut. (Fig. 100.)

The habit of some Testacea to adhere to floating wood is proved by their fixing themselves to the bottoms of ships. By this mode of conveyance *Mytilus polymorphus*, previously known only in the Danube and

Wolga, may have been brought to the Commercial Docks in the Thames, and to Hamburgh, where the species is now domiciled. But Mr. Gray suggests that as the animal is known to have the faculty of living for a very long time out of water, it is more probable that it was brought in Russian timber, than borne uninjured through the salt water at the bottom of a vessel.*

A lobster (*Astacus marinus*) was lately taken alive covered with living mussels (*Mytilus edulis*)†; and a large female crab (*Cancer pagurus*), covered with oysters, and bearing also *Anomia ephippium*, and Actiniæ, was taken in April, 1832, off the English coast. The oysters, seven in number, include individuals of six years' growth, and the two largest are four inches long and three inches and a half broad. Both the crab and the oysters were seen alive by Mr. Robert Brown.‡

Fig. 100.

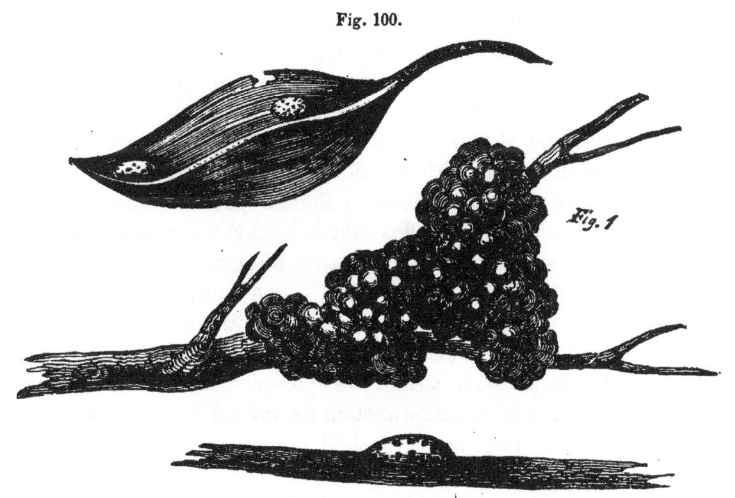

Eggs of Freshwater mollusks.

Fig. 1. Eggs of *Ampullaria ovata* (a fluviatile species) fixed to a small sprig which had fallen into the water.
Fig. 2. Eggs of *Planorbis albus*, attached to a dead leaf lying under water.
Fig. 3. Eggs of the common Limneus (*L. vulgaris*), adhering to a dead stick under water.

From this example we learn the manner in which oysters may be diffused over every part of the sea where the crab wanders; and if they are at length carried to a spot where there is nothing but fine mud, the foundation of a new oyster-bank may be laid on the death of the crab. In this instance the oysters survived the crab many days, and were killed at last only by long exposure to the air.

* Phil. Trans. 1835, p. 303.

† The specimen is preserved in the Museum of the Zool. Soc. of London.

‡ This specimen is in the collection of my friend Mr. Broderip, who observes, that this crab, which was apparently in perfect health, could not have cast her shell for six years, whereas some naturalists have stated that the species moults annually, without limiting the moulting period to the early stages of the growth of the animal.

Geographical Distribution and Migrations of Zoophytes.

Zoophytes are very imperfectly known; but there can be little doubt that each maritime region possesses species peculiar to itself. The Madrepores, or lamelliferous Polyparia, are found in their fullest development only in the tropical seas of Polynesia and the East and West Indies; and this family is represented only by a few species in our seas. The zoophytes of the Mediterranean, according to Ehrenberg, differ almost entirely from those of the Red Sea, although only seventy miles distant. Out of 120 species of Anthozoa, only two are common to both seas.* Péron and Lesueur, after studying the Holothuriæ, Medusæ, and other congeners of delicate and changeable forms, came to the conclusion that each kind has its place of residence determined by the temperature necessary to support its existence. Thus, for example, they found the abode of *Pyrosoma Atlantica* to be confined to one particular region of the Atlantic Ocean.†

Let us now inquire how the transportation of zoophytes from one part of the globe to another is effected. Many of them, as in the families Flustra and Sertularia, attach themselves to sea-weed, and are occasionally drifted along with it. Many fix themselves to the shells of Mollusca, and are thus borne along by them to short distances. Others, like some species of sea-pens, float about in the ocean, and are usually believed to possess powers of spontaneous motion. But the most frequent mode of transportation consists in the buoyancy of their eggs, or certain small vesicles, which are detached, and are capable of becoming the foundation of a new colony. These gems, as they are called, have, in many instances, a locomotive power of their own, by which they proceed in a determinate direction for several days after separation from the parent. They are propelled by means of numerous short threads or *ciliæ*, which are in constant and rapid vibration; and, when thus supported in the water, they may be borne along by currents to a great distance.

That some zoophytes adhere to floating bodies, is proved by their being found attached to the bottoms of ships, like certain Testacea before alluded to.

Geographical Distribution and Migrations of Insects.

Before I conclude this sketch of the manner in which the habitable parts of the earth are shared out among particular assemblages of organic beings, I must offer a few remarks on insects, which, by their numbers and the variety of their powers and instincts, exert a prodigious influence in the economy of animate nature. As a large portion of these minute creatures are strictly dependent for their subsistence on certain species of vegetables, the entomological provinces must coincide in considerable degree with the botanical.

* Quart. Journ. Geol. Soc., vol. iv. p. 336.
† Voy. aux Terres Australes, tom. i. p. 492.

All the insects, says Latreille, brought from the eastern parts of Asia and China, whatever be their latitude and temperature, are distinct from those of Europe and of Africa. The insects of the United States, although often approaching very close to our own, are, with very few exceptions, specifically distinguishable by some characters. In South America, the equinoctial lands of New Granada and Peru on the one side, and of Guiana on the other, contain for the most part distinct groups; the Andes forming the division, and interposing a narrow line of severe cold between climates otherwise very similar.*

Migratory instincts.—Nearly all the insects of the United States and Canada, differ specifically from the European; while those of Greenland appear to be in a great measure identical with our own. Some insects are very local; while a few, on the contrary, are common to remote countries, between which the torrid zone and the ocean intervene. Thus our painted lady butterfly (*Vanessa cardui*) re-appears at the Cape of Good Hope and in New Holland and Japan with scarcely a varying streak.† The same species is said to be one of the few insects which are universally dispersed over the earth, being found in Europe, Asia, Africa, and America; and its wide range is the more interesting, because it seems explained by its migratory instinct, seconded, no doubt, by a capacity, enjoyed by few species, of enduring a great diversity of temperature.

A vast swarm of this species, forming a column from ten to fifteen feet broad, was, a few years since, observed in the Canton de Vaud; they traversed the country with great rapidity from north to south, all flying onwards in regular order, close together, and not turning from their course on the approach of other objects. Professor Bonelli, of Turin, observed, in March of the same year, a similar swarm of the same species, also directing their flight from north to south, in Piedmont, in such immense numbers that at night the flowers were literally covered with them. They had been traced from Coni, Raconi, Susa, &c. A similar flight at the end of the last century is recorded by M. Louch in the Memoirs of the Academy of Turin. The fact is the more worthy of notice, because the caterpillars of this butterfly are not gregarious, but solitary from the moment that they are hatched; and this instinct remains dormant, while generation after generation passes away, till it suddenly displays itself in full energy when their numbers happen to be in excess.

Not only peculiar species, but certain types, distinguish particular countries; and there are groups, observes Kirby, which represent each other in distant regions, whether in their form, their functions, or in both. Thus the honey and wax of Europe, Asia, and Africa, are in each case prepared by bees congenerous with our common hive-bee (*Apis*, Latr.); while, in America, this genus is nowhere indigenous, but is replaced by Melipona, Trigona, and Euglossa; and in New Holland by

* Géographie Générale des Insectes et des Arachnides. Mém. du Mus. d'Hist. Nat., tom. iii.

† Kirby and Spence, vol. iv. p. 487; and other authors.

a still different but undescribed type. *The European bee (*Apis mellifica*), although not a native of the new world, is now established both in North and South America. It was introduced into the United States by some of the early settlers, and has since overspread the vast forests of the interior, building hives in the decayed trunks of trees. "The Indians," says Irving, "consider them as the harbinger of the white man, as the buffalo is of the red man, and say that in proportion as the bee advances the Indian and the buffalo retire. It is said," continues the same writer, "that the wild bee is seldom to be met with at any great distance from the frontier, and that they have always been the heralds of civilization, preceding it as it advanced from the Atlantic borders. Some of the ancient settlers of the west even pretend to give the very year when the honey-bee first crossed the Mississippi." † The same species is now also naturalized in Van Diemen's Land and New Zealand.

As almost all insects are winged, they can readily spread themselves wherever their progress is not opposed by uncongenial climates, or by seas, mountains, and other physical impediments; and these barriers they can sometimes surmount by abandoning themselves to violent winds, which, as I before stated, when speaking of the dispersion of seeds (p. 618.), may in a few hours carry them to very considerable distances. On the Andes some sphinxes and flies have been observed by Humboldt, at the height of 19,180 feet above the sea, and which appeared to him to have been involuntarily carried into these regions by ascending currents of air. ‡

White mentions a remarkable shower of aphides which seem to have emigrated, with an east wind, from the great hop plantations of Kent and Sussex, and blackened the shrubs and vegetables where they alighted at Selbourne, spreading at the same time in great clouds all along the vale from Farnham to Alton. These aphides are sometimes accompanied by vast numbers of the common lady-bird (*Coccinella septem-punctata*), which feed upon them. §

It is remarkable, says Kirby, that many of the insects which are occasionally observed to emigrate, as, for instance, the Libellulæ, Coccinellæ, Carabi, Cicadæ, &c. are not usually social insects; but seem to congregate, like swallows, merely for the purpose of emigration. ‖ Here, therefore, we have an example of an instinct developing itself on certain rare emergencies, causing unsocial species to become gregarious and to venture sometimes even to cross the ocean.

The armies of locusts which darken the air in Africa, and traverse the globe from Turkey to our southern counties in England, are well known to all. When the western gales sweep over the Pampas they

* Kirby and Spence, vol. iv. p. 497.
† Washington Irving's Tour in the Prairies, ch. ix.
‡ Malte-Brun, vol. v. p. 379.
§ Kirby and Spence, vol. ii. p. 9. 1817.
‖ Kirby and Spence, vol. ii. p. 12. 1817.

bear along with them myriads of insects of various kinds. As a proof of the manner in which species may be thus diffused, I may mention that when the Creole frigate was lying in the outer roads off Buenos Ayres, in 1819, at the distance of six miles from the land, her decks and rigging were suddenly covered by thousands of flies and grains of sand. The sides of the vessel had just received a fresh coat of paint, to which the insects adhered in such numbers as to spot and disfigure the vessel, and to render it necessary partially to renew the paint.* Captain W. H. Smyth was obliged to repaint his vessel, the Adventure, in the Mediterranean, from the same cause. He was on his way from Malta to Tripoli, when a southern wind blowing from the coast of Africa, then one hundred miles distant, drove such myriads of flies upon the fresh paint, that not the smallest point was left unoccupied by insects.

To the southward of the river Plate, off Cape St. Antonio, and at the distance of fifty miles from land, several large dragon-flies alighted on the Adventure frigate, during Captain King's late expedition to the Straits of Magellan. If the wind abates when insects are thus crossing the sea, the most delicate species are not necessarily drowned; for many can repose without sinking on the water. The slender long-legged tipulæ have been seen standing on the surface of the sea, when driven out far from our coast, and took wing immediately on being approached.† Exotic beetles are sometimes thrown on our shore, which revive after having been long drenched in salt water; and the periodical appearance of some conspicuous butterflies amongst us, after being unseen some for five others for fifty years, has been ascribed, not without probability, to the agency of the winds.

Inundations of rivers, observes Kirby, if they happen at any season except in the depths of winter, always carry down a number of insects, floating on the surface of bits of stick, weeds, &c.; so that when the waters subside, the entomologist may generally reap a plentiful harvest. In the dissemination, moreover, of these minute beings, as in that of plants, the larger animals play their part. Insects are, in numberless instances, borne along in the coats of animals, or the feathers of birds; and the eggs of some species are capable, like seeds, of resisting the digestive powers of the stomach, and after they are swallowed with herbage, may be ejected again unharmed in the dung.

Geographical Distribution and Diffusion of Man.

I have reserved for the last some observations on the range and diffusion of the human species over the earth, and the influence of man in spreading other animals and plants, especially the terrestrial.

Many naturalists have amused themselves in speculating on the probable birth-place of mankind, the point from which, if we assume the whole human race to have descended from a single pair, the tide of emigration

* I am indebted to Lieutenant Graves, R.N., for this information

† I state this fact on the authority of my friend, Mr. John Curtis.

must originally have proceeded. It has been always a favorite conjecture, that this birth-place was situated within or near the tropics, where perpetual summer reigns, and where fruits, herbs, and roots are plentifully supplied throughout the year. The climate of these regions, it has been said, is suited to a being born without any covering, and who had not yet acquired the arts of building habitations or providing clothes.

Progress of Human Population.—"The hunter state," it has been argued, "which Montesquieu placed the first, was probably only the second stage to which mankind arrived; since so many arts must have been invented to catch a salmon, or a deer, that society could no longer have been in its infancy when they came into use."* When regions where the spontaneous fruits of the earth abound became overpeopled, men would naturally diffuse themselves over the neighboring parts of the temperate zone; but a considerable time would probably elapse before this event took place; and it is possible, as a writer before cited observes, that in the interval before the multiplication of their numbers and their increasing wants had compelled them to emigrate, some arts to take animals were invented, but far inferior to what we see practised at this day among savages. As their habitations gradually advanced into the temperate zone, the new difficulties they had to encounter would call forth by degrees the spirit of invention, and the probability of such inventions always rises with the number of people involved in the same necessity.†

A distinguished modern writer, who coincides for the most part in the views above mentioned, has introduced one of the persons in his second dialogue, as objecting to the theory of the human race having gradually advanced from a savage to a civilized state, on the ground that "the first man must have inevitably been destroyed by the elements or devoured by savage beasts, so infinitely his superiors in physical force."‡ He then contends against the difficulty here started by various arguments, all of which were, perhaps, superfluous; for if a philosopher is pleased to indulge in conjectures on this subject, why should he not assign, as the original seat of man, some one of those large islands within the tropics, which are as free from large beasts of prey as Van Diemen's Land or Australia? Here man may have remained for a period, peculiar to a single island, just as some of the large anthropomorphous species are now limited to one island within the tropics. In such a situation, the new-born race might have lived in security, though far more helpless than the New Holland savages, and might have found abundance of vegetable food. Colonies may afterwards have been sent forth from this mother country, and then the peopling of the earth may have proceeded according to the hypothesis before alluded to.

To form a probable conjecture respecting the country from whence the early civilization of India was derived, has been found almost as difficult as to determine the original birth-place of the human race. That the dawn of oriental civilization did not arise within the limits of the tropics,

* Brand's Select Dissert. from the Amœn. Acad., vol. i. p. 118. † Ibid
‡ Sir H. Davy, Consolations in Travel, p. 74

is the conclusion to which Baron William von Humboldt has come after much patient research into "the diversities of the structure of language and their influence on the mental development of the human race." According to him the ancient Zend country from whence the spread of knowledge and the arts has been traced in a south-easterly direction, lay to the north-west of the upper Indus.*

As to the time of the first appearance of man upon the earth, if we are to judge from the discordance of opinion amongst celebrated chronologers, not even a rude approximation has yet been made towards determining a point of so much interest. The problem seems hitherto to have baffled the curiosity of the antiquary, if possible, more completely than the fixing on a geographical site for the original habitation of the ancestors of the human race. The Chevalier Bunsen, in his elaborate and philosophical work on Ancient Egypt,† has satisfied not a few of the learned, by an appeal to monumental inscriptions still extant, that the successive dynasties of kings may be traced back without a break, to Menes, and that the date of his reign would correspond with the year 3640 B. C. He supposes at the same time, what is most reasonable, that the Egyptian people must have existed for a long period (probably at least for five centuries), in their earlier and less settled state, before they reached the point of civilization at which Menes consolidated them into a great and united empire. This would carry us back to upwards of 4000 years B. C., or to an epoch coincident with that commonly set down for the creation of the world in accordance with computations founded on the combined ages of the successive antediluvian patriarchs. It follows that the same epoch of Menes is anterior by a great many centuries to the most ancient of the dates usually fixed upon for the Mosaic deluge. The fact that no record or tradition of any great and overwhelming flood has been detected in the mythology, or monumental annals of the Egyptians, will suggest many reflections to a geologist who has weighed well the evidence we possess of a variety of partial deluges which have happened in districts not free like Egypt, for the last 3000 years, from earthquakes and other causes of great aqueous catastrophes. The tales and legends of calamitous floods preserved in Greece, Asia Minor, the southern shores of the Baltic, China, Peru, and Chili, have, as we have seen, been all of them handed down to us by the inhabitants of regions in which the operation of natural causes in modern times, and the recurrence of a succession of disastrous floods, afford us data for interpreting the meaning of the obscure traditions of an illiterate age.‡

In his learned treatise on ancient chronology, Dr. Hales has selected, from a much greater number, a list of no less than 120 authors, all of whom give a different period for the epoch of the creation of the world,

* W. von Humboldt, "On the Kawi Language," &c. cited in Cosmos. Introduction.

† Egypten's Stelle, &c. Egypt restored to her Place in Universal History, by C. C. J. Bunsen. 1845.

‡ For Grecian and Asiatic deluges, see above, p. 356.; Cimbrian, p. 331., Chinese, p. 7. Peruvian, p. 502.; Chilian or Araucanian deluge, p. 500.

the extreme range of difference between them amounting to no less than 3268 years. It appears that even amongst authorities, who in England are generally regarded as orthodox, there is a variance, not of years or of one or two centuries, but of upwards of a millennium, according as they have preferred to follow the Hebrew, or the Samaritan, or the Greek versions of the Mosaic writings. Can we then wonder that they who decipher the monuments of Egypt, or the geologist who interprets the earth's autobiography, should arrive at views respecting the date of an ancient empire, or the age of our planet, irreconcileable with every one of these numerous and conflicting chronologies? The want of agreement amongst the learned in regard to the probable date of the deluge of Noah is a source of far greater perplexity and confusion than our extreme uncertainty as to the epoch of the creation,—the deluge being a comparatively modern event, from which the repeopling of the earth and the history of the present races of mankind is made to begin.

Naturalists have long felt that to render probable the received opinion that all the leading varieties of the human family have originally sprung from a single pair, (a doctrine against which there appears to me to be no sound objection,) a much greater lapse of time is required for the slow and gradual formation of races, (such as the Caucasian, Mongolian, and Negro,) than is embraced in any of the popular systems of chronology. The existence of two of those marked varieties above mentioned can be traced back 3000 years before the present time, or to the painting of pictures, preserved in the tombs or on the walls of buried temples in Egypt. In these we behold the Negro and Caucasian physiognomies portrayed as faithfully and in as strong contrast as if the likenesses of those races had been taken yesterday. When we consider therefore the extreme slowness of the changes, which climate and other modifying causes have produced in modern times, we must allow for a vast series of antecedent ages, in the course of which the long-continued influence of similar external circumstances gave rise to peculiarities, probably increased in many successive generations, until they were fixed by hereditary transmission. The characteristic forms and features thus acquired by certain tribes, may have been afterwards diffused by migration from a few centres over wide continental spaces. The theory, therefore, that all the races of man have come from one common stock receives support from every investigation which forces us to expand our ideas of the duration of past time, or which multiplies the number of years that have passed away since the origin of man. Hitherto, geology has neither enlarged nor circumscribed the "human period;" but simply proved that in the history of animated nature it is comparatively modern, or the last of a long series of antecedent epochs, in each of which the earth has been successively peopled by distinct species of animals and plants.

In an early stage of society the necessity of hunting acts as a principle of repulsion, causing men to spread with the greatest rapidity over a country, until the whole is covered with scattered settlements. It has been calculated that eight hundred acres of hunting-ground produce only as much food as half an acre of arable land. When the game has been in a great measure exhausted, and a state of pasturage succeeds, the several hunter tribes, being already scattered, may multiply in a short time into the greatest number which the pastoral state is capable of sustaining. The necessity, says Brand, thus imposed upon the two savage states, of dispersing themselves far and wide over the country, affords a reason why, at a very early period, the worst parts of the earth may have become inhabited.

But this reason, it may be said, is only applicable in as far as regards the peopling of a continuous continent; whereas the smallest islands, however remote from continents, have almost always been found inhabited by man. St. Helena, it is true, afforded an exception; for when that island was discovered in 1501, it was only inhabited by sea-fowl, and occasionally by seals and turtles, and was covered with a forest of trees and shrubs, all of species peculiar to it, with one or two exceptions, and which seem to have been expressly created for this remote and insulated spot.*

The islands also of Mauritius, Bourbon, Pitcairns, and Juan Fernandez, and those of the Galapagos archipelago, one of which is seventy miles long, were inhabited when first discovered, and, what is more remarkable than all, the Falkland Islands, which together are 120 miles in length by 60 in breadth, and abounding in food fit for the support of man.

Drifting of canoes to vast distances.—But very few of the numerous coral islets and volcanoes of the vast Pacific, capable of sustaining a few families of men, have been found untenanted; and we have, therefore, to inquire whence and by what means, if all the members of the great human family have had one common source, could those savages have migrated. Cook, Forster, and others, have remarked that parties of savages in their canoes must have often lost their way, and must have been driven on distant shores, where they were forced to remain, deprived both of the means and of the requisite intelligence for returning to their own country. Thus Captain Cook found on the island of Wateoo three inhabitants of Otaheite, who had been drifted thither in a canoe, although the distance between the two isles is 550 miles. In 1696, two canoes, containing thirty persons, who had left Ancorso, were thrown by contrary winds and storms on the island of Samar, one of the Philippines, at a distance of 800 miles. In 1721, two canoes, one of which contained twenty-four, and the other six persons, men, women, and children, were drifted from an island called Farroilep to the island of Guaham, one of the Marians, a distance of 200 miles.†

Kotzebue, when investigating the Coral Isles of Radack, at the eastern extremity of the Caroline Isles, became acquainted with a person of the name of Kadu, who was a native of Ulea, an isle 1500 miles distant, from which he had been drifted with a party. Kadu and three of his

* See p. 615.

† Malte-Brun's Geography, vol. iii. p. 419.

countrymen one day left Ulea in a sailing boat, when a violent storm arose, and drove them out of their course: they drifted about the open sea for eight months, according to their reckoning by the moon, making a knot on a cord at every new moon. Being expert fishermen, they subsisted entirely on the produce of the sea; and when the rain fell, laid in as much fresh water as they had vessels to contain it. "Kadu," says Kotzebue, "who was the best diver, frequently went down to the bottom of the sea, where it is well known that the water is not so salt, with a cocoa-nut shell, with only a small opening." * When these unfortunate men reached the isles of Radack, every hope and almost every feeling had died within them; their sail had long been destroyed, their canoe had long been the sport of winds and waves, and they were picked up by the inhabitants of Aur in a state of insensibility; but by the hospitable care of those islanders they soon recovered, and were restored to perfect health. †

Captain Beechey, in his voyage to the Pacific, fell in with some natives of the Coral Islands, who had in a similar manner been carried to a great distance from their native country. They had embarked, to the number of 150 souls, in three double canoes, from Anaa, or Chain Island, situated about three hundred miles to the eastward of Otaheite. They were overtaken by the monsoon, which dispersed the canoes; and after driving them about the ocean, left them becalmed, so that a great number of persons perished. Two of the canoes were never heard of; but the other was drifted from one uninhabited island to another, at each of which the voyagers obtained a few provisions; and at length, after having wandered for a distance of 600 miles, they were found and carried to their home in the Blossom. ‡

Mr. Crawfurd informs me that there are several well-authenticated accounts of canoes having been drifted from Sumatra to Madagascar, and by such causes a portion of the Malayan language, with some useful plants, have been transferred to that island, which is principally peopled by negroes.

The space traversed in some of these instances was so great, that similar accidents might suffice to transport canoes from various parts of Africa to the shores of South America, or from Spain to the Azores, and thence to North America; so that man, even in a rude state of society, is liable to be scattered involuntarily by the winds and waves over the globe, in a manner singularly analogous to that in which many plants and animals are diffused. We ought not, then, to wonder, that during the ages required for some tribes of the human race to attain that advanced stage of civilization which empowers the navigator to cross the ocean in all directions with security, the whole earth should

* Chamisso states that the water which they brought up was cooler, and *in their opinion*, less salt. It is difficult to conceive its being fresher near the bottom, except where submarine springs may happen to rise.

† Kotzebue's Voyage, 1815–1818. Quarterly Review, vol. xxvi. p. 361.

‡ Narrative of a Voyage to the Pacific, &c., in the years 1825, 1826, 1827, 1828, p. 170.

have become the abode of rude tribes of hunters and fishers. Were the whole of mankind now cut off, with the exception of one family, inhabiting the old or new continent, or Australia, or even some coral islet of the Pacific, we might expect their descendants, though they should never become more enlightened than the South Sea Islanders or the Esquimaux, to spread in the course of ages over the whole earth, diffused partly by the tendency of population to increase, in a limited district, beyond the means of subsistence, and partly by the accidental drifting of canoes by tides and currents to distant shores.

Involuntary Influence of Man in diffusing Animals and Plants.

Many of the general remarks which have been made respecting the influence of man in spreading or in checking the diffusion of plants apply equally to his relations with the animal kingdom. On a future occasion I shall be led to speak of the instrumentality of our species in naturalizing useful animals and plants in new regions, when explaining my views of the effects which the spreading and increase of certain species exert in the extirpation of others. At present I shall confine myself to a few remarks on the involuntary aid which man lends to the dissemination of species.

In the mammiferous class our influence is chiefly displayed in increasing the number of quadrupeds which are serviceable to us, and in exterminating or reducing the number of those which are noxious.

Sometimes, however, we unintentionally promote the multiplication of inimical species, as when we introduced the rat, which was not indigenous in the new world, into all parts of America. They have been conveyed over in ships, and now infest a great multitude of islands and parts of that continent. In like manner the Norway rat (*Mus decumanus*) has been imported into England, where it plunders our property in ships and houses.

Among birds, the house sparrow may be cited as a species known to have extended its range with the tillage of the soil. During the last century it has spread gradually over Asiatic Russia towards the north and east, always following the progress of cultivation. It made its first appearance on the Irtisch in Tobolsk, soon after the Russians had ploughed the land. It came in 1735 up the Obi to Beresow, and four years after to Naryn, about fifteen degrees of longitude farther east. In 1710, it had been seen in the higher parts of the coast of the Lena, in the government of Irkutzk. In all these places it is now common, but is not yet found in the uncultivated regions of Kamtschatka.*

The great viper (*Fer de lance*), a species no less venomous than the rattlesnake, which now ravages Martinique and St. Lucia, was accidentally introduced by man, and exists in no other part of the West Indies.

Many parasitic insects which attack our persons, and some of which are supposed to be peculiar to our species, have been carried into all parts

* Gloger, Abänd. der Vögel, p. 103.; Pallas, Zoog. Rosso-Asiat., tom. ii. p. 197.

of the earth, and have as high a claim as man to a *universal* geographical distribution.

A great variety of insects have been transported in ships from one country to another, especially in warmer latitudes. The European house-fly has been introduced in this way into all the South Sea Islands. Notwithstanding the coldness of our climate in England we have been unable to prevent the cockroach (*Blatta orientalis*) from entering and diffusing itself in our ovens and kneading troughs, and availing itself of the artificial warmth which we afford. It is well known also, that beetles, and many other kinds of ligniperdous insects, have been introduced into Great Britain in timber; especially several North American species. "The commercial relations," says Malte-Brun*, "between France and India have transported from the latter country the aphis, which destroys the apple tree, and two sorts of Neuroptera, the *Lucifuga* and *Flavicola*, mostly confined to Provence and the neighbourhood of Bourdeaux, where they devour the timber in the houses and naval arsenals."

Among mollusks we may mention the *Teredo navalis*, which is a native of equatorial seas, but which, by adhering to the bottom of ships, was transported to Holland, where it has been most destructive to vessels and piles. The same species has also become naturalized in England, and other countries enjoying an extensive commerce. *Bulimus undatus*, a land species of considerable size, native of Jamaica and other West Indian islands, has been imported, adhering to tropical timber, into Liverpool; and, as I learn from Mr. Broderip, is now naturalized in the woods near that town.

In all these and innumerable other instances we may regard the involuntary agency of man as strictly analogous to that of the inferior animals. Like them, we unconsciously contribute to extend or limit the geographical range and numbers of certain species, in obedience to general rules in the economy of nature, which are for the most part beyond our control.

* Syst. of Geog., vol. viii. p. 169.

CHAPTER XL.

THEORIES RESPECTING THE ORIGINAL INTRODUCTION OF SPECIES.

Proposal of an hypothesis on this subject—Supposed centres or foci of creation—Why distinct provinces of animals and plants have not become more blended together—Brocchi's speculations on the loss of species—Stations of plants and animals—Causes on which they depend—Stations of plants how affected by animals—Equilibrium in the number of species how preserved—Peculiar efficacy of insects in this task—Rapidity with which certain insects multiply or decrease in numbers—Effect of omnivorous animals in preserving the equilibrium of species—Reciprocal influence of aquatic and terrestrial species on each other.

Theory of Linnæus.—It would be superfluous to examine the various attempts which were made to explain the phenomena of the distribution of species alluded to in the preceding chapters, in the infancy of the sciences of botany, zoology, and physical geography. The theories or rather conjectures then indulged now stand refuted by a simple statement of facts; and if Linnæus were living he would be the first to renounce the notions which he promulgated. For he imagined the habitable world to have been for a certain time limited to one small tract, the only portion of the earth's surface that was as yet laid bare by the subsidence of the primæval ocean. In this fertile spot he supposed the originals of all the species of plants which exist on this globe to have been congregated together with the first ancestors of all animals and of the human race. "In quâ commodè habitaverint animalia omnia, et vegetabilia lætè germinaverint." In order to accommodate the various habitudes of so many creatures, and to provide a diversity of climate suited to their several natures, the tract in which the creation took place was supposed to have been situated in some warm region of the earth, but to have contained a lofty mountain range, on the heights and in the declivities of which were to be found all temperatures and every climate, from that of the torrid to that of the frozen zone.*

That there never was a universal ocean since the planet was inhabited, or, rather, since the oldest groups of strata yet known to contain organic remains were formed, is proved by the presence of terrestrial plants or by indications of shores in all the older formations; and if this conclusion was not established, yet no geologist could deny that, since the first small portion of the earth was laid dry, there have been many entire changes in the species of plants and animals inhabiting the land.

But, without dwelling on the above and other refuted theories, let us inquire whether some hypothesis cannot be substituted as simple as that of Linnæus, to which the phenomena now ascertained in regard to the distribution both of aquatic and terrestrial species may be referred.

* De terrâ habitabili incremento; also Prichard, Phys. Hist. of Mankind, vol. i. p. 17., where the hypotheses of different naturalists are enumerated.

The following may, perhaps, be reconcileable with known facts:—Each species may have had its origin in a single pair, or individual, where an individual was sufficient, and species may have been created in succession at such times and in such places as to enable them to multiply and endure for an appointed period, and occupy an appointed space on the globe.

In order to explain this theory, let us suppose every living thing to be destroyed in the western hemisphere, both on the land and in the ocean, and permission to be given to man to people this great desert, by transporting into it animals and plants from the eastern hemisphere, a strict prohibition being enforced against introducing two original stocks of the same species.

Now it is easy to show that the result of such a mode of colonizing would correspond exactly, so far as regards the grouping of animals and plants, with that now observed throughout the globe. In the first place, it would be necessary for naturalists, before they imported species into particular localities, to study attentively the climate and other physical conditions of each spot. It would be no less requisite to introduce the different species in succession, so that each plant and animal might have time and opportunity to multiply before the species destined to prey upon it was admitted. Many herbs and shrubs, for example, must spread far and wide before the sheep, the deer, and the goat could be allowed to enter, lest they should devour and annihilate the original stocks of many plants, and then perish themselves for want of food. The above-mentioned herbivorous animals in their turn must be permitted to make considerable progress before the entrance of the first pair of wolves or lions. Insects must be allowed to swarm before the swallow could be permitted to skim through the air, and feast on thousands at one repast.

It is evident that, however equally in this case our original stocks were distributed over the whole surface of land and water, there would nevertheless arise distinct botanical and zoological provinces, for there are a great many natural barriers which oppose common obstacles to the advance of a variety of species. Thus, for example, almost all the animals and plants naturalized by us, towards the extremity of South America, would be unable to spread beyond a certain limit, towards the east, west, and south; because they would be stopped by the ocean, and a few of them only would succeed in reaching the cooler latitudes of the northern hemisphere, because they would be incapable of bearing the heat of the tropics, through which they must pass. In the course of ages, undoubtedly, exceptions would arise, and some species might become common to the temperate and polar regions, or both sides of the equator; for I have before shown that the powers of diffusion conferred on some classes are very great. But we might confidently predict that these exceptions would never become so numerous as to invalidate the general rule.

Some of the plants and animals transplanted by us to the coast of Chili and Peru would never be able to cross the Andes, so as to reach the eastern plains; nor, for a similar reason, would those first established in the Pampas, or the valleys of the Amazon and the Orinoco, ever arrive at the shores of the Pacific.

In the ocean an analogous state of things would prevail; for there, also, climate would exert a great influence in limiting the range of species, and the land would stop the migrations of aquatic tribes as effectually as the sea arrests the dispersion of the terrestrial. As certain birds, insects, and the seeds of plants, can never cross the direction of prevailing winds, so currents form natural barriers to the dissemination of many oceanic races. A line of shoals may be as impassable to deep-water species, as are the Alps and the Andes to plants and animals peculiar to plains; while deep abysses may prove insuperable obstacles to the migrations of the inhabitants of shallow waters.

Supposed centres, or foci, of creation.—It is worthy of observation, that one effect of the introduction of single pairs of each species must be the confined range of certain groups in spots, which, like small islands, or solitary inland lakes, have few means of interchanging their inhabitants with adjoining regions. Now this congregating in a small space of many peculiar species, would give an appearance of *centres* or *foci* of creation, as they have been termed, as if they were favourite points where the creative energy has been in greater action than in others, and where the numbers of peculiar organic beings have consequently become more considerable.

I do not mean to call in question the soundness of the inferences of some botanists, as to the former existence of certain limited spots whence species of plants have been propagated, radiating, as it were, in all directions from a common centre. On the contrary, I conceive these phenomena to be the necessary consequences of the plan of nature before suggested, operating during the successive mutations of the surface, some of which the geologist can prove to have taken place subsequently to the period when many species now existing were created. In order to exemplify how this arrangement of plants may have been produced, let us imagine that, about three centuries before the discovery of St. Helena (itself of submarine volcanic origin), a multitude of new islands had been thrown up in the surrounding sea, and that these had each become clothed with plants emigrating from St. Helena, in the same manner as the wild plants of Campania have diffused themselves over Monte Nuovo. Whenever the first botanist investigated the new archipelago, he would, in all probability, find a different assemblage of plants in each of the islands of recent formation; but in St. Helena itself, he would meet with individuals of every species, belonging to all parts of the archipelago, and some, in addition, peculiar to itself, viz., those which had not been able to obtain a passage into any one of the surrounding new-formed lands. In this case it might be truly said that the original island was the primitive focus, or centre, of a certain type of vegetation; whereas, in the surrounding islands, there would be a smaller number of species, yet all belonging to the same group.

But this peculiar distribution of plants would not warrant the conclusion that, in the space occupied by St. Helena, there had been a greater exertion of creative power than in the spaces of equal area occupied by the new adjacent lands; because, within the period in which St. Helena had acquired its peculiar vegetation, each of the spots supposed to be subsequently converted into land may have been the birth-place of a great number of *marine* animals and plants, which may have had time to scatter themselves far and wide over the southern Atlantic.

Why distinct provinces not more blended.—Perhaps it may be objected to some parts of the foregoing train of reasoning, that during the lapse of past ages, especially during many partial revolutions of the globe of comparatively modern date, different zoological and botanical provinces ought to have become more confounded and blended together—that the distribution of species approaches too nearly to what might have been expected, if animals and plants had been introduced into the globe when its physical geography had already assumed the features which it now wears; whereas we know that, in certain districts, considerable geographical changes have taken place since species identical with those now in being were created.

Brocchi's speculations on loss of species.—These and many kindred topics cannot be fully discussed until we have considered, not merely the general laws which may regulate the first introduction of species, but those which may limit their *duration* on the earth. Brocchi remarked, when hazarding some interesting conjectures respecting "the loss of species," that a modern naturalist had no small assurance, who declared "that individuals alone were capable of destruction, and that species were so perpetuated that nature could not annihilate them, so long as the planet lasted, or at least that nothing less than the shock of a comet, or some similar disaster, could put an end to their existence."* The Italian geologist, on the contrary, had satisfied himself that many species of Testacea, which formerly inhabited the Mediterranean, had become extinct, although a great number of others, which had been the contemporaries of those lost races, still survived. He came to the opinion that about half the species which peopled the waters when the Subapennine strata were deposited had gone out of existence; and in this inference he does not appear to have been far wrong.

But, instead of seeking a solution of this problem, like some other geologists of his time, in a violent and general catastrophe, Brocchi endeavoured to imagine some regular and constant law by which species might be made to disappear from the earth gradually and in succession. The death, he suggested, of a species might depend, like that of individuals, on certain peculiarities of constitution conferred upon them at their birth; and as the longevity of the one depends on a certain force of vitality, which, after a period, grows weaker and weaker, so the duration of the other may be governed by the quantity of prolific power

* Necker, Phytozool. Philosoph. p. 21.; Brocchi, Conch. Foss. Subap., tome i. p. 229.

bestowed upon the species which, after a season, may decline in energy, so that the fecundity and multiplication of individuals may be gradually lessened from century to century, "until that fatal term arrives when the embryo, incapable of extending and developing itself, abandons, almost at the instant of its formation, the slender principle of life by which it was scarcely animated,—and so all dies with it."

Now we may coincide in opinion with the Italian naturalist, as to the gradual extinction of species one after another, by the operation of regular and constant causes, without admitting an inherent principle of deterioration in their physiological attributes. We might concede, "that many species are on the decline, and that the day is not far distant when they will cease to exist;" yet deem it consistent with what we know of the nature of organic beings, to believe that the last individuals of each species retain their prolific powers in their full intensity.

Brocchi has himself speculated on the share which a change of climate may have had in rendering the Mediterranean unfit for the habitation of certain Testacea, which still continued to thrive in the Indian Ocean, and of others which were now only represented by analogous forms within the tropics. He must also have been aware that other extrinsic causes, such as the progress of human population, or the increase of some one of the inferior animals, might gradually lead to the extirpation of a particular species, although its fecundity might remain to the last unimpaired. If, therefore, amid the vicissitudes of the animate and inanimate world, there are known causes capable of bringing about the decline and extirpation of species, it became him thoroughly to investigate the full extent to which these might operate, before he speculated on any cause of so purely hypothetical a kind as "the diminution of the prolific virtue."

If it could have been shown that some wild plant had insensibly dwindled away and died out, as sometimes happens to cultivated varieties propagated by cuttings, even though climate, soil, and every other circumstance, should continue identically the same—if any animal had perished while the physical condition of the earth, and the number and force of its foes, with every other extrinsic cause, remain unaltered, then might we have some ground for suspecting that the infirmities of age creep on as naturally on species as upon individuals. But, in the absence of such observations, let us turn to another class of facts, and examine attentively the circumstances which determine the *stations* of particular animals and plants, and perhaps we shall discover, in the vicissitudes to which these stations are exposed, a cause fully adequate to explain the phenomena under consideration.

Stations of plants and animals.—Stations comprehend all the circumstances, whether relating to the animate or inanimate world, which determine whether a given plant or animal can exist in a given place; so that if it be shown that stations can become essentially modified by the influence of known causes, it will follow that species, as well as individuals, are mortal.

Every naturalist is familiar with the fact, that although in a particular country, such as Great Britain, there may be more than three thousand species of plants, ten thousand insects, and a great variety in each of the other classes; yet there will not be more than a hundred, perhaps not half that number, inhabiting any given locality. There may be no want of space in the supposed tract: it may be a large mountain, or an extensive moor, or a great river plain, containing room enough for individuals of every species in our island; yet the spot will be occupied by a few to the exclusion of many, and these few are enabled, throughout long periods, to maintain their ground successfully against every intruder, notwithstanding the facilities which species enjoy, by virtue of their power of diffusion, of invading adjacent territories.

The principal causes which enable a certain assemblage of plants thus to maintain their ground against all others depend, as is well known, on the relations between the physiological nature of each species, and the climate, exposure, soil, and other physical conditions of the locality. Some plants live only on rocks, others in meadows, a third class in marshes. Of the latter, some delight in a fresh-water morass,—others in salt marshes, where their roots may copiously absorb saline particles. Some prefer an alpine region in a warm latitude, where, during the heat of summer, they are constantly irrigated by the cool waters of melting snows. To others loose sand, so fatal to the generality of species, affords the most proper station. The *Carex arenaria* and the *Elymus arenarius* acquire their full vigor on a sandy dune, obtaining an ascendancy over the very plants which in a stiff clay would immediately stifle them.

Where the soil of a district is of so peculiar a nature that it is extremely favorable to certain species, and agrees ill with every other, the former get exclusive possession of the ground, and, as in the case of heaths, live in societies. In like manner the bog moss (*Sphagnum*) is fully developed in peaty swamps, and becomes, like the heath, in the language of botanists, a social plant. Such monopolies, however, are not common, for they are checked by various causes. Not only are many species endowed with equal powers to obtain and keep possession of similar stations, but each plant, for reasons not fully explained by the physiologist, has the property of rendering the soil where it has grown less fitted for the support of other individuals of its own species, or even other species of the same family. Yet the same spot, so far from being impoverished, is improved, for plants of *another* family. Oaks, for example, render the soil more fertile for the fir tribe, and firs prepare the soil for oaks. Every agriculturist feels the force of this law of the organic world, and regulates accordingly the rotation of his crops.

Equilibrium in the number of species, how preserved.—"All the plants of a given country," says De Candolle, in his usual spirited style, "are at war one with another. The first which establish themselves by chance in a particular spot tend, by the mere occupancy of space, to exclude other species—the greater choke the smaller; the longest livers replace those which last for a shorter period; the more prolific gradually make themselves masters of the ground, which species multiplying more slowly would otherwise fill."

In this continual strife it is not always the resources of the plant itself which enable it to maintain or extend its ground. Its success depends, in a great measure, on the number of its foes or allies among the animals and plants inhabiting the same region. Thus, for example, a herb which loves the shade may multiply, if some tree with spreading boughs and dense foliage flourish in the neighborhood. Another, which, if unassisted, would be overpowered by the rank growth of some hardy competitor, is secure because its leaves are unpalatable to cattle; which, on the other hand, annually crop down its antagonist, and rarely suffer it to ripen its seed.

Oftentimes we see some herb which has flowered in the midst of a thorny shrub, when all the other individuals of the same species, in the open fields around, are eaten down, and cannot bring their seed to maturity. In this case, the shrub has lent his armor of spines and prickles to protect the defenceless herb against the mouths of the cattle, and thus a few individuals which occupied, perhaps, the most unfavorable station in regard to exposure, soil, and other circumstances, may, nevertheless, by the aid of an ally, become the principal source whereby the winds are supplied with seeds which perpetuate the species throughout the surrounding tract. Thus, in the New Forest in Hampshire, the young oaks which are not consumed by the deer, or uprooted by the swine, are indebted to the holly for their escape.

In the above examples we see one plant shielding another from the attacks of animals; but instances are, perhaps, still more numerous, where some animal defends a plant against the enmity of some other subject of the vegetable kingdom.

Scarcely any beast, observes a Swedish naturalist, will touch the nettle, but fifty different kinds of insects are fed by it.* Some of these seize upon the root, others upon the stem; some eat the leaves, others devour the seeds and flowers; but for this multitude of enemies, the nettle (*Urtica dioica*), which is now found in all the four quarters of the globe, would annihilate a great number of plants. Linnæus tells us, in his "Tour in Scania," that goats were turned into an island which abounded with the *Agrostis arundinacea*, where they perished by famine; but horses which followed them grew fat on the same plant. The goat, also, he says, thrives on the meadow-sweet and water-hemlock, plants which are injurious to cattle.†

Agency of insects.—Every plant, observes Wilcke, has its proper insect allotted to it to curb its luxuriancy, and to prevent it from multiplying to the exclusion of others. "Thus grass in meadows sometimes flourishes so as to exclude all other plants; here the Phalæna graminis (*Bombyx gram.*), with her numerous progeny, finds a well-spread table;

* Amœn. Acad. vol. vi. p. 17. § 12.

† Ibid. vol. vii. p. 409.

they multiply in immense numbers, and the farmer, for some years, laments the failure of his crop; but the grass being consumed, the moths die with hunger, or remove to another place. Now the quantity of grass being greatly diminished, the other plants, which were before choked by it, spring up, and the ground becomes variegated with a multitude of different species of flowers. Had not nature given a commission to this minister for that purpose, the grass would destroy a great number of species of vegetables, of which the equilibrium is now kept up."*

In the above passage allusion is made to the ravages committed in 1740, and the two following years, in many provinces of Sweden, by a most destructive insect. The same moth is said never to touch the fox-tail grass, so that it may be classed as a most active ally and benefactor of that species, and as peculiarly instrumental in preserving it in its present abundance.† A discovery of Rolander, cited in the treatise of Wilcke above mentioned, affords a good illustration of the checks and counter-checks which nature has appointed to preserve the balance of power among species. "The *Phalæna strobilella* has the fir cone assigned to it to deposit its eggs upon; the young caterpillars coming out of the shell consume the cone and superfluous seed; but, lest the destruction should be too general, the *Ichneumon strobilellæ* lays its eggs in the caterpillar, inserting its long tail in the openings of the cone till it touches the included insect, for its body is too large to enter. Thus it fixes its minute egg upon the caterpillar, which being hatched, destroys it."‡

Entomologists enumerate many parallel cases where insects, appropriated to certain plants, are kept down by other insects, and these again by parasites expressly appointed to prey on them.§ Few perhaps are in the habit of duly appreciating the extent to which insects are active in preserving the balance of species among plants, and thus regulating indirectly the relative numbers of many of the higher orders of terrestrial animals.

The peculiarity of their agency consists in their power of suddenly multiplying their numbers to a degree which could only be accomplished in a considerable lapse of time in any of the larger animals, and then as instantaneously relapsing, without the intervention of any violent disturbing cause, into their former insignificance.

If, for the sake of employing, on different but rare occasions, a power of many hundred horses, we were under the necessity of feeding all these animals at great cost in the intervals when their services were not required, we should greatly admire the invention of a machine, such as the steam-engine, which was capable at any moment of exerting the same degree of strength without any consumption of food during periods of inaction. The same kind of admiration is strongly excited when we contemplate the powers of insect life, in the creation of which the Author

* Amœn. Acad., vol. vi. p. 17. § 11, 12.
† Kirby and Spence, vol. i. p. 178.
‡ Amœn. Acad., vol. vi. p. 26. § 14.
§ Kirby and Spence, vol. iv. p. 218.

of nature has been so prodigal. A scanty number of minute individuals, to be detected only by careful research, are ready in a few days, weeks, or months, to give birth to myriads, which may repress any degree of monopoly in another species, or remove nuisances, such as dead carcases, which might taint the air. But no sooner has the destroying commission been executed than the gigantic power becomes dormant—each of the mighty host soon reaches the term of its transient existence, and the season arrives when the whole species passes naturally into the egg, and thence into the larva and pupa state. In this defenceless condition it may be destroyed either by the elements, or by the augmentation of some of its numerous foes which may prey upon it in the early stages of its transformation: or it often happens that in the following year the season proves unfavorable to the hatching of the eggs or the development of the pupæ.

Thus the swarming myriads depart which may have covered the vegetation like the aphides, or darkened the air like locusts. In almost every season there are some species which in this manner put forth their strength, and then, like Milton's spirits, which thronged the spacious hall, " reduce to smallest forms their shapes immense"—

> ———— So thick the aëry crowd
> Swarm'd and were straiten'd; till the signal given,
> Behold a wonder! they but now who seem'd
> In bigness to surpass earth's giant sons,
> Now less than smallest dwarfs.

A few examples will illustrate the mode in which this force operates. It is well known that, among the countless species of the insect creation, some feed on animal, others on vegetable matter; and upon considering a catalogue of eight thousand British Insects and Arachnidæ, Mr. Kirby found that these two divisions were nearly a counterpoise to each other, the carnivorous being somewhat preponderant. There are also distinct species, some appointed to consume living, others dead or putrid animal and vegetable substances. One female, of *Musca carnaria*, will give birth to twenty thousand young; and the larvæ of many flesh-flies devour so much food in twenty-four hours, and grow so quickly, as to increase their weight two hundred-fold! In five days after being hatched they arrive at their full growth and size, so that there was ground, says Kirby, for the assertion of Linnæus, that three flies of *M. vomitoria* could devour a dead horse as quickly as a lion*; and another Swedish naturalist remarks, that so great are the powers of propagation of a single species even of the smallest insects, that each can commit, when required, more ravages than the elephant.†

Next to locusts, the aphides, perhaps, exert the greatest power over the vegetable world, and, like them, are so metimes so numerous as to darken the air. The multiplication of these little creatures is without

* Kirby and Spence, vol. i. p. 250. † Wilcke, Amœn. Acad. c. ii.

parallel, and almost every plant has its peculiar species. Reaumur has proved that in five generations one aphis may be the progenitor of 5,904,900,000 descendants; and it is supposed that in one year there may be twenty generations.* Mr. Curtis observes that, as among caterpillars we find some that are constantly and unalterably attached to one or more particular species of plants, and others that feed indiscriminately on most sorts of herbage, so it is precisely with the aphides: some are particular, others more general feeders; and as they resemble other insects in this respect, so they do also in being more abundant in some years than in others.† In 1793 they were the chief, and in 1798 the sole, cause of the failure of the hops. In 1794, a season almost unparalleled for drought, the hop was perfectly free from them; while peas and beans, especially the former, suffered very much from their depredations.

The ravages of the caterpillars of some of our smaller moths afford a good illustration of the temporary increase of a species. The oak-trees of a considerable wood have been stripped of their leaves as bare as in winter by the caterpillars of a small green moth (*Tortrix viridana*), which has been observed the year following not to abound.‡ The silver Y moth (*Plusia gamma*), although one of our common species, is not dreaded by us for its devastations; but legions of their caterpillars have at times created alarm in France, as in 1735. Reaumur observes that the female moth lays about four hundred eggs; so that if twenty caterpillars were distributed in a garden, and all lived through the winter and became moths in the succeeding May, the eggs laid by these, if half of them were female and all fertile, would in the next generation produce 800,000 caterpillars.§ A modern writer, therefore, justly observes that, did not Providence put causes in operation to keep them in due bounds, the caterpillars of this moth alone, leaving out of consideration the two thousand other British species, might soon destroy more than half of our vegetation.‖

In the latter part of the last century an ant most destructive to the sugar-cane (*Formica saccharivora*), appeared in such infinite hosts in the island of Granada, as to put a stop to the cultivation of that vegetable. Their numbers were incredible. The plantations and roads were filled with them; many domestic quadrupeds, together with rats, mice, and reptiles, and even birds, perished in consequence of this plague. It was not till 1780 that they were at length annihilated by torrents of rain, which accompanied a dreadful hurricane.¶

Devastations caused by locusts.—We may conclude by mentioning some instances of the devastations of locusts in various countries. Among other parts of Africa, Cyrenaica has been at different periods infested by myriads of these creatures, which have consumed nearly every green thing. The effect of the havoc committed by them may be estimated

* Kirby and Spence, vol. i. p. 174. † Trans. Linn. Soc., vol. vi.
‡ Lib. Ent. Know., Insect Trans., p. 203. See Haworth, Lep.
§ Reaumur, ii. 337. ‖ Lib. Ent. Know., Insect Trans., p. 212.
¶ Kirby and Spence, vol. i. p. 183. Castle, Phil. Trans., xxx. 346.

by the famine they occasioned. St. Augustin mentions a plague of this kind in Africa, which destroyed no less than 800,000 men in the kingdom of Massinissa alone, and many more upon the territories bordering upon the sea. It is also related, that in the year 591 an infinite army of locusts migrated from Africa into Italy; and, after grievously ravaging the country, were cast into the sea, when there arose a pestilence from their stench, which carried off nearly a million of men and beasts.

In the Venetian territory, also, in 1748, more than thirty thousand persons are said to have perished in a famine occasioned by this scourge; and other instances are recorded of their devastations in France, Spain, Italy, Germany, &c. In different parts of Russia also, Hungary, and Poland, in Arabia and India, and other countries, their visitations have been periodically experienced. Although they have a preference for certain plants, yet, when these are consumed, they will attack almost all the remainder. In the accounts of the invasion of locusts, the statements which appear most marvellous relate to the prodigious mass of matter which encumbers the sea wherever they are blown into it, and the pestilence arising from its putrefaction. Their dead bodies are said to have been, in some places, heaped one upon another, to the depth of four feet, in Russia, Poland and Lithuania; and when, in Southern Africa, they were driven into the sea, by a north-west wind, they formed, says Barrow, along the shore, for fifty miles, a bank three or four feet high.* But when we consider that forests are stripped of their foliage, and the earth of its green garment for thousands of square miles, it may well be supposed that the volume of animal matter produced may equal that of great herds of quadrupeds and flights of large birds suddenly precipitated into the sea.

The occurrence of such events, at certain intervals, in hot countries, like the severe winters and damp summers returning after a series of years in the temperate zone, may affect the proportional numbers of almost all classes of animals and plants, and probably prove fatal to the existence of many which would otherwise thrive there; while, on the contrary, the same occurrences can scarcely fail to be favorable to certain species which, if deprived of such aid, might not maintain their ground.

Although it may usually be remarked that the extraordinary increase of some one species is immediately followed and checked by the multiplication of another, yet this does not always happen; partly because many species feed in common on the same kinds of food, and partly because many kinds of food are often consumed indifferently by one and the same species. In the former case, where a variety of different animals have precisely the same taste, as, for example, when many insectivorous birds and reptiles devour alike some particular fly or beetle, the unusual numbers of these insects may cause only a slight and almost imperceptible augmentation of each of these species of bird and reptile.

* Travels in Africa, p. 257. Kirby and Spence, vol. i. p. 215.

In the other instances, where one animal preys on others of almost every class, as for example, where our English buzzards devour not only small quadrupeds, as rabbits and field-mice, but also birds, frogs, lizards, and insects, the profusion of any one of these last may cause all such general feeders to subsist more exclusively upon the species thus in excess, by which means the balance may be restored.

Agency of omnivorous animals.—The number of species which are nearly omnivorous is considerable; and although every animal has, perhaps, a predilection for some one description of food rather than another, yet some are not even confined to one of the great kingdoms of the organic world. Thus, when the raccoon of the West Indies can procure neither fowls, fish, snails, nor insects, it will attack the sugar-canes, and devour various kinds of grain. The civets, when animal food is scarce, maintain themselves on fruits and roots.

Numerous birds, which feed indiscriminately on insects and plants, are perhaps more instrumental than any other of the terrestrial tribes in preserving a constant equilibrium between the relative numbers of different classes of animals and vegetables. If the insects become very numerous and devour the plants, these birds will immediately derive a larger portion of their subsistence from insects, just as the Arabians, Syrians, and Hottentots feed on locusts, when the locusts devour their crops.

Reciprocal influence of aquatic and terrestrial species.—The intimate relation of the inhabitants of the water to those of the land, and the influence exerted by each on the relative number of species, must not be overlooked amongst the complicated causes which determine the existence of animals and plants in certain regions. A large portion of the amphibious quadrupeds and reptiles prey partly on aquatic plants and animals, and in part on terrestrial; and a deficiency of one kind of prey causes them to have immediate recourse to the other. The voracity of certain insects, as the dragon-fly, for example, is confined to the water during one stage of their transformations, and in their perfect state to the air. Innumerable water-birds, both of rivers and seas, derive in like manner their food indifferently from either element; so that the abundance or scarcity of prey in one induces them either to forsake or more constantly to haunt the other. Thus an intimate connection between the state of the animate creation in a lake or river, and in the adjoining dry land, is maintained; or between a continent, with its lakes and rivers, and the ocean. It is well known that many birds migrate, during stormy seasons, from the sea-shore into the interior, in search of food; while others, on the contrary, urged by like wants, forsake their inland haunts, and live on substances rejected by the tide.

The migration of fish into rivers during the spawning season supplies another link of the same kind. Suppose the salmon to be reduced in numbers by some marine foes, as by seals and grampuses, the consequence must often be, that in the course of a few years the otters at the distance of several hundred miles inland will be lessened in number from the scarcity of fish. On the other hand, if there be a dearth of food for the young fry of the salmon in rivers and estuaries, so that few return to the sea, the sand eels and other marine species, which are usually kept down by the salmon, will swarm in greater profusion.

It is unnecessary to accumulate a greater number of illustrations in order to prove that the stations of different plants and animals depend on a great complication of circumstances,—on an immense variety of relations in the state of the inanimate worlds. Every plant requires a certain climate, soil, and other conditions, and often the aid of many animals, in order to maintain its ground. Many animals feed on certain plants, being often restricted to a small number, and sometimes to one only; other members of the animal kingdom feed on plant-eating species, and thus become dependent on the conditions of the *stations* not only of their prey, but of the plants consumed by them.

Having duly reflected on the nature and extent of these mutual relations in the different parts of the organic and inorganic worlds, we may next proceed to examine the results which may be anticipated from the fluctuations now continually in progress in the state of the earth's surface, and in the geographical distribution of its living productions.

CHAPTER XLI.

EXTINCTION OF SPECIES.—CHANGES IN THE STATIONS OF ANIMALS.

Extension of the range of one species alters the condition of many others—The first appearance of a new species causes the chief disturbance—Changes known to have resulted from the advance of human population—Whether man increases the productive powers of the earth—Indigenous quadrupeds and birds extirpated in Great Britain—Extinction of the dodo—Rapid propagation of domestic quadrupeds in America—Power of exterminating species no prerogative of man—Concluding remarks

WE have seen that the stations of animals and plants depend not merely on the influence of external agents in the inanimate world, and the relations of that influence to the structure and habits of each species, but also on the state of the contemporary living beings which inhabit the same part of the globe. In other words, the possibility of the existence of a certain species in a given place, or of its thriving more or less therein, is determined not merely by temperature, humidity, soil, elevation, and other circumstances of the like kind; but also by the existence or non-existence, the abundance or scarcity, of a particular assemblage of other plants and animals in the same region.

If it be shown that both these classes of circumstances, whether relating to the animate or inanimate creation, are perpetually changing, it will follow that species are subject to incessant vicissitudes; and if the result of these mutations, in the course of ages, be so great as materially to affect the general condition of *stations*, it will follow that the successive destruction of species must now be part of the regular and constant order of nature.

Extension of the range of one species alters the condition of the others.—It will be desirable, first, to consider the effects which every extension of the numbers or geographical range of one species must produce on the condition of others inhabiting the same regions. When the necessary consequences of such extensions have been fully explained, the reader will be prepared to appreciate the important influence which slight modifications in the physical geography of the globe may exert on the condition of organic beings.

In the first place, it is clear that when any region is stocked with as great a variety of animals and plants as the productive powers of that region will enable it to support, the addition of any new species, or the *permanent* numerical increase of one previously established, must always be attended either by the local extermination or the numerical decrease of some other species.

There may undoubtedly be considerable fluctuations from year to year, and the equilibrium may be again restored without any permanent alteration; for, in particular seasons, a greater supply of heat, humidity, or other causes, may augment the total quantity of vegetable produce, in which case all the animals subsisting on vegetable food, and others which prey on them, may multiply without any one species giving way: but whilst the aggregate quantity of vegetable produce remains unaltered, the progressive increase of one animal or plant implies the decline of another.

All agriculturists and gardeners are familiar with the fact that when weeds intrude themselves into the space appropriated to cultivated species, the latter are starved in their growth, or stifled. If we abandon for a short time a field or garden, a host of indigenous plants,

> The darnel, hemlock, and rank fumitory,

pour in and obtain the mastery, extirpating the exotics, or putting an end to the monopoly of some native plants.

If we inclose a park, and stock it with as many deer as the herbage will support, we cannot add sheep without lessening the number of the deer; nor can other herbivorous species be subsequently introduced, unless the individuals of each species in the park become fewer in proportion.

So, if there be an island where leopards are the only beasts of prey, and the lion, tiger, and hyæna afterwards enter, the leopards, if they stand their ground, will be reduced in number. If the locusts then arrive and swarm greatly, they may deprive a large number of plant-eating animals of their food, and thereby cause a famine, not only among them, but among the beasts of prey: certain species perhaps, which had the weakest footing in the island, may thus be annihilated.

We have seen how many distinct geographical provinces there are of aquatic and terrestrial species, and how great are the powers of migration conferred on different classes, whereby the inhabitants of one region may be enabled from time to time to invade another, and do actually so migrate and diffuse themselves over new countries. Now, although our knowledge of the history of the animate creation dates from so recent a period, that we can scarcely trace the advance or decline of any animal or plant, except in those cases where the influence of man has intervened; yet we can easily conceive what must happen when some new colony of wild animals or plants enters a region for the first time, and succeeds in establishing itself.

Supposed effects of the first entrance of the polar bear into Iceland.—Let us consider how great are the devastations committed at certain periods by the Greenland bears, when they are drifted to the shores of Iceland in considerable numbers on the ice. These periodical invasions are formidable even to man; so that when the bears arrive, the inhabitants collect together, and go in pursuit of them with fire-arms—each native who slays one being rewarded by the King of Denmark. The Danes of old, when they landed in their marauding expeditions upon our coast, hardly excited more alarm, nor did our islanders muster more promptly for the defence of their lives and property against the common enemy, than the modern Icelanders against these formidable brutes. It often happens, says Henderson, that the natives are pursued by the bear when he has been long at sea, and when his natural ferocity has been heightened by the keenness of hunger; if unarmed, it is frequently by stratagem only that they make their escape.*

Let us cast our thoughts back to the period when the first polar bears reached Iceland, before it was colonized by the Norwegians in 874: we may imagine the breaking up of an immense barrier of ice like that which, in 1816 and the following year, disappeared from the east coast of Greenland, which it had surrounded for four centuries. By the aid of such means of transportation a great number of these quadrupeds might effect a landing at the same time, and the havoc which they would make among the species previously settled in the island would be terrific. The deer, foxes, seals, and even birds, on which these animals sometimes prey, would be soon thinned down.

But this would be a part only, and probably an insignificant portion, of the aggregate amount of change brought about by the new invader. The plants on which the deer fed, being less consumed in consequence of the lessened numbers of that herbivorous species, would soon supply more food to several insects, and probably to some terrestrial testacea, so that the latter would gain ground. The increase of these would furnish other insects and birds with food, so that the numbers of these last would be augmented. The diminution of the seals would afford a respite to some fish which they had persecuted; and these fish, in their turn, would then

* Journal of a Residence in Iceland, p. 276.

multiply and press upon their peculiar prey. Many water-fowls, the eggs and young of which are devoured by foxes, would increase when the foxes were thinned down by the bears; and the fish on which the water-fowls subsisted would then, in their turn, be less numerous. Thus the numerical proportions of a great number of the inhabitants, both of the land and sea, might be permanently altered by the settling of one new species in the region; and the changes caused indirectly would ramify through all classes of the living creation, and be almost endless.

An actual illustration of what we have here only proposed hypothetically, is in some degree afforded by the selection of small islands by the eider duck for its residence during the season of incubation, its nest being seldom if ever found on the shores of the main land, or even of a large island. The Icelanders are so well aware of this, that they have expended a great deal of labor in forming artificial islands, by separating from the main land certain promontories, joined to it by narrow isthmuses. This insular position is necessary to guard against the destruction of the eggs and young birds, by foxes, dogs, and other animals. One year, says Hooker, it happened that, in the small island of Vidoe, adjoining the coast of Iceland, a fox got over *upon the ice*, and caused great alarm, as an immense number of ducks were then sitting on their eggs or young ones. It was long before he was taken, which was at last, however, effected by bringing another fox to the island, and fastening it by a string near the haunt of the former, by which he was allured within shot of the hunter.*

The first appearance of a new species causes the chief disturbance.—It is usually the first appearance of an animal or plant, in a region to which it was previously a stranger, that gives rise to the chief alteration; since, after a time, an equilibrium is again established. But it must require ages before such a new adjustment of the relative forces of so many conflicting agents can be definitely settled. The causes in simultaneous action are so numerous, that they admit of an almost infinite number of combinations; and it is necessary that all these should have occurred once before the total amount of change, capable of flowing from any new disturbing force, can be estimated.

Thus, for example, suppose that once in two centuries a frost of unusual intensity, or a volcanic eruption of great violence accompanied by floods from the melting of glaciers, should occur in Iceland; or an epidemic disease, fatal to the larger number of individuals of some one species, and not affecting others,—these, and a variety of other contingencies, all of which may occur at once, or at periods separated by different intervals of time, ought to happen before it would be possible for us to declare what ultimate alteration the presence of any new comer, such as the bear before mentioned, might occasion in the animal population of the isle.

Every new condition in the state of the organic or inorganic creation, a new animal or plant, an additional snow-clad mountain, any permanent change, however slight in comparison to the whole, gives rise to a

* Tour in Iceland, vol. i. p. 64, 2nd edit.

new order of things, and may make a material change in regard to some one or more species. Yet a swarm of locusts, or a frost of extreme intensity, or an epidemic disease, may pass away without any great apparent derangement; no species may be lost, and all may soon recover their former relative numbers, because the same scourges may have visited the region again and again, at preceding periods. Every plant that was incapable of resisting such a degree of cold, every animal which was exposed to be entirely cut off by an epidemic or by famine caused by the consumption of vegetation by the locusts, may have perished already, so that the subsequent recurrence of similar catastrophes is attended only by a temporary change.

Changes caused by Man

We are best acquainted with the mutations brought about by the progress of human population, and the growth of plants and animals favored by man. To these, therefore, we should in the first instance turn our attention. If we conclude, from the concurrent testimony of history and of the evidence yielded by geological data, that man is, comparatively speaking, of very modern origin, we must at once perceive how great a revolution in the state of the animate world the increase of the human race, considered merely as consumers of a certain quantity of organic matter, must necessarily cause.

Whether man increases the productive powers of the earth.—It may perhaps, be said, that man has, in some degree, compensated for the appropriation to himself of so much food, by artificially improving the natural productiveness of soils, by irrigation, manure, and a judicious intermixture of mineral ingredients conveyed from different localities. But it admits of reasonable doubt whether, upon the whole, we fertilize or impoverish the lands which we occupy. This assertion may seem startling to many; because they are so much in the habit of regarding the sterility or productiveness of land in relation to the wants of man, and not as regards the organic world generally. It is difficult, at first, to conceive, if a morass is converted into arable land, and made to yield a crop of grain, even of moderate abundance, that we have not improved the capabilities of the habitable surface—that we have not empowered it to support a larger quantity of organic life. In such cases, however, a tract, before of no utility to man, may be reclaimed, and become of high agricultural importance, though it may, nevertheless, yield a scantier vegetation. If a lake be drained, and turned into a meadow, the space will provide sustenance to man, and many terrestrial animals serviceable to him, but not, perhaps, so much food as it previously yielded to the aquatic races.

If the pestiferous Pontine marshes were drained, and covered with corn, like the plains of the Po, they might, perhaps, feed a smaller number of animals than they do now; for these morasses are filled with herds of buffaloes and swine, and they swarm with birds, reptiles, and insects.

The felling of dense and lofty forests, which covered, even within the records of history, a considerable space on the globe, now tenanted by civilized man, must generally have lessened the amount of vegetable food throughout the space where these woods grew. We must also take into our account the area covered by towns, and a still larger surface occupied by roads.

If we force the soil to bear extraordinary crops one year, we are, perhaps, compelled to let it lie fallow the next. But nothing so much counterbalances the fertilizing effects of human art as the extensive cultivation of foreign herbs and shrubs, which, although they are often more nutritious to man, seldom thrive with the same rank luxuriance as the native plants of a district. Man is, in truth, continually striving to diminish the natural diversity of the *stations* of animals and plants in every country, and to reduce them all to a small number fitted for species of economical use. He may succeed perfectly in attaining his object, even though the vegetation be comparatively meagre, and the total amount of animal life be greatly lessened.

Spix and Martius have given a lively description of the incredible number of insects which lay waste the crops in Brazil, besides swarms of monkeys, flocks of parrots, and other birds, as well as the paca, agouti, and wild swine. They describe the torment which the planter and the naturalist suffer from the musquitoes, and the devastation of the ants and blattæ; they speak of the dangers to which they were exposed from the jaguar, the poisonous serpents, crocodiles, scorpions, centipedes, and spiders. But with the increasing population and cultivation of the country, say these naturalists, these evils will gradually diminish; when the inhabitants have cut down the woods, drained the marshes, made roads in all directions, and founded villages and towns, man will, by degrees, triumph over the rank vegetation and the noxious animals, and all the elements will second and amply recompense his activity.*

The number of human beings now peopling the earth is supposed to amount to eight hundred millions, so that we may easily understand how great a number of beasts of prey, birds, and animals of every class, this prodigious population must have displaced, independently of the still more important consequences which have followed from the derangement brought about by man in the relative numerical strength of particular species.

Indigenous quadrupeds and birds extirpated in Great Britain.—Let us make some inquiries into the extent of the influence which the progress of society has exerted during the last seven or eight centuries, in altering the distribution of indigenous British animals. Dr. Fleming has prosecuted this inquiry with his usual zeal and ability; and in a memoir on the subject has enumerated the best-authenticated examples of the decrease or extirpation of certain species during a period when

* Travels in Brazil, vol. i. p. 260.

our population has made the most rapid advances. I shall offer a brief outline of his results.*

The stag, as well as the fallow deer and the roe, were formerly so abundant in our island, that, according to Lesley, from five hundred to a thousand were sometimes slain at a hunting match; but the native races would already have been extinguished, had they not been carefully preserved in certain forests. The otter, the marten, and the polecat, were also in sufficient numbers to be pursued for the sake of their fur; but they have now been reduced within very narrow bounds. The wild cat and fox have also been sacrificed throughout the greater part of the country, for the security of the poultry-yard or the fold. Badgers have been expelled from nearly every district, which at former periods they inhabited.

Besides these, which have been driven out from their favorite haunts, and everywhere reduced in number, there are some which have been wholly extirpated; such as the ancient breed of indigenous horses, and the wild boar; of the wild oxen a few remains are still preserved in some of the old English parks. The beaver, which is eagerly sought after for its fur, had become scarce at the close of the ninth century; and, by the twelfth century, was only to be met with, according to Giraldus de Barri, in one river in Wales, and another in Scotland. The wolf, once so much dreaded by our ancestors, is said to have maintained its ground in Ireland so late as the beginning of the eighteenth century (1710), though it had been extirpated in Scotland thirty years before, and in England at a much earlier period. The bear, which, in Wales, was regarded as a beast of the chase equal to the hare or the boar†, only perished, as a native of Scotland, in the year 1057.‡

Many native birds of prey have also been the subjects of unremitting persecution. The eagles, larger hawks, and ravens, have disappeared from the more cultivated districts. The haunts of the mallard, the snipe, the redshank, and the bittern, have been drained equally with the summer dwellings of the lapwing and the curlew. But these species still linger in some portion of the British isles; whereas the larger capercailzies or wood grouse, formerly natives of the pine-forests of Ireland and Scotland, have been destroyed within the last sixty years. The egret and the crane, which appear to have been formerly very common in Scotland, are now only occasional visitants.§

The bustard (*Otis tarda*), observes Graves, in his British Ornithology‖, "was formerly seen in the downs and heaths of various parts of our island, in flocks of forty or fifty birds; whereas it is now a circumstance of rare occurrence to meet with a single individual." Bewick also remarks, "that they were formerly more common in this island than at present; they are now found only in the open counties of the

* Ed. Phil. Journ., No. xxii. p. 287. Oct. 1824.

† Ray. Syn. Quad., p. 214.

‡ Fleming, Ed. Phil. Journ., No. xxii. p. 295.

§ Fleming, ibid., p. 292.

‖ Vol. iii. London, 1821.

south and east—in the plains of Wiltshire, Dorsetshire, and some parts of Yorkshire."* In the few years that have elapsed since Bewick wrote, this bird has entirely disappeared from Wiltshire and Dorsetshire.

These changes, it may be observed, are derived from very imperfect memorials, and relate only to the larger and more conspicuous animals inhabiting a small spot on the globe; but they cannot fail to exalt our conception of the enormous revolutions which, in the course of several thousand years, the whole human species must have effected.

Extinction of the dodo.—The kangaroo and the emu are retreating rapidly before the progress of colonization in Australia; and it scarcely admits of doubt, that the general cultivation of that country must lead to the extirpation of both. The most striking example of the loss, even within the last two centuries, of a remarkable species, is that of the dodo—a bird first seen by the Dutch, when they landed on the Isle of France, at that time uninhabited, immediately after the discovery of the passage to the East Indies by the Cape of Good Hope. It was of a large size, and singular form; its wings short, like those of an ostrich, and wholly incapable of sustaining its heavy body, even for a short flight. In its general appearance it differed from the ostrich, cassowary, or any known bird.†

Many naturalists gave figures of the dodo after the commencement of the seventeenth century; and there is a painting of it in the British Museum, which is said to have been taken from a living individual. Beneath the painting is a leg, in a fine state of preservation, which ornithologists are agreed cannot belong to any other known bird. In the museum at Oxford, also, there is a foot and a head in an imperfect state.

In spite of the most active search, during the last century, no information respecting the dodo was obtained, and some authors have gone so far as to pretend that it never existed; but a great mass of satisfactory evidence in favor of its recent existence has now been collected by Mr. Broderip,‡ and by Mr. Strickland and Dr. Melville. Mr. Strickland, agreeing with Professor Reinhardt, of Copenhagen, in referring the dodo to the Columbidæ, calls it a "vulture-like frugivorous pigeon." It appears, also, that another short-winged bird of the same order, called "The Solitaire," inhabited the small island of Rodrigues, 300 miles east of the Mauritius, and has been exterminated by man, as have one or two different but allied birds of the Isle of Bourbon.§

* Land Birds, vol. i. p. 316. ed. 1821.

† Some have complained that inscriptions on tomb-stones convey no general information, except that individuals were born and died, accidents which must happen alike to all men. But the death of a *species* is so remarkable an event in natural history that it deserves commemoration, and it is with no small interest that we learn, from the archives of the University of Oxford, the exact day and year when the remains of the last specimen of the dodo, which had been permitted to rot in the Ashmolean Museum, were cast away. The relics, we are told, were "a musæo subducta, annuente vice-cancellario aliisque curatoribus, ad ea lustranda convocatis, die Januarii 8vo, A.D. 1755." Zool. Journ. No. 12. p. 559. 1828.

‡ Penny Cyclopædia, "Dodo." 1837.

§ Messrs. Strickland and Melville on "the Dodo and its Kindred." London, 1848.

Rapid propagation of domestic quadrupeds over the American continent.—Next to the direct agency of man, his indirect influence in multiplying the numbers of large herbivorous quadrupeds of domesticated races may be regarded as one of the most obvious causes of the extermination of species. On this, and on several other grounds, the introduction of the horse, ox, and other mammalia, into America, and their rapid propagation over that continent within the last three centuries, is a fact of great importance in natural history. The extraordinary herds of wild cattle and horses which overran the plains of South America sprung from a very few pairs first carried over by the Spaniards; and they prove that the wide geographical range of large species in great continents does not necessarily imply that they have existed there from remote periods.

Humboldt observes, in his Travels, on the authority of Azzara, that it is believed there exist, in the Pampas of Buenos Ayres, twelve million cows and three million horses, without comprising, in this enumeration, the cattle that have no acknowledged proprietor. In the Llanos of Caraccas, the rich hateros, or proprietors of pastoral farms, are entirely ignorant of the number of cattle they possess. The young are branded with a mark peculiar to each herd, and some of the most wealthy owners mark as many as fourteen thousand a year.* In the northern plains, from the Orinoco to the lake of Maraycabo, M. Depons reckoned that 1,200,000 oxen, 180,000 horses, and 90,000 mules, wandered at large.† In some parts of the valley of the Mississippi, especially in the country of the Osage Indians, wild horses are immensely numerous.

The establishment of black cattle in America dates from Columbus's second voyage to St. Domingo. They there multiplied rapidly; and that island presently became a kind of nursery from which these animals were successively transported to various parts of the continental coast, and from thence into the interior. Notwithstanding these numerous exportations, in twenty-seven years after the discovery of the island, herds of four thousand head, as we learn from Oviedo, were not uncommon, and there were even some that amounted to eight thousand. In 1587, the number of hides exported from St. Domingo alone, according to Acosta's report, was 35,444; and in the same year there were exported 64,350 from the ports of New Spain. This was in the sixty-fifth year after the taking of Mexico, previous to which event the Spaniards, who came into that country, had not been able to engage in anything else than war.‡ Every one is aware that these animals are now established throughout the American continent from Canada to the Straits of Magellan.

The ass has thriven very generally in the New World; and we learn from Ulloa, that in Quito they ran wild, and multiplied in amazing numbers, so as to become a nuisance. They grazed together in herds,

* Pers. Nar. vol. iv.
† Quarterly Review, vol. xxi. p. 335.
‡ Ibid.

and when attacked defended themselves with their mouths. If a horse happened to stray into the places where they fed, they all fell upon him, and did not cease biting and kicking till they left him dead.*

The first hogs were carried to America by Columbus, and established in the Island of St. Domingo the year following its discovery, in November, 1493. In succeeding years they were introduced into other places where the Spaniards settled; and, in the space of half a century, they were found established in the New World, from the latitude of 25° north, to the 40th degree of south latitude. Sheep, also, and goats have multiplied enormously in the New World, as have also the cat and the rat; which last, as before stated, has been imported unintentionally in ships. The dogs introduced by man which have at different periods become wild in America, hunted in packs, like the wolf and the jackall, destroying not only hogs, but the calves and foals of the wild cattle and horses.

Ulloa in his voyage, and Buffon on the authority of old writers, relate a fact which illustrates very clearly the principle before explained, of the check which the increase of one animal necessarily offers to that of another. The Spaniards had introduced goats into the Island of Juan Fernandez, where they became so prolific as to furnish the pirates who infested those seas with provisions. In order to cut off this resource from the buccaneers, a number of dogs were turned loose into the island; and so numerous did they become in their turn, that they destroyed the goats in every accessible part, after which the number of the wild dogs again decreased.†

Increase of rein-deer imported into Iceland.—As an example of the rapidity with which a large tract may become peopled by the offspring of a single pair of quadrupeds, it may be mentioned that in the year 1773 thirteen rein-deer were exported from Norway, only three of which reached Iceland. These were turned loose into the mountains of Guldbringè Syssel, where they multiplied so greatly, in the course of forty years, that it was not uncommon to meet with herds, consisting of from forty to one hundred, in various districts.

The rein-deer, observes a modern writer, is in Lapland a loser by his connexion with man, but Iceland will be this creature's paradise. There is, in the interior, a tract which Sir. G. Mackenzie computes at not less than forty thousand square miles, without a single human habitation, and almost entirely unknown to the natives themselves. There are no wolves: the Icelanders will keep out the bears; and the rein-deer, being almost unmolested by man, will have no enemy whatever, unless it has brought with it its own tormenting gad-fly.‡

Besides the quadrupeds before enumerated, our domestic fowls have also succeeded in the West Indies and America, where they have the

* Ulloa's Voyage. Wood's Zoog. vol. i. p. 9.
† Buffon, vol. v. p. 100. Ulloa's Voyage, vol. ii. p. 220.
‡ Travels in Iceland in 1810, p. 342.

common fowl, the goose, the duck, the peacock, the pigeon, and the guinea-fowl. As these were often taken suddenly from the temperate to very hot regions, they were not reared at first without much difficulty: but after a few generations, they became familiarized to the climate, which, in many cases, approached much nearer than that of Europe to the temperature of their original native countries.

The fact of so many millions of wild and tame individuals of our domestic species, almost all of them the largest quadrupeds and birds, having been propagated throughout the new continent within the short period that has elapsed since the discovery of America, while no appreciable improvement can have been made in the productive powers of that vast continent, affords abundant evidence of the extraordinary changes which accompany the diffusion and progressive advancement of the human race over the globe. That it should have remained for us to witness such mighty revolutions is a proof, even if there was no other evidence, that the entrance of man into the planet is, comparatively speaking, of extremely modern date, and that the effects of his agency are only beginning to be felt.

Population which the globe is capable of supporting.—A modern writer has estimated, that there are in America upwards of four million square miles of useful soil, each capable of supporting 200 persons; and nearly six million, each mile capable of supporting 490 persons.* If this conjecture be true, it will follow, as that author observes, that if the natural resources of America were fully developed, it would afford sustenance to five times as great a number of inhabitants as the entire mass of human beings existing at present upon the globe. The new continent, he thinks, though less than half the size of the old, contains an equal quantity of useful soil, and much more than an equal amount of productive power. Be this as it may, we may safely conclude that the amount of human population now existing constitutes but a small proportion of that which the globe is capable of supporting, or which it is destined to sustain at no distant period, by the rapid progress of society, especially in America, Australia, and certain parts of the old continent.

Power of exterminating species no prerogative of man.—But if we reflect that many millions of square miles of the most fertile land, occupied originally by a boundless variety of animal and vegetable forms, have been already brought under the dominion of man, and compelled, in a great measure, to yield nourishment to him, and to a limited number of plants and animals which he has caused to increase, we must at once be convinced, that the annihilation of a multitude of species has already been effected, and will continue to go on hereafter, in certain regions, in a still more rapid ratio, as the colonies of highly civilized nations spread themselves over unoccupied lands.

Yet, if we wield the sword of extermination as we advance, we have no reason to repine at the havoc committed, nor to fancy, with the Scottish

* Maclaren, art. America, Encyc. Brit.

poet, that "we violate the social union of nature;" or, complain, with the melancholy Jacques, that we

> Are mere usurpers, tyrants, and what's worse,
> To fright the animals and to kill them up
> In their assign'd and native dwelling-place.

We have only to reflect, that in thus obtaining possession of the earth by conquest, and defending our acquisitions by force, we exercise no exclusive prerogative. Every species which has spread itself from a small point over a wide area must, in like manner, have marked its progress by the diminution or the entire extirpation of some other, and must maintain its ground by a successful struggle against the encroachments of other plants and animals. That minute parasitic plant, called "the rust" in wheat, has, like the Hessian fly, the locust, and the aphis, caused famines ere now amongst the "lords of the creation." The most insignificant and diminutive species, whether in the animal or vegetable kingdom, have each slaughtered their thousands, as they disseminated themselves over the globe, as well as the lion, when first it spread itself over the tropical regions of Africa.

Concluding remarks.—Although we have as yet considered one class only of the causes (the organic) by which species may become exterminated, yet it cannot but appear evident that the continued action of these alone, throughout myriads of future ages, must work an entire change in the state of the organic creation, not merely on the continents and islands, where the power of man is chiefly exerted, but in the great ocean, where his control is almost unknown. The mind is prepared by the contemplation of such future revolutions to look for the signs of others, of an analogous nature, in the monuments of the past. Instead of being astonished at the proofs there manifested of endless mutations in the animate world, they will appear to one who has thought profoundly on the fluctuations now in progress, to afford evidence in favor of the uniformity of the system, unless, indeed, we are precluded from speaking of *uniformity* when we characterize a principle of endless variation.

CHAPTER XLII.

EXTINCTION OF SPECIES.—INFLUENCE OF INORGANIC CAUSES.

Powers of diffusion indispensable, that each species may maintain its ground—How changes in physical geography affect the distribution of species—Rate of the change of species due to this cause cannot be uniform—Every change in the physical geography of large regions tends to the extinction of species—Effects of a general alteration of climate on the migration of species—Gradual refrigeration would cause species in the northern and southern hemispheres to become distinct—Elevation of temperature the reverse—Effects on the condition of species which must result from inorganic changes inconsistent with the theory of transmutation.

Powers of diffusion indispensable, that each species may maintain its ground.—HAVING shown in the last chapter, how considerably the numerical increase or the extension of the geographical range of any one species must derange the numbers and distribution of others, let us now direct our attention to the influence which the inorganic causes described in the second book are continually exerting on the habitations of species.

So great is the instability of the earth's surface, that if nature were not continually engaged in the task of sowing seeds and colonizing animals, the depopulation of a certain portion of the habitable sea and land would in a few years be considerable. Whenever a river transports sediment into a lake or sea, so as materially to diminish its depth, the aquatic animals and plants which delight in deep water are expelled: the tract, however, is not allowed to remain useless; but is soon peopled by species which require more light and heat, and thrive where the water is shallow. Every addition made to the land by the encroachment of the delta of a river banishes many subaqueous species from their native abodes; but the new-formed plain is not permitted to lie unoccupied, being instantly covered with terrestrial vegetation. The ocean devours continuous lines of sea-coasts, and precipitates forests or rich pasture land into the waves: but this space is not lost to the animate creation; for shells and sea-weeds soon adhere to the new-made cliffs, and numerous fish people the channel which the current has scooped out for itself. No sooner has a volcanic island been thrown up than some lichens begin to grow upon it, and it is sometimes clothed with verdure while smoke and ashes are still occasionally thrown from the crater. The cocoa, pandanus, and mangrove take root upon the coral reef before it has fairly risen above the waves. The burning stream of lava that descends from Etna rolls through the stately forest, and converts to ashes every tree and herb which stands in its way; but the black strip of land thus desolated is covered again in the course of time, with oaks, pines, and chestnuts, as luxuriant as those which the fiery torrent swept away.

Every flood and landslip, every wave which a hurricane or earthquake throws upon the shore, every shower of volcanic dust and ashes which buries a country far and wide to the depth of many feet, every advance of the sand-flood, every conversion of salt water into fresh when rivers alter their main channel of discharge, every permanent variation in the rise or fall of tides in an estuary—these and countless other causes displace, in the course of a few centuries, certain plants and animals from stations which they previously occupied. If, therefore, the Author of nature had not been prodigal of those numerous contrivances, before alluded to, for spreading all classes of organic beings over the earth—if he had not ordained that the fluctuations of the animate and inanimate creation should be in perfect harmony with each other, it is evident that considerable spaces, now the most habitable on the globe, would soon be as devoid of life as are the Alpine snows, or the dark abysses of the ocean, or the moving sands of the Sahara.

The powers, then, of migration and diffusion conferred on animals and plants are indispensable to enable them to maintain their ground, and would be necessary, even though it were never intended that a species should gradually extend its geographical range. But a facility of shifting their quarters being once given, it cannot fail to happen that the inhabitants of one province should occasionally penetrate into some other; since the strongest of those barriers which I before described as separating distinct regions are all liable to be thrown down, one after the other, during the vicissitudes of the earth's surface.

How changes in physical Geography affect the distribution of species.—The numbers and distribution of particular species are affected in two ways, by changes in the physical geography of the earth:—First, these changes promote or retard the migrations of species; secondly, they alter the physical conditions of the localities which species inhabit. If the ocean should gradually wear its way through an isthmus, like that of Suez, it would open a passage for the intermixture of the aquatic tribes of two seas previously disjoined, and would, at the same time, close a free communication which the terrestrial plants and animals of two continents had before enjoyed. These would be, perhaps, the most important consequences, in regard to the distribution of species, which would result from the breach made by the sea in such a spot; but there would be others of a distinct nature, such as the conversion of a certain tract of land, which formed the isthmus, into sea. This space, previously occupied by terrestrial plants and animals, would be immediately delivered over to the aquatic; a local revolution which might have happened in innumerable other parts of the globe, without being attended by any alteration in the blending together of species of two distinct provinces.

Rate of change of species cannot be uniform.—This observation leads me to point out one of the most interesting conclusions to which we are led by the contemplation of the vicissitudes of the inanimate world in relation to those of the animate. It is clear that, if the agency of inorganic causes be uniform, as I have supposed, they must operate very irregularly on the state of organic beings, so that the rate according to which these will change in particular regions will not be equal in equal periods of time.

I am not about to advocate the doctrine of general catastrophes recurring at certain intervals, as in the ancient Oriental cosmogonies, nor do I doubt that, if very considerable periods of equal duration could be compared one with another, the rate of change in the living, as well as in the inorganic world, might be nearly uniform; but if we regard each of the causes separately, which we know to be at present the most instrumental in remodelling the state of the surface, we shall find that we must expect each to be in action for thousands of years, without producing any extensive alterations in the habitable surface, and then to give rise, during a very brief period, to important revolutions.

Illustration derived from subsidences.—I shall illustrate this principle by a few of the most remarkable examples which present themselves. In the course of the last century, as we have seen, a considerable number of instances are recorded of the solid surface, whether covered by water or not, having been permanently sunk or upraised by subterranean movements. Most of these convulsions are only accompanied by temporary fluctuations in the state of limited districts, and a continued repetition of these events for thousands of years might not produce any decided change in the state of many of those great zoological or botanical provinces of which I have sketched the boundaries.

When, for example, large parts of the ocean and even of inland seas are a thousand fathoms or upwards in depth, it is a matter of no moment to the animate creation that vast tracts should be heaved up many fathoms at certain intervals, or should subside to the same amount. Neither can any material revolution be produced in South America either in the terrestrial or the marine plants or animals by a series of shocks on the coast of Chili, each of which, like that of Penco, in 1751, should uplift the coast about twenty-five feet. Nor if the ground sinks fifty feet at a time, as in the harbor of Port Royal, in Jamaica, in 1692, will such alterations of level work any general fluctuations in the state of organic beings inhabiting the West Indian Islands, or the Carribean Sea.

It is only when the subterranean powers, by shifting gradually the points where their principal force is developed, happen to strike upon some particular region where a slight change of level immediately affects the distribution of land and water, or the state of the climate, or the barriers between distinct groups of species over extensive areas, that the rate of fluctuation becomes accelerated, and may, in the course of a few years or centuries, work mightier changes than had been experienced in myriads of antecedent years.

Thus, for example, a repetition of subsidences causing the narrow isthmus of Panama to sink down a few hundred feet, would, in a few centuries, bring about a great revolution in the state of the animate creation in the western hemisphere. Thousands of aquatic species would pass, for the first time, from the Caribbean Sea into the Pacific; and thousands of others, before peculiar to the Pacific Ocean, would make their way into the Caribbean Sea, the Gulf of Mexico, and the Atlantic. A considerable modification would probably be occasioned by the same event in the direction or volume of the Gulf stream, and thereby the temperature of the sea and the contiguous lands might be altered as far as the influence of that current extends. A change of climate might thus be produced in the ocean from Florida to Spitzbergen, and in many countries of North America, Europe, and Greenland. Not merely the heat, but the quantity of rain which falls, would be altered in certain districts, so that many species would be excluded from tracts where they before flourished: others would be reduced in number; and some would thrive more and multiply. The seeds also and the fruits of plants would no longer be drifted in precisely the same directions, nor the eggs of aquatic animals; neither would species be any longer impeded in their migrations towards particular stations before shut out from them by their inability to cross the mighty current.

Let us take another example from a part of the globe which is at present liable to suffer by earthquakes, namely, the low sandy tract which intervenes between the sea of Azof and the Caspian. If there should occur a sinking down to a trifling amount, and such ravines should be formed as might be produced by a few earthquakes, not more considerable than have fallen within our limited observation during the last 150 years, the waters of the Sea of Azof would pour rapidly into the Caspian, which, according to the measurements lately made by the Academy of St. Petersburg, is 84 feet below the level of the Black Sea.* The Sea of Azof would immediately borrow from the Black Sea, that sea again from the Mediterranean, and the Mediterranean from the Atlantic, so that an inexhaustible current would pour down into the low tracts of Asia bordering the Caspian, by which all the sandy salt steppes adjacent to that sea would be inundated. An area of several thousand square leagues, now below the level of the Mediterranean, would be converted from land into sea.

Illustration derived from the elevation of land.—Let us next imagine a few cases of the elevation of land of small extent at certain critical points, as, for example, in the shallowest part of the Straits of Gibraltar, where the deepest soundings from the African to the European side give only 220 fathoms. In proportion as this submarine barrier of rock was upheaved, the whole channel would be contracted in width and depth, and the volume of water which the current constantly flowing from the Atlantic pours into the Mediterranean would be lessened. But the loss of the inland sea by evaporation would remain the same; so that being no longer able to draw on the ocean for a supply sufficient to restore its equilibrium, it must sink, and leave dry a certain portion of land around

* See a note on this subject, chap. x. p. 157.

its borders. The current which now flows constantly out of the Black Sea into the Mediterranean would then rush in more rapidly, and the level of the Mediterranean would be thereby prevented from falling so low; but the level of the Black Sea would, for the same reason, sink; so that when, by a continued series of elevatory movements, the Straits of Gibraltar had become completely closed up, we might expect large and level sandy steppes to surround both the Black Sea and Mediterranean, like those occurring at present on the skirts of the Caspian and the Lake of Aral. The geographical range of hundreds of aquatic species would be thereby circumscribed, and that of hundreds of terrestrial plants and animals extended.

A line of submarine volcanos crossing the channel of some strait, and gradually choking it up with ashes and lava, might produce a new barrier as effectually as a series of earthquakes; especially if thermal springs, charged with carbonate of lime, silica, and other mineral ingredients, should promote the rapid multiplication of corals and shells, and cement them together with solid matter precipitated during the intervals between eruptions. Suppose in this manner a stoppage to be caused of the Bahama channel between the bank of that name and the coast of Florida. This insignificant revolution, confined to a mere spot in the bottom of the ocean, would, by diverting the main current of the Gulf stream, give rise to extensive changes in the climate and distribution of animals and plants inhabiting the northern hemisphere.

Illustration from the formation of new islands.—A repetition of elevatory movements of earthquakes might continue over an area as extensive as Europe, for thousands of ages, at the bottom of the ocean, in certain regions, and produce no visible effects; whereas, if they should operate in some shallow parts of the Pacific, amid the coral archipelagos, they would soon give birth to a new continent. Hundreds of volcanic islands may be thrown up, and become covered with vegetation, without causing more than local fluctuations in the animate world; but if a chain like the Aleutian archipelago, or the Kurile Isles, run for a distance of many hundred miles, so as to form an almost uninterrupted communication between two continents, or two distant islands, the migrations of plants, birds, insects, and even of some quadrupeds, may cause, in a short time, an extraordinary series of revolutions tending to augment the range of some animals and plants, and to limit that of others. A new archipelago might be formed in the Mediterranean, the Bay of Biscay, and a thousand other places, and might produce less important events than one rock which should rise up between Australia and Java, so placed that winds and currents might cause an interchange of the plants, insects, and birds.

From the wearing through of an isthmus.—If we turn from the igneous to the aqueous agents, we find the same tendency to an irregular rate of change, naturally connected with the strictest uniformity in the energy of those causes. When the sea, for example, gradually encroaches upon both sides of a narrow isthmus, as that of Sleswick, separating the

North Sea from the Baltic, where, as before stated, the cliffs on both the opposite coasts are wasting away*, no material alteration results for thousands of years, save only that there is a progressive conversion of a small strip of land into water. A few feet only, or a few yards, are annually removed; but if, at last, the partition should be broken down, and the tides of the ocean should enter by a direct passage into the inland sea, instead of going by a circuitous route through the Cattegat, a body of salt water would sweep up as far as the Gulfs of Bothnia and Finland, the waters of which are now brackish, or almost fresh; and this revolution would be attended by the local annihilation of many species.

Similar consequences must have resulted on a small scale, when the sea opened its way through the Isthmus of Staveren in the thirteenth century, forming a union between an inland lake and the ocean, and opening, in the course of one century, a shallow strait, more than half as wide as the narrowest part of that which divides England from France.

Changes in physical geography which must occasion extinction of species.—It will almost seem superfluous, after I have thus traced the important modifications in the condition of living beings which flow from changes of trifling extent, to argue that entire revolutions might be brought about, if the climate and physical geography of the whole globe were greatly altered. It has been stated, that species are in general local, some being confined to extremely small spots, and depending for their existence on a combination of causes, which, if they are to be met with elsewhere, occur only in some very remote region. Hence it must happen that, when the nature of these localities is changed, the species will perish; for it will rarely happen that the cause which alters the character of the district will afford new facilities to the species to establish itself elsewhere.

African deserts.—If we attribute the origin of a great part of the desert of Africa to the gradual progress of moving sands driven eastward by the westerly winds, we may safely infer that a variety of species must have been annihilated by this cause alone. The sand-flood has been inundating, from time immemorial, some of the rich lands on the west of the Nile; and we have only to multiply this effect a sufficient number of times in order to understand how, in the lapse of ages, a whole group of terrestrial animals and plants may become extinct.

The African desert, without including Bornou and Darfour, extends, according to the calculation of Humboldt, over 194,000 square leagues; an area nearly three times as great as that of France. In a small portion of so vast a space, we may infer from analogy that there were many peculiar species of plants and animals which must have been banished by the sand, and their habitations invaded by the camel, and by birds and insects formed for the arid sands.

There is evidently nothing in the nature of the catastrophe to favor the escape of the former inhabitants to some adjoining province; nothing

* See above, p. 317.

to weaken, in the bordering lands, that powerful barrier against emigration—pre-occupancy. Nor, even if the exclusion of a certain group of species from a given tract were compensated by an extension of their range over a new country, would that circumstance tend to the conservation of species in general; for the extirpation would merely then be transferred to the region so invaded. If it be imagined, for example, that the aboriginal quadrupeds, birds, and other animals of Africa, emigrated in consequence of the advance of drift-sand, and colonized Arabia, the indigenous Arabian species must have given way before them, and have been reduced in number or destroyed.

Let us next suppose that, in some central or more elevated parts of the great African desert, the upheaving power of subterranean movements should be exerted throughout an immense series of ages, accompanied, at certain intervals, by volcanic eruptions, such as gave rise at once, in 1755, to a mountain 1600 feet high, on the Mexican plateau. When the continued repetition of these events had caused a mountain-chain, it is obvious that a complete transformation in the state of the climate would be brought about throughout a vast area.

We may imagine the summits of the new chain to rise so high as to be covered, like Mount Atlas, for several thousand feet, with snow, during a great part of the year. The melting of these snows, during the greatest heat, would cause the rivers to swell in the season when the greatest drought now prevails; the waters, moreover, derived from this source, would always be of lower temperature than the surrounding atmosphere, and would thus contribute to cool the climate. During the numerous earthquakes and volcanic eruptions supposed to accompany the gradual formation of the chain, there would be many floods caused by the bursting of temporary lakes, and by the melting of snows by lava. These inundations might deposit alluvial matter far and wide over the original sands, as the country assumed varied shapes, and was modified again and again by the moving power from below, and the aqueous erosion of the surface above. At length the Sahara might be fertilized, irrigated by rivers and streamlets intersecting it in every direction, and covered by jungle and morasses; so that the animals and plants which now people Northern Africa would disappear, and the region would gradually become fitted for the reception of a population of species perfectly dissimilar in their forms, habits, and organization.

There are always some peculiar and characteristic features in the physical geography of each large division of the globe; and on these peculiarities the state of animal and vegetable life is dependent. If, therefore, we admit incessant fluctuations in the physical geography, we must, at the same time, concede the successive extinction of terrestrial and aquatic species to be part of the economy of our system. When some great class of *stations* is in excess in certain latitudes, as, for example, in wide savannahs, arid sands, lofty mountains, or inland seas, we find a corresponding development of species adapted for such circumstances. In North America, where there is a chain of vast inland lakes of fresh water, we find an extraordinary abundance and variety of aquatic birds, fresh-water fish, testacea, and small amphibious reptiles, fitted for such a climate. The greater part of these would perish if the lakes were destroyed,—an event that might be brought about by some of the least of those important revolutions contemplated in geology. It might happen that no fresh-water lakes of corresponding magnitude might then exist on the globe; or that, if they occurred elsewhere, they might be situated in New Holland, Southern Africa. Eastern Asia, or some region so distant as to be quite inaccessible to the North American species; or they might be situated within the tropics, in a climate uninhabitable by creatures fitted for a temperate zone; or, finally, we may presume that they would be pre-occupied by *indigenous* tribes.

A vivid description has been given by Mr. Darwin and Sir W. Parish of the great droughts which have sometimes visited the Pampas of South America, for three or four years in succession, during which an incredible number of wild animals, cattle, horses, and birds, have perished from want of food and water. Several hundred thousand animals were drowned in the Parana alone, having rushed into the river to drink, and being too much exhausted by hunger to escape.* Such droughts are often attended in South America and other hot climates by wide-spreading conflagrations, caused by lightning, which fires the dried grass and brushwood. Thus quadrupeds, birds, insects, and other creatures, are destroyed by myriads. How many species, both of the animal and vegetable world, which once flourished in the country between the valley of the Parana and the Straits of Magellan, may not have been annihilated, since the first drought or first conflagration began!

To pursue this train of reasoning farther is unnecessary; the geologist has only to reflect on what has been said of the habitations and stations of organic beings in general, and to consider them in relation to those effects which were contemplated in the second book, as resulting from the igneous and aqueous causes now in action, and he will immediately perceive that, amidst the vicissitudes of the earth's surface, species cannot be immortal, but must perish, one after the other, like the individuals which compose them. There is no possibility of escaping from this conclusion, without resorting to some hypothesis as violent as that of Lamarck, who imagined, as we have before seen, that species are each of them endowed with indefinite powers of modifying their organization, in conformity to the endless changes of circumstances to which they are exposed.

Effects of a general Alteration in Climate on the Distribution of Species.

Some of the effects which must attend every general alteration of *climate* are sufficiently peculiar to claim a separate consideration before concluding the present chapter.

* Darwin's Journal, p. 156., 2d ed. p. 133. Sir W. Parish, Buenos Ayres, &c. p. 371. and 151.

I have before stated that, during seasons of extraordinary severity, many northern birds, and in some countries many quadrupeds, migrate southwards. If these cold seasons were to become frequent, in consequence of a gradual and general refrigeration of the atmosphere, such migrations would be more and more regular, until, at length, many animals, now confined to the arctic regions, would become the tenants of the temperate zone; while the inhabitants of the temperate zone would approach nearer to the equator. At the same time, many species previously established on high mountains would begin to descend, in every latitude, towards the middle regions; and those which were confined to the flanks of mountains would make their way into the plains. Analogous changes would also take place in the vegetable kingdom.

If, on the contrary, the heat of the atmosphere be on the increase, the plants and animals of low grounds would ascend to higher levels, the equatorial species would migrate into the temperate zone, and those of the temperate into the arctic circle.

But although some species might thus be preserved, every great change of climate must be fatal to many which can find no place of retreat when their original habitations become unfit for them. For if the general temperature be on the rise, then there is no cooler region whither the polar species can take refuge; if it be on the decline, then the animals and plants previously established between the tropics have no resource. Suppose the general heat of the atmosphere to increase, so that even the arctic region became too warm for the musk-ox and rein-deer, it is clear that they must perish; so if the torrid zone should lose so much of its heat, by the progressive refrigeration of the earth's surface, as to be an unfit habitation for apes, boas, bamboos, and palms, these tribes of animals and plants, or, at least, most of the species now belonging to them, would become extinct, for there would be no warmer latitudes for their reception.

It will follow, therefore, that as often as the climates of the globe are passing from the extreme of heat to that of cold—from the summer to the winter of the great year before alluded to*—the migratory movement will be directed constantly from the poles towards the equator; and for this reason the species inhabiting parallel latitudes, in the northern and southern hemispheres, must become widely different. For I assume, on grounds before explained, that the original stock of each species is introduced into one spot of the earth only, and, consequently, no species can be at once indigenous in the arctic and antarctic circles.

But when, on the contrary, a series of changes in the physical geography of the globe, or any other supposed cause, occasions an elevation of the general temperature,—when there is a passage from the winter to one of the vernal or summer seasons of the great cycle of climate,—then the order of the migratory movement is inverted. The different species of animals and plants direct their course from the equator towards the poles;

* See above, chap. vii. p. 112.

and the northern and southern hemispheres may become peopled to a certain limited extent by identical species.

I say limited, because we cannot speculate on the entire transposition of a group of animals and plants from tropical to polar latitudes, or the reverse, as a probable or even possible event. We may believe the mean annual temperature of one zone to be transferable to another, but we know that the same climate cannot be so transferred. Whatever be the general temperature of the earth's surface, comparative equability of heat will characterize the tropical regions; while great periodical variations will belong to the temperate, and still more to the polar latitudes. These, and many other peculiarities connected with heat and light, depend on fixed astronomical causes, such as the motion of the earth and its position in relation to the sun, and not on those fluctuations of its surface which may influence the general temperature.

Among many obstacles to such extensive transference of habitations, we must not forget the immense lapse of time required, according to the hypothesis before suggested, to bring about a considerable change in climate. During a period so vast, the other cause of extirpation, before enumerated, would exert so powerful an influence as to prevent all, save a very few hardy species, from passing from equatorial to polar regions, or from the tropics to the pole.*

But the power of accommodation to new circumstances is great in certain species, and might enable many to pass from one zone to another, if the mean annual heat of the atmosphere and the ocean were greatly altered. To the marine tribes, especially, such a passage would be possible; for they are less impeded in their migrations by barriers of land, than are the terrestrial by the ocean. Add to this, that the temperature of the ocean is much more uniform than that of the atmosphere investing the land; so that we may easily suppose that most of the testacea, fish, and other classes, might pass from the equatorial into the temperate regions, if the mean temperature of those regions were transposed, although a second expatriation of these species of tropical origin into the arctic and antarctic circles would probably be impossible.

Let us now consider more particularly the effect of vicissitudes of climate in causing one species to give way before the increasing numbers of some other.

When temperature forms the barrier which arrests the progress of an animal or plant in a particular direction, the individuals are fewer and less vigorous as they approach the extreme confines of the geographical range of the species. But these stragglers are ready to multiply rapidly on the slightest increase or diminution of heat that may be favorable to them, just as particular insects increase during a hot summer, and certain plants and animals gain ground after a series of congenial seasons.

In almost every district, especially if it be mountainous, there are a

* See above, chaps. vi. vii. and viii

variety of species the limits of whose habitations are conterminous, some being unable to proceed farther without encountering too much heat, others too much cold. Individuals, which are thus on the borders of the regions proper to their respective species, are like the outposts of hostile armies, ready to profit by every slight change of circumstances in their favor, and to advance upon the ground occupied by their neighbors and opponents.

The proximity of distinct climates produced by the inequalities of the earth's surface, brings species possessing very different constitutions into such immediate contact, that their naturalizations are very speedy whenever opportunities of advancing present themselves. Many insects and plants, for example, are common to low plains within the arctic circle, and to lofty mountains in Scotland and other parts of Europe. If the climate, therefore, of the polar regions were transferred to our own latitudes, the species in question would immediately descend from these elevated stations to overrun the low grounds. Invasions of this kind, attended by the expulsion of the pre-occupants, are almost instantaneous, because the change of temperature not only places the one species in a more favorable position, but renders the others sickly and almost incapable of defence.

These changes inconsistent with the theory of transmutation.—Lamarck, when speculating on the transmutation of species, supposed every modification in organization and instinct to be brought about slowly and insensibly in an indefinite lapse of ages. But he does not appear to have sufficiently considered how much every alteration in the physical condition of the habitable surface changes the relations of a great number of coexisting species, and that some of these would be ready instantly to avail themselves of the slightest change in their favor, and to multiply to the injury of others. Even if we thought it possible that the palm or the elephant, which now flourish in equatorial regions, could ever learn to bear the variable seasons of our temperate zone, or the rigors of an arctic winter, we might with no less confidence affirm, that they must perish before they had time to become habituated to such new circumstances. That they would be displaced by other species as often as the climate varied, may be inferred from the data before explained respecting the local extermination of species produced by the multiplication of others.

Suppose the climate of the highest part of the woody zone of Etna to be transferred to the sea-shore of the base of the mountain, no botanist would anticipate that the olive, lemon-tree, and prickly pear (*Cactus Opuntia*) would be able to contend with the oak and chestnut, which would begin forthwith to descend to a lower level; or that these last would be able to stand their ground against the pine, which would also, in the space of a few years, begin to occupy a lower position. We might form some kind of estimate of the time which might be required for the migrations of these plants; whereas we have no data for concluding that any number of thousands of years would be sufficient for one step in the pretended metamorphosis of one species into another, possessing distinct attributes and qualities.

This argument is applicable not merely to *climate*, but to any other cause of mutation. However slowly a lake may be converted into a marsh, or a marsh into a meadow, it is evident that before the lacustrine plants can acquire the power of living in marshes, or the marsh-plants of living in a less humid soil, other species, already existing in the region, and fitted for these several stations, will intrude and keep possession of the ground. So, if a tract of salt water becomes fresh by passing through every intermediate degree of brackishness, still the marine mollusks will never be permitted to be gradually metamorphosed into fluviatile species; because long before any such transformation can take place by slow and insensible degrees, other tribes, already formed to delight in brackish or fresh water, will avail themselves of the change in the fluid, and will, each in their turn, monopolize the space.

It is idle, therefore, to dispute about the abstract possibility of the conversion of one species into another, when there are known causes so much more active in their nature, which must always intervene and prevent the actual accomplishment of such conversions. A faint image of the certain doom of a species less fitted to struggle with some new condition in a region which it previously inhabited, and where it has to contend with a more vigorous species, is presented by the extirpation of savage tribes of men by the advancing colony of some civilized nation. In this case the contest is merely between two different *races*—two varieties, moreover, of a species which exceeds all others in its aptitude to accommodate its habits to the most extraordinary variations of circumstances. Yet few future events are more certain than the speedy extermination of the Indians of North America and the savages of New Holland in the course of a few centuries, when these tribes will be remembered only in poetry or history.

Concluding remarks.—We often hear astonishment expressed at the disappearance from the earth in times comparatively modern of many small as well as large animals, the remains of which have been found in a fossil state, under circumstances implying that neither any great geographical revolution, nor the exterminating influence of man has intervened to account for their extinction. But in all such cases we should inquire whether we are sufficiently acquainted with the numerous and complicated conditions on which the perpetuation of each species depends, to entitle us to wonder if it should be suddenly cut off.

Mr. Darwin, when calling attention to the fact that the horse, megatherium, megalonyx, and many contemporary Mammalia, had perished in South America after that continent had acquired its present configuration, and when, if we may judge by the Testacea, the climate very nearly resembled the present, observes, "that in the living creation one species is often extremely rare in a given region, while another of the same genus and with closely allied habits is exceedingly common. A zoologist familiar with such phenomena, if asked to explain them, usually replies, that some slight difference in climate, food, or the number of its enemies, must determine the relative strength of the two species in question, although we may be unable to point out the precise manner of the action of the check. We are, therefore, driven to the conclusion, that causes generally quite inappreciable by us determine whether a given species shall be abundant or scanty in numbers. Why, then, should we feel astonishment if the rarity is occasionally carried a step farther,—to extinction?" *

CHAPTER XLIII.

EXTINCTION AND CREATION OF SPECIES.

Theory of the successive extinction of species consistent with a limited geographical distribution—Opinions of botanists respecting the centres from which plants have been diffused—Whether there are grounds for inferring that the loss, from time to time, of certain animals and plants, is compensated by the introduction of new species?—Whether any evidence of such new creations could be expected within the historical era?—The question whether the existing species have been created in succession must be decided by geological monuments.

Successive Extinction of Species consistent with their limited Geographical Distribution.

IN the preceding chapters I have pointed out the strict dependence of each species of animal and plant on certain physical conditions in the state of the earth's surface, and on the number and attributes of other organic beings inhabiting the same region. I have also endeavored to show that all these conditions are in a state of continual fluctuation, the igneous and aqueous agents remodelling, from time to time, the physical geography of the globe, and the migrations of species causing new relations to spring up successively between different organic beings. I have deduced as a corollary, that the species existing at any particular period, must, in the course of ages, become extinct one after the other. "They must die out," to borrow an emphatical expression from Buffon, "because Time fights against them."

If the views which I have taken are just, there will be no difficulty in explaining why the habitations of so many species are now restrained within exceedingly narrow limits. Every local revolution, such as those contemplated in the preceding chapter, tends to circumscribe the range of some species, while it enlarges that of others; and if we are

* Journ. of Nat. Hist. &c. 2d edit., 1845, p. 175; also Lyell's 2d Visit to the United States, vol. i. p. 351.

led to infer that new species originate in one spot only, each must require time to diffuse itself over a wide area. It will follow, therefore, from the adoption of this hypothesis, that the recent origin of some species, and the high antiquity of others, are equally consistent with the general fact of their limited distribution; some being local, because they have not existed long enough to admit of their wide dissemination; others, because circumstances in the animate or inanimate world have occurred to restrict the range which they may once have obtained. As a general rule, however, species, common to many distant provinces, or those now found to inhabit very distant parts of the globe, are to be regarded as the most ancient. Numerically speaking, they may not perhaps be largely represented, but their wide diffusion shows that they have had a long time to spread themselves, and have been able to survive many important revolutions in physical geography.

After so much evidence has been brought to light by the geologist, of land and sea having changed places in various regions since the existing species were in being, we can feel no surprise that the zoologist and botanist have hitherto found it difficult to refer the geographical distribution of species to any clear and determinate principles, since they have usually speculated on the phenomena, upon the assumption that the physical geography of the globe had undergone no material alteration since the introduction of the species now living. So long as this assumption was made, the facts relating to the geography of plants and animals appeared capricious in the extreme, and by many the subject was pronounced to be so full of mystery and anomalies, that the establishment of a satisfactory theory was hopeless.*

Centres from which plants have been diffused.—Some botanists conceived, in accordance with the hypothesis of Wildenow, that mountains were the centres of creation from which the plants now inhabiting large continents have radiated; to which De Candolle and others, with much reason, objected, that mountains, on the contrary, are often the barriers between two provinces of distinct vegetation. The geologist who is acquainted with the extensive modifications which the surface of the

* This and the preceding chapter, on the causes of extinction of species and their present geographical distribution, are reprinted almost verbatim from the original edition of the second volume of "The Principles," published in January, 1832. It was I believe the first attempt to point out how former changes in the geography and local climate of many parts of the globe must be taken into account when we endeavor to explain the actual provinces of plants and animals, the changes alluded to having been proved by geological evidence to be subsequent to the creation of a great proportion of the species now living, and these having been, according to the view which I advocated, introduced in succession, and not all at one geological epoch. In my third volume, published in May, 1833, I announced my conviction that the greater part of the existing Fauna and Flora of Sicily were older than the mountains, plains, and rivers, which the same species of animals and plants now inhabit. (Prin. of Geol., vol. iii. ch. ix.; repeated in Elements of Geol., 2d edit., vol. i. p. 297.) This line of reasoning has since been ably followed up and elucidated by Professor E. Forbes in an excellent paper (published in 1846) already alluded to. (See page 86.)

earth has undergone in very recent geological epochs, may be able, perhaps, to reconcile both these theories in their application to different regions.

A lofty range of mountains, which is so ancient as to date from a period when the species of animals and plants differed from those now living, will naturally form a barrier between contiguous provinces; but a chain which has been raised, in great part, within the epoch of existing species, and around which new lands have arisen from the sea within that period, will be a centre of peculiar vegetation.

"In France," observes De Candolle, "the Alps and Cevennes prevent a great number of the plants of the south from spreading themselves to the northward; but it has been remarked that some species have made their way through the gorges of these chains, and are found on their northern sides, principally in those places where they are lower and more interrupted."* Now the chains here alluded to have probably been of considerable height ever since the era when the existing vegetation began to appear, and were it not for the deep fissures which divide them, they might have caused much more abrupt terminations to the extension of distinct assemblages of species.

Parts of the Italian peninsula, on the other hand, have gained a considerable portion of their present height since a majority of the marine species now inhabiting the Mediterranean, and probably, also, since the terrestrial plants of the same region were in being. Large tracts of land have been added, both on the Adriatic and Mediterranean side, to what originally constituted a much narrower range of mountains, if not a chain of islands running nearly north and south, like Corsica and Sardinia. It may therefore be presumed that the Apennines have been a centre whence species have diffused themselves over the contiguous *lower* and *newer* regions. In this and all analogous situations, the doctrine of Wildenow, that species have radiated from the mountains as from centres, may be well founded.

Introduction of New Species.

If the reader should infer, from the facts laid before him in the preceding chapters, that the successive extinction of animals and plants may be part of the constant and regular course of nature, he will naturally inquire whether there are any means provided for the repair of these losses? Is it part of the economy of our system that the habitable globe should, to a certain extent, become depopulated both in the ocean and on the land; or that the variety of species should diminish until some new era arrives when a new and extraordinary effort of creative energy is to be displayed? Or is it possible that new species can be called into being from time to time, and yet that so astonishing a phenomenon can escape the observation of naturalists?

* Essai Elémentaire, &c. p. 46.

Humboldt has characterized these subjects as among the mysteries which natural science cannot reach; and he observes that the investigation of the origin of beings does not belong to zoological or botanical geography. To geology, however, these topics do strictly appertain; and this science is chiefly interested in inquiries into the state of the animate creation as it now exists, with a view of pointing out its relations to antecedent periods when its condition was different.

Before offering any hypothesis towards the solution of so difficult a problem, let us consider what kind of evidence we ought to expect, in the present state of science, of the first appearance of new animals or plants, if we could imagine the successive creation of species to constitute, like their gradual extinction, a regular part of the economy of nature.

In the first place it is obviously more easy to prove that a species, once numerously represented in a given district, has ceased to be, than that some other which did not pre-exist has made its appearance—assuming always, for reasons before stated, that single stocks only of each animal and plant are originally created, and that individuals of new species do not suddenly start up in many different places at once.

So imperfect has the science of natural history remained down to our own times, that, within the memory of persons now living, the numbers of known animals and plants have been doubled, or even quadrupled, in many classes. New and often conspicuous species are annually discovered in parts of the old continent, long inhabited by the most civilized nations. Conscious, therefore, of the limited extent of our information, we always infer, when such discoveries are made, that the beings in question had previously eluded our research; or had at least existed elsewhere, and only migrated at a recent period into the territories where we now find them. It is difficult, even in contemplation, to anticipate the time when we shall be entitled to make any other hypothesis in regard to all the marine tribes, and to by far the greater number of the terrestrial;—such as birds, which possess such unlimited powers of migration; insects, which, besides the variability of each species in number, are also so capable of being diffused to vast distances; and cryptogamous plants, to which, as to many other classes, both of the animal and vegetable kingdom, similar observations are applicable.

What kind of evidence of new creations could be expected?—What kind of proofs, therefore, could we reasonably expect to find of the origin at a particular period of a new species?

Perhaps it may be said in reply that, within the last two or three centuries, some forest tree or new quadruped might have been observed to appear suddenly in those parts of England or France which had been most thoroughly investigated;—that naturalists might have been able to show that no such living being inhabited any other region of the globe, and that there was no tradition of anything similar having before been observed in the district where it had made its appearance.

Now, although this objection may seem plausible, yet its force will be found to depend entirely on the rate of fluctuation which we suppose to prevail in the animate world, and on the proportion which such conspicuous subjects of the animal and vegetable kingdoms bear to those which are less known and escape our observation. There are, perhaps, more than a million species of plants and animals, exclusive of the microscopic and infusory animalcules, now inhabiting the terraqueous globe. The terrestrial plants may amount, says De Candolle, to somewhere between 110,000 and 120,000;* but the data on which this conjecture is founded are considered by many botanists to be vague and unsatisfactory. Sprengel only enumerated, in 1827, about 31,000 known phænogamous, and 6000 cryptogamous plants; but that naturalist omitted many, perhaps 7000 phænogamous, and 1000 cryptogamous species. Mr. Lindley, in a letter to the author in 1836, expressed his opinion that it would be rash to speculate on the existence of more than 80,000 phænogamous, and 10,000 cryptogamous plants. "If we take," he says, in a letter to the author on this subject, "37,000 as the number of published phænogamous species, and then add, for the undiscovered species in Asia and New Holland, 15,000, in Africa 10,000, and in America 18,000, we have 80,000 species; and if 7000 be the number of published cryptogamous plants, and we allow 3000 for the undiscovered species (making 10,000), there would then be, on the whole, 90,000 species." But since that period one catalogue, as I learn from Dr. J. Hooker, contains a list of the names of 78,000 phænogamous plants which had been published before 1841.

It was supposed by Linnæus that there were four or five species of insects in the world for each phænogamous plant: but if we may judge from the relative proportion of the two classes in Great Britain, the number of insects must be still greater; for the total number of British insects, "according to the last census," is about 12,500;† whereas there are only 1500 phænogamous plants indigenous to our island. As the insects are much more numerous in hot countries than in our temperate latitudes, it seems difficult to avoid the conclusion that there are more than half a million species in the world.

The number of known mammifers, when Temminck wrote, exceeded 800, and Mr. Waterhouse informs me that more than 1200 are now (1850) ascertained to exist. Baron Cuvier estimated the amount of known fishes at 6000; and Mr. G. Gray, in his "Genera of Birds," enumerates 8000 species. We have still to add the reptiles, and all the invertebrated animals, exclusive of insects. It remains, in a great degree, mere matter of conjecture what proportion the aquatic tribes may bear to the denizens of the land; but the habitable surface beneath the waters can hardly be estimated at less than double that of the continents and islands, even admitting that a very considerable area is destitute of life, in consequence of great depth, cold, darkness, and other circumstances. In the late polar expedition it was found that, in some regions, as in Baffin's Bay, there were marine animals inhabiting the bottom at great depths, where the temperature of the water was below the freezing point. That there is

* Geog. des Plantes. Dict. des Sci.

† See Catalogue of Brit. Insects, by John Curtis, Esq.

life at much greater profundities in warmer regions may be confidently inferred.

The ocean teems with life—the class of *Polyps* alone are conjectured by Lamarck to be as strong in individuals as insects. Every tropical reef is described as covered with Corals and Sponges, and swarming with Crustacea, Echini, and Testacea; while almost every tide-washed rock in the world is carpeted with Fuci, and supports some Corallines, Actiniæ, and Mollusca. There are innumerable forms in the seas of the warmer zones, which have scarcely begun to attract the attention of the naturalist; and there are parasitic animals without number, three or four of which are sometimes appropriated to one genus, as to the whale (*Balæna*), for example. Even though we concede, therefore, that the geographical range of marine species is more extensive in general than that of the terrestrial (the temperature of the sea being more uniform, and the land impeding less the migrations of the oceanic than the ocean those of the terrestrial species), yet it seems probable that the aquatic tribes far exceed in number the inhabitants of the land.

Without insisting on this point, it may be safe to assume, that, exclusive of microscopic beings, there are between one and two millions of species now inhabiting the terraqueous globe; so that if only one of these were to become extinct annually, and one new one were to be every year called into being, much more than a million of years might be required to bring about a complete revolution in organic life.

I am not hazarding at present any hypothesis as to the probable rate of change; but none will deny that when the *annual* birth and the *annual* death of one species on the globe is proposed as a mere speculation, this at least is to imagine no slight degree of instability in the animate creation. If we divide the surface of the earth into twenty regions of equal area, one of these might comprehend a space of land and water about equal in dimensions to Europe, and might contain a twentieth part of the million of species which may be assumed to exist in the animal kingdom. In this region one species only would, according to the rate of mortality before assumed, perish in twenty years, or only five out of fifty thousand in the course of a century. But as a considerable proportion of the whole would belong to the aquatic classes, with which we have a very imperfect acquaintance, we must exclude them from our consideration; and if they constitute half of the entire number, then one species only might be lost in forty years among the terrestrial tribes. Now the Mammalia, whether terrestrial or aquatic, bear so small a proportion to other classes of animals, forming less, perhaps, than one thousandth part of the whole, that if the longevity of species in the different orders were equal, a vast period must elapse before it would come to the turn of this conspicuous class to lose one of their number. If one species only of the whole animal kingdom died out in forty years, no more than one mammifer might disappear in 40,000 years in a region of the dimensions of Europe.

It is easy, therefore, to see, that in a small portion of such an area, in countries, for example, of the size of England and France, periods of much greater duration must elapse before it would be possible to authenticate the first appearance of one of the larger plants and animals, assuming the annual birth and death of one species to be the rate of vicissitude in the animate creation throughout the world.

The observations of naturalists upon living species may, in the course of future centuries, accumulate positive data, from which an insight into the laws which govern this part of our terrestrial system may be derived; but, in the present deficiency of historical records, we have traced up the subject to that point where geological monuments alone are capable of leading us on to the discovery of ulterior truths. To these, therefore, we must appeal, carefully examining the strata of recent formation wherein the remains of *living* species, both animal and vegetable, are known to occur. We must study these strata in strict reference to their chronological order, as deduced from their superposition, and other relations. From these sources we may learn which of the species, now our contemporaries, have survived the greatest revolutions of the earth's surface; which of them have co-existed with the greatest number of animals and plants now extinct; and which have made their appearance only when the animate world had nearly attained its present condition.

From such data we may be enabled to infer, whether species have been called into existence in succession, or all at one period; whether singly, or by groups simultaneously; whether the antiquity of man be as high as that of any of the inferior beings which now share the planet with him, or whether the human species is one of the most recent of the whole.

To some of these questions we can even now return a satisfactory answer; and with regard to the rest, we have some data to guide conjecture, and to enable us to speculate with advantage: but in order to be fully qualified to enter upon such discussions the reader must study the ample body of materials amassed by the industry of modern geologists.

CHAPTER XLIV.

EFFECTS PRODUCED BY THE POWERS OF VITALITY ON THE STATE OF THE EARTH'S SURFACE.

Modifications in physical geography caused by organic beings—Why the vegetable soil does not augment in thickness—The theory, that vegetation is an antagonist power counterbalancing the degradation caused by running water untenable—Conservative influence of vegetation—Rain diminished by felling of forests—Distribution of American forests dependent on direction of predominant winds—Influence of man in modifying the physical geography of the globe.

THE second branch of our inquiry, respecting changes of the organic world, relates to the processes by which the remains of animals and plants become fossil, or, to speak still more generally, to all the effects produced by the powers of vitality on the surface and shell of the earth.

Before entering on the principal division of this subject, the imbedding and preservation of animal and vegetable remains, I shall offer a few remarks on the superficial modifications caused directly by the agency of organic beings, as when the growth of certain plants covers the slope of a mountain with peat, or converts a swamp into dry land; or when vegetation prevents the soil, in certain localities, from being washed away by running water.

In considering alterations of this kind, brought about in the physical geography of particular tracts, we are too apt to think exclusively of that part of the earth's surface which has emerged from beneath the waters, and with which alone, as terrestrial beings, we are familiar. Here the direct power of animals and plants to cause any important variation is, of necessity, very limited, except in checking the progress of that decay of which the land is the chief theatre. But if we extend our views, and instead of contemplating the dry land, consider that larger portion which is assigned to the aquatic tribes, we discover the great influence of the living creation, in imparting varieties of conformation to the solid exterior which the agency of inanimate causes alone could not produce.

Thus, when timber is floated into the sea, it is often drifted to vast distances, and subsides in spots where there might have been no deposit, at that time and place, if the earth had not been tenanted by living beings. If, therefore, in the course of ages, a hill of wood, or lignite, be thus formed in the subaqueous regions, a change in the submarine geography may be said to have resulted from the action of organic powers. So in regard to the growth of coral reefs; it is probable that a large portion of the matter of which they are composed is supplied by mineral springs, which often rise up at the bottom of the sea, and which, on land, abound throughout volcanic regions hundreds of leagues in extent. The matter thus constantly given out could not go on accumulating for ever in the waters, but would be precipitated in the abysses of the sea, even if there were no polyps and testacea; but these animals arrest and secrete the carbonate of lime on the summits of submarine mountains, and form reefs many hundred feet in thickness, and hundreds of miles in length, where, but for them, none might ever have existed.

Why the vegetable soil does not augment in thickness.—If no such voluminous masses are formed on the land, it is not from the want of solid matter in the structure of terrestrial animals and plants; but merely because, as I have so often stated, the continents are those parts of the globe where accessions of matter can scarcely ever take place—where, on the contrary, the most solid parts already formed are, each in their turn, exposed to gradual degradation. The quantity of timber and vegetable matter which grows in a tropical forest in the course of a century is enormous, and multitudes of animal skeletons are scattered there during the same period, besides innumerable land shells and other organic substances. The aggregate of these materials, therefore, might constitute a mass greater in volume than that which is produced in any coral-reef during the same lapse of years; but, although this process should continue on the land for ever, no mountains of wood or bone would be seen stretching far and wide over the country, or pushing out bold promontories into the sea. The whole solid mass is either devoured by animals, or decomposes, as does a portion of the rock and soil on which the animals and plants are supported.

The waste of the strata themselves, accompanied by the decomposition of their organic remains, and the setting free of their alkaline ingredients, is one source from whence running water and the atmosphere may derive the materials which are absorbed by the roots and leaves of plants. Another source is the passage into a gaseous form of even the hardest parts of animals and plants which die and putrefy in the air, where they are soon resolved into the elements of which they are composed: and while a portion of these constituents is volatilized, the rest is taken up by rain-water, and sinks into the earth, or flows towards the sea; so that they enter again and again into the composition of different organic beings.

The principal elements found in plants are hydrogen, carbon, and oxygen; so that water and the atmosphere contain all of them, either in their own composition or in solution.* The constant supply of these elements is maintained not only by the putrefaction of animal and vegetable substances, and the decay of rocks, but also by the copious evolution of carbonic acid and other gases from volcanoes and

* See some good remarks on the Formation of Soils, Bakewell's Geology, chap. xviii

mineral springs, and by the effects of ordinary evaporation, whereby aqueous vapors are made to rise from the ocean, and to circulate round the globe.

It is well known, that when two gases of different specific gravity are brought into contact, even though the heavier be the lowermost, they soon become uniformly diffused by mutual absorption through the whole space which they occupy. By virtue of this law, the heavy carbonic acid finds its way upwards through the lighter air of the atmosphere, and conveys nourishment to the lichen which covers the mountain top.

If the quantity of food consumed by terrestrial animals, and the elements imbibed by the roots and leaves of plants, were derived entirely from that supply of hydrogen, carbon, oxygen, nitrogen, and other elements, given out into the atmosphere and the waters by the putrescence of organic substances, then we might imagine that the vegetable mould would, after a series of years, neither gain nor lose a single particle by the action of organic beings; and this conclusion is not far from the truth; but the operation which renovates the vegetable and animal mould is by no means so simple as that here supposed. Thousands of carcases of terrestrial animals are floated down, every century, into the sea; and, together with forests of drift-timber, are imbedded in subaqueous deposits, where their elements are imprisoned in solid strata, and may there remain locked up throughout whole geological epochs before they again become subservient to the purposes of life.

On the other hand, fresh supplies are derived by the atmosphere and by running water, as before stated, from the disintegration of rocks and their organic contents, and through the agency of mineral springs from the interior of the earth, from whence all the elements before mentioned, which enter principally into the composition of animals and vegetables, are continually evolved. Even nitrogen is found, by chemists, to be contained very generally in the waters of mineral springs.

Vegetation not an antagonist power counterbalancing the action of running water.—If we suppose that the copious supply from the nether regions, by springs and volcanic vents, of carbonic acid and other gases, together with the decomposition of rocks, may be just sufficient to counterbalance that loss of matter which, having already served for the nourishment of animals and plants, is annually carried down in organized forms, and buried in subaqueous strata, we concede the utmost that is consistent with probability. An opinion, however, has been expressed, that the processes of vegetable life, by absorbing various gases from the atmosphere, cause so large a mass of solid matter to accumulate on the surface of the land, that this mass alone may constitute a great counterpoise to all the matter transported to lower levels by the aqueous agents of decay. "Torrents and rivers," it is said—"the waves of the sea and marine currents—act upon lines only; but the power of vegetation to absorb the elastic and non-elastic fluids circulating round the earth, extends over the whole surface of the continents. By the silent but universal action of this great antagonist power, the spoliation and waste caused by running water on the land, and by the movements of the ocean, are neutralized, and even counterbalanced."*

In opposition to these views, I conceive that we shall form a juster estimate of the influence of vegetation, if we consider it as being in a slight degree conservative, and capable of retarding the waste of land, but not of acting as an antagonist power. The vegetable mould is seldom more than a few feet in thickness, and frequently does not exceed a few inches; and we by no means find that its volume is more considerable on those parts of our continents which we can prove, by geological data, to have been elevated at more ancient periods, and where, consequently, there has been the greatest time for the accumulation of vegetable matter, produced throughout successive zoological epochs. On the contrary, these higher and older regions are more frequently denuded, so as to expose the bare rock to the action of the sun and air.

We find in the torrid zone, where the growth of plants is most rank and luxurious, that accessions of matter due to their agency are by no means the most conspicuous. Indeed it is in these latitudes, where the vegetation is most active, that, for reasons to be explained in the next chapter, even those superficial peat mosses are unknown which cover a large area in some parts of our temperate zone. If the operation of animal and vegetable life could restore to the general surface of the continents a portion of the elements of those disintegrated rocks of which such enormous masses are swept down annually into the sea, the effects would long ere this have constituted one of the most striking features in the structure and composition of our continents. All the great steppes and table-lands of the world, where the action of running water is feeble, would have become the grand repositories of organic matter, accumulated without that intermixture of earthy sediment which so generally characterizes the subaqueous strata.

I have already stated that, in the known operation of the *igneous* causes, a real antagonist power is found, which may counterbalance the levelling action of running water (p. 563); and there seems no good reason for presuming that the upheaving and depressing force of earthquakes, together with the ejection of matter by volcanoes, may not be fully adequate to restore that inequality of the surface which rivers and the waves and currents of the ocean annually tend to lessen. If a counterpoise be derived from this source, the quantity and elevation of land above the sea may for ever remain the same, in spite of the action of the aqueous causes, which, if thus counteracted, may never be able to reduce the surface of the earth more nearly to a state of equilibrium than that which it has now attained; and, on the other hand, the force of the aqueous agents themselves might thus continue for ever unimpaired.

Conservative influence of vegetation.—If, then, vegetation cannot act as an antagonist power amid the mighty agents of change which are always modifying the surface of the globe, let us next inquire how far

* See Professor Sedgwick's Anniversary Address to the Geological Society, Feb. 1831, p. 24.

its influence is conservative,—how far it may retard the levelling effects of running water, which it cannot oppose, much less counterbalance.

It is well known that a covering of herbage and shrubs may protect a loose soil from being carried away by rain, or even by the ordinary action of a river, and may prevent hills of loose sand from being blown away by the wind; for the roots bind together the separate particles into a firm mass, and the leaves intercept the rain-water, so that it dries up gradually, instead of flowing off in a mass and with great velocity. The old Italian hydrographers make frequent mention of the increased degradation which has followed the clearing away of natural woods in several parts of Italy. A remarkable example was afforded in the Upper Val d' Arno, in Tuscany, on the removal of the woods clothing the steep declivities of the hills by which that valley is bounded. When the ancient forest laws were abolished by the Grand Duke Joseph, during the last century, a considerable tract of surface in the Cassentina (the Clausentinium of the Romans) was denuded, and immediately the quantity of sand and soil washed down into the Arno increased enormously. Frisi, alluding to such occurrences, observes, that as soon as the bushes and plants were removed, the waters flowed off more rapidly, and, in the manner of floods, swept away the vegetable soil.*

This effect of vegetation is of high interest to the geologist, when he is considering the formation of those valleys which have been principally due to the action of rivers. The spaces intervening between valleys, whether they be flat or ridgy, when covered with vegetation, may scarcely undergo the slightest waste, as the surface may be protected by the green sward of grass; and this may be renewed, in the manner before described, from elements derived from rain-water and the atmosphere. Hence, while the river is continually bearing down matter in the alluvial plain, and undermining the cliffs on each side of every valley, the height of the intervening rising grounds may remain stationary.

In this manner, a cone of loose scoriæ, sand, and ashes, such as Monte Nuovo, may, when it has once become densely clothed with herbage and shrubs, suffer scarcely any further dilapidation; and the perfect state of the cones of hundreds of extinct volcanoes in France, the Neapolitan territory, Sicily, and elsewhere, may prove nothing whatever, either as to their relative or absolute antiquity. We may be enabled to infer, from the integrity of such conical hills of incoherent materials, that no flood can have passed over the countries where they are situated, since their formation; but the atmospheric action alone, in spots where there happen to be no torrents, and where the surface was clothed with vegetation, could scarcely in any lapse of ages have destroyed them.

During a tour in Spain, in 1830, I was surprised to see a district of gently undulating ground in Catalonia, consisting of red and gray sandstone, and in some parts of red marl, almost entirely denuded of herbage; while the roots of the pines, holm oaks, and some other trees, were half

* Treatise on Rivers and Torrents, p. 5. Garston's translation.

exposed, as if the soil had been washed away by a flood. Such is the state of the forests, for example, between Oristo and Vich, and near San Lorenzo. But, being overtaken by a violent thunder-storm, in the month of August, I saw the whole surface, even the highest levels of some flat-topped hills, streaming with mud, while on every declivity the devastation of torrents was terrific. The peculiarities in the physiognomy of the district were at once explained; and I was taught that, in speculating on the greater effects which the direct action of rain may once have produced on the surface of certain parts of England, we need not revert to periods when the heat of the climate was *tropical.*

In the torrid zone the degradation of land is generally more rapid; but the waste is by no means proportioned to the superior quantity of rain or the suddenness of its fall, the transporting power of water being counteracted by a greater luxuriance of vegetation. A geologist who is no stranger to tropical countries observes, that the softer rocks would speedily be washed away in such regions, if the numerous roots of plants were not matted together in such a manner as to produce considerable resistance to the destructive power of the rains. The parasitical and creeping plants also entwine in every possible direction, so as to render the forests nearly impervious, and the trees possess forms and leaves best calculated to shoot off the heavy rains; which, when they have thus been broken in their fall, are quickly absorbed by the ground beneath, or, when thrown into the drainage depressions, give rise to furious torrents.*

Influence of Man in modifying the Physical Geography of the Globe.

Before concluding this chapter, I shall offer a few observations on the influence of man in modifying the physical geography of the globe; for we must class his agency among the powers of organic nature.

Felling of forests.—The felling of forests has been attended, in many countries, by a diminution of rain, as in Barbadoes and Jamaica.† For in tropical countries, where the quantity of aqueous vapor in the atmosphere is great, but where, on the other hand, the direct rays of the sun are most powerful, any impediment to the free circulation of air, or any screen which shades the earth from the solar rays, becomes a source of humidity; and wherever dampness and cold have begun to be generated by such causes, the condensation of vapor continues. The leaves, moreover, of all plants are alembics, and some of those in the torrid zone have the remarkable property of distilling water, thus contributing to prevent the earth from becoming parched up.

Distribution of the American forests.—There can be no doubt then, that the state of the climate, especially the humidity of the atmosphere, influences vegetation, and that, in its turn, vegetation re-acts upon the

* De la Beche, Geol. Man., p. 184., 1st ed.

† Phil. Trans., vol. ii. p. 294.

climate: but some writers seem to have attributed too much importance to the influence of forests, particularly those of America, as if they were the primary cause of the moisture of the climate.

The theory of a modern author on this subject "that forests exist in those parts of America only where the predominant winds carry with them a considerable quantity of moisture from the ocean," seems far more rational. In all countries, he says, "having a summer heat exceeding 70°, the presence or absence of natural woods, and their greater or less luxuriance, may be taken as a measure of the amount of humidity, and of the fertility of the soil. Short and heavy rains in a warm country will produce grass, which, having its roots near to the surface, springs up in a few days, and withers when the moisture is exhausted; but transitory rains, however heavy, will not nourish trees; because, after the surface is saturated, the remainder of the water runs off, and the moisture lodged in the soil neither sinks deep enough, nor is in sufficient quantity, to furnish the giants of the forests with the necessary sustenance. It may be assumed that twenty inches of rain falling moderately or at intervals, will leave a greater permanent supply in the soil than forty inches falling, as it sometimes does in the torrid zone, in as many hours."*

"In all regions," he continues, "where ranges of mountains intercept the course of the constant or predominant winds, the country on the windward side of the mountains will be moist, and that on the leeward dry; and hence parched deserts will generally be found on the west side of countries within the tropics, and on the east side of those beyond them, the prevailing winds in these cases being generally in opposite directions. On this principle, the position of forests in North and South America may be explained. Thus, for example, in the region within the thirtieth parallel, the moisture swept up by the trade-wind from the Atlantic is precipitated in part upon the mountains of Brazil, which are but low, and so distributed as to extend far into the interior. The portion which remains is borne westward, and, losing a little as it proceeds, is at length arrested by the Andes, where it falls down in showers on their summits. The aërial current, now deprived of all the humidity with which it can part, arrives in a state of complete exsiccation at Peru, where consequently no rain falls. But in the region of America, beyond the thirtieth parallel, the Andes serve as a screen to intercept the moisture brought by the prevailing winds from the Pacific Ocean: rains are copious on their summits, and in Chili on their *western* declivities; but none falls on the plains to the *eastward*, except occasionally when the wind blows from the Atlantic."†

I have been more particular in explaining these views, because they appear to place in a true light the dependence of vegetation on climate, the humidity being increased, and more uniformly diffused throughout the year, by the gradual spreading of wood.

* Maclaren, art. America, Encyc. Britannica.

† Maclaren, art. America, Encyc. Britannica, where the position of the American forests, in accordance with this theory, is laid down in a map.

It has been affirmed, that formerly, when France and England were covered with wood, Europe was much colder than at present; that the winters in Italy were longer, and that the Seine, and many other rivers, froze more regularly every winter than now. M. Arago, in an essay on this subject, has endeavored to show, by tables of observations on the congelation of the Rhine, Danube, Rhone, Po, Seine, and other rivers, at different periods, that there is no reason to believe the cold to have been in general more intense in ancient times.* He admits, however, that the climate of Tuscany has been so far modified, by the removal of wood, as that the winters are less cold; but the summers also, he contends, are less hot than of old; and the summers, according to him, were formerly hotter in France than in our own times. His evidence is derived chiefly from documents showing that wine was made three centuries ago in the Vivarais and several other provinces, at an earlier season, at greater elevations, and in higher latitudes, than are now found suitable to the vine.

There seems little doubt that in the United States of North America the rapid *clearing* of the country has rendered the winters less severe and the summers less hot; in other words, the extreme temperatures of January and July have been observed from year to year to approach somewhat nearer to each other. Whether in this case, or in France, the *mean* temperature has been raised, seems by no means as yet decided; but there is no doubt that the climate has become, as Buffon would have said, "less excessive."

I have before shown, when treating of the excavation of new estuaries in Holland by inroads of the ocean, as also of the changes on our own coasts, that although the conversion of sea into land by artificial labors may be great, yet it must always be in subordination to the power of the tides and currents, or to the great movements which alter the relative level of the land and sea, (Chap. XX.) If, in addition to the assistance obtained by parliamentary grants for defending Dunwich from the waves, all the resources of Europe had been directed to the same end, the existence of that port might perhaps have been prolonged for several centuries (p. 310.) But in the mean time, the current would have continued to sweep away portions from the adjoining cliffs on each side, giving to the whole line of coast its present form, until at length the town, projecting as a narrow promontory, must have become exposed to the irresistible fury of the waves.

It is scarcely necessary to observe, that the control which man can obtain over the igneous agents is less even than that which he may exert over the aqueous. He cannot modify the upheaving or depressing force of earthquakes, or the periods or degree of violence of volcanic eruptions; and on these causes the inequalities of the earth's surface, and, consequently, the shape of the sea and land, appear mainly to depend. The utmost that man can hope to effect in this respect is occasionally to divert the course of a lava-stream, and to prevent the burning matter, for

* Annuaire du Bureau des Long. 1834.

a season, from overwhelming a city, or some other of the proudest works of human industry.

If all the nations of the earth should attempt to quarry away the lava which flowed during one eruption from the Icelandic volcanoes in 1783, and the two following years, and should attempt to consign it to the deepest abysses of the ocean, they might toil for thousands of years and not accomplish their task. Yet the matter borne down to the sea by two great rivers, the Ganges and Burrampooter, in each quarter of a century, probably equals in weight and volume the mass of Icelandic lava produced by that great eruption (p. 282). So insignificant is the aggregate force exerted by man, when contrasted with the ordinary operations of aqueous or igneous agents in the natural world.

No application, perhaps, of human skill and labor tends so greatly to vary the state of the habitable surface, as that employed in the drainage of lakes and marshes, since not only the *stations* of many animals and plants, but the general climate of a district, may thus be modified. It is also a kind of alteration to which it is difficult to find anything analogous in the agency of inferior beings; for we ought always, before we decide that any part of the influence of man is novel and anomalous, carefully to consider the powers of all other animated agents which may be limited or superseded by him.* Many who have reasoned on these subjects seem to have forgotten that the human race often succeeds to the discharge of functions previously fulfilled by other species. Suppose the growth of some of the larger terrestrial plants, or, in other words, the extent of forest, to be diminished by man, and the climate to be thereby modified, it does not follow that this kind of innovation is unprecedented. It is a change in the state of vegetation, and such may often have been the result of the appearance of new species upon the earth. The multiplication, for example, of certain insects in parts of Germany, during the last century, destroyed more trees than man, perhaps, could have felled during an equal period.

It would be rash, however, to affirm that the power of man to modify the surface may not differ in kind or degree from that of other living beings; although the problem is certainly more complex than many who have speculated on such topics have imagined. If land be raised from the sea, the greatest alteration in its physical condition, which could ever arise from the influence of organic beings, would probably be produced by the first immigration of terrestrial plants, whereby the new tract would become covered with vegetation. The change next in importance would seem to be when animals first enter, and modify the proportionate numbers of certain species of plants. If there be any anomaly in the intervention of man, in farther varying the rela-

* Since this was written I have seen in New Brunswick (1852) a lake formed by beavers who had thrown a dam, consisting of stakes, stones, and mud, across the course of a small streamlet, between Dorchester and the Portage south of the Peticodiac river. The beavers have since been extirpated by man, but the lake remains, and musk rats have taken possession of the shallow parts of the lake to build their habitations in them.

tive numbers in the vegetable kingdom, it may not so much consist in the kind or absolute quantity of alteration, as in the circumstance that a *single species*, in this case, would exert, by its superior power and universal distribution, an influence equal to that of hundreds of other terrestrial animals.

If we inquire whether man, by his direct power, or by the changes which he may give rise to indirectly, tends, upon the whole, to lessen or increase the inequalities of the earth's surface, we shall incline, perhaps, to the opinion that he is a levelling agent. In mining operations he conveys upwards a certain quantity of materials from the bowels of the earth; but, on the other hand, much rock is taken annually from the land, in the shape of ballast, and afterwards thrown into the sea, and by this means, in spite of prohibitory laws, many harbors, in various parts of the world, have been blocked up. We rarely transport heavy materials to higher levels, and our pyramids and cities are chiefly constructed of stone brought down from more elevated situations. By ploughing up thousands of square miles, and exposing a surface for part of the year to the action of the elements, we assist the abrading force of rain, and diminish the conservative effects of vegetation.

CHAPTER XLV.

INCLOSING OF FOSSILS IN PEAT, BROWN SAND, AND VOLCANIC EJECTIONS.

Division of the subject—Imbedding of organic remains in deposits on emerged land—Growth of peat—Site of ancient forests in Europe now occupied by peat—Bog iron-ore—Preservation of animal substances in peat—Miring of quadrupeds—Bursting of the Solway moss—Great Dismal Swamp—Imbedding of organic bodies and human remains in blown sand—Moving sands of African deserts—De Luc on their recent origin—Buried temple of Ipsambul—Dried carcases in the sands—Towns overwhelmed by sand-floods—Imbedding of organic and other remains in volcanic formations on the land.

Division of the subject.—The next subject of inquiry is the mode in which the remains of animals and plants become fossil, or are buried in the earth by natural causes. M. Constant Prevost has observed, that the effects of geological causes are divisible into two great classes; those produced during the submersion of land beneath the waters, and those which take place after its emersion. Agreeably to this classification, I shall consider, first, in what manner animal and vegetable remains become included and preserved in deposits on emerged land, or that part of the surface which is not *permanently* covered by water, whether of seas or lakes; secondly, the manner in which organic remains become imbedded in subaqueous deposits.

Under the first division, I shall treat of the following topics:—1st, the growth of peat, and the preservation of vegetable and animal remains therein;—2dly, the burying of organic remains in blown sand;—3dly, of the same in the ejections and alluviums of volcanoes;—4thly, in alluviums generally, and in the ruins of landslips;—5thly, in the mud and stalagmite of caves and fissures.

Growth of Peat, and Preservation of Vegetable and Animal Remains therein.

The generation of peat, when not completely under water, is confined to moist situations, where the temperature is low, and where vegetables may decompose without putrefying. It may consist of any of the numerous plants which are capable of growing in such *stations;* but a species of moss (*Sphagnum*) constitutes a considerable part of the peat found in marshes of the north of Europe; this plant having the property of throwing up new shoots in its upper part, while its lower extremities are decaying.* Reeds, rushes, and other aquatic plants may usually be traced in peat; and their organization is often so entire that there is no difficulty in discriminating the distinct species.

* For a catalogue of plants which form peat, see Rev. Dr. Rennie's Essays on Peat, p. 171; and Dr. MacCulloch's Western Isles, vol. i. p. 129.

Analysis of peat.—In general, says Sir H. Davy, one hundred parts of dry peat contain from sixty to ninety-nine parts of matter destructible by fire; and the residuum consists of earths usually of the same kind as the substratum of clay, marl, gravel, or rock, on which they are found, together with oxide of iron. "The peat of the chalk counties of England," observes the same writer, "contains much gypsum: but I have found very little in any specimens from Ireland or Scotland, and in general these peats contain very little saline matter." * From the researches of Dr. MacCulloch, it appears that peat is intermediate between simple vegetable matter and lignite, the conversion of peat to lignite being gradual, and being brought about by a prolonged action of water. †

Peat abundant in cold and humid climates.—Peat is sometimes formed on a declivity in mountainous regions, where there is much moisture; but in such situations it rarely, if ever, exceeds four feet in thickness. In bogs, and in low grounds into which alluvial peat is drifted, it is found forty feet thick, and upwards; but in such cases it generally owes one half of its volume to the water which it contains. It has seldom, if ever, been discovered within the tropics; and it rarely occurs in the valleys, even in the south of France and Spain. It abounds more and more, in proportion as we advance farther from the equator, and becomes not only more frequent but more inflammable in northern latitudes.‡

The same phenomenon is repeated in the southern hemisphere. No peat is found in Brazil, nor even in the swampy parts of the country drained by the La Plata on the east side of South America, or in the island of Chiloe on the west; yet when we reach the 45th degree of latitude and examine the Chonos Archipelago or the Falkland Islands, and Tierra del Fuego, we meet with an abundant growth of this substance. Almost all plants contribute here by their decay to the production of peat, even the grasses; but it is a singular fact, says Mr. Darwin, as contrasted with what occurs in Europe, that no kind of moss enters into the composition of the South American peat, which is formed by many plants, but chiefly by that called by Brown *Astelia pumila*.§

I learn from Dr. Forchhammer (1849) that water charged with vegetable matter in solution does not throw down a deposit of peat in countries where the mean temperature of the year is above 43° or 44° Fahrenheit. Frost causes the precipitation of such peaty matter, but in warm climates the attraction of the carbon for the oxygen of the air mechanically mixed with the water increases with the increasing temperature, and the dissolved vegetable matter or humic acid (which is organic matter in a progressive state of decomposition) being converted into carbonic acid, rises and is absorbed into the atmosphere, and thus disappears.

* Irish Bog Reports, p. 209.
† System of Geology, vol. ii. p. 353.
‡ Rev. Dr. Rennie on Peat, p. 260.
§ Darwin's Journal, p. 349.; 2d ed. p. 287.

Extent of surface covered by peat.—There is a vast extent of surface in Europe covered with peat, which, in Ireland, is said to extend over a tenth of the whole island. One of the mosses on the Shannon is described as being fifty miles long, by two or three broad; and the great marsh of Montoire, near the mouth of the Loire, is mentioned, by Blavier, as being more than fifty leagues in circumference. It is a curious and well-ascertained fact, that many of these mosses of the north of Europe occupy the place of forests of pine and oak, which have, many of them, disappeared within the historical era. Such changes are brought about by the fall of trees and the stagnation of water, caused by their trunks and branches obstructing the free drainage of the atmospheric waters, and giving rise to a marsh. In a warm climate, such decayed timber would immediately be removed by insects, or by putrefaction; but, in the cold temperature now prevailing in our latitudes, many examples are recorded of marshes originating in this source. Thus, in Mar forest, in Aberdeenshire, large trunks of Scotch fir, which had fallen from age and decay, were soon immured in peat, formed partly out of their perishing leaves and branches, and in part from the growth of other plants. We also learn, that the overthrow of a forest by a storm, about the middle of the seventeenth century, gave rise to a peat-moss near Lochbroom, in Ross-shire, where, in less than half a century after the fall of the trees, the inhabitants dug peat.* Dr. Walker mentions a similar change, when, in the year 1756, the whole wood of Drumlanrig in Dumfries-shire was overset by the wind. Such events explain the occurrence, both in Britain and on the Continent, of mosses where the trees are all broken within two or three feet of the original surface, and where their trunks all lie in the same direction.†

It may however be suggested in these cases, that the soil had become exhausted for trees, and that, on the principle of that natural rotation which prevails in the vegetable world, one set of plants died out and another succeeded. It is certainly a remarkable fact that in the Danish islands, and in Jutland and Holstein, fir wood of various species, especially Scotch fir, is found at the bottom of the peat-mosses, although it is well ascertained that for the last five centuries no Coniferæ have grown wild in these countries; the coniferous trees which now flourish there having been all planted towards the close of the last century.

Nothing is more common than the occurrence of buried trees at the bottom of the Irish peat-mosses, as also in most of those of England, France, and Holland; and they have been so often observed with parts of their trunks standing erect, and with their roots fixed to the subsoil, that no doubt can be entertained of their having generally grown on the spot. They consist, for the most part, of the fir, the oak, and the birch: where the subsoil is clay, the remains of oak are the most abundant; where sand is the substratum, fir prevails. In the marsh of Curragh, in the Isle of Man, vast trees are discovered standing firm on

* Rennie's Essays on Peat, p. 65. † Ibid. p. 30.

their roots, though at the depth of eighteen or twenty feet below the surface. Some naturalists have desired to refer the imbedding of timber in peat-mosses to aqueous transportation, since rivers are well known to float wood into lakes; but the facts above mentioned show that, in numerous instances, such an hypothesis is inadmissible. It has, moreover, been observed, that in Scotland, as also in many parts of the Continent, the largest trees are found in those peat-mosses which lie in the least elevated regions, and that the trees are proportionally smaller in those which lie at higher levels; from which fact De Luc and Walker have both inferred that the trees grew on the spot, for they would naturally attain a greater size in lower and warmer levels. The leaves, also, and fruits of each species, are continually found immersed in the moss along with the parent trees; as, for example, the leaves and acorns of the oak, the cones and leaves of the fir, and the nuts of the hazel.

Recent origin of some peat-mosses.—In Hatfield moss, in Yorkshire, which appears clearly to have been a forest eighteen hundred years ago, fir-trees have been found ninety feet long, and sold for masts and keels of ships; oaks have also been discovered there above one hundred feet long. The dimensions of an oak from this moss are given in the Philosophical Transactions, No. 275, which must have been larger than any tree now existing in the British dominions.

In the same moss of Hatfield, as well as in that of Kincardine, in Scotland, and several others, Roman roads have been found covered to the depth of eight feet by peat. All the coins, axes, arms, and other utensils found in British and French mosses, are also Roman; so that a considerable portion of the peat in European peat-bogs is evidently not more ancient than the age of Julius Cæsar. Nor can any vestiges of the ancient forests described by that general, along the line of the great Roman way in Britain, be discovered, except in the ruined trunks of trees in peat.

De Luc ascertained that the very sites of the aboriginal forests of Hercinia, Semana, Ardennes, and several others, are now occupied by mosses and fens; and a great part of these changes have, with much probability, been attributed to the strict orders given by Severus, and other emperors, to destroy all the wood in the conquered provinces. Several of the British forests, however, which are now mosses, were cut at different periods, by order of the English parliament, because they harbored wolves or outlaws. Thus the Welsh woods were cut and burned, in the reign of Edward I.; as were many of those in Ireland, by Henry II., to prevent the natives from harboring in them, and harassing his troops.

It is curious to reflect that considerable tracts have, by these accidents, been permanently sterilized, and that, during a period when civilization has been making great progress, large areas in Europe have, by human agency, been rendered less capable of administering to the wants of man. Rennie observes,* with truth, that in those regions

* Essays on Peat, &c., p. 74.

alone which the Roman eagle never reached—in the remote circles of the German empire, in Poland and Prussia, and still more in Norway, Sweden, and the vast empire of Russia—can we see what Europe was before it yielded to the power of Rome. Desolation now reigns where stately forests of pine and oak once flourished, such as might now have supplied all the navies of Europe with timber.

Sources of bog iron-ore.—At the bottom of peat-mosses there is sometimes found a cake, or "pan," as it is termed, of oxide of iron, and the frequency of bog iron-ore is familiar to the mineralogist. The oak, which is so often dyed black in peat, owes its color to the same metal. From what source the iron is derived has often been a subject of discussion, until the discoveries of Ehrenberg seem at length to have removed the difficulty. He had observed in the marshes about Berlin a substance of a deep ochre yellow passing into red, which covered the bottom of the ditches, and which, where it had become dry after the evaporation of the water, appeared exactly like oxide of iron. But under the microscope it was found to consist of slender articulated threads or plates, partly siliceous and partly ferruginous, of what he considered an animalcule, *Gaillonella ferruginea*, but which most naturalists now regard as a plant.* There can be little doubt, therefore, that bog iron-ore consists of an aggregate of millions of these organic bodies invisible to the naked eye.†

Fig. 101.

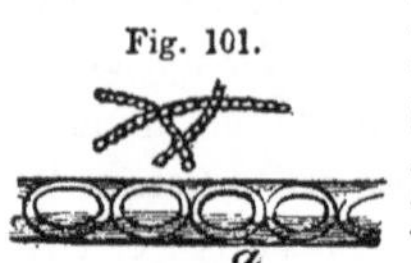

Gaillonella ferruginea.
a. 2000 times magnifie

Preservation of animal substances in peat.—One interesting circumstance attending the history of peat mosses is the high state of preservation of animal substances buried in them for periods of many years. In June, 1747, the body of a woman was found six feet deep, in a peat-moor in the Isle of Axholm, in Lincolnshire. The antique sandals on her feet afforded evidence of her having been buried there for many ages: yet her nails, hair, and skin, are described as having shown hardly any marks of decay. On the estate of the Earl of Moira, in Ireland, a human body was dug up, a foot deep in gravel, covered with eleven feet of moss; the body was completely clothed and the garments seemed all to be made of hair. Before the use of wool was known in that country the clothing of the inhabitants was made of hair, so that it would appear that this body had been buried at that early period; yet it was fresh and unimpaired.‡ In the Philosophical Transactions we find an example recorded of the bodies of two persons having been buried in moist peat, in Derbyshire, in 1674, about a yard deep, which were examined twenty-eight years and nine months afterwards; "the color of their skin was fair and natural, their flesh soft as that of persons newly dead."§

Among other analogous facts we may mention, that in digging a pit

* See above, p. 388, note.
† Ehrenberg, Taylor's Scientific Mem. vol. i. part iii. p. 402.
‡ Dr. Rennie, on Peat, p. 521; where several other instances are referred to.
§ Phil. Trans., vol. xxxviii. 1734.

for a well near Dulverton, in Somersetshire, many pigs were found in various postures, still entire. Their shape was well preserved, the skin, which retained the hair, having assumed a dry, membranous appearance. Their whole substance was converted into a white, friable, laminated, inodorous, and tasteless substance; but which, when exposed to heat, emitted an odor precisely similar to broiled bacon.*

Cause of the antiseptic property of peat.—We naturally ask whence peat derives this antiseptic property? It has been attributed by some to the carbonic and gallic acids which issue from decayed wood, as also to the presence of charred wood in the lowest strata of many peat-mosses, for charcoal is a powerful antiseptic, and capable of purifying water already putrid. Vegetable gums and resins also may operate in the same way.†

The tannin occasionally present in peat is the produce, says Dr. MacCulloch, of tormentilla, and some other plants; but the quantity he thinks too small, and its occurrence too casual, to give rise to effects of any importance. He hints that the soft parts of animal bodies, preserved in peat-bogs, may have been converted into adipocire by the action of water merely; an explanation which appears clearly applicable to some of the cases above enumerated.‡

Miring of quadrupeds.—The manner, however, in which peat contributes to preserve, for indefinite periods, the harder parts of terrestrial animals, is a subject of more immediate interest to the geologist. There are two ways in which animals become occasionally buried in the peat of marshy grounds; they either sink down into the semifluid mud, underlying a turfy surface upon which they have rashly ventured, or, at other times, as we shall see in the sequel, a bog "bursts," and animals may be involved in the peaty alluvium.

In the extensive bogs of Newfoundland, cattle are sometimes found buried with only their heads and necks above ground; and after having remained for days in this situation, they have been drawn out by ropes and saved. In Scotland, also, cattle venturing on the "quaking moss" are often mired, or "laired," as it is termed; and in Ireland, Mr. King asserts that the number of cattle which are lost in sloughs is quite incredible.§

Solway moss.—The description given of the Solway moss will serve to illustrate the general character of these boggy grounds. That moss, observes Gilpin, is a flat area, about seven miles in circumference, situated on the western confines of England and Scotland. Its surface is covered with grass and rushes, presenting a dry crust and a fair appearance; but it shakes under the least pressure, the bottom being unsound and semifluid. The adventurous passenger, therefore, who sometimes in dry seasons traverses this perilous waste, to save a few miles, picks his cautious way over the rushy tussocks as they appear before him, for here the soil

* Dr. Rennie, on Peat, &c., p. 521.
† Ibid. p. 531.
‡ Syst. of Geol. vol. ii. pp. 340—346.
§ Phil. Trans. vol. xv. p. 949.

is firmest. If his foot slip, or if he venture to desert this mark of security, it is possible he may never more be heard of.

"At the battle of Solway, in the time of Henry VIII. (1542), when the Scotch army, commanded by Oliver Sinclair, was routed, an unfortunate troop of horse, driven by their fears, plunged into this morass, which instantly closed upon them. The tale was traditional, but it is now authenticated; a man and horse, in complete armor, having been found by peat-diggers, in the place where it was always supposed the affair had happened. The skeleton of each was well preserved, and the different parts of the armor easily distinguished."*

The same moss, on the 16th of December, 1772, having been filled like a great sponge with water during heavy rains, swelled to an unusual height above the surrounding country, and then burst. The turfy covering seemed for a time to act like the skin of a bladder retaining the fluid within, till it forced a passage for itself, when a stream of black half-consolidated mud began at first to creep over the plain, resembling, in the rate of its progress, an ordinary lava-current. No lives were lost, but the deluge totally overwhelmed some cottages, and covered 400 acres. The highest parts of the original moss subsided to the depth of about twenty-five feet; and the height of the moss, on the lowest parts of the country which it invaded, was at least fifteen feet.

Bursting of a peat-moss in Ireland.—A recent inundation in Sligo (January, 1831), affords another example of this phenomenon. After a sudden thaw of snow, the bog between Bloomfield and Geevah gave way; and a black deluge, carrying with it the contents of a hundred acres of bog, took the direction of a small stream and rolled on with the violence of a torrent, sweeping along heath, timber, mud, and stones, and overwhelming many meadows and arable land. On passing through some boggy land, the flood swept out a wide and deep ravine, and part of the road leading from Bloomfield to St. James's Well was completely carried away from below the foundation for the breadth of 200 yards.

Great Dismal Swamp.—I have described, in my Travels in North America,† an extensive swamp or morass, forty miles long from north to south, and twenty-five wide, between the towns of Norfolk in Virginia, and Weldon in North Carolina. It is called the "Great Dismal," and has somewhat the appearance of an inundated river-plain covered with aquatic trees and shrubs, the soil being as black as that of a peat bog. It is higher on all sides except one than the surrounding country, towards which it sends forth streams of water to the north, east, and south, receiving a supply from the west only. In its centre it rises 12 feet above the flat region which bounds it. The soil, to the depth of 15 feet, is formed of vegetable matter without any admixture of earthy particles, and offers an exception to a general rule before alluded to, namely, that such peaty accumulations scarcely ever occur so far south as lat. 36°, or in any region where the summer heat is so great as in Virginia. In dig-

* Gilpin, Observ. on Picturesque Beauty, &c., 1772.
† Travels, &c., in 1841, 1842, vol. i. p. 143.

ging canals through the morass for the purpose of obtaining timber, much of the black soil has been thrown out from time to time, and exposed to the sun and air, in which case it soon rots away so that nothing remains behind, showing clearly that it owes its preservation to the shade afforded by a luxuriant vegetation and to the constant evaporation of the spongy soil by which the air is cooled during the hot months. The surface of the bog is carpeted with mosses, and densely covered with ferns and reeds, above which many evergreen shrubs and trees flourish, especially the White Cedar (*Cupressus thyoides*), which stands firmly supported by its long tap roots in the softest parts of the quagmire. Over the whole the deciduous cypress (*Taxodium distichum*) is seen to tower with its spreading top, in full leaf in the season when the sun's rays are hottest, and when, if not intercepted by a screen of foliage, they might soon cause the fallen leaves and dead plants of the preceding autumn to decompose, instead of adding their contributions to the peaty mass. On the surface of the wide morass lie innumerable trunks of large and tall trees, while thousands of others, blown down by the winds, are buried at various depths in the black mire below. They remind the geologist of the prostrate position of large stems of Sigillaria and Lepidodendron, converted into coal in ancient carboniferous rocks.

Bones of herbivorous quadrupeds in peat.—The antlers of large and full-grown stags are amongst the most common and conspicuous remains of animals in peat. They are not horns which have been shed; for portions of the skull are found attached, proving that the whole animal perished. Bones of the ox, hog, horse, sheep, and other herbivorous animals, also occur. M. Morren has discovered in the peat of Flanders the bones of otters and beavers*; but no remains have been met with belonging to those extinct quadrupeds, of which the living congeners inhabit warmer latitudes, such as the elephant, rhinoceros, hippopotamus, hyæna, and tiger, though these are so common in superficial deposits of silt, mud, sand, or stalactite, in various districts throughout Great Britain. Their absence seems to imply that they had ceased to live before the atmosphere of this part of the world acquired that cold and humid character which favors the growth of peat.

Remains of ships, &c., in peat mosses.—From the facts before mentioned, that mosses occasionally burst, and descend in a fluid state to lower levels, it will readily be seen that lakes and arms of the sea may occasionally become the receptacles of drift peat. Of this, accordingly, there are numerous examples; and hence the alternations of clay and sand with different deposits of peat so frequent on some coasts, as on those of the Baltic and German Ocean. We are informed by Deguer, that remains of ships, nautical instruments, and oars, have been found in many of the Dutch mosses; and Gerard, in his History of the Valley of the Somme, mentions that in the lowest tier of that moss was found a boat loaded with bricks, proving that these mosses were at one period

* Bulletin de la Soc. Géol. de France, tom. ii. p. 26.

navigable lakes and arms of the sea, as were also many mosses on the coast of Picardy, Zealand, and Friesland, from which soda and salt are procured.* The canoes, stone hatchets, and stone arrow-heads found in peat in different parts of Great Britain, lead to similar conclusions.

Imbedding of human and other remains, and works of Art, in Blown Sand.

The drifting of sand may next be considered among the causes capable of preserving organic remains and works of art on the emerged land.

African Sands.—The sands of the African deserts have been driven by the west winds over part of the arable land of Egypt, on the western bank of the Nile, in those places where valleys open into the plain, or where there are gorges through the Libyan mountains. By similar sand-drifts the ruins of ancient cities have been buried between the temple of Jupiter Ammon and Nubia. M. G. A. De Luc attempted to infer the recent origin of our continents, from the fact that these moving sands have arrived only in modern times at the fertile plains of the Nile. The same scourge, he said, would have afflicted Egypt for ages anterior to the times of history, had the continents risen above the level of the sea several hundred centuries before our era.† But the author proceeded in this, as in all his other chronological computations, on a multitude of gratuitous assumptions. He ought, in the first place, to have demonstrated that the whole continent of Africa was raised above the level of the sea at one period; for unless this point was established, the region from whence the sands began to move might have been the last addition made to Africa, and the commencement of the sand-flood might have been long posterior to the laying dry of the greater portion of that continent. That the different parts of Europe were not all elevated at one time is now generally admitted. De Luc should also have pointed out the depth of drift sand in various parts of the great Libyan deserts, and have shown whether any valleys of large dimensions had been filled up—how long these may have arrested the progress of the sands, and how far the flood had upon the whole advanced since the times of history.

We have seen that Sir J. G. Wilkinson is of opinion that, while the sand-drift is making aggressions at certain points upon the fertile soil of Egypt, the alluvial deposit of the Nile is advancing very generally upon the desert; and that, upon the whole, the balance is greatly in favor of the fertilizing mud.‡

No mode of interment can be conceived more favorable to the conservation of monuments for indefinite periods than that now so common in the region immediately westward of the Nile. The sand which surrounded and filled the great temple of Ipsambul, first discovered by

* Dr. Rennie, Essays on Peat Moss, p. 205.
† M. G. A. De Luc, Mercure de France, Sept. 1809.
‡ See p. 262.

Burckhardt, and afterwards partially uncovered by Belzoni and Beechey, was so fine as to resemble a fluid when put in motion. Neither the features of the colossal figures, nor the color of the stucco with which some were covered, nor the paintings on the walls, had received any injury from being enveloped for ages in this dry impalpable dust.*

At some future period, perhaps when the pyramids shall have perished, the action of the sea, or an earthquake, may lay open to the day some of these buried temples. Or we may suppose the desert to remain undisturbed, and changes in the surrounding sea and land to modify the climate and the direction of the prevailing winds, so that these may then waft away the Libyan sands as gradually as they once brought them to those regions. Thus, many a town and temple of higher antiquity than Thebes or Memphis may reappear in their original antiquity, and a part of the gloom which overhangs the history of the earlier nations be dispelled.

Whole caravans are said to have been overwhelmed by the Libyan sands; and Burckhardt informs us that "after passing the Akaba near the head of the Red Sea, the bones of dead camels are the only guides of the pilgrim through the wastes of sand."—"We did not see," says Captain Lyon, speaking of a plain near the Soudah mountains, in Northern Africa, "the least appearance of vegetation; but observed many skeletons of animals, which had died of fatigue on the desert, and occasionally the grave of some human being. All these bodies were so dried by the heat of the sun, that putrefaction appears not to have taken place after death. In recently expired animals I could not perceive the slightest offensive smell; and in those long dead, the skin with the hair on it remained unbroken and perfect, although so brittle as to break with a slight blow. The sand-winds never cause these carcases to change their places; for, in a short time, a slight mound is formed round them, and they become stationary."†

Towns overwhelmed by sand floods.—The burying of several towns and villages in England, France, and Jutland, by blown sand, is on record; thus, for example, near St. Pol de Leon, in Brittany, a whole village was completely buried beneath drift sand, so that nothing was seen but the spire of the church.‡ In Jutland marine shells adhering to sea-weed are sometimes blown by the violence of the wind to the height of 100 feet, and buried in similar hills of sand.

In Suffolk, in the year 1688, part of Downham was overwhelmed by sands which had broken loose about 100 years before, from a warren five miles to the south-west. This sand had, in the course of a century, travelled five miles, and covered more than 1000 acres of land.§ A considerable tract of cultivated land on the north coast of Cornwall has been inundated by drift sand, forming hills several hundred feet above the

* Stratton, Ed. Phil. Journ., No. v. p. 62.

† Travels in North Africa in the Years 1818, 1819, and 1820, p. 83.

‡ Mém. de l'Acad des Sci. de Paris, 1772. See also the case of the buried church of Eccles, above, p. 306.

§ Phil. Trans., vol. ii. p. 722.

level of the sea, and composed of comminuted marine shells, in which some terrestrial shells are enclosed entire. By the shifting of these sands the ruins of ancient buildings have been discovered; and in some cases where wells have been bored to a great depth, distinct strata, separated by a vegetable crust, are visible. In some places, as at New Quay, large masses have become sufficiently indurated to be used for architectural purposes. The lapidification, which is still in progress, appears to be due to oxide of iron held in solution by the water which percolates the sand.*

Imbedding of Organic and other Remains in Volcanic Formations on the Land.

I have in some degree anticipated the subject of this section in former chapters, when speaking of the buried cities around Naples, and those on the flanks of Etna (pp. 385. 400.). From the facts referred to, it appeared that the preservation of human remains and works of art is frequently due to the descent of floods caused by the copious rains which accompany eruptions. These aqueous lavas, as they are called in Campania, flow with great rapidity, and in 1822 surprised and suffocated, as was stated, seven persons in the villages of St. Sebastian and Massa, on the flanks of Vesuvius.

In the tuffs, moreover, or solidified mud, deposited by these aqueous lavas, impressions of leaves and of trees have been observed. Some of those, formed after the eruption of Vesuvius in 1822, are now preserved in the museum at Naples.

Lava itself may become indirectly the means of preserving terrestrial remains, by overflowing beds of ashes, pumice, and ejected matter, which may have been showered down upon animals and plants, or upon human remains. Few substances are better non-conductors of heat than volcanic dust and scoriæ, so that a bed of such materials is rarely melted by a superimposed lava-current. After consolidation, the lava affords secure protection to the lighter and more removable mass below, in which the organic relics may be enveloped. The Herculanean tuffs containing the rolls of papyrus, of which the characters are still legible, have, as was before remarked, been for ages covered by lava.

Another mode by which lava may tend to the conservation of imbedded remains, at least of works of human art, is by its overflowing them when it is not intensely heated, in which case they sometimes suffer little or no injury.

Thus when the Etnean lava-current of 1669 covered fourteen towns and villages, and part of the city of Catania, it did not melt down a great number of statues and other articles in the vaults of Catania; and at the depth of thirty-five feet in the same current, on the site of Mompiliere, one of the buried towns, the bell of a church and some statues were found uninjured (p. 401.).

* Boase on Submersion of Part of the Mount's Bay, &c., Trans. Roy. Geol. Soc. of Cornwall, vol. ii. p. 140.

We read of several buried cities in Central India, and among others of Oujein (or Oojain) which about fifty years before the Christian era was the seat of empire, of art, and of learning; but which in the time of the Rajah Vicramaditya, was overwhelmed, according to tradition, together with more than eighty other large towns in the provinces of Malwa and Bagur, "by a shower of earth." The city which now bears the name is situated a mile to the southward of the ancient town. On digging on the spot where the latter is supposed to have stood, to the depth of fifteen or eighteen feet, there are frequently discovered, says Mr. Hunter, entire brick walls, pillars of stone, and pieces of wood of an extraordinary hardness, besides utensils of various kinds, ancient coins, and occasionally buried wheat in a state resembling charcoal.*

The soil which covers Oujein is described as "being of an ash-gray color, with minute specks of black sand."† And the "shower of earth," said to have "fallen from heaven," has been attributed by some travellers to volcanic agency. There are, however, no active volcanoes in Central India, the nearest to Oujein being Denodur hill near Bhooj, the capital of Cutch, 300 geographical miles distant, if indeed that hill has ever poured out lava in historical times, which is doubted by many.‡ The latest writers on Oujein avow their suspicion that the supposed "catastrophe" was nothing more than the political decline and final abandonment of a great city which, like Nineveh or Babylon, and many an ancient seat of empire in the East, after losing its importance as a metropolis, became a heap of ruins. The rapidity with which the sun-dried bricks, of which even the most splendid oriental palaces are often constructed, crumble down when exposed to rain and sun, and are converted into mounds of ordinary earth and clay, is well known. According to Captain Dangerfield, trap tuff and columnar basalt constitute the rocks in the environs of Oujein§, and the volcanic nature of these formations, from which the materials of the bricks were originally derived, may have led to the idea of the city having been overwhelmed by a volcanic eruption.

* Narrative of Journey from Agra to Oujein, Asiatic Researches, vol. vi. p. 36.
† Asiatic Journal, vol. ix. p. 35. ‡ See above, p. 460.
§ Sir J. Malcolm's Central India. Appendix, No. 2. p. 324.

CHAPTER XLVI.

BURYING OF FOSSILS IN ALLUVIAL DEPOSITS AND IN CAVES.

Fossils in alluvium—Effects of sudden inundations—Terrestrial animals most abundantly preserved in alluvium where earthquakes prevail—Marine alluvium—Buried town—Effects of Landslips—Organic remains in fissures and caves—Form and dimensions of caverns—their probable origin—Closed basins and subterranean rivers of the Morea—Katavothra—Formation of breccias with red cement—Human remains imbedded in Morea—Intermixture, in caves of South of France and elsewhere, of human remains and bones of extinct quadrupeds, no proof of former co-existence of man with those lost species.

Fossils in alluvium.—THE next subject for our consideration, according to the division before proposed, is the embedding of organic bodies in alluvium.

The gravel, sand, and mud in the bed of a river does not often contain any animal or vegetable remains; for the whole mass is so continually shifting its place, and the attrition of the various parts is so great, that even the hardest rocks contained in it are, at length, ground down to powder. But when sand and sediment are suddenly swept by a flood, and then let fall upon the land, such an alluvium may envelop trees or the remains of animals, which, in this manner, are often permanently preserved. In the mud and sand produced by the floods in Scotland, in 1829, the dead and mutilated bodies of hares, rabbits, moles, mice, partridges, and even the bodies of men, were found partially buried.* But in these and similar cases one flood usually effaces the memorials left by another, and there is rarely a sufficient depth of undisturbed transported matter, in any one spot, to preserve the organic remains for ages from destruction.

Where earthquakes prevail, and the levels of a country are changed from time to time, the remains of animals may more easily be inhumed and protected from disintegration. Portions of plains, loaded with alluvial accumulations by transient floods, may be gradually upraised; and, if any organic remains have been imbedded in the transported materials, they may, after such elevation, be placed beyond the reach of the erosive power of streams. In districts where the drainage is repeatedly deranged by subterranean movements, every fissure, every hollow caused by the sinking in of land, becomes a depository of organic and inorganic substances, hurried along by transient floods.

Marine alluvium.—In May, 1787, a dreadful inundation of the sea was caused at Coringa, Ingeram, and other places, on the coast of Coromandel, in the East Indies, by a hurricane blowing from the N. E., which raised the waters so that they rolled inland to the distance of about twenty miles from the shore, swept away many villages, drowned more

* Sir T. D. Lauder, Bart., on Floods in Morayshire, Aug. 1839, p. 177.

than 10,000 people, and left the country covered with marine mud, on which the carcasses of about 100,000 head of cattle were strewed. An old tradition of the natives of a similar flood, said to have happened about a century before, was, till this event, regarded as fabulous by the European settlers.* The same coast of Coromandel was, so late as May, 1832, the scene of another catastrophe of the same kind; and when the inundation subsided, several vessels were seen grounded in the fields of the low country about Coringa.

Many of the storms termed hurricanes have evidently been connected with submarine earthquakes, as is shown by the atmospheric phenomena attendant on them, and by the sounds heard in the ground and the odors emitted. Such were the circumstances which accompanied the swell of the sea in Jamaica, in 1780, when a great wave desolated the western coast, and bursting upon Savanna la Mar, swept away the whole town in an instant, so that not a vestige of man, beast, or habitation, was seen upon the surface.†

Houses and works of art in alluvial deposits.—A very ancient subterranean town, apparently of Hindoo origin, was discovered in India in 1833, in digging the Doab canal. Its site is north of Saharunpore, near the town of Behat, and seventeen feet below the present surface of the country. More than 170 coins of silver and copper have already been found, and many articles in metal and earthenware. The overlying deposit consisted of about five feet of river sand, with a substratum about twelve feet thick of red alluvial clay. In the neighborhood are several rivers and torrents, which descend from the mountains charged with vast quantities of mud, sand, and shingle; and within the memory of persons now living the modern Behat has been threatened by an inundation, which, after retreating, left the neighboring country strewed over with a superficial covering of sand several feet thick. In sinking wells in the environs, masses of shingle and boulders have been reached resembling those now in the river-channels of the same district, under a deposit of thirty feet of reddish loam. Captain Cautley, therefore, who directed the excavations, supposes that the matter discharged by torrents has gradually raised the whole country skirting the base of the lower hills; and that the ancient town, having been originally built in a hollow, was submerged by floods, and covered over with sediment seventeen feet in thickness.‡

We are informed, by M. Boblaye, that in the Morea, the formation termed céramique, consisting of pottery, tiles, and bricks, intermixed with various works of art, enters so largely into the alluvium and vegetable soil upon the plains of Greece, and into hard and crystalline breccias which have been formed at the foot of declivities, that it constitutes an important stratum which might, in the absence of zoological characters, serve to mark our epoch in a most indestructible manner.§

* Dodsley's Ann. Regist, 1788.
† Edwards, Hist. of West Indies, vol. i. p. 235, ed. 1801
‡ Journ. of Asiat. Soc., Nos. xxv. and xxix., 1834.
§ Ann. des Sci. Nat. tom. xxii. p. 117, Feb. 1831.

Landslips.—The landslip, by suddenly precipitating large masses of rock and soil into a valley, overwhelms a multitude of animals, and sometimes buries permanently whole villages, with their inhabitants and large herds of cattle. Thus three villages, with their entire population, were covered, when the mountain of Piz fell in 1772, in the district of Treviso, in the state of Venice,* and part of Mount Grenier, south of Chambery, in Savoy, which fell down in the year 1248, buried five parishes, including the town and church of St. André, the ruins occupying an extent of about nine square miles.†

The number of lives lost by the slide of the Rossberg, in Switzerland, in 1806, was estimated at more than 800, a great number of the bodies, as well as several villages and scattered houses, being buried deep under mud and rock. In the same country, several hundred cottages, with eighteen of their inhabitants and a great number of cows, goats, and sheep, were victims to the sudden fall of a bed of stones, thirty yards deep, which descended from the summits of the Diablerets in Vallais. In the year 1618, a portion of Mount Conto fell, in the county of Chiavenna, in Switzerland, and buried the town of Pleurs with all its inhabitants, to the number of 2430.

It is unnecessary to multiply examples of similar local catastrophes, which however numerous they may have been in mountainous parts of Europe, within the historical period, have been, nevertheless, of rare occurrence when compared to events of the same kind which have taken place in regions convulsed by earthquakes. It is then that enormous masses of rock and earth, even in comparatively low and level countries, are detached from the sides of valleys, and cast down into the river courses, and often so unexpectedly that they overwhelm, even in the daytime, every living thing upon the plains.

Preservation of Organic Remains in Fissures and Caves.

In the history of earthquakes it was shown that many hundreds of new fissures and chasms had opened in certain regions during the last 150 years, some of which are described as being of unfathomable depth. We also perceive that mountain masses have been violently fractured and dislocated, during their rise above the level of the sea; and thus we may account for the existence of many cavities in the interior of the earth by the simple agency of earthquakes; but there are some caverns, especially in limestone rocks, which, although usually, if not always, connected with rents, are nevertheless of such forms, and dimensions, alternately expanding into spacious chambers, and then contracting again into narrow passages, that it is difficult to conceive that they can owe their origin to the mere fracturing and displacement of solid masses.

In the limestone of Kentucky, in the basin of Green River, one of the

* Malte-Brun's Geog., vol. i. p. 435.
† Bakewell, Travels in the Tarentaise, vol. i. p. 201.

tributaries of the Ohio, a line of underground cavities has been traced in one direction for a distance of ten miles, without any termination; and one of the chambers, of which there are many, all connected by narrow tunnels, is no less than ten acres in area and 150 feet in its greatest height. Besides the principal series of "antres vast," there are a great many lateral embranchments not yet explored.*

The cavernous structure here alluded to is not altogether confined to calcareous rocks; for it has lately been observed in micaceous and argillaceous schist in the Grecian island of Thermia (Cythnos of the ancients), one of the Cyclades. Here also spacious halls, with rounded and irregular walls, are connected together by narrow passages or tunnels, and there are many lateral branches which have no outlet. A current of water has evidently at some period flowed through the whole, and left a muddy deposit of bluish clay upon the floor; but the erosive action of the stream cannot be supposed to have given rise to the excavations in the first instance. M. Virlet suggests that fissures were first caused by earthquakes, and that these fissures became the chimneys or vents for the disengagement of gas, generated below by volcanic heat. Gases, he observes, such as the muriatic, sulphuric, fluoric, and others, might, if raised to a high temperature, alter and decompose the rocks which they traverse. There are signs of the former action of such vapors in rents of the micaceous schist of Thermia, and thermal springs now issue from the grottoes of that island. We may suppose that afterwards the elements of the decomposed rocks were gradually removed in a state of solution by mineral waters; a theory which, according to M. Virlet, is confirmed by the effect of heated gases which escape from rents in the isthmus of Corinth, and which have greatly altered and corroded the hard siliceous and jaspideous rocks.†

When we reflect on the quantity of carbonate of lime annually poured out by mineral waters, we are prepared to admit that large cavities must, in the course of ages, be formed at considerable depths below the surface in calcareous rocks.‡ These rocks, it will be remembered, are at once more soluble, more permeable, and more fragile, than any others, at least all the compact varieties are very easily broken by the movements of earthquakes, which would produce only flexures in argillaceous strata. Fissures once formed in limestone are not liable, as in many other formations, to become closed up by impervious clayey matter, and hence a stream of acidulous water might for ages obtain a free and unobstructed passage.§

Morea.—Nothing is more common in limestone districts than the engulfment of rivers, which after holding a subterranean course for many miles escape again by some new outlet. As they are usually charged

* Nahum Ward, Trans. of Antiq. Soc. of Massachusetts. Holmes's United States, p. 438.

† Bull. de la Soc. Géol. de France, tom. ii. p. 329.

‡ See above, p. 240.

§ See remarks by M. Boblaye, Ann. des Mines, 3me série, tom. iv with fine sediment, and often with sand and pebbles where they enter, whereas they are usually pure and limpid where they flow out again, they must deposit much matter in empty spaces in the interior of the earth. In addition to the materials thus introduced, stalagmite, or carbonate of lime, drops from the roofs of caverns, and in this mixture the bones of animals washed in by rivers are often entombed. In this manner we may account for those bony breccias which we often find in caves, some of which are of high antiquity while others are very recent and in daily progress. In no district are engulfed streams more conspicuous than in the Morea, where the phenomena attending them have been lately studied and described in great detail by M. Boblaye and his fellow-laborers of the French expedition to Greece.* Their account is peculiarly interesting to geologists, because it throws light on the red osseous breccias containing the bones of extinct quadrupeds which are so common in almost all the countries bordering the Mediterranean. It appears that the numerous caverns of the Morea occur in a compact limestone, of the age of the English chalk, immediately below which are arenaceous strata referred to the period of our greensand. In the more elevated districts of that peninsula there are many deep land-locked valleys, or basins, closed round on all sides by mountains of fissured and cavernous limestone. The year is divided almost as distinctly as between the tropics into a rainy season, which lasts upwards of four months, and a season of drought of nearly eight months' duration. When the torrents are swollen by the rains, they rush from surrounding heights into the inclosed basins; but, instead of giving rise to lakes, as would be the case in most other countries, they are received into gulfs or chasms, called by the Greeks "Katavothra," and which correspond to what are termed "swallow-holes" in the north of England. The water of these torrents is charged with pebbles and red ochreous earth, resembling precisely the well-known cement of the osseous breccias of the Mediterranean. It dissolves in acids with effervescence, and leaves a residue of hydrated oxide of iron, granular iron, impalpable grains of silex, and small crystals of quartz. Soil of the same description abounds everywhere on the surface of the decomposing limestone in Greece, that rock containing in it much siliceous and ferruginous matter.

Many of the Katavothra being insufficient to give passage to all the water in the rainy season, a temporary lake is formed round the mouth of the chasm, which then becomes still farther obstructed by pebbles, sand, and red mud, thrown down from the turbid waters. The lake being thus raised, its waters generally escape through other openings, at higher levels, around the borders of the plain, constituting the bottom of the closed basin.

In some places, as at Kavaros and Tripolitza, where the principal discharge is by a gulf in the middle of the plain, nothing can be seen over the opening in summer, when the lake dries up, but a deposit of red

* Ann. des Mines, 3me série, tom. iv., 1833.

mud, cracked in all directions. But the Katavothron is more commonly situated at the foot of the surrounding escarpment of limestone; and in that case there is sometimes room enough to allow a person to enter, in summer, and even to penetrate far into the interior. Within is seen a suite of chambers, communicating with each other by narrow passages; and M. Virlet relates, that in one instance he observed, near the entrance, human bones imbedded in recent red mud, mingled with the remains of plants and animals of species now inhabiting the Morea. It is not wonderful, he says, that the bones of man should be met with in such receptacles; for so murderous have been the late wars in Greece, that skeletons are often seen lying exposed on the surface of the country.*

In summer, when no water is flowing into the Katavothron, its mouth, half closed up with red mud, is masked by a vigorous vegetation, which is cherished by the moisture of the place. It is then the favorite hiding-place and den of foxes and jackals; so that the same cavity serves at one season of the year for the habitation of carnivorous beasts, and at another as the channel of an engulfed river. Near the mouth of one chasm, M. Boblaye and his companions saw the carcass of a horse, in part devoured, the size of which seemed to have prevented the jackals from dragging it in: the marks of their teeth were observed on the bones, and it was evident that the floods of the ensuing winter would wash in whatsoever might remain of the skeleton.

It has been stated that the waters of all these torrents of the Morea are turbid where they are engulfed; but when they come out again, often at the distance of many leagues, they are perfectly clear and limpid, being only charged occasionally with a slight quantity of calcareous sand. The points of efflux are usually near the sea-shores of the Morea, but sometimes they are submarine; and when this is the case, the sands are seen to boil up for a considerable space, and the surface of the sea, in calm weather, swells in large convex waves. It is curious to reflect, that when this discharge fails in seasons of drought, the pressure of the sea may force its salt waters into subterraneous caverns, and carry in marine sand and shells, to be mingled with ossiferous mud, and the remains of terrestrial animals.

In general, however, the efflux of water at these inferior openings is surprisingly uniform. It seems, therefore, that the large caverns in the interior must serve as reservoirs, and that the water escapes gradually from them, in consequence of the smallness of the rents and passages by which they communicate with the surface.

The phenomena above described are not confined to the Morea, but occur in Greece generally, and in those parts of Italy, Spain, Asia Minor, and Syria, where the formations of the Morea extend. The Copaic lake in Bœotia has no outlet, except by underground channels; and hence we can explain those traditional and historical accounts of its having gained on the surrounding plains and overflowed towns, as such floods

* Bull. de la Soc. Géol. de France, tom. iii. p. 223.

must have happened whenever the outlet was partially choked up by mud, gravel, or the subsidence of rocks, caused by earthquakes. When speaking of the numerous fissures in the limestone of Greece, M. Boblaye reminds us of the famous earthquake of 469 B.C., when, as we learn from Cicero, Plutarch, Strabo, and Pliny, Sparta was laid in ruins, part of the summit of Mount Taygetus torn off, and numerous gulfs and fissures caused in the rocks of Laconia.

During the great earthquake of 1693, in Sicily, several thousand people were at once entombed in the ruins of caverns in limestone, at Sortino Vecchio; and, at the same time, a large stream, which had issued for ages from one of the grottoes below that town, changed suddenly its subterranean course, and came out from the mouth of a cave lower down the valley, where no water had previously flowed. To this new point the ancient water-mills were transferred, as I learnt when I visited the spot in 1829.

When the courses of engulfed rivers are thus liable to change, from time to time, by alterations in the levels of a country, and by the rending and shattering of mountain masses, we must suppose that the dens of wild beasts will sometimes be inundated by subterranean floods, and their carcasses buried under heaps of alluvium. The bones, moreover, of individuals which have died in the recesses of caves, or of animals which have been carried in for prey, may be drifted along, and mixed up with mud, sand, and fragments of rocks, so as to form osseous breccias.

In 1833 I had an opportunity of examining the celebrated caves of Franconia, and among others that of Rabenstein, newly discovered. Their general form, and the nature and arrangement of their contents, appeared to me to agree perfectly with the notion of their having once served as the channels of subterranean rivers. This mode of accounting for the introduction of transported matter into the Franconian and other caves, filled up as they often are even to their roofs with osseous breccia, was long ago proposed by M. C. Prevost,* and seems at length to be very generally adopted. But I do not doubt that bears inhabited some of the German caves, or that the cavern of Kirkdale, in Yorkshire, was once the den of hyænas. The abundance of bony dung, associated with hyænas' bones, has been pointed out by Dr. Buckland, and with reason, as confirmatory of this opinion.

The same author observed in every cave examined by him in Germany, that deposits of mud and sand, with or without rolled pebbles and angular fragments of rock, were covered over with a *single* crust of stalagmite.† In the English caves he remarked a similar absence of *alterations* of alluvium and stalagmite. But Dr. Schmerling has discovered in a cavern at Chockier, about two leagues from Liège, three distinct beds of stalagmite, and between each of them a mass of breccia, and mud mixed with quartz pebbles, and in the three deposits the bones of extinct quadrupeds.‡

* Mém. de la Soc. d'Hist. Nat. de Paris, tom. iv.
† Reliquiæ Diluvianæ, p. 108.
‡ Journ. de Géol., tom. i. p. 286. July, 1830.

This exception does not invalidate the generality of the phenomenon pointed out by Dr. Buckland, one cause of which may perhaps be this, that if several floods pass at different intervals of time through a subterranean passage, the last, if it has power to drift along fragments of rock, will also tear up any alternating stalagmitic and alluvial beds that may have been previously formed. Another cause may be, that a particular line of caverns will rarely be so situated, in relation to the lowest levels of a country, as to become, at two distinct epochs, the receptacle of engulfed rivers; and if this should happen, some of the caves, or at least the tunnels of communication, may at the first period be entirely choked up with transported matter, so as not to allow the subsequent passage of water in the same direction.

As the same chasms may remain open throughout periods of indefinite duration, the species inhabiting a country may in the meantime be greatly changed, and thus the remains of animals belonging to very different epochs may become mingled together in a common tomb. For this reason it is often difficult to separate the monuments of the human epoch from those relating to periods long antecedent, and it was not without great care and skill that Dr. Buckland was enabled to guard against such anachronisms in his investigations of several of the English caves. He mentions that human skeletons were found in the cave of Wokey Hole, near Wells, in the Mendips, dispersed through reddish mud and clay, and some of them united by stalagmite into a firm osseous breccia. "The spot on which they lie is within reach of the highest floods of the adjacent river, and the mud in which they are buried is evidently fluviatile." *

In speaking of the cave of Paviland on the coast of Glamorganshire the same author states that the entire mass through which bones were dispersed appeared to have been disturbed by ancient diggings, so that the remains of extinct animals had become mixed with recent bones and shells. In the same cave was a human skeleton, and the remains of recent testacea of eatable species, which may have been carried in by man.

In several caverns on the banks of the Meuse, near Liège, Dr. Schmerling has found human bones in the same mud and breccia with those of the elephant, rhinoceros, bear, and other quadrupeds of extinct species. He has observed none of the dung of any of these animals: and from this circumstance, and the appearance of the mud and pebbles, he concludes that these caverns were never inhabited by wild beasts, but washed in by a current of water. As the human skulls and bones were in fragments, and no entire skeleton had been found, he does not believe that these caves were places of sepulture, but that the human remains were washed in at the same time as the bones of extinct quadrupeds, and that these lost species of mammalia co-existed on the earth with man.

Caverns in the south of France.—Similar associations in the south of France, of human bones and works of art, with remains of extinct quadrupeds, have induced other geologists to maintain that man was an

* Reliquiæ Diluvianæ, p. 165.

inhabitant of that part of Europe before the rhinoceros, hyæna, tiger, and many fossil species disappeared. I may first mention the cavern of Bize, in the department of Aude, where M. Marcel de Serres met with a small number of human bones mixed with those of extinct animals and with land shells. They occur in a calcareous stony mass, bound together by a cement of stalagmite. On examining the same caverns, M. Tournal found not only in these calcareous beds, but also in a black mud which overlies a red osseous mud, several human teeth, together with broken angular fragments of a rude kind of pottery, and also recent marine and terrestrial shells. The teeth preserve their enamel; but the fangs are so much altered as to adhere strongly when applied to the tongue. Of the terrestrial shells thus associated with the bones and pottery, the most common are *Cyclostoma elegans*, *Bulimus decollatus*, *Helix nemoralis*, and *H. nitida*. Among the marine are found *Pecten jacobæus*, *Mytilus edulis*, and *Natica mille-punctata*, all of them eatable kinds, and which may have been brought there for food. Bones were found in the same mass belonging to three new species of deer, the brown bear (*Ursus arctoïdeus*), and the wild bull (*Bos urus*), formerly a native of Germany.*

In the same parts of France, M. de Christol has found in caverns in a tertiary limestone at Poudres and Souvignargues, two leagues north of Lunel-viel, in the department of Herault, human bones and pottery confusedly mixed with remains of the rhinoceros, bear, hyæna, and other terrestrial mammifers. They were imbedded in alluvial mud, of the solidity of calcareous tufa, and containing some flint pebbles and fragments of the limestone of the country. Beneath this mixed accumulation, which sometimes attained a thickness of thirteen feet, is the original floor of the cavern, about a foot thick, covered with bones and the dung of animals (*album græcum*), in a sandy and tufaceous cement.

The human bones in these caverns of Pondres and Souvignargues were found, upon a careful analysis, to have parted with their animal matter to as great a degree as those of the hyæna which accompany them, and are equally brittle, and adhere as strongly to the tongue.

In order to compare the degree of alteration of these bones with those known to be of high antiquity, M. Marcel de Serres and M. Ballard, chemists of Montpelier, procured some from a Gaulish sarcophagus, in the plain of Lunel, supposed to have been buried for fourteen or fifteen centuries at least. In these the cellular tissue was empty, but they were more solid than fresh bones. They did not adhere to the tongue in the same manner as those of the caverns of Bize and Pondres, yet they had lost at least three fourths of their original animal matter.

The superior solidity of the Gaulish bones to those in a fresh skeleton is a fact in perfect accordance with the observations made by Dr. Mantell on bones taken from a Saxon tumulus near Lewes.

M. Tessier has also described a cavern near Mialet, in the department

* M. Marcel de Serres, Géognosie des Terrains Tertiaires, p. 64. Introduction.

of Gard, where the remains of the bear and other animals were mingled confusedly with human bones, coarse pottery, teeth pierced for amulets, pointed fragments of bone, bracelets of bronze, and a Roman urn. Part of this deposit reached to the roof of the cavity, and adhered firmly to it. The author suggests that the exterior portion of the grotto may at one period have been a den of bears, and that afterwards the aboriginal inhabitants of the country took possession of it either for a dwelling or a burial-place, and left there the coarse pottery, amulets, and pointed pieces of bone. At a third period the Romans may have used the cavern as a place of sepulture or concealment, and to them may have belonged the urn and bracelets of metal. If we then suppose the course of the neighboring river to be impeded by some temporary cause, a flood would be occasioned, which, rushing into the open grotto, may have washed all the remains into the interior caves and tunnels, heaping the whole confusedly together.*

In the controversy which has arisen on this subject, MM. Marcel de Serres, De Christol, Tournal, and others, have contended, that the phenomena of this and other caverns in the south of France prove that the fossil rhinoceros, hyæna, bear, and several other lost species, were once contemporaneous inhabitants of the country, together with man; while M. Desnoyers has supported the opposite opinion. The flint hatchets and arrow-heads, he says, and the pointed bones and coarse pottery of many French and English caves, agree precisely in character with those found in the tumuli, and under the dolmens (rude altars of unhewn stone) of the primitive inhabitants of Gaul, Britain, and Germany. The human bones, therefore, in the caves which are associated with such fabricated objects, must belong not to antediluvian periods, but to a people in the same stage of civilization as those who constructed the tumuli and altars.

In the Gaulish monuments we find, together with the objects of industry above mentioned, the bones of wild and domestic animals of species now inhabiting Europe, particularly of deer, sheep, wild-boars, dogs, horses, and oxen. This fact has been ascertained in Quercy, and other provinces; and it is supposed by antiquaries that the animals in question were placed beneath the Celtic altars in memory of sacrifices offered to the Gaulish divinity Hesus, and in the tombs to commemorate funeral repasts, and also from a supposition prevalent among savage nations, which induces them to lay up provisions for the manes of the dead in a future life. But in none of these ancient monuments have any bones been found of the elephant, rhinoceros, hyæna, tiger, and other quadrupeds, such as are found in caves, as might certainly have been expected had these species continued to flourish at the time that this part of Gaul was inhabited by man.†

We are also reminded by M. Desnoyers of a passage in Florus, in which it is related that Cæsar ordered the caves into which the Aquitanian

* Bull. de la Soc. Géol. de France, tom. ii. pp. 56—63.

† Desnoyers, Bull. de la Soc. Géol. de France, tom. ii. p. 252.

Gauls had retreated to be closed up.* It is also on record, that so late as the eighth century, the Aquitanians defended themselves in caverns against King Pepin. As many of these caverns, therefore, may have served in succession as temples and habitations, as places of sepulture, concealment, or defence, it is easy to conceive that human bones, and those of animals, in osseous breccias of much older date, may have been swept away together, by inundations, and then buried in one promiscuous heap.

It is not on the evidence of such intermixtures that we ought readily to admit either the high antiquity of the human race, or the recent date of certain lost species of quadrupeds.

Among the various modes in which the bones of animals become preserved, independently of the agency of land floods and engulfed rivers, I may mention that open fissures often serve as natural pitfalls in which herbivorous animals perish. This may happen the more readily when they are chased by beasts of prey, or when surprised while carelessly browsing on the shrubs which so often overgrow and conceal the edges of fissures.†

During the excavations recently made near Behat in India, the bones of two deer were found at the bottom of an ancient well which had been filled up with alluvial loam. Their horns were broken to pieces, but the jaw bones and other parts of the skeleton remained tolerably perfect. "Their presence," says Captain Cautley, "is easily accounted for, as a great number of these and other animals are constantly lost in galloping over the jungles and among the high grass by falling into deserted wells."‡

Above the village of Selside, near Ingleborough in Yorkshire, a chasm of enormous but unknown depth occurs in the scar-limestone, a member of the carboniferous series. "The chasm," says Professor Sedgwick, "is surrounded by grassy shelving banks, and many animals, tempted towards its brink, have fallen down and perished in it. The approach of cattle is now prevented by a strong lofty wall; but there can be no doubt that, during the last two or three thousand years, great masses of bony breccia must have accumulated in the lower parts of the great fissure, which probably descends through the whole thickness of the scar-limestone, to the depth of perhaps five or six hundred feet."§

When any of these natural pit-falls happen to communicate with lines of subterranean caverns, the bones, earth, and breccia, may sink by their own weight, or be washed into the vaults below.

At the north extremity of the rock of Gibraltar are perpendicular fissures, on the ledges of which a number of hawks nestle and rear their young in the breeding season. They throw down from their nests the bones of small birds, mice, and other animals, on which they feed, and

* Hist. Rom. Epit., lib. iii. c. 10.
† Buckland, Reliquiæ Diluvianæ, p. 25.
‡ See above, pp. 730, 731.
§ On the Lake Mountains of North of England, Geol. Soc. Jan. 5, 1831.

these are gradually united into a breccia of angular fragments of the decomposing limestone with a cement of red earth.

At the pass of Escrinet in France, on the northern escarpment of the Coiron hills, near Aubenas, I have seen a breccia in the act of forming. Small pieces of disintegrating limestone are transported, during heavy rains, by a streamlet, to the foot of the declivity, where land shells are very abundant. The shells and pieces of stone soon become cemented together by stalagmite into a compact mass, and the talus thus formed is in one place fifty feet deep, and five hundred yards wide. So firmly is the lowest portion consolidated, that it is quarried for mill-stones.

Recent stalagmitic limestone of Cuba.—One of the most singular examples of the recent growth of stalagmitic limestone in caves and fissures is that described by Mr. R. C. Taylor, as observable on the north-east part of the island of Cuba.* The country there is composed of a white marble, in which are numerous cavities, partially filled with a calcareous deposit of a brick-red color. In this red deposit are shells, or often the hollow casts of shells, chiefly referable to eight or nine species of land snails, a few scattered bones of quadrupeds, and, what is still more singular, marine univalve shells, often at the height of many hundred, or even one thousand feet above the sea. The following explanation is given of the gradual increase of this deposit. Land snails of the genera Helix, Cyclostoma, Pupa, and Clausilia, retire into the caves, the floors of which are strewed with myriads of their dead and unoccupied shells, at the same time that water infiltered through the mountain throws down carbonate of lime, enveloping the shells, together with fragments of the white limestone which occasionally falls from the roof. Multitudes of bats resort to the caves; and their dung, which is of a bright red color, (probably derived from the berries on which they feed,) imparts its red hue to the mass. Sometimes also the Hutia, or great Indian rat of the island, dies and leaves its bones in the caves. "At certain seasons the soldier-crabs resort to the sea-shore, and then return from their pilgrimage, each carrying with them, or rather dragging, the shell of some marine univalve for many a weary mile. They may be traced even at the distance of eight or ten miles from the shore, on the summit of mountains 1200 feet high, like the pilgrims of the olden times, each bearing his shell to denote the character and extent of his wanderings." By this means several species of marine testacea of the genera Trochus, Turbo, Littorina, and Monodonta, are conveyed into inland caverns, and enter into the composition of the newly formed rock.

* Notes on Geol. of Cuba, 1836, Phil. Mag., July, 1837.

CHAPTER XLVII.

IMBEDDING OF ORGANIC REMAINS IN SUBAQUEOUS DEPOSITS.

Division of the subject—Imbedding of terrestrial animals and plants—Increased specific gravity of wood sunk to great depths in the sea—Drift-timber of the Mackenzie in Slave Lake and Polar Sea—Floating trees in the Mississippi—in the Gulf Stream—on the coast of Iceland, Spitzbergen, and Labrador—Submarine forests—Example on coast of Hampshire—Mineralization of plants—Imbedding of marine plants—of insects—of reptiles—Bones of birds why rare—Imbedding of terrestrial quadrupeds by river floods—Skeletons in recent shell marl—Imbedding of mammiferous remains in marine strata.

Division of the subject.—Having treated of the imbedding of organic remains in deposits formed upon the land, I shall next consider the including of the same in deposits formed under water.

It will be convenient to divide this branch of our subject into three parts; considering, first, the various modes whereby the relics of *terrestrial* species may be buried in subaqueous formations; secondly, the modes whereby animals and plants inhabiting *fresh water* may be so entombed; thirdly, how *marine* species may become preserved in new strata.

The phenomena above enumerated demand a fuller share of attention than those previously examined, since the deposits which originate upon dry land are insignificant in thickness, superficial extent, and durability, when contrasted with those of subaqueous origin. At the same time, the study of the latter is beset with greater difficulties; for we are here concerned with the results of processes much farther removed from the sphere of ordinary observation. There is, indeed, no circumstance which so seriously impedes the acquisition of just views in our science as an habitual disregard of the important fact, that the reproductive effects of the principal agents of change are confined to another element—to that larger portion of the globe, from which by our very organization we are almost entirely excluded.*

Imbedding of Terrestrial Plants.

When a tree falls into a river from the undermining of the banks or from being washed in by a torrent or flood, it floats on the surface, not because the woody portion is specifically lighter than water, but because it is full of pores containing air. When soaked for a considerable time, the water makes its way into these pores, and the wood becomes *water-logged* and sinks. The time required for this process varies in different woods; but several kinds may be drifted to great distances, sometimes across the ocean, before they lose their buoyancy.

* See above, p. 67.

Wood sunk to a great depth in the sea.—If wood be sunk to vast depths in the sea, it may be impregnated with water suddenly. Captain Scoresby informs us, in his Account of the Arctic Regions, that on one occasion a whale, on being harpooned, ran out all the lines in the boat, which it then dragged under water, to the depth of several thousand feet, the men having just time to escape to a piece of ice. When the fish returned to the surface "to blow," it was struck a second time, and soon afterwards killed. The moment it expired it began to sink,—an unusual circumstance, which was found to be caused by the weight of the sunken boat, which still remained attached to it. By means of harpoons and ropes the fish was prevented from sinking, until it was released from the weight by connecting a rope to the lines of the attached boat, which was no sooner done than the fish rose again to the surface. The sunken boat was then hauled up with great labor; for so heavy was it, that although before the accident it would have been buoyant when full of water, yet it now required a boat at each end to keep it from sinking. "When it was hoisted into the ship, the paint came off the wood in large sheets; and the planks, which were of wainscot, were as completely soaked in every pore as if they had lain at the bottom of the sea since the flood! A wooden apparatus that accompanied the boat in its progress through the deep, consisting chiefly of a piece of thick deal, about fifteen inches square, happened to fall overboard, and, though it originally consisted of the lightest fir, sank in the water like a stone. The boat was rendered useless; even the wood of which it was built, on being offered to the cook for fuel, was tried and rejected as incombustible."*

Captain Scoresby found that, by sinking pieces of fir, elm, ash, &c., to the depth of four thousand and sometimes six thousand feet, they became impregnated with sea-water, and when drawn up again, after immersion for an hour, would no longer float. The effect of this impregnation was to increase the dimensions as well as the specific gravity of the wood, every solid inch having increased one-twentieth in size and twenty-one twenty-fifths in weight.†

Drift-wood of the Mackenzie River.—When timber is drifted down by a river, it is often arrested by lakes; and, becoming water-logged, it may sink and be imbedded in lacustrine strata, if any be there forming; sometimes a portion floats on till it reaches the sea. In the course of the Mackenzie River we have an example of vast accumulations of vegetable matter now in progress under both these circumstances.

In Slave Lake in particular, which vies in dimensions with some of the great fresh-water seas of Canada, the quantity of drift-timber brought down annually is enormous. "As the trees," says Dr. Richardson, "retain their roots, which are often loaded with earth and stones, they readily sink, especially when water-soaked; and, accumulating in the eddies, form shoals, which ultimately augment into islands. A thicket of small willows covers the new-formed island as soon as it appears

* Account of the Arctic Regions, vol. ii. p. 193. † Ibid. p. 202.

above water, and their fibrous roots serve to bind the whole firmly together. Sections of these islands are annually made by the river, assisted by the frost; and it is interesting to study the diversity of appearances they present, according to their different ages. The trunks of the trees gradually decay until they are converted into a blackish brown substance resembling peat, but which still retains more or less of the fibrous structure of the wood; and layers of this often alternate with layers of clay and sand, the whole being penetrated, to the depth of four or five yards or more, by the long fibrous roots of the willows. A deposition of this kind, with the aid of a little infiltration of bituminous matter, would produce an excellent imitation of coal, with vegetable impressions of the willow-roots. What appeared most remarkable was the horizontal slaty structure that the old alluvial banks presented, or the *regular curve* that the strata assumed from unequal subsidence.

"It was in the rivers only that we could observe sections of these deposits; but the same operation goes on, on a much more magnificent scale, in the lakes. A shoal of many miles in extent is formed on the south side of Athabasca Lake, by the drift-timber and vegetable debris brought down by the Elk River; and the Slave Lake itself must in process of time be filled up by matters daily conveyed into it from Slave River. Vast quantities of drift-timber are buried under the sand at the mouth of the river, and enormous piles of it are accumulated on the shores of every part of the lake." *

The banks of the Mackenzie display almost everywhere horizontal beds of wood coal, alternating with bituminous clay, gravel, sand, and friable sandstone; sections, in short, of such deposits as are now evidently forming at the bottom of the lakes which it traverses.

Notwithstanding the vast forests intercepted by the lakes, a still greater mass of drift-wood is found where the Mackenzie reaches the sea, in a latitude where no wood grows at present except a few stunted willows. At the mouths of the river the alluvial matter has formed a barrier of islands and shoals, where we may expect a great formation of coal at some distant period.

The abundance of floating timber on the Mackenzie is owing, as Dr. Richardson informs me, to the direction and to the length of the course of this river, which runs from south to north, so that the sources of the stream lie in much warmer latitudes than its mouths. In the country, therefore, where the sources are situated, the frost breaks up at an earlier season, while yet the waters in the lower part of its course are ice-bound. Hence the current of water, rushing down northward, reaches a point where the thaw has not begun, and, finding the channel of the river blocked up with ice, it overflows the banks, sweeping through forests of pines, and carrying away thousands of uprooted trees.

Drift-timber on coasts of Iceland, Spitzbergen, &c.—The ancient forests of Iceland, observes Malte-Brun, have been improvidently exhausted;

* Dr. Richardson's Geognost. Obs. on Capt. Franklin's Polar Expedition.

but, although the Icelander can obtain no timber from the land, he is supplied with it abundantly by the ocean. An immense quantity of thick trunks of pines, firs, and other trees, are thrown upon the northern coast of the island, especially upon the North Cape and Cape Langaness, and are then carried by the waves along these two promontories to other parts of the coast, so as to afford sufficiency of wood for fuel and for constructing boats. Timber is also carried to the shores of Labrador and Greenland; and Crantz assures us that the masses of floating wood thrown by the waves upon the island of John de Mayen often equal the whole of that island in extent.*

In a similar manner the bays of Spitzbergen are filled with drift-wood, which accumulates also upon those parts of the coast of Siberia that are exposed to the east, consisting of larch trees, pines, Siberian cedars, firs, and Pernambuco and Campeachy woods. These trunks appear to have been swept away by the great rivers of Asia and America. Some of them are brought from the Gulf of Mexico by the Bahama stream; while others are hurried forward by the current which, to the north of Siberia, constantly sets in from east to west. Some of these trees have been deprived of their bark by friction, but are in such a state of preservation as to form excellent building timber.† Parts of the branches and almost all the roots remain fixed to the pines which have been drifted into the North Sea, into latitudes too cold for the growth of such timber, but the trunks are usually barked.

The leaves and lighter parts of plants are seldom carried out to sea, in any part of the globe, except during tropical hurricanes among islands, and during the agitations of the atmosphere which sometimes accompany earthquakes and volcanic eruptions.

Comparative number of living and fossilized species of plants.—It will appear from these observations that, although the remains of terrestrial vegetation, borne down by aqueous causes from the land, are chiefly deposited at the bottom of lakes or at the mouths of rivers, yet a considerable quantity is drifted about in all directions by currents, and may become imbedded in any *marine* formation, or may sink down, when water-logged, to the bottom of unfathomable abysses, and there accumulate without intermixture with other substances.

It may be asked whether we have any data for inferring that the remains of a considerable proportion of the existing species of plants will be permanently preserved, so as to be hereafter recognizable, supposing the strata now in progress to be at some future period upraised? To this inquiry it may be answered, that there are no reasons for expecting that more than a small number of the plants now flourishing in the globe will become fossilized; since the entire habitations of a great number of them are remote from lakes and seas, and even where they grow near to large bodies of water, the circumstances are quite accidental and partial which

* Malte-Brun, Geog., vol. v. part 1. p. 112.—Brantz, Hist. of Greenland, tom. i. pp. 53, 54.

† Olafsen, Voyage to Iceland, tom. i.—Malte-Brun's Geog., vol. v. part i. p. 112.

favor the imbedding and conservation of vegetable remains. Suppose, for example, that the species of plants inhabiting the hydrographical basin of the Rhine, or that region, extending from the Alps to the sea, which is watered by the Rhine and its numerous tributaries, to be about 2500 in number, exclusive of the cryptogamic class. This estimate is by no means exaggerated; yet if a geologist could explore the deposits which have resulted from the sediment of the Rhine in the Lake of Constance, and off the coast of Holland, he could scarcely expect to obtain from the recent strata the leaves, wood, and seeds of *fifty* species in such a state of preservation as to enable a botanist to determine their specific characters with certainty.

Those naturalists, therefore, who infer that the ancient flora of the globe was, at certain periods, less varied than now, merely because they have as yet discovered only a few hundred fossil species of a particular epoch, while they can enumerate more than one hundred thousand living ones, are reasoning on a false basis, and their standard of comparison is not the same in the two cases.

Submarine forests on coast of Hants.—We have already seen that the submarine position of several forests, or the remains of trees standing in a vertical position on the British shores, has been due, in some instances, to the subsidence of land.* There are some cases which require a different explanation. My friend, Mr. Charles Harris, discovered, in 1831, evident traces of a fir-wood beneath the mean level of the sea, at Bournmouth, in Hampshire, the formation having been laid open during a low spring tide. It is composed of peat and wood, and is situated between the beach and a bar of sand about 200 yards off, and extends fifty yards along the shore. It also lies in the direct line of the Bournmouth Valley, from the termination of which it is separated by 200 yards of shingle and drift-sand. Down the valley flows a large brook, traversing near its mouth a considerable tract of rough, boggy, and heathy ground, which produces a few birch-trees, and a great abundance of the *Myrica gale.* Seventy-six rings of annual growth were counted in a transverse section of one of the buried fir-trees, which was fourteen inches in diameter. Besides the stumps and roots of fir, pieces of alder and birch are found in the peat; and it is a curious fact, that a part of many of the trees have been converted into iron pyrites. The peat rests on pebbly strata, precisely similar to the sand and pebbles occurring on the adjoining heaths.

As the sea is encroaching on this shore, we may suppose that at some former period the Bourne Valley extended farther, and that its extremity consisted, as at present, of boggy ground, partly clothed with fir-trees. The bog rested on that bed of pebbles which we now see below the peat; and the sea, in its progressive encroachments, eventually laid bare, at low water, the sandy foundations; upon which a stream of fresh water, rushing through the sand at the fall of the tides, carried out loose sand with it. The super-stratum of vegetable matter, being matted and

* See above, pp. 308 and 323.

bound together by the roots of trees, remained; but being undermined, sank down below the level of the sea, and then the waves washed sand and shingle over it. In support of this hypothesis, it may be observed, that small streams of fresh water often pass under the sands of the sea-beach, so that they may be crossed dry-shod; and the water is seen, at the point where it issues, to carry out sand and even pebbles.

Mineralization of plants.—Although the botanist and chemist have as yet been unable to explain fully the manner in which wood becomes petrified, it is nevertheless ascertained that, under favorable circumstances, the lapidifying process is now continually going on. A piece of wood was lately procured by Mr. Stokes, from an ancient Roman aqueduct in Westphalia, in which some portions were converted into spindle-shaped bodies, consisting of carbonate of lime, while the rest of the wood remained in a comparatively unchanged state.* It appears that in some cases the most perishable, in others the most durable, portions of plants are preserved, variations which doubtless depend on the time when the mineral matter was supplied. If introduced immediately, on the first commencement of decomposition, then the most destructible parts are lapidified, while the more durable do not waste away till afterwards, when the supply has failed, and so never become petrified. The converse of these circumstances gives rise to exactly opposite results.

Professor Göppert, of Breslau, has instituted a series of curious experiments, in which he has succeeded in producing some very remarkable imitations of fossil petrifactions. He placed recent ferns between soft layers of clay, dried these in the shade, and then slowly and gradually heated them, till they were red-hot. The result was the production of so perfect a counterpart of fossil plants as might have deceived an experienced geologist. According to the different degrees of heat applied, the plants were obtained in a brown or perfectly carbonized condition; and sometimes, but more rarely, they were in a black shining state, adhering closely to the layer of clay. If the red heat was sustained until all the organic matter was burnt up, only an impression of the plant remained.

The same chemist steeped plants in a moderately strong solution of sulphate of iron, and left them immersed in it for several days, until they were thoroughly soaked in the liquid. They were then dried, and kept heated until they would no longer shrink in volume, and until every trace of organic matter had dissappeared. On cooling them he found that the oxide formed by this process had taken the form of the plants. A variety of other experiments were made by steeping animal and vegetable substances in siliceous, calcareous, and metallic solutions, and all tended to prove that the mineralization of organic bodies can be carried much farther in a short time than had been previously supposed.†

* Geol. Trans., second series, vol. v. p. 212.

† Göppert, Poggendorff's Annalen der Physik und Chemie, vol. xxxviii. part iv., Leipsic, 1836. See also Lyell's Manual of Geol., p. 40.

Imbedding of the Remains of Insects.

I have observed the elytra and other parts of beetles in a band of fissile clay, separating two beds of recent shell-marl, in the Loch of Kinnordy in Forfarshire. Amongst these, Mr. Curtis recognized *Elator lineatus* and *Atopa cervina*, species still living in Scotland. These, as well as other remains which accompanied them, appear to belong to terrestrial, not aquatic species, and must have been carried down in muddy water during an inundation. In the lacustrine peat of the same locality, the elytra of beetles are not uncommon; but in the deposits of drained lakes generally, and in the silt of our estuaries, the relics of this class of the animal kingdom are rare. In the blue clay of very modern origin of Lewes levels, Dr. Mantell has found the Indusia, or cases of the larvæ of Phryganea, in abundance, with minute shells belonging to the genera Planorbis, Limnea, &c., adhering to them.*

When speaking of the migrations of insects, I pointed out that an immense number are floated into lakes and seas by rivers, or blown by winds far from the land; but they are so buoyant that we can only suppose them, under very peculiar circumstances, to sink to the bottom before they are either devoured by insectivorous animals or decomposed.

Remains of Reptiles.

As the bodies of several crocodiles were found in the mud brought down to the sea by the river inundation which attended an earthquake in Java, in the year 1699, we may imagine that extraordinary floods of mud may stifle many individuals of the shoals of alligators and other reptiles which frequent lakes and the deltas of rivers in tropical climates. Thousands of frogs were found leaping about among the wreck, carried into the sea by the inundations in Morayshire, in 1829;† and it is evident that whenever a sea-cliff is undermined, or land is swept by other violent causes into the sea, land reptiles may be carried in.

Remains of Birds.

We might have anticipated that the imbedding of the remains of birds in new strata would be of very rare occurrence; for their powers of flight insure them against perishing by numerous casualties to which quadrupeds are exposed during floods; and if they chance to be drowned, or to die when swimming on the water, it will scarcely ever happen that they will be submerged so as to become preserved in sedimentary deposits. In consequence of the hollow tubular structure of their bones and the quantity of their feathers, they are extremely light in proportion to their volume; so that when first killed they do not sink to the bottom like

* Trans. Geol. Soc., vol. iii. part i. p. 201, second series.
† Sir T. D. Lauder's Account, 2d. ed., p. 312.

quadrupeds, but float on the surface until the carcass either rots away or is devoured by predaceous animals. To these causes we may ascribe the absence of any vestige of the bones of birds in the recent marl formations of Scotland; although these lakes, until the moment when they were artificially drained, were frequented by a great abundance of waterfowl.

Imbedding of Terrestrial Quadrupeds.

River inundations recur in most climates at very irregular intervals, and expend their fury on those rich alluvial plains where herds of herbivorous quadrupeds congregate together. These animals are often surprised; and, being unable to stem the current, are hurried along until they are drowned, when they sink at first immediately to the bottom. Here their bodies are drifted along, together with sediment, into lakes or seas, and may then be covered by a mass of mud, sand, and pebbles, thrown down upon them. If there be no sediment superimposed, the gases generated by putrefaction usually cause the bodies to rise again to the surface about the ninth, or at latest the fourteenth day. The pressure of a thin covering of mud would not be sufficient to retain them at the bottom; for we see the putrid carcasses of dogs and cats, even in rivers, floating with considerable weights attached to them, and in sea-water they would be still more buoyant.

Where the body is so buried in drift sand, or mud accumulated upon it, as never to rise again, the skeleton may be preserved entire; but if it comes again to the surface while in the process of putrefaction, the bones commonly fall piecemeal from the floating carcass, and may in that case be scattered at random over the bottom of the lake, estuary, or sea; so that a jaw may afterwards be found in one place, a rib in another, a humerus in a third—all included, perhaps, in a matrix of fine materials, where there may be evidence of slight transporting power in the current, or even of none, but simply of some chemical precipitate.

A large number of the bodies of drowned animals, if they float into the sea or a lake, especially in hot climates, are instantly devoured by sharks, alligators, and other carnivorous beasts, which may have power to digest even the bones; but during extraordinary floods, when the greatest number of land animals are destroyed, the waters are commonly so turbid, especially at the bottom of the channel, that even aquatic species are compelled to escape into some retreat where there is clearer water, lest they should be stifled. For this reason, as well as the rapidity of sedimentary deposition at such seasons, the probability of carcasses becoming permanently imbedded is considerable.

Flood in the Solway Firth, 1794.—One of the most memorable floods of modern date, in our island, is that which visited part of the southern borders of Scotland, on the 24th of January, 1794, and which spread particular devastation over the country adjoining the Solway Firth.

We learn from the account of Captain Napier, that the heavy rains had swollen every stream which entered the Firth of Solway; so that the inundation not only carried away a great number of cattle and sheep, but many of the herdsmen and shepherds, washing down their bodies into the estuary. After the storm, when the flood subsided, an extraordinary spectacle was seen on a large sand-bank called "the beds of Esk," where there is a meeting of the tidal waters, and where heavy bodies are usually left stranded after great floods. On this single bank were found collected together the bodies of 9 black cattle, 3 horses, 1840 sheep, 45 dogs, 180 hares, besides a great number of smaller animals, and, mingled with the rest, the corpses of two men and one woman.*

Floods in Scotland, 1829.—In those more recent floods in Scotland, in August, 1829, whereby a fertile district on the east coast became a scene of dreadful desolation, a vast number of animals and plants were washed from the land, and found scattered about after the storm, around the mouths of the principal rivers. An eye-witness thus describes the scene which presented itself at the mouth of the Spey, in Morayshire:—"For several miles along the beach crowds were employed in endeavoring to save the wood and other wreck with which the heavy-rolling tide was loaded; whilst the margin of the sea was strewed with the carcasses of domestic animals, and with millions of dead hares and rabbits." †

Savannahs of South America.—We are informed by Humboldt, that during the periodical swellings of the large rivers in South America great numbers of quadrupeds are annually drowned. Of the wild horses, for example, which graze in immense troops in the savannahs, thousands are said to perish when the river Apure, a tributary of the Orinoco, is swollen, before they have time to reach the rising ground of the Llanos. The mares, during the season of high water, may be seen, followed by their colts, swimming about and feeding on the grass, of which the top alone waves above the waters. In this state they are pursued by crocodiles; and their thighs frequently bear the prints of the teeth of these carnivorous reptiles. "Such is the pliability," observes the celebrated traveller, "of the organization of the animals which man has subjected to his sway, that horses, cows, and other species of European origin, lead, for a time, an amphibious life, surrounded by crocodiles, water-serpents, and manatees. When the rivers return again into their beds, they roam in the savannah, which is then spread over with a fine odoriferous grass, and enjoy, as in their native climate, the renewed vegetation of spring." ‡

Floods of the Parana.—The great number of animals which are drowned in seasons of drought in the tributaries of the Plata, was before mentioned. Sir W. Parish states, that the Parana, flowing from the mountains of Brazil to the estuary of the Plata, is liable to great floods,

* Treatise on Practical Store Farming, p. 25.
† Sir T. D. Lauder's Floods in Morayshire, 1829; and above, p. 196.
‡ Humboldt's Pers. Nar., vol. iv. p. 394.

and during one of these, in the year 1812, vast quantities of cattle were carried away, "and when the waters began to subside, and the islands which they had covered became again visible, the whole atmosphere for a time was poisoned by the effluvia from the innumerable carcasses of skunks, capybaras, tigers, and other wild beasts which had been drowned."*

Floods of the Ganges.—We find it continually stated, by those who describe the Ganges and Burrampooter, that these rivers carry before them, during the flood season, not only floats of reeds and timber, but dead bodies of men, deer, and oxen.†

In Java, 1699.—I have already referred to the effects of a flood which attended an earthquake in Java in 1699, when the turbid waters of the Batavian river destroyed all the fish except the carp; and when drowned buffaloes, tigers, rhinoceroses, deer, apes, and other wild beasts, were brought down to the sea-coast by the current, with several crocodiles which had been stifled in the mud. (See above, p. 503.)

On the western side of the same island, in the territory of Galongoon, in the Regencies, a more recent volcanic eruption (that of 1822, before described) (see above, p. 431) was attended by a flood, during which the river Tandoi bore down hundreds of carcasses of rhinoceroses and buffaloes, and swept away more than one hundred men and women from a multitude assembled on its banks to celebrate a festival. Whether the bodies reached the sea, or were deposited, with drift matter, in some large intervening alluvial plains, we are not informed.‡

Sumatra.—"On the coast of Orissa," says Heynes, "I have seen tigers and whole herds of black cattle carried along by what are called freshes, and trees of immense size."§

In Virginia, 1771.—I might enumerate a great number of local deluges that have swept through the fertile lands bordering on large rivers, especially in tropical countries, but I should surpass the limits assigned to this work. I may observe, however, that the destruction of the islands, in rivers, is often attended with great loss of lives. Thus when the principal river in Virginia rose, in 1771, to the height of twenty-five feet above its ordinary level, it swept entirely away Elk Island, on which were seven hundred head of quadrupeds,—horses, oxen, sheep, and hogs,—and nearly one hundred houses.‖

The reader will gather, from what was before said respecting the deposition of sediment by aqueous causes, that the greater number of the remains of quadrupeds drifted away by rivers must be intercepted by lakes before they reach the sea, or buried in freshwater formations near the mouths of rivers. If they are carried still farther, the probabilities are increased of their rising to the surface in a state of putrefaction, and, in that case, of being there devoured by aquatic beasts of prey, or of

* Buenos Ayres and La Plata, p. 187. † Malte-Brun's Geog., vol. iii. p. 22.

‡ This account I had from Mr. Baumhauer, Director-General of Finances in Java.

§ Tracts on India, p. 397. ‖ Scots Mag., vol. xxxiii.

subsiding into some spots whither no sediment is conveyed, and, consequently, where every vestige of them will, in the course of time, disappear.

Skeletons of animals in recent shell-marl, Scotland.—In some instances, the skeletons of quadrupeds are met with abundantly in recent shell-marls in Scotland, where we cannot suppose them to have been imbedded by the action of rivers or floods. They all belong to species which now inhabit, or are known to have been indigenous in Scotland. The remains of several hundred skeletons have been procured within the last century from five or six small lakes in Forfarshire, where shell-marl has been worked. Those of the stag (*Cervus Elaphas*) are most numerous; and if the others be arranged in the order of their relative abundance, they will nearly follow thus—the ox, the boar, the horse, the sheep, the dog, the hare, the fox, the wolf, and the cat. The beaver seems extremely rare; but it has been found in the shell-marl of Loch Marlie, in Perthshire, and in the parish of Edrom, in Berwickshire.

In the greater part of these lake-deposits there are no signs of floods; and the expanse of water was originally so confined, that the smallest of the above-mentioned quadrupeds could have crossed, by swimming from one shore to the other. Deer, and such species as take readily to the water, may often have been mired in trying to land, where the bottom was soft and quaggy, and in their efforts to escape may have plunged deeper into the marly bottom. Some individuals, I suspect, of different species, have fallen in when crossing the frozen surface in winter; for nothing can be more treacherous than the ice when covered with snow, in consequence of the springs, which are numerous, and which, retaining always an equal temperature, cause the ice, in certain spots, to be extremely thin, while in every other part of the lake it is strong enough to bear the heaviest weights.

Mammiferous remains in marine strata.—As the bones of mammalia are often so abundantly preserved in peat, and such lakes as have just been described, the encroachments of a sea upon a coast may sometimes throw down the imbedded skeletons, so that they may be carried away by tides and currents, and entombed in submarine formations. Some of the smaller quadrupeds, also, which burrow in the ground, as well as reptiles and every species of plant, are liable to be cast down into the waves by this cause, which must not be overlooked, although probably of comparatively small importance amongst the numerous agents whereby terrestrial organic remains are included in submarine strata.

During the great earthquake of Conception in 1835, some cattle, which were standing on the steep sides of the island of Quiriquina, were rolled by the shock into the sea, while on a low island at the head of the Bay of Conception seventy animals were washed off by a great wave and drowned.*

* Darwin's Journal, p. 372. 2d ed., 1845, p. 304.

CHAPTER XLVIII.

IMBEDDING OF THE REMAINS OF MAN AND HIS WORKS IN SUBAQUEOUS STRATA.

Drifting of human bodies to the sea by river inundations—Destruction of bridges and houses—Loss of lives by shipwreck—How human corpses may be preserved in recent deposits—Number of wrecked vessels—Fossil skeletons of men—Fossil canoes, ships, and works of art—Chemical changes which metallic articles have undergone after long submergence—Imbedding of cities and forests in subaqueous strata by subsidence—Earthquake of Cutch in 1819—Buried Temples of Cashmere—Berkeley's arguments for the recent date of the creation of man—Concluding remarks.

I SHALL now proceed to inquire in what manner the mortal remains of man and the works of his hands may be permanently preserved in subaqueous strata. Of the many hundred million human beings which perish in the course of every century on the land, every vestige is usually destroyed in the course of a few thousand years; but of the smaller number that perish in the waters, a certain proportion must be entombed under circumstances that may enable parts of them to endure throughout entire geological epochs.

The bodies of men, together with those of the inferior animals, are occasionally washed down during river inundations into seas and lakes. (See pp. 726—728.) Belzoni witnessed a flood on the Nile in September, 1818, where, although the river rose only three feet and a half above its ordinary level, several villages, with some hundreds of men, women, and children, were swept away.* It was before mentioned that a rise of six feet of water in the Ganges, in 1763, was attended with a much greater loss of lives. (See above, p. 278.)

In the year 1771, when the inundations in the north of England appear to have equalled the floods of Morayshire in 1829, a great number of houses and their inhabitants were swept away by the rivers Tyne, Can, Wear, Tees, and Greta; and no less than twenty-one bridges were destroyed in the courses of these rivers. At the village of Bywell the flood tore the dead bodies and coffins out of the churchyard, and bore them away, together with many of the living inhabitants. During the same tempest an immense number of cattle, horses, and sheep, were also transported to the sea, while the whole coast was covered with the wreck of ships. Four centuries before (in 1338), the same district had been visited by a similar continuance of heavy rains, followed by disastrous floods, and it is not improbable that these catastrophes may recur periodically, though at uncertain intervals. As the population increases, and

* Narrative of Discovery in Egypt, &c., London, 1820.

buildings and bridges are multiplied, we must expect the loss of lives and property to augment.*

Fossilization of human bodies in the bed of the sea.—If to the hundreds of human bodies committed to the deep in the way of ordinary burial we add those of individuals lost by shipwrecks, we shall find that in the course of a single year, a great number of human remains are consigned to the subaqueous regions. I shall hereafter advert to a calculation by which it appears that more than five hundred *British* vessels alone, averaging each a burthen of about 120 tons, are wrecked, and sink to the bottom, *annually*. Of these the crews for the most part escape, although it sometimes happens that all perish. In one great naval action several thousand individuals sometimes share a watery grave.

Many of these corpses are instantly devoured by predaceous fish, sometimes before they reach the bottom; still more frequently when they rise again to the surface, and float in a state of putrefaction. Many decompose on the floor of the ocean, where no sediment is thrown down upon them; but if they fall upon a reef where corals and shells are becoming agglutinated into a solid rock, or subside where the delta of a river is advancing, they may be preserved for an incalculable series of ages.

Often at the distance of a few hundred feet from a coral reef, where wrecks are not unfrequent, there are no soundings at the depth of many hundred fathoms. Canoes, merchant vessels, and ships of war, may have sunk and have been enveloped, in such situations, in calcareous sand and breccia, detached by the breakers from the summit of a submarine mountain. Should a volcanic eruption happen to cover such remains with ashes and sand, and a current of lava be afterwards poured over them, the ships and human skeletons might remain uninjured beneath the superincumbent mass, like the houses and works of art in the subterranean cities of Campania. Already many human remains may have been thus preserved beneath formations more than a thousand feet in thickness; for, in some volcanic archipelagoes, a period of thirty or forty centuries might well be supposed sufficient for such an accumulation. It was stated, that at the distance of about forty miles from the base of the delta of the Ganges there is an elliptical space about fifteen miles in diameter, where soundings of from 100 to 300 fathoms sometimes fail to reach the bottom. (See above, p. 279.) As during the flood season the quantity of mud and sand poured by the great rivers into the Bay of Bengal is so great that the sea only recovers its transparency at the distance of sixty miles from the coast, this depression must be gradually shoaling, especially as during the monsoons, the sea loaded with mud and sand, is beaten back in that direction towards the delta. Now, if a ship or human body sink to the bottom in such a spot, it is by no means improbable that it may become buried under a depth of a thousand feet of sediment in the same number of years.

* Scots Mag., vol. xxxiii., 1771.

Even on that part of the floor of the ocean to which no accession of drift matter is carried (a part which probably constitutes, at any given period, by far the larger proportion of the whole submarine area), there are circumstances accompanying a wreck which favor the conservation of skeletons. For when the vessel fills suddenly with water, especially in the night, many persons are drowned between decks and in their cabins, so that their bodies are prevented from rising again to the surface. The vessel often strikes upon an uneven bottom, and is overturned; in which case the ballast, consisting of sand, shingle, and rock, or the cargo, frequently composed of heavy and durable materials, may be thrown down upon the carcasses. In the case of ships of war, cannon, shot, and other warlike stores, may press down with their weight the timbers of the vessel as they decay, and beneath these and the metallic substances the bones of man may be preserved.

Number of wrecked vessels.—When we reflect on the number of curious monuments consigned to the bed of the ocean in the course of every naval war from the earliest times, our conceptions are greatly raised respecting the multiplicity of lasting memorials which man is leaving of his labors. During our last great struggle with France, thirty-two of our ships of the line went to the bottom in the space of twenty-two years, besides seven 50-gun ships, eighty-six frigates, and a multitude of smaller vessels. The navies of the other European powers, France, Holland, Spain, and Denmark, were almost annihilated during the same period, so that the aggregate of their losses must have many times exceeded that of Great Britain. In every one of these ships were batteries of cannon constructed of iron or brass, whereof a great number had the dates and places of their manufacture inscribed upon them in letters cast in metal. In each there were coins of copper, silver, and often many of gold, capable of serving as valuable historical monuments; in each were an infinite variety of instruments of the arts of war and peace; many formed of materials, such as glass and earthenware, capable of lasting for indefinite ages when once removed from the mechanical action of the waves, and buried under a mass of matter which may exclude the corroding action of sea-water. The quantity, moreover, of timber which is conveyed from the land to the bed of the sea by the sinking of ships of a large size is enormous, for it is computed that 2000 tons of wood are required for the building of one 74-gun ship; and reckoning fifty oaks of 100 years growth to the acre, it would require forty acres of oak forest to build one of these vessels.*

It would be an error to imagine that the fury of war is more conducive than the peaceful spirit of commercial enterprise to the accumulation of wrecked vessels in the bed of the sea. From an examination of Lloyd's lists, from the year 1793 to the commencement of 1829, Captain W. H. Smyth ascertained that the number of *British vessels* alone lost during that period amounted on an average to no less than one and a half *daily;*

* Quart. Journ. of Agricult., No. ix. p. 433.

an extent of loss which would hardly have been anticipated, although we learn from Moreau's tables that the number of merchant vessels employed at one time, in the navigation of England and Scotland, amounts to about twenty thousand, having one with another a mean burthen of 120 tons.* My friend, Mr. J. L. Prevost, also informs me that on inspecting Lloyd's list for the years 1829, 1830, and 1831, he finds that no less than 1953 vessels were lost in those three years, their average tonnage being about 150 tons, or in all nearly 300,000 tons, being at the enormous rate of 100,000 tons annually of the merchant vessels of one nation only. This increased loss arises, I presume, from increasing activity in commerce.

Out of 551 ships of the royal navy lost to the country during the period above mentioned, only 160 were taken or destroyed by the enemy, the rest having either stranded or foundered, or having been burnt by accident; a striking proof that the dangers of our naval warfare, however great, may be far exceeded by the storm, the shoal, the lee-shore, and all the other perils of the deep.†

Durable nature of many of their contents.—Millions of silver dollars and other coins have been sometimes submerged in a single ship, and on these, when they happen to be enveloped in a matrix capable of protecting them from chemical changes, much information of historical interest will remain inscribed, and endure for periods as indefinite as have the delicate markings of zoophytes or lapidified plants in some of the ancient secondary rocks. In almost every large ship, moreover, there are some precious stones set in seals, and other articles of use and ornament composed of the hardest substances in nature, on which letters and various images are carved—engravings which they may retain when included in subaqueous strata, as long as a crystal preserves its natural form.

It was, therefore, a splendid boast, that the deeds of the English chivalry at Agincourt made Henry's chronicle

> —— as rich with praise
> As is the ooze and bottom of the deep
> With sunken wreck and sumless treasuries

for it is probable that a greater number of monuments of the skill and industry of man will, in the course of ages, be collected together in the bed of the ocean, than will exist at any one time on the surface of the continents.

If our species be of as recent a date as is generally supposed, it will be vain to seek for the remains of man and the works of his hands imbedded in submarine strata, except in those regions where violent earthquakes are frequent, and the alterations of relative level so great, that the bed of the sea may have been converted into land within the historical era. We need not despair, however, of the discovery of such monu-

* Cæsar Moreau's Tables of the Navigation of Great Britain.
† I give these results on the authority of Captain W. H. Smyth, R. N.

ments, when those regions which have been peopled by man from the earliest ages, and which are at the same time the principal theatres of volcanic action, shall be examined by the joint skill of the antiquary and geologist.

Power of human remains to resist decay.—There can be no doubt that human remains are as capable of resisting decay as are the harder parts of the inferior animals; and I have already cited the remark of Cuvier, that "in ancient fields of battle the bones of men have suffered as little decomposition as those of horses which were buried in the same grave." (See above, p. 147.) In the delta of the Ganges bones of men have been found in digging a well at the depth of ninety feet;* but as that river frequently shifts its course and fills up its ancient channels, we are not called upon to suppose that these bodies are of extremely high antiquity, or that they were buried when that part of the surrounding delta where they occur was first gained from the sea.

Fossil skeletons of men.—Several skeletons of men, more or less mutilated, have been found in the West Indies, on the north-west coast of the main land of Guadaloupe, in a kind of rock which is known to be forming daily, and which consists of minute fragments of shells and corals, incrusted with a calcareous cement resembling travertin, by which also the different grains are bound together. The lens shows that some of the fragments of coral composing this stone still retain the same red color which is seen in the reefs of living coral which surround the island. The shells belong to species of the neighboring sea intermixed with some terrestrial kinds which now live on the island, and among them is the *Bulimus Gaudaloupensis* of Férussac. The human skeletons still retain some of their animal matter, and all their phosphate of lime. One of them, of which the head is wanting, may now be seen in the British Museum, and another in the Royal Cabinet at Paris. According to M. König, the rock in which the former is inclosed is harder under the mason's saw and chisel than statuary marble. It is described as forming a kind of glacis, probably an indurated beach, which slants from the steep cliffs of the island to the sea, and is nearly all submerged at high tide.

Similar formations are in progress in the whole of the West Indian archipelago, and they have greatly extended the plain of Cayes in St. Domingo, where fragments of vases and other human works have been found at a depth of twenty feet. In digging wells also near Catania, in Sicily, tools have been discovered in a rock somewhat similar.

Buried ships, canoes, and works of art.—When a vessel is stranded in shallow water, it usually becomes the nucleus of a sand-bank, as has been exemplified in several of our harbors, and this circumstance tends greatly to its preservation. Between the years 1780 and 1790 a vessel from Purbeck, laden with three hundred tons of stone, struck on a shoal off the entrance of Poole harbor and foundered; the crew were saved,

* Von Hoff, vol. i. p. 379.

but the vessel and cargo remain to this day at the bottom. Since that period the shoal at the entrance of the harbor has so extended itself in a westerly direction towards Peveril Point in Purbeck, that the navigable channel is thrown a mile nearer that point.* The cause is obvious; the tidal current deposits the sediment with which it is charged around any object which checks its velocity. Matter also drifted along the bottom is arrested by any obstacle, and accumulates round it, just as the African sand-winds, before described, raise a small hillock over the carcass of every dead camel exposed on the surface of the desert.

I before alluded to an ancient Dutch vessel, discovered in the deserted channel of the river Rother in Sussex, of which the oak wood was much blackened, but its texture unchanged. (See above, p. 316.) The interior was filled with fluviatile silt, as was also the case in regard to a vessel discovered in a former bed of the Mersey, and another disinterred where the St. Katherine Docks are excavated in the alluvial plain of the Thames. In like manner many ships have been found preserved entire in modern strata, formed by the silting up of estuaries along the southern shores of the Baltic, especially in Pomerania. Between Bromberg and Nakel, for example, a vessel and two anchors in a very perfect state were dug up far from the sea.†

Several vessels have been lately detected half buried in the delta of the Indus, in the numerous deserted branches of that river, far from where the stream now flows. One of these found near Vikkar in Sinde, was 400 tons in burthen, old fashioned, and pierced for fourteen guns, and in a region where it had been matter of dispute whether the Indus had ever been navigable by large vessels.‡

At the mouth of a river in Nova Scotia, a schooner of thirty-two tons, laden with live stock, was lying with her side to the tide, when the bore, or tidal wave, which rises there about ten feet in perpendicular height, rushed into the estuary, and overturned the vessel, so that it instantly disappeared. After the tide had ebbed, the schooner was so totally buried in the sand, that the taffrel or upper rail over the stern was alone visible.§ We are informed by Leigh that, on draining Martin Meer, a lake eighteen miles in circumference, in Lancashire, a bed of marl was laid dry, wherein no fewer than eight canoes were found imbedded. In figure and dimensions they were not unlike those now used in America. In a morass about nine miles distant from this Meer a whetstone and an axe of mixed metal were dug up.‖ In Ayrshire, also, three canoes were found in Loch Doon some few years ago; and during the year 1831 four others, each hewn out of separate oak trees. They were twenty-three feet in length, two and a half in depth, and nearly four feet in breadth at the stern. In the mud which filled one of them was found a

* This account I received from the Honorable and Rev. Charles Harris.
† Von Hoff, vol. i. p. 368.
‡ Lieut. Carless, Geograph. Journ., vol. viii. p. 338.
§ Silliman's Geol Lectures, p. 78, who cites Penn.
‖ Leigh's Lancashire, p. 17, A. D. 1700.

war-club of oak and a stone battle-axe. A canoe of oak was also found in 1820, in peat overlying the shell-marl of the Loch of Kinnordy, in Forfarshire.*

Manner in which ships may be preserved in a deep sea.—It is extremely possible that the submerged woodwork of ships which have sunk where the sea is two or three miles deep has undergone greater chemical changes in an equal space of time, than in the cases above mentioned; for the experiments of Scoresby show that wood may at certain depths be impregnated in a single hour with salt water, so that its specific gravity is entirely altered. It may often happen that hot springs, charged with carbonate of lime, silex, and other mineral ingredients, may issue at great depths, in which case every pore of the vegetable tissue may be injected with the lapidifying liquid, whether calcareous or siliceous, before the smallest decay commences. The conversion, also, of wood into lignite is probably more rapid under enormous pressure. But the change of the timber into lignite or coal would not prevent the original form of a ship from being distinguished; for as we find, in strata of the carboniferous era, the bark of the hollow reed-like trees converted into coal, and the central cavity filled with sandstone, so might we trace the outline of a ship in coal; while in the indurated mnd, sandstone, or limestone, filling the interior, we might discover instruments of human art, ballast consisting of rocks foreign to the rest of the stratum, and other contents of the ship.

Submerged metallic substances.—Many of the metallic substances which fall into the waters probably lose, in the course of ages, the forms artificially imparted to them; but under certain circumstances these may be preserved for indefinite periods. The cannon enclosed in a calcareous rock, drawn up from the delta of the Rhone, which is now in the museum at Montpellier, might probably have endured as long as the calcareous matrix; but even if the metallic matter had been removed, and had entered into new combinations, still a mould of its original shape would have been left, corresponding to those impressions of shells which we see in rocks, from which all the carbonate of lime has been subtracted. About the year 1776, says Mr. King, some fishermen, sweeping for anchors in the Gulf stream (a part of the sea near the Downs), drew up a very curious old swivel gun, nearly eight feet in length. The barrel, which was about five feet long, was of brass; but the handle by which it was traversed was about three feet in length, and the swivel and pivot on which it turned were of iron. Around these latter were formed incrustations of sand converted into a kind of stone, of exceedingly strong texture and firmness; whereas round the barrel of the gun, except where it was near adjoining to the iron, there were no such incrustations, the greater part of it being clean, and in good condition, just as if it had still continued in use. In the incrusting stone, adhering to it on the outside, were a number of shells and corallines, "just as they are often found in a fossil

* Geol. Trans., second series, vol. ii. p. 87.

state." These were all so strongly attached, that it required as much force to separate them from the matrix "as to break a fragment off any hard rock."*

In the year 1745, continues the same writer, the Fox man-of-war was stranded on the coast of East Lothian, and went to pieces. About thirty-three years afterwards a violent storm laid bare a part of the wreck, and threw up near the place several masses, "consisting of iron, ropes, and balls," covered over with ochreous sand, concreted and hardened into a kind of stone. The substance of the rope was very little altered. The consolidated sand retained perfect impressions of parts of an iron ring, "just as impressions of extraneous fossil bodies are found in various kinds of strata."†

After a storm in the year 1824, which occasioned a considerable shifting of the sands near St. Andrew's, in Scotland, a gun-barrel of ancient construction was found, which is conjectured to have belonged to one of the wrecked vessels of the Spanish Armada. It is now in the museum of the Antiquarian Society of Scotland, and is incrusted over by a thin coating of sand, the grains of which are cemented by brown ferruginous matter. Attached to this coating are fragments of various shells, as of the common cardium, mya, &c.

Many other examples are recorded of iron instruments taken up from the bed of the sea near the British coast, incased by a thick coating of conglomerate, consisting of pebbles and sand, cemented by oxide of iron.

Dr. Davy describes a bronze helmet, of the antique Grecian form, taken up in 1825, from a shallow part of the sea, between the citadel of Corfu and the village of Castrades. Both the interior and exterior of the helmet were partially incrusted with shells, and a deposit of carbonate of lime. The surface generally, both under the incrustation, and where freed from it, was of a variegated color, mottled with spots of green, dirty white, and red. On minute inspection with a lens, the green and red patches proved to consist of crystals of the red oxide and carbonate of copper, and the dirty white chiefly of oxide of tin.

The mineralizing process, says Dr. Davy, which has produced these new combinations, has, in general, penetrated very little into the substance of the helmet. The incrustation and rust removed, the metal is found bright beneath; in some places considerably corroded, in others very slightly. It proves, on analysis, to be copper, alloyed with 18·5 per cent. of tin. Its color is that of our common brass, and it possesses a considerable degree of flexibility.

"It is a curious question," he adds, "how the crystals were formed in the helmet, and on the adhering calcareous deposit. There being no reason to suppose deposition from solution, are we not under the necessity of inferring, that the mineralizing process depends on a small motion and separation of the particles of the original compound? This motion may have been due to the operation of electro-chemical powers which may have separated the different metals of the alloy.*

Effects of the Subsidence of Land, in imbedding Cities and Forests in subaqueous Strata.

We have hitherto considered the transportation of plants and animals from the land by *aqueous* agents, and their inhumation in lacustrine or submarine deposits, and we may now inquire what tendency the subsidence of tracts of land may have to produce analogous effects. Several examples of the sinking down of buildings, and portions of towns near the shore, to various depths beneath the level of the sea during subterranean movements, were before enumerated in treating of the changes brought about by *inorganic* causes. The events alluded to were comprised within a brief portion of the historical period, and confined to a small number of the regions of active volcanoes. Yet these authentic facts, relating merely to the last century and a half, gave indications of considerable changes in the physical geography of the globe, and we are not to suppose that these were the only spots throughout the surrounding land and sea which suffered similar depressions.

If, during the short period since South America has been colonized by Europeans, we have proof of alterations of level at the three principal ports on the western shores, Callao, Valparaiso, and Conception,† we cannot for a moment suspect that these cities, so distant from each other, have been selected as the peculiar points where the desolating power of the earthquake has expended its chief fury. On considering how small is the area occupied by the seaports of this disturbed region—points where alone each slight change of the relative level of the sea and land can be recognized,—and reflecting on the proofs in our possession of the local revolutions that have happened on the site of each port, within the last century and a half,—our conceptions must be greatly exalted respecting the magnitude of the alterations which the country between the Andes and the sea may have undergone, even in the course of the last six thousand years.

Cutch earthquake.—The manner in which a large extent of surface may be submerged, so that the terrestrial plants and animals may be imbedded in subaqueous strata, cannot be better illustrated than by the earthquake of Cutch, in 1819, before alluded to (p. 460). It is stated, that, for some years after that earthquake, the withered tamarisks and other shrubs protruded their tops above the waves, in parts of the lagoon formed by subsidence, on the site of the village of Sindree and its environs; but, after the flood of 1826, they were seen no longer. Every geologist will at once perceive, that forests sunk by such subterranean movements may become imbedded in subaqueous deposits, both fluviatile

* Phil. Trans., 1826, part ii. p. 55. † See above, pp. 453. 457. 499. 501.

and marine, and the trees may still remain erect, or sometimes the roots and part of the trunks may continue in their original position, while the current may have broken off, or levelled with the ground, their upper stems and branches.

Buildings how preserved under water.—Some of the buildings which have at different times subsided beneath the level of the sea have been immediately covered up to a certain extent with strata of volcanic matter showered down upon them. Such was the case at Tomboro in Sumbawa, in the present century, and at the site of the Temple of Serapis, in the environs of Puzzuoli, probably about the 12th century. The entrance of a river charged with sediment in the vicinity may still more frequently occasion the rapid envelopment of buildings in regularly stratified formations. But if no foreign matter be introduced, the buildings, when once removed to a depth where the action of the waves is insensible, and where no great current happens to flow, may last for indefinite periods, and be as durable as the floor of the ocean itself, which may often be composed of the very same materials. There is no reason to doubt the tradition mentioned by the classic writers, that the submerged Grecian towns of Bura and Helice were seen under water; and it has been already mentioned that different eye-witnesses have observed the houses of Port Royal, at the bottom of the sea, at intervals of 88, 101, and 143 years after the convulsion of 1692. (p. 505.)

Buried temples of Cashmere.—The celebrated valley of Cashmere (or Kashmir) in India, situated at the southern foot of the Himâlaya range, is about 60 miles in length, and 20 in breadth, surrounded by mountains which rise abruptly from the plain to the height of about 5000 feet. In the cliffs of the river Jelam and its tributaries, which traverse this beautiful valley, strata consisting of fine clay, sand, soft sandstone, pebbles, and conglomerate are exposed to view. They contain freshwater shells, of the genera Lymneus, Paludina, and Cyrena, with land shells, all of recent species, and are precisely such deposits as would be formed if the whole valley were now converted into a great lake, and if the numerous rivers and torrents descending from the surrounding mountains were allowed sufficient time to fill up the lake-basin with fine sediment and gravel. Fragments of pottery met with at the depth of 40 and 50 feet in this lacustrine formation show that the upper part of it at least has accumulated within the human epoch.

Dr. Thomas Thomson, who visited Cashmere in 1848, observes that several of the lakes which still exist in the great valley, such as that near the town of Cashmere, five miles in diameter, and some others, are deeper than the adjoining river-channels, and may have been formed by subsidence during the numerous earthquakes which have convulsed that region in the course of the last 2000 years. It is also probable that the freshwater strata seen to extend far and wide over the whole of Cashmere originated not in one continuous sheet of water once occupying the entire valley, but in many lakes of limited area, formed and filled in succession. Among other proofs of such lake-basins of moderate dimensions having once existed and having been converted into land at different periods, Dr. Thomson mentions that the ruins of Avantipura, not far from the modern village of that name, stand on an older fresh-water deposit at the base of the mountains, and terminate abruptly towards the plain in a straight line, such as admits of no other explanation than by supposing that the advance of the town in that direction was arrested by a lake, now drained or represented only by a marsh. In that neighborhood, as very generally throughout Cashmere, the rivers run in channels or alluvial flats, bounded by cliffs of lacustrine strata, horizontally stratified, and these strata form low table-lands from 20 to 50 feet high between the different watercourses. On a table-land of this kind near Avantipura, portions of two buried temples are seen, which have been partially explored by Major Cunningham, who, in 1847, discovered that in one of the buildings a magnificent colonnade of seventy-four pillars is preserved underground. He exposed to view three of the pillars in a cavity still open. All the architectural decorations below the level of the soil are as perfect and fresh-looking as when first executed. The spacious quadrangle must have been silted up gradually at first, for some unsightly alterations, not in accordance with the general plan and style of architecture, were detected, evidently of subsequent date, and such as could only have been required when the water and sediment had already gained a certain height in the interior of the temple.

This edifice is supposed to have been erected about the year 850 of our era, and was certainly submerged before the year 1416, when the Mahomedan king, Sikandar, called Butshikan or the idol-breaker, destroyed all the images of Hindoo temples in Cashmere. Ferishta the historian particularly alludes to Sikandar having demolished every Cashmerian temple save one, dedicated to Mahadéva, which escaped "in consequence of its foundations being below the neighboring water." The unharmed condition of the human-headed birds and other images in the buried edifice near Avantipura leaves no doubt that they escaped the fury of the iconoclast by being under water, and perhaps silted up before the date of his conquest.*

Berkeley's arguments for the recent date of the creation of man.— I cannot conclude this chapter without recalling to the reader's mind a memorable passage written by Bishop Berkeley a century ago, in which he inferred, on grounds which may be termed strictly geological, the recent date of the creation of man. "To any one," says he, "who considers that on digging into the earth, such quantities of shells, and in some places, bones and horns of animals, are found sound and entire, after having lain there in all probability some thousands of years; it should seem probable that guns, medals, and implements in metal or stone, might have lasted entire, buried under ground forty or fifty

* Thomson's Western Himalaya and Thibet, p. 292. London, 1852. Cunningham, vol. xvii. Journ. Asiat. Soc. Bengal, pp. 241. 277.

thousand years, if the world had been so old. How comes it then to pass that no remains are found, no antiquities of those numerous ages preceding the Scripture accounts of time; that no fragments of buildings, no public monuments, no intaglios, cameos, statues, basso-relievos, medals, inscriptions, utensils, or artificial works of any kind, are ever discovered, which may bear testimony to the existence of those mighty empires, those successions of monarchs, heroes, and demi-gods, for so many thousand years? Let us look forward and suppose ten or twenty thousand years to come, during which time we will suppose that plagues, famine, wars, and *earthquakes* shall have made great havoc in the world, is it not highly probable that at the end of such a period, pillars, vases, and statues now in being, of granite, or porphyry, or jasper (stones of such hardness as we know them to have lasted two thousand years above ground, without any considerable alteration), would bear record of these and past ages? Or that some of our current coins might then be dug up, or old walls and the foundations of buildings show themselves, as well as the shells and stones of *the primeval world*, which are preserved down to our times."*

That many signs of the agency of man would have lasted at least as long as "the shells of the primeval world," had our race been so ancient, we may feel as fully persuaded as Berkeley; and we may anticipate with confidence that many edifices and implements of human workmanship and the skeletons of men, and casts of the human form, will continue to exist when a great part of the present mountains, continents, and seas have disappeared. Assuming the future duration of the planet to be indefinitely protracted, we can foresee no limit to the perpetuation of some of the memorials of man, which are continually entombed in the bowels of the earth or in the bed of the ocean, unless we carry forward our views to a period sufficient to allow the various causes of change, both igneous and aqueous, to remodel more than once the entire crust of the earth. *One* complete revolution will be inadequate to efface every monument of our existence; for many works of art might enter again and again into the formations of successive eras, and escape obliteration even though the very rocks in which they had been for ages imbedded were destroyed, just as pebbles included in the conglomerates of one epoch often contain the organized remains of beings which flourished during a prior era.

Yet it is no less true, as a late distinguished philosopher has declared, "that none of the works of a mortal being can be eternal."† They are in the first place wrested from the hands of man, and lost as far as regards their subserviency to his use, by the instrumentality of those very causes which place them in situations where they are enabled to endure for indefinite periods. And even when they have been included in rocky strata, when they have been made to enter as it were into the solid frame-

* Alciphron, or the Minute Philosopher, vol. ii. pp. 84, 85., 1732.

† Davy, Consolations in Travel, p. 276.

work of the globe itself, they must nevertheless eventually perish; for every year some portion of the earth's crust is shattered by earthquakes, or melted by volcanic fire, or ground to dust by the moving waters on the surface. "The river of Lethe," as Bacon eloquently remarks, "runneth as well above ground as below."*

CHAPTER XLIX.

IMBEDDING OF AQUATIC SPECIES IN SUBAQUEOUS STRATA.

Inhumation of fresh water plants and animals—Shell marl—Fossilized seed-vessels and stems of chara—Recent deposits in American lakes—Freshwater species drifted into seas and estuaries—Lewes levels—Alternations of marine and freshwater strata, how caused—Imbedding of marine plants and animals—Cetacea stranded on our shores—Littoral and estuary Testacea swept into the deep sea—Burrowing shells—Living Testacea found at considerable depths—Blending of organic remains of different ages.

HAVING treated of the imbedding of terrestrial plants and animals, and of human remains, in deposits now forming beneath the waters, I come next to consider in what manner *aquatic* species may be entombed in strata formed in their own element.

Freshwater plants and animals.—The remains of species belonging to those genera of the animal and vegetable kingdoms which are more or less exclusively confined to fresh water are for the most part preserved in the beds of lakes or estuaries, but they are oftentimes swept down by rivers into the sea, and there intermingled with the exuviæ of marine races. The phenomena attending their inhumation in lacustrine deposits are sometimes revealed to our observation by the drainage of small lakes, such as are those in Scotland, which have been laid dry for the sake of obtaining shell marl for agricultural uses.

In these recent formations, as seen in Forfarshire, two or three beds of calcareous marl are sometimes observed separated from each other by layers of drift peat, sand, or fissile clay. The marl often consists almost entirely of an aggregate of shells of the genera Limnea, Planorbis, Valvata, and Cyclas, of species now existing in Scotland. A considerable proportion of the Testacea appear to have died very young, and few of the shells are of a size which indicates their having attained a state of maturity. The shells are sometimes entirely decomposed, forming a pulverulent marl; sometimes in a state of good preservation. They are frequently intermixed with stems of Charæ and other aquatic vegetables, the whole being matted together and compressed, forming laminæ often as thin as paper.

* Essay on the Vicissitude of Things.

Fossilized seed-vessels and stems of Chara.—As the Chara is an aquatic plant which occurs frequently fossil in formations of different eras, and is often of much importance to the geologist in characterizing entire groups of strata, I shall describe the manner in which I have found the recent species in a petrified state. They occur in a marl-lake in Forfarshire, inclosed in nodules, and sometimes in a continuous stratum of a kind of travertin.

Fig. 102.

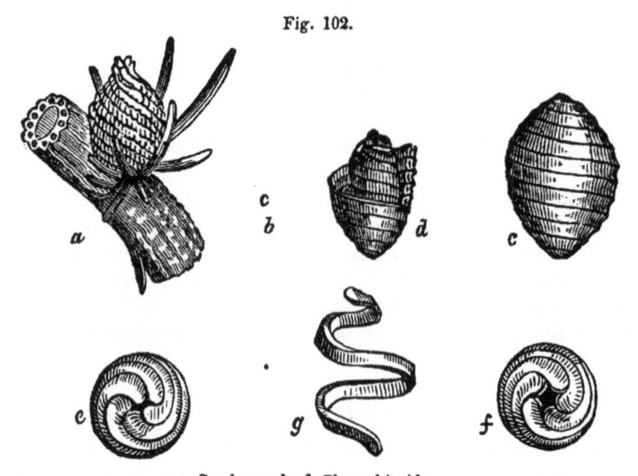

Seed-vessel of Chara hispida.

a, Part of the stem with the seed-vessel attached. Magnified.
b, Natural size of the seed vessel.
c, Integument of the Gyrogonite, or petrified seed-vessel of *Chara hispida*, found in the Scotch marl-lakes. Magnified.
d, Section showing the nut within the integument.
e, Lower end of the integument to which the stem was attached.
f, Upper end of the integument to which the stigmata were attached.
g, One of the spiral valves of *c*.

The seed-vessel of these plants is remarkably tough and hard, and consists of a membranous nut covered by an integument (*d*, fig. 102.) both of which are spirally striated or ribbed. The integument is composed of five spiral valves, of a quadrangular form (*g*). In *Chara hispida*, which abounds in the lakes of Forfarshire, and which has become fossil in the Bakie Loch, each of the spiral valves of the seed-vessel turns rather more than twice round the circumference, the whole together making between ten and eleven rings. The number of these rings differs greatly in different species, but in the same appears to be very constant.

The stems of Charæ occur fossil in the Scotch marl in great abundance. In some species, as in *Chara hispida*, the plant when living contains so much carbonate of lime in its vegetable organization, independently of calcareous incrustation, that it effervesces strongly with acids when dry. The stems of *Chara hispida* are longitudinally striated, with a tendency to be spiral. These striæ, as appears to be the case with all Charæ, turn always like the worm of a screw from right to left, while those of the seed-vessel wind round in a contrary direction. A cross section of the stem exhibits a curious structure, for it is composed of a large tube surrounded by smaller tubes (fig. 103., *b*, *c*) as is seen in some extinct as well as recent species. In the stems of several species, however, there is only a single tube.*

Fig. 103.

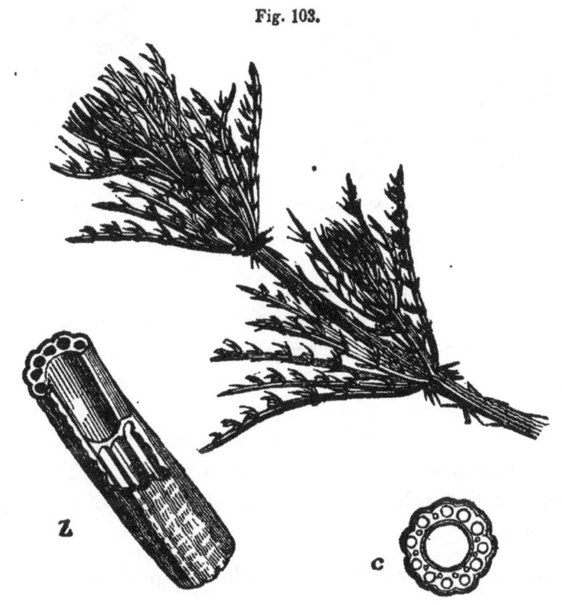

Stem and branches of Chara hispida.

a, Stem and branches of the natural size.
b, Section of the stem magnified.
c, Showing the central tube surrounded by two rings of smaller tubes.

The valves of a small animal called cypris (*C. ornata?* Lam.) occur completely fossilized, like the stems of Charæ, in the Scotch travertin above mentioned. The same cypris inhabits the lakes and ponds of England, where, together with many other species, it is not uncommon. Although extremely minute, they are visible to the naked eye, and may be observed in great numbers, swimming swiftly through the waters of our stagnant pools and ditches. The antennæ, at the end of which are fine pencils of hair, are the principal organs for swimming, and are moved with great rapidity. The animal resides within two small valves, not unlike those of a bivalve shell, and moults its integuments annually, which the conchiferous mollusks do not. The cast-off shells, resembling thin scales, and occurring in countless myriads in many ancient freshwater marls, impart to them a divisional structure, like that so frequently derived from plates of mica.

The recent strata of lacustrine origin above alluded to are of very small extent, but analogous deposits on the grandest scale are forming in the great Canadian lakes, as in Lakes Superior and Huron, where beds

* On Freshwater Marl, &c. By C. Lyell. Geol. Trans., vol. ii., second series, p. 73.

Fig. 104.

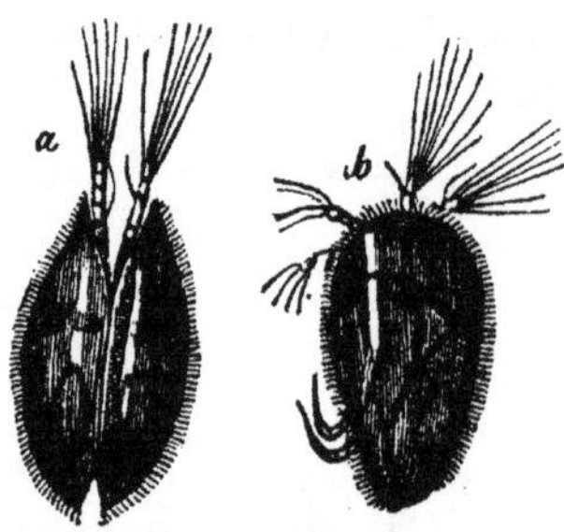

Cypris unifasciata, a living species, greatly magnified.
a, Upper part. *b*, Side view of the same.

Fig. 105

Cypris vidua, a living species, greatly magnified.*

of sand and clay are seen inclosing shells of existing species.† The Chara also plays the same part in the subaqueous vegetation of North America as in Europe. I observed along the borders of several fresh-water lakes in the state of New York a luxuriant crop of this plant in clear water of moderate depth, rendering the bottom as verdant as a grassy meadow. Here, therefore, we may expect some of the tough seed vessels to be preserved in mud, just as we detect them fossil in the Eocene strata of Hampshire, or in the neighborhood of Paris, and many other countries.

Imbedding of freshwater Species in Estuary and Marine Deposits.

In Lewes levels.—We have sometimes an opportunity of examining the deposits which within the historical period have silted up some of our estuaries; and excavations made for wells and other purposes, where the sea has been finally excluded, enable us to observe the state of the organic remains in these tracts. The valley of the Ouze between Newhaven and Lewes is one of several estuaries from which the sea has retired within the last seven or eight centuries; and here, as appears from the researches of Dr. Mantell, strata thirty feet and upwards in thickness have accumulated. At the top, beneath the vegetable soil, is a bed of peat about five feet thick, inclosing many trunks of trees. Next below is a stratum of blue clay containing freshwater shells of about nine species, such as now inhabit the district. Intermixed with these was observed the skeleton of a deer. Lower down, the layers of blue clay contain, with the above-mentioned freshwater shells, several marine species well known on our coast. In the lowest beds, often at the depth of thirty-six feet, these marine Testacea occur without the slightest intermixture of fluviatile species, and amongst them the skull of the narwal, or sea unicorn (*Monodon monoceros*), has been detected. Underneath all these deposits is a bed of pipe-clay, derived from the subjacent chalk.‡

* See Desmarest's Crustacea, pl. 55.
† Dr. Bigsby, Journ. of Science, &c. No. xxxvii. pp. 262, 263.
‡ Mantell, Geol. of Sussex, p. 285; also Catalogue of Org. Rem., Geol. Trans., vol. iii. part i. p. 201., 2nd series.

If we had no historical information respecting the former existence of an inlet of the sea in this valley and of its gradual obliteration, the inspection of the section above described would show, as clearly as a written chronicle, the following sequence of events. First, there was a salt-water estuary peopled for many years by species of marine Testacea identical with those now living, and into which some of the larger Cetacea occasionally entered. Secondly, the inlet grew shallower, and the water became brackish, or alternately salt and fresh, so that the remains of freshwater and marine shells were mingled in the blue argillaceous sediment of its bottom. Thirdly, the shoaling continued until the river-water prevailed, so that it was no longer habitable by marine Testacea, but fitted only for the abode of fluviatile species and aquatic insects. Fourthly, a peaty swamp or morass was formed, where some trees grew, or perhaps were drifted during floods, and where terrestrial quadrupeds were mired. Finally, the soil being flooded by the river only at distant intervals, became a verdant meadow.

In delta of Ganges. and Indus.—It was before stated, that on the sea-coast, in the delta of the Ganges, there are eight great openings, each of which has evidently, at some ancient period, served in its turn as the principal channel of discharge.* As the base of the delta is 200 miles in length, it must happen that, as often as the great volume of river-water is thrown into the sea by a new mouth, the sea will at one point be converted from salt to fresh, and at another from fresh to salt; for, with the exception of those parts where the principal discharge takes place, the salt water not only washes the base of the delta, but enters far into every creek and lagoon. It is evident, then, that repeated alternations of beds containing freshwater shells, with others filled with marine exuviæ, may here be formed. It has also been shown by artesian borings at Calcutta (see above, p. 267), that the delta once extended much farther than now into the gulf, and that the river is only recovering from the sea the ground which had been lost by subsidence at some former period. Analogous phenomeua must sometimes be occasioned by such alternate elevation and depression as has occurred in modern times in the delta of the Indus.† But the subterranean movements affect but a small number of the deltas formed at one period on the globe; whereas the silting up of some of the arms of great rivers and the opening of others, and the consequent variation of the points where the chief volume of their waters is discharged into the sea, are phenomena common to almost every delta.

The variety of species of Testacea contained in the recent calcareous marl of Scotland, before mentioned, is very small, but the abundance of individuals extremely great, a circumstance very characteristic of freshwater formations in general, as compared to marine; for in the latter, as is seen on sea-beaches, coral-reefs, or in the bottom of the seas examined by dredging, wherever the individual shells are exceedingly numerous, there rarely fails to be a vast variety of species.

* Page 276.

† Page 460.

Imbedding of the Remains of Marine Plants and Animals.

Marine plants.—The large banks of drift sea-weed which occur on each side of the equator in the Atlantic, Pacific, and Indian oceans, were before alluded to.* These, when they subside, may often produce considerable beds of vegetable matter. In Holland, sub-marine peat is derived from Fuci, and on parts of our own coast from *Zostera marina.* In places where Algæ do not generate peat, they may nevertheless leave traces of their form imprinted on argillaceous and calcareous mud, as they are usually very tough in their texture.

Sea-weeds are often cast up in such abundance on our shores during heavy gales, that we cannot doubt that occasionally vast numbers of them are imbedded in littoral deposits now in progress. We learn from the researches of Dr. Forchhammer, that besides supplying in common with land plants the materials of coal, the Algæ must give rise to important chemical changes in the composition of strata in which they are imbedded. These plants always contain sulphuric acid, and sometimes in as large a quantity as 8½ per cent., combined with potash: magnesia also and phosphoric acid are constant ingredients. Whenever large masses of sea-weeds putrefy in contact with ferruginous clay, sulphuret of iron, or iron pyrites, is formed by the union of the sulphur of the plants with the iron of the clay; while the potash, released from its union with the clay (*i. e.* silicate of alumina), forms with it a peculiar compound. Many of the mineral characteristics of ancient rocks, especially the alum slates, and the pyrites which occur in clay slate, and the fragments of anthracite in marine Silurian strata, may be explained by the decompositiou of fucoids or sea-weeds.†

Imbedding of cetacea.—It is not uncommon for the larger Cetacea, which can float only in a considerable depth of water, to be carried during storms or high tides into estuaries, or upon low shores, where, upon the retiring of high water, they are stranded. Thus a narwal (*Monodon monoceros*) was found on the beach near Boston in Lincolnshire, in the year 1800, the whole of its body buried in the mud. A fisherman going to his boat saw the horn, and tried to pull it out, when the animal began to stir itself.‡ An individual of the common whale (*Balæna mysticetus*), which measured seventy feet, came ashore near Peterhead, in 1682. Many individuals of the genus Balænoptera have met the same fate. It will be sufficient to refer to those cast on shore near Burnt Island, and at Alloa, recorded by Sibbald and Neill. The other individual mentioned by Sibbald, as having come ashore at Boyne, in Banffshire, was probably a razor-back. Of the genus Catodon (*Cachalot*), Ray mentions a large one stranded on the west coast of Holland in 1598, and the fact is also commemorated in a Dutch engraving of the time of much merit. Sibbald, too, records that a herd of Cachalots, upwards of 100 in number,

* Page 599. † Forchhammer, Report British Assoc. 1844.
‡ Fleming's Brit. Animals, p. 37; in which work other cases are enumerated.

were found stranded at Cairston, in Orkney. The dead bodies of the larger Cetacea are sometimes found floating on the surface of the waters, as was the case with the immense whale exhibited in London in 1831. And the carcase of a sea-cow or Lamantine (*Halicora*) was, in 1785, cast ashore near Leith.

To some accident of this kind we may refer the position of the skeleton of a whale, seventy-three feet long, which was found at Airthrey, on the Forth, near Stirling, imbedded in clay twenty feet higher than the surface of the highest tide of the river Forth at the present day. From the situation of the Roman station and causeways at a small distance from the spot, it is concluded that the whale must have been stranded there at a period prior to the Christian era.*

Other fossil remains of this class have also been found in estuaries known to have been silted up in recent times, one example of which has been already mentioned near Lewes, in Sussex.

Marine reptiles.—Some singular fossils have lately been discovered in the Island of Ascension, in a stone said to be continually forming on the beach, where the waves threw up small rounded fragments of shells and corals, which, in the course of time, become firmly agglutinated together, and constitute a stone used largely for building and making lime. In a quarry on the N. W. side of the island, about 100 yards from the sea, some fossil eggs of turtles have been discovered in the hard rock thus formed. The eggs must have been nearly hatched at the time when they perished; for the bones of the young turtle are seen in the interior, with their shape fully developed, the interstices between the bones being entirely filled with grains of sand, which are cemented together, so that when the egg-shells are removed perfect casts of their form remain in stone. In the single specimen here figured (fig. 106), which is only five inches in its longest diameter, no less than seven eggs are preserved.‡

Fig. 106.

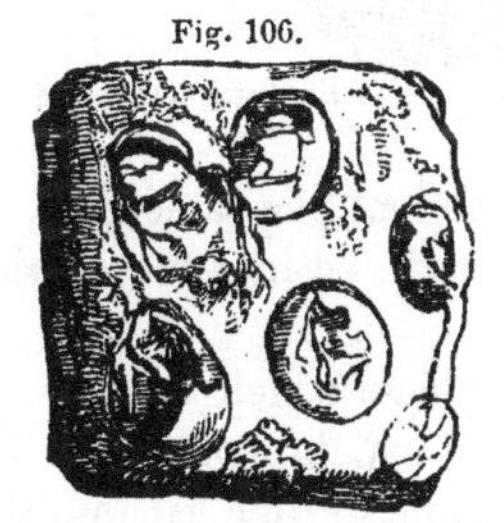

Fossil eggs of turtles from the Island of Ascension.†

To explain the state in which they occur fossil, it seems necessary to suppose that after the eggs were almost hatched in the warm sand, a great wave threw upon them so much more sand as to prevent the rays of the sun from penetrating, so that the yolk was chilled and deprived of

* Quart. Journ. of Lit. Sci., &c., No. xv., p. 172. Oct. 1819.

† This specimen has been presented by Mr. Lonsdale to the Geological Society of London.

‡ The most conspicuous of the bones represented within the shell in fig. 107, appear to be the clavicle and coracoid bone. They are hollow; and for this reason resemble, at first sight, the bones of birds rather than of reptiles; for the latter have no medullary cavity. Prof. Owen, of the College of Surgeons, in order to elucidate this point, dissected for me a very young turtle, and found that the exterior portion only of the bones was ossified, the interior being still filled with cartilage. This cartilage soon dried up and shrank to a mere thread upon the evaporation of the spirits of wine in which the specimen had been preserved, so that in a short time the bones became as empty as those of birds.

Fig. 107.

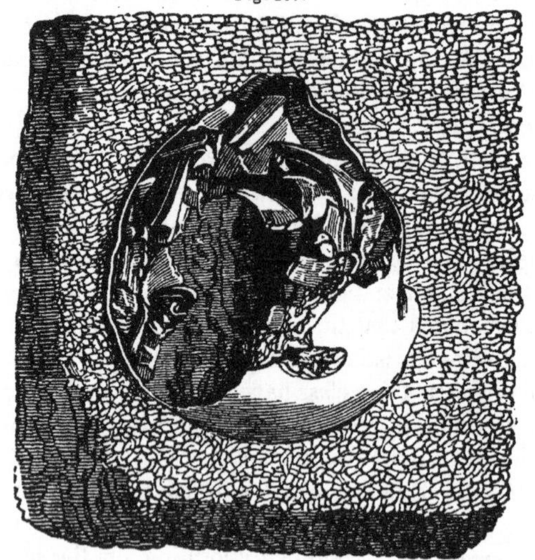

One of the eggs in fig. 106, of the natural size, showing the bones of the fœtus which had been nearly hatched.

vitality. The shells were, perhaps, slightly broken at the same time, so that small grains of sand might gradually be introduced into the interior by water as it percolated through the beach.

Marine testacea.—The aquatic animals and plants which inhabit an estuary are liable, like the trees and land animals which people the alluvial plains of a great river, to be swept from time to time far into the deep; for as a river is perpetually shifting its course, and undermining a portion of its banks with the forests which cover them, so the marine current alters its direction from time to time, and bears away the banks of sand and mud against which it turns its force. These banks may consist in great measure of shells peculiar to shallow and sometimes brackish water, which may have been accumulating for centuries, until at length they are carried away and spread out along the bottom of the sea, at a depth at which they could not have lived and multiplied. Thus littoral and estuary shells are more frequently liable even than freshwater species, to be intermixed with the exuviæ of pelagic tribes.

After the storm of February 4, 1831, when several vessels were wrecked in the estuary of the Forth, the current was directed against a bed of oysters with such force, that great heaps of them were thrown *alive* upon the beach, and remained above high-water mark. I collected many of these oysters, as also the common eatable whelks (*Buccina*), thrown up with them, and observed that, although still living, their shells were worn by the long attrition of sand which had passed over them as they lay in their native bed, and which had evidently not resulted from the mere action of the tempest by which they were cast ashore.

From these facts we learn that the union of the two parts of a bivalve shell does not prove that it has not been transported to a distance; and when we find shells worn, and with all their prominent parts rubbed off, they may still have been imbedded where they grew.

Burrowing shells.—It sometimes appears extraordinary, when we observe the violence of the breakers on our coast, and see the strength of the current in removing cliffs, and sweeping out new channels, that many tender and fragile shells should inhabit the sea in the immediate vicinity of this turmoil. But a great number of the bivalve Testacea, and many also of the turbinated univalves, burrow in sand or mud. The Solen and the Cardium, for example, which are usually found in shallow water near the shore, pierce through a soft bottom without injury to their shells; and the Pholas can drill a cavity through mud of considerable hardness. The species of these and many other tribes can sink, when alarmed, with considerable rapidity, often to the depth of several feet, and can also penetrate upwards again to the surface, if a mass of matter be heaped upon them. The hurricane, therefore, may expend its fury in vain, and may sweep away even the upper part of banks of sand or mud, or may roll pebbles over them, and yet these Testacea may remain below secure and uninjured.

Shells become fossil at considerable depths.—I have already stated that, at the depth of 950 fathoms, between Gibraltar and Ceuta, Captain Smith found a gravelly bottom, with fragments of broken shells, carried thither probably from the comparatively shallow parts of the neighboring straits, through which a powerful current flows. Beds of shelly sand might here, in the course of ages, be accumulated several thousand feet thick. But, without the aid of the drifting power of a current, shells may accumulate in the spot where they live and die, at great depths from the surface, if sediment be thrown down upon them; for even in our own colder latitudes, the depths at which living marine animals abound is very considerable. Captain Vidal ascertained, by soundings made off Tory Island, on the northwest coast of Ireland, that Crustacea, Star-fish, and Testacea occurred at various depths between fifty and one hundred fathoms; and he drew up Dentalia from the mud of Galway Bay, in 230 and 240 *fathoms* water.

The same hydrographer discovered on the Rockhall Bank large quantities of shells at depths varying from 45 to 190 fathoms. The shells were for the most part pulverized, and evidently recent, as they retained their colors. In the same region a bed of fish bones was observed extending for two miles along the bottom of the sea in eighty and ninety fathoms water. At the eastern extremity also of Rockhall Bank, fish-bones were met with, mingled with pieces of fresh shell, at the depth of 235 fathoms.

Analogous formations are in progress in the submarine tracts extending from the Shetland Isles to the north of Ireland, wherever soundings can be procured. A continuous deposit of sand and mud, replete with broken and entire shells, Echini, &c., has been traced for upwards of twenty miles to the eastward of the Faroe Islands, usually at the depth of from forty to one hundred fathoms. In one part of this tract (lat.

61° 50′, long. 6° 30′) fish-bones occur in extraordinary profusion, so that the lead cannot be drawn up without some vertebræ being attached. This "bone bed," as it was called by our surveyors, is three miles and a half in length, and forty-five fathoms under water, and contains a few shells intermingled with the bones.

In the British seas, the shells and other organic remains lie in soft mud or loose sand and gravel; whereas, in the bed of the Adriatic, Donati found them frequently inclosed in stone of recent origin. This is precisely the difference in character which we might have expected to exist between the British marine formations now in progress and those of the Adriatic; for calcareous and other mineral springs abound in the Mediterranean and lands adjoining, while they are almost entirely wanting in our own country. I have already adverted to the eight regions of different depths in the Ægean Sea, each characterized by a peculiar assemblage of shells, which have been described by Professor E. Forbes, who explored them by dredging. (See above, p. 649.)

During his survey of the west coast of Africa, Captain Sir E. Belcher found, by frequent soundings between the twenty-third and twentieth degrees of north latitude, that the bottom of the sea, at the depth of from twenty to about fifty fathoms, consists of sand with a great intermixture of shells, often entire, but sometimes finely comminuted. Between the eleventh and ninth degrees of north latitude, on the same coast, at soundings varying from twenty to about eighty fathoms, he brought up abundance of corals and shells mixed with sand. These also were in some parts entire, and in others worn and broken.

In all these cases, it is only necessary that there should be some depositiou of sedimentary matter, however minute, such as may be supplied by rivers draining a continent, or currents preying on a line of cliffs, in order that stratified formations, hundreds of feet in thickness, and replete with organic remains, should result in the course of ages.

But although some deposits may thus extend continuously for a thousand miles or more near certain coasts, the greater part of the bed of the ocean, remote from continents and islands, may very probably receive, at the same time, no new accessions of drift matter, all sediment being intercepted by intervening hollows, in which a marine current must clear its waters as thoroughly as a turbid river in a lake. Erroneous theories in geology may be formed not only from overlooking the great extent of simultaneous deposits now in progress, but also from the assumption that such formations may be universal or coextensive with the bed of the ocean.

We frequently observe, on the sea beach, very perfect specimens of fossil shells, quite detached from their matrix, which have been washed out of older formations, constituting the sea-cliffs. They may be all of extinct species, like the Eocene freshwater and marine shells strewed over the shores of Hampshire, yet when they become mingled with the shells of the present period, and buried in the same deposits of mud and sand, they would appear, if upraised and examined by future geologists, to have been all of the same age. That such intermixture and blending of organic remains of different ages have actually taken place in former times, is unquestionable, though the occurrence appears to be very local and exceptional. It is, however, a class of accidents more likely than almost any other to lead to serious anachronisms in geological chronology.

CHAPTER L.

FORMATION OF CORAL REEFS.

Growth of coral chiefly confined to tropical regions—Principal genera of coral-building zoophytes—Their rate of growth—Seldom flourish at greater depths than twenty fathoms—Atolls or annular reefs with lagoons—Maldive Isles—Origin of the circular form—Coral reefs not based on submerged volcanic craters—Mr. Darwin's theory of subsidence in explanation of atolls, encircling and barrier reefs—Why the windward side of atolls highest—Subsidence explains why all atolls are nearly on one level—Alternate areas of elevation and subsidence—Origin of openings into the lagoons—Size of atolls and barrier reefs—Objection to the theory of subsidence considered—Composition, structure, and stratified arrangement of rocks now forming in coral reefs—Lime, whence derived—Supposed increase of calcareous matter in modern epochs controverted—Concluding remarks.

THE powers of the organic creation in modifying the form and structure of the earth's crust, are most conspicuously displayed in the labors of the coral animals. We may compare the operation of these zoophytes in the ocean, to the effects produced on a smaller scale upon the land by the plants which generate peat. In the case of the Sphagnum, the upper part vegetates while the lower part is entering into a mineral mass, in which the traces of organization remain when life has entirely ceased. In corals, in like manner, the more durable materials of the generation that has passed away serve as the foundation on which the living animals continue to rear a similar structure.

The stony part of the lamelliform zoophyte may be likened to an internal skeleton; for it is always more or less surrounded by a soft animal substance capable of expanding itself; yet, when alarmed, it has the power of contracting and drawing itself almost entirely into the cells and hollows of the hard coral. Although oftentimes beautifully colored in their own element, the soft parts become when taken from the sea nothing more in appearance than a brown slime spread over the stony nucleus.*

The growth of those corals which form reefs of solid stone is entirely confined to the warmer regions of the globe, rarely extending beyond the tropics above two or three degrees, except under peculiar circumstances,

* Ehrenberg, Nat. und Bild. der Coralleninseln, &c., Berlin, 1834.

as in the Bermuda Islands, in lat. 32° N., where the Atlantic is warmed by the Gulf stream. The Pacific Ocean, throughout a space comprehended between the thirtieth parallels of latitude on each side of the equator, is extremely productive of coral; as also are the Arabian and Persian Gulfs. Coral is also abundant in the sea between the coast of Malabar and the island of Madagascar. Flinders describes a reef of coral on the east coast of New Holland as having a length of nearly 1000 miles, and as being in one part unbroken for a distance of 350 miles. Some groups of coral islands in the Pacific are from 1100 to 1200 miles in length, by 300 or 400 in breadth, as the Dangerous Archipelago, for example, and that called Radack by Kotzebue; but the islands within these spaces are always small points, and often very thinly sown.

Fig. 108.

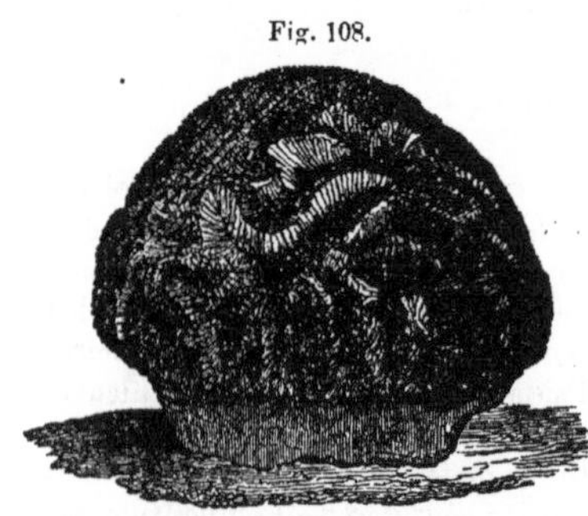

Meandrina labyrinthica, Lam.

Of the numerous species of zoophytes which are engaged in the production of coral banks, some of the most common belong to the Lamarckian genera Astrea, Porites, Madrepora, Millepora, Caryophyllia, and Meandrina.

Rate of the growth of Coral.—Very different opinions have been entertained in regard to the rate at which coral reefs increase. In Captain Beechey's late expedition to the Pacific, no positive information could be obtained of any channel having been filled up within a given period; and it seems established, that several reefs had remained for more than half a century, at about the same depth from the surface.

Ehrenberg also questions the fact of channels and harbors having been closed up in the Red Sea by the rapid increase of coral limestone. He supposes the notion to have arisen from the circumstance of havens having been occasionally filled up in some places with coral sand, in others with large quantities of ballast of coral rock thrown down from vessels.

The natives of the Bermuda Islands point out certain corals now

Genera of Zoophytes most common in coral reefs.

Fig. 109.

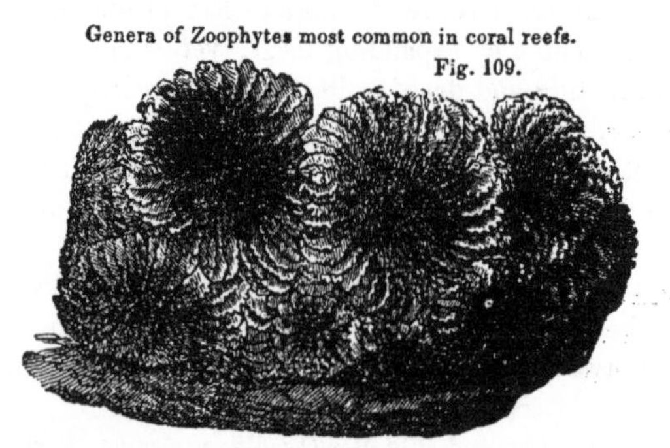

Astrea dipsacca, Lam.

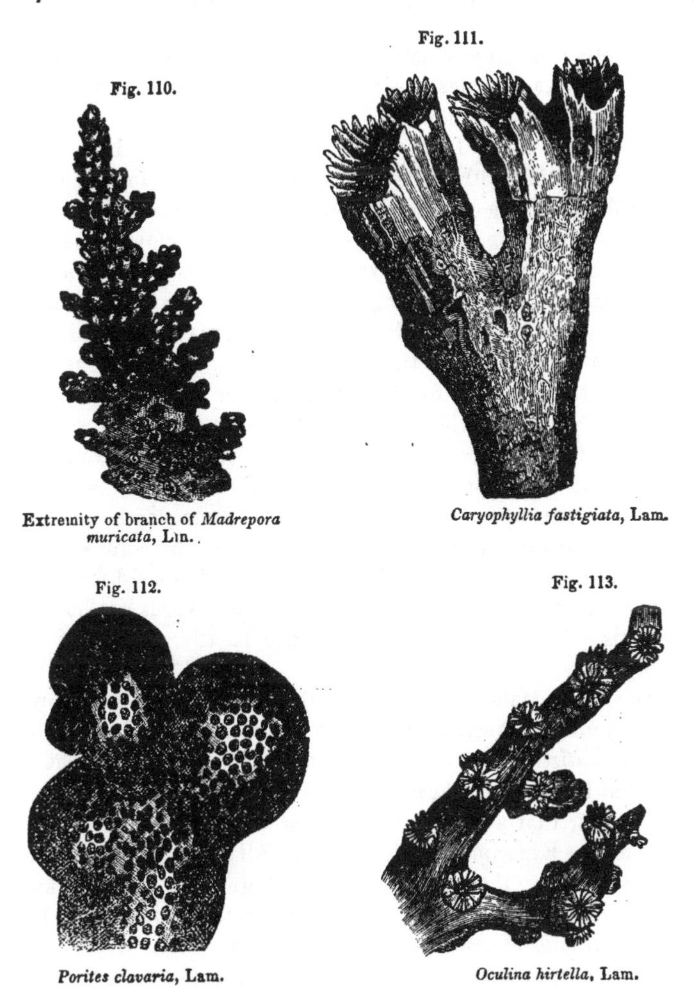

Fig. 110.

Extremity of branch of *Madrepora muricata*, Lin.

Fig. 111.

Caryophyllia fastigiata, Lam.

Fig. 112.

Porites clavaria, Lam.

Fig. 113.

Oculina hirtella, Lam.

growing in the sea, which, according to tradition, have been living in the same spots for centuries. It is supposed that some of them may vie in age with the most ancient trees of Europe. Ehrenberg also observed single corals of the genera Meandrina and Favia, having a globular form, from six to nine feet in diameter, "which must (he says) be of immense antiquity, probably several thousand years old, so that Pharaoh may have looked upon these same individuals in the Red Sea."* They certainly imply, as he remarks, that the reef on which they grow has increased at a very slow rate. After collecting more than 100 species, he found none of them covered with parasitic zoophytes, nor any instance of a living coral growing on another living coral. To this repulsive power which they exert whilst living, against all others of their own class, we owe the beautiful symmetry of some large Meandri-

* See Ehrenberg's work above cited, p. 751.

næ, and other species which adorn our museums. Yet Balani and Serpulæ can attach themselves to living corals, and holes are excavated in them by saxicavous mollusca.

At the island called Taaopoto, in the South Pacific, the anchor of a ship, wrecked about 50 years before, was observed in seven fathoms water, still preserving its original form, but entirely incrusted by coral.* This fact would seem to imply a slow rate of augmentation; but to form a correct estimate of the average rate must be very difficult, since it must vary not only according to the species of coral, but according to the circumstances under which each species may be placed; such, for example, as the depth from the surface, the quantity of light, the temperature of the water, its freedom from sand or mud, or the absence or presence of breakers, which is favorable to the growth of some kinds and is fatal to that of others. It should also be observed that the apparent stationary condition of some coral reefs, which according to Beechey have remained for centuries at the same depth under water, may be due to subsidence, the upward growth of the coral having been just sufficient to keep pace with the sinking of the solid foundation on which the zoophytes have built. We shall afterwards see how far this hypothesis is borne out by other evidence in the regions of annular reefs or atolls.

In one of the Maldive islands a coral reef, which, within a few years, existed on an islet bearing cocoa-nut trees, was found by Lieutenant Prentice, "*entirely covered with live coral and madrepore.*" The natives stated that the islet had been washed away by a change in the currents, and it is clear that a coating of growing coral had been formed in a short time.† Experiments, also, of Dr. Allan, on the east coast of Madagascar, prove the possibility of coral growing to a thickness of three feet in about half a year,‡ so that the rate of increase may, under favorable circumstances, be very far from slow.

It must not be supposed that the calcareous masses termed coral reefs are exclusively the work of zoophytes: a great variety of shells, and, among them, some of the largest and heaviest of known species, contribute to augment the mass. In the South Pacific, great beds of oysters, mussels, *Pinnæ marinæ*, *Chamæ* (or *Tridacnæ*), and other shells, cover in profusion almost every reef; and on the beach of coral islands are seen the shells of echini and broken fragments of crustaceous animals. Large shoals of fish are also discernible through the clear blue water, and their teeth and hard palates cannot fail to be often preserved although their soft cartilaginous bones may decay.

It was the opinion of the German naturalist Forster, in 1780, after his voyage round the world with Captain Cook, that coral animals had the power of building up steep and almost perpendicular walls from great depths in the sea, a notion afterwards adopted by Captain Flinders and

* Stutchbury, West of England Journal, No. i. p. 49.
† Darwin's Coral Reefs, p. 77.
‡ Ibid. 78.

others; but it is now very generally believed that these zoophytes cannot live in water of great depths.

Mr. Darwin has come to the conclusion, that those species which are most effective in the construction of reefs, rarely flourish at a greater depth than 20 fathoms, or 120 feet. In some lagoons, however, where the water is but little agitated, there are, according to Kotzebue, beds of living coral in 25 fathoms of water, or 150 feet; but these may perhaps have begun to live in shallower water, and may have been carried downwards by the subsidence of the reef. There are also various species of zoophytes, and among them some which are provided with calcareous as well as horny stems, which live in much deeper water, even in some cases to a depth of 180 fathoms; but these do not appear to give origin to stony reefs.

There is every variety of form in coral reefs, but the most remarkable and numerous in the Pacific consist of circular or oval strips of dry land, enclosing a shallow lake or lagoon of still water, in which zoophytes and mollusca abound. These annular reefs just raise themselves above the level of the sea, and are surrounded by a deep and often unfathomable ocean.

In the annexed cut (fig. 114), one of these circular islands is represented, just rising above the waves, covered with the cocoa-nut and other trees, and inclosing within a lagoon of tranquil water.

Fig. 114.

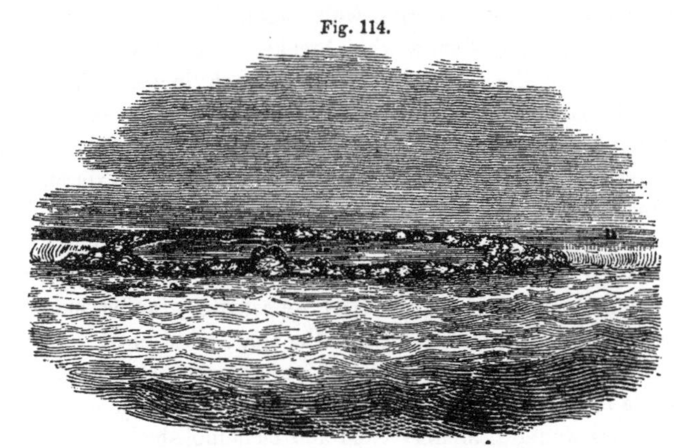

View of Whitsunday Island. (Capt. Beechey.)*

The accompanying section will enable the reader to comprehend the usual form of such islands. (Fig. 115.)

Fig. 115.

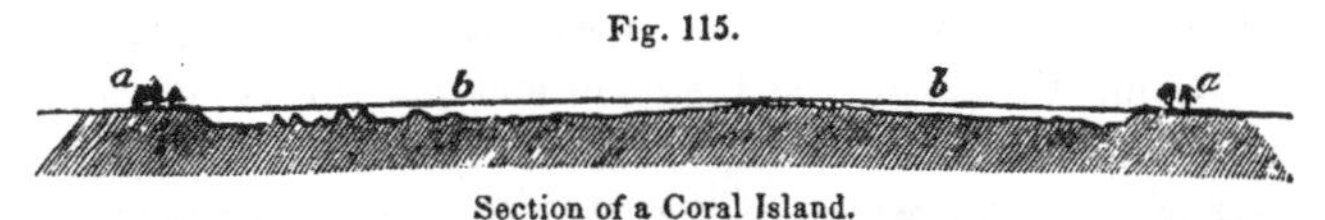

Section of a Coral Island.

a, a, Habitable part of the island, consisting of a strip of coral, inclosing the lagoon.
b, b, The lagoon.

* Voyage to the Pacific, &c. in 1825–28.

The subjoined cut (fig. 116.) exhibits a small part of the section of a coral island on a larger scale.

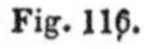

Section of part of a Coral Island.

a, b, Habitable part of the island.
b, c, Slope of the side of the island, plunging at an angle of forty-five to the depth of fifteen hundred feet.
c, c, Part of the lagoon.
d, d, Knolls of coral in the lagoon, with overhanging masses of coral resembling the capitals of columns.

Of thirty-two of these coral islands visited by Beechey in his voyage to the Pacific, twenty-nine had lagoons in their centres. The largest was 30 miles in diameter, and the smallest less than a mile. All were increasing their dimensions by the active operations of the lithophytes, which appeared to be gradually extending and bringing the immersed parts of their structure to the surface. The scene presented by these annular reefs is equally striking for its singularity and beauty. A strip of land a few hundred yards wide is covered by lofty cocoa-nut trees, above which is the blue vault of heaven. This band of verdure is bounded by a beach of glittering white sand, the outer margin of which is encircled with a ring of snow-white breakers, beyond which are the dark heaving waters of the ocean. The inner beach incloses the still clear water of the lagoon, resting in its greater part on white sand, and when illuminated by a vertical sun, of a most vivid green.* Certain species of zoophytes abound most in the lagoon, others on the exterior margin, where there is a great surf. "The ocean," says Mr. Darwin, "throwing its breakers on these outer shores, appears an invincible enemy, yet we see it resisted and even conquered by means which at first seem most weak and inefficient. No periods of repose are granted, and the long swell caused by the steady action of the trade wind never ceases. The breakers exceed in violence those of our temperate regions, and it is impossible to behold them without feeling a conviction that rocks of granite or quartz would ultimately yield and be demolished by such irresistible forces. Yet these low insignificant coral islets stand and are victorious, for here another power, as antagonist to the former, takes part in the contest. The organic forces separate the atoms of carbonate of lime one by one from the foaming breakers, and unite them into a symmetrical structure; myriads of architects are at work night and day, month after month, and we see their soft and gelatinous bodies through the agency of the vital laws conquering the great mechanical power of the waves of an ocean, which neither the art of man, nor the inanimate works of nature could successfully resist." †

* Darwin's Journal, &c., p. 540, and new edit., of 1845, p. 453.
† Darwin's Journal, &c., pp. 547, 548., and 2d edit., of 1845, p. 460.

As the coral animals require to be continually immersed in salt water, they cannot raise themselves by their own efforts, above the level of the lowest tides. The manner in which the reefs are converted into islands above the level of the sea is thus described by Chamisso, a naturalist, who accompanied Kotzebue in his voyages:—"When the reef," says he, "is of such a height that it remains almost dry at low water the corals leave off building. Above this line a continuous mass of solid stone is seen composed of the shells of mollusks and echini, with their broken-off prickles and fragments of coral, united by calcareous sand, produced by the pulverization of shells. The heat of the sun often penetrates the mass of stone when it is dry, so that it splits in many places, and the force of the waves is thereby enabled to separate and lift blocks of coral, frequently six feet long and three or four in thickness, and throw them upon the reef, by which means the ridge becomes at length so high that it is covered only during some seasons of the year by the spring tides. After this the calcareous sand lies undisturbed, and offers to the seeds of trees and plants cast upon it by the waves a soil upon which they rapidly grow, to overshadow its dazzling white surface. Entire trunks of trees, which are carried by the rivers from other countries and islands, find here, at length, a resting-place after their long wanderings: with these come some small animals such as insects and lizards, as the first inhabitants. Even before the trees form a wood, the sea-birds nestle here; stray land-birds take refuge in the bushes; and, at a much later period, when the work has been long since completed, man appears and builds his hut on the fruitful soil."*

In the above description the solid stone is stated to consist of shell and coral, united by sand; but masses of very compact limestone are also found even in the uppermost and newest parts of the reef, such as could only have been produced by chemical precipitation. Professor Agassiz also informs me that his observations on the Florida reefs (which confirm Darwin's theory of atolls to be mentioned in the sequel) have convinced him, that large blocks are loosened, not by shrinkage in the sun's heat, as Chamisso imagined, but by innumerable perforations of lithodomi and other boring testacea.

The carbonate of lime may have been principally derived from the decomposition of corals and testacea; for when the animal matter undergoes putrefaction, the calcareous residuum must be set free under circumstances very favorable to precipitation, especially when there are other calcareous substances, such as shells and corals, on which it may be deposited. Thus organic bodies may be inclosed in a solid cement, and become portions of rocky masses.†

The width of the circular strip of dead coral forming the islands explored by Captain Beechey, exceeded in no instance half a mile from the usual wash of the sea to the edge of the lagoon, and, in general, was only about three or four hundred yards.‡ The depth of the lagoons is

* Kotzebue's Voy., 1815–18, vol. iii. pp. 331–333.

† Stutchbury, West of Eng. Journ., No. i. p. 50. ‡ Captain Beechey, part i. p. 188.

various; in some, entered by Captain Beechey, it was from twenty to thirty-eight fathoms.

The two other peculiarities which are most characteristic of the annular reef or atoll are first, that the strip of dead coral is invariably highest on the windward side, and secondly, that there is very generally an opening at some point in the reef affording a narrow passage, often of considerable depth, from the sea into the lagoon.

Fig. 117.

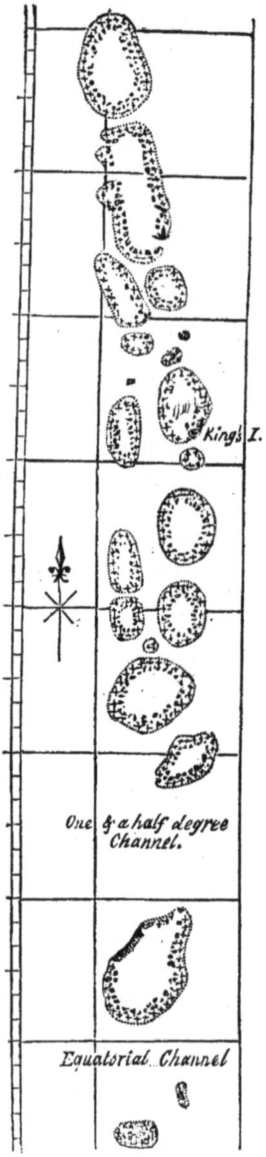

King's I.

One & a half degree Channel.

Equatorial Channel

Maldive and Laccadive Isles.—The chain of reefs and islets called the Maldives (see fig. 117.), situated in the Indian Ocean, to the south-west of Malabar, forms a chain 470 geographical miles in length, running due north and south, with an average breadth of about 50 miles. It is composed throughout of a series of circular assemblages of islets, all formed of coral, the larger groups being from forty to ninety miles in their longest diameter. Captain Horsburgh, whose chart of these islands is subjoined, states, that outside of each circle or *atoll*, as it is termed, there are coral reefs sometimes extending to the distance of two or three miles, beyond which there are no soundings at immense depths. But in the centre of each atoll there is a lagoon from fifteen to forty-nine fathoms deep. In the channels between the atolls no soundings can usually be obtained at the depth of 150 or even 250 fathoms, but during Captain Moresby's survey, soundings were struck at 150 and 200 fathoms, the only instances as yet known of the bottom having been reached, either in the Indian or Pacific oceans, in a space intervening between two separate and well characterized atolls.

The singularity in the form of the atolls of this archipelago consists in their being made up, not of one continuous circular reef but of a ring of small coral islets sometimes more than a hundred in number, each of which is a miniature atoll in itself; in other words, a ring-shaped strip of coral surrounding a lagoon of salt water. To account for the origin of these, Mr. Darwin supposes the larger annular reef to have been broken up into a number of fragments, each of which acquired its peculiar configurations under the influence of causes similar to those to which the structure of the parent atoll has been due. Many of the minor rings are no less than three, and even five miles in diameter, and some are situated in the midst of the principal lagoon; but this happens only in cases where the sea can enter freely through breaches in the outer or marginal reef.

The rocks of the Maldives are composed of sandstone formed of broken shells and corals, such as may be obtained in a loose state from the beach, and which is seen when exposed for a few days to the air to become hardened. The sandstone is sometimes observed to be an aggregate of broken shells, corals, pieces of wood, and shells of the cocoa-nut.*

The Laccadive islands run in the same line with the Maldives, on the north, as do the isles of the Chagos Archipelago, on the south; so that these may be continuations of the same chain of submerged mountains, crested in a similar manner by coral limestones.

Origin of the circular form—not volcanic.—The circular and oval shape of so many reefs, each having a lagoon in the centre, and being surrounded on all sides by a deep ocean, naturally suggested the idea that they were nothing more than the crests of submarine volcanic craters overgrown by coral; and this theory I myself advocated in the earlier editions of this work. Although I am now about to show that it must be abandoned, it may still be instructive to point out the grounds on which it was formerly embraced. In the first place, it had been remarked that there were many active volcanoes in the coral region of the Pacific, and that in some places, as in Gambier's group, rocks composed of porous lava rise up in a lagoon bordered by a circular reef, just as the two cones of eruption called the Kamenis have made their appearance in the times of history within the circular gulf of Santorin.† It was also observed that, as in S. Shetland, Barren Island, and others of volcanic origin, there is one narrow breach in the walls of the outer cone by which ships may enter a circular gulf, so in like manner there is often a single deep passage leading into the lagoon of a coral island, the lagoon itself seeming to represent the hollow or gulf just as the ring of dry coral recals to our minds the rim of a volcanic crater. More lately, indeed, Mr. Darwin has shown that the numerous volcanic craters of the Galapagos Archipelago in the Pacific have all of them their southern sides the lowest, or in many cases quite broken down, so that if they were submerged and incrusted with coral, they would resemble true atolls in shape.‡

Another argument which I adduced when formerly defending this doctrine was derived from Ehrenberg's statement, that some banks of coral in the Red Sea were square, while many others were ribbon-like strips, with flat tops, and without lagoons. Since, therefore, all the genera and many of the species of zoophytes in the Red Sea agreed with those which elsewhere construct lagoon islands, it followed that the stone-making

* Captain Moresby on the Maldives, Journ. Roy. Geograph. Soc., vol. v. part ii. p. 400.

† See above, p. 442.

‡ Darwin, Volcanic Islands, p. 113.

zoophytes are not guided by their own instinct in the formation of annular reefs, but that this peculiar shape and the position of such reefs in the midst of a deep ocean must depend on the outline of the submarine bottom, which resembles nothing else in nature but the crater of a lofty submerged volcanic cone. The enormous size, it is true, of some atolls, made it necessary for me to ascribe to the craters of many submarine volcanoes a magnitude which was startling, and which had often been appealed to as a serious objection to the volcanic theory. That so many of them were of the same height, or just level with the water, did not present a difficulty so long as we remained ignorant of the fact that the reef-building species do not grow at greater depths than twenty-five fathoms.

May be explained by subsidence.—Mr. Darwin, after examining a variety of coral formations in different parts of the globe, was induced to reject the opinion that their shape represented the form of the original bottom. Instead of admitting that the ring of dead coral rested on a circular or oval ridge of rock, or that the lagoon corresponded to a pre-existing cavity, he advanced a new opinion, which must, at first sight, seem paradoxical in the extreme; namely, that the lagoon is precisely in the place once occupied by the highest part of a mountainous island, or, in other cases, by the top of a shoal.

The following is a brief sketch of the facts and arguments in favor of this new view:—Besides those rings of dry coral which enclose lagoons, there are others having a similar form and structure which encircle lofty islands. Of the latter kind is Vanikoro, (see map, fig. 39, p. 351,) celebrated on account of the shipwreck of La Peyrouse, where the coral reef runs at the distance of two or three miles from the shore, the channel between it and the land having a general depth of between 200 and 300 feet. This channel, therefore, is analogous to a lagoon, but with an island standing in the middle like a picture in its frame. In like manner in Tahiti we see a mountainous land, with everywhere round its margin a lake or zone of smooth salt water, separated from the ocean by an encircling reef of coral, on which a line of breakers is always foaming. So also New Caledonia, a long narrow island east of New Holland, in which the rocks are granitic, is surrounded by a reef which runs for a length of 400 miles. This reef encompasses not only the island itself, but a ridge of rocks which are prolonged in the same direction beneath the sea. No one, therefore, will contend for a moment that in this case the corals are based upon the rim of a volcanic crater, in the middle of which stands a mountain or island of granite.

The great barrier reef, already mentioned as running parallel to the north-east coast of Australia for nearly 1000 miles, is another most remarkable example of a long strip of coral running parallel to a coast. Its distance from the mainland varies from twenty to seventy miles, and the depth of the great arm of the sea thus enclosed is usually between ten and twenty fathoms, but towards one end from forty to sixty. This great reef would extend much farther, according to Mr. Jukes, if the growth of coral were not prevented off the shores of New Guinea by a muddy bottom, caused by rivers charged with sediment which flow from the southern coast of that great island.*

Two classes of reefs, therefore, have now been considered; first, the atoll, and, secondly, the encircling and barrier reef, all agreeing perfectly in structure, and the sole difference lying in the absence in the case of the atoll of all land, and in the others the presence of land bounded either by an encircling or a barrier reef. But there is still a third class of reefs, called by Mr. Darwin "fringing reefs," which approach much nearer the land than those of the encircling and barrier class, and which indeed so nearly touched the coast as to leave nothing in the intervening space resembling a lagoon. "That these reefs are not attached quite close to the shore appears to be the result of two causes; first, that the water immediately adjoining the beach is rendered turbid by the surf, and therefore injurious to all zoophytes; and, secondly, that the larger and efficient kinds only flourish on the outer edge amidst the breakers of the open sea." †

It will at once be conceded that there is so much analogy between the form and position of the strip of coral in the atoll, and in the encircling and barrier reef, that no explanation can be satisfactory which does not include the whole. If we turn in the first place to the encircling and barrier reefs, and endeavor to explain how the zoophytes could have found a bottom on which to begin to build, we are met at once with a great difficulty. It is a general fact, long since remarked by Dampier, that high land and deep seas go together. In other words, steep mountains coming down abruptly to the sea-shore are generally continued with the same slope beneath the water. But where the reef, as at *b* and *c* (fig. 118), is distant several miles from a steep coast, a line drawn per-

Fig. 118.

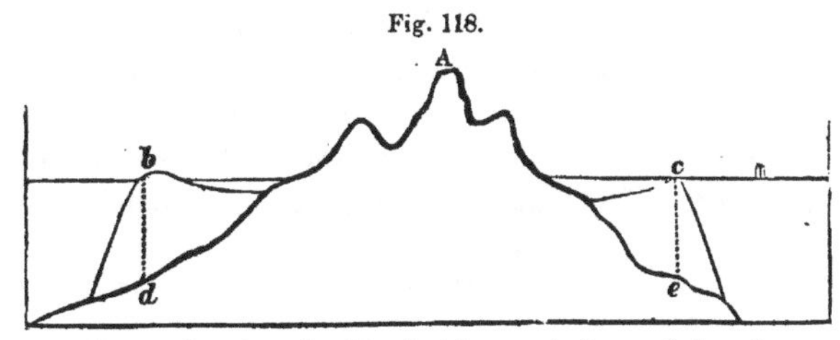

Supposed section of an island with an encircling reef of coral.

A, The island.
b, *c*, Highest points of the encircling reef between which and the coast is seen a space occupied by still water.

pendicularly downwards from its outer edges *b c* to the fundamental rock *d e*, must descend to a depth exceeding by several thousand feet the limits at which the efficient stone-building corals can exist, for we have seen that they cease to grow in water which is more than 120 feet deep. That the original rock immediately beneath the points *b c* is actually as far from the surface as *d e*, is not merely inferred from Dampier's rule,

* Quart. Journ. Geol. Soc 4. XCIII.

† Darwin's Journal, p. 557. 2d edit. chap. 20, and Coral Islands, chapters 1, 2, 3.

but confirmed by the fact, that, immediately outside the reef, soundings are either not met with at all, or only at enormous depths. In short, the ocean is as deep there as might have been anticipated in the neighborhood of a bold coast; and it is obviously the presence of the coral alone which has given rise to the anomalous existence of shallow water on the reef and between it and the land.

After studying in minute detail all the phenomena above described, Mr. Darwin has offered in explanation a theory now very generally adopted. The coral-forming polypi, he states, begin to build in water of a moderate depth, and while they are yet at work, the bottom of the sea subsides gradually, so that the foundation of their edifice is carried downwards at the same time that they are raising the superstructure. If, therefore, the rate of subsidence be not too rapid, the growing coral will continue to build up to the surface; the mass always gaining in height above its original base, but remaining in other respects in the same position. Not so with the land; each inch lost is irreclaimably gone; as it sinks, the water gains foot by foot on the shore, till in many cases the highest peak of the original island disappears. What was before land is then occupied by the lagoon, the position of the encircling coral remaining unaltered, with the exception of a slight contraction of its dimensions.

In this manner are encircling reefs and atolls produced; and in confirmation of his views Mr. Darwin has pointed out examples which illustrate every intermediate state, from that of lofty islands, such as Otaheite, encircled by coral, to that of Gambier's group, where a few peaks only of land rise out of a lagoon, and lastly, to the perfect atoll, having a lagoon several hundred feet deep, surrounded by a reef rising steeply from an unfathomed ocean.

If we embrace these views, it is clear, that in regions of growing coral a similar subsidence must give rise to barrier reefs along the shores of a continent. Thus suppose A (fig. 119), to represent the north-east portion of Australia, and *b c* the ancient level of the sea, when the coral reef *d* was formed. If the land sink so that it is submerged more and

Fig. 119.

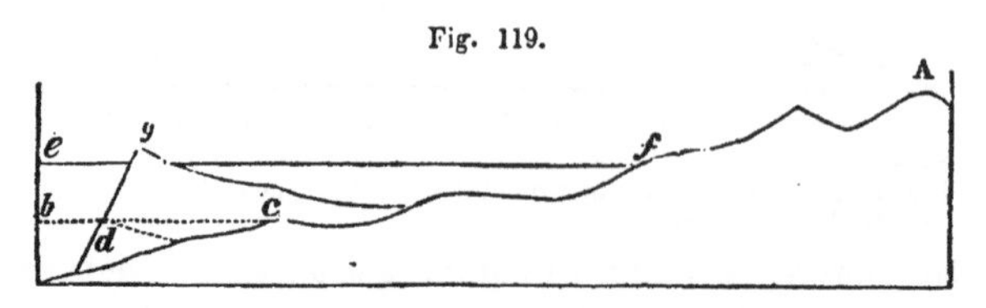

more, the sea must at length stand at the level *e f*, the reef in the mean time having been enlarged and raised to the point *g*. The distance between the shore *f*, and the barrier reef *g*, is now much greater than originally between the shore *c* and the reef *d*, and the longer the subsidence continues the farther will the coast of the mainland recede.

When the first edition of this work appeared in 1831, several years before Mr. Darwin had investigated the facts on which his theory is founded, I had come to the opinion that the land was subsiding at the bottom of those parts of the Pacific where atolls are numerous, although I failed to perceive that such a subsidence, if conceded, would equally solve the enigma as to the form both of annular and barrier reefs.

I shall cite the passage referred to, as published by me in 1831 :—"It is a remarkable circumstance that there should be so vast an area in Eastern Oceanica, studded with minute islands, without one single spot where there is a wider extent of land than belongs to such islands as Otaheite, Owhyhee, and a few others, which either have been or are still the seats of active volcanoes. If an equilibrium only were maintained between the upheaving and depressing force of earthquakes, large islands would very soon be formed in the Pacific; for, in that case, the growth of limestone, the flowing of lava, and the ejection of volcanic ashes, would combine with the upheaving force to form new land.

"Suppose a shoal, 600 miles in length, to sink fifteen feet, and then to remain unmoved for a thousand years; during that interval the growing coral may again approach the surface. Then let the mass be re-elevated fifteen feet, so that the original reef is restored to its former position: in this case, the new coral formed since the first subsidence will constitute an island 600 miles long. An analogous result would have occurred if a lava-current fifteen feet thick had overflowed the submerged reef. The absence, therefore, of more extensive tracts of land in the Pacific, seems to show that the amount of subsidence by earthquakes exceeds, in that quarter of the globe, at present, the elevation due to the same cause."*

Another proof also of subsidence derived from the structure of atolls, was pointed out by me in the following passage in all former editions. "The low coral islands of the Pacific," says Captain Beechey, "follow one general rule in having their windward side higher and more perfect than the other. At Gambia and Matilda islands this inequality is very conspicuous, the weather side of both being wooded, and of the former inhabited, while the other sides are from twenty to thirty feet under water; where, however, they may be perceived to be equally *narrow* and well defined. It is on the leeward side also that the entrances into the lagoons occur; and although they may sometimes be situated on a side that runs in the direction of the wind, as at Bow Island, yet there are none to windward." These observations of Captain Beechey accord with those which Captain Horsburgh and other hydrographers have made in regard to the coral islands of other seas. From this fortunate circumstance ships can enter and sail out with ease; whereas if the narrow inlets were to windward, vessels which once entered might not succeed for months in making their way out again. The well-known security of many of these harbors depends entirely on this fortunate peculiarity in their structure.

"In what manner is this singular conformation to be accounted for? The action of the waves is seen to be the cause of the superior elevation

* See Principles of Geology, 1st edit., vol. ii. p. 296.

of some reefs on their windward sides, where sand and large masses of coral rock are thrown up by the breakers; but there is a variety of cases where this cause alone is inadequate to solve the problem; for reefs submerged at considerable depths, where the movements of the sea cannot exert much power, have, nevertheless, the same conformation, the leeward being much lower than the windward side.*

"I am informed by Captain King, that, on examining the reefs called Rowley Shoals, which lie off the north-west coast of Australia, where the east and west monsoons prevail alternately, he found the open side of one crescent-shaped reef, the Impérieuse, turned to the east, and of another, the Mermaid, turned to the west; while a third oval reef, of the same group, was entirely submerged. This want of conformity is exactly what we should expect, where the winds vary periodically.

"It seems impossible to refer the phenomenon now under consideration to any original uniformity in the configuration of submarine volcanoes, on the summits of which we may suppose the coral reefs to grow; for although it is very common for craters to be broken down on one side only, we cannot imagine any cause that should breach them all in the same direction. But the difficulty will, perhaps, be removed, if we call in another part of the volcanic agency—subsidence by earthquakes. Suppose the windward barrier to have been raised by the mechanical action of the waves to the height of two or three yards above the wall on the leeward side, and then the whole island to sink down a few fathoms, the appearances described would then be presented by the submerged reef. A repetition of such operations, by the alternate elevation and depression of the same mass (an hypothesis strictly conformable to analogy), might produce still greater inequality in the two sides, especially as the violent efflux of the tide has probably a strong tendency to check the accumulation of the more tender corals on the leeward reef; while the action of the breakers contributes to raise the windward barrier."†

Previously to my adverting to the signs above enumerated of a downward movement in the bed of the ocean, Dr. MacCulloch, Captain Beechey, and many other writers, had shown that masses of recent coral had been laid dry at various heights above the sea-level, both in the Red Sea, the islands of the Pacific, and in the East and West Indies. After describing thirty-two coral islands in the Pacific, Captain Beechey mentioned that they were all formed of living coral except one, which, although of coral formation, was raised about seventy or eighty feet above the level of the sea, and was encompassed by a reef of living coral. It is called Elizabeth or Henderson's Island, and is five miles in length by one in breadth. It has a flat surface, and, on all sides, except the north, is bounded by perpendicular cliffs about fifty feet high, composed entirely of dead coral, more or less porous, honey-combed at the surface, and har-

* Voyage to the Pacific, &c., p. 189.
† See Principles of Geology, 1st ed., 1832, vol. ii. p. 293.

dening into a compact calcareous mass, which possesses the fracture of secondary limestone, and has a species of millepore interspersed through it. These cliffs are considerably undermined by the action of the waves, and some of them appear on the eve of precipitating their superincum-

Fig. 120.

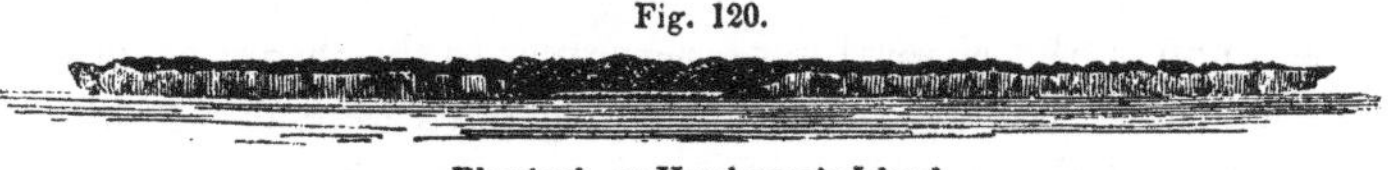

Elizabeth, or Henderson's Island.

bent weight into the sea. Those which are less injured in this way present no alternate ridges or indication of the different levels which the sea might have occupied at different periods; but a smooth surface, as if the island, which has probably been raised by volcanic agency, had been forced up by one great subterraneous convulsion.* At the distance of a few hundred yards from this island, no bottom could be gained with 200 fathoms of line.

It will be seen, from the annexed sketch, communicated to me by Lieutenant Smith, of the Blossom, that the trees came down to the beach towards the centre of the island; a break at first sight resembling the openings which usually lead into lagoons; but the trees stand on a steep slope, and no hollow of an ancient lagoon was perceived.

Beechey also remarks, that the surface of Henderson's Island is flat, and that in Queen Charlotte's Island, one of the same group, but under water, there was no lagoon, the coral having grown up everywhere to one level. The probable cause of this obliteration of the central basin or lagoon will be considered in the sequel.

That the bed of the Pacific and Indian oceans, where atolls are frequent, must have been sinking for ages, might be inferred, says Mr. Darwin, from simply reflecting on two facts; first, that the efficient coral-building zoophytes do not flourish in the ocean at a greater depth than 120 feet; and, secondly, that there are spaces occupying areas of many hundred thousand square miles, where all the islands consist of coral, and yet none of which rise to a greater height than may be accounted for by the action of the winds and waves on broken and triturated coral. Were we to take for granted that the floor of the ocean had remained stationary from the time when the coral began to grow, we should be compelled to assume that an incredible number of submarine mountains of vast height (for the ocean is always deep, and often unfathomable between the different atolls) had all come to within 120 feet of the surface, and yet no one mountain had risen above water. But no sooner do we admit the theory of subsidence, than this great difficulty vanishes. However varied may have been the altitude of different islands, or the separate peaks of particular mountain-chains, all may have been reduced to one uniform level by the gradual submergence of the loftiest points, and the additions made to the calcareous cappings of the less elevated summits as they subsided to great depths.

* Beechey's Voyage to the Pacific, &c., p. 46.

Openings into the lagoons.—In the general description of atolls and encircling reefs, it was mentioned that there is almost always a deep narrow passage opening into the lagoon, or into the still water between the reef and the shore, which is kept open by the efflux of the sea as the tide goes down.

The origin of this channel must, according to the theory of subsidence before explained, be traced back to causes which were in action during the existence of the encircling reef, and when an island or mountain-top rose within it, for such a reef precedes the atoll in the order of formation. Now in those islands in the Pacific, which are large enough to feed small rivers, there is generally an opening or channel in the surrounding coral reef at the point where the stream of fresh water enters the sea. The depth of these channels rarely exceeds twenty-five feet; and they may be attributed, says Captain Beechey, to the aversion of the lithophytes to fresh water, and to the probable absence of the mineral matter of which they construct their habitations.*

Mr. Darwin, however, has shown, that mud at the bottom of river-courses is far more influential than the freshness of the water in preventing the growth of the polypi, for the walls which inclose the openings are perpendicular, and do not slant off gradually, as would be the case, if the nature of the element presented the only obstacle to the increase of the coral-building animals.

When a breach has thus been made in the reef, it will be prevented from closing up by the efflux of the sea at low tides; for it is sufficient that a reef should rise a few feet above low-water mark to cause the waters to collect in the lagoon at high tide, and when the sea falls, to rush out at one or more points where the reef happens to be lowest or weakest. This event is strictly analogous to that witnessed in our estuaries, where a body of salt water accumulated during the flow issues with great velocity at the ebb of the tide, and scours out or keeps open a deep passage through the bar, which is almost always formed at the mouth of a river. At first there are probably many openings, but the growth of the coral tends to obstruct all those which do not serve as the principal channels of discharge; so that their number is gradually reduced to a few, and often finally to one. The fact observed universally, that the principal opening fronts a considerable valley in the encircled island, between the shores of which and the outer reef there is often deep water, scarcely leaves any doubt as to the real origin of the channel in all those countless atolls where the nucleus of land has vanished.

Size of atolls and barrier reefs.—In regard to the dimensions of atolls, it was stated that some of the smallest observed by Beechey in the Pacific were only a mile in diameter. If their external slope under water equals upon an average an angle of 45°, then would such an atoll at the depth of half a mile, or 2640 feet, have a diameter of two miles. Hence it would appear that there must be a tendency in every atoll to

* Voyage to the Pacific, &c., p. 194.

grow smaller, except in those cases where oscillations of level enlarge the base on which the coral grows by throwing down a talus of detrital matter all round the original cone of limestone.

Bow Island is described by Captain Beechey as seventy miles in circumference, and thirty in its greatest diameter, but we have seen that some of the Maldives are much larger.

As the shore of an island or continent which is subsiding will recede from a coral reef at a slow or rapid rate according as the surface of the land has a steep or gentle slope, we cannot measure the thickness of the coral by its distance from the coast; yet, as a general rule, those reefs which are farthest from the land imply the greatest amount of subsidence. We learn from Flinders, that the barrier reef of north-eastern Australia is in some places seventy miles from the mainland, and it should seem that a calcareous formation is there in progress 1000 miles long from north to south, with a breadth varying from twenty to seventy miles. It may not, indeed, be continuous over this vast area, for doubtless innumerable islands have been submerged one after another between the reef and mainland, like some which still remain, as, for example, Murray's Islands, lat. 9° 54′ S. We are also told that some parts of the gulf inclosed within a barrier are 400 feet deep, so that the efficient rock-building corals cannot be growing there, and in other parts of it islands appear encircled by reefs.

It will follow as one of the consequences of the theory already explained that, provided the bottom of the sea does not sink too fast to allow the zoophytes to build upwards at the same pace, the thickness of coral will be great in proportion to the rapidity of subsidence, so that if one area sinks two feet while another sinks one, the mass of coral in the first area will be double that in the second. But the downward movement must in general have been very slow and uniform, or where intermittent, must have consisted of a great number of depressions, each of slight amount, otherwise the bottom of the sea would have been carried down faster than the corals could build upwards, and the island or continent would be permanently submerged, having reached a depth of 120 or 150 feet, at which the effective reef-constructing zoophytes cease to live. If, then, the subsidence required to account for all the existing atolls must have amounted to three or four thousand feet, or even sometimes more, we are brought to the conclusion that there has been a *slow* and *gradual* sinking to this enormous extent. Such an inference is perfectly in harmony with views which the grand scale of denudation, everywhere observable in the older rocks, has led geologists to adopt in reference to upward movements. They must also have been gradual and continuous throughout indefinite ages to allow the waves and currents of the ocean to operate with adequate power.

The map constructed by Mr. Darwin to display at one view the geographical position of all the coral reefs throughout the globe is of the highest geological interest (see above, p. 351.), leading to splendid generalizations, when we have once embraced the theory that all atolls and barrier reefs indicate recent subsidence, while the presence of fringing reefs proves the land to be stationary or rising. These two classes of coral formations are depicted by different colors; and one of the striking facts brought to light by the same classification of coral formations is the absence of active volcanoes in the areas of subsidence, and their frequent presence in the areas of elevation. The only supposed exception to this remarkable coincidence at the time when Mr. Darwin wrote, in 1842, was the volcano of Torres Strait, at the northern point of Australia, placed on the borders of an area of subsidence; but it has been since proved that this volcano has no existence.

We see, therefore, an evident connection, first, between the bursting forth every now and then of volcanic matter through rents and fissures, and the expansion or forcing outwards of the earth's crust, and, secondly, between a dormant and less energetic development of subterranean heat, and an amount of subsidence sufficiently great to cause mountains to disappear over the broad face of the ocean, leaving only small and scattered lagoon islands, or groups of atolls, to indicate the spots where those mountains once stood.

On a review of the differently-colored reefs on the map alluded to, it will be seen that there are large spaces in which upheaval, and others in which depression prevails, and these are placed alternately, while there are a few smaller areas where movements of oscillation occur. Thus if we commence with the western shores of South America, between the summit of the Andes and the Pacific (a region of earthquakes and active volcanoes), we find signs of recent elevation, not attested indeed by coral formations, which are wanting there, but by upraised banks of marine shells. Then proceeding westward, we traverse a deep ocean without islands, until we come to a band of *atolls* and encircled islands, including the Dangerous and Society archipelagoes, and constituting an area of subsidence more than 4000 miles long and 600 broad. Still farther, in the same direction, we reach the chain of islands to which the New Hebrides, Salomon, and New Ireland belong, where fringing reefs and masses of elevated coral indicate another area of upheaval. Again, to the westward of the New Hebrides we meet with the encircling reef of New Caledonia and the great Australian barrier, implying a second area of subsidence.

The only objection deserving attention which has hitherto been advanced against the theory of atolls, as before explained (p. 759.), is that proposed by Mr. Maclaren.* "On the outside," he observes, "of coral reefs very highly inclined, no bottom is sometimes found with a line of 2000 or 3000 feet, and this is by no means a rare case. It follows that the reef ought to have this thickness; and Mr. Darwin's diagrams show that he understood it so. Now, if such masses of coral exist under the sea, they ought somewhere to be found on *terra firma;* for there is evidence that all the lands yet visited by geologists, have been at one

* Scotsman, Nov. 1842, and Jameson's Edin. Journ. of Science, 1843.

time submerged. But neither in the great volcanic chain, extending from Sumatra to Japan, nor in the West Indies, nor in any other region yet explored, has a bed or formation of coral even 500 feet thick been discovered, so far as we know."

When considering this objection, it is evident that the first question we have to deal with is, whether geologists have not already discovered calcareous masses of the required thickness and structure, or precisely such as the upheaval of atolls might be expected to expose to view? We are called upon, in short, to make up our minds both as to the internal composition of the rocks that must result from the growth of corals, whether in lagoon islands or barrier reefs, and the external shape which the reefs would retain when upraised gradually to a vast height,—a task by no means so easy as some may imagine. If the reader has pictured to himself large masses of entire corals, piled one upon another, for a thickness of several thousand feet, he unquestionably mistakes altogether the nature of the accumulations now in progress. In the first place, the strata at present forming very extensively over the bottom of the ocean, within such barrier reefs as those of Australia and New Caledonia, are known to consist chiefly of horizontal layers of calcareous sediment, while here and there an intermixture must occur of the detritus of granitic and other rocks brought down by rivers from the adjoining lands, or washed from sea-cliffs by the waves and currents. Secondly, in regard to atolls, the stone-making polypifers grow most luxuriantly on the outer edge of the island, to a thickness of a few feet only. Beyond this margin broken pieces of coral and calcareous sand are strewed by the breakers over a steep seaward slope, and as the subsidence continues the next coating of live coral does not grow vertically over the first layer, but on a narrow annular space within it, the reef, as was before stated (p. 761), constantly contracting its dimensions as it sinks. Thirdly, within the lagoon the accumulation of calcareous matter is chiefly sedimentary, a kind of chalky mud derived from the decay of the softer corallines, with a mixture of calcareous sand swept by the winds and waves from the surrounding circular reef. Here and there, but only in partial clumps, are found living corals, which grow in the middle of the lagoon, and mixed with fine mud and sand, a great variety of shells, and fragments of testacea and echinoderms.

We owe to Lieutenant Nelson the discovery that in the Bermudas the calcareous mud resulting from the decomposition of the softer corallines is absolutely undistinguishable when dried from the ordinary white chalk of Europe,* and this mud is carried to great distances by currents, and spread far and wide over the floor of the ocean. We also have opportunities of seeing in upraised atolls, such as Elizabeth Island, Tonga, and Hapai, which rise above the level of the sea to heights varying from ten to eighty feet, that the rocks of which they consist do not differ in structure or in the state of preservation of their included zoophytes and

* Trans. Geol. Soc., London, 2d series, vol. v.

shells from some of the oldest limestones known to the geologist. Captain Beechey remarks that the dead coral in Elizabeth Island is more or less porous and honeycombed at the surface, and hardening into a com pact rock which has the fracture of *secondary limestone.* *

The island of Pulo Nias, off Sumatra (see Map, fig. 39. p. 351), which is about 3000 feet high, is described by Dr. Jack as being overspread by coral and large shells of the *Chama* (*Tridacna*) *gigas*, which rest on quartzose and arenaceous rocks, at various levels from the sea-coast to the summit of the highest hills.

The cliffs of the island of Timor in the Indian Ocean are composed, says Mr. Jukes, of a raised coral reef abounding in *Astræa*, *Meandrina*, and *Porites*, with shells of *Strombus*, *Conus*, *Nerita*, *Arca*, *Pecten*, *Venus*, and *Lucina*. On a ledge about 150 feet above the sea, a Tridacna (or large clam shell), two feet across, was found bedded in the rock with closed valves, just as they are often seen in barrier reefs. This formation in the islands of Sandlewood, Sumbawa, Madura, and Java, where it is exposed in sea cliffs, was found to be from 200 to 300 feet thick, and is believed to ascend to much greater heights in the interior. It has usually the form of a "chalk-like" rock, white when broken, but in the weathered surface turning nearly black. †

It appears, therefore, premature to assert that there are no recent coral formations uplifted to great heights, for we are only beginning to be acquainted with the geological structure of the rocks of equatorial regions. Some of the upraised islands, such as Elizabeth and Queen Charlotte, in the Pacific, although placed in regions of atolls, are described by Captain Beechey and others as flat-topped, and exhibiting no traces of lagoons. In explanation of the fact, we may presume that after they had been sinking for ages, the descending movement was relaxed; and while it was in the course of being converted into an ascending one, the ground remained for a long season almost stationary, in which case the corals within the lagoon would build up to the surface, and reach the level already attained by those on the margin of the reef. In this manner the lagoon would be effaced, and the island acquire a flat summit.

It may, however, be thought strange that many examples have not been noticed of fringing reefs uplifted above the level of the sea. Mr. Darwin, indeed, cites one instance where the reef preserved, on dry land in the Mauritius, its peculiar moat-like structure; but they ought, he says, to be of rare occurrence, for in the case of atolls or of barrier or fringing reefs, the characteristic outline must usually be destroyed by denudation as soon as a reef begins to rise; since it is immediately exposed to the action of the breakers, and the large and conspicuous corals on the outer rim of the atoll or barrier are the first to be destroyed and to fall to the bottom of vertical and undermined cliffs. After slow and continued

* Beechey's Voyage, vol. i. p. 45.
† Paper read to Brit. Assoc., Southampton, 1846.

upheaval a wreck alone can remain of the original reef. If, therefore, says Mr. Darwin, "at some period as far in futurity as the secondary rocks are in the past, the bed of the Pacific with its atolls and barrier reefs should be converted into a continent, we may conceive that scarcely any or none of the existing reefs would be preserved, but only widely spread strata of calcareous matter derived from their wear and tear." *

When it is urged in support of the objection before stated (p. 767), that the theory of atolls by subsidence implies the accumulation of calcareous formations 2000 or 3000 feet thick, it must be conceded that this estimate of the minimum density of the deposits is by no means exaggerated. On the contrary, when we consider that the space over which atolls are scattered in Polynesia and the Indian oceans may be compared to the whole continent of Asia, we cannot but infer from analogy that the differences in level in so vast an area have amounted, antecedently to subsidence, to 5000 or even a greater number of feet. Whatever was the difference in height between the loftiest and lowest of the original mountains or mountainous islands on which the different atolls are based, that difference must represent the thickness of coral which has now reduced all of them to one level. Flinders, therefore, by no means exaggerated the volume of the limestone, which he conceived to have been the work of coral animals; he was merely mistaken as to the manner in which they were enabled to build reefs in an unfathomed ocean.

But is it reasonable to expect, after the waste caused by denudation, that calcareous masses, gradually upheaved in an open sea, should retain such vast thicknesses? Or may not the limestones of the cretaceous and oolitic epochs, which attain in the Alps and Pyrenees a density of 3000 or 4000 feet, and are in great part made up of coralline and shelly matter, present us with a true geological counterpart of the recent coral reefs of equatorial seas?

Before we attach serious importance to arguments founded on negative evidence, and opposed to a theory which so admirably explains a great variety of complicated phenomena, we ought to remember that the upheaval to the height of 4000 feet of atolls in which the coralline limestone would be 4000 feet thick, implies, first, a slow subsidence of 4000 feet, and, secondly, an elevation of the same amount. Even if the reverse or ascending movement began the instant the downward one ceased, we must allow a great lapse of ages for the accomplishment of the whole operation. We must also assume that at the commencement of the period in question, the equatorial regions were as fitted as now for the support of reef-building zoophytes. This postulate would demand the continuance of a complicated variety of conditions throughout a much longer period than they are usually persistent in one place.

To show the difficulty of speculating on the permanence of the geographical and climatal circumstances requisite for the growth of reef-

* Letter to Mr. Maclaren, Scotsman, 1843.

building corals, we have only to state the fact that there are no reefs in the Atlantic, off the west coast of Africa, nor among the islands of the Gulf of Guinea, nor in St. Helena, Ascension, the Cape Verdes, or St. Paul's. With the exception of Bermuda, there is not a single coral reef in the central expanse of the Atlantic, although in some parts the waves, as at Ascension, are charged to excess with calcareous matter. This capricious distribution of coral reefs is probably owing to the absence of fit stations for the reef-building polypifers, other organic beings in those regions obtaining in the great struggle for existence a mastery over them. Their absence, in whatever manner it be accounted for, should put us on our guard against expecting upraised reefs at all former geological epochs, similar to those now in progress.

Lime, whence derived.—Dr. Maculloch, in his system of Geology, vol. i. p. 219, expressed himself in favor of the theory of some of the earlier geologists, that all limestones have originated in organized substances. If we examine, he says, the quantity of limestone in the primary strata, it will be found to bear a much smaller proportion to the siliceous and argillaceous rocks than in the secondary; and this may have some connexion with the rarity of testaceous animals in the ancient ocean. He farther infers, that in consequence of the operations of animals, "the quantity of calcareous earth deposited in the form of mud or stone is always increasing; and that as the secondary series far exceeds the primary in this respect, so a third series may hereafter arise from the depths of the sea, which may exceed the last in the proportion of its calcareous strata."

If these propositions went no farther than to suggest that every particle of lime that now enters into the crust of the globe, may possibly in its turn have been subservient to the purposes of life, by entering into the composition of organized bodies, I should not deem the speculation improbable; but, when it is hinted that lime may be an animal product combined by the powers of vitality from some simple elements, I can discover no sufficient grounds for such an hypothesis, and many facts militate against it.

If a large pond be made in almost any soil, and filled with rain water, it may usually become tenanted by testacea; for carbonate of lime is almost universally diffused in small quantities. But if no calcareous matter be supplied by waters flowing from the surrounding high grounds, or by springs, no tufa or shell-marl are formed. The thin shells of one generation of mollusks decompose, so that their elements afford nutriment to the succeeding races; and it is only where a stream enters a lake, which may introduce a fresh supply of calcareous matter, or where the lake is fed by springs, that shells accumulate and form marl.

All the lakes in Forfarshire which have produced deposits of shell-marl have been the sites of springs, which still evolve much carbonic acid, and a small quantity of carbonate of lime. But there is no marl in Loch Fithie, near Forfar, where there are *no springs*, although that lake is surrounded by these calcareous deposits, and although, in every other respect, the site is favorable to the accumulation of aquatic testacea.

We find those Charæ which secrete the largest quantity of calcareous matter in their stems to abound near springs impregnated with carbonate of lime. We know that, if the common hen be deprived altogether of calcareous nutriment, the shells of her eggs will become of too slight a consistency to protect the contents; and some birds eat chalk greedily during the breeding season.

If, on the other hand, we turn to the phenomena of inorganic nature, we observe that, in volcanic countries, there is an enormous evolution of carbonic acid, either free, in a gaseous form, or mixed with water; and the springs of such districts are usually impregnated with carbonate of lime in great abundance. No one who has travelled in Tuscany, through the region of extinct volcanos and its confines, or who has seen the map constructed by Targioni (1827), to show the principal sites of mineral springs, can doubt, for a moment, that if this territory was submerged beneath the sea, it might supply materials for the most extensive coral reefs. The importance of these springs is not to be estimated by the magnitude of the rocks which they have thrown down on the slanting sides of hills, although of these alone large cities might be built, nor by a coating of travertin that covers the soil in some districts for miles in length. The greater part of the calcareous matter passes down in a state of solution to the sea, and in all countries the rivers which flow from chalk and other marly and calcareous rocks carry down vast quantities of lime into the ocean. Lime is also one of the component parts of augite and other volcanic and hypogene minerals, and when these decompose is set free, and may then find its way in a state of solution to the sea.

The lime, therefore, contained generally in sea water, and secreted so plentifully by the testacea and corals of the Pacific, may have been derived either from springs rising up in the bed of the ocean, or from rivers fed by calcareous springs, or impregnated with lime derived from disintegrated rocks, both volcanic and hypogene. If this be admitted, the greater proportion of limestone in the more modern formations as compared to the most ancient, will be explained, for springs in general hold no argillaceous, and but a small quantity of siliceous matter in solution, but they are continually subtracting calcareous matter from the inferior rocks. The constant transfer, therefore, of carbonate of lime from the lower or older portions of the earth's crust to the surface, must cause at all periods and throughout an indefinite succession of geological epochs, a preponderance of calcareous matter in the newer as contrasted with the older formations.

THE END.

CONCLUDING REMARKS.

In the concluding chapters of the first book, I examined in detail a great variety of arguments which have been adduced to prove the distinctness of the state of the earth's crust at remote and recent epochs. Among other supposed proofs of this distinctness, the dearth of calcareous matter, in the ancient rocks above adverted to, might have been considered. But it would have been endless to enumerate all the objections urged against those geologists who represent the course of nature at the earliest periods as resembling in all essential circumstances the state of things now established. We have seen that, in opposition to this doctrine, a strong desire has been manifested to discover in the ancient rocks the signs of an epoch when the planet was uninhabited, and when its surface was in a chaotic condition and uninhabitable. The opposite opinion, indeed, that the oldest of the rocks now visible may be the last monuments of an antecedent era in which living beings may already have peopled the land and water, has been declared to be equivalent to the assumption that there never was a beginning to the present order of things.

With equal justice might an astronomer be accused of asserting that the works of creation extended throughout *infinite* space, because he refuses to take for granted that the remotest stars now seen in the heavens are on the utmost verge of the material universe. Every improvement of the telescope has brought thousands of new worlds into view; and it would, therefore, be rash and unphilosophical to imagine that we already survey the whole extent of the vast scheme, or that it will ever be brought within the sphere of human observation.

But no argument can be drawn from such premises in favor of the infinity of the space that has been filled with worlds; and if the material universe has any limits, it then follows, that it must occupy a minute and infinitesimal point in infinite space.

So if, in tracing back the earth's history, we arrive at the monuments of events which may have happened millions of ages before our times, and if we still find no decided evidence of a commencement, yet the arguments from analogy in support of the probability of a beginning remain unshaken; and if the past duration of the earth be finite, then the aggregate of geological epochs, however numerous, must constitute a mere moment of the past, a mere infinitesimal portion of eternity.

It has been argued, that, as the different states of the earth's surface, and the different species by which it has been inhabited have all had their origin, and many of them their termination, so the entire series may have commenced at a certain period. It has also been urged, that, as we admit the creation of man to have occurred at a comparatively modern epoch—as we concede the astonishing fact of the first introduction of a moral and intellectual being—so also we may conceive the first creation of the planet itself.

I am far from denying the weight of this reasoning from analogy; but, although it may strengthen our conviction, that the present system of change has not gone on from eternity, it cannot warrant us in presuming that we shall be permitted to behold the signs of the earth's origin, or the evidences of the first introduction into it of organic beings. We aspire in vain to assign limits to the works of creation in *space*, whether we examine the starry heavens, or that world of minute animalcules which is revealed to us by the microscope. We are prepared, therefore, to find that in *time* also the confines of the universe lie beyond the reach of mortal ken. But in whatever direction we pursue our researches, whether in time or space, we discover everywhere the clear proofs of a Creative Intelligence, and of His foresight, wisdom, and power.

As geologists, we learn that it is not only the present condition of the globe which has been suited to the accommodation of myriads of living creatures, but that many former states also have been adapted to the organization and habits of prior races of beings. The disposition of the seas, continents, and islands, and the climates, have varied; the species likewise have been changed; and yet they have all been so modelled, on types analogous to those of existing plants and animals, as to indicate, throughout, a perfect harmony of design and unity of purpose. To assume that the evidence of the beginning or end of so vast a scheme lies within the reach of our philosophical inquiries, or even of our speculations, appears to be inconsistent with a just estimate of the relations which subsist between the finite powers of man and the attributes of an Infinite and Eternal Being.

GLOSSARY

OF GEOLOGICAL AND OTHER SCIENTIFIC TERMS USED IN THIS WORK.

ACEPHALOUS. The Acephala are that division of molluscous animals which, like the oyster and scallop, are without heads. The class Acephala of Cuvier comprehends many genera of animals with bivalve shells, and a few which are devoid of shells. *Etym.*, α, *a*, without, and κεφαλη, *cephale*, the head.

ACIDULOUS. Slightly acid.

ACROGENS. One of five classes into which all plants may be divided; it includes such flowerless ones as grow from the top only, and whose stems consequently do not increase materially in bulk, as Mosses, Ferns, Lycopodiums, Equisetums, &c. The trunk of a tree fern is a good example. They are also called Acrobrya. *Etym.*, ακρον, *acron*, the top, and γενεσις, *genesis*, increase.

ADIPOCIRE. A substance apparently intermediate between fat and wax, into which dead animal matter is converted when buried in the earth, and in a certain stage of decomposition. *Etym.*, *adeps*, fat, and *cera*, wax.

ALBITE. See "Felspar."

ALEMBIC. An apparatus for distilling.

ALGÆ. An order or division of the cryptogamic class of plants. The whole of the sea-weeds are comprehended under this division, and the application of the term in this work is to marine plants. *Etym.*, *alga*, sea-weed.

ALLUVIAL. The adjective of alluvium, which see.

ALLUVION. Synonymous with alluvium, which see.

ALLUVIUM. Earth, sand, gravel, stones, and other transported matter which has been washed away and thrown down by rivers, floods, or other causes upon land not *permanently* submerged beneath the waters of lakes or seas. *Etym.*, *alluo*, to wash upon, or *alluvio*, an inundation.

ALUM-STONE, ALUMEN, ALUMINOUS. Alum is the base of pure clay, and strata of clay are often met with containing much iron pyrites. When the latter substance decomposes, sulphuric acid is produced, which unites with the aluminous earth of the clay to form sulphate of alumine, or common alum. Where manufactories are established for obtaining the alum, the indurated beds of clay employed are called Alum-stone.

AMMONITE. An extinct and very numerous genus of the order of molluscous animals called Cephalopoda, allied to the modern genus Nautilus, which inhabited a chambered shell, curved like a coiled snake. Species of it are found in all geological periods of the secondary strata; but they have not been seen in the tertiary beds. They are named from their resemblance to the horns on the statues of Jupiter Ammon.

AMORPHOUS. Bodies devoid of regular form. *Etym.*, α, *a*, without, and μορφη, *morphe*, form.

AMYGDALOID. One of the forms of the Trap-rocks, in which agates and simple minerals appear to be scattered like almonds in a cake. *Etym.*, αμυγδαλα, *amygdala*, an almond.

ANALCIME. A simple mineral of the Zeolite family, also called Cubizite, of frequent occurrence in the Trap-rocks.

ANALOGUE. A body that resembles or corresponds with another body. A recent shell of the same species as a fossil shell is the analogue of the latter.

ANGIOSPERMS. A term applied to all flowering plants in which the ovules are inclosed in an ovary, and the seeds in a pericarp or covering, as in all flowering plants except those mentioned under gymnosperms and gymnogens, which see. *Etym.*, αγγος, *angos*, a vessel, and σπερμα, a seed.

ANOPLOTHERIUM. A fossil extinct quadruped belonging to the order Pachydermata, resembling a pig. It has received its name because the animal must have been singularly wanting in means of defence, from the form of its teeth and the absence of claws, hoofs, and horns. *Etym.*, ανοπλος, *anoplos*, unarmed, and θηριον, *therion*, a wild beast.

ANTAGONIST POWER. Two powers in nature, the action of the one counteracting that of the other, by which a kind of equilibrium or balance is maintained, and the destructive effect prevented that would be produced by one operating without a check.

ANTENNÆ. The articulated horns with which the heads of insects are invariably furnished.

ANTHRACITE. A shining substance like black-lead; a species of mineral charcoal. *Etym.*, ανθραξ, *anthrax*, coal.

ANTHRACOTHERIUM. A name given to an extinct quadruped, supposed to belong to the Pachydermata, the bones of which were first found in lignite and coal of the tertiary strata. *Etym.*, ανθραξ, *anthrax*, coal, and θηριον, *therion*, wild beast.

ANTHROPOMORPHOUS. Having a form resembling the human. *Etym.*, ανθρωπος, *anthropos*, a man, and μορφη, *morphe*, form.

ANTISEPTIC. Substances which prevent corruption in animal and vegetable matter, as common salt does, are said to be antiseptic. *Etym.*, αντι, *anti*, against, and σηπω, *sepo*, to putrefy.

ARENACEOUS. Sandy. *Etym.*, *arena*, sand.

ARGILLACEOUS. Clayey, composed of clay. *Etym.*, *argilla*, clay.

ARRAGONITE. A simple mineral, a variety of carbonate of lime, so called from having been first found in Aragon in Spain.

ATOLLS. Coral islands of an annular form, or consisting of a circular strip or ring of coral surrounding a central lagoon.

AUGITE. A simple mineral of a dark green, or black color, which forms a constituent part of many varieties of volcanic rocks. Name applied by Pliny to a particular mineral, from the Greek αυγη, *auge*, lustre.

AVALANCHES. Masses of snow which, being detached from great heights in the Alps, acquire enormous bulk by fresh accumulations as they descend; and when they fall into the valleys below often cause great destruction. They are also called *lavanges* and *lavanches* in the dialects of Switzerland.

BASALT. One of the most common varieties of the Trap-rocks. It is a dark green or black stone, composed of augite and felspar, very compact in texture, and of considerable hardness, often found in regular pillars of three or more sides called basaltic columns. Remarkable examples of this kind are seen at the Giant's Causeway, in Ireland, and at Fingal's Cave, in Staffa, one of the Hebrides. The term is used by Pliny, and is said to come from *basal*, an Æthiopian word signifying iron. The rock often contains much iron.

"BASIN" of Paris, "BASIN" of London. Deposits lying in a hollow or trough, formed of older rocks; sometimes used in geology almost synonymously with "formations," to express the deposits lying in a certain cavity or depression in older rocks.

BELEMNITE. An extinct genus of the order of molluscous animals called Cephalopoda, having a long, straight, and chambered conical shell. *Etym.*, βελεμνον, *belemnon*, a dart.

BITUMEN. Mineral pitch, of which the tar-like substance which is often seen to ooze out of the Newcastle coal when on the fire, and which makes it cake, is a good example. *Etym.*, *bitumen*, pitch.

BITUMINOUS SHALE. An argillaceous shale, much impregnated with bitumen, which is very common in the Coal Measures.

BLENDE. A metallic ore, a compound of the metal zinc with sulphur. It is often found in brown shining crystals; hence its name among the German miners, from the word *blenden*, to dazzle.

BLUFFS. High banks presenting a precipitous front to the sea or a river. A term used in the United States of North America.

BOTRYOIDAL. Resembling a bunch of Grapes. *Etym.*, βοτρυς, *botrys*, a bunch of grapes, and ειδος, *eidos*, form.

BOULDERS. A provincial term for large rounded blocks of stone lying on the surface of the ground, or sometimes imbedded in loose soil, different in composition from the rocks in their vicinity, and which have been therefore transported from a distance.

BRECCIA. A rock composed of angular fragments connected together by lime or other mineral substance. An Italian term.

CALO SINTER. A German name for the deposits from springs holding carbonate of lime in solution—petrifying springs. *Etym.*, *kalk*, lime, and *sintern*, to drop.

CALCAIRE GROSSIER. An extensive stratum, or rather series of strata, found in the Paris Basin, belonging to the Eocene tertiary period. *Etym.*, *calcaire*, limestone, and *grossier*, coarse.

CALCAREOUS ROCK. Limestone. *Etym.*, *calx*, lime.

CALCAREOUS SPAR. Crystallized carbonate of lime.

CARBON. An undecomposed inflammable substance, one of the simple elementary bodies. Charcoal is almost entirely composed of it. *Etym.*, *carbo*, coal.

CARBONATE OF LIME. Lime combines with great avidity with carbonic acid, a gaseous acid only obtained fluid when united with water,—and all combinations of it with other substances are called *Carbonates*. All limestones are carbonates of lime, and quicklime is obtained by driving off the carbonic acid by heat.

CARBONATED SPRINGS. Springs of water, containing carbonic acid gas. They are very common, especially in volcanic countries; and sometimes contain so much gas, that if a little sugar be thrown into the water it effervesces like soda-water.

CARBONIC ACID GAS. A natural gas which often issues from the ground, especially in volcanic countries. *Etym.*, *carbo*, coal; because the gas is obtained by the slow burning of charcoal.

CARBONIFEROUS. A term usually applied, in a technical sense, to an ancient group of secondary strata; but any bed containing coal may be said to be carboniferous. *Etym.*, *carbo*, coal, and *fero*, to bear.

CATACLYSM. A deluge. *Etym.*, κατακλυζω, *catacluzo*, to deluge.

CEPHALOPODA. A class of molluscous animals, having their organs of motion arranged round their head. *Etym.*, κεφαλη, *cephale*, head, and ποδα, *poda*, feet.

CETACEA. An order of vertebrated mammiferous animals inhabiting the sea. The whale, dolphin, and narwal are examples. *Etym.*, *cete*, whale.

CHALCEDONY. A siliceous simple mineral, uncrystallized. Agates are partly composed of chalcedony.

CHALK. A white earthy limestone, the uppermost of the secondary series of strata.

CHERT. A siliceous mineral, nearly allied to chalcedony and flint, but less homogeneous and simple in texture. A gradual passage from chert to limestone is not uncommon.

CHLORITIC SAND. Sand colored green by an admixture of the simple mineral chlorite. *Etym.*, χλωρυς, *chlorus*, green.

CLEAVAGE. Certain rocks, usually called Slate-rocks, may be cleaved into an indefinite number of thin laminæ which are parallel to each other, but which are generally not parallel to the planes of the true strata or layers of deposition. The planes of cleavage, therefore, are distinguishable from those of stratification.

CLINKSTONE, called also phonolite, a felspathic rock of the trap family, usually fissile. It is sonorous when struck with a hammer, whence its name.

COAL FORMATION. This term is generally understood to mean the same as the Coal Measures, or Carboniferous group.

COLEOPTERA. An order of insects (Beetles) which have four wings, the upper pair being crustaceous and forming a shield. *Etym.*, κολεος, *coleos*, a sheath, and πτερον, *pteron*, a wing.

CONFORMABLE. When the planes of one set of strata are generally parallel to those of another set which are in contact, they are said to be conformable. Thus the

Fig. 98.

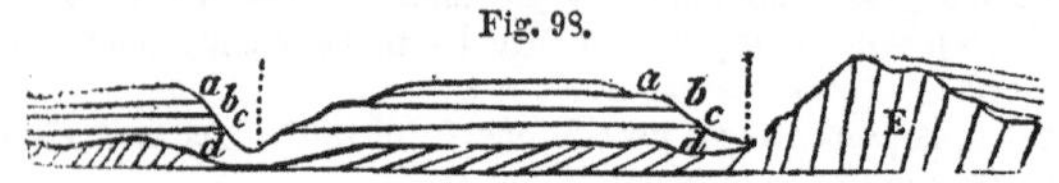

set *a*, *b*, Fig. 98, rest conformably on the inferior set *c*, *d*; but *c*, *d* rest unconformably on E.

CONGENERS. Species which belong to the same genus.

CONGLOMERATE, or PUDDINGSTONE. Rounded water-worn fragments of rock or pebbles, cemented together by another mineral substance, which may be of a siliceous, calcareous, or argillaceous nature. *Etym.*, *con*, together, *glomero*, to heap.

CONIFERÆ. An order of plants, all of which have disks in their wood fibres, by which they are recognized in a fossil state. Their ovules are naked (see GYMNOGENS). Most of the northern kinds bear the seeds in cones; but the yew does not, nor do a host of tropical and south temperate species. *Etym.*, *conus*, a cone, and *fero*, to bear.

COSMOGONY, COSMOLOGY. Words synonymous in meaning, applied to speculations respecting the first origin or mode of creation of the earth. *Etym.*, κοσμος, *kosmos*, the world, and γονη, *gonee*, generation, or λογος, *logos*, discourse.

CRAG. A provincial name in Norfolk and Suffolk for certain tertiary deposits usually composed of sand with shells, belonging to the Older Pliocene period.

CRATER. The circular cavity at the summit of a volcano, from which the volcanic matter is ejected. *Etym.*, *crater*, a great cup or bowl.

CRETACEOUS. Belonging to chalk. *Etym.*, *creta*, chalk.

CROP OUT. A miner's or mineral surveyor's term, to express the rising up or exposure at the surface of a stratum or series of strata.

CRUST OF THE EARTH. See "Earth's crust."

CRUSTACEOUS. Animals having a shelly coating or crust which they cast periodically. Crabs, shrimps, and lobsters are examples.

CRYPTOGAMIC. Asexual, flowerless, or Acotyledonous plants; a term applied to half the vegetable kingdom in contradistinction to Phænogamic, sexual, or flowering plants. It includes Fungi, Sea-weeds, Lichens, Mosses, Ferns, &c., which have no obvious flowers, and no cotyledons (seed-lobes) to their spores or seeds. *Etym.*, κρυπτος, *cruptos*, concealed, and γαμος, *gamos*, marriage.

CRYSTALS. Simple minerals are frequently found in regular forms, with facets like the drops of cut glass of chandeliers. Quartz being often met with in rocks in such forms, and beautifully transparent like ice, was called *rock-crystal*, κρυσταλλος, *crystallos*, being Greek for ice. Hence the regular *forms* of other minerals are called crystals, whether they be clear or opake.

CRYSTALLIZED. A mineral which is found in regular forms or crystals is said to be crystallized.

CRYSTALLINE. The internal texture which regular crystals exhibit when broken, or a confused assemblage of ill-defined crystals. Loaf-sugar and statuary-marble have a *crystalline* texture. Sugar-candy and calcareous spar are crystallized.

CUPRIFEROUS. Copper-bearing. *Etym.*, *cuprum*, copper, and *fero*, to bear.

CYCADEÆ. A small and very anomalous order of flowering plants, chiefly found in Mexico, the East Indian Islands, South Africa, and Australia. They are Gymnogens as to ovules, and neither Exogens nor Endogens in the wood of their short, simple, or branched trunks, and they have dicotyledonous seeds. The leaves are pinnated (like those of cocoa-nut palms), and when young are rolled inwards as in Ferns. The wood fibres are curiously perforated, and marked, by which they are recognized in a fossil state as well as by the trunk and foliage, and the cones, which contain the male flowers. The term is derived from κυκας, *cycas*, a name applied by the ancient Greek naturalist Threophrastus to a palm.

CYPERACEÆ. A tribe of plants answering to the English sedges; they are distinguished from grasses by their stems being solid, and generally triangular, instead of being hollow and round. Together with *Gramineæ* they constitute what writers on botanical geography often call *glumaceæ*.

DEBACLE. A great rush of waters, which, breaking down all opposing barriers, carries forward the broken fragments of rocks, and spreads them in its course. *Etym.*, *débacler*, French, to unbar, to break up as a river does at the cessation of a long-continued frost.

DELTA. When a great river, before it enters the sea, divides into separate streams, they often diverge and form two sides of a triangle, the sea being the base. The land included by the three lines, and which is invariably alluvial, was first called, in the case of the Nile a delta, from its resemblance to the letter of the Greek alphabet which goes by that name Δ. Geologists apply the term to alluvial land formed by a river at its mouth, without reference to its precise shape.

DENUDATION. The carrying away by the action of running water of a portion of the solid materials of the land, by which inferior rocks are laid bare. *Etym.*, *denudo*, to lay bare.

DEOXIDIZED, DEOXIDATED. Deprived of oxygen. Disunited from oxygen.

DESICCATION. The art of drying up. *Etym.*, *desicco*, to dry up.

DETRITUS. Matter worn or rubbed off from rocks. *Etym.*, *de*, from, and *tero*, to rub.

DICOTYLEDONOUS. A grand division of the vegetable kingdom, founded on the plant having two *cotyledons*, or seed-lobes. *Etym.*, δις, *dis*, double, and κοτυληδον, cotyledon.

DIKES. When a mass of the unstratified or igneous rocks, such as granite, trap, and lava, appears as if injected into a rent in the stratified rocks, cutting across the strata, it forms a dike. They are sometimes seen running along the ground, and projecting, like a wall, from the softer strata on both sides of them having wasted away; whence they were first called in the north of England and in Scotland *dikes*, a provincial name for wall. It is not easy to draw the line between dikes and veins. The former are generally of larger dimensions, and have their sides parallel for considerable distances; while veins have generally many ramifications, and these often thin away into slender threads.

DILUVIUM. Those accumulations of gravel and loose materials, which, by some geologists, are said to have been produced by the action of a diluvian wave or deluge sweeping over the surface of the earth. *Etym.*, *diluvium*, deluge.

DIP. When a stratum does not lie horizontally, but is inclined, it is said to *dip* towards some point of the compass, and the angle it makes with the horizon is called the angle of dip or inclination.

DIPTERA. An order of insects, comprising those which have only two wings. *Etym.*, δις, *dis*, double, and πτερον, *pteron*, wing.

DOLERITE. One of the varieties of the Trap-rocks, composed of augite and felspar.

DOLOMITE. A crystalline limestone, containing magnesia as a constituent part. Named after the French geologist Dolomieu.

DUNES. Low hills of blown sand that skirt the shores of Holland, England, Spain, and other countries.

EARTH'S CRUST. Such superficial parts of our planet as are accessible to human observation.

ECPYROSIS. A Greek term for a destruction by fire.

ELYTRA. The wing-sheaths, or upper crustaceous membranes, which form the superior wings in the tribe of beetles. They cover the body, and protect the true membranous wing. *Etym.*, ελυτρον, *elytron*, a sheath.

ENDOGENS. A class of flowering plants, whose stems present no distinction of wood, pith, and bark. The wood is disposed in bundles, placed nearer the axis than those of the previous year, as in palm trunks. This class answers to the Monocotyledones of Jussieu. *Etym.*, ενδον, *endon*, within, and γενεσις, *genesis*, increase.

ENTOMOSTRACA. Cuvier's second section of Crustacea; so called from their relationship to insects. *Etym.*, εντομα, *entoma*, insects.

EOCENE. A name given to the lowest division of the tertiary strata, containing an extremely small percentage of living species amongst its fossil shells, which indicate the first commencement or dawn of the existing state of the animate creation. *Etym.*, ηως, *eos*, aurora or the dawn, and καινος, *kainos*, recent.

ESCARPMENT. The abrupt face of a ridge of high land. *Etym.*, *escarper*, French, to cut steep.

ESTUARIES. Inlets of the land, which are entered both by rivers and the tides of the sea. Thus we have the estuaries of the Thames, Severn, Tay, &c. *Etym.*, *æstus*, the tide.

EXOGENS. A class of flowering plants whose stems have bark, wood, and pith. The bark is increased by layers deposited within the previously formed layers and the wood of layers or rings placed outside of those of the previous year. This class answers to the Dicotyledones of Jussieu, and includes all common English trees except pines, &c. (See GYMNOGENS.) *Etym.*, ἐξω, *exo*, outside, γενεσις, *genesis*, increase.

EXPERIMENTUM CRUCIS. A decisive experiment, so called, because, like a cross or direction-post, it directs men to true knowledge; or, as some explain it, because it is a kind of torture whereby the nature of the thing is extorted, as it were, by violence.

EXUVIÆ. Properly speaking, the transient parts of certain animals which they put off or lay down to assume new ones, as serpents and caterpillars shift their skins; but in geology it refers not only to the cast-off coverings of animals, but to fossil shells and other remains which animals have left in the strata of the earth. *Etym.*, *exuere*, to put off or divest.

FALUNS. A French provincial name for some tertiary strata abounding in shells in Touraine, which resemble in lithological characters the "Crag" of Norfolk and Suffolk.

FAULT, in the language of miners, is the sudden interruption of the continuity of strata in the same plane, accompanied by a crack or fissure, varying in width from a mere line to several feet, which is generally filled with broken stone, clay, &c.

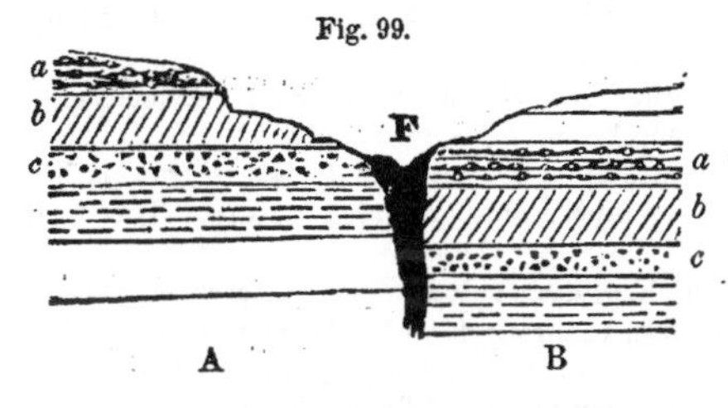

Fig. 99.

The strata, *a*, *b*, *c*, &c., must at one time have been continuous; but a fracture having taken place at the fault F, either by the upheaving of the portion A, or the sinking of the portion B, the strata were so displaced that the bed *a* in B is many feet lower than the same bed *a* in the portion A.

FAUNA. The various kinds of animals peculiar to a country constitute its FAUNA, as the various kinds of plants constitute its FLORA. The term is derived from the FAUNI, or rural deities, in Roman Mythology.

FELSPAR. A simple mineral, which, next to quartz, constitutes the chief material of rocks. The white angular portions in granite are felspar. This mineral always contains some alkali in its composition. In *common felspar* the alkali is potash; in another variety, called Albite or Cleavlandite, it is soda. Glassy felspar is a term applied when the crystals have a considerable degree of transparency. *Compact* felspar is a name of more vague signification. The substance so called appears to contain both potash and soda.

FELSPATHIC. Of or belonging to felspar.

FERRUGINOUS. Any thing containing iron. *Etym.*, *ferrum*, iron.

FISSILE, easily cleft, dividing readily into an indefinite number of parallel laminæ, like slates.

FLOETZ ROCKS. A German term applied to the secondary strata by the geologists of that country, because these rocks were supposed to occur most frequently in flat horizontal beds. *Etym.*, *flotz*, a layer or stratum.

FLORA. The various kinds of trees and plants found in any country constitute the FLORA of that country in the language of botanists.

FLUVIATILE. Belonging to a river. *Etym.*, *fluvius*, a river.

FORAMINIFERA. A name given by D'Orbigny to a family of microscopic shells. Their different chambers are united by a small perforation or *foramen*. Recent observation has shown that some at least are not Cephalopoda, as D'Orbigny supposed.

FORMATION. A group, whether of alluvial deposits, sedimentary strata, or igneous rocks, referred to a common origin or period.

FOSSIL. All minerals were once called fossils, but geologists now use the word only to express the remains of animals and plants found buried in the earth. *Etym.*, *fossilis*, any thing that may be dug out of the earth.

FOSSILIFEROUS. Containing organic remains.

GALENA. A metallic ore, a compound of lead and sulphur. It has often the appearance of highly polished lead. *Etym.*, γαλεω, *galeo*, to shine.

GARNET. A simple mineral, generally of a deep red color, crystallized; most commonly met with in mica slate, but also in granite and other igneous rocks.

GASTEROPODS. A division of the Testacea, in which, as in the limpet, the foot is attached to the body. *Etym.*, γαστηρ, *gaster*, belly, and ποδα, *poda*, feet.

GAULT. A provincial name in the east of England for a series of beds of clay and marl, the geological position of which is between the Upper and Lower Greensand.

GAVIAL. A kind of crocodile found in India.

GEM, or GEMMULE, from the Latin *gemma*, a bud. The term, applied to zoophytes, means a young animal not confined within an envelope or egg.

GEOLOGY, GEOGNOSY. Both mean the same thing; but with an unnecessary degree of refinement in terms, it has been proposed to call our description of the structure of the earth *geognosy* (*Etym.*, γεα, *gea*, earth, and γινωσκω, *ginosco*, to know), and our theoretical speculations as to its formation *geology* (*Etym.*, γεα, and λογος, *logos*, a discourse).

GLACIER. Vast accumulations of ice and hardened snow in the Alps and other lofty mountains. *Etym.*, *glace*, French for ice.

GLACIS. A term borrowed from the language of fortification, where it means an easy insensible slope or declivity, less steep than a *talus*, which see.

GNEISS. A stratified primary rock, composed of the same materials as granite, but having usually a larger proportion of mica and a laminated texture. The word is a German miner's term.

GRAMINEÆ. The order of plants to which grasses belong. *Etym.*, *gramen*, grass.

GRANITE. An unstratified or igneous rock, generally found inferior to or associated with the oldest of the stratified rocks, and sometimes penetrating them in the form of dikes and veins. It is usually composed of three simple minerals, felspar, quartz, and mica, and derives its name from having a coarse *granular* structure; *granum*, Latin for grain. Waterloo bridge, and the paving-stones in the carriage-way of the London streets, afford good examples of the most common varieties of granite.

GREENSAND. Beds of sand, sandstone, limestone, belonging to the Cretaceous Period. The name is given to these beds because they often, but not always, contain an abundance of green earth or chlorite scattered through the substance of the sandstone, limestone, &c.

GREENSTONE. A variety of trap, composed of hornblende and felspar.

GREYWACKÉ. *Grauwacke*, a German name, generally adopted by geologists for some of the most ancient fossiliferous strata. The rock is very often of a gray color; hence the name, *grau*, being German for gray, and *wacke*, being a provincial miner's term.

GRIT. A provincial name for a coarse-grained sandstone.

GYMNOSPERMOUS. *Etym.*, γυμνος, *gymnos*, naked, and σπερμα, *sperma*, a seed. (See GYMNOGENS.)

GYMNOGENS. A class of flowering plants, in which the ovules are not inclosed in an ovary. They are also called *gymnosperms*, the seeds in like manner not being inclosed in a pericarp. It includes all *Coniferæ*, as pine, fir, juniper, cypress, yew, cedar, &c., and *Cycadeæ*. All are Dicotyledonous (a few have many cotyledons), and all Exogenous, except *Cycas*, the growth of which is anomalous. The term is applied in contradistinction to *Angiosperms*, which see. *Etym.*, γυμνος, naked, and γενεσις, increase.

GYPSUM. A mineral composed of lime and sulphuric acid, hence called also *sulphate of lime*. Plaster and stucco are obtained by exposing gypsum to a strong heat. It is found so abundantly near Paris, that plaster of Paris is a common term in this country for the white powder of which casts are made. The term is used by Pliny for a stone used for the same purposes by the ancients. The derivation is unknown.

GYPSEOUS, of or belonging to gypsum.

GYROGONITES. Bodies found in freshwater deposits, originally supposed to be microscopic shells, but subsequently discovered to be seed-vessels of freshwater plants of the genus *Chara*. See above p. 742. *Etym.*, γυρος, *gyros*, curved, and γονος, *gonos*, seed, on account of their external structure.

HEMIPTERA. An order of insects, so called from a peculiarity in their wings, the superior being coriaceous at the base and membranous at the apex, ἡμισυ, *hemisu*, half, and πτερον, *pteron*, wing.

HORNBLENDE. A simple mineral of a dark green or black color, which enters largely into the composition of several varieties of the Trap-Rocks.

HORNSTONE. A siliceous mineral substance, sometimes approaching nearly to flint, or common quartz. It has a conchoidal fracture, and is infusible, which distinguishes it from compact felspar.

HUMERUS. The bone of the upper arm.

HYDROPHYTES. Plants which grow in water. *Etym.*, ὑδωρ, *hydor*, water, and φυτον, *phyton*, plant.

HYPOGENE ROCKS. Those rocks which are *nether-formed*, or which have not assumed their present form and structure at the surface, such as granite, gneiss, &c. The term, which includes both the plutonic and metamorphic rocks, is substituted for *primary*, because some members of both these classes, such as granite and gneiss, are posterior to many secondary or fossiliferous rocks. *Etym.*, ὑπο, *hypo*, under, and γινομαι, *ginomai*, to be formed or produced.

ICEBERG. Great masses of ice, often the size of hills, which float in the polar and adjacent seas. *Etym.*, ice, and *berg*, German for hill.

ICHTHYOSAURUS. A gigantic fossil marine reptile, allied in part of its structure to a fish. *Etym.*, ιχθυς, *ichthus*, a fish, and σαυρα, *saura*, a lizard.

IGNEOUS ROCKS. All rocks, such as lava, trap, and granite, known or supposed to have been melted by volcanic heat.

INCANDESCENT. White hot—having a more intense degree of heat than red heat.

INDUCTION. A consequence, inference, or general principle drawn from a number of particular facts or phenomena. The inductive philosophy, says Mr. Whewell, has been rightly described as a science which ascends from particular facts to general principles, and then descends again from these general principles to particular applications.

INFUSORY ANIMALCULES. Minute living creatures found in many *infusions*; and the term *infusori* has been given to all such animalcules, whether found in infusions or in stagnant water, vinegar, &c.

INSPISSATED. Thickened. *Etym.*, *spissus*, thick.

INVERTEBRATED ANIMALS. Animals which are not furnished with a back-bone. For a further explanation, see "Vertebrated Animals."

ISOTHERMAL. Such zones or divisions of the land, ocean, or atmosphere, which have an equal degree of mean annual warmth, are said to be isothermal, from ισος, *isos*, equal, and θερμη, *therme*, heat.

JOINTS. Fissures or lines of parting in rocks, often at right angles to the planes of stratification. The partings which divide columnar basalt into prisms are joints.

JURA LIMESTONE. The limestones belonging to the Oolite Group constitute the chief part of the mountains of Jura between France and Switzerland; and hence the geologists of the Continent have given the name to the group.

KEUPER. A German name for a member of the Upper New Red Sandstone.

KIMMERIDGE CLAY. A thick bed of clay, constituting a member of the Oolite Group. So called because it is found well developed at Kimmeridge, in the Isle of Purbeck, Dorsetshire.

LACUSTRINE. Belonging to a lake. *Etym.*, *lacus*, a lake.

LAMANTINE. A living species of the herbivorous Cetacea or whale tribe which inhabits the mouth of rivers on the coasts of Africa and South America: the sea-cow.

LAMELLIFEROUS. Having a structure consisting of thin plates or leaves like paper. *Etym.*, *lamella*, the diminutive of *lamina*, plate, and *fero*, to bear.

LAMINÆ. Latin for plates; used in geology for the smaller layers of which a stratum is frequently composed.

LANDSLIP. A portion of land that has slid down in consequence of disturbance by an earthquake, or from being undermined by water washing away the lower beds which supported it.

LAPIDIFICATION. Lapidifying process. Conversion into stone. *Etym.*, *lapis*, stone, and *fio*, to make.

LAPILLI. Small volcanic cinders. *Lapillus*, a little stone.

LAVA. The stone which flows in a melted state from a volcano.

LEPIDODENDRON, a genus of fossil plants of the Coal Measures, intermediate in character between the Lycopodiums and coniferous plants.

LEUCITE. A simple mineral found in volcanic rocks, crystallized, and of a white color. *Etym.*, λευκος, *leucos*, white.

LIAS. A provincial name for an argillaceous limestone, characterized together with its associated beds by peculiar fossils, and forming a particular group of strata, interposed between the Oolite and the New Red Sandstone.

LIGNIPERDOUS. A term applied to insects which destroy wood. *Etym.*, *lignum*, wood, and *perdo*, to destroy.

LIGNITE. Wood converted into a kind of coal. *Etym.*, *lignum*, wood.

LITHODOMI. Molluscous animals which form holes in the solid rocks in which they lodge themselves. The holes are not perforated mechanically, but the rock appears to be dissolved. *Etym.*, λιθος, *lithos*, stone, and δεμω, *demo*, to build.

LITHOGENOUS POLYPS. Animals which form coral.

LITHOGRAPHIC STONE. A slaty compact limestone, of a yellowish color and fine grain, used in lithography, which is the art of drawing upon and printing from stone *Etym.*, λιθος, *lithos*, stone, and γραφο, *grapho*, to write.

LITHOIDAL. Having a stony structure.

LITHOLOGICAL. A term expressing the stony structure or character of a mineral mass. We speak of the lithological character of a stratum as distinguished from its zoological character. *Etym.*, λιθος, *lithos*, stone, and λογος, *logos*, discourse.

LITHOPHAGI. Molluscous animals which form holes in solid stones. See "Lithodomi." *Etym.*, λιθος, *lithos*, stone, and φαγειν, *phagein*, to eat.

LITHOPHITES. The animals which form Stone-coral.

LITTORAL. Belonging to the shore. *Etym.*, *littus*, the shore.

LOAM. A mixture of sand and clay.

LOPHIODON. A genus of extinct quadrupeds, allied to the tapir, named from eminences on the teeth.

LYCOPODIACEÆ. Plants of an inferior degree of organization to Coniferæ, some of which they very much resemble in foliage, but all recent species are infinitely smaller. Many of the fossil species are as gigantic as recent Coniferæ. Their mode of reproduction is analogous to that of ferns. In English they are called club-mosses, generally found in mountainous heaths in the north of England.

LYDIAN STONE. Flinty slate; a kind of quartz or flint, allied to Hornstone, but of a grayish black color.

MACIGNO. In Italy this term has been applied to a siliceous sandstone sometimes containing calcareous grains, mica, &c.

MADREPORE. A genus of corals, but generally applied to all the corals distinguished by superficial star-shaped cavities. There are several fossil species.

MAGNESIAN LIMESTONE. An extensive series of beds, the geological position of which is immediately above the Coal Measures; so called, because the limestone, the principal member of the series, contains much of the earth magnesia as a constituent part.

MAMMIFEROUS. Mammifers. Animals which give suck to their young. To this class all the warm-blooded quadrupeds, and the Cetacea, or whales, belong. *Etym.*, *mamma*, a breast, *fero*, to bear.

MAMMILLARY. A surface which is studded over with rounded projections. *Etym.*, *mammilla*, a little breast or pap.

MAMMOTH. An extinct species of the elephant (*E. primigenius*), of which the fossil bones are frequently met with in various countries. The name is of Tartar origin, and is used in Siberia for animals that burrow under ground.

MANATI. One of the Cetacea, the sea-cow, or lamantine (*Trichechus manatus*, Lin.)

MARL. A mixture of clay and lime; usually soft, but sometimes hard, in which case it is called indurated marl.

MARSUPIAL ANIMALS. A tribe of quadrupeds having a sack or pouch under the belly, in which they carry their young. The kangaroo is a well-known example. *Etym.*, *marsupium*, a purse.

MASTODON. A genus of fossil extinct quadrupeds allied to the elephants; so called from the form of the hind teeth or grinders, which have their surface covered with conical mammillary crests. *Etym.*, *μαστος*, *mastos*, pap, and *οδων*, *odon*, tooth.

MATRIX. If a simple mineral or shell, in place of being detached, be still fixed in a portion of rock, it is said to be in its matrix. *Matrix*, womb.

MECHANICAL ORIGIN, ROCKS OF. Rocks composed of sand, pebbles, or fragments, are so called to distinguish them from those of a uniform crystalline texture, which are of chemical origin.

MEDUSÆ. A genus of marine radiated animals, without shells; so called, because their organs of motion spread out like the snaky hair of the fabulous Medusa.

MEGALOSAURUS. A fossil gigantic amphibious animal of the saurian or lizard and crocodile tribe. *Etym.*, *μεγαλη*, *megale*, great, and *σαυρα*, *saura*, lizard.

MEGATHERIUM. A fossil extinct quadruped, resembling a gigantic sloth. *Etym.*, *μεγα*, *mega*, great, and *θηριον*, *therion*, wild beast.

MELASTOMA. A genus of MELASTOMACEA, an order of exotic plants of the evergreen tree and shrubby kinds. *Etym.*, *μελας*, *melas*, black, and *στομα*, *stoma*, mouth; because the fruit of one of these species stains the lips.

MESOTYPE. A simple mineral, white, and needle-shaped, one of the Zeolite family, frequently met with in the Trap-rocks.

METAMORPHIC ROCKS. A stratified division of hypogene rocks, highly crystalline, such as gneiss and mica-schist, and so named because they have been *altered* by plutonic action. *Etym.*, *μετα*, *meta*, trans, and *μορφη*, *morphe*, form.

MICA. A simple mineral, having a shining silvery surface, and capable of being split into very thin elastic leaves or scales. It is often called *talc* in common life; but mineralogists apply the term talc to a different mineral. The brilliant scales in granite are mica. *Etym.*, *mico*, to shine.

MICA-SLATE, MICA-SCHIST, MICACEOUS SCHISTUS. One of the metamorphic or crystalline stratified rocks of the hypogene class, which is characterized by being composed of a large proportion of mica united with quartz.

MIOCENE. A division of tertiary strata intervening between the Eocene and Pliocene formations; so called, because a minority of its fossil shells are referable to living species. *Etym.*, *μειων*, *meion*, less, and *καινος*, *kainos*, recent.

MOLASSE. A provincial name for a soft green sandstone, associated with marl and conglomerates, belonging to the Miocene Tertiary Period, extensively developed in the lower country of Switzerland. *Etym.*, French, *molle*, soft.

MOLLUSCA, MOLLUSCOUS ANIMALS. Animals, such as shell-fish, which, being devoid of bones, have soft bodies. *Etym.*, *mollis*, soft.

MONAD. The smallest of visible animalcules, spoken of by Buffon and his followers as constituting the elementary molecules of organic beings.

MONITOR. An animal of the saurian or lizard tribe, species of which are found in both the fossil and recent state.

MONOCOTYLEDONOUS. A grand division of the vegetable kingdom (including palms, grasses, Lilaceæ, &c.), founded on the plant having only one *cotyledon*, or seed-lobe. *Etym.*, *μονος*, *monos*, single.

MORAINE, a Swiss term for the débris of rocks brought into valleys by glaciers. See p. 228.

MOSCHUS. A quadruped resembling the chamois or mountain goat, from which the perfume musk is obtained.

MOUNTAIN LIMESTONE, OR CARBONIFEROUS LIMESTONE. A series of limestone strata of marine origin, usually forming the lowest member of the Coal Measures.

MOYA. A term applied in South America to mud poured out from volcanoes during eruptions.

MULTILOCULAR. Many-chambered; a term applied to those shells which, like the nautilus, ammonite, and others, are divided into many compartments. *Etym.*, *multus*, many, and *loculus*, a partition.

Muriate of Soda. The scientific name for common culinary salt, because it is composed of muriatic acid and the alkali soda.

Musaceæ. A family of tropical monocotyledonous plants, including the banana and plantains.

Muschelkalk. A limestone, belonging to the Upper New Red Sandstone group. Its position is between the Magnesian Limestone and the Lias. This formation has not yet been found in England, and the German name is adopted by English geologists. The word means shell limestone. *Etym.*, *muschel*, shell, and *kalkstein*, limestone.

Naphtha. A very thin, volatile, inflammable, and fluid mineral substance, of which there are springs in many countries, particularly in volcanic districts.

Nenuphar. A yellow water-lily. P. 618.

New Red Sandstone. A formation so named, because it consists chiefly of sandy and argillaceous strata, the predominant color of which is brick-red, but containing portions which are of a greenish-gray. These occur often in spots and stripes, so that the series has sometimes been called the variegated sandstone. This formation is divided into the Upper New Red in which the Muschelkalk is included, and the Lower New Red, of which the Magnesian Limestone is a member.

Nodule. A rounded irregular-shaped lump or mass. *Etym.*, diminutive of *nodus*, knot.

Normal Groups. Groups of certain rocks taken as a rule or standard. *Etym.*, *norma*, rule or pattern.

Nucleus. A solid central piece, around which other matter is collected. The word is Latin for kernel.

Nummulites. An extinct genus of the order of molluscous animals, called Cephalopoda, of a thin lenticular shape, internally divided into small chambers. *Etym.*, *nummus*, Latin for money, and λιθος, *lithos*, stone, from its resemblance to a coin.

Obsidian. A volcanic product, or species of lava, very like common green bottle glass, which is almost black in large masses, but semi-transparent in thin fragments. Pumice-stone is obsidian in a frothy state; produced, most probably, by water that was contained in or had access to the melted stone, and converted into steam. There are very often portions in masses of solid obsidian, which are partially converted into pumice.

Ochre. A yellow powder, a combination of some earth with oxide of iron.

Ogygian Deluge. A great inundation mentioned in fabulous history, supposed to have taken place in the reign of Ogyges in Attica, whose death is fixed in Blair's Chronological Tables in the year 1764 before Christ. See p. 341.

Old Red Sandstone. A formation immediately below the Carboniferous Group. The term Devonian has been recently proposed for strata of this age, because in Devonshire they are largely developed, and contain many organic remains.

Oligoclase. A mineral of the felspar family.

Olivine. An olive-colored, semi-transparent, simple mineral, very often occurring in the form of grains and of crystals in basalt and lava.

Oolite, Oolitic. A limestone; so named because it is composed of rounded particles like the roe or eggs of a fish. The name is also applied to a large group of strata, characterized by peculiar fossils, in which limestone of this texture occurs. *Etym.*, ωον, *oon*, egg, and λιθος, *lithos*, stone.

Opalized Wood. Wood petrified by siliceous earth, and acquiring a structure similar to the simple mineral called opal.

Ophidious Reptiles. Vertebrated animals, such as snakes and serpents. *Etym.*, οφις, *ophis*, a serpent.

Organic Remains. The remains of animals and plants (*organized* bodies) found in a fossil state.

Orthocerata or Orthoceræ. An extinct genus of the order of molluscous animals, called Cephalopoda, that inhabited a long-chambered conical shell, like a straight horn. *Etym.*, ορθος, *orthos*, straight, and κερας, *ceras*, horn.

OSSEOUS BRECCIA. The cemented mass of fragments of bones of extinct animals found in caverns and fissures. *Osseous* is a Latin adjective, signifying bony.

OSTEOLOGY. That division of anatomy which treats of the bones; from *οστεον*, *osteon*, bone, and λογος, *logos*, a discourse.

OUTLIERS. When a portion of a stratum occurs at some distance, detached from the general mass of the formation to which it belongs, some practical mineral surveyors call it an *outlier*, and the term is adopted in geological language.

OVATE. The shape of an egg. *Etym.*, *ovum*, egg.

OVIPOSITING. The laying of eggs.

OXIDE. The combination of a metal with oxygen; rust is oxide of iron.

OXYGEN. One of the constituent parts of the air of the atmosphere; that part which supports life. For a farther explanation of the word, consult elementary works on chemistry.

PACHYDERMATA. An order of quadrupeds, including the elephant, rhinoceros, horse, pig, &c., distinguished by having thick skins. *Etym.*, παχυς, *pachus*, thick, and δερμα, *derma*, skin, or hide.

PACHYDERMATOUS. Belonging to Pachydermata.

PALÆOTHERIUM, PALEOTHERE. A fossil extinct quadruped, belonging to the order Pachydermata, resembling a pig, or tapir, but of great size. *Etym.*, παλαιος, *palaios*, ancient, and θηριον, *therion*, wild beast.

PALEONTOLOGY. The science which treats of fossil remains, both animal and vegetable. *Etym.*, παλαιος, *palaios*, ancient, οντα, *onta*, beings, and λογος, *logos*, a discourse.

PELAGIAN, PELAGIC. Belonging to the *deep* sea. *Etym.*, *pelagus*, sea.

PEPERINO. An Italian name for a particular kind of volcanic rock, formed like tuff, by the cementing together of volcanic sand, cinders, or scoriæ, &c.

PETROLEUM. A liquid mineral pitch, so called because it is seen to ooze like oil out of the rock. *Etym.*, *petra*, rock, and *oleum*, oil.

PHÆNOGAMOUS or PHANEROGAMIC PLANTS. A name given by Linnæus to those plants in which the reproductive organs are apparent. *Etym.*, φανερος, *phaneros*, evident, or φαινω, *phaino*, to show, and γαμος, *gamos*, marriage.

PHLEGRÆAN FIELDS. Campi Phlegræi, or "the Burnt Fields." The country around Naples, so named by the Greeks, from the traces of igneous action everywhere visible.

PHONOLITE. See "Clinkstone."

PHRYGANEA. A genus of four-winged insects, the larvæ of which, called caddis-worms, are used by anglers as a bait.

PHYSICS. The department of science which treats of the properties of natural bodies, laws of motion, &c.; sometimes called natural philosophy and mechanical philosophy. *Etym.*, φυσις, *physis*, nature.

PHYTOLOGY, PHYTOLOGICAL. The department of science which relates to plants—synonymous with botany and botanical. *Etym.*, φυτον, *phyton*, plant, and λογος, *logos*, discourse.

PHYTOPHAGOUS. Plant-eating. *Etym.*, φυτον, *phyton*, plant, and φαγειν, *phagein*, to eat.

PISOLITE. A stone possessing a structure like an agglutination of peas. *Etym.*, πισον, *pison*, pea, and λιθος, *lithos*, stone.

PISTIA. P. 618. The plant mentioned by Malte-Brun is probably the *Pistia Stratiotes*, a floating plant, related to English duckweed, but very much larger.

PIT COAL. Ordinary coal; called so, because it is obtained by sinking pits in the ground.

PITCHSTONE. A rock of a uniform texture, belonging to the unstratified and volcanic classes, which has an unctuous appearance like indurated pitch.

PLASTIC CLAY. One of the beds of the Eocene Tertiary Period; so called, because it is used for making pottery. The formation to which this name is applied is a series of beds chiefly sands, with which the clay is associated. *Etym.*, πλασσω, *plasso*, to form or fashion.

PLESIOSAURUS. A fossil extinct amphibious animal, resembling the saurian, or lizard and crocodile tribe. *Etym.*, πλησιον, *plesion*, near to, and σαυρα, *saura*, a lizard.

PLIOCENE, OLDER and NEWER. Two divisions of the Tertiary Period which are the most modern, and of which the largest part of the fossil shells are of recent species. *Etym.*, πλειων, *pleion*, more, and καινος, *kainos*, recent.

PLUTONIC ACTION. The influence of volcanic heat and other subterranean causes under pressure.

PLUTONIC ROCKS. Granite, porphyry, and other igneous rocks supposed to have consolidated from a melted state at a great depth from the surface.

POLYPARIA. CORALS. A numerous class of invertebrated animals, belonging to the great division called Radiata.

PORPHYRY. An unstratified or igneous rock. The term is as old as the time of Pliny, and was applied to a red rock with small, angular, white bodies diffused through it, which are crystallized felspar, brought from Egypt. The term is hence applied to every species of unstratified rock in which detached crystals or felspar or some other mineral are diffused through a base of other mineral composition. *Etym.*, πορφυρα, *porphyra*, purple.

PORTLAND LIMESTONE, PORTLAND BEDS. A series of limestone strata, belonging to the upper part of the Oolite Group, found chiefly in England in the Island of Portland on the coast of Dorsetshire. The great supply of the building-stone used in London is from these quarries.

POZZUOLANA. Volcanic ashes, largely used as mortar for buildings, similar in nature to what is called in this country Roman cement. It gets its name from Puzzuoli, a town in the Bay of Naples, from which it is shipped in large quantities to all parts of the Mediterranean.

PRECIPITATE. Substances which, having been dissolved in a fluid, are separated from it by combining chemically and forming a solid, which falls to the bottom of the fluid. This process is the opposite to that of chemical solution.

PRODUCTA. An extinct genus of fossil bivalve shells occurring only in the older secondary rocks. It is closely allied to the living genus Terebratula.

PTERODACTYL. A flying reptile: species of this genus have been found in the Oolite and Muschelkalk. Some of the finger-joints are lengthened, so as to serve as the expansors of a membranous wing. Hence the name *wing-fingered*. *Etym.*, πτερον, *pteron*, a wing, and δακτυλος, *dactylos*, a finger.

PUBESCENCE. The soft hairy down on insects. *Etym.*, *pubesco*, the first growth of the beard.

PUDDINGSTONE. See "Conglomerate."

PUMICE. A light spongy lava, chiefly felspathic, of a white color, produced by gases or watery vapor getting access to the particular kind of glassy lava called obsidian, when in a state of fusion; it may be called the froth of melted volcanic glass. The word comes from the Latin name of the stone, *pumex*.

PURBECK LIMESTONE, PURBECK BEDS. Limestone strata, belonging to the Wealden Group, which intervenes between the Greensand and the Oolite.

PYRITES. (Iron.) A compound of sulphur and iron, found usually in yellow shining crystals like brass, and in almost every rock, stratified and unstratified. The shining metallic bodies so often seen in common roofing slate are a familiar example of the mineral. The word is Greek, and comes from πυρ, *pyr*, fire; because under particular circumstances, the stone produces spontaneous heat, and even inflammation.

PYROMETER. An instrument for measuring intense degrees of heat.

QUADRUMANA. The order of mammiferous animals to which apes belong. *Etym.*, *quadrus*, a derivative of the Latin word for the number four, and *manus*, hand, the four feet of those animals being in some degree usable as hands.

QUA-QUA-VERSAL DIP. The dip of beds to all points of the compass around a centre, as in the case of beds of lava round the crater of a volcano. *Etym.*, *quâ-quâ-versum*, on every side.

QUARTZ. A German provincial term, universally adopted in scientific language for a simple mineral composed of pure silex, or earth of flints: rock-crystal is an example.

QUARTZITE or QUARTZ ROCK. An aggregate of grains of quartz, sometimes passing into compact quartz.

RED MARL. A term often applied to the New Red Sandstone.

RETICULATE. A structure of cross lines, like a net, is said to be reticulated, from *rete*, a net.

ROCK SALT. Common culinary salt, or muriate of soda, found in vast solid masses or beds, in different formations, extensively in the New Red Sandstone formation, as in Cheshire; and it is then called *rock*-salt.

RUBBLE. A term applied by quarry-men to the upper fragmentary and decomposed portion of a mass of stone.

RUMINANTIA. Animals which ruminate or chew the cud, such as the ox, deer, &c. *Etym.*, the Latin verb *rumino*, meaning the same thing.

SACCHAROID, SACCHARINE. When a stone has a texture resembling that of loaf-sugar. *Etym.*, σακχαρ, *sacchar*, sugar, and ειδος, *eidos*, form.

SALIENT ANGLE. In a zigzag line *a a* are the salient angles, *b b* the re-entering angles. *Etym.*, *salire*, to leap or bound forward.

Fig. 100.

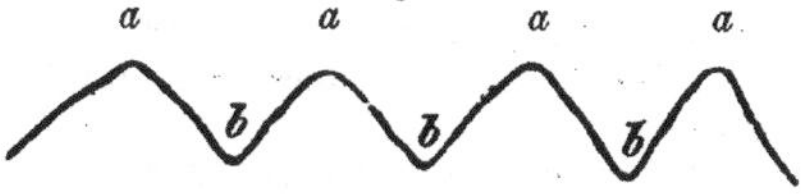

SALT SPRINGS. Springs of water containing a large quantity of common salt. They are very abundant in Cheshire and Worcestershire, and culinary salt is obtained from them by mere evaporation.

SANDSTONE. Any stone which is composed of an agglutination of grains of sand, whether calcareous, siliceous, or of any other mineral nature.

SAURIAN. Any animal belonging to the lizard tribe. *Etym.*, σαυρα, *saura*, a lizard.

SAXICAVOUS. Hollowing out stone.

SCHIST is often used as synonymous with slate; but it may be very useful to distinguish between a schistose and a slaty structure. The hypogene or primary *schists*, as they are termed, such as gneiss, mica-schist, and others, cannot be split into an indefinite number of parallel laminæ like rocks which have a true slaty cleavage. The uneven schistose layers of mica-schist and gneiss are probably layers of deposition, which have assumed a crystalline texture. See "Cleavage." *Etym.*, *schistus*, adj. Latin, that which may be split.

SCHISTOSE ROCKS. See "Schist."

SCORIÆ. Volcanic cinders. The word is Latin for cinders.

SEAMS. Thin layers which separate two strata of greater magnitude.

SECONDARY STRATA. An extensive series of the stratified rocks which compose the crust of the globe, with certain characters in common, which distinguish them from another series below them called *primary*, and from a third series above them called *tertiary*.

SECULAR REFRIGERATION. The periodical cooling and consolidation of the globe from a supposed original state of fluidity from heat. *Sæculum*, age or period.

SEDIMENTARY ROCKS are those which have been formed by their materials having been thrown down from a state of suspension or solution in water.

SELENITE. Crystallized gypsum, or sulphate of lime—a simple mineral.

SEPTARIA. Flattened balls of stone, generally a kind of iron-stone, which, on being split, are seen to be separated in their interior into irregular masses. *Etym.*, *septa*, inclosures.

SERPENTINE. A rock usually containing much magnesian earth, for the most part unstratified, but sometimes appearing to be an altered or metamorphic stratified rock. Its name is derived from frequently presenting contrasts of color, like the skin of some serpents.

SHALE. A provincial term, adopted by geologists, to express an indurated slaty clay. *Etym.*, German *schalen*, to peel, to split.

SHELL MARL. A deposit of clay, peat, and other substances mixed with shells, which collects at the bottom of lakes.

SHINGLE. The loose and completely water-worn gravel on the sea-shore.

SILEX. The name of one of the pure earths, being the Latin word for *flint*, which is wholly composed of that earth. French geologists have applied it as a generic name for all minerals composed entirely of that earth, of which there are many of different external forms.

SILICA. One of the pure earths. *Etym.*, *silex*, flint, because found in that mineral.

SILICATE. A chemical compound of silica and another substance, such as silicate of iron. Consult elementary works on chemistry.

SILICEOUS. Of or belonging to the earth of flint. *Etym.*, *silex*, which see. A siliceous rock is one mainly composed of silex.

SILICIFIED. Any substance that is petrified or mineralized by *siliceous* earth.

SILT. The more comminuted sand, clay, and earth, which is transported by running water. It is often accumulated by currents in banks. Thus the mouth of a river is silted up when its entrance into the sea is impeded by such accumulation of loose materials.

SIMPLE MINERAL. Individual mineral substances, as distinguished from rocks, which last are usually an aggregation of simple minerals. They are not simple in regard to their nature; for when subjected to chemical analysis, they are found to consist of a variety of different substances. Pyrites is a simple mineral in the sense we use the term, but it is a chemical compound of sulphur and iron.

SINTER, CALCAREOUS OR SILICEOUS. A German name for a rock precipitated from mineral waters. *Etym.*, *sintern*, to drop.

SLATE. See "Cleavage" and "Schist."

SOLFATARA. A volcanic vent from which sulphur, sulphureous, watery, and acid vapors and gases are emitted.

SPORULES. The reproductory corpuscula (minute bodies) of cryptogamic plants. *Etym.*, *σπορα*, *spora*, a seed.

STALACTITE. When water holding lime in solution deposits it as it drops from the roof of a cavern, long rods of stone hang down like icicles, and these are called *stalactites*. *Etym.*, *σταλαζω*, *stalazo*, to drop.

STALAGMITE. When water holding lime in solution drops on the floor of a cavern, the water evaporating leaves a crust composed of layers of limestone: such a crust is called *stalagmite*, from *σταλαγμα*, *stalagma*, a drop, in opposition to *stalactite*, which see.

STATICAL FIGURE. The figure which results from the equilibrium of forces. From *στατος*, *statos*, stable, or standing still.

STERNUM. The breast-bone, or the flat bone occupying the front of the chest.

STILBITE. A crystallized simple mineral, usually white, one of the Zeolite family, frequently included in the mass of the Trap-rocks.

STRATIFIED. Rocks arranged in the form of *strata*, which see.

STRATIFICATION. An arrangement of rocks in *strata*, which see.

STRATA, STRATUM. The term stratum, derived from the Latin verb *struo*, to strew or lay out, means a bed or mass of matter spread out over a certain surface by the action of water, or in some cases by wind. The deposition of successive layers of sand and gravel in the bed of a river, or in a canal, affords a perfect illustration both of the form and origin of stratification. A large portion of the masses constituting the earth's crust are thus stratified, the successive strata of a given rock preserving a general parallelism to each other; but the planes of stratification not being perfectly parallel throughout a great extent like the planes of *cleavage*, which see.

STRIKE. The direction or line of bearing of strata, which is always at right angles to their prevailing dip.

STUFAS. Jets of steam issuing from fissures in volcanic regions at a temperature often above the boiling point.

SUBAPENNINES. Low hills which skirt or lie at the foot of the great chain of the Apennines in Italy. The term Subapennine is applied geologically to a series of strata of the Older Pliocene Period.

SYENITE. A kind of granite; so called, because it was brought from Syene in Egypt.

TALUS. When fragments are broken off by the action of the weather from the face of a steep rock, as they accumulate at its foot, they form a sloping heap, called a talus. The term is borrowed from the language of fortification, where *talus* means the outside of a wall of which the thickness is diminished by degrees, as it rises in height, to make it the firmer.

TARSI. The feet in insects, which are articulated, and formed of five or a less number of joints.

TERTIARY STRATA. A series of sedimentary rocks, with characters which distinguish them from two other great series of strata—the secondary and primary—which lie *beneath* them.

TESTACEA. Molluscous animals, having a shelly covering. *Etym.*, *testa*, a shell, such as snails, whelks, oysters, &c.

THALLOGENS. A class of flowerless plants including all those that have no defined axis, stem, or leaves; as Lichens, Seaweeds, and Fungi. *Etym.*, *θαλλος*, *thallos*, a branch, and *γενεσις*, *genesis*, increase.

THERMAL. Hot. *Etym.*, *θερμος*, *thermos*, hot.

THERMO-ELECTRICITY. Electricity developed by heat.

THIN OUT. When a stratum, in the course of its prolongation in any direction, becomes gradually less in thickness, the two surfaces approach nearer and nearer; and when at last they meet, the stratum is said to thin out or disappear.

TRACHYTE. A variety of lava essentially composed of glassy felspar, and frequently having detached crystals of felspar in the base or body of the stone, giving it the structure of porphyry. It sometimes contains hornblende and augite; and when these last predominate, the trachyte passes into the varieties of trap, called Greenstone, Basalt, Dolorite, &c. The term is derived from *τραχυς*, *trachus*, rough, because the rock has a peculiar rough feel.

TRAP and TRAPPEAN ROCKS. Volcanic rocks composed of felspar, augite, and hornblende. The various proportions and state of aggregation of these simple minerals, and differences in external forms, give rise to varieties, which have received distinct appellations, such as Basalt, Amygdaloid, Dolorite, Greenstone, and others. The term is derived from *trappa*, a Swedish word for stair, because the rocks of this class sometimes occur in large tabular masses, rising one above another like steps.

TRAVERTIN. A white concretionary limestone, usually hard and semi-crystalline, deposited from the water of springs holding lime in solution.—*Etym.* This stone was called by the ancients Lapis Tiburtinus, the stone being formed in great quantity by the river Anio, at Tibur, near Rome. Some suppose travertin to be an abbreviation of trasterverino from transtiburtinus.

TRIPOLI. The name of a powder used for polishing metals and stones, first imported from Tripoli, which, as well as a certain kind of siliceous stone of the same name, has been lately found to be composed of the flinty cases of Infusoria.

TROPHI, of Insects. Organs which form the mouth, consisting of an upper and under lip, and comprising the parts called mandibles, maxillæ, and palpi.

TUFA, CALCAREOUS. A porous rock deposited by calcareous waters on their exposure to the air, and usually containing portions of plants and other organic substances incrusted with carbonate of lime. The more solid form of the same deposit is called "travertin," into which it passes.

TUFA, VOLCANIC. See "Tuff."

TUFACEOUS. A rock with the texture of tuff, or tufa, which see.

TUFF, or TUFA VOLCANIC. An Italian name for a variety of volcanic rock of an earthy texture, seldom very compact, and composed of an agglutination of fragments of scoriæ and loose materials ejected from a volcano.

TURBINATED. Shells which have a spiral or screw-form structure. *Etym.*, *turbinatus*, made like a top.

TURRILITE. An extinct genus of chambered shells, allied to the Ammonites, having the siphuncle near the dorsal margin.

UNCONFORMABLE. See "Conformable."

UNOXIDIZED, UNOXIDATED. Not combined with oxygen.

VEINS, MINERAL. Cracks in rocks filled up by substances different from the rock, which may either be earthy or metallic. Veins are sometimes many yards wide;

and they ramify or branch off into innumerable smaller parts, often as slender as threads, like the veins in an animal, hence their name.

VERTEBRATED ANIMALS. A great division of the animal kingdom, including all those which are furnished with a back-bone, as the mammalia, birds, reptiles, and fishes. The separate joints of the back-bone are called *vertebræ*, from the Latin verb *verto*, to turn.

VESICLE. A small, circular, inclosed space, like a little bladder. *Etym.*, diminutive of *vesica*, Latin for a bladder.

VITRIFICATION. The conversion of a body into glass by heat.

VOLCANIC BOMBS. Volcanoes throw out sometimes detached masses of melted lava, which, as they fall, assume rounded forms (like bomb-shells), and are often elongated into a pear-shape.

VOLCANIC FOCI. The subterranean centres of action in volcanoes, where the heat is supposed to be in the highest degree of energy.

WACKE. A rock nearly allied to basalt, of which it may be regarded as a soft and earthy variety.

WARP. The deposit of muddy waters, artificially introduced into low lands. See p. 326.

ZEOLITE. A family of simple minerals, including stilbite, mesotype, analcime, and some others, usually found in the trap or volcanic rocks. Some of the most common varieties swell or boil up when exposed to the blow-pipe, and hence the name of ζεω, *zeo*, to boil, and λιθος, *lithos*, stone.

ZOOPHITES. Corals, sponges, and other aquatic animals allied to them; so called because, while they are the habitation of animals, they are fixed to the ground, and have the form of plants. *Etym.*, ζωον, *zoon*, animal, and φυτον, *phyton*, plant.

INDEX.

A.

B.

C.

E.

F.

G.

H.

I.

N.

O.

P.

Q.

R.

S.

T.

U.

V.

W.

X

Y.

Z.

ND - #0028 - 281123 - C0 - 229/152/46 [48] - CB - 9780656337972 - Gloss Lamination